# POPULATION GENETICS, MOLECULAR EVOLUTION, AND THE NEUTRAL THEORY

Photo by William B. Provine

MOTOO KIMURA

Jacques E. Schoutens, PhD.

# POPULATION GENETICS, MOLECULAR EVOLUTION, AND THE NEUTRAL THEORY

*Selected Papers*

❖

MOTOO KIMURA

*Edited and with Introductory Essays by*
NAOYUKI TAKAHATA

*With a Foreword by*
JAMES F. CROW

THE UNIVERSITY OF CHICAGO PRESS
*Chicago and London*

Motoo Kimura, former head of the Department of Population Genetics at the National Institute of Genetics, Mishima, Japan, is a member of the Japan Academy, a foreign associate of the U.S. National Academy of Sciences, and an honorary member of the Genetical Society of Great Britain. He is the author of *The Neutral Theory of Molecular Evolution.* Naoyuki Takahata is professor in the Department of Genetics at The Graduate University for Advanced Studies, Mishima, Japan.

The University of Chicago Press, Chicago 60637
The University of Chicago Press, Ltd., London

Printed in the United States of America
03 02 01 00 99 98 97 96 95 94     5 4 3 2 1

ISBN (cloth): 0-226-43562-8
ISBN (paper): 0-226-43563-6

Library of Congress Cataloging-in-Publication Data

Kimura, Motoo, 1924–
Population genetics, molecular evolution, and the neutral theory : selected papers / Motoo Kimura ; edited and with introductory essays by Naoyuki Takahata ; with a foreword by James F. Crow.
p. cm.
Includes bibliographical references.
ISBN 0-226-43562-8. — ISBN 0-226-43563-6 (pbk.)
1. Population genetics. 2. Molecular evolution. I. Title.
QH455.K54 1994
575.1′5—dc20 94-16056
CIP

♾ The paper used in this publication meets the minimum requirements of the American National Standard for Information Sciences—Permanence of Paper for Printed Library Materials, ANSI Z39.48-1984.

# CONTENTS

# FOREWORD

It was in the late summer of 1953, a chance encounter that changed two lives. During a Genetics Society meeting in Madison, Wisconsin, I was walking down the corridor of the Wisconsin Union Building and encountered a Japanese person who appeared not quite sure of his way. In his then somewhat halting English he said that his name was Kimura. I think it is correct to say that the total number of American geneticists then familiar with Kimura's name could be counted on one hand; but I was one, having learned of his work from Newton Morton, a student of mine who had been in Japan.

I quickly learned that Kimura's given name, Motoo, is pronounced *Mo-toe,* not *Mo-two.* The extra *o* is part of the original Japanese kanji, but had the unfortunate effect of leading most westerners to mispronounce it.

Motoo was exceedingly eager to meet Sewall Wright, who happened to be in town to attend the same meeting, so I introduced them. I also learned that Kimura had a manuscript with him, a paper he had written on board ship while crossing the Pacific. I realized that this was a major contribution and quickly agreed to consider it for *Genetics,* of which I was then assistant editor. The paper (number 5 in this collection) was reviewed by Sewall Wright, who praised it effusively, quite unusual for Wright.

Kimura was at that time beginning graduate work at Iowa State College. He had hoped to study in Chicago with Wright, who because of his imminent retirement was not accepting students and recommended that Kimura study with J. L. Lush at Iowa State. Kimura found the extensive course requirements there and the research emphasis (subdivision of epistatic variance) not to his liking and during the year wrote to see if he could come to Wisconsin. I was delighted, of course, and the next summer he moved to Madison. By another wonderful coincidence, Dr. Wright was also planning to move to Madison after his retirement from the University of Chicago, and he arrived a few months after Motoo. Kimura's dream of studying with Wright had come true.

So the chance encounter changed Kimura's life by facilitating his work in stochastic processes and his association with Sewall Wright. My life was changed by the beginning of a continuing partnership that led to several joint papers and a book.

Kimura's work falls into two categories. First is his research in theoretical population genetics. Second is his neutral, random drift theory of molecular evolution. To most biologists the neutral theory is his major contribution, and indeed it has revolutionized the way we think about molecular evolution. Certainly this is what Kimura is best known for. Yet population geneticists—the insiders, the pros—are, if anything, more impressed by the power, originality, and ingenuity of his theoretical work. His capacity to manipulate diffusion equations so as to answer difficult and important biological questions is a source of wonder to all who have associated with him.

In his first few months at Wisconsin Kimura completed a now-famous paper giving the complete process of random genetic drift with a diffusion model (number 1). Wright communicated this to the Proceedings of the National Academy of Sciences USA. During the year Kimura wrote two other papers and was invited to speak at the 1955 Cold Spring Harbor Symposium. There he met a number of leading population geneticists, who were interested in seeing this young phenomenon from Japan. After Kimura's talk, Wright rose to comment that only those who had tried to solve these problems could fully appreciate Kimura's great accomplishment.

After receiving his Ph.D. from the University of Wisconsin, Kimura returned to Japan and continued to produce one important paper after another. The topics included population structure (the famous stepping-stone model), variable selection intensity (with the concept of quasi-fixation), genetics of quantitative characters (including the remarkable property of quasi-linkage equilibrium, rendering epistatic variance components almost irrelevant for evolutionary studies), inbreeding systems (showing that mating of least-related individuals is not the best way to conserve heterozygosity in the long run), reversibility of changes by random drift (with the surprising result that favored and unfavored genes with the same absolute fitness are fixed at the same rate), and many more that are represented in this volume. Most interesting from the standpoint of history of science is his exploiting the Kolmogorov backward equation to study such things as the probability of fixation of a mutant gene and the time required. Experimental science regularly uses existing theory, but in this case the theory and its later use were by the same person. Little did we know it at the time, but Kimura's theoretical studies turned out to be precisely what was needed for development of the neutral theory—a remarkable instance of evolutionary preadaptation.

Kimura first presented his neutral theory of molecular evolution in 1968. This proposal was blasphemy to evolutionists accustomed to thinking of natural selection as being the sole directive force in evolution. The theory was initially laughed out of court, but over the years the derision has changed to interest and respect. The theory can be stated in two forms, a strong one and weak one. The weak form asserts that the great majority of changes at the DNA level are the result of mutation and random drift. The strong form asserts that this is also true of amino acid changes. The weak form is hardly controversial anymore, but the strong form is still a matter of debate. Kimura has stated his argument forcefully and convincingly, as several papers in this volume attest. Regardless of the ultimate answer as to how much amino acid change is neutral, it is beyond question that random processes must taken into account. Needless to say, the neutral theory does not contradict the Darwinian theory as the basis of changes in form and function. The detailed reconciliation of molecular and phenotypic changes is a major agenda item for future research.

Population genetics was founded and completely dominated in its early years by the great trinity, J. B. S. Haldane, R. A. Fisher, and Sewall Wright. Their creativity laid the foundation for later work that led to the more rigorous, model-based subject that is modern population genetics. Two names stand out in this transition. One is Gustave Malécot, whose major contributions were written in French and secluded in obscure journals. Until recently his deepest work has been largely unknown to English- and Japanese-speaking geneticists. The other name, of course, is Kimura.

How does one contrast these two successors to the great trinity? Although they sometimes independently solved the same problems, their viewpoints and much of their work are complementary. Malécot is a theorists's theorist. Kimura is a biologists's theorist. Whereas Malécot has been interested in the systematic development of mathematical theory, Kimura has constantly had his eye out for biologically interesting problems. He solved problems, but he also formulated them. His role in our understanding of evolution is seminal, and would be even without his contribution to molecular evolution. The neutral theory is icing on an already delicious cake.

This book provides a chance to read Kimura's greatest papers, and to admire his range and depth, his ingenuity in problem solving, and his originality and daring in forming hypotheses. Takahata has performed a most welcome service not only by setting the background for the papers reprinted here, but also by pointing out their relation to other work and to subsequent developments.

James F. Crow

# EDITOR'S PREFACE

The topics that Motoo Kimura has treated during the past forty years are diverse. Except for a few topics such as those in evolutionary stable strategies and applied quantitative genetics, he has indeed covered most that have been raised in modern evolutionary biology. He heralded a rebirth of the mathematical theory of population genetics, opened up the whole topic of retrospective properties of diffusion models, and revolutionized the way we think about molecular evolution.

The neutral theory of molecular evolution, for which Kimura is best known, is simple, intuitively understandable, and testable. Nevertheless, to understand the concept correctly, it is necessary to understand the underlying logic in quantitative terms. It is also important to recognize what the theory has actually demanded. The theory has bifurcated molecular evolution and phenotypic evolution, but the molecular basis of phenotypic evolution has been left unsolved. In the last decade, molecular evolutionary study has become popular without much connection to population genetics. This situation is ironic to Kimura, for molecular evolutionary study has been propelled by the neutral theory, which in turn grew on theoretical population genetics. Under these circumstances of population genetics, molecular evolution, and the neutral theory, I believe, there are many things to be learned from Kimura's writing.

When writing this, I first tried to make comments on each paper focusing on the prospective rather than the retrospective value. However, since the same topic sometimes appears repeatedly in several papers, I thought it sensible to group them. This grouping obviously does not follow the chronological order and may result in some inconvenience for readers who use this volume as a reference. Furthermore, such grouping is neither completely unambiguous nor mutually exclusive. Nevertheless, I have taken this compromise of grouping not only to save space but also to emphasize the prospective value of Kimura's work. In my comments, the fifty-seven papers reproduced in this volume are classified into eighteen groups, as in the Contents.

The papers in this volume are facsimiles of reprints, a few of which contain minor, handwritten corrections made by Kimura.

## Acknowledgments

Thanks are due to Drs. C. F. Aquadro, C. J. Basten, J. F. Crow, M. Kimura, J. Klein, T. Nagylaki, M. Nei, T. Ohta, W. B. Provine, Y. Satta, M. Slatkin, H. Tachida, and F. Tajima, who gave helpful advice and valuable comments on early versions of this manuscript. I am especially thankful to Dr. Motoo Kimura for his comments, which minimized my misunderstanding about his work.

Naoyuki Takahata

# PART ONE

# Random Gene Frequency Drift

## Introduction

In the early days when Kimura started to work at the National Institute of Genetics, he presented his preliminary results in the annual report of the Institute. The 1951 annual report contains Kimura's formulas for the moments of the gene frequency distribution under random genetic drift. The drift occurs in every population when gametes unite in no predetermined way in the process of reproduction and plays an important role in shaping the genetic architecture of populations. At that time, he had obtained only the first four eigenvalues, but he made a clear distinction between fixed and unfixed probabilities. For instance, if $p$ is the initial allele frequency and $t$ is time in generations, the probability of fixation of this allele for large $t$ was given as

$$p - 3p(1 - p)\left(1 - \frac{1}{2N_e}\right)^t$$

where $N_e$ is the effective population number. Robertson (1952) also gave the exact formulas for the first four moments.

Soon after coming to the University of Wisconsin, Kimura obtained the complete solution by the diffusion approximation method (no. 1). Wright was sufficiently impressed by this work to communicate it to the Proceedings of the National Academy of Sciences USA. Kimura's figures showing the distribution of the unfixed classes are often reproduced in textbooks. Kimura was the first to obtain the complete solution, but it is interesting that the approximation obtained by Malécot in 1944 for the probability density for large $t$

$$\phi(p,x,t) \approx 6p(1 - p)e^{-t/2Ne}$$

is the same as the leading term in Kimura's solution (see Nagylaki 1989a). Most interesting is the rate of steady decay. The rate is $1/2N_e$ per generation and after $2N_e$ generations the distribution becomes almost flat, as earlier found by Fisher (1922, 1930) and Wright (1931).

It may be noted that the complete set of eigenvalues $-i(i-1)/4N_e$ ($i = 1, 2, \ldots$) of the diffusion equation is the same as that in the coalescent process of Kingman (1982a,b), a death process which describes the family relationships of all the genes at a locus in a randomly mating population. For genealogical processes of neutral genes, see Hudson (1983), Tajima (1983, 1989a), Tavaré (1984), and Watterson (1984), while for those of selected genes, see Takahata (1991a) and references therein.

The complete description of gene frequency changes under the joint effects of random sampling drift, selection, and other evolutionary forces is often very difficult to obtain. Even for drift and selection alone, there are only a few successful cases. Kimura (no. 2) derived the complete solution of the distribution of gene frequency for the case where genic selection and drift interact. The solution is expressed in terms of the Gegenbauer polynomials. The analysis of the smallest eigenvalue that determines the rate of steady decay indicates that selection predominates if $N_e s \geq 1$ where $s$ is the selection coefficient. In the second through the fourth sections, Kimura studied the steady distribution of gene frequency due to random fluctuation of selection intensity with and without mutation and random sampling drift. The main conclusion about fluctuating selection intensity was that the effect is less important in a small population, because selection in any form is ineffective in a relatively monomorphic population. The paper ends up with two notes on cyclic changes in selection intensity and multidimensional diffusion equations. It also includes Malécot's comment on isolation by distance. Throughout, it is clear that Kimura was convinced of the stochastic nature of evolutionary processes.

Kimura no. 3 is an extensive review of stochastic processes in genetics. Of five topics treated, the first two are somewhat different from the rest. These are concerned with random assortment of a replicated genetic entity in cell divisions. When there are multiple copies of a genetic element within a cell, and the replicated elements are distributed randomly into the daughter cells, there will be random segregation of these copies. The formulation assumes that each element (chromosome) replicates exactly but the doubled elements are randomly partitioned. This stochastic process is similar to genetic drift caused by random sampling of gametes and is suitable for describing the stochastic process of organelle genes in multicellular organisms (Birky 1983, 1991; Takahata 1985) and plasmids in bacteria (Levin and Stewart 1976). Applying the formulation to *Paramecium*, which was considered to have a polyploid macronucleus, Kimura, following the work of Sonneborn, assumed that senescence or aging occurs due to random loss of chromosomes of any one type. In the remaining article, the rate of steady decay of the gene frequency distribution for unfixed classes was provided under the joint effects of sampling drift and selection with an arbitrary degree of dominance (See chapter 8 in Crow and Kimura 1970).

A diffusion equation for describing Brownian motion was invented independently by A. Einstein and M. von Smoluchowski. In genetics, Fisher (1922) first investigated random sampling drift (then called the *Hagedoorn effect*) by a diffusion equation similar to that for Brownian motion (Wright 1931). These preceded Kolmogorov's (1931) rigorous mathematical foundation of diffusion processes. Diffusion theory has played a significant role in developing the modern theory of population genetics. A stochastic process arising from reproduction in a finite population interacting with other evolutionary forces is often very difficult to treat by a Markov chain. As remarked by Ewens (1979, p. 140), Kimura (nos. 1, 2, and 3; Kimura 1955, 1956) heralded a rebirth of the mathematical theory of population genetics.

When Kimura was an assistant in Kihara's laboratory in Kyoto University, he was greatly influenced by Wright's 1931 paper (Kimura 1985a) and dreamed of the application of the diffusion theory to significant problems in population genetics. He quickly mastered the theory, recognized its usefulness, and solved many difficult stochastic problems by means of diffusion approximations. Kimura (no. 4) introduced geneticists to the Kolmogorov forward and backward equations in an elementary way and at the same time demonstrated their power. Diffusion equations in population genetics have singularities at the boundaries which require caution. It is surprising to see his intuition and ability in manipulating such delicate equations, even though Feller's (1951, 1952) pioneer work was by then available (see also chapter 15 in Karlin and Taylor 1981). We should, however, mention one error in Kimura no. 4. It occurs in the general formula for obtaining the probability of joint fixation of mutant genes involving two or more loci under epistatic interaction in fitness (this part is omitted, but see p. 430 in Crow and Kimura 1970). Kimura still regrets this, often asking his colleagues if we can get the correct formula or even an approximate answer. So far no one has succeeded (but see Rutledge 1970 for an approximate solution).

# SOLUTION OF A PROCESS OF RANDOM GENETIC DRIFT WITH A CONTINUOUS MODEL*

BY MOTOO KIMURA

DEPARTMENT OF GENETICS,† UNIVERSITY OF WISCONSIN

*Communicated by Sewall Wright, November 15, 1954*

The problem of random genetic drift in finite populations due to random sampling of gametes in reproduction was first treated mathematically by R. A. Fisher,[1] using a differential equation. Fisher's general method was appropriate, but owing to the omission of a term in the equation, his result for the rate of decay of variance was only half large enough. The correct solution for the state of steady decay was first supplied by S. Wright,[2] using the method of path coefficients and an integral equation.

Later Fisher[3] corrected his results and also elaborated the terminal part of the distribution in the statistical equilibrium by his method of functional equations.

In all these works, however, it was assumed that a state of steady decay had been attained, but nothing was known about the complete solution which might show how the process finally leads to the state of steady decay. The present writer,[4] by calculating the moments of the distribution and with the help of the Fokker-Planck equation, obtained a solution which assumed an infinite series under the continuous model, showing that the process approaches asymptotically the state of steady decay. At that time, however, only the first few coefficients in the terms of the series could be determined. Pursuing the pıoblem further, he arrived at the complete solution, which will be reported here. After obtaining these results, the writeı recently discovered the work of S. Goldberg.[5] In his unpublished thesis Goldberg solved the diffusion equation for the gene-frequency distribution in a finite population when recurrent mutations occur. His solutions have a direct connection with the frequency distribution of unfixed classes in the case of pure random genetic drift, as will be seen below.

Consider a random mating population of $N$ diploid parents. Let $A$ and $A'$ be a pair of alleles with frequencies $x$ and $1 - x$, respectively. In order to single out the effect of random drift, we shall assume an idealized situation in which selection, migration, and mutation are absent and generations do not overlap. The process of the change in gene frequencies is most adequately described by giving the frequency distribution $f(x, t)$ at the $t$th generation, where $x$ takes on a series of discrete values: $0, 1/2N, 2/2N, \ldots, 1 - 1/2N, 1$. For fairly large $N$, however, $x$ can be treated as a continuous variable without serious error.

First, we shall derive the moment formula which is useful heuristically. Let $x_t$ be the gene frequency in the $t$th generation, and let $\delta x_t$ be the amount of change due to random sampling of gametes per generation, such that

$$x_{t+1} = x_t + \delta x_t. \tag{1}$$

Let $\mu'^{(t+1)}_n = E(x^n_{t+1})$ be the $n$th moment of the distribution about 0 in the $(t + 1)$th generation. We write the expectation of $x^n_{t+1}$ in terms of $(x_t + \delta x_t)^n$ in two steps: (1) taking expectation for the random change $\delta x_t$, which will be denoted by $E_\delta$, and (2) taking the expectation for the existing distribution, which will be denoted by $E_\phi$.

Noting that $E_\delta(\delta x_t) = 0$, $E_\delta(\delta x_t)^2 = x_t(1 - x_t)/2N$, etc.,

$$\begin{aligned}\mu'^{(t+1)}_n &= E(x_t + \delta x_t)^n \\ &= E_\phi\{x^n_t + \tbinom{n}{1}x^{n-1}_t E_\delta(\delta x) + \tbinom{n}{2}x^{n-2}_t E_\delta(\delta x_t)^2 + \ldots\} \\ &= E_\phi\left\{x^n_t + \frac{n(n-1)}{2}x^{n-2}_t\frac{x_t(1-x_t)}{2N} + O\left(\frac{1}{N^2}\right)\right\}.\end{aligned} \tag{2}$$

The intrinsic assumption in the continuous model is that the effective size $N$ is sufficiently large so that terms of order $1/N^2$ and higher can be omitted without serious error.

Thus

$$\mu'^{(t+1)}_n = \left\{1 - \frac{n(n-1)}{4N}\right\}\mu'^{(t)}_n + \frac{n(n-1)}{4N}\mu'^{(t)}_{n-1}.$$

For large $N$, the moments change very slowly per generation, and we can replace the above equation by the system of differential equations:

$$\frac{d\mu'^{(t)}_n}{dt} = -\frac{n(n-1)}{4N}\{\mu'^{(t)}_n - \mu'^{(t)}_{n-1}\} \quad (n = 1, 2, 3, \ldots). \tag{3}$$

If the population starts from the gene frequency $p$ $(0 < p < 1)$, $\mu'^{(0)}_n = p^n$, and we can obtain the $n$th moment as the solution of (3),

$$\mu'^{(t)}_n = p + \sum_{i=1}^{\infty}(2i+1)pq(-1)^i F(1-i, i+2, 2, p) \times \frac{(n-1)\ldots(n-i)}{(n+1)\ldots(n+i)}e^{-[i(i+1)/4N]t}, \tag{4}$$

where $F(1 - i, i + 2, 2, p)$ is the hypergeometric function and $q = 1 - p$. For finite $n$ the series is finite. Putting $n = 1, 2, 3$, and 4, it will be seen that, for large $N$, the resulting formulas give a very good approximation to the exact moment formulas obtained by A. Robertson[6] (p. 205).

The probability $f(1, t)$ of the gene $A$ being fixed in the population by the $t$th generation can be obtained by using the relation

$$f(1, t) = \lim_{n \to \infty} \sum_{x=0}^{1} x^n f(x, t) = \lim_{n \to \infty} \mu'^{(t)}_n.$$

The resulting series,

$$f(1, t) = p + \sum_{i=1}^{\infty} (2i + 1)pq(-1)^i F(1 - i, i + 2, 2, p)e^{-[i(i+1)/4N]t}, \tag{5}$$

is now an infinite series whose convergence must be examined. It is convenient here to introduce the Gegenbauer polynomial $T^1_{i-1}(z)$ which is related to the hypergeometric function by

$$T^1_{i-1}(z) = \frac{i(i+1)}{2} F\left(i + 2, 1 - i, 2, \frac{1 - z}{2}\right).$$

The properties of this function have been thoroughly studied (see Morse and Feshbach[7] [pp. 782–783]). Using this relation and putting $p = (1 - r)/2$, $(-1 < r < 1)$, we obtain

$$f(1, t) = p + \sum_{i=1}^{\infty} (-1)^i \frac{(2i + 1)}{2i(i + 1)} (1 - r^2)\, T^1_{i-1}(r)e^{-[i(i+1)/4N]t}, \tag{6}$$

where $T_0^1(r) = 1$, $T_1^1(r) = 3r$, $T_2^1(r) = {}^3/_2(5r^2 - 1)$, $T_3^1(r) = {}^5/_2(7r^3 - 3r)$, etc. Here, if we use the recurrence relation: $(2i + 1)(1 - r^2)T^1_{i-1}(r) = i(i + 1)P_{i-1}(r) - i(i + 1)P_{i+1}(r)$, the above formula becomes

$$f(1, t) = p + \sum_{i=1}^{\infty} \frac{(-1)^i}{2} \{P_{i-1}(r) - P_{i+1}(r)\}e^{-[i(i+1)/4N]t}, \tag{7}$$

where $P_n(r)$ represents a Legendre polynomial: $P_0 = 1$, $P_1 = r$, $P_2 = {}^1/_2(3r^2 - 1)$, $P_3 = {}^1/_2(5r^3 - 3r)$, etc. For $t = 0$, the partial sum of the first $n$ terms of equation (7) is $(-1)^{n-1}(P_{n-1} - P_n)/2$ $(n \geq 3)$. By using a proper integral expression (see later part), it can easily be shown that, if $-1 < r < 1$, the partial sum tends to zero as $n$ goes to infinity. For $t > 0$, the series (7) is uniformly convergent and tends to $p$ as $t \to \infty$. The probability of the gene $A'$ being fixed in the population by the $t$th generation $f(0, t)$ can be obtained by replacing $p$ with $q$ and $r$ with $-r$. If we note that $P_n(-r) = (-1)^n P_n(r)$, the frequency of the fixed classes is seen to be

$$f(1, t) + f(0, t) = 1 - \sum_{j=0}^{\infty} \{P_{2j}(r) - P_{2j+2}(r)\}\ e^{-[(2j+1)(2j+2)/4N]t}, \tag{8}$$

which is 0 when $t = 0$ and tends to 1 as $t \to \infty$.

Let us now consider the probability distribution of unfixed classes. Let $\phi(x, t)$ be the probability density that the gene frequency in the $t$th generation is between $x$ and $x + dx$ $(0 < x < 1)$. It has been shown that, under the assumption of the continuous model, $\phi(x, t)$ satisfies the following partial differential equation:

$$\frac{\partial \phi}{\partial t} = \frac{1}{4N} \frac{\partial^2}{\partial X^2} \{x(1 - x)\phi\}, \tag{9}$$

which is a Fokker-Planck equation for the case of the random drift.[4, 8] This equation has singularities at the boundaries, and no arbitrary conditions can be imposed there.[9] But the moment formula which can be obtained from equations (4) and (5) by calculating,

$$\mu'^{(t)}_n - 1^n f(1, t) = \int_0^1 x^n \phi(x, t)\, dx, \tag{10}$$

suggests that equation (9) must have the solution of the form

$$\sum_{i=1}^{\infty} C_i X_i(x) e^{-[i(i+1)/4N]t},$$

where $C_i$ are constants and $X_i(x)$ are functions of $x$ only. In order to solve equation (9), if we put $\phi \propto X_i(x) \exp\{-i(i + 1)t/4N\}$ $(i = 1, 2, 3, \ldots)$, we obtain the hypergeometric equation

$$x(1 - x)\frac{d^2X_i}{dx^2} + 2(1 - 2x)\frac{dX_i}{dx} - (1 - i)(i + 2)X_i = 0,$$

or, putting $x = (1 - z)/2$ such that $z = 1 - 2x$ $(-1 < z < 1)$, we obtain the Gegenbauer equation:

$$(z^2 - 1)\frac{d^2X_i}{dz^2} + 4z\frac{dX_i}{dz} - (i - 1)(i + 2)X_i = 0. \tag{11}$$

From the comparison of the results obtained from equation (10), it can be found that a Gegenbauer polynomial $X_i = T^1_{i-1}(z)$ is a pertinent solution.

Thus

$$\phi(x, t) = \sum_{i=1}^{\infty} C_i T^1_{i-1}(z) e^{-[i(i+1)/4N]t} \tag{12}$$

The coefficients $C_i$ can be determined from the initial condition that the population starts from the gene frequency $p$. Mathematically,

$$\delta(x - p) = \sum_{i=1}^{\infty} C_i T^1_{i-1}(z), \tag{13}$$

where $\delta(x)$ represents the delta function. Multiplying $(1 - z^2)T^1_{i-1}(z)$ on both sides of equation (13) and using the orthogonal property,

$$\int_{-1}^{1} (1 - z^2) T_m{}^1(z) T^1_{i-1}(z)\, dz = \delta_{m,\, i-1} \frac{2(i + 1)i}{(2i + 1)}, \tag{14}$$

where $m$ in Kronecker's notation represents zero or a positive integer, we obtain

$$2\{1 - (1 - 2p)^2\} T^1_{i-1}(1 - 2p) = C_i \frac{2(i + 1)i}{(2i + 1)}$$

or

$$C_i = 4pq \frac{(2i + 1)}{i(i + 1)} T^1_{i-1}(1 - 2p).$$

Thus the formal solution is

$$\phi(x, t) = \sum_{i=1}^{\infty} \frac{(2i+1)(1-r^2)}{i(i+1)} T_{i-1}^1(r) T_{i-1}^1(z) e^{-[i(i+1)/4N]t} \quad (15)$$

or, in terms of the hypergeometric function,

$$\phi(x, t) = \sum_{i=1}^{\infty} pqi(i+1)(2i+1)F(1-i, i+2, 2, p) \times F(1-i, i+2, 2, x)e^{-[i(i+1)/4N]t}. \quad (15)'$$

For $t > 0$, the series is uniformly convergent for $x$ and $p$, since the exponential term approachs zero rapidly. It is interesting that this solution agrees with the "absorbing barrier solution" of Goldberg[5] by putting $\alpha = \beta = 0$ in his formula.

The probability that both $A$ and $A'$ coexist in the population in the $t$th generation ($\Omega_t$) is easily obtained from equation (15), by noting that $dP_i(z)/dz = T_{i-1}^1(z)$ and $P_n(1) = 1$:

$$\Omega_t = \int_0^1 \phi(x, t)\, dx = \int_{-1}^1 \phi(x, t) \frac{dz}{2} = \sum_{m=1}^{\infty} \frac{(4m-1)(1-r^2)}{(2m-1)2m} T_{2m-2}^1(r)\, e^{-[(2m-1)2m/4N]t}. \quad (16)$$

For $t > 0$, the series is easily seen to be convergent, and, as $t \to \infty$, $\Omega_t$ goes to zero. For $t = 0$, we must show that this series converges to 1. Let $\Omega_{0,n}$ be a partial sum of the first $n$ terms; then, by the recurrence relation $(4m-1)(1-r^2)T_{2m-2}^1(r)/(2m-1)2m = P_{2m-2}(r) - P_{2m}(r)$, we have $\Omega_{0,n} = 1 - P_{2n}(r)$. By using an integral expression of $P_n$, i.e., $P_n(z) = (1/\pi)\int_0^\pi \{z + \sqrt{z^2-1}\cos t\}^n dt$, we can show that, for $|r| < 1$ $P_{2n}(r) \to 0$ as $n \to \infty$. For

$$|P_{2n}(r)| \le \frac{1}{\pi}\int_0^\pi |r + \sqrt{r^2-1}\cos t|^{2n} di = \frac{1}{\pi}\int_0^\pi \{r^2 + (1-r^2)\cos^2 t\}^n dt \to 0 \quad (n \to \infty).$$

Furthermore, from (16),

$$\Omega_t = \sum_{j=0}^{\infty} \{P_{2j}(r) - P_{2j+2}(r)\} e^{-[(2j+1)(2j+2)/4N]t}. \quad (17)$$

Consequently, it can be seen that, from equations (8) and (17), $f(1, t) + \Omega_t + f(0, t) = 1$, as it should be.

The processes of the change in the distribution of the unfixed classes when the population starts from $p = 0.5$ and $p = 0.1$ are illustrated in Figures 1 and 2, respectively. In Figure 1 it will be seen that after $2N$ generations the distribution curve becomes almost flat, and the genes are still unfixed in about 50 per cent of the cases. In Figure 2, the initial gene frequency is assumed to be 10 per cent, and it takes $4N$ or $5N$ generations before the distribution curve becomes practically flat. By that time, however, the genes are fixed in more than 90 per cent of the cases, and the simplest asymptotic formula $Ce^{-(1/2N)t}$ may not be useful as in the case of $p = 0.5$.

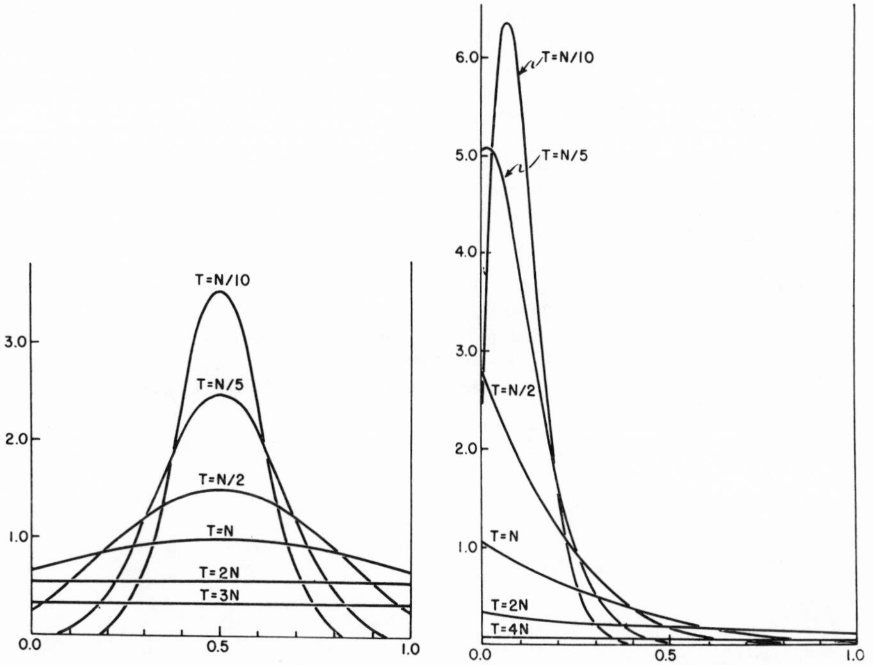


FIGS. 1–2.—The processes of the change in the probability distribution of heterallelic classes, due to random sampling of gametes in reproduction. It is assumed that the population starts from the gene frequency 0.5 in Fig. 1 (left) and 0.1 in Fig. 2 (right). $t$ = time in generation; $N$ = effective size of the population; abscissa is gene frequency; ordinate is probability density.

The probability of heterozygosis is calculated by equation (15):

$$H_t = \int_0^1 2x(1 - x)\phi(x, t)\, dx = \sum_{i=1}^{\infty} pq \frac{(2i + 1)}{i(i + 1)} T_{i-1}^{1} (1 - 2p) \times \int_{-1}^{1} (1 - z^2) T_{i-1}^{1}(z)\, e^{-[i(i+1)/4N]t}\, dz.$$

By virtue of equation (14) (put $m = 0$), the last integral is 0 except for $i = 1$. Hence

$$H_t = pq \cdot \frac{3}{2} \cdot 1 \cdot \frac{4}{3} \cdot e^{-(1/2N)t} = 2pqe^{-(1/2N)t} = H_0 e^{-(1/2N)t}, \tag{18}$$

showing that the heterozygosis decreases exactly at the rate of $1/(2N)$ per generation. This is readily confirmed by a simple calculation: Let $p$ be the frequency of $A$ in the population, where the frequency of the heterozygotes is $2p(1 - p)$. The amount of heterozygosis to be expected after one generation of random sampling of the gametes is

$$E\{2(p + \delta p)(1 - p - \delta p)\} = 2p(1 - p) - 2E(\delta p)^2 = 2p(1 - p) - 2\frac{p(1 - p)}{2N} = \left(1 - \frac{1}{2N}\right) 2p(1 - p),$$

as was to be shown.

The author expresses his appreciation to Dr. James F. Crow for valuable help during the course of this work. Thanks are also due to Dr. E. R. Immel for his helpful suggestions.

* Contribution No. 84 of the National Institute of Genetics, Mishima-shi, Japan.

† Contribution No. 570. This work was supported by a grant from the University Research Committee from funds supplied by the Wisconsin Alumni Research Foundation.

[1] R. A. Fisher, *Proc. Roy. Soc. Edinburgh,* **42,** 321–341, 1922.

[2] S. Wright, *Genetics,* **16,** 97–159, 1931.

[3] R. A. Fisher, *The Genetical Theory of Natural Selection* (Oxford, 1930).

[4] M. Kimura, *Genetics,* **39,** 280–295, 1954.

[5] S. Goldberg, Ph.D. thesis, Cornell University, 1950.

[6] A. Robertson, *Genetics,* **37,** 189–207, 1952.

[7] M. Morse and H. Feshbach, *Methods of Theoretical Physics* (New York, 1953).

[8] S. Wright, these PROCEEDINGS, **31,** 382–389, 1945.

[9] W. Feller, *Proc. Second Berkeley Symposium on Math. Stat. and Prob., Univ. of California,* pp. 227–246, 1951.

# PAPER TWO

# STOCHASTIC PROCESSES AND DISTRIBUTION OF GENE FREQUENCIES UNDER NATURAL SELECTION[1]

MOTOO KIMURA

University of Wisconsin, Madison, Wisconsin

## INTRODUCTION

Evolution is a stochastic process of change in gene frequencies in natural populations. Since the populations making up a species consist of many individuals and since evolution extends over enormous periods of time, laws which govern the process of change are inevitably "statistical." In this sense the genetical theory of evolution, as R. A. Fisher (1922) suggests, is comparable to the theory of gases. This analogy can be pushed further: Instead of considering populations as aggregates of genes, we find it more convenient to consider populations as aggregates of gene frequencies (or ratios). This is similar to the situation in physics where the specification of the population of velocities is sometimes more useful than that of a population of particles (Fisher, 1953). As far as I know, this fruitful idea was first incorporated into the theory of population genetics by Fisher in his 1922 paper, which led to a later elaboration (Fisher, 1930a).

The deductive theory of genetics of natural populations has been greatly advanced since then by S. Wright in his numerous papers starting in 1931. The problem of steady state distribution of gene frequencies has been solved under more and more general conditions. Out of his investigations he has proposed a new thesis that the most favorable condition for rapid evolution is the subdivision of the population into numerous partially isolated local groups. The essential idea is that this population structure provides stochastic differentiation of local groups which is the basis of intergroup selection. This view, though it has been accepted by many evolutionists (Dobzhansky, 1951; Haldane, 1949b; Muller, 1949) is a controversial one, having been criticized by Fisher's school (see for example Sheppard, 1954). One of the criticisms came from the study of the isolated population of the moth *Panaxia dominula* by Fisher and Ford (1947). They generalized their findings on the *medionigra* gene to the statement that natural populations in general are affected by selective action varying from time to time in direction and intensity, and of sufficient magnitude to cause fluctuating variation in all gene-ratios. Thus they (Fisher and Ford, 1950) consider that the claim for ascribing a special evolutionary advantage to small isolated communities due to fluctuations in gene ratios had better be dropped.

The controversy on this problem (Fisher and Ford, 1947, 1950; Wright, 1948, 1951a) is an intriguing one and I have started to study the logical consequences of the effect of random fluctuation of selection intensities, some results of which have already been published (Kimura, 1954). In the present paper I shall first consider two stochastic processes: First, the process of genic selection under random drift due to small population number and second, the process of change in gene frequency due to random fluctuation in selection intensity. In the latter case, we consider a very large population. We then see the effect of random fluctuations of selection intensity on the probability of fixation of an advantageous mutant gene.

The main part of this paper will be devoted to the study of steady state distributions of gene frequencies under the joint effect of systematic (mutation, migration and selection) and random fluctuation effects (random sampling of gametes in finite populations and random fluctuation of selection intensity).

The final part concerns the general equation for the stochastic process of change in gene frequencies, taking account of an arbitrary number of alleles as well as of loci. Though the treatment of intergroup selection is beyond the scope of the present paper, one method of attacking this problem will be suggested.

## THE PARTIAL DIFFERENTIAL EQUATION METHOD

Usually evolution is a slow process in our time scale so that the rate of change in gene frequency per generation is small. This fact enables us to treat the process as a continuous stochastic process[2] with good approximation. Also we introduce the assumption that the probability distribution of gene frequencies in the next generation depends on the gene frequency of the present generation but not on the history in the previous generations which have led to the present gene frequency. For the study of this continuous Markov process one of the most powerful methods uses the Kolmogorov equation (Kolmogorov, 1931). For our purpose, the

[1] Paper No. 600 from the Department of Genetics, University of Wisconsin. Also Contribution No. 114 of the National Institute of Genetics, Mishima-shi, Japan.

[2] We call the stochastic process $\{x_t\}$ continuous if the random variable $x_t$ changes in such way that for any given $\delta > 0$, the probability that $x_t$ changes more than $\delta$ for any given time interval $(t, t + \Delta t)$ is $o(\Delta t)$, that is, of the smaller order of magnitude than $\Delta t$.

equation will be written in the following form, assuming a pair of alleles at a single locus.

$$\frac{\partial\phi(x, p; t)}{\partial t} = \tfrac{1}{2}\frac{\partial^2}{\partial x^2}\{V_{\delta x}\phi(x, p; t)\} - \frac{\partial}{\partial x}\{M_{\delta x}\phi(x, p; t)\} \quad (1)$$

In this equation x is the gene frequency at the $t^{th}$ generation and the function $\phi$ (x, p; t) denotes the density of the conditional probability that the frequency lies between x and $x + dx$ at the $t^{th}$ generation given that the initial gene frequency was p at $t = 0$. $\delta x$ denotes the rate of change of the gene frequency per generation and $V_{\delta x}$ and $M_{\delta x}$ respectively denote the variance and the mean of $\delta x$.

In many mathematical writings the term corresponding to $V_{\delta x}$ is defined by means of the second moment of change in the random variable per infinitesimal time interval (see for example Feller, 1951a). Mathematically, the definition through the variance and the one through the second moment should be the same. However, for our purpose in population genetics, the variance definition has definite advantages. The reason for this together with a simple derivation of (1) is given in Appendix 1. This type of partial differential equation is also called the Fokker-Planck equation by physicists and seems to have been known for a long time (*cf.* Chandrasekhar, 1943). In the field of population genetics, we owe the application of this equation to S. Wright, who introduced it in 1945 (Wright, 1945), though the simplest type of heat diffusion equation had been used already by Fisher (1922). The equation (1), when applied to concrete genetical problems, almost always has singularities at the boundaries. This fact has attracted the attention of some mathematicians and the nature of the equation has been studied for the case where $M_{\delta x}$ is linear in x, that is, in the case of mutation (Feller, 1951a; Goldberg, 1950).

With selection, $M_{\delta x}$ is at least quadratic and the general solution may be expected to be very difficult. Fortunately, in the simplest case of genic selection with random drift, the exact solution has been obtained, as will be shown in the next section.

## Genic Selection with Random Drift

We consider a randomly mating population and let N be the effective number of reproducing individuals.[3] We assume a pair of alleles $A_1$ and $A_2$ with frequencies x and 1—x. Selection intensity or fitness is measured by the Malthusian parameter, that is to say the rate of geometric growth, throughout this paper. Let $m_{11}$, $m_{12}$ and $m_{22}$ be the fitness of the three genotypes $A_1A_1$, $A_1A_2$ and $A_2A_2$:

| *Genotypes* | *Frequencies* | *Fitness* |
|---|---|---|
| $A_1A_1$ | $x^2$ | $m_{11}$ |
| $A_1A_2$ | $2x(1-x)$ | $m_{12}$ |
| $A_2A_2$ | $(1-x)^2$ | $m_{22}$ |

With no dominance, $m_{11}-m_{12} = m_{12}-m_{22}$ which we shall denote by s. In this case s is the average excess in fitness associated with the gene $A_1$ substitution (Fisher, 1930b, 1941) and for small s it is approximately equal to the selection coefficient of $A_1$ in the usual sense.

With this definition, the rate of change in the frequency of $A_1$ becomes exactly

$$\frac{dx}{dt} = sx(1-x). \quad (2)$$

Thus under the assumption of slow change, we can write the amount of change per generation as

$$\delta x = sx(1-x) + \delta' x \quad (3)$$

where $\delta' x$ is the amount of change due to random sampling of gametes per generation. In this section we assume that s is constant. Then

$$M_{\delta x} = sx(1-x),\ V_{\delta x} = \frac{x(1-x)}{2N},$$

and the partial differential equation (1) becomes

$$\frac{\partial\phi}{\partial t} = \frac{1}{4N}\frac{\partial^2}{\partial x^2}\{x(1-x)\phi\} - s\frac{\partial}{\partial x}\{x(1-x)\phi\}, \quad (0<x<1), \quad (4)$$

which has singularities at the boundaries; $x = 0$ and 1.

Recently the same equation has been used to analyze the gene frequency change in very small experimental populations of *Drosophila melanogaster* (Wright and Kerr, 1954). In that work Wright devised an ingenious method of his own to study the process of steady decay. The present method is different and in some ways more formal.

We seek a solution of the form

$$\phi = e^{2cx}Ve^{-\lambda t}$$

where V is a function of x only and $c = Ns$.

If we substitute this in (4), we have

$$x(1-x)\frac{d^2V}{dx^2} + 2(1-2x)\frac{dV}{dx} - \{2+4c^2x(1-x)-4N\lambda\}V=0 \quad (5)$$

Then by the substitution

$$x = \frac{1-z}{2},$$

(5) becomes

$$(1-z^2)\frac{d^2V}{dz^2} - 4z\frac{dV}{dz} + \{(4N\lambda-2-c^2)+c^2z^2\}V=0 \quad (6)$$

3 We assume that N is constant. For our present purpose, this will not be as drastic an assumption as Feller (1951a) has stressed. One useful model will be to assume that N changes fortuitously from generation to generation with mean $\bar{N}$ and with relatively small variance $V_N$. In this case N should be replaced by $\bar{N} = (V_N/N)$ in the following treatment (*cf.* Kimura, 1955b).

This type of differential equation is known as the oblate spheroidal equation. We want here the solutions which are finite at the singularities, $z = \pm 1$. Such a solution has been studied by Stratton and others (1941) and is expressed in the form:

$$V_{1k}^{(1)}(z) = \sum_{n=0,1}' f_n^k T_n^1(z), \qquad (7)$$

where $k = 0, 1, 2, \ldots$ In this formula the $f_n^k$'s are constants and $T_n^1(z)$'s are the Gegenbauer polynomials:

$$T_0^1(z) = 1, T_1^1(z) = 3z,$$

$$T_2^1(z) = \frac{3}{2}(5z^2-1), T_3^1(z) = \frac{5}{2}(7z^3-3z),$$

$$T_4^1(z) = \frac{15}{8}(21z^4-14z^2+1), \text{etc. In general}$$

$$T_n^1(z) = \frac{(n+1)(n+2)}{2} F(-n, n+3, 2, \frac{1-z}{2})$$

where F denotes the hypergeometric function. The primed summation in (7) is over even values of n if k is even, odd values of n if k is odd.

The desired solution of (4) is given by summing up $V_{1k}^{(1)}(z)$ for all possible values of k, after having multiplied through by $e^{2cx-\lambda_k t}$, where $\lambda_k$ is the $k^{th}$ eigenvalue.

$$\phi(x, p; t) = \sum_{k=0}^{\infty} C_k e^{-\lambda_k t+2cx} V_{1k}^{(1)}(z). \qquad (8)$$

In this formula the $C_k$'s are determined by the initial condition that the population starts from the gene frequency p at the $0^{th}$ generation, that is,

$$\phi(x, p; 0) = \delta(x-p).$$

Since the details will be given elsewhere, only the results are given here.

$$C_k = \frac{(1-r^2)e^{-c(1-r)}V_{1k}^{(1)}(r)}{\sum_{n=0,1}' \frac{(n+1)(n+2)}{(2n+3)}(f_n^k)^2}, \qquad (9)$$

where $r = 1-2p$ and $c = Ns$. The solution (8) with coefficients given by (9) gives the probability distribution of unfixed classes in the $t^{th}$ generation.

As t increases, the exponential terms in (8) decrease in absolute value very rapidly, and for large t only the first few terms are important. The numerical values of the first few eigenvalues $\lambda_0$, $\lambda_1$ and $\lambda_2$ can be obtained from the tables of the separation constants $(B_{1,k})$ which are listed in the book of Stratton *et al.* (1941). This is done by using the relation

$$4N\lambda_k = c^2 - B_{1,k}$$

Among them, the smallest eigenvalue $\lambda_0$ gives the final rate of decay of the frequency distribution curve (for unfixed classes); that is,

$$\lim_{t\to\infty} \frac{1}{\phi}\frac{d\phi}{dt} = -\lambda_0,$$

and has special importance. It is interesting to note that $\lambda_0$ is equivalent to Wright's K (Wright and Kerr, 1954).

For small values of c, the eigenvalues can be expanded into a power series in c. Using the method of continued fraction expansion due to A. H. Wilson (1928; see also Bateman, 1944, pp. 442-443), we obtain the following formula for $\lambda_0$:

$$4N\lambda_0 = 2 + \frac{2^2}{5}c^2 - \frac{2^2}{5^3\cdot 7}c^4 - \frac{2^3}{3\cdot 5^5\cdot 7}c^6 - \frac{2^2\cdot 31}{5^6\cdot 7^3\cdot 11}c^8 - \frac{17507389}{2\cdot 3^4\cdot 5^9\cdot 7^4\cdot 11\cdot 13}c^{10} -$$

or in terms of $2c = 2Ns$, $2N\lambda_0 = 2NK$ is given by

$$2N\lambda_0 = 1 + \frac{(2Ns)^2}{10} - \frac{(2Ns)^4}{7{,}000} - \frac{(2Ns)^6}{1{,}050{,}000} - 4.108\times 10^{-9}(2Ns)^8 - 7.869\times 10^{-11}(2Ns)^{10} - \cdots (10)$$

It is impressive to see that Wright's empirical formula (Wright and Kerr, 1954, p. 236) is exact up to the coefficients of the order of $10^{-9}$. For the purpose of numerical calculation the above power series is not suitable except for small values of c. According to Stratton *et al.* (*loc. cit*), convergence of the power series is usually slow and it proves unsatisfactory for values of $c^2 > 1$. However in the present case ($m = 1$, $l = 0$ in the spheroidal function), (10) gives the right answer to 3 significant figures even for $c = 3$. This agrees with the statement of Wright and Kerr (1954). Figure 1 gives the relation between c and $2N\lambda_0$. More exact figures are listed in Table 1.

The eigenfunctions $V_{1k}^{(1)}(z)$ corresponding to $\lambda_k$'s are given by (7). The coefficients $f_n^k$ corresponding to the first three eigenvalues are found in

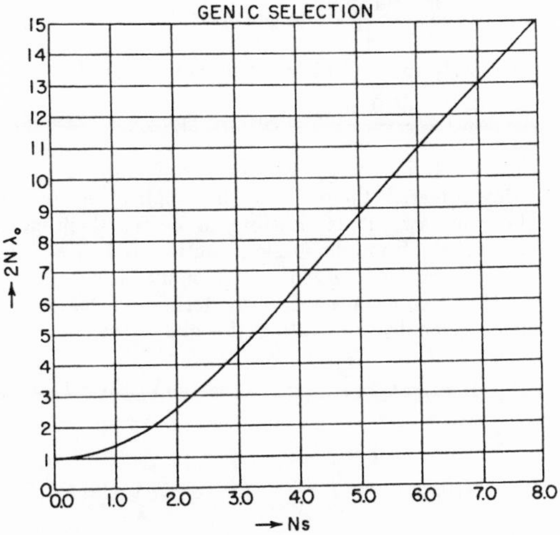


FIGURE 1. Relation between rate of steady decay ($\lambda_0$) and intensity of selection (s) in the process of genic selection in finite populations. N is effective size of population.

TABLE 1. RELATION BETWEEN c (= Ns) AND $2N\lambda_0$

The main part (c= 0.0 ~ 5.0) of this table is made from the table of separation constants ($B_{1,k}$) of oblate spheroidal function by Stratton *et al.* (1941), using the relation $4N\lambda_0 = c^2 - B_{1,0}$. The remaining part is calculated from the unpublished oblate eigen values $t_{11}(g)$ kindly supplied by Dr. P. M. Morse of the Massachusetts Institute of Technology.

| c | $2N\lambda_0$ | c | $2N\lambda_0$ |
|---|---|---|---|
| 0.0 | 1.00000 | 3.5 | 5.431835 |
| 0.5 | 1.099855 | 4.0 | 6.54540 |
| 1.0 | 1.397655 | 4.5 | 7.661215 |
| 1.5 | 1.887710 | 5.0 | 8.753305 |
| 2.0 | 2.559275 | 6.0 | 10.857286 |
| 2.5 | 3.39445 | 7.0 | 12.899831 |
| 3.0 | 4.36529 | 8.0 | 14.919894 |

the tables of Stratton *et al.* (pp. 116, 118 and 120). It will be noted here that for c = 0, all the formulae given above reduce to the ones for the case of pure random drift (*cf.* Kimura, 1955a).

The eigenfunction $V_{10}^{(1)}(z)$ corresponding to the smallest eigenvalue $\lambda_0$ is of special significance, since it gives the frequency distribution of unfixed classes at the state of steady decay, when it is multiplied by $e^{c(1-z)}$. It is expressed by

$$V_{10}(z) = f_0{}^0T_0{}^1(z) + f_2{}^0T_2{}^1(z) + f_4{}^0T_4{}^1(z) + \ldots \quad (11)$$

Figure 2 illustrates the distribution of unfixed classes for some values of c. The area under each curve is adjusted so that it is unity. The case c = 1.7 corresponds to the case experimentally studied by Wright and Kerr (*loc. cit*) and the present result agrees quite well with theirs (including the rate of decay).

It is also possible to calculate the rate of fixation at each terminal class: If we denote by f(1, p, t) the probability that $A_1$ has become fixed by the t^th^ generation, then the rate of fixation of $A_1$ is

$$\frac{df(1, p, t)}{dt} = \phi(1, p; t)/(4N). \quad (12)$$

Similarly the rate of loss (or fixation of $A_2$) is

$$\frac{df(0, p, t)}{dt} = \phi(0, p; t)/(4N). \quad (13)$$

These relations have been used by Wright (*loc. cit*) who inferred them from the application of the Poisson law. It is possible to derive them more rigorously from the basic equation (4) itself and the derivation is given in Appendix II.

The frequencies of both terminal classes are then given by integrating the above formulae:

$$f(1, p, t) = e^{2c} \sum_{k=0}^{\infty} \frac{C_k}{4N\lambda_k} (1 - e^{-\lambda_k t}) V_{1k}^{(1)}(-1), \quad (14)$$

and

$$f(0, p, t) = \sum_{k=0}^{\infty} \frac{C_k}{4N\lambda_k} (1 - e^{-\lambda_k t}) V_{1k}^{(1)}(1) \quad (15)$$

where

$$V_{1k}^{(1)}(-1) = \frac{1}{2} \sum_{n=0,1}' (-1)^n (n+1)(n+2) f_n{}^k,$$

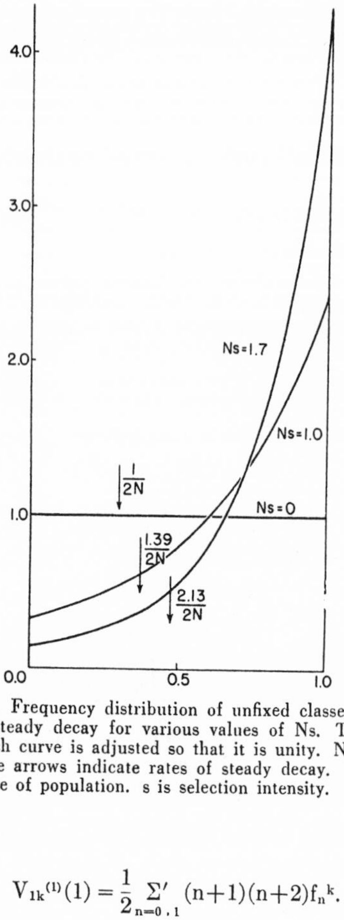


FIGURE 2. Frequency distribution of unfixed classes at the state of steady decay for various values of Ns. The area under each curve is adjusted so that it is unity. Numerals beside the arrows indicate rates of steady decay. N is effective size of population. s is selection intensity.

and

$$V_{1k}^{(1)}(1) = \frac{1}{2} \sum_{n=0,1}' (n+1)(n+2) f_n{}^k.$$

Note that

$$V_{1k}^{(1)}(-1) = V_{1k}^{(1)}(1) \text{ for } k = 0, 2, 4, \ldots$$

RANDOM FLUCTUATION OF SELECTION INTENSITY

We assume genic selection as before but consider a random mating population so large that the effect of random sampling of gametes is negligible. Then

$$\delta x = sx(1-x).$$

We now assume that s is a random variable and fluctuates from generation to generation with mean $\bar{s}$ and variance $V_s$. Therefore

$$M_{\delta x} = \bar{s}x(1-x), \; V_{\delta x} = V_s x^2 (1-x)^2$$

and the differential equation (1) becomes

$$\frac{\partial \phi}{\partial t} = \frac{V_s}{2} \frac{\partial^2}{\partial x^2} \{x^2(1-x)^2 \phi\} - \bar{s} \frac{\partial}{\partial x} \{x(1-x)\phi\}, \; (0<x<1). \quad (16)$$

The transformations

$$\begin{cases} \phi = 2x^{\frac{1+s_1}{2}-2}(1-x)^{\frac{1+s_1}{2}-2}e^{-\lambda t_1}U \\ x = \frac{1}{2}\{1+\tanh(\xi/2)\} \end{cases}$$

where $t_1 = (tV_s)/2$, $s_1 = (2\bar{s})/V_s$, reduce (16) to

$$\frac{d^2U}{d\xi^2} + \left\{\lambda - \frac{1+s_1^2}{4} - \frac{s_1}{2}\tanh\left(\frac{\xi}{2}\right)\right\}U = 0, \quad (-\infty < \xi < \infty). \qquad (17)$$

This has the following two independent solutions denoted by $U_+$ and $U_-$:

$$U_+ = \left(\frac{e^\xi}{1+e^\xi}\right)^a \left(\frac{1}{1+e^\xi}\right)^b \times F\left(a+b, a+b+1, 1+2a, \frac{e^\xi}{1+e^\xi}\right)$$

$$U_- = \left(\frac{e^\xi}{1+e^\xi}\right)^{-a} \left(\frac{1}{1+e^\xi}\right)^b \times F\left(-a+b, a+b+1, 1-2a, \frac{e^\xi}{1+e^\xi}\right)$$

where

$$a = \sqrt{\left(\frac{1-s_1}{2}\right)^2 - \lambda} \text{ and } b = \sqrt{\left(\frac{1+s_1}{2}\right)^2 - \lambda}.$$

Unfortunately these solutions are too complicated for obtaining the required solution of (16) for arbitrary values of $V_s$ and $\bar{s}$.

However a few special cases deserve mention. For $\bar{s} = 0$ (16) can be solved exactly. This is a case where the gene is neutral in the sense of long term average. The rather interesting behavior of the process of change in gene frequency is given in a previous paper (Kimura, 1954). The solution in this case is

$$\phi(x, p; t) = \frac{1}{\sqrt{2\pi V_s t}} \times \exp\left\{-\frac{V_s}{8}t - \frac{\left[\log\frac{x(1-p)}{(1-x)p}\right]^2}{2V_s t}\right\}\frac{[p(1-p)]^{1/2}}{[x(1-x)]^{3/2}}, \qquad (18)$$

where p is the initial gene frequency. As time elapses the gene frequency shifts toward either terminal of the distribution ($x = 0$ or $1$) indefinitely and accumulates in the neighborhood just short of fixation or loss but never becomes fixed or lost completely (quasifixation, quasiloss).

On the other hand if $V_s$ is much smaller than $\bar{s}$ we have a situation very near to the deterministic process of genic selection in an infinite population. The approximate solution is given by

$$\phi(x, p; t) = \frac{1}{\sqrt{2\pi V_s t}} \times \exp\left\{-\frac{\left[\log\frac{x(1-p)}{p(1-x)} - \bar{s}t\right]^2}{2V_s t}\right\}\frac{1}{x(1-x)}. \qquad (19)$$

This is equivalent to one dimensional random walk with definite drift toward the right (if $\bar{s}>0$ or the left (if $\bar{s}<0$) if we measure the gene frequency by the logit scale, that is, $\xi = \log[x/(1-x)]$ (*cf.* Fisher, 1930b). This result will be very useful in practice since it can be applied to any case where the order of $\sigma_s$ ($=\sqrt{V_s}$) does not exceed that of $|\bar{s}|$ and also $|\bar{s}|$ is small. Figure 3 shows the process of change on the logit scale when $\bar{s} = 0.1$ and $V_s = 0.0025$ ($\sigma_s = 0.05$).

Based on this solution (19), we can prove that

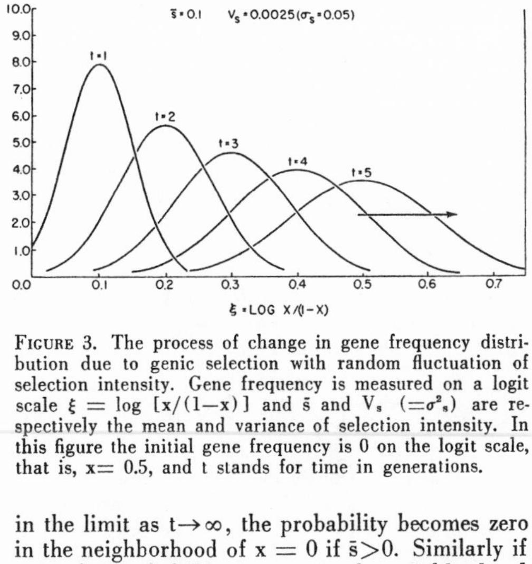


FIGURE 3. The process of change in gene frequency distribution due to genic selection with random fluctuation of selection intensity. Gene frequency is measured on a logit scale $\xi = \log[x/(1-x)]$ and $\bar{s}$ and $V_s$ ($=\sigma^2_s$) are respectively the mean and variance of selection intensity. In this figure the initial gene frequency is 0 on the logit scale, that is, $x = 0.5$, and t stands for time in generations.

in the limit as $t \to \infty$, the probability becomes zero in the neighborhood of $x = 0$ if $\bar{s}>0$. Similarly if $\bar{s}<0$ the probability is zero in the neighborhood of $x = 1$. However we can not give much weight to this conclusion since the argument is based on the approximate solution.

The interesting question now is: When $\bar{s}$ and $V_s$ are of the same order of magnitude, or $\bar{s}(>0)$ is much smaller than $V_s$, is quasifixation still possible on the side of the disadvantageous class? Intuition seems to deny the possibility. However, one argument which seems to be against this intuition will be given here: By the logit transformation, the gene frequency is expressed by a point on the entire real axis. As x changes from 0 to 1, the transformed gene frequency $\xi$ changes continuously from $-\infty$ to $\infty$. If $\delta\xi$ be the amount of change in $\xi$ per generation, this can be expanded in powers of $\delta x$;

$$\delta\xi = \frac{1}{x(1-x)}\delta x + \frac{2x-1}{2x^2(1-x)^2}(\delta x)^2 + \ldots$$

or using the relation,

$$\delta x = sx(1-x),$$

$$\delta\xi = s + \frac{(2x-1)}{2}s^2 + \ldots$$

Therefore

$$M_{\delta\xi} = \bar{s} + \frac{2x-1}{2}V_s \text{ and } V_{\delta\xi} = V_s$$

with good approximation. Thus for $\bar{s} \gg V_s$, we have a simple one dimensional random walk as mentioned above. However if $V_s \gg \bar{s} > 0$, $M_{\delta\xi}$ can be negative for small values of x (ie.. for large negative value of $\xi$) and pushes the gene frequency toward the left side. Then if the initial gene frequency is very low, it might lead to accumulation of probability in the neighborhood of 0. At any rate, until a rigorous demonstration is made, this will be left as an open question.

## The Chance of Fixation of a Single Mutant Gene

Suppose a new gene $A'$ has appeared in an infinitely large population and a gamete carrying this gene has selective advantage s over a gamete with the original gene. Fisher (1930a) and Haldane (1927) studied the case with constant s. Here we assume that s is a random variable with probability distribution f(s).

Let P(z) be the probability generating function of the number of the descendant gene $A'$ in the next ($1^{st}$) generation. Then

$$P(z) = \int_{-\infty}^{\infty} e^{(1+s)(z-1)}f(s)ds. \tag{20}$$

By a well known property of the branching process (*cf.* Bartlett, 1949; Feller, 1951b), the probability generating function of the second generation, $P_2(z)$, is given by

$$P_2(z) = P[P(z)].$$

In general, the probability generating function of the $n^{th}$ generation is

$$P_n(z) = P_{n-1}(P(z)), \tag{21}$$

where $P_1(z) \equiv P(z)$. By using (21) we can follow the probability distribution of the number of surviving mutant genes from generation to generation. The important theorem on the ultimate fate of the mutant gene is that if $P'(1) = 1 + \bar{s} \leqq 1$, the extinction of the gene is almost certain, whereas if $P'(1) > 1$, there is a finite probability $\epsilon$ that $A'$ will be established in the population. The latter probability can be calculated from the theorem that the chance of ultimate extinction which we will call $\zeta$ $(=1-\epsilon)$ is obtained as the minimum root of

$$P(\zeta) = \zeta. \tag{22}$$

For the proof of these results the reader is referred to Feller's book (1951b, p. 226). Thus it is easy to see that if $\bar{s} \leqq 0$, extinction is almost certain, whereas if $\bar{s} > 0$, the gene has a finite chance of fixation. For small $\bar{s}$, this chance is given approximately by

$$\varepsilon = \frac{2\bar{s}}{1+V_s} \tag{23}$$

If there is no random fluctuation in selective advantage, $V_s = 0$ and the above result agrees with the previously known one (Haldane, 1927; Fisher, 1930a; Wright, 1931). Thus random fluctuation of selection intensity tends to reduce the probability of fixation of an advantageous gene, but the amount of reduction is very small and may even be negligible.

## Relative Importance of the Population Size and the Random Fluctuation of Selection Intensity in Determining the Gene Frequency Distribution at the Steady State

So far we have studied separately the factors which will cause the stochastic processes of change in gene frequency. In nature not only these factors but also systematic factors such as recurrent mutation, migration and the directed pressure of selection will act simultaneously. If the condition persists for a sufficiently long period, the balance between all these factors will ultimately lead to a stable distribution of gene frequencies which we will denote by $\phi(x)$; namely

$$\phi(x, p; t) \rightarrow \phi(x), \text{ as } t \rightarrow \infty. \tag{24}$$

The important point here is that the final distribution is independent of the initial gene frequency p. This is understandable since any population which has been carried to the fixed (homallelic) states by chance will sooner or later return to the unfixed (heterallelic) states either by mutation or by migration, thus creating a continual shuffling of the gene frequency distribution.

At this steady state we have

$$\tfrac{1}{2}\frac{\partial}{\partial x}\{V_{\delta x}\phi\} - M_{\delta x}\phi = 0,$$

as shown in Appendix I.

Here we assume that $M_{\delta x}$ and $V_{\delta x}$ do not depend on t. This leads to

$$\phi(x) = \frac{C}{V_{\delta x}} e^{2\int \frac{M_{\delta x}}{V_{\delta x}}dx}, \tag{25}$$

where C is determined such that $\int_0^1 \phi(x)\,dx = 1$. The critical point or points of the curve can be obtained by solving the equation;

$$2M_{\delta x} - \frac{d}{dx}(V_{\delta x}) = 0 \tag{26}$$

The important formula (25) giving the steady state distribution was first derived by Wright (1938) using an entirely different method. In the language of the theory of probability (24) represents the ergodic property of the process under consideration. Thus $\phi(x)$ gives not only the probability distribution of gene frequencies in different populations at a given moment (spatial distribution) but also the probability distribution of gene frequencies which will be realized in the same population over a long period of time (see Wright, 1939).

Let us consider now a randomly mating population with effective size N. We denote by u the mutation rate of gene $A_1$ to its allele $A_2$ and by v the mutation rate in the reverse direction. The effect of migration[4] has the same mathematical expression as that of mutation; that is, if $\alpha$ is the rate of migration and if $x_I$ is the frequency of the gene $A_1$ among the immigrants, we can use $\alpha(1-x_I)$ instead of u, and $\alpha x_I$ instead of v to express the effect of migration. Therefore in the following discussion, unless otherwise stated, we shall use u and v to include mutation and migration, that is, to mean $u+\alpha(1-x_I)$ and $v+\alpha x_I$ respectively.

As to the mode of selection we consider two cases in this section: The case of no dominance and the case of complete dominance.

### 1. *Case of no dominance*

We express the selective advantage of $A_1$ over $A_2$ by s, a random variable with mean $\bar{s}$ and variance $V_s$. Under this assumption the amount of change in the frequency of $A_1$ per generation is

$$\delta x = sx(1-x) - ux + v(1-x) + \delta' x.$$

Therefore

$$M_{\delta x} = \bar{s}x(1-x) - ux + v(1-x)$$

and

$$V_{\delta x} = V_s x^2(1-x)^2 + \frac{x(1-x)}{2N}$$

By substituting these into (25) we obtain

$$\phi(x) = Cx^{4Nv-1}(1-x)^{4Nu-1}(\lambda_1 - x)^{4NA-1} \times (x-\lambda_2)^{4NB-1} \qquad (27)$$

where

$$A = -\{(\bar{s}/2NV_s) + u\lambda_1 - v\lambda_2\}/(\lambda_1 - \lambda_2)$$

and

$$B = \{(\bar{s}/2NV_s) - v\lambda_1 + u\lambda_2\}/(\lambda_1 - \lambda_2).$$

Here $\lambda_1$ and $\lambda_2$ are the positive and the negative roots of the quadratic equation:

$$\lambda^2 - \lambda - \frac{1}{2NV_s} = 0$$

that is,

$$\lambda_1 = \frac{1+\sqrt{1+(2/NV_s)}}{2} \qquad (>1)$$

and

$$\lambda_2 = \frac{1-\sqrt{1+(2/NV_s)}}{2} \qquad (<0)$$

In Table 2, numerical values of $\lambda_1$ and $\lambda_2$ are listed for a few values of $2NV_s$.

TABLE 2. NUMERICAL VALUES OF $\lambda_1$ AND $\lambda_2$ IN THE FORMULA (27) FOR SOME VALUES OF $2NV_s$.

| $2NV_s$ | $\lambda_1$ | $\lambda_2$ |
|---|---|---|
| 0.1 | 3.701562 | −2.701562 |
| 1.0 | 1.618034 | −0.618034 |
| 10.0 | 1.091607 | −0.091607 |
| 100.0 | 1.009902 | −0.009902 |

In order to compare the relative effect due to the two different random factors, we will study the symmetrical case where $\bar{s} = 0$ and $u = v$ in detail. With these assumptions (27) reduces to

$$\phi(x) = C[x(1-x)]^{4Nu-1}[(\lambda_1 - x)(x-\lambda_2)]^{-4Nu-1}.$$

If the population size is small such that 4Nu is much less than 1, the genes are fixed most of the time and the distribution curves for unfixed classes are U-shaped, as shown in Figure 4. Here we assume $N = 10^3$. If the variance of s is $0.1/(2N)$ or $5 \times 10^{-5}$ the effect of the fluctuation is so small that the curve is indistinguishable from the one with no fluctuation when drawn on the graph. With $V_s = 10/(2N)$ or $5 \times 10^{-3}$ the effect is still not striking; the frequencies of subterminal classes merely rise by about 33 per cent as compared with the case of $V_s = 0$. With $V_s = 100/(2N)$ or 0.05, the subterminal classes become about 2.2 times as high as the case with $V_s = 0$. $V_s = 0.05$ is a high fluctuation since $\sigma_s = \sqrt{V_s} \doteqdot 0.2236$. Thus we might say that for small populations the effect of random fluctuation of selection is rather unimportant.

Figure 5 shows the distribution of gene frequencies in intermediate populations in which one mutation appears every two generations on the average $(4Nu = 1)$. This means that if we assume the mutation rate of one in one hundred thousand $(10^{-5})$, N should be 25000, that is, 25 times as large as before. In such populations, if there is no fluctuation, the flat distribution will be realized, that is, all the heterallelic classes are equally prob-

[4] With the island model, $x_I$ may be equated to the mean gene frequency of the species $(\bar{x})$. With the model of continuous distribution the situation is much more complicated (*cf.* Wright, 1951b). With the stepping stone model of linear type, $2\epsilon(1-r)(1-\bar{x})$ and $2\epsilon(1-r)\bar{x}$ should be substituted for u and v, where r represents the correlation coefficient between the gene frequencies of two adjacent subgroups and is a positive root of the quadratic equation;

$$\epsilon r^2 + (2-3\epsilon)r + (4\epsilon - 2) = 0.$$

Here $\epsilon$ is the net rate of exchange of individual between two adjacent groups per generation (*cf.* Kimura, 1953a).

able. Here the effect of the fluctuation may not be negligible: With $V_s = 10/(2N)$ or $2 \times 10^{-4}$, the distribution curve becomes U-shaped and the heights of the subterminal classes are about 4.9 times that with $V_s = 0$. With $V_s = 100/(2N)$ or 0.002, the distribution resembles that of a very small population ($4Nu \approx 0$), the frequency of the subterminal classes rising about 48 times as high as when $V_s = 0$.

In Figure 6 the population size is assumed to be twice as large as before ($4Nu = 2$), so that one new mutation is expected per generation. With a mutation rate of $u = 10^{-5}$, N should be 50,000. In this case it can be shown that if $V_s$ does not exceed $1/N$ or $2 \times 10^{-5}$, the distribution curve is unimodal. On the other hand, if $V_s$ exceeds this value, the curve becomes bimodal. As will be seen in the figures, if $V_s = 10/2N$ or $10^{-4}$, the gene frequencies giving the two modes of the distribution are about 0.053 and 0.947 respectively. With $V_s = 100/2N$ or $10^{-3}$, the modal gene frequencies are about 0.5 per cent and 99.5 per cent, and the modal classes are about 104 times as frequent as the class with 50 per cent gene frequency. Therefore in the last case the distribution curve looks as if it were U-shaped on the figure. However the class frequencies fall sharply outside of these modes and the frequencies of terminal classes are zero.

Generally, if 4Nu is larger than one, the following rules hold: (i) If $V_s \leqq 4u-(1/N)$, the curve is unimodal with the maximum at $x = 0.5$; (ii) If $V_s > 4u-(1/N)$, the curve is bimodal and the gene frequencies which give these modes are given by

$$x_{max_1} = (1-\sqrt{D})/2 \text{ and } x_{max_2} = (1+\sqrt{D})/2,$$

where $D = 1 - (4Nu-1)/(NV_s)$. Therefore for very large populations ($uN \gg 1$), $x_{max_1} \rightarrow u/V_s$ and $x_{max_2} \rightarrow 1-(u/V_s)$ as $V_s$ increases ($V_s \gg u$). How-

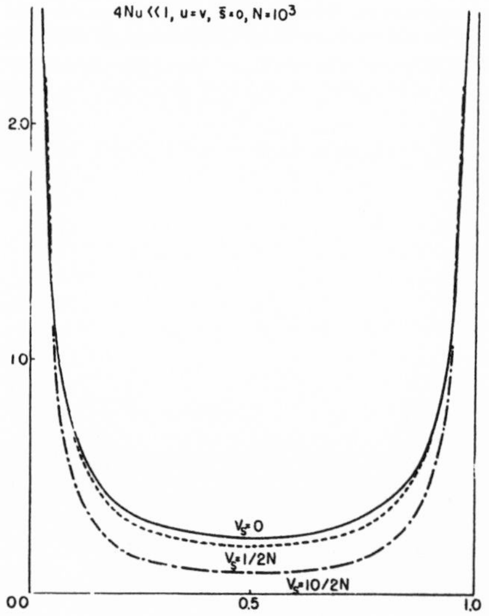


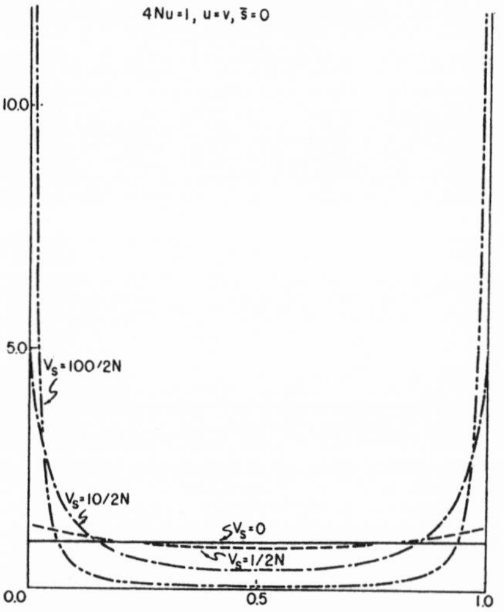


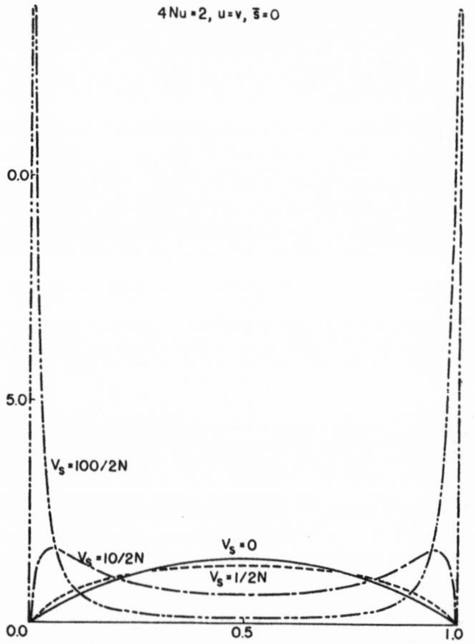


FIGURES 4-6. Effects of random fluctuation of selection intensity on the probability distribution of gene frequencies at steady state. In these figures it is assumed that there is no dominance between a pair of alleles which are neutral on the average ($\bar{s} = 0$). Also mutation rates in both directions are equal ($u = v$). N stands for effective size of population and $V_s$ is the variance of selection intensity. Figure 4 shows the situation in a small population and Figure 6 that in a large population, while Figure 5 shows the intermediate situation. Abscissa; gene frequency (x). Ordinate; relative probability.

ever the frequencies of the terminal classes are always zero. The behavior of the curve when N becomes infinitely large will be discussed later.

### 2. *Case of complete dominance*

Here we assume that $A_1$ is completely recessive to $A_2$, and let s be the selective advantage of the recessive. In terms of Malthusian parameters of fitness $s = m_{11} - m_{12}$ and $m_{12} - m_{22} = 0$. The change in the frequency of $A_1$ per generation $(\delta x)$, the mean $(M_{\delta x})$, and the variance $(V_{\delta x})$ are given as follows:

$$\delta x = -ux + v(1-x) + sx^2(1-x) + \delta' x$$

$$M_{\delta x} = -ux + v(1-x) + \bar{s}x^2(1-x)$$

$$V_{\delta x} = V_s x^4(1-x)^2 + x(1-x)/(2N).$$

The application of Wright's formula (25) gives the following distribution function:

$$\phi(x) = Cx^{4Nv-1}(1-x)^{4Nu-1}(\lambda_1 - x)^{4NA-1}(x-\lambda_2)^{4NB-1} \times [(x-\alpha)^2+\beta^2]^{4ND-1}\exp\left\{8NE\tan^{-1}\left(\frac{\beta}{x-\alpha}\right)\right\} \quad (28)$$

where

$$A = \{-v + (u+v)\lambda_1 + (c\bar{s}/\lambda_1)\}/(3-4\lambda_1),$$

$$B = \{-v + (u+v)\lambda_2 + (c\bar{s}/\lambda_2)\}/(3-4\lambda_2),$$

$$D = (ab + a'b')/(b^2 + b'^2),$$

$$E = (a'b - ab')/(b^2 + b'^2)$$

and

$$a = -v + (u + v - \lambda_1\lambda_2\bar{s})\alpha,$$

$$a' = (u + v + \lambda_1\lambda_2\bar{s})\beta$$

$$b = 3 - 4\alpha,\ b' = -4\beta,\ c = 1/(2NV_s).$$

In the above formula $\tan^{-1}$ takes only its principal values and the definition by the following integral will be convenient:

$$\tan^{-1}x = \int_0^x (1+t^2)^{-1}dt$$

$\lambda_1$, $\lambda_2$ and $\alpha \pm \beta i$ are the four roots of the quartic equation;

$$x^4 - x^3 - 1/(2NV_s) = 0,$$

of which $\lambda_1$ is real and larger than 1, $\lambda_2$ is negative and $\alpha \pm \beta i$ are both imaginary. Table 3 gives the numerical values of these roots for some values of $2NV_s$.

In order to show the effect of random fluctuation in populations of different sizes, Figure 7 and Figure 8 are constructed assuming $\bar{s} = 0$ and $u = v$. In Figure 7, the population is assumed to be so small $(N = 10^3)$ that 4Nu is negligible as compared with 1 $(4Nu \ll 1)$. If we compare three curves corresponding to $V_s = 0$, $V_s = 10/2N = 0.005$ and $V_s = 100/(2N) = 0.05$, we notice as in the case of no dominance that the effect of random fluctuation in selection intensity is rather unimportant. In fact even with $V_s = 0.05$, the height of the subterminal class rises only by about 63 per cent as compared with the case of $V_s = 0$. Also from the graphs it will be recognized that dominance makes the distribution asymmetrical even if $\bar{s} = 0$. In small populations the asymmetry is less. For example with $V_s = 100/(2N)$ the heights of two subterminal classes $x = 1/2N = 0.0005$ and $x = 1 - (1/2N) = 0.9995$ are respectively about 199.5 and 190.0.

In Figure 8, we consider the much larger popula-

TABLE 3. NUMERICAL VALUES OF $\lambda_1$, $\lambda_2$, $\alpha$ AND $\beta$ IN THE FORMULA (28) FOR SOME VALUES OF $2NV_s$.

| $2NV_s$ | $\lambda_1$ | $\lambda_2$ | $\alpha$ | $\beta$ |
|---|---|---|---|---|
| 0.1 | 2.0920904 | −1.5723723 | 0.2401409 | 1.7269238 |
| 1.0 | 1.3802775 | −0.8191725 | 0.2194474 | 0.9144736 |
| 10.0 | 1.0794947 | −0.4135803 | 0.1670427 | 0.4428115 |
| 100.0 | 1.0097141 | −0.2025941 | 0.0964400 | 0.1989478 |

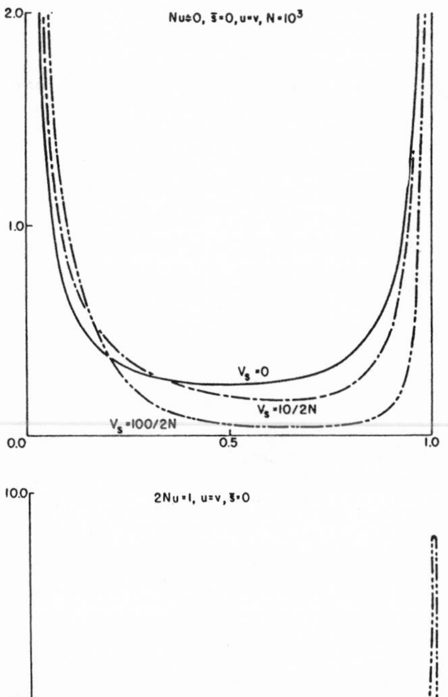


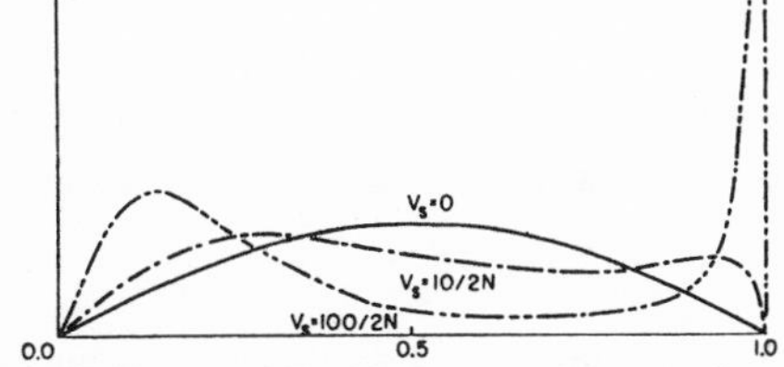


FIGURES 7 and 8. Graphs showing the probability distribution of gene frequencies in the case of complete dominance. Figures 7 and 8 show the situation in a small and large population respectively. Abscissa; gene frequency. Ordinate; relative probability.

tion in which one new mutation is expected per generation ($2Nu = 1$). The three curves correspond to $V_s = 0$, $V_s = 10/2N = 10^{-4}$ and $V_s = 100/2N = 10^{-3}$. As will be seen, a remarkable asymmetry starts to appear as $V_s$ becomes large and there is a very sharp peak in the neighborhood of fixation ($x = 1$). More precise analysis of this sort of asymmetry will be made later for infinitely large populations.

## The Effect of Random Fluctuation of Selection Intensity in Infinitely Large Populations

As a model of a very large population, we consider an infinitely large population where random sampling of gametes plays no role. This allows us to single out the effect of random fluctuation of selection intensity on the distribution of gene frequencies. First we will consider the simplest case of no dominance, followed by the case of complete dominance. Finally the case of an arbitrary degree of dominance will be considered to study heterotic genes.

### 1. *Case of no dominance*

Let $A_1$ have the selective advantage s over $A_2$. With u and v as defined above, we have

$$\delta x = -ux + v(1-x) + sx(1-x),$$
$$M_{\delta x} = -ux + v(1-x) + \bar{s}x(1-x)$$

and

$$V_{\delta x} = V_s x^2(1-x)^2.$$

Thus the steady state distribution is given by

$$\phi(x) = C \frac{e^{-\frac{2}{V_s}\left(\frac{v}{x}+\frac{u}{1-x}\right)}}{x^2(1-x)^2}\left(\frac{x}{1-x}\right)^{-2\left(\frac{-\bar{s}+u-v}{V_s}\right)} \qquad (29)$$

If we consider migration rather than mutation, u and v should be replaced by $m(1-x_I)$ and $mx_I$ respectively. The resulting formula agrees with the one obtained by Wright (1948, p. 292). In (29) the constant C is chosen such that $\int\phi(x)dx = 1$. If the mutation rates are much smaller than $\bar{s}, > 0$, the gene frequencies cluster around their equilibrium position;

$$\hat{x} \doteqdot 1-(u/\bar{s})$$

as $Vs \rightarrow 0$. Under such conditions it can be shown that C approaches

$$\sqrt{\bar{s}/(\pi V_s)}\ (u/\bar{s})^{1+(2s/V_s)} e^{2s/V_s}\ (V_s \rightarrow 0).$$

On the other hand if $\bar{s} = 0$ and $u = v$, (29) reduces to the simple form:

$$\phi(x) = Ce^{\frac{-2u}{V_s}\frac{1}{x(1-x)}} \Big/ x^2(1-x)^2.$$

In such cases, if $V_s \leqq 4u$, the curve is unimodal, while if $V_s > 4u$, the curve is bimodal. Since u will be a very small quantity, even the small fluctuation in s will act to carry the distribution toward both termini.

### 2. *Case of complete dominance*

Let $-s$ be the selective value of the recessive relative to the dominant:

| *Genotype* | *Frequencies* | *Relative fitness* |
|---|---|---|
| $A_1A_1$ | $x^2$ | $-s$ |
| $A_2-$ | $1-x^2$ | 0 |

Then

$$M_{\delta x} = ux + v(1-x) - \bar{s}x^2(1-x)$$
$$V_{\delta x} = V_s x^4(1-x)^2$$

and the steady state distribution is given by

$$\phi(x) = \frac{C}{x^4(1-x)^2} e^{-\frac{1}{V_s}\psi(x)} \qquad (30)$$

where

$$\psi(x) = \frac{2v}{3x^3} + \frac{v-u}{x^2} + \frac{2v-4u-2\bar{s}}{x} + \frac{2u}{1-x} + (2v-6u-2\bar{s})\log\frac{1-x}{x}.$$

First consider a recessive deleterious gene. In this case $\bar{s}$ ($>0$) will be much larger than the mutation rates u, v and if $V_s$ is very small ($V_s \rightarrow 0$), the distribution clusters around the equilibrium gene frequency x which can be obtained by solving

$$M_{\delta x} = 0,$$

giving $\hat{x} = \sqrt{v/\bar{s}}$ approximately. Under this condition[5]

$$\phi(x) \rightarrow \sqrt{\frac{-M'_{\delta\hat{x}}\hat{x}^2(1-\hat{x})}{\pi V_s\ x^4(1-x)^2}}\ e^{-\frac{1}{V_s}\left\{\psi(x)-\psi(\hat{x})\right\}}, \qquad (31)$$

as $V_s \rightarrow 0$. In this formula

$$-M'_{\delta\hat{x}} = u+v+\bar{s}(2\hat{x}-3\hat{x}^2).$$

Figure 9 shows the case where $u = v = 10^{-5}$ and $\bar{s} = 10^{-3}$. Three curves corresponding to $V_s = 4 \times 10^{-5}$, $V_s = 10^{-5}$, and $V_s = (1/4) \times 10^{-5}$ are drawn, each of which centers around the equilibrium position $\hat{x} = 0.09463$.

Next we consider the case of $\bar{s} = 0$ and $u = v$. In this case $\psi(x)$ in (30) becomes

$$u\left\{\frac{2}{3x^3} - \frac{2}{x} + \frac{2}{1-x} - 4\log\frac{1-x}{x}\right\}$$

Figure 10 shows the distributions for $u/V_s = 1$, 0.05 and 0.01. As expected the curve is unimodal when $V_s$ is small but becomes bimodal as $V_s$ increases. In the latter case, the distribution curve be-

[5] This can be derived by using the following result which is due to Laplace: Let $f(x)$ and $\phi(x)$ be continuous in $[a, b]$ and suppose that $\phi(x)$ has only 1 minimum at a point $\xi$ lying inside $[a, b]$, then

$$\int_a^b f(x)e^{-n\phi(x)}dx \sim \sqrt{\frac{2\pi}{n}}\frac{f(\xi)}{\sqrt{\phi''(\xi)}}e^{-n\phi(\xi)}\ (n\rightarrow\infty)$$

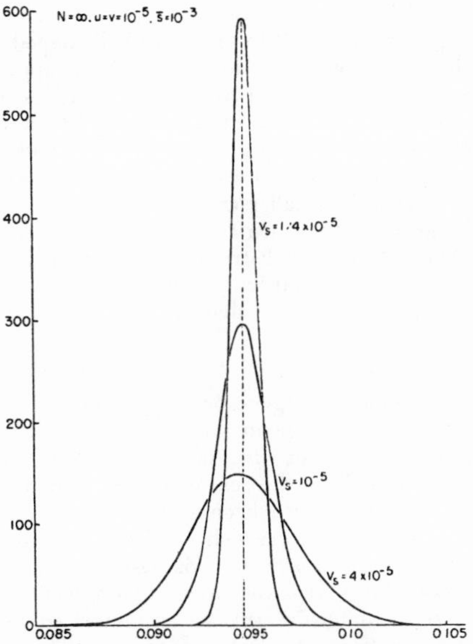


FIGURE 9. Probability distributions of frequencies of a recessive deleterious gene in an infinitely large population. u and v are mutation rates in both directions and $\bar{s}$ and $V_s$ are respectively the mean and variance of selection intensity. Abscissa is gene frequency, and ordinate is probability density.

comes remarkably asymmetrical in spite of $\bar{s} = 0$. This phenomenon was briefly mentioned, but will be discussed here in more detail. It may be interesting to note that the genes are neutral in the sense of long term average in this case.

Figure 11 is intended to show the location of maxima and minima in the distribution curve. The figure is so constructed that the perpendicular projections on the x-axis of the intersecting points between the lines and the curve give the gene frequencies corresponding to the maxima and minima. If the line intersects the curve only once, the distribution curve is unimodal and has only one maximum. On the other hand if the line intersects the curve in three points the distribution has two maxima and one minimum. Thus it will be seen that if $V_s$ is less than about $u/(0.07)$, the curve is unimodal, while it becomes bimodal if $V_s$ exceeds this value. From this graph the asymmetry of the distribution is quite clear. This asymmetry becomes more extreme as $V_s$ gets larger.

### 3. *General case of any degree of dominance and application to heterotic genes*

Let $-s$ and $-t$ be the fitness of two homozygous genotypes relative to that of the heterozygotes; that is, $m_{11} - m_{12} = -s$ and $m_{22} - m_{12} = -t$. Here s and t are arbitrary and therefore we can represent any degree of dominance.

From the following table;

| *Genotypes* | *Frequency* | *Relative fitness* |
|---|---|---|
| $A_1A_1$ | $x^2$ | $-s$ |
| $A_1A_2$ | $2x(1-x)$ | $0$ |
| $A_2A_2$ | $(1-x)^2$ | $-t$ |

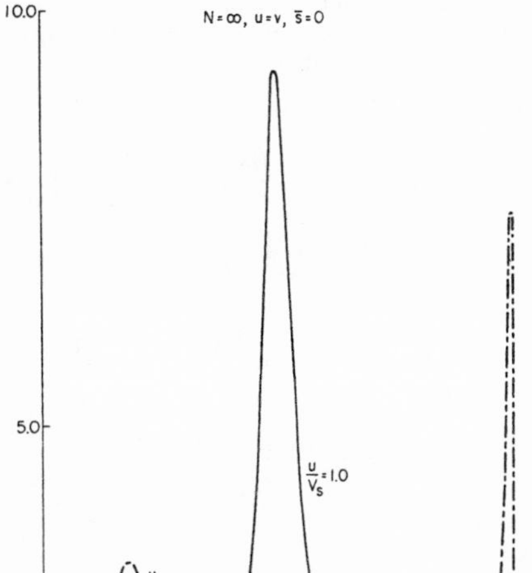

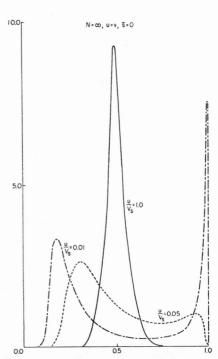

FIGURE 10. Probability distributions of frequencies of recessive genes in an infinitely large population. In this figure it is assumed that the gene is neutral on the average ($\bar{s} = 0$) and the mutation rates in both directions are equal ($u = v$). Note the asymmetry of distribution despite $\bar{s} = 0$. Abscissa is gene frequency and ordinate is probability density.

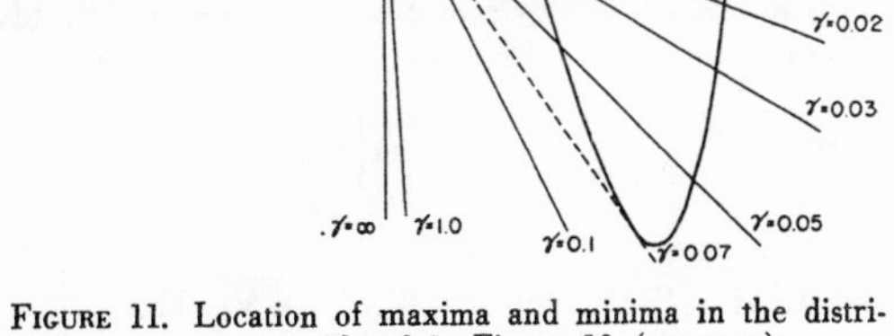

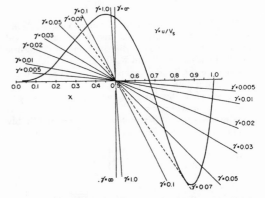

FIGURE 11. Location of maxima and minima in the distribution curve as considered in Figure 10 (see text).

we have

$$M_{\delta x} = -ux + v(1-x) - \bar{s}x^2(1-x) + \bar{t}x(1-x)^2$$

and

$$V_{\delta x} = V_s x^4(1-x)^2 - 2W_{st}x^3(1-x)^3 + V_t x^2(1-x)^4.$$

In this case we must consider not only the variances of s and t but also their covariance which we denote by $W_{st}$. This makes the distribution formula much more complicated.

$$\phi(x) = Cx^{A-2}(1-x)^{B-2}[(x-\alpha)^2+\beta^2]^{D-1}$$
$$\times \exp\left\{-2E\tan^{-1}\left(\frac{x-\alpha}{\beta}\right) - \frac{2u}{V_s}\left(\frac{1}{1-x}\right) - \frac{2v}{V_t}\left(\frac{1}{x}\right)\right\}, \qquad (32)$$

where

$$A = \frac{2}{V_t}\left\{\bar{t} - u + \frac{2W_{st}+3V_t}{V_t}v\right\}$$

$$B = \frac{2}{V_s}\left\{\bar{s} - v + \frac{2W_{st}+3V_s}{V_s}u\right\}$$

$$D = -\frac{\bar{s}-\bar{t}}{V_sV_t} - \frac{V_tV_{s+t}-(V_t-V_s)^2}{V_s^2V_t}u - \frac{V_sV_{s+t}-(V_t-V_s)^2}{V_sV_t^2}v$$

$$E = \frac{1}{\sqrt{V_tV_s-W_{st}^2}}\left\{\left(1+\frac{W_{st}}{V_s}\right)\bar{s} - \left(1+\frac{W_{st}}{V_t}\right)\bar{t} + \frac{(W_{st}+V_t)V_sV_{s+t}-2(V_tV_s-W_{st}^2)(V_t-V_s)}{V_s^2V_t}u - \frac{(V_s+W_{st})V_tV_{s+t}+2(V_tV_s-W_{st}^2)(V_t-V_s)}{V_t^2V_s}v\right\}$$

$$(V_tV_s - W_{st}^2 \neq 0),$$

and

$$\alpha = (W_{st}+V_t)/V_{s+t}, \quad \beta = \sqrt{V_tV_s - W_{st}^2}/V_{s+t}$$

$$V_{s+t} = V_s + 2W_{st} + V_t.$$

Here $\tan^{-1}$ takes only its principal values. As an important application of the above formula we will consider the frequency distribution of heterotic genes, in which $\bar{s}>0$ and $\bar{t}>0$. If the variances of s and t are small the distribution clusters around the equilibrium point $\hat{x} = \bar{t}/(\bar{s}+\bar{t})$.

To simplify the formula, we consider the symmetrical case where $\bar{s} = \bar{t}$, $V_s = V_t$ and $u = v$ and also assume that the mutation rates are negligibly small. Then (32) reduces to

$$\phi(x) = C\left\{\frac{1}{2(1+\rho)} - x(1-x)\right\}^{-s_1-1}\{x(1-x)\}^{s_1-2}, \qquad (33)$$

where $s_1 = 2\bar{s}/V_s$ and $\rho = W_{st}/\sqrt{V_sV_t}$. Thus $\rho$ represents the correlation coefficient between s and t. Figure 12 shows the case of $\bar{s}/V_s = 1$. The curve changes from the flat to the sharp bell shaped one as $\rho$ changes from $-1.0$ to $+0.9$. As $\rho$ approaches $+1$, the frequency distribution condenses to the equilibrium point $\hat{x} = 1/2$. Figure 13 shows the case of $\bar{s}/V_s = 10$. In this case, the distribution has a stronger tendency to condense around $\hat{x}$ as is expected from the smaller variance of s.

In general there are two factors which increase the tendency for the distribution to condense around $\hat{x}$: (1) smaller variances that is, $V_s$ and $V_t \rightarrow 0$; and (2) larger correlation, that is, $\rho \rightarrow 1$.

## Note on Cyclic Changes in Selection Intensity

Cyclic change in gene or chromosome frequency due to cyclic change in selection intensity is often observed with "adaptive polymorphism." A well known example is the seasonal change in the proportion of different chromosome types in *Drosophila pseudo-obscura* which has been studied extensively by Dobzhansky and his associates (*cf.* Dobzhansky, 1951). It is very likely that a cyclic change in selection intensity will be accompanied by random fluctuations of selection intensity. The general treatment of the resulting process is mathematically very difficult. The formula for the steady state distribution (25) cannot be applied to such a system. In this paper we will consider a very simple model of cyclic change with no random fluctuation.

Let the relative fitness of three genotypes $A_1A_1$, $A_1A_2$ and $A_2A_2$ be sp(t), 0 and −sp(t) respectively, where p(t) is a periodic function of t. Assume for convenience that $s>0$. Thus

$$\frac{dx}{dt} = sp(t)x(1-x). \qquad (34)$$

This is equivalent to genic selection with selection intensity as a function of time.

First we consider the case where the number of generations (m) per cycle is large and $p(t) = \sin(2\pi t/m)$. Then from (34) we have the result that the gene frequency on a logit scale at the $t^{th}$ generation ($\xi_t$) is given by

$$\xi_t = \xi_o + \frac{sm}{2\pi}\{1-\cos(2\pi t/m)\}. \qquad (35)$$

This shows that $\xi_t$ oscillates around the mean value $\xi_o + (sm)/(2\pi)$. From this we can show that the maxima and the minima of the frequency of $A_1A_1$ are reached always m/4 generations later than those of the fitness of $A_1A_1$.

This may indicate that, with a cyclic change in environmental conditions, the mode of adaptation due to genetic "coadaptation" mechanism is less efficient than that due to the direct change of phenotypes.

Next we consider the contrasting case where one cycle consists of two generations and $p(t) = (-1)^t$. This was suggested by R. C. Lewontin (personal communication). In this case we must

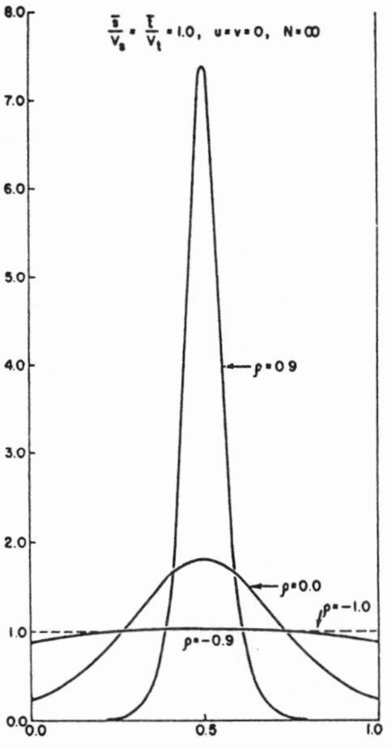


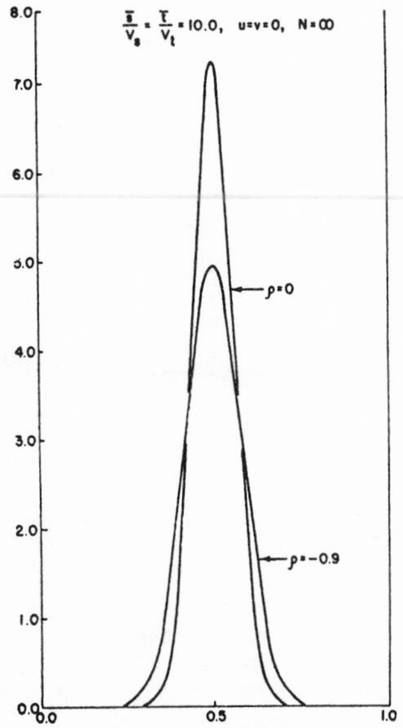


FIGURES 12 and 13. Probability distributions of frequencies of heterotic genes in an infinitely large population. In these figures the fitness of the two homozygotes relative to that of the heterozygotes is assumed to be $-s$ and $-t$. $\bar{s}$, $\bar{t}$ and $V_s$, $V_t$ are respectively means and variances of s and t, and $\rho$ stands for the correlation coefficient between these two variables. In both figures completely symmetrical situations are assumed.

use discrete generation time ($t = 0, 1, 2$, etc.) and (34) must be replaced by the finite difference equation

$$\frac{\Delta x}{\Delta t} = s(-1)^t x(1-x), \quad (s>0) \qquad (36)$$

In this case though the gene frequency fluctuates from generation to generation there is a definite drift toward fixation or loss. This becomes evident if we observe the change of the frequency only at even generations; $\tau = 2t$ ($\tau = 0, 2, 4, \ldots$). If $\Delta_2$ denotes the change in two generations;

$$\frac{\Delta_2 x}{\Delta \tau} = -s^3x^2(1-x)\{1-2x-sx(1-x)\}. \qquad (37)$$

For small s, the behavior of x at the even generations will be obtained by treating (37) as if it were a differential equation. Generally, it is easily seen from (37) that if x is small (more precisely less than about $1/2 - s/8$), $\Delta_2 x/\Delta\tau$ is always negative and since the fluctuation per generation becomes smaller as x decreases (*cf.* (36)) the gene will ultimately be lost ($x \to 0$). On the other hand if x is near 1, the gene will ultimately be fixed ($x \to 1$).

### GENERAL EQUATION FOR THE STOCHASTIC PROCESSES AND THE DISTRIBUTION OF GENE FREQUENCIES

So far we have considered the simplest case of one locus with a pair of alleles. The fitness of an individual, however, often depends not only on the sum of the average effect of each gene but also on the interaction between them. The partial differential equation (1) can be extended to cover such cases: Let $x_i$ ($i = 1, 2, \ldots, K$) be the gene frequency in the $i^{th}$ locus and let $\phi(x_1, x_2, \ldots, x_K, t)$ be the probability density that the $i^{th}$ gene frequency lies between $x_i$ and $x_i + dx_i$ ($i = 1, 2, \ldots K$) at the $t^{th}$ generation, then

$$\frac{\partial\phi}{\partial t} = \frac{1}{2}\sum_{i=1}^{k}\frac{\partial^2}{\partial x_i^2}(V_{\delta x_i}\phi) + \sum_{i>j}\frac{\partial^2}{\partial x_i \partial x_j}(W_{\delta x_i \delta x_j}\phi)$$

$$-\sum_{i=1}^{k}\frac{\partial}{\partial x_i}(M_{\delta x_i}\phi), \quad (i, j = 1, 2, \ldots, K), \qquad (38)$$

where $\delta x_i$ is the change in $x_i$ per generation. In this equation $M_{\delta x_i}$ and $V_{\delta x_i}$ are respectively the mean and the variance of $\delta x_i$, and $W_{\delta x_i \delta x_j}$ is the covariance between $\delta x_i$ and $\delta x_j$. These quantities are in general functions of all gene frequencies as well as of t. The equation (38) corresponds to the Kolomogorov equation in K variables (*cf.* Kimura, 1953b). If the random fluctuation of the gene frequencies is due solely to random sampling of gametes, all the covariances are zero[6] and (38) reduces to the following form under random mating;

[6] Strictly speaking, if there is linkage between genes, this may not hold at the start. However, the effect of the linkage will disappear rather rapidly in the population and the covariance approaches zero, except for very close linkage.

$$\frac{\partial\phi}{\partial t}=\frac{1}{2}\sum_{i=1}^{k}\frac{\partial^2}{\partial x_i^2}\left(\frac{x_i(1-x_i)}{2N}\phi\right)$$
$$-\sum_{i=1}^{k}\frac{\partial}{\partial x_i}\left[\left(\frac{x_i(1-x_i)}{2}\frac{\partial\bar{m}}{\partial x_i}\right.\right.$$
$$\left.\left.-u_ix_i-v_i(1-x_i)\right)\phi\right], \qquad (39)$$

where $\bar{m}$ is the mean fitness in the population (assuming each m is constant), and $u_i$ and $v_i$ are the mutation rates of the $i^{th}$ gene to and from its allele. If there is a steady state distribution, which we will denote by $\varphi(x_1, x_2, \ldots, x_K)$, it should be obtained by setting $\partial\phi/\partial t = 0$ in (39). It is not difficult to verify that Wright's formula (1937) written in our terminology;

$$\phi(x_1, x_2, \ldots, x_K)=Ce^{2N\bar{m}}\prod_{i=1}^{k}x_i^{4Nv_i-1}(1-x_i)^{4Nu_i-1} \qquad (40)$$

satisfies this condition. However if there are random fluctuations of selection intensities the covariance term in the differential equation (38) does not vanish in general and Wright's formula (40) cannot be applied.

In natural populations, each locus may contain multiple alleles. The extension to such cases may also be made (Crow and Kimura, 1955): If we consider K loci with $n_i$ alleles at the $i^{th}$ locus, we have

$$\frac{\partial\phi}{\partial t}=L(\phi), \qquad (41)$$

where

$$L(\phi)=\frac{1}{2}\sum_{i=1}^{k}\sum_{j=1}^{n_i-1}\frac{\partial^2}{\partial x_{ij}^2}(V_{\delta x_{ij}}\phi)$$
$$+\frac{1}{2}\sum_{(i,j)\neq(i',j')}\frac{\partial^2}{\partial x_{ij}\partial x_{i'j'}}(W_{\delta x_{ij}\delta x_{i'j'}}\phi)-\sum_{i=1}^{k}\sum_{j=1}^{n_i-1}(M_{\delta x_{ij}}\phi),$$

and $x_{ij}$ ($i = 1, \ldots K$, $j = 1, \ldots (n_i-1)$), is the frequency of $j^{th}$ allele at the $i^{th}$ locus. The summation in the second term is over all different genes, that is, for all i and j except for the case of $(i, j) = (i', j')$. ($i, i' = 1, \ldots K$, $j, j' = 1, \ldots (n_i-1)$).

If the random fluctuation is due solely to random sampling of gametes

$$V_{\delta x_{ij}}=x_{ij}(1-x_{ij})/(2N) \text{ and}$$
$$W_{\delta x_{ij}\delta x_{i'j'}}=\begin{cases}0 \text{ for } i\neq i'\\ -x_{ij}x_{ij}'/(2N) \text{ for } i=i',\ j\neq j'.\end{cases}$$

If we assume selection and migration but disregard mutation, the rate of which may differ from allele to allele, the most general formula so far obtained for the steady state distribution is

$$\phi(x_{ij}; s)=Ce^{2N\bar{m}}\prod_{i=1}^{k}\prod_{j=1}^{n_i}x_{ij}^{4N\alpha\bar{x}_{ij}-1}, \qquad (42)$$

in our terminology, where $\bar{m}$ is the mean fitness in the population, $\alpha$ is the rate of migration and $\bar{x}_{ij}$'s are the gene frequencies of the immigrants (Wright, 1949). As in (40), this formula should satisfy the condition

$$L(\phi) = 0,$$

though the demonstration seems to be less simple.

The differential equation (41), when applied to a single locus proves to be a very powerful tool to solve the problem of random drift for a multi-allelic system (Kimura, 1956).

## CONCLUSION

From the standpoint of population genetics the most elementary process in evolution is the change in gene frequency, and according to Wright (1949) we can classify the mode of change into three categories: (1) Systematic change due to recurrent mutation, migration and selection; (2) random fluctuations due to accidents of sampling as well as due to random fluctuations in the systematic pressure; among the latter, random fluctuation in selection intensity may be the most important; (3) change due to events which are unique in the life of the species, for example, non-recurrent mutation, non-recurrent hybridization, etc.

Therefore it will be expected that the process of change in gene frequency in natural populations is generally stochastic.

However, the problem arises as to how important is the role played by random fluctuations. First the importance of the role played by accidents of sampling must depend on the prevailing population structure (especially "effective size," N) and sensitivity of mutant genes to natural selection. At present our knowledge of either factor is discouragingly small. Common sense tells us that populations are usually so large that accidents of sampling are negligible. However, the splitting of the population due to geographical barriers, isolation by distance due to restricted locomotion, and other causes will tend to make effective size smaller (Wright, 1939, 1940, 1951b). Thus verv small or very large effective size will be rare and the distribution curve for effective N may tend to be bell-shaped. As to the distribution of selection intensitv of mutant genes in the population, Haldane (1949) considered that there was reason to think that the distribution was U-shaped, if measured on a scale from zero (for complete lethal) to one (for normal). This will be apparent if we notice that cross-fertilizing populations contain a number of lethals and semi-lethals and if we also assume that natural populations contain a great variety of isoalleles which are very nearly neutral. As shown by the foregoing treatment of genic selection with random genetic drift, if Ns is (approximately) less than 1, even fixation of an unfavorable allele could occur with appreciable change leading to random differentiation of local populations. Accumulation of this process acting on isoalleles may create the great diversity of genetic backgrounds among local groups.

For the genes for which Ns is much larger selection will exert decisive control over the change in gene frequency. However the selection intensity in general could hardly remain constant over extensive time and space. The mode of change through time will be classified into two categories. (1) There may be cyclic fluctuations with various lengths of period and amplitudes. Wright (1935) considered the significance of oscillations of the position of the optima of quantitative characters and assumed that even trivial oscillations would be enough to carry the population from one adaptive peak to another (see also Wright, 1951a, p. 454). The importance of the long-term fluctuations was also stressed by Sheppard (1954) who considered that secular fluctuations would be more important than short-term ones. (2) There may be random fluctuation of selective force from generation to generation without any specified rule. Since the biological systems in nature have complicated interdependent structure, any random fluctuation in one part will inevitably be propagated to the other. This will especially be true for populations of higher animals. Actually cyclic and random fluctuations will not be separable in practice; wherever there is a cyclic fluctuation, random fluctuation will accompany it. Moreover if we consider subdivided population structure as a whole for a long enough time, the "local conditions are so varied and unpredictable," as Muller (1949) stated, "an element of indeterminacy is thereby introduced."

On pages 38-42 of this paper we have considered the relative importance of the population size and random fluctuations of selection intensity on the determination of gene frequency distribution at the steady state. This shows that for a small population the effect of random fluctuations is rather small, while for a large one, the effect is significant as is also demonstrated on pages 42-44.

As far as I know, there seems to be little disagreement with the thesis that a subdivided population structure is the most favorable condition for rapid evolution, though there are divergent opinions about the underlying mechanism; Wright (since 1931) emphasizes the proper balance between directed and random processes leading to the maximum response to "ecologic opportunity" (see Wright, 1950). Fisher and Ford (1950), and also Sheppard (1954) consider that what is important is the variety of environmental conditions to which the colonies are exposed but not the random process due to accidents of sampling. Mayr (1954) claims that drastic change of selective value for many genes which will be realized on the altered genetic background or "genetic environment" of a newly founded peripherally isolated population is the key to rapid evolution.

The most attractive point in Wright's theory is the creative role ascribed to intergroup selection (Wright, 1945, 1949). As in the case of intragroup selection there must be various ways by which intergroup selection can occur. Wright (1951b) considers the excessive migration from the spot in the continuum where the favorable genotype has happened to be established. In what follows we shall consider a different model in which the total population is subdivided into small and mutually competing subgroups, each of which has finite probability of splitting and multiplying into two or more groups as well as of being exterminated. Under such a circumstance the process of change in gene frequency of the given population is no longer Markovian and the above partial differential equations cannot be applied. However the following extension is possible: Let C be the rate of increase of the number of subgroups with given genic constitution due to intergroup competition. Generally C is a function of $\mathbf{x}$ and t, where $\mathbf{x}$ is the vector representing all the gene freqeuncies as one quantity. Let $\psi(\mathbf{x}, t)$ be the quantity proportional to the probability of the group whose gene frequency is $\mathbf{x}$ at time t. Then $\psi$ will satisfy the equation

$$\frac{\partial \psi}{\partial t} = L(\psi) + C\psi, \qquad (43)$$

where L is the differential operator as given in (41). C in this sense measures the fitness of the subgroup. It will be convenient to treat the fitness in this level as entirely different from the fitness of an individual in a given subgroup.

## Appendix I

### *Elementary derivation of the Fokker-Planck equation applied to population genetics*

Let $y = \phi(x, t)$ represent the curve for probability distribution of gene frequencies at time t. As shown in Figure 14, the distribution is approximated by histograms, each column having width h. We represent the gene frequency of each class by its middle point. Consider the class with gene frequency x. For sufficiently small h, the area of the column $\phi(x, t)h$ gives the probability that the population has gene frequency $x \pm (1/2)h$.

After time interval $\Delta t$, the population with gene frequency x will move to another class due to systematic as well as to random changes. However, by virtue of the assumption of continuous stochastic process (*cf.* footnote 2), we make this movement sufficiently small by taking $\Delta t$ small, so that consideration of two adjacent classes is sufficient. Let $m(x)\Delta t$ be the probability that the population moves to the higher class $(x + h)$ by systematic pressure. Let $v(x)\Delta t$ be the probability that it moves outside the class by random fluctuation, half of the time to the left class (x—h) and the other half to the right class (x + h). Greater displacements have a smaller order of probability ($o(\Delta t)$) and can be neglected.

Thus the probability that the population will have gene frequency $x \pm (\frac{1}{2})h$ after $\Delta t$ is obtained by considering the exchange of gene frequencies between these adjacent classes.

$$\phi(x, t+\Delta t)h = \phi(x, t)h - \{v(x)+m(x)\}\Delta t\phi(x, t)h + \{v(x-h)/2\}\Delta t\phi(x-h, t)h + \{v(x+h)/2\} \times \Delta t\phi(x+h, t)h + m(x-h)\Delta t\phi(x-h, t)h \qquad (44)$$

The second term in the right side is the amount of loss due to movement to other classes, the third term is the contribution from the left class, the fourth by the right class both due to random change and the final term is the contribution from the left class due to systematic change.

If we denote by $\sigma^2(x,t)\Delta t$ the variance of the change in x per $\Delta t$ due to random change,

$$\sigma^2(x, t)\Delta t = h^2(v(x)/2)\Delta t + (-h)^2(v(x)/2)\Delta t$$

or

$$\sigma^2(x, t) = h^2 v(x).$$

Similarly if $M(x, t)\Delta t$ be the mean change in x per $\Delta t$,

$$M(x,t)\Delta t = hm(x)\Delta t \text{ or } M(x,t) = m(x)h.$$

Substituting these in the right side of (44) and dividing both sides by $\Delta t \cdot h$, we have after some rearrangement,

$$\frac{\phi(x, t+\Delta t) - \phi(x, t)}{\Delta t} = \frac{1}{2}\frac{\dfrac{\sigma^2(x+h, t)\phi(x+h, t) - \sigma^2(x, t)\phi(x, t)}{h} - \dfrac{\sigma^2(x, t)\phi(x, t) - \sigma^2(x-h, t)\phi(x-h, t)}{h}}{h} - \frac{M(x, t)\phi(x, t) - M(x-h, t)\phi(x-h, t)}{h}$$

Taking the limit $\Delta t \to 0$, $h \to 0$, we have

$$\frac{\partial\phi(x,t)}{\partial t} = \frac{1}{2}\frac{\partial^2}{\partial x^2}\left(\sigma^2(x, t)\phi(x, t)\right) - \frac{\partial}{\partial x}\left(M(x, t)\,\phi(x, t)\right) \qquad (45)$$

In the above derivation we have assumed that the systematic pressure pushes the gene frequency toward the right, but no essential change is required for the argument if it pushes toward the left, in which case we simply use $m(x+h)\Delta t\phi(x+h,t)h$ as the last term of (44).

Next we want to show that if the steady state distribution $y = \phi(x)$ exists, it satisfies

$$\frac{1}{2}\frac{\partial}{\partial x}\left\{V_{\delta x}\phi(x)\right\} - M_{\delta x}\phi(x) = 0.$$

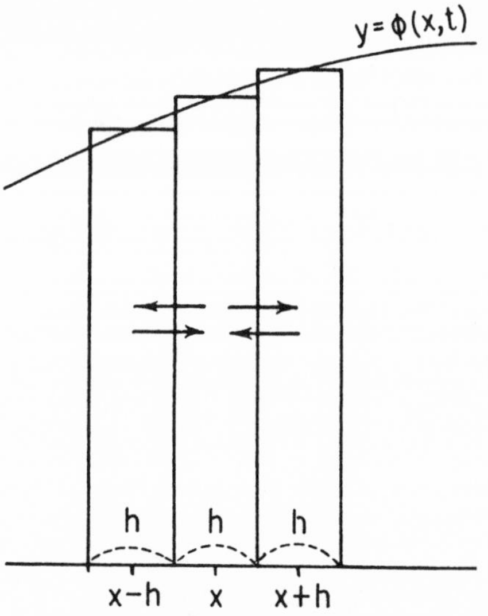


FIGURE 14.

Here $V_{\delta x}$ and $M_{\delta x}$ must be independent of t. Consider the net flow of probability, $P(x)\Delta t$, across the point $x+(h/2)$, per $\Delta t$. Under the assumption of small change per infinitesimal time, we may consider only the exchange of frequencies between the class with gene frequency x and that of x + h.

By the similar argument as above

$$P(x)\Delta t = m(x)\Delta t\phi(x)h + \{v(x)/2\}\Delta t\phi(x)h - \{v(x+h)/2\}\Delta t\phi(x+h)h$$

or

$$P(x) = m(x)h\phi(x) - \frac{1}{2}\left\{v(x+h)\phi(x+h) - v(x)\phi(x)\right\}h.$$

Substitute $m(x)h = M_{\delta x}$ and $v(x)h = V_{\delta x}/h$ and let $h \to 0$. Then we have

$$P(x) = M_{\delta x}\phi(x) - \frac{1}{2}\frac{\partial}{\partial x}\left\{V_{\delta x}\phi(x)\right\}.$$

At the steady state, there is no change in the distribution curve and the net flow of probability must be zero at any point on [0, 1]. Thus the condition $P(x) = 0$ gives the required relation.

For a more rigorous and less intuitive derivation of the Fokker-Planck equation, the reader may refer to some standard works on stochastic processes.

Usually they start by defining the quantity;

$$\mu_n(x, t, \Delta t) = \int (x_1 - x)^n p(x_1, t+\Delta t | x,t)dt,$$

where $p(x_1,t + \Delta t \mid x, t)$ is the conditional probability that the random variable changes to $x_1$ after time interval $\Delta t$ given that it is x at time t. It is then assumed that

$$\lim_{\Delta t \to 0} \{\mu_1(x, t, \Delta t)/\Delta t\} = M(x, t) \text{ and}$$

$$\lim_{\Delta t \to 0} \{\mu_2(x, t, \Delta t)/\Delta t\} = \sigma^2(x, t)$$

exist and

$$\lim_{\Delta t \to 0} \{\mu_n(x, t, \Delta t)/\Delta t\} = 0 \text{ for } n \geq 3.$$

Therefore it seems as if the second moment of $\delta x$, that is, $E(\delta x)^2 = V_{\delta x} + (M_{\delta x})^2$, should be used for $\sigma^2(x,t)$ which is defined through the mean square of the infinitesimal change relative to the original position (x). Mathematically, the definition through the second moment is equivalent to the one through variance, since the latter turns out to be

$$\lim_{\Delta t \to 0} \frac{1}{\Delta t} \int (x_1 - \bar{x}_1)^2 p(x_1, t+\Delta t | x, t)dx_1$$

$$= \lim_{\Delta t \to 0} \frac{\mu_2(x, t, \Delta t) - \mu_1^2(x, t, \Delta t)}{\Delta t} =$$

$$\lim_{\Delta t \to 0} \frac{1}{\Delta t} \{ \sigma^2(x, t)\Delta t - (M(x, t)\Delta t)^2\} = \sigma^2(x, t).$$

However, when applied to population genetics, to use $V_{\delta x}$ for $\sigma^2(x,t)$ has definite advantages over using $V_{\delta x} + (M_{\delta x})^2$. This comes from the fact that $\sigma^2(x,t)\Delta t$ can only be obtained for finite $\Delta t$, for example, one generation. It would be highly desirable for (1) or (45) to give the deterministic change when no random fluctuation is involved ($V_{\delta x} = 0$), that is if x at time t satisfies the first order equation:

$$\frac{dx_t}{dt} = M_{\delta x}. \qquad (46)$$

If we stick to the definition through the second moment, $\sigma^2(x,t)$ should be replaced by $(M_{\delta x})^2$, a quantity which, though small, is not strictly zero. On the other hand if we use on the variance definition, we must put $\sigma^2(x,t) = 0$. We can demonstrate that the latter is the adequate one. For, putting $\sigma^2(x,t) = 0$, (1) becomes

$$\frac{\partial \phi}{\partial t} = -\frac{\partial}{\partial x}\{M_{\delta x}\phi\},$$

which is easily changed into Lagrange form by putting $M_{\delta x}\phi = \psi$;

$$\frac{\partial \psi}{\partial t} + M_{\delta x}\frac{\partial \psi}{\partial x} = 0.$$

This yields the subsidiary equations

$$\frac{d\psi}{0} = \frac{dt}{1} = \frac{dx}{M_{\delta x}}$$

Thus $\psi = M_{\delta x}\phi = C_1$ and $x = x_t + C_2$ where $x_t$ satisfies (46), and $C_1$ and $C_2$ are arbitrary constants. Then it follows that $\phi$ should be expressed in the form: $\phi(x,t) = f(x-x_t)$ where f is an arbitrary function. If the gene frequency at time $t = 0$ is $x_0$, we have $\phi(x,0) = \delta(x-x_0)$ as the initial condition, giving

$$\phi(x, t) = \delta(x - x_t), \qquad (47)$$

where $\delta$ is Dirac's delta function. (47) is equivalent to (46).

The above elementary method can readily be extended to derive the bivariate Fokker-Planck equation. Let the gene frequencies in two loci be $x_1$ and $x_2$ ($0 \leq x_1 \leq 1$, $0 \leq x_2 \leq 1$). Here we must consider the distribution surface; $y = \phi(x_1, x_2; t)$. The area under this surface is approximated by a three-dimensional histogram standing on the $x_1$— $x_2$ plane, which is divided into squares with side of length h. As shown in Figure 15, we must consider exchanges of frequencies between 6 surrounding classes, and the exchanges between diagonal directions give the covariance term ($W_{\delta x_1 \delta x_2}$).

## Appendix II

*Derivation of the formulae for the rate of fixation and of loss (12) and (13)*

As shown in (8) the solution is written in the following form:

$$\phi(x, p; t) = \sum_{k=0}^{\infty} C_k U_k e^{-\lambda_k t} \qquad (48)$$

where $U_k(x)$'s are continuous and finite on [0, 1].

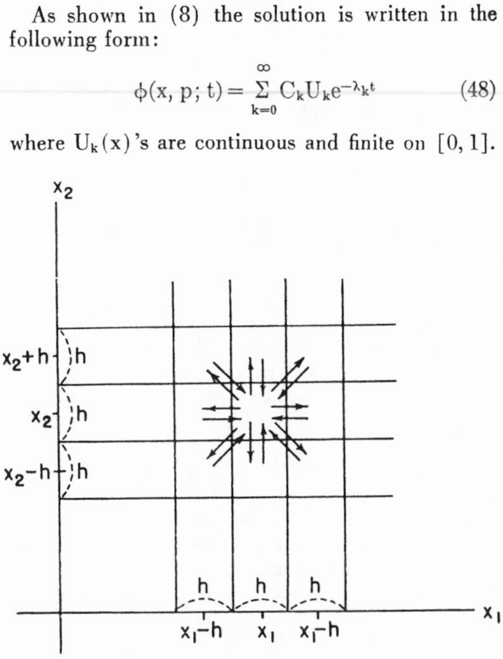


Figure 15.

The rate of decrease of the frequency of total unfixed classes at time t is

$$-\frac{\partial}{\partial t}\int_0^1 \phi(x, p; t)dx$$

which is the rate of increase of the fixed (homallelic) classes. From (48) this becomes

$$-\frac{\partial}{\partial t}\int_0^1 \phi dx = \sum_{k=o}^{\infty} C_k \int_0^1 \lambda_k U_k(x) e^{-\lambda_k t} \quad (49)$$

since the series is uniformly convergent for $t > 0$. Now from (4) and (48), we have

$$-4N\lambda_k U_k(x) = \{x(1-x)U_k(x)\}'' -4Ns\{x(1-x)U_k(x)\}'$$

and if we substitute this in (49)

$$-\frac{\partial}{\partial t}\int_0^1 \phi dx = -\frac{1}{4N}\sum_{k=0}^{\infty} C_k \int_0^1 \Big[\{x(1-x)U_k(x)\}'' -4Ns\{x(1-x)U_k(x)\}'\Big] e^{-\lambda_k t}dx$$

$$= -\frac{1}{4N}\sum_{k=0}^{\infty} C_k \Big[\{x(1-x)U_k(x)\}' -4Nsx(1-x)U_k(x)\Big]_0^1 e^{-\lambda_k t}$$

$$= -\frac{1}{4N}\sum_{k=0}^{\infty} C_k \Big[x(1-x)U_k'(x) +(1-2x)U_k(x)\Big]_0^1 e^{-\lambda_k t}$$

$$= \frac{1}{4N}\left\{\sum_{k=0}^{\infty} C_k U_k(1) e^{-\lambda_k t} + \sum_{k=0}^{\infty} C_k U(0) e^{-\lambda_k t}\right\}$$

$$= \{\phi(1, p; t) + \phi(0, p; t)\}/(4N).$$

Since $\phi(x,p,t)$ is continuous on the closed interval [0, 1], the subterminal classes have a sufficiently flat distribution at the neighborhoods of 0 and 1 under the continuous model. Thus the rate of fixation and that of loss (fixation to $x = 0$) must be in the ratio $\phi(1,p,t) : \phi(0,p,t)$, which means that they are $\phi(1,p,t)/(4N)$ and $\phi(0,p,t)/(4N)$ respectively.

## SUMMARY


The purpose of the present paper is to discuss the stochastic processes and the distributions of gene frequencies under natural selection, with special reference to recent controversies on the bearing of random genetic drift and selection on evolution.

First we considered the process of genic selection in finite populations, that is, the process of genic selection under random drift. The complete solution has been worked out in terms of spheroidal wave functions and the behavior at the state of steady decay has been illustrated using some examples. The relation between the rate of decay (fixation plus loss) and the joint effect of selection and population size has been tabulated (Table 1, see also Figs. 1 and 2). It will be seen from these results that if Ns is less than about 1, there is a considerable chance for fixation of a nonadaptive gene.

Second we studied the process due to random fluctuation of selective force from generation to generation. Though a complete solution has not been obtained for an arbitrary selective value s, two contrasting cases have been analyzed successfully: First is the process leading to "quasifixation" and "quasiloss" when the gene is neutral in the sense of long-term average ($\bar{s} = 0$).

Second is the case where the mean selective value $\bar{s}$ is much larger than the variance of s (Fig. 3). The subtle interplay between deterministic pressure and the random effect in selection which is expected at the intermediate situation is left as an open question.

Then the chance of fixation of a favorable mutant gene when there is random fluctuation of selective value was studied. This has led to the conclusion that for a semidominant gene with mean selective advantage $\bar{s}$, $>0$, the chance of ultimate fixation is approximately $2\bar{s}/(1 + V_s)$, where $V_s$ is the variance of s. Thus the random fluctuation of selection intensity tends to reduce the probability of fixation, but the amount of reduction may be generally very small or even negligible.

The central part of this paper is devoted to the analysis of the relative importance of effective population number and random fluctuation of selection intensity in determining the gene frequency distributions at the steady state. The distributions have been illustrated for various population sizes as well as for various modes and intensity of selection (see Figs. 4-14). Through this analysis it has been shown that for small populations the random fluctuation of selection intensity is rather unimportant, while for a large one it plays a prominent role.

Becaues of its mathematical difficulty, the analysis of effect of cyclic change in selection intensity could not be carried out extensively, but two contrasting cases studied showed some interesting properties.

Throughout this paper I have tried to develop the technique of consistent application of the method of partial differential equations (Fokker-Planck or Kolmogorov equation) to population genetics. A simple and elementary derivation of the equation is presented with the hope that this will help our understanding of the meaning of this important equation (Appendix I). The equation can be extended to cover the general case of an arbitrary number of loci each with an arbitrary number of alleles, and the relation of this equation with the most general formula for the steady state gene frequency distribution given by Wright (1949) was discussed. These differential equations should have a fundamental meaning for the study of the stochastic processes and the steady state distribution in population genetics. However, if we take intergroup selection into consideration, the process of change involved is not Markovian and the Kolmogorov equation cannot be applied. The extension of the method of the partial differential equation to such a situation is discussed briefly.


ACKNOWLEDGMENT

The writer expresses his thanks to Dr. James F. Crow for his valuable help in the course of this work. Thanks are also due to Dr. E. R. Immel who read the manuscript and gave criticism and helpful suggestions. Finally the writer expresses his indebtedness to Dr. Philip M. Morse of Massachusetts Institute of Technology for useful information on spheroidal wave functions.


## REFERENCES


BARTLETT, M. S., 1949, Some evolutionary stochastic processes. J. R. Statist. Soc. (B) *11:* 211-229.

BATEMAN, H., 1944, Partial Differential Equations of Mathematical Physics. Dover, New York.

CHANDRASEKHAR, S., 1943, Stochastic problems in physics and astronomy. Rev. Mod. Physics *15:* 1-89.

CROW, J. F., and KIMURA, M., 1955, Some genetic problems in natural populations. Proc. Third Berkeley Symp. Math. Statist. Prob. Univ. Calif. Press (in press).

DOBZHANSKY, TH., 1951, Genetics and the Origin of Species. Third Edition. New York, Columbia Univ. Press.

FELLER, W., 1951a, Diffusion processes in genetics. Proc. Second Berkeley Symp. Math. Statist. Prob. Univ. Calif. Press. pp. 227-246.

1951b, An Introduction to Probability Theory and its Applications. John Wiley, New York.

FISHER, R. A., 1922, On the dominance ratio. Proc. Roy. Soc. Edinb. *42:* 321-341.

1930a, The distribution of gene ratios for rare mutations. Proc. Roy. Soc. Edinb. *50:* 204-219.

1930b, The Genetical Theory of Natural Selection. Oxford, Clarendon Press.

1941, Average excess and average effect of a gene substitution. Ann. Eugen. *11:* 53-63.

1953, Population genetics. Proc. Roy. Soc. B. *141:* 510-523.

FISHER, R. A., and FORD, E. B., 1947, The spread of a gene in natural conditions in a colony of the moth *Panaxia dominula* L. Heredity *1:* 143-174.

1950, The "Sewall Wright Effect." Heredity *4:* 117-119.

GOLDBERG, S., 1950, On a Singular Diffusion Equation. Ph.D. Thesis, Cornell University. Unpub.

HALDANE, J. B. S., 1927, A mathematical theory of natural and artificial selection, Part V: Selection and mutation. Proc. Camb. Phil. Soc. *23:* 838-844.

1949a, Some statistical problems arising in genetics. J. R. Statist. Soc. (B) *11:* 1-9.

1949b, Human evolution: Past and future. In: Genetics, Paleontology, and Evolution, pp. 405-418. Princeton Univ. Press.

KIMURA, M. 1953a, "Stepping-Stone" Model of Population. Ann. Rep. Nat. Inst. Gen. Japan *3:* 62-63.

1953b, On simultaneous distribution of gene frequencies in populations. Ann. Rep. Nat. Inst. Gen. Japan *3:* 65-66.

1954, Process leading to quasi-fixation of genes in natural populations due to random fluctuation of selection intensities. Genetics *39:* 280-295.

1955a, Solution of a process of random genetic drift with a continuous model. Proc. Nat. Acad. Sci. Wash. *41:* 144-150.

1955b, Random genetic drift in multi-allelic locus. (in press).

1956, Random genetic drift in a tri-allelic locus: Exact solution with a continuous model. (in press).

KOLMOGOROV, A., 1931, Über die analytischen Methoden in der Wahrscheinlichkeitsrechnung. Math. Ann. *104:* 415-458.

MAYR, E., 1954, Change of genetic environment and evolution. In: Evolution as a Process, pp. 157-180. London, George Allen and Unwin.

MULLER, H. J., 1949, Reintegration of the Symposium on Genetics, Paleontology, and Evolution. In: Genetics, Paleontology and Evolution, pp. 421-445. Princeton Univ. Press.

SHEPPARD, P. M., 1954, Evolution in bisexually reproducing organisms. In: Evolution as a Process, pp. 201-218. London, George Allen and Unwin.

STRATTON, J. A., MORSE, P. M., CHU, L. J., and HUTNER, R. A., 1941, Elliptic Cylinder and Spheroidal Wave Functions. New York, John Wiley.

WILSON, A. H., 1928, A generalised spheroidal wave equation. Proc. Roy. Soc. (A) *118:* 617-635.

WRIGHT, S., 1931, Evolution in Mendelian populations. Genetics *16:* 97-159.

1935, Evolution in populations in approximate equilibrium. J. Genet. *30:* 257-266.

1937, The distribution of gene frequencies in populations. Proc. Nat. Acad. Sci. Wash. *23:* 307-320.

1938, The distribution of gene frequencies under irreversible mutation. Proc. Nat. Acad. Sci. Wash. *24:* 253-259.

1939, Statistical genetics in relation to evolution. Exposés de Biométrie et de Statistique Biologique 13, Actualités Sci. Industr. *802.*

1940, Breeding structure of populations in relation to speciation. Amer. Nat. *74:* 232-248.

1945, Tempo and Mode in Evolution: A critical review. Ecology *26:* 415-419.

1948, On the roles of directed and random changes in gene frequency in the genetics of populations. Evolution *2:* 279-294.

1949, Adaptation and selection. In: Genetics, Paleontology, and Evolution, pp. 365-389. Princeton, Princeton Univ. Press.

1950, Population structure as a factor in evolution. In: Moderne Biologie, pp. 275-287. Berlin.

1951a, Fisher and Ford on "The Sewall Wright Effect." Amer. Scient. *39:* 452-479.

1951b, The genetical structure of populations. Ann. Eugen. *15:* 323-354.

WRIGHT, S., and KERR, W. E., 1954, Experimental studies of the distribution of gene frequencies in very small populations of *Drosophila melanogaster.* II. Bar. Evolution *8:* 225-240.


## DISCUSSION

KEMPTHORNE: I wonder if Mr. Kimura's treatment can be extended to deal with the case of interactions between genes both within and between loci. Also it would be nice if one could do something about non-random mating, either consanguineous or assortative.

KIMURA: The general equation, $\partial\phi/\partial t = L(\phi)$, where L is the differential operator as given in (41), can include any degree of interaction between genes at the same locus as well as different loci. In that case $V_{\delta xij}$, $W_{\delta xij;x_{i'j'}}$ and $M_{\delta x_{ij}}$ are not only functions of $x_{ij}$ alone but also functions of frequencies of all other genes and also functions of time t. The trouble here is the great difficulty involved in solving the equation for complicated systems of gene interaction. Though random mating is assumed in this equation, the method can be extended to include any system of mating. In that case we will have to specify a population as an aggregate of zygotic frequencies rather than an aggregate of gene frequencies alone.

MALÉCOT: The following remarks concern the decrease of relationship with distance.

Consider a population extending over a large area. Assuming that there is no selection on the gene *a* and its alleles and that the mutation pressure is the same over the whole area, the general mean of the frequency q of gene *a* over the whole area will be, when a steady state is reached, a constant $\bar{q}$, depending only on the mutation pressure. But the local frequencies q in small parts of the area will fluctuate around $\bar{q}$. To study these fluctuations, Dr. Wright uses the "coefficient of inbreeding" (correlation between uniting gametes, or probability that two homologous loci of an individual come from the same locus of a common ancestor). I shall call it $f_0$. A rigorous calculation of $f_0$ needs a previous study of a more general coefficient, the coefficient of relationship between two individuals of the same generation $F_n$ located in places $\alpha$ and $\beta$ separated by a distance x; this coefficient is, by definition, the probability that two homologous loci of the two individuals come from the same locus of a common ancestor; in the simplest case of "homogeneous" dispersal of individuals, this coefficient is a function of x only, $f_n(x)$. If we assume random mating, the "coefficient of inbreeding" $f_0$ is the same as $f_n(0)$. To calculate $f_n(x)$, it is necessary to know the law of dispersal of individuals. If one were to assume that each parent of an individual I of $F_n$ born in place $\alpha$ has a probability $g(\alpha,\mu)\,d\mu$ to be born in place u, it would be easy to write the probability for each ancestor of I to be born in each possible place, and therefore the probability for two individuals I and J of $F_n$ to have a common ancestor (a probability which is a function of the density $\delta$ of individuals, that is, of their number per unit area, or length). But it is better to have a formula relating $f_n(x)$ and $f_{n-1}(x)$. For mere simplification of writing we shall suppose that the dispersal is only along one dimension, for instance along a straight line.

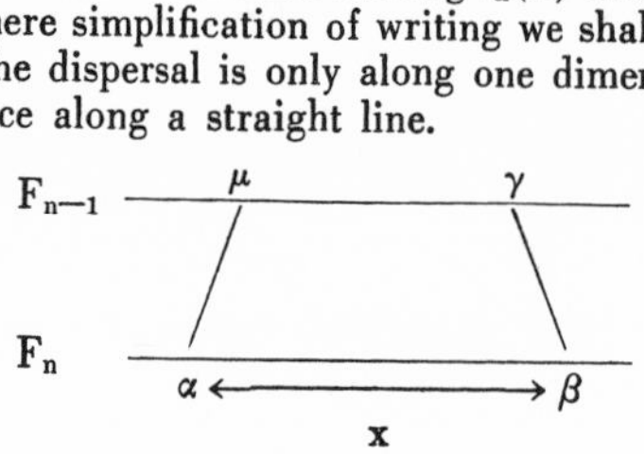


Where $x = \beta - \alpha$, $\alpha$ and $\beta$ being the abscissas of I and J. Let us introduce the "moments of dispersal":

$$\int (\mu-\alpha) g(\alpha, \mu) d\mu = M_1$$

$$\int (\mu-\alpha)^2 g(\alpha, \mu) d\mu = M_2, \text{ etc.}$$

$\mu$ being the abscissa, at birth, of a parent of individual I, himself of abscissa $\alpha$ at birth. The same hypotheses as in the derivation of the diffusion equation of Kolmogoroff allow us to write, if u is the mean mutation rate and $M_2 - M_1^2 = \sigma^2$ the "variance of dispersal":

$$(1)\quad f_n(x) = (1-u)^2\left[ f_{n-1}(x) + \sigma^2 \frac{\partial^2 f_{n-1}}{\partial x^2} + \frac{1-f_0}{2\delta} \int g(\alpha, \mu) g(\beta, \mu) d\mu \right]$$

When n tends to infinity, $f_n(x)$ tends to a limit $f(x)$ which, when neglecting $u^2$, satisfies the equation:

$$(2)\quad 2uf(x) - \sigma^2 \frac{\partial^2 f}{\partial x^2} = (1-u)^2 \frac{1-f_0}{2\delta} \int g(\alpha,\mu) g(\beta,\mu) d\mu$$

The second member (which corresponds in (1) to the case where the two individuals in $\alpha$ and $\beta$ have the same locus from the same parent in place $\mu$ in $F_{n-1}$) becomes negligible when the distance $\beta - \alpha = x$ is sufficiently larger than $\sigma$. When $x \gg \sigma$, $f(x)$ is given by equation (2) without the second member; $f(x) = K e^{-\sqrt{\frac{2u}{\sigma^2}}\, x}$ (since it must be bounded). So the coefficient of relationship decreases exponentially when the distance x increases. It is easy to show that the same formula applies to the coefficient of correlation $r(x)$ between the frequencies q in places at distance x [coefficient which may be estimated by

$$\frac{\sum_{\alpha} (q_\alpha - \bar{q})(q_{\alpha+x} - \bar{q})}{\sum_{\alpha} (q_\alpha - \bar{q})^2},$$

the summation being over places $\alpha$ where $q_a$ is empirically determined]. I have shown that the formula remains valid for dispersal in two dimensions or more, and also for a population clustered in regular groups, the variance of dispersal being calculated from the probabilities of escaping from one group to another.

But, when $x/\sigma$ is not large, the second member of formula (2) cannot be neglected; applying to (2) a Fourier transform, one obtains (when dispersal is in one dimension and if $g(\alpha,\mu)$ is a normal density of probability):

$$(3)\quad f(x) = \frac{1-f_0}{2\pi\delta} \int_{-\infty}^{\infty} \frac{e^{-\sigma^2 t^2}}{\sigma^2 t^2 + 2u} e^{-itx} dt.$$

In particular, putting $x = 0$ and solving in $f(0)$, one obtains:

$$f_0 = \frac{1}{1+4\sigma\delta\sqrt{2u}}. \tag{4}$$

In case of dispersal in two dimensions, a different formula is found:

$$f_0 = \frac{1}{1+8\pi\delta\sigma^2(-1/\log 2u)}. \tag{5}$$

In the denominator of both formulas, one finds what Dr. Wright calls the "effective number" $N = 4\pi\delta\sigma^2$ (or $N = 4\sigma\delta$), the number of individuals in a circle (or a segment) of radius $2\sigma$; but the mutation rate does not play the same part in (4) and (5). So the coefficient of inbreeding $f_0$ is much influenced by the number of dimensions; on the contrary, the decrease with distance of the coefficient of relationship or of correlation is approximately the same, $e^{-\sqrt{\frac{2u}{\sigma^2}}x}$ in all cases; this latter formula also remains valid for a quantitative character determined by additive gene effects (it is remarkable that this formula does not depend upon density $\delta$, whereas $f_0$ depends on it). It seems that this exponential decrease could be tested experimentally in some populations, as Dr. Lamotte did in *Cepaea nemoralis*. The measure of decrease with distance gives an estimation of $u/\sigma^2$, and hence of $u$; there is indeed for $u$ a risk of being overestimated if there is some "microselection" causing the coefficient of correlation to fall off more quickly than if mutation alone were operating. Perhaps this fact could be taken in account in the formula (naturally, if there is a "cline" in geographical selection the formula becomes meaningless).

FREIRE-MAIA: Using the Catholic marriage records, the mean inbreeding coefficient has been determined in 17 South American dioceses (15 in Brazil, 1 in Argentina and 1 in Uruguay). They are the following: 0.0083, 0.0045, 0.0029, 0.0025, 0.0023, 0.0019, 0.0018, 0.0017, 0.0013, 0.0011, 0.0010, 0.0010, 0.0008, 0.0006 (Montevideo and surrounding localities), 0.0005 (Buenos Aires), 0.0005 and 0.0004. In the parishes, the highest inbreeding coefficients are 0.0156, 0.0152, 0.0137, 0.0126, 0.0122, 0.0118, 0.0114 and 0.0106, found in localities of a northern and a central state. The lowest inbreeding rates have been found in southern states.

A negative correlation was found, as expected, between inbreeding coefficient and exogamy index, being in this case the ratio of the frequency of marriages between persons born in different localities by the frequency of marriages between persons born in the same locality.

It is interesting to note that the highest exogamy indices so far discovered (19.83 and 15.39) have not been found in the big cities studied, but in some localities of the northern part of the state of Paraná. This region has undergone since about 15 years ago a quite rapid development caused mainly by the new and extensive coffee plantations. To explain the high exogamy indices found there it is sufficient to say that in one of the localities studied, about 87 per cent of the marrying people in 1953-1954 were born outside the state.

# SOME PROBLEMS OF STOCHASTIC PROCESSES IN GENETICS[1]

MOTOO KIMURA[2]

*Department of Genetics, University of Wisconsin*

**Summary.** In genetics, stochastic processes arise at all levels of organization ranging from subunits of the gene to natural populations. Types of stochastic processes involved are also diverse. In the present paper, the following five topics have been selected for mathematical discussion and new results are presented: (1) Random assortment of subunits of a gene. (2) Senescence in paramecium due to random assortment of chromosomes. (3) Process of natural selection in a finite population (interaction between selection and random genetic drift). (4) Chance of fixation of mutant genes. (5) Population structure and evolution. Finally it is pointed out that new mathematical techniques will be needed for a satisfactory treatment of Wright's theory of evolution.

"Elles n'auroient dû leur premiere origine qu'à quelques productions fortuites, dans lesquelles les parties élémentaires n'auroient pas retenu l'ordre qu'elles tenoient dans les animaux peres & meres: chaque degré d'erreur auroit fait une nouvelle espece: . . .

Des moyens différents des moyens ordinaires que la Nature emploie pour la production des animaux, loin d'être des objections contre ce systême, lui sont indifférents, ou lui seroient plutôt favorables."[3]

—Maupertuis (Oeuvres, 1756)

**1. Introduction.** These words, written two centuries ago, foreshadow the stochastic nature of genetic and evolutionary processes. Actually, stochastic processes are found in all levels of organization with which genetics is concerned, in the gene, the cell, the organism, and the population.

The types of stochastic processes involved are also diverse. Of special importance is the Markov process, which Kolmogorov [1] called stochastic definite; the exact treatment of regular systems of inbreeding is a typical example of a

Received March 26, 1956; revised May 14, 1957.

[1] Paper No. 680 from the Department of Genetics, University of Wisconsin. This work was supported by a grant from the University Research Committee from funds supplied by the Wisconsin Alumi Research Foundation. Parts of this paper were presented to the New York Meeting of the Institute of Mathematical Statistics, December 27, 1955, at the session "Probability and Statistics in Genetics."

[2] On leave from the National Institute of Genetics, Mishima-shi, Japan. This paper also constitutes Contribution No. 148 of the National Institute of Genetics.

[3] "They [i.e. species] would have owed their first origin only to certain fortuitous productions, in which the elementary particles have not retained the arrangement that they had in the father and mother animals: each grade of error would have made a new species. . . .

"As to methods [e.g. parthenogenesis, fission] different from the ordinary methods that Nature employes for the production of animals, far from being objections to this theory, they are indifferent to it, or rather would be favorable to it."

finite Markov chain, though workers in this field have seldom used such a terminology [2] [3]. The fate of an individual mutant gene appearing in a population may best be studied by the theory of branching processes. The probability distribution of gene frequencies in natural populations is important in the mathematical theory of evolution developed by Fisher and Wright. It contains many difficult problems of continuous stochastic processes in which the Kolmogorov equation plays a fundamental role [4] [5] [6] [7].

In the present paper a few topics will be selected from various levels of organization ranging from subunits of the gene to natural populations.

**2. Random assortment of subunits in chromosome division.** The idea that each gene is composed of a number of subunits is a natural one, since analogous situations are familiar in physics and chemistry. If the subunits are of two or more different kinds, they will be sorted out in the process of chromosome division.

In fact, such a model was proposed more than three decades ago to explain the high mutability of the so-called mutable genes. Unfortunately, precise experimental results in a few higher organisms apparently contradicted this model [8]. Later, however, as investigations of the finer structure of the chromosome have developed, a multiple-strand structure has been revealed and this encouraged the formulation of the same type of model again. Matsuura and Suto [9], in order to explain certain irregular segregations in maize, assumed that each chromosome contains 8 strands and that mutation may affect any one of the 8 gene replicates. A similar model was used by Auerbach [10] to explain the occurrence of mosaics in the offspring of *Drosophila* males treated with mustard gas. More recently, Friedrich-Freksa and Kaudewitz [11] carried out an interesting experiment with *Amoeba proteus* treated with radioactive $P^{32}$, in which they assumed that sorting-out of the radiation-damaged strands or subunits causes death to the organism in later generations.

Let us consider a model in which each chromosome consists of $n$ subunits and suppose that a mutation has occurred in one of them. The subunits duplicate to produce $2n$ which separate at random into two groups of $n$ subunits to form the daughter chromosomes. Thus the total number of subunits per chromosome is kept constant, but the number of the mutant subunits may change from generation to generation due to random segregation. We follow a single line of descent obtained by selecting randomly one of the pair of daughter chromosomes in each generation. We shall designate by $E_i (i = 0, 1, \cdots, n)$ the state in which a given chromosome contains exactly $i$ mutant subunits. Let $a_i^{(t)}$ be the probability that the chromosome is in the state $E_i$ at the $t$th generation (assuming that the mutation occurred at $t = 0$). In the present model, the transition probabilities $p_{j|i} \equiv \Pr\{E_j \leftarrow E_i\} (i, j = 0, 1, 2, \cdots, n)$ are given by

$$p_{j|i} = \binom{2i}{j}\binom{2n-2i}{n-j} \Big/ \binom{2n}{n}. \tag{2.1}$$

Thus the probability chain will be expressed by

$$a^{(t)} = Pa^{(t-1)}, \tag{2.2}$$

in which $a^{(t)}$ is a column vector whose $i$th element is $a_i^{(t)}$ and $P$ is an $n \times n$ matrix whose element in the $i$th row, $j$th column is $p_{j|i}$. Obviously $E_0$ and $E_n$ are absorbing barriers and the remaining states $E_1, \cdots, E_{n-1}$ are all transient. Eigenvalues of the matrix $P$, which satisfy $|P - \lambda I| = 0$, are

$$\lambda_r = 2^r \binom{2n-r}{n-r} \Big/ \binom{2n}{n} \qquad (r = 0, 1, \cdots, n). \tag{2.3}$$

This can be shown by following a procedure similar to that which Feller [5] gave in an appendix of his paper (p. 244), noting that a non-trivial set of $y_j$'s which satisfy

$$\sum_{j=0}^{n} p_{j|i} y_j = \lambda_r y_i$$

can be written in the form

$$y_j = \sum_{\nu=0}^{r} c_\nu j^{(\nu)} \qquad (c_\nu \text{ is a constant})$$

and

$$\sum_{j=0}^{n} p_{j|i} j^{(\nu)} = (2i)^{(\nu)} \binom{2n-\nu}{n-\nu} \Big/ \binom{2n}{n},$$

where $j^{(\nu)} = j(j-1) \cdots (j - \nu + 1)$ is the factorial of degree $\nu$.

Though the general expressions for eigenvectors of $P$ and its transpose $P'$ corresponding to these eigenvalues do not seem easily obtainable, we can obtain numerical results for small values of $n$. Thus we can construct formulae giving probabilities of various states at a given generation.

I have worked out the cases of $n = 2, 3, 4, 5, 6, 8$, and 16, the details of which will be published elsewhere.

We are particularly interested in the probability ($d^{(t)}$) with which a mutant chromosome (i.e., a chromosome containing only mutant subunits) first appears by the sorting out process in a given generation $t(t = 1, 2, 3, \cdots)$. This is obtained as $d_n^{(t)} = a_n^{(t)} - a_n^{(t-1)}$ for the case of $n$ subunits. For $n = 2, 4, 8$ and 16, we have

$$d_2^{(t)} = \frac{1}{6}\left(\frac{2}{3}\right)^{t-1},$$

$$d_4^{(t)} = \frac{1}{3080}\left\{195\left(\frac{6}{7}\right)^{t-1} - 330\left(\frac{4}{7}\right)^{t-1} + 135\left(\frac{16}{70}\right)^{t-1}\right\},$$

$$d_8^{(t)} = \frac{1}{12870}\left\{(242.0)\left(\frac{14}{15}\right)^{t-1} - (709.1)\left(\frac{12}{15}\right)^{t-1} + (904.5)\left(\frac{120}{195}\right)^{t-1} - (686.9)\left(\frac{80}{195}\right)^{t-1} + \cdots\right\},$$

TABLE 1

| $n$ | $t_{max}$ | $d_{max}$ (%) |
|---|---|---|
| 2 | 1 | 16.667 |
| 4 | 5 | 2.287 |
| 8 | 15 | 0.481 |
| 16 | 35 | 0.111 |

$$d_{16}^{(t)} = 0.0051626\left(\frac{30}{31}\right)^{t-1} - 0.0193172\left(\frac{28}{31}\right)^{t-1} + 0.0358431\left(\frac{728}{899}\right)^{t-1} - 0.0457009\left(\frac{624}{899}\right)^{t-1} + 0.0451643\left(\frac{13728}{24273}\right)^{t-1} + \cdots.$$

If the mutant chromosome causes death to the Ameba as in the model of Friedrich-Freksa *et al.* [11], $d^{(t)}$ gives the probability of a death at the $t$th generation.

It may be expected on intuitive grounds that the larger the number of subunits, the later will be the appearance of the mutant chromosome. One way of expressing this tendency is to calculate the value of $t$ which maximizes $d_n^{(t)}$. Table 1 shows these values and corresponding values of $d$.

Next we shall consider the situation in which $n$ is very large. The proportion of mutant subunits $x = i/n$ $(0 \leqq x \leqq 1)$ may be treated as a continuous variable with good approximation. Let $\phi(x, t)$ be the probability density of $x$ at time $t$ measured in generations. If $\delta x$ is the amount of change in $x$ per generation,

$$E(\delta x) = 0, \qquad E(\delta x)^2 = x(1 - x)/(2n - 1) \qquad \text{and, for } k \geqq 3,$$

$E(\delta x)^k$ is $o(1/n)$. Therefore we can use the following differential equation to obtain $\phi(x, t)$ (see [16]):

$$\frac{\partial\phi(x, t)}{\partial t} = \frac{1}{2(2n - 1)}\frac{\partial^2}{\partial x^2}\{x(1 - x)\phi(x, t)\} \qquad (0 < x < 1), \tag{2.4}$$

with the initial condition

$$\phi(x, 0) = \boldsymbol{\delta}\left(x - \frac{1}{n}\right), \tag{2.5}$$

where $\boldsymbol{\delta}$ is the Dirac function. The singular equation (2.4) is equivalent to the one describing the process of random genetic drift in natural populations if we put $2n - 1 = 2N$, $N$ being the effective population size. The complete solution of this equation has been worked out (see Section 4). The points $x = 0$ and $x = 1$ act as absorbing barriers and the rate of fixation at $x = 1$ is given by $\phi(1, t)/4N$ which reduces to

$$d_n^{(t)} = \frac{1}{4Nn}\sum_{i=1}^{\infty}(-1)^{i-1}i(i + 1)(2i + 1)e^{-i(i+1)t/4N}$$

The value of $t$ giving $d_n^{(t)}$ maximum is obtained by solving

$$1 - 15e^{-2r} + 84e^{-5r} - 300e^{-9r} + 825e^{-14r} - \cdots = 0,$$

where $r = t/2N$. The required root of this transcendental equation can be obtained numerically; $r = 1.2940 \cdots$.

From this we can derive the following remarkable asymptotic relations ($n \to \infty$):

$$\text{(2.6)} \qquad \begin{cases} t_{\max} \sim 2.59n, \\ d_{\max} \sim 1.08/(4n^2). \end{cases}$$

Namely, $t_{\max}$ will be proportional to the number of subunits and $d_{\max}$ will be inversely proportional to the square of that number.

Finally it is desirable to consider the cases in which the initial number of mutant subunits is more than one, that is, some number, say $np$, where $p$ is the initial proportion of the subunits ($0 < p < 1$). The limiting probability of absorption at $n$, starting at $i$, is the $i$th component of an eigenvector corresponding to $\lambda = 1$, and from the third formula after equation (2.3) it follows that $p = i/n$ is the required probability. The total frequency of mutant chromosomes in the $t$th generation will be expressed in the form:

$$\text{(2.7)} \qquad f_n^{(t)}(p) = \sum_{\tau=0}^{t} d_n^{(\tau)} = p + \sum_{i=2}^{n} C_i(p, n)\lambda_i^t,$$

where the $C_i(p, n)$'s are functions of $p$ and $n$ but not of $t$. If the mutant chromosome changes the fitness of its possesser, this type of formula must be applied with caution. Generally $d_n^{(t)}$ should be used as a basis of comparing expectation with observed results.

For very large $n$, (2.7) should approach

$$\text{(2.8)} \qquad p + \sum_{i=1}^{\infty} (2i + 1)pq(-1)^i F(1 - i, i + 2, 2, p)e^{-[i(i+1)t/4N]},$$

where $q = 1 - p$ and $F$ designates the hypergeometric function (see (5.3) in Section 5).

The experimental data of Friedrich-Freksa [11] appear to agree with the model for $n = 16$.

**3. Senescence in Paramecium.** It has been known to biologists for a long time that if cultures of the protozoon, *Paramecium*, are kept under exclusive asexual reproduction, they lose vigor and eventually die. This phenomenon is known as senescence or aging of paramecium and in fact is one of the old problems in biology. Recently Dr. T. M. Sonneborn has made extensive studies of this phenomenon and discussed a hypothesis that the aging is due to an accumulation of chromosome aberrations in the macronucleus (cf. e.g. Sonneborn and Schneller [12]). Following the suggestion of Dr. J. Lederberg, I have tried to work out the logical consequences of the stochastic model involved.

The macronucleus is considered to be polyploid, consisting of, say, $m$ chromosome sets each with $n$ chromosomes. As in a polyploid nucleus of higher plants, the various chromosomes are mixed at random inside the nucleus[4]. If we designate the chromosomes of a set by successive letters $A, B, C, \cdots, N$ and designate sets by subscripts, then the normal constituent of the nucleus will be expressed in the form:

$$m\begin{cases} \overbrace{\begin{matrix} A_1 & B_1 & C_1 \cdots N_1 \\ A_2 & B_2 & C_2 \cdots N_2 \\ \vdots & \vdots & \vdots \quad\;\; \vdots \\ A_m & B_m & C_m \cdots N_m \end{matrix}}^{n} \end{cases}$$

We designate the total number of chromosomes by $M(= mn)$.[5]

On this model we assume that at the division of the macronucleus, each chromosome duplicates itself followed by the random distribution of chromosomes into two groups of equal number to form the daughter macronuclei. The death by aging is assumed to occur whenever chromosomes of any one type are lost entirely from the nucleus by chance. Various states of the nucleus will be expressed as $n$ dimensional vectors. Here we have a hierarchical structure of absorbing barriers, and a direct attack on the problem may seem extremely difficult. However, because of the symmetry of the model, we can find an easier approach. Suppose we start from an individual with normal macronucleus, and each generation takes one of the daughters to continue the lineage. Our purpose is to calculate the probability that all the $n$ chromosome types coexist in the individual at the $t$th generation. Since the process of loss of any type of chromosome is irreversible, we can treat the problem as if all possible chromosome constituents are viable and then remove unsuitable parts afterwards.

Let us fix our attention on the $t$th generation. We designate by $P_1$ the probability that all the chromosomes except those of one specific type have been lost by that time, by $P_2$ the probability that all but 2 specific types have been lost and that these two coexist. Generally $P_i$ will be defined in a similar way. Since we can classify $n$ chromosome types into two alternative groups like $A$ vs. non $A$, $A$ or $B$ vs. neither $A$ nor $B$, etc.,

$$\phi_1 = P_1,$$

$$\phi_2 = \binom{2}{1} P_1 + P_2,$$

$$\phi_3 = \binom{3}{1} P_1 + \binom{3}{2} P_2 + \binom{3}{3} P_3, \text{ etc.,}$$

[4] This model is essentially different from the one considered by Kimball and Householder [13] to explain the delayed lethal effect of radiation.

[5] $n$ in this section has a different meaning from that of the previous section. Generally the same symbol in different sections may not have the same meaning.

up to

$$\phi_{n-1} = \binom{n-1}{1} P_1 + \binom{n-1}{2} P_2 + \cdots + \binom{n-1}{n-1} P_{n-1},$$

where $\phi_i$ represents $f_M^{(t)}(i/n)$ in (2.7) $(i = 1, \cdots, n-1)$. For example, $\phi_2$ is the probability that all the chromosomes except $A$ or $B$ or both have been lost and this is a sum of the probabilities that all but $A$ have been lost $(P_1)$, that all but $B$ have been lost $(P_1)$, and that $A$ and $B$ coexist but all others have been lost $(P_2)$. It is convenient to consider the above relations as a linear transformation of $P_i$'s into $\phi_i$'s with an $(n-1) \times (n-1)$ matrix whose element in the $i$th row and $j$th column is $\binom{i}{j}$. The inverse transformation can be shown to have a matrix whose element in the $i$th row and $j$th column is $(-1)^{i+j}\binom{i}{j}$. Let $\Omega_t (\equiv P_n)$ be the probability that all the chromosome types coexist in the macronucleus at the $t$th generation. Since $\sum_{i=1}^{n} \binom{n}{i} P_i = 1$,

$$\Omega_t = 1 - \sum_{i=1}^{n-1} \binom{n}{i} P_i = 1 - \sum_{j=1}^{n-1} (-1)^{n-1+j} \binom{n}{j} \phi_j .$$

Substituting $\phi_j = f_M^{(t)}(j/n)$ from (2.7) and noting that

$$\sum_{j=1}^{n-1} (-1)^{n-1+j} \binom{n}{j} (j/n) = 1,$$

we have

$$\Omega_t = \sum_{j=1}^{n-1} \sum_{i=2}^{M} (-1)^{n+j} \binom{n}{j} C_i\left(\frac{j}{n}, M\right) \lambda_i^t ,$$

where

$$\lambda_i = 2^i \binom{2M-i}{M-i} \Big/ \binom{2M}{M}.$$

According to Sonneborn, the usual strains of *Paramecium* have a chromosome number of $n \doteqdot 41$, but also there are strains with $n \doteqdot 35$ and 50. There is a good reason to believe that the macronucleus is at least 100 ploid $(m = 100)$. Thus the total number of chromosomes $M$ in the macronucleus would be of the order of 5000. This fact will enable us to use the asymptotic formula for $f_M^{(t)}(j/n)$ given in (2.8).

$$\Omega_t = \sum_{j=1}^{n-1} (-1)^{n+j} \binom{n}{j} \sum_{i=1}^{\infty} (2i+1)(j/n)(1-j/n)(-1)^i$$
$$\times F(1-i, i+2, 2, j/n) e^{-[i(i+1)t/4M]}$$

or if we put

$$\alpha_i \equiv \sum_{j=1}^{n-1} (-1)^j \binom{n}{j} (j/n)(1-j/n) F(1-i, i+2, 2, j/n)$$

$$\Omega_t = \sum_{i=1}^{\infty} (2i + 1)(-1)^{n+i} \alpha_i e^{-[i(i+1)t/4M]}. \tag{3.1}$$

It can be shown that $\alpha_i = 0$ for $i < n - 1$. For $i \geqq n - 1$

$$\alpha_{n-1} = -\frac{(2n-2)!}{(n-1)!n^n}, \qquad \alpha_n = 0,$$

and in general, writing $i = (n - 1) + \nu (\nu \geqq 0)$,

$$\alpha_{n-1+\nu} = \frac{-(n-1)!}{(n+\nu)!n^{n-1}} \sum_{\mu=0}^{\nu} (-1)^{\mu} \binom{\nu+n-2}{\mu+n-2} \frac{[\mu+\nu+2(n-1)]!}{(\mu+n-1)!n^{\mu}} \times \mathfrak{S}_{\mu+n-1}^{n-1},$$

where $\mathfrak{S}_{\mu+n-1}^{n-1}$ is Stirling's number of the second kind defined by

$$j^n = \sum_{\nu=1}^{n} \mathfrak{S}_n^{\nu} j^{(\nu)}$$

(see [14]). Examination of the absolute values of $\alpha_{n-1+\nu}$ at $\nu = 1, 2, \cdots$ (small values of $\nu$) suggests that they are at most of the order of $1/n$ relative to that of $\alpha_{n-1}$. This enables us to write down the following asymptotic formula for small $\Omega_t$:

$$\Omega_t \sim \frac{(2n-1)!}{(n-1)!} \left(\frac{1}{n}\right)^n e^{-(n-1)t/4m} \qquad (t \to \infty); \tag{3.2}$$

or applying Stirling's formula for $n!$,

$$\Omega_t \sim \frac{1}{\sqrt{2}} \exp\left\{(2 \log 2 - 1)n - \left(\frac{n-1}{4m}\right) t\right\}. \tag{3.2'}$$

Formula (3.2) can also be derived by a different method, by using the multivariate Kolmogorov forward equation. To reach a given small probability of survival ($\Omega$), the approximate number of generations required will be given by

$$\hat{t} = \frac{4m}{n-1} (0.39n - 0.35 - 2.3 \log_{10} \Omega).$$

In the case of $m = 100$, $n = 41$, we have $\hat{t} = 156.4 - 23 \log_{10} \Omega$ and the generations giving 99%, 99.9% and 99.99% deaths are respectively about 202, 225, and 248 generations. This agrees reasonably well with the finding of Sonneborn that under exclusively asexual reproduction, many of the lines die before 200 fissions and almost all die before 324.

A slightly modified model was suggested to the author by Dr. J. Lederberg: After chromosomes have reduplicated themselves in the macronucleus, they are distributed into two daughter nuclei in such a way that each chromosome has an independent and equal chance of going to either daughter. This differs from the previous model in that the total number of chromosomes per cell does not remain constant. This leads to the following asymptotic formula for the probability of survival:

$$\Omega_t \sim (1 - e^{-4m/(t+4)})^n \qquad (t \to \infty).$$

For $n = 41$, $m = 100$, the number of generations for 99%, 99.9% and 99.99% deaths are respectively about 175, 210, 246, rather similar to the previous model.

Also these models allow predictions for time of death of a lineage derived from repeated regeneration from a small fraction of the macronucleus and for segregation of recessive factors, thus permitting two additional independent tests of the models by comparison with data.

**4. Process of natural selection in finite population (interaction between selection and random genetic drift).** From the standpoint of population genetics, the most elementary step in evolution is the change in gene frequency, especially the one due to natural selection. It may not be difficult to imagine that the process of change is not entirely deterministic, since there exist various factors which introduce an element of indeterminacy into the process, among which random sampling of gametes due to finite population size is of special interest. Let $A$ and $A'$ be a pair of alleles whose frequencies are respectively $x$ and $1 - x$ in the population. In natural populations, the number of individuals is usually large and there may be overlapping of generations, so that gene frequency and time parameter ($t$) may be treated as continuous variables with advantage. We shall designate by $\phi(x, p; t)$ the probability density that the gene frequency lies between $x$ and $x + dx$ at the $t$th generation given that the initial gene frequency is $p$ at $t = 0$.

The simplest situation is obviously that of pure random genetic drift in which no mutation, selection, or migration is involved. The gene frequency changes randomly from generation to generation due to random sampling of gametes in reproduction. In this case if $N$ is the number of reproducing individuals in the random mating population, $\phi$ satisfies the following partial differential equation [15], [16].

$$\frac{\partial \phi}{\partial t} = \frac{1}{4N} \frac{\partial^2}{\partial x^2} \{x(1 - x)\phi\} \qquad (0 < x < 1), \tag{4.1}$$

with the initial condition

$$\phi(x, p; 0) = \delta(x - p),$$

where $\delta$ represents Dirac's delta function. Equation (4.1) is a special case of the Kolmogorov forward (or Fokker-Planck) equation, and its pertinent solution is given by

$$\phi(x, p; t) = \sum_{i=1}^{\infty} \frac{(2i + 1)(1 - r^2)}{i(i + 1)} T_{i-1}^1(r) T_{i-1}^1(z) e^{-i(i+1)t/4N}, \tag{4.2}$$

where $r = 1 - 2p$, $z = 1 - 2x$ and $T_{i-1}^1(r)$ is the Gegenbauer Polynomial defined by

$$T_{i-1}^1(r) = \frac{i(i + 1)}{2} F\left(i + 2, 1 - i, 2, \frac{1 - r}{2}\right).$$

Boundaries $x = 0$ and $x = 1$ act as absorbing barriers and as the number of generations increases, the probability distribution of the classes in which $A$ and $A'$ coexist ("heterallelic," or unfixed, classes) approaches a definite form and decays at the constant rate of $1/2N$. The process ultimately leads to complete fixation or loss of one of the alleles.

When linear pressures (mutation, migration) are involved, the problem becomes a little more complicated. A thoroughgoing analysis of the solutions of the differential equations in this case has been made by Goldberg [17]. The present author also obtained the pertinent solution[6] by studying the law of change in the moments of the distribution [6]. Malécot [4] [18] studied interesting problems of migration and decrease of correlation with distance in the case of no selection.

For the evolution of the genetic system, however, natural selection which acts on mutant genes will be of utmost importance. The simplest situation here is genic selection in which no dominance exists. Suppose gene $A$ has selective advantage $s$ over $A'$, measured in Malthusian parameters [19], that is to say, the rate of geometric growth. The partial differential equation now becomes

$$\frac{\partial \phi}{\partial t} = \frac{1}{4N} \frac{\partial^2}{\partial x^2} \{x(1 - x)\phi\} - s \frac{\partial}{\partial x} \{x(1 - x)\phi\} \qquad (0 < x < 1), \tag{4.3}$$

with the same initial condition as before. Recently this equation was used by Wright and Kerr [20] in connection with their selection experiment in very small populations. The state of steady decay of the heterallelic classes was successfully analysed by Wright. The complete solution of the above equation, which reduces to that of pure random drift for $s = 0$, is given in terms of oblate spheroidal functions studied by Stratton and others [21]:

$$\phi(x, p; t) = \sum_{k=0}^{\infty} C_k e^{-\lambda_k t + 2cx} V_{1k}^{(1)}(z),$$

where $c = Ns$ and $z = 1 - 2x$. The spheroidal function $V_{1k}^{(1)}(z)$ is expressed as a series of Gegenbauer polynomials:

$$V_{1k}^{(1)}(z) = \sum_{n=0,1}' f_n^k \, T_n^1(z),$$

where $f_n^k$'s are constants, and primed summation is over even values of $n$ if $k$ is even, odd values of $n$ if $k$ is odd. For details of the solution see [7]. The boundaries $x = 0$ and $x = 1$ act as absorbing barriers as in the preceding cases and the gene $A$ will ultimately be fixed in the population or completely lost from it The probability of fixation is larger, the larger the value of $s$.

[6] Strictly speaking, the existing solution which treats boundaries as reflecting barriers is not entirely satisfactory, because for small populations boundaries should act as elastic barriers.

At the state of steady decay, the probability distribution $\phi$ decreases in value at a constant rate $\lambda_0$ :

$$\lim_{t\to\infty} \frac{1}{\phi}\frac{d\phi}{dt} = -\lambda_0 .$$

For small values of $c$, we can expand $\lambda_0$ into a power series in $c$.

$$(4.4) \qquad 2N\lambda_0 = 1 + \frac{2}{5}c^2 - \frac{2}{5^3\cdot 7}c^4 - \frac{2^2}{3\cdot 5^5\cdot 7}c^6 - \frac{2\cdot 31}{5^6\cdot 7^3\cdot 11}c^8 - \cdots .$$

This suggests that genic selection increases the rate of decay as compared with the case of no selection ($c = 0$), at least when $c$ is small. Values of $\lambda_0$ for larger values of $c$ will be found in the above reference [7] in which values of $2N\lambda_0$ up to $Ns = 8$ have been studied.

Very often, however, there is some dominance between alleles, and usually "complete" dominance. The main purpose of this section is to develop a method to analyse this situation.

Let us suppose that $A$ is dominant over $A'$ and the dominant genotypes $AA$ and $AA'$ have selective advantage $s$, measured in Malthusian parameters, over the homozygous recessive ($A'A'$). The differential equation for the probability distribution $\phi$ is

$$(4.5) \qquad \frac{\partial\phi}{\partial t} = \frac{1}{4N}\frac{\partial^2}{\partial x^2}\{x(1-x)\phi\} - \frac{\partial}{\partial x}\{sx(1-x)^2\phi\} \qquad (0 < x < 1),$$

with the initial condition

$$\phi(x, p, 0) = \delta(x - p).$$

If we apply the transformation

$$\begin{cases} \phi = e^{-\lambda t}e^{2cx(1-x/2)}w, \\ x = (1-z)/2, \end{cases}$$

to (4.5), we obtain the following ordinary differential equation:

$$(4.6) \quad (1-z^2)w'' - 4zw' + \left[\Lambda - 2 - \frac{c}{2}(z^2-1) + \frac{c^2}{4}(z^2-1)(1+z)^2\right]w = 0,$$

in which $\Lambda = 4N\lambda$ and $c = Ns$. We note that for the case of no selection ($c = 0$) the pertinent solution is the Gegenbauer polynomial. So we try to expand the solution into a series of Gegenbauer polynomials, which are known to form a complete orthogonal system in the interval $[-1, 1]$. Let

$$w = \sum_{n=0}^{\infty} d_n T_n^1(z),$$

in which the $d_n$'s are constants. If we substitute this into (4.6) and use repeatedly the recurrence relation,

$$zT_n^1(z) = \frac{n+2}{2n+3}T_{n-1}^1(z) + \frac{n+1}{2n+3}T_{n+1}^1(z) \qquad (\text{set } T_{-1}^1(z) \equiv 0)$$

we obtain a 9-term recursion formula for the $d_n$'s. Now we expand $\Lambda$ and $d_n$'s into power series of $c$:

$$\begin{aligned}
\Lambda &= k_0 + k_1 c + k_2 c^2 + k_3 c^3 + \cdots , \\
d_1 &= (\alpha_1^1 c + \alpha_2^1 c^2 + \alpha_3^1 c^3 + \cdots)\, d_0 , \\
d_2 &= (\alpha_1^2 c + \alpha_2^2 c^2 + \alpha_3^2 c^3 + \cdots)\, d_0 , \\
d_3 &= (\alpha_2^3 c^2 + \alpha_3^3 c^3 + \alpha_4^3 c^4 + \cdots)\, d_0 , \\
d_4 &= (\alpha_2^4 c^2 + \alpha_3^4 c^3 + \cdots)\, d_0 , \\
d_5 &= (\alpha_3^5 c^3 + \cdots)\, d_0 , \text{ etc.,}
\end{aligned}$$

and substitute these into the recursion formula. By picking out coefficients of equal powers of $c$, we can determine the $k$'s and $\alpha$'s, by means of which the eigenvalue $\lambda$ (or $\Lambda$) and the eigenfunction $w$ are expressed. The most important information is the smallest eigenvalue ($\lambda_0$) which gives the "rate of decay," and the corresponding eigenfunction. To get $\lambda_0$, we set $k_0 = 2$, since for $c = 0$, $\Lambda(= 4N\lambda)$ should be 2, as shown in the previous treatment of pure random drift in which the final rate of decay is $1/2N$.

Though the calculation involved is quite tedious, we can obtain the desired coefficients step by step. For the smallest eigenvalue $\lambda_0$ we have:

$$(4.7) \qquad 2N\lambda_0 = 1 - \tfrac{1}{5}c + \frac{199}{2\cdot 5^3\cdot 7}c^2 + \frac{17}{2\cdot 5^5\cdot 7}c^3 - \frac{23\cdot 41\cdot 29599}{2^3\cdot 3^3\cdot 5^6\cdot 7^3\cdot 11}c^4 \cdots .$$

The coefficients of the eigenfunction are:

$$\alpha_1^1 = 0, \qquad \alpha_2^1 = -\frac{1}{2\cdot 3\cdot 7}, \qquad \alpha_3^1 = \frac{11}{3^4\cdot 5^2\cdot 7},$$

$$\alpha_1^2 = -\frac{1}{2\cdot 3\cdot 5^2}, \qquad \alpha_2^2 = \frac{13}{3^2\cdot 5^4}, \qquad \alpha_3^2 \doteqdot -7.31 \times 10^{-5}, \cdots ,$$

$$\alpha_2^3 = \frac{1}{2\cdot 3^2\cdot 5\cdot 7}, \cdots ,$$

$$\alpha_2^4 \doteqdot 2.49 \times 10^{-4}, \text{ etc.}$$

The same method may be applied to get similar expansions for other eigenvalues and eigenfunctions.

The shape of the distribution curve at the state of steady decay is given by

$$(4.8) \qquad \phi(x) = e^{2cx(1-x/2)} w_0 .$$

It will be convenient to adjust $d_0$ so that $\int_0^1 \phi(x)\, dx = 1$ (fixed classes excluded). The rate of fixation and loss of the gene $A$ per generation at this state is given by $\phi(0)/4N$ and $\phi(1)/4N$ and therefore

$$(4.9) \qquad 4N\lambda_0 = \phi(1) + \phi(0).$$

This can be derived from (4.5) noting that $\phi(x)$ is finite at the boundaries ($x = 0$ and $x = 1$), as shown for the case of no dominance in [7].

TABLE 2

| $x$ | $2Ns = 1$ | $2Ns = -1$ |
|---|---|---|
| 0.0 | 0.688 | 1.389 |
| 0.1 | 0.764 | 1.251 |
| 0.2 | 0.838 | 1.142 |
| 0.3 | 0.910 | 1.056 |
| 0.4 | 0.977 | 0.990 |
| 0.5 | 1.037 | 0.940 |
| 0.6 | 1.088 | 0.903 |
| 0.7 | 1.128 | 0.879 |
| 0.8 | 1.155 | 0.865 |
| 0.9 | 1.168 | 0.860 |
| 1.0 | 1.166 | 0.866 |

A numerical example will be given here. For weak selection favoring dominants; $2Ns = 2c = 1$, we get

$$2N\lambda_0 = 0.928$$

and

$$w_0 = d_0[T_0^1(z) - 0.0058T_1^1(z) - 0.0028T_2^1(z) + 0.0004T_3^1(z) + 0.00006T_4^1(z) + \cdots],$$

in which $T_0(z) = 1$, $T_1(z) = 3z$, $T_2(z) = \frac{3}{2}(5z^2 - 1)$, $T_3(z) = \frac{5}{2}(7z^3 - 3z)$, $\cdots$. Values of $\phi(x)$ at 0, 0.1, 0.2, $\cdots$ and 1 are listed in Table 2. They are adjusted by Simpson's rule so that the area under the curve is unity. $\phi(1) + \phi(0)$ comes out 1.855, while $4N\lambda_0$ is 1.856. The agreement is satisfactory for this level of approximation. As a second example, we assume weak selection against the dominants: $2Ns = 2c = -1$. $2N\lambda_0$ is 1.128 and values of $\phi(x)$ are given in Table 2. In this case $\phi(1) + \phi(0)$ comes out 2.254 while $4N\lambda_0$ is 2.256. Again the agreement is satisfactory.

The above treatment leading to the power series expansion of eigenvalues and of coefficients of eigenfunctions is rather heuristic. For the more rigorous treatment of the problem, further investigation of these series will be required.

As to uniqueness of the solutions of the type of singular partial differential equations considered in this section, an investigation could presumably be based on Section 23 of Feller's paper [22].

The most remarkable fact suggested by the above analysis seems to be that as compared with the case of pure random drift, selection toward dominants ($s > 0$) decreases the final rate of decay, while selection against dominants ($s < 0$) increases it. At least for weak selection the above results follow from (4.7), since the most influential term $-\frac{1}{5}c$ is negative if $c(= Ns)$ is positive and positive if $c$ is negative.

For this continuous treatment to be applicable, the population number $N$

TABLE 3

| Classes | Frequencies % |
|---|---|
| $7A + 1A'$ | 14.80 |
| $6A + 2A'$ | 16.48 |
| $5A + 3A'$ | 16.32 |
| $4A + 4A'$ | 15.51 |
| $3A + 5A'$ | 14.28 |
| $2A + 6A'$ | 12.64 |
| $1A + 7A'$ | 9.97 |
| Total.................................. | 100.00 |

should be fairly large so that $1/N$ is negligible as compared with 1. If the population is extremely small, we must treat the problem by the methods of finite Markov chains. The transition probability that the number of $A$ genes in the population becomes $j$ in the next generation, given that it is $i$ in the present generation will be given by

$$p_{j|i} = \binom{2N}{j} x'^j (1 - x')^{2N-j} \qquad (i, j = 0, 1, \cdots, 2N),$$

where $x' = x + \delta x$, in which $x = i/(2N)$, and $\delta x$ is the change of gene frequency by selection per generation and is $sx(1 - x)^2$ if $s$ is small. The rate of decay of the unfixed classes and their limiting distribution may be obtained by iteration. For example, if $N = 4$, $2Ns = 1$, the limiting form of the distribution (fixed classes excluded) becomes as follows (Table 3), with rate of decay ($\lambda_0$) 11.875%, giving $2N\lambda_0 = 0.9500$. If there is no selection ($s = 0$), it turns out that the rate of decay becomes $1/2N = 0.125$ or 12.5%. Note that with selection for dominants, the rate of decay is smaller.

**5. Chance of fixation of mutant genes.** In any large natural population, gene mutations may be occurring in each generation. Most of the mutant genes are likely to be deleterious but a few of them may turn out to be advantageous. Such advantageous mutant genes have a tendency to increase their frequencies in later generations thus having a positive chance of establishing themselves even in a very large population. Because of its importance in evolutionary genetics, the probability of fixation of mutant genes has been studied by Fisher [23], Haldane [24] and Wright [25] [26]. However, due to mathematical difficulties involved, so far only a few cases have been successfully worked out.

In this section I will try to present the solution under quite general conditions and will show that the previous results are obtained as special cases.

We will designate the selective advantage of the mutant homozygote ($AA$) by $s$ and that of the heterozygote ($AA'$) by $sh$. Let $u(p, t)$ be the conditional probability that the mutant gene reaches fixation by the $t$th generation, given that its initial frequency is $p$. Under the assumption of a continuous model and random mating it is possible to show that $u(p, t)$ satisfies the following partial differential equation:

$$\frac{\partial u}{\partial t} = \frac{p(1-p)}{4N}\frac{\partial^2 u}{\partial p^2} + sp(1-p)\{h + (1-2h)p\}\frac{\partial u}{\partial p}. \tag{5.1}$$

Here we have inevitably the following boundary conditions:

$$u(0, t) = 0, \qquad u(1, t) = 1. \tag{5.2}$$

For the special case of neutral genes ($s = 0$), the pertinent solution is

$$u(p, t) = p + \sum_{i=1}^{\infty} (2i+1)pq(-1)^i F(1-i, i+2, 2, p)e^{-[i(i+1)t/4N]}, \tag{5.3}$$

which agrees exactly with the results obtained by the study of moments [16]. Usually the process of evolution extends over an enormous period of time and hence the probability of ultimate fixation will be of special importance. We will designate such probability by $u(p)$ which is defined by

$$u(p) = \lim_{t \to \infty} u(p, t).$$

For the neutral mutant gene, $u(p) = p$. If $\nu$ is the initial number of mutant genes, $u(p) = \nu/2N$ for this case and hence the probability of fixation per mutant gene is $1/2N$.

For the general case the probability may be obtained by setting $\partial u/\partial t = 0$ in (5.1). This leads to

$$u(p) = \int_0^p e^{-2cDx(1-x)-2cx}\,dx \Big/ \int_0^1 e^{-2cDx(1-x)-2cx}\,dx, \tag{5.4}$$

where $c = Ns$ and $D = 2h - 1$.

The rate of approach to the ultimate state of complete fixation or loss may be given by the smallest eigenvalue $\lambda_0$ of equation (5.1). For a small value of $c$, we can expand $\lambda_0$ into a power series in $c$ as follows:

$$2N\lambda_0 = 1 + K_1c + K_2c^2 + K_3c^3 + K_4c^4 + \cdots, \tag{5.5}$$

where

$$K_1 = -\tfrac{1}{5}D, \qquad K_2 = \frac{1}{2\cdot 5} + \frac{2^2\cdot 3}{5^3\cdot 7}D^2, \qquad K_3 = \frac{1}{2\cdot 5^3\cdot 7}D - \frac{2^2}{5^5\cdot 7}D^3,$$

$$K_4 = -\frac{1}{2^3\cdot 5^3\cdot 7} - \frac{7^3}{2\cdot 3^3\cdot 5^5}D^2 - \frac{2^2\cdot 3^5}{5^6\cdot 7^3\cdot 11}D^4, \text{ etc.}$$

It should be noted that for the case of no dominance $D = 0$ and the above series (5.5) agrees with (4.4) provided that $2s$ is used instead of $s$ to express the selective advantage of the homozygous mutants. For the case of complete dominance, $D = 1$ or $-1$ according as the mutant gene is either dominant or recessive. In the former case of $D = 1$, (5.5) agrees with (4.7). Returning to formula (5.4), we will consider a few cases of special importance in evolution. To obtain the chance of fixation of individual mutant gene denoted by $u$ we may put $p = 1/2N$. For the case of no dominance ($D = 0$), we have

$$u = (1 - e^{-s})/(1 - e^{-2c}),$$

or denoting the selective advantage of the homozygote by $2s$,

$$u = (1 - e^{-2s})/(1 - e^{-4Ns}).$$

Thus for a slightly advantageous mutant gene we may write

$$u = 2s/(1 - e^{-4Ns}) \tag{5.6}$$

with good approximation. The result agrees with Fisher [23] who used the method of branching processes and also with Wright [25] who used the method of integral equations. For a large $N$ this chance is very close to $2s$ as given by Haldane [24]. For a slightly disadvantageous mutant gene ($s < 0$), we have

$$u = 2s'/(e^{4Ns'} - 1), \tag{5.7}$$

where $s' = -s$. The chance is not negligible if $Ns'$ is small. The result agrees with that obtained by Wright [25].

For the completely dominant gene ($D = 1$) with small selective advantage $s(s > 0)$ we may use the formula $u = 2s$ unless $Ns$ is small.

The case of a completely recessive mutant gene ($D = -1$) with small selective advantages $s$ ($s > 0$) in the homozygous state is of special interest. Haldane [24] estimated the chance of fixation as of the order of $\sqrt{s/N}$ using the method of branching processes and Wright [26] estimated it as of the order of $\sqrt{s/2N}$ by his method of integral equations. Our formula (5.4) gives

$$u = \sqrt{2s/(\pi N)} \tag{5.8}$$

as the best simple approximation for a large $N$. Since

$$\sqrt{2s/\pi N} = \sqrt{2/\pi}\sqrt{s/N} = \sqrt{4/\pi}\sqrt{s/2N},$$

it may readily be seen that our result lies between those of Haldane and Wright. Furthermore it is interesting to note that Wright [26] obtained numerically the formula $1.1(s/2N)^{\frac{1}{2}}$ as the average chance of fixation for values of $s$ ranging from $4/2N$ to $64/2N$. The factor 1.1 is indeed very close to $\sqrt{4/\pi}$ which is $1.128 \cdots$.

Finally our general formula (5.4) allows us to calculate the chance of fixation of a nearly recessive mutant gene with selective advantage $s$ ($s > 0$) in the homozygous state. Namely for $0 < h \ll 1$, we may have

$$u = e^{-2Nsh^2/(1-2h)} \sqrt{\frac{2s(1-2h)}{\pi N}} \Big/ \{1 - 2\Phi(\sqrt{4Nsh^2/(1-2h)})\} \tag{5.9}$$

as a good approximation, unless $2Ns$ is small. Here $\Phi(x)$ stands for the error function

$$\Phi(x) = (1/\sqrt{2\pi}) \int_0^x e^{-x^2/2}\, dx.$$

As an example consider a case with $N = 10^3$ and $s = 10^{-1}$. If the mutant gene is completely recessive ($h = 0$), $u \approx 0.8 \times 10^{-2}$. With slight phenotypic effect of $h = 0.01$ in the heterozygote $u \approx 0.9 \times 10^{-2}$, while with

$$h = 0.1, \qquad u \approx 2.3 \times 10^{-2}.$$

**6. Population structure and evolution.** So far we have considered the process of change in gene frequency in an isolated population in which mating is random and the number of individuals remains constant through generations. This may be an over-simplification for the study of evolutionary processes in general, since most species in nature may have a much more complicated breeding structure. Unfortunately this immediately brings us baffling problems, for the solution of which new techniques will be required.

First let us suppose that a species is subdivided into numerous isolated colonies, each of which may receive, from time to time, migrants taken as random samples from the whole population. Mating is assumed to be random within each colony. Following Wright [27] we will call this the "island model." The model may be realistic to describe a species inhabiting an archipelago such as the Galapagos Islands studied by Darwin. The number of individuals may fluctuate from generation to generation not only due to fluctuation in environmental conditions but also due to change in the genetic make up of each colony which in turn is influenced by the population number. If the number of reproducing individuals per colony is small, say less than 100, and if isolation is so severe that less than one migrant is expected per thousand generations, the chance of disadvantageous mutant genes reaching fixation may be considerable, as suggested by (5.7), and accumulation of such genes will lead to extinction of colonies. We would like to know then what is the chance that an isolated colony becomes extinct before a migrant comes in to start a new colony. What is the joint distribution of the population number and the gene frequency among colonies at the steady state? These questions may have to be answered before we reach conclusions on the optimum structure of populations for the evolution of a species.

Next we will consider the continuum model of a population. The model is realistic for representing a species inhabiting a wide range with more or less uniform density. Here the whole population can not be a random mating unit since a tendency toward "isolation by distance" may arise due to limitation in the locomotive ability of the organism [27]. In the course of time advantageous mutant genes may arise with exceedingly low rate in various spots in the continuum and these will spread into the population. If the local fluctuation of gene frequencies is negligible, the process of spread will be very similar to diffusion of physical particles in a medium, except here that differential rate of multiplication is involved among particles.

Let $x(\mu, \nu)$ be the relative frequency of a mutant gene denoted by $A$ at a point $(\mu, \nu)$ in the continuum with rectangular coordinate system. The process of spread of the advantageous dominant gene may be described by the equation

$$\frac{\partial x}{\partial t} = m\nabla^2 x + sx(1 - x)^2, \tag{6.1}$$

where $m$ represents locomotive ability of an individual and corresponds to a diffusion constant in physical systems, $\nabla^2$ denotes the two-dimensional Laplace operator $(\partial^2/\partial\mu^2 + \partial^2/\partial\nu^2)$, and $s$ is the selective advantage of the dominant gene $A$ to its allele $A'$. The simplest situation is that $s$ is constant throughout the continuum. The mutant gene will spread in the form of concentric circles from the point of origin which we may take as $(0, 0)$. Introducing the polar coordinates $(r, \theta)$ and assuming that $\partial^2 x/\partial^2\theta = 0$, (6.1) becomes

$$\frac{\partial x}{\partial \tau} = \frac{\partial^2 x}{\partial r^2} + \frac{1}{r}\frac{\partial x}{\partial r} + \alpha x(1 - x)^2, \tag{6.2}$$

where $\tau = mt$ and $\alpha = s/m$. At an early stage when the frequency of $A$ is still low, the distribution may be approximated by

$$x(r, \tau) = x_0 e^{\alpha\tau - r^2/4\tau} r/2\tau,$$

where $x_0$ is the initial frequency of $A$ at the origin. Beyond this stage, however, we face a difficult problem of solving a non-linear diffusion equation.

The problem of steady state distribution is worthwhile to investigate if the mutant gene is advantageous within a closed region but disadvantageous outside, as in the case of melanic genes in many lepidopteran species which in recent years have increased their relative frequencies in a spectacular fashion in many industrial areas but remain in low frequencies in rural districts—a phenomenon known as "industrial melanism" [28].

Real mathematical difficulties arise, however, when we take random fluctuation of local gene frequencies into consideration. The fluctuation may be due to random sampling of gametes in reproduction or due to random fluctuation of selection intensities brought about by chance fluctuation of local environments. Notable contributions have been made by Wright [27] [29] [30] and Malécot [4] [18] for the case of no selection, but more important cases involving selection are yet to be worked out.

Such studies should be indispensable for our understanding of the process of speciation and also of the mechanism of evolution in general.

In his theories of evolution Wright [31] put forward an important concept of "balance," especially of balance between directional factors such as selection, mutation, and migration and undirectional or stochastic factors such as random sampling of gametes and random fluctuation of environmental conditions. It appears that new methods of stochastic processes will be needed for a satisfactory treatment of Wright's theory of evolution.

**7. Acknowledgement.** The author expresses his appreciation to Dr. James F. Crow for valuable help during the course of this work. Dr. E. R. Immel read the manuscript and offered helpful suggestions. Thanks are also due to the Numerical Analysis Laboratory of the University of Wisconsin for computing some of the eigenvectors in the Section 2. Finally the author wishes to express his indebtness to Professor Sewall Wright for guidance and help in many ways.

REFERENCES

[1] A. Kolmogorov, "Über die analytischen Methoden in der Wahrscheinlichkeitsrechnung," *Math. Ann.*, Vol. 104 (1931), pp. 415–458.
[2] M. S. Bartlett and J. B. S. Haldane, "The theory of inbreeding in autotetraploids," *J. Genetics*, Vol. 24 (1934), pp. 175–180.
[3] R. A. Fisher, *The Theory of Inbreeding*, Oliver and Boyd, Ltd., 1949.
[4] G. Malécot, *Le Mathematiques de L'Hérédité*, Masson et Cie, Paris, 1948.
[5] W. Feller, "Diffusion processes in genetics," *Proceedings of the Second Berkeley Symposium on Mathematical Statistics and Probability*, University of California Press, 1951, pp. 227–246.
[6] J. F. Crow and M. Kimura, "Some genetic problems in natural populations," *Proceedings of the Third Berkeley Symposium on Mathematical Statistics and Probability*, Vol. 4 (1956), pp. 1–22.
[7] M. Kimura, "Stochastic processes and distribution of gene frequencies under natural selection," *Cold Spring Harbor Symposium*, Vol. 20 (1955), pp. 33–55.
[8] M. Demerec, "Behavior of two mutable genes of *Delphinium ajacis*," *J. Genetics*, Vol. 24 (1931), pp. 179–193.
[9] H. Matsuura and T. Suto, "Genic analysis in maize V. mode of calico character" (in Japanese), *Japanese J. Genetics*, Vol. 23 (1948), p. 31.
[10] Ch. Auerbach, "Problems in chemical mutagenesis," *Cold Spring Harbor Symposium*, Vol. 16 (1951), pp. 199–213.
[11] Von H. Friedrich-Freksa und F. Kaudewitz, "Letale Spätfolgen nach Einbau von $^{32}P$ in *Amoeba proteus* und ihre Deutung durch genetische Untereinheiten," *Z. Naturforsch.*, Vol. 86 (1953), pp. 343–355.
[12] T. M. Sonneborn and M. V. Schneller, "Genetic consequences of aging in variety 4 of *Paramecium aurelia*," *Records Genetics Soc. America*, Vol. 24 (1955), p. 596.
[13] A. W. Kimball and A. S. Householder, "A stochastic model for the selection of macronuclear units in paramecium growth," *Biometics*, Vol. 10 (1954), pp. 361–374.
[14] Ch. Jordan, *Calculus of Finite Differences*, New York, 1950.
[15] S. Wright, "The differential equation of the distribution of gene frequencies," *Proc. Nat. Acad. Sci.*, Vol. 31 (1945), pp. 382–389.
[16] M. Kimura, "Solution of a process of random genetic drift with a continuous model," *Proc. Nat. Acad. Sci.*, Vol. 41 (1955), pp. 144–150.
[17] S. Goldberg, "On a Singular Diffusion Equation," Ph.D. Thesis (Cornell University), 1950, unpublished.
[18] G. Malécot, "Un traitement stochastique des problèmes linéaires [mutation, linkage, migration] en Génétique de Population," *Sciences Mathématiques et Astronomie XIV, Annales de l'Université de Lyon* (1951), pp. 79–117.
[19] R. A. Fisher, *The Genetical Theory of Natural Selection*, Clarendon Press, 1930.
[20] S. Wright and W. E. Kerr, "Experimental studies of the distribution of gene frequencies in very small populations of *Drosophila melanogaster*, II. Bar," *Evolution*, Vol. 8 (1954), pp. 225–240.
[21] J. A. Stratton, P. M. Morse, L. J. Chu, and R. A. Hutner, *Elliptic Cylinder and Spheroidal Wave Functions*, John Wiley & Sons, New York, 1941.
[22] W. Feller, "The parabolic differential equations and the associated semi-group of transformations," *Ann. Mathematics*, Vol. 55 (1952), pp. 468–519.

[23] R. A. Fisher, "The distribution of gene ratios for rare mutations," *Proc. Roy. Soc. Edinburgh*, Vol. 50, Pt. II (1930), pp. 205–220.

[24] J. B. S. Haldane, "A mathematical theory of natural and artificial selection. Part V: Selection and mutation," *Proc. Cambridge Philos. Soc.*, Vol. 23 (1927), pp. 838–844.

[25] S. Wright, "Evolution in Mendelian populations," *Genetics*, Vol. 16 (1931), pp. 97–159.

[26] S. Wright, "Statistical genetics and evolution," *Bull. Amer. Math. Soc.*, Vol. 48 (1942), pp. 223–246.

[27] S. Wright, "Isolation by distance," *Genetics*, Vol. 28 (1943), pp. 114–138.

[28] H. B. D. Kettlewell, "Selection experiments on industrial melanism in the *Lepidoptera*," *Heredity*, Vol. 9 (1955) pp. 323–342.

[29] S. Wright, "Isolation by distance under diverse systems of mating," *Genetics*, Vol. 31 (1946), pp. 39–59.

[30] S. Wright, "The genetical structure of populations," *Ann. Eugenics*, Vol. 15 (1951), pp. 323–354.

[31] S. Wright, "Population structure as a factor in evolution," in *Moderne Biologie*, Berlin, 1950, pp. 275–287.

# DIFFUSION MODELS IN POPULATION GENETICS

MOTOO KIMURA, *National Institute of Genetics, Mishima, Japan*

## CONTENTS



Received in revised form 10 June 1964. Contribution No. 453 from the Nationa Institute of Genetics.

## I. Problems and Methods of Population Genetics

### 1. Introduction

Population genetics is that branch of genetics, whose object is the study of the genetical make-up of natural populations. By investigating the laws which govern the genetic structure of natural populations, we intend to clarify the mechanism of evolution.

In a natural population of sexually reproducing species, with only a hundred loci segregating, the number of possible genotypes may be practically infinite, and the genotype of each individual is quite likely to be unique in the entire history of the species. Thus, as an aggregate of individual genotypes, a population is an enormously complicated system, sometimes too complicated to be treated theoretically. On the other hand, in any reasonably large population, the relative proportion of an allele (a particular form of a gene) within the population changes almost continuously with time. This is because, unlike genotypes, each gene reproduces its own kind with complete fidelity except for the very rare event of mutation. As pointed out by Fisher (1953), "the frequencies with which the different genotypes occur define the gene ratios characteristic of the population, so that it is often convenient to consider a natural population not so much as an aggregate of living individuals as an aggregate of gene ratios. Such a change of viewpoint is similar to that familiar in the theory of gases, where the specification of the population of velocities is often more useful than that of a population of particles." This line of investigation was initiated by Fisher (1922) and later elaborated by him (Fisher, 1930), Haldane (cf. 1932) and especially by Wright (1931, and later publications).

In the present paper, I shall review the theoretical works on population genetics, treating the changes of gene frequencies as stochastic processes, and describing these especially by the use of diffusion equations. Since I started my work in this field as a geneticist, the mathematical sophistication of my approach has been rather limited; I cannot escape from this limitation in the present paper, but I hope it will stimulate mathematicians to work in this fascinating field. Indeed, there is much to be done in the refinement and extension of the mathematical methods involved, as is shown by the works of Feller (1951, 1952) and Moran (cf. 1962).

### 2. Changes of gene frequencies as stochastic processes

From the standpoint of population genetics, the most elementary step in evolution is the change of gene frequencies. Here, gene frequencies mean the proportions of genes in a population. The simplest mathematical approach to this problem is to regard the process of change as deterministic. Such an approach was first used extensively by Haldane in his series of papers starting in 1924 (Haldane, 1924). Strictly speaking, it applies only if a population is infinitely large and is placed in an environment which remains constant or changes in a deterministic way. There are many circumstances in which this is sufficiently realistic as a first approximation. Furthermore, because of its simplicity, this approach is still the most useful, and is often the only manageable one for many problems. In nature, however, the process of change may not be quite deterministic, because of the existence of factors which produce random fluctuation in gene frequencies, of which two different types may be recognized (Wright, 1949). One is the random sampling of gametes in reproduction. The process of change in gene frequency which is due solely to this factor is often called *drift*. However, the term drift in this context is hardly adequate unless the prefix *random* is also attached, and it may be called the random drift in the narrow sense. This factor becomes prominent in a small population. The other type consists of random fluctuations in what Wright called the systematic evolutionary pressures, of which random fluctuation in selection intensity may be especially important. These two types of factors introduce a random element into the process of change in gene frequencies (random drift in the wide sense). Thus, in the present review, we will regard the process of change as a stochastic process, where this means the mathematical formulation of a chance event evolving in time.

So far as I know, this line of investigation was initiated by Fisher (1922). In his paper, Fisher considered the random sampling of gametes as the factor causing random fluctuation in gene frequency and, assuming no selection, he investigated its effect (*Hagedoorn Effect*) on the decrease of variability in a species. Though his treatment was restricted to a quite simple situation, the paper was important in that he introduced the method of partial differential equations in the study of gene frequency distributions in a population. This method, if properly extended, is equivalent to the approach which makes use of the Fokker-Planck equation, later introduced by Wright (1945). Here Fisher used the transformed gene frequency rather than the gene frequency itself. Also he suggested the method of functional equations to study the probability of fixation of an individual mutant gene. As in many other pioneering works, this 1922 paper was not finished and contained some minor errors and ambiguity. Later a more complete treatment was presented (Fisher, 1930) in which the errors were amended and the results were greatly extended. In my opinion this is one of the most beautiful papers ever written on the mathematical theory of population genetics.

In 1931, Wright published his now classical paper "Evolution in Mendelian populations" in which he studied similar problems by his method of integral equations (Wright, 1931). Since then, Wright has published a number of important papers on the probability distribution of gene frequencies and on the role of random processes in evolution. He has emphasized the importance of proper balance between the directed and the random processes in relation to population structure in evolution (cf. Wright, 1950).

These two authors, as Feller (1951) has remarked, have studied individual problems with great ingenuity, with the result that many limiting probability distributions have been worked out. However, the problem of constructing a model for the entire process of change in the gene frequencies has not been dealt with by these authors.

In the field of the mathematical theory of probability, progress in our knowledge of stochastic processes has been quite extensive since Kolmogorov's fundamental paper (Kolmogorov, 1931). This is doubtless a result of the growing need for the stochastic treatment of problems in diverse fields of modern science. It is not surprising therefore that pioneering attempts to construct a model for the entire process of change in gene frequencies were made by mathematicians. Malécot (1948) who considered "evolution of the probability law in the course of time", especially for the case of mutation pressure and random sampling of gametes, sketched a method by which the solution might be obtained. Goldberg (1950) in his unpublished thesis studied the same case and succeeded in obtaining two solutions of the diffusion equation involved.

From the mathematical standpoint, various types of stochastic processes arise in population genetics. Of special importance is the Markov process which Kolmogorov called stochastically definite. The process of change in gene frequencies in a very small population consisting of a few individuals with non-overlapping generations will most appropriately be treated as a finite Markov chain. The fate of an individual mutant gene in a large population can be treated by the theory of branching processes.

Generally, however, the processes of organic evolution in nature are very slow and the number of individuals involved per generation is very large, so that they may be treated with advantage as a continuous Markov process in space (gene frequency) and time, as will be explained in the next section. Here the Fokker-Planck equation (Kolmogorov forward equation) plays a fundamental role. Using this approach, the problem of constructing a model for the entire process of change in gene frequencies starting from an arbitrary initial frequency was solved for several genetically interesting cases by the present author (Kimura, 1954, 1955 a, b, c, 1956 a, b, 1957, Crow and Kimura, 1956). Also, it has been shown by the author (Kimura, 1957, 1962) that the Kolmogorov backward equation may be used to obtain the probability of fixation of mutant genes in a population.

Recently, the stochastic theory of gene frequency change was used by Robertson (1960) in his theory of limits in artificial selection. He also studied the problem of selection for heterozygotes in small populations (Robertson, 1962) based on the mathematical work of Miller (1962) who developed a powerful method for evaluating eigenvalues of the Fokker-Planck equation involved.

Usually, the Fokker-Planck equations which appear in population genetics have singularities at the boundaries and a deep mathematical investigation of these was carried out by Feller (1952) who clarified the nature of boundaries by using semi-group theory.

The diffusion equation approach has been used extensively in the study of gene frequency change because of its extreme usefulness. But it is an approximation based on rather intuitive arguments. Therefore, to investigate the conditions under which such approximation may be valid, is an important task for mathematicians; Moran (1958 a and b) introduced two population models, with overlapping and non-overlapping generations for the study of gene frequency distribution in populations. Watterson (1962) obtained sufficient conditions, concerning the change of gene frequency per unit length of time, under which the diffusion approximation is valid, even if the gene frequency does not necessarily form a Markov process. Recently he applied these conditions to unify Moran's two models (Watterson, 1964). Moran's models have also been investigated by Karlin and McGregor (1962). An important contribution was made by Moran (1961) to the problem of gene fixation in a finite population. Though he was able to treat only a very simple genetical situation, his rigorous treatment is important in giving a case where the diffusion approximation introduced by the author (Kimura, 1957) can be checked.

Accuracy of the diffusion approximation method may also be checked numerically by high-speed computer when the population number is very small; a recent study of Ewens (1963) seems to indicate that the approximation is quite good for populations of reasonable size. Furthermore (Ewens, 1964), formulae for the leading terms of the corrections to diffusion approximations can be found.

**3. The partial differential equation method**

A natural population which plays a significant role as an evolutionary unit should consist of a large number of individuals, and the gene frequencies for these behave practically as continuous variables. Also any change of gene frequencies must in general be very slow by our ordinary time scale. There are certain cases in which relatively rapid changes were observed in polymorphic characters, such as the spread of industrial melanism in moths. However, the typical rate of evolution shown in fossil records is of the order of one-tenth of a darwin, one darwin standing for the rate of change with a factor of $e\ (=2.71\cdots)$ per million years (Haldane, 1949); this suggests that the change in gene frequencies involved must be correspondingly slow.

For these reasons, the process of change in gene frequency may be treated as a continuous stochastic process; this means roughly that as the time interval becomes smaller, so also does the amount of change in gene frequency $x$ during that interval. More strictly, the process is called a continuous stochastic process if for any given positive value $\varepsilon$, the probability that the change in $x$ during the time interval $(t, t+\delta t)$ exceeds $\varepsilon$ is $o(\delta t)$, i.e. an infinitesimal of higher order than $\delta t$. We will assume also that change in gene frequencies is Markovian, that is, the probability distribution of gene frequencies at a given moment $t$ depends on the gene frequencies at a preceding time $t_0 (t_0 < t)$ but not on the previous history which has led to the gene frequencies at $t_0$.

For the study of this continuous Markov process, one of the most powerful methods available makes use of the Kolmogorov equations (Kolmogorov, 1931). We will first derive the Kolmogorov forward equation as applied to population genetics. Throughout this article, we will assume, unless otherwise stated, a diploid population consisting of a fixed number $N$ of individuals in each generation. Thus, there are $2N$ genes at each locus.

Consider a pair of alleles $A_1$ and $A_2$ with respective frequencies $x$ and $1-x$. Let $\phi(p,x;t)$ be the conditional probability density that the gene frequency is $x$ at time $t$, given that the initial frequency is $p$ at time $t=0$. This gives the transition probability that the gene frequency moves from $p$ to $x$ after time $t$. With $p$ fixed, $\phi(p,x;t)$ determines a frequency distribution such that when $1/(2N)$ is substituted for $dx$, $\phi(p,x;t)dx$ gives an approximation to the frequency of the class with gene frequency $x$ $(0<x<1)$ at time $t$, which when expressed in terms of generations ,we may also roughly refer to as the $t$th generation; this frequency distribution may be denoted by

$$f(x,t) = \phi(p,x;t)\,\frac{1}{2N} \qquad (0<x<1). \tag{3.1}$$

When $p$ is fixed, it will often be omitted so that $\phi(p,x;t)$ is written as $\phi(x,t)$. Also it should be noted that the above relation (3.1) holds only for *unfixed* classes, i.e. for $0<x<1$. Frequencies of classes with $x=0$ or 1 have to be treated separately.

Let $g(\delta x, x; \delta t, t)$ be the probability density that the gene frequency changes from $x$ to $x+\delta x$ during the time interval $(t, t+\delta t)$. Using this probability density, the assumption of a continuous stochastic process may be expressed as

$$\int_{|\delta x|>\varepsilon} g(\delta x, x; \delta t, t)\, d(\delta x) = o(\delta t), \quad (\delta t \to 0), \tag{3.2}$$

where $\varepsilon$ is some arbitrary preassigned positive value.

Furthermore, for the process of change in gene frequency, we have

$$\phi(p,x;t+\delta t) = \int \phi(p, x-\delta x; t)\, g(\delta x, x-\delta x; \delta t, t)\, d(\delta x), \tag{3.3}$$

where the integral on the right is taken over all possible values of $\delta x$. The above relation is a natural consequence of the assumption that the process is Markovian. The probability that the gene frequency is $x$ at time $t + \delta t$ is the sum total of the probabilities of cases in which the gene frequency is $x - \delta x$ at time $t$, and the gene frequency increases by $\delta x$ during the subsequent time interval $(t, t + \delta t)$, with $\delta x$ taking all possible values. Actually, the above relation is a special form of the Kolmogorov-Chapman equation in the theory of stochastic processes. In this expression, $\delta x$ may take any value such that $x - \delta x$ lies between 0 and 1, exclusive of the end points. However, because of (3.2), only values in the range $|\delta x| < \varepsilon$ are of any significance.

Expanding the integrand on the right side of (3.3) in terms of $\delta x$, we have

$$\phi(p, x - \delta x; t)\, g(\delta x, x - \delta x; \delta t, t)$$

$$= \phi g - \delta x \frac{\partial}{\partial x}(\phi g) + \frac{(\delta x)^2}{2!}\frac{\partial^2}{\partial x^2}(\phi g) - \frac{(\delta x)^3}{3!}\frac{\partial^3}{\partial x^3}(\phi g) + \cdots,$$

where $\phi$ and $g$ stand for $\phi(p,x;t)$ and $g(\delta x, x; \delta t, t)$ respectively. Thus (3.3) may be expressed as

$$\phi(p, x; t + \delta t) = \phi \int g d(\delta x) - \frac{\partial}{\partial x}\left\{\phi \int (\delta x) g d(\delta x)\right\} + \frac{1}{2}\frac{\partial^2}{\partial x^2}\left\{\phi \int (\delta x)^2 g d(\delta x)\right\} - \cdots. \tag{3.4}$$

Here we have assumed that the orders of summation, integration and differentiation may be interchanged freely.

Noting that

$$\int g d(\delta x) = 1,$$

and transferring the first term on the right side of (3.4) to the left, we have, after dividing both sides by $\delta t$,

$$\frac{\phi(p, x; t + \delta t) - \phi(p, x; t)}{\delta t} \tag{3.5}$$

$$= -\frac{\partial}{\partial x}\left\{\phi(p,x;t)\frac{1}{\delta t}\int(\delta x)\,g(\delta x,x;\delta t,t)\,d(\delta x)\right\}$$

$$+\frac{1}{2}\frac{\partial^2}{\partial x^2}\left\{\phi(p,x;t)\frac{1}{\delta t}\int(\delta x)^2 g(\delta x,x:\delta t,t)\,d(\delta x)\right\}$$

$$-\frac{1}{3!}\frac{\partial^3}{\partial x^3}\left\{\phi(p,x;t)\frac{1}{\delta t}\int(\delta x)^3 g(\delta x,x;\delta t,t)\,d(\delta x)\right\}$$

$$+\cdots.$$

Let

$$\lim_{\delta t\to 0}\frac{1}{\delta t}\int(\delta x)\,g(\delta x,x;\delta t,t)\,d(\delta x) = M(x,t), \tag{3.6}$$

$$\lim_{\delta t\to 0}\frac{1}{\delta t}\int(\delta x)^2 g(\delta x,x;\delta t,t)\,d(\delta x) = V(x,t), \tag{3.7}$$

and assume that

$$\lim_{\delta t\to 0}\frac{1}{\delta t}\int(\delta x)^n g(\delta x,x;\delta t,t)\,d(\delta x) = 0 \tag{3.8}$$

for $n \geqq 3$. Then we have

$$\frac{\partial\phi(p,x;t)}{\partial t} = \frac{1}{2}\frac{\partial^2}{\partial x^2}\{V(x,t)\phi(p,x;t)\} - \frac{\partial}{\partial x}\{M(x,t)\phi(p,x;t)\} \tag{3.9}$$

where $M(x,t)$ and $V(x,t)$ refer to the first and second moments of $\delta x$ during the infinitesimal time interval $(t, t+\delta t)$.

In practice, however, quantities such as mutation rates, rate of migration, intensity of selection, and effect of random sampling of gametes which determine $\delta x$ are all measured with one generation as a time unit and the limiting rate with $\delta t \to 0$ can only be obtained by extrapolation. So we will replace $M(x,t)$ and $V(x,t)$ in the above equation by $M_{\delta x}$ and $V_{\delta x}$, the mean and variance of the change in gene frequency per generation ($\delta t$ corresponding to one generation). Thus we obtain

$$\frac{\partial\phi}{\partial t} = \frac{1}{2}\frac{\partial^2}{\partial x^2}(V_{\delta x}\phi) - \frac{\partial}{\partial x}(M_{\delta x}\phi). \tag{3.10}$$

Such an equation, as given in (3.9), is called the Kolmogorov forward equation by mathematicians. It is also called the Fokker-Planck equation by physicists. Actually, Fokker derived the steady state form in 1914 and Planck (1917) later extended it to a quite general form, though rigorous mathematical formuations were first given by Kolmogorov (1931).

The above derivation leading to equation (3.9), is rather formal. More rigorous derivations may be found in the mathematical literature, such as Kolmogorov's paper (1931). On the other hand, the above derivation may be too formal for most biologists to see the physical meaning of the terms involved. A less rigorous but very elementary derivation of the equation was devised by the author, based on the geometrical interpretation of the process involved (Kimura, 1955 c). It was shown that the first and second terms on the right side of (3.10) give the rates of change in the probability distribution due respectively to random fluctuation and systematic pressures. It was also pointed out that the variance

$$E\{(\delta x)^2\} - \{E(\delta x)\}^2$$

rather than the second moment, i.e. $E\{(\delta x)^2\}$ should be used for $V_{\delta x}$ in (3.10), when the equation is applied to actual population genetics problems. This is based on the consideration that (3.10) should give the deterministic process correctly when there is no random fluctuation, i.e. in the limit when $V_{\delta x} = 0$.

Since the gene frequency $x$ lies between 0 and 1 in general, the process of change in gene frequency in a population through time is represented as the stochastic movement of a point $x$ on the closed real interval $[0,1]$. The equation (3.10) can describe this movement at least on the open interval $(0,1)$. We will now show that

$$P(x,t) = -\frac{1}{2}\frac{\partial}{\partial x}\{V_{\delta x}\phi(x,t)\} + M_{\delta x}\phi(x,t) \tag{3.11}$$

which satisfies

$$\frac{\partial\phi(x,t)}{\partial t} = -\frac{\partial P(x,t)}{\partial x} \tag{3.12}$$

represents the rate (per generation) of net flow of probability across the point $x$. First, let us consider the amount of probability which flows over the point $x$ in a positive direction during the time interval of length $\delta t$. The contribution of the class with gene frequency $\xi$ to this is (see Figure 1):

$$\phi(p,\xi;t)d\xi \int_{\delta\xi > x-\xi} g(\delta\xi,\xi;\delta t,t)\,d(\delta\xi) \qquad (\xi < x),$$

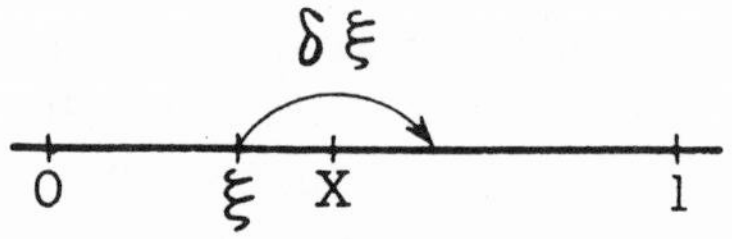


**Figure 1**
**Probability flux across the point $x$.**

and the total amount of probability flow in the positive direction, denoted by $P_+(x)\delta t$, is the sum of all contributions from the class at the left of $x$:

$$P_+(x)\delta t = \int_{\xi<x} \phi(p,\xi;t)d\xi \int_{\delta\xi>x-\xi} g(\delta\xi,\xi;\delta t,t)d(\delta\xi)$$

$$= \int_{\delta\xi>0} d(\delta\xi) \int_{x-\delta\xi}^{x} \phi(p,\xi;t)\,g(\delta\xi,\xi;\delta t,t)d\xi.$$

Let $\xi = x+\eta$ and expand the integrand in terms of $\eta$, or $\xi - x$; we have

$$\phi(p,\xi;t)\,g(\delta\xi,\xi;\delta t,t) = \phi g + \eta\frac{\partial(\phi g)}{\partial x} + \cdots,$$

where $\phi$ and $g$ respectively denote $\phi(p,x;t)$ and $g(\delta\xi,x;\delta t,t)$.

Thus

(3.13)
$$P_+(x)\delta t = \int_{\delta\xi>0} d(\delta\xi)\int_{-\delta\xi}^{0}\left\{\phi g + \eta\frac{\partial(\phi g)}{\partial x} + \cdots\right\}d\eta$$

$$= \int_{\delta\xi>0}\left\{\delta\xi\cdot\phi g - \frac{(\delta\xi)^2}{2}\frac{\partial(\phi g)}{\partial x} + \cdots\right\}d(\delta\xi).$$

Similarly, the total amount of probability which flows over the point $x$ in the opposite direction is

$$P_-(x)\delta t = \int_{\xi>x} \phi(p,\xi;t)d\xi \int_{\delta\xi<x-\xi} g(\delta\xi,\xi;\delta t,t)d(\delta\xi)$$

(3.14)
$$= \int_{\delta\xi<0} d(\delta\xi)\int_{x}^{x-\delta\xi} \phi(p,\xi;t)\,g(\delta\xi,\xi;\delta t,t)d\xi$$

$$= \int_{\delta\xi<0} d(\delta\xi)\int_{0}^{-\delta\xi}\left\{\phi g + \eta\frac{\partial(\phi g)}{\partial x} + \cdots\right\}d\eta$$

$$= \int_{\delta\xi<0}\left\{(-\delta\xi)\phi g + \frac{(\delta\xi)^2}{2}\frac{\partial(\phi g)}{\partial x} + \cdots\right\}d(\delta\xi).$$

Thus the net amount of probability which flows past the point $x$ during the time interval $(t, t+\delta t)$ is

$$
\begin{aligned}
P(x,t)\delta t &= P_+(x)\delta t - P_-(x)\delta t \\
&= \int (\delta\xi)\phi g d(\delta\xi) - \frac{1}{2}\int (\delta\xi)^2 \frac{\partial(\phi g)}{\partial x}\, d(\delta\xi) + \cdots \\
&= \phi \int \delta\xi g(\delta\xi, x; dt, t)\, d(\delta\xi) \\
&\quad - \frac{1}{2}\frac{\partial}{\partial x}\left\{\phi \int (\delta\xi)^2 g(\delta\xi, x; \delta t, t)\, d(\delta\xi)\right\} + \cdots
\end{aligned}
$$

and if we note (3.6), (3.7) and (3.8), we obtain, in the limit as $\delta t \to 0$,

$$
P(x,t) = M(x,t)\phi(p,x;t) - \frac{1}{2}\frac{\partial}{\partial x}\{V(x,t)\phi(p,x;t)\}. \tag{3.15}
$$

In terms of a generation as the unit of time, the above equation becomes (3.11) as was to be shown.

In equations (3.10) and (3.11), $M_{\delta x}$ and $V_{\delta x}$ are in general functions of both $x$ and $t$, but in most of our applications to population genetics, they are functions of $x$ only, and independent of the time parameter $t$. Actually in the present paper, we shall be concerned only with the cases where $M_{\delta x}$ and $V_{\delta x}$ are functions of $x$ but independent of $t$, namely where the process is time homogeneous. However, except for the special case of being identically zero, they can never take constant values, as might be the case in many diffusion problems in physics.

As stated already, our fundamental equation (3.10) can describe the "movement" of the point $x$ representing the gene frequency of a population on the open interval (0,1) and as noted in (3.1), $\phi(x,t)dx$ with $dx = 1/(2N)$ gives the approximate frequency of the class with gene frequency $x$ for $0 < x < 1$. The equation by itself cannot give the rates of change in the relative frequencies of terminal classes. However, these rates can be obtained by utilizing the established relation (3.11); we use the fact that $-P(0,t)$ and $P(1,t)$ respectively represent the rates at which the probability flows into the classes $x = 0$ and $x = 1$, from the open interval (0,1). In the special but important case in which the change in frequencies of these terminal classes ($x = 0$ and $x = 1$) is solely due to such an inflow of the probability, i.e. when boundaries act as "absorbing barriers", we have

$$
\frac{df(0,t)}{dt} = -P(0,t), \tag{3.16}
$$

$$
\frac{df(1,t)}{dt} = P(1,t), \tag{3.17}
$$

where $f(0,t)$ and $f(1,t)$ are respectively the frequencies of classes with $x = 0$ and $x = 1$ at the $t$th generation.

Now, if $\phi(x,t)$ and its first derivative with respect to $x$ are finite at $x = 0$ and if $V_{\delta x}$ and $M_{\delta x}$ vanish there, then

$$-P(0,t) = \frac{1}{2}\left[\frac{d}{dx}V_{\delta x}\right]_{x=0}\phi(0,t).$$

In particular, if the random fluctuation is due solely to random sampling of gametes, $V_{\delta x} = x(1-x)/(2N)$ and $[dV_{\delta x}/dx]_{x=0} = 1/(2N)$. Therefore

$$\frac{df(0,t)}{dt} = \frac{1}{2}\phi(0,t)\frac{1}{2N}. \tag{3.18}$$

The right-hand side of the above equation is approximately equal to half the relative frequency of the subterminal class with $x = 1/(2N)$. This is because, for a large value of $N$, $\phi(0,t)/(2N)$ must be very near to $f(1/2N,t)$ unless $|\partial\phi/\partial x|$ is very large at the neighborhood of $x = 0$. It should be noted here that if the effective size ($N_e$) is different from the actual size ($N$) of the population, we must put $V_{\delta x} = x(1-x)/(2N_e)$ and therefore the right-hand side of (3.18) must be multiplied by the factor of $N/N_e$. Similarly, we have

$$\frac{df(1,t)}{dt} = \frac{1}{2}\phi(1,t)\frac{1}{2N}. \tag{3.19}$$

The relation (3.11) is also useful in deriving the probability distribution of gene frequencies in the steady state when the distribution curve reaches constancy in form. The distribution in this state may either be obtained from

$$P(x,t) = 0 \quad \text{(stable distribution)} \tag{3.20}$$

or from

$$P(x,t) = \text{constant} \quad \text{(steady flux)} \tag{3.21}$$

depending on the circumstances.

The above arguments may readily be extended to the cases of two or more random variables. For the case of two random variables such as appear in the tri-allelic system having three alleles, $A_1$, $A_2$ and $A_3$ with respective frequencies $x_1$, $x_2$ and $x_3(=1-x_1-x_2)$, the corresponding differential equation becomes

$$\frac{\partial\phi}{\partial t} = \frac{1}{2}\frac{\partial^2}{\partial x_1^2}\left\{V_{\delta x_1}\phi\right\} + \frac{1}{2}\frac{\partial^2}{\partial x_2^2}\left\{V_{\delta x_2}\phi\right\} + \frac{\partial^2}{\partial x_1 \partial x_2}\{W_{\delta x_1 \delta x_2}\phi\} - \frac{\partial}{\partial x_1}\{M_{\delta x_1}\phi\} - \frac{\partial}{\partial x_2}\{M_{\delta x_2}\phi\} \tag{3.22}$$

where $\phi = \phi(p_1, p_2, x_1, x_2; t)$ gives the probability density that the frequencies of $A_1$ and $A_2$ become $x_1$ and $x_2$ at the $t$th generation given that their frequencies are $p_1$ and $p_2$ at $t=0$. In the above equation, $W_{\delta x_1 \delta x_2}$ stands for the covariance between $\delta x_1$ and $\delta x_2$, where $\delta x_1$ and $\delta x_2$ stand respectively for the rates of change of $x_1$ and $x_2$ per generation. Also their mean and variance $M_{\delta x_i}$ and $V_{\delta x_i}$ $(i=1,2)$ are in general functions of $x_1$ and $x_2$, as well as of $t$. More generally, for the case of $n$ segregating loci each having a pair of alleles, if $x^{(i)}$ is the frequency of an allele at the $i$th locus $(i=1,2,\cdots,n)$ then the Kolmogorov forward equation becomes

$$\text{(3.23)} \qquad \begin{aligned} \frac{\partial \phi}{\partial t} = \frac{1}{2}\sum_i \frac{\partial^2}{\partial^2 x^{(i)}}\{V_{\delta x^{(i)}}\phi\} &+ \sum_{i<j} \frac{\partial^2}{\partial x^{(i)}\partial x^{(j)}}\{W_{\delta x^{(i)}\delta x^{(j)}}\phi\} \\ &- \sum_i \frac{\partial}{\partial x^{(i)}}\{M_{\delta x^{(i)}}\phi\}. \end{aligned}$$

where $\phi=\phi(x^{(1)},\cdots,x^{(n)};t)$ is the probability density that gene frequencies are $x^{(1)} \sim x^{(1)}+dx^{(1)},\cdots,x^{(n)} \sim x^{(n)}+dx^{(n)}$ in the $t$th generation. In the above equation $\delta x^{(i)}$ is the rate of change of $x^{(i)}$ per generation and $M$, $V$ and $W$ respectively denote the mean, variance and covariance of the $\delta x^{(i)}$'s.

The equations for the probability flux in the multivariate case may be derived as in the case of a single variable. Here, I will merely present the equations for the case of two independent random variables, $x_1$ and $x_2$:

$$\text{(3.24)} \qquad \begin{cases} P(x_1|x_2;t) = -\dfrac{1}{2}\dfrac{\partial}{\partial x_1}\{V_{\delta x_1}\phi\} - \dfrac{1}{2}\dfrac{\partial}{\partial x_2}\{W_{\delta x_1 \delta x_2}\phi\} + M_{\delta x_1}\phi \\ Q(x_2|x_1;t) = -\dfrac{1}{2}\dfrac{\partial}{\partial x_1}\{W_{\delta x_1 \delta x_2}\phi\} - \dfrac{1}{2}\dfrac{\partial}{\partial x_2}\{V_{\delta x_2}\phi\} + M_{\delta x_2}\phi, \end{cases}$$

where $P(x_1|x_2;t)$ and $Q(x_2|x_1;t)$ are respectively the fluxes at point $(x_1,x_2)$ along the $x_1$ and $x_2$ axes. In terms of these quantities, (3.22) is expressed in the form:

$$\frac{\partial \phi}{\partial t} = -\frac{\partial P(x_1|x_2;t)}{\partial x_1} - \frac{\partial Q(x_2|x_1;t)}{\partial x_2}.$$

It should be noted here that the existence of a stable gene frequency distribution at equilibrium does not necessarily mean that

$$\text{(3.25)} \qquad P(x_1|x_2;t) = Q(x_2|x_1;t) = 0.$$

For example, in a locus with three alleles, $A_1$, $A_2$ and $A_3$, if genes mutate only in the sequence $A_1 \to A_2 \to A_3 \to A_1 \cdots$, a stable non-trivial distribution may be realized under a cyclic flow of probability.

So far, we have treated gene frequencies after $t$ generations as random variables and initial gene frequencies as fixed. For example, in the expression $\phi(p,x;t)$, $x$ is considered as a random variable and $p$ is assumed fixed. This means that we have considered the process of change in gene frequency in the forward direction in time.

On the other hand, we may regard $x$ as fixed and consider $p$ as a random variable. Namely, we reverse the time sequence and view the process retrospectively. In order to make our argument simpler, we will assume in what follows, that the process is time homogeneous. That is, if $x(t_1)$ and $x(t_2)$ are respectively frequencies of a gene at times $t_1$ and $t_2$ $(t_1 < t_2)$ then the probability distribution of $x(t_2)$ given $x(t_1)$, which in general should be a function of $t_1$ and $t_2$ separately, depends only on the difference $t_1 - t_2$. Then, we have

$$\phi(p,x;t+\delta t) = \int g(\delta p, p; \delta t)\,\phi(p+\delta p, x; t)\,d(\delta p). \tag{3.26}$$

The above equation which is a counterpart of (3.3), contains $g$ as a function of three variables only, i.e. $\delta p$, $p$ and $\delta t$. This is because the probability that the gene frequency changes from $p$ to $p + \delta p$ during the time interval of length $\delta t$ is the same for any $t$ (generation) due to the assumption of time homogeneity. Expanding $\phi\ (p+\delta p, x; t)$ on the right-hand side of the above equation in terms of $\delta p$ and using relations (3.6), (3.7) and (3.8) we obtain

$$\frac{\partial \phi(p,x;t)}{\partial t} = \frac{V(x)}{2}\,\frac{\partial^2 \phi(p,x;t)}{\partial p^2} + M(x)\,\frac{\partial \phi(p,x;t)}{\partial p}, \tag{3.27}$$

or in terms of one generation as a unit of time, we have

$$\frac{\partial \phi}{\partial t} = \frac{V_{\delta p}}{2}\,\frac{\partial^2 \phi}{\partial p^2} + M_{\delta p}\,\frac{\partial \phi}{\partial p}. \tag{3.28}$$

Note here that the initial gene frequency $p$ is the variable and $x$ is assumed to be constant. Mathematically, equation (3.27) is the Kolmogorov backward equation as applied to the time homogeneous case, and it is the adjoint form of (3.9).

When $x = 1$, $\phi$ in (3.28) gives the probability that the gene whose initial frequency was $p$ becomes fixed in the population by the $t$th generation. We will denote this probability by $u(p,t)$, for which we have

$$\frac{\partial u(p,t)}{\partial t} = \frac{V_{\delta p}}{2}\,\frac{\partial^2 u(p,t)}{\partial p^2} + M_{\delta p}\,\frac{\partial u(p,t)}{\partial p}. \tag{3.29}$$

The probability of fixation by a given time $t$ will then be obtained by solving the above equation with boundary conditions

$$u(0,t) = 0,\ u(1,t) = 1. \tag{3.30}$$

In the present paper we will be especially interested in the ultimate probability of fixation defined by

(3.31) $$u(p) = \lim_{t\to\infty} u(p,t).$$

For this probability,

$$\frac{\partial u}{\partial t} = 0$$

and $u(p)$ satisfies the ordinary differential equation

(3.32) $$\frac{V_{\delta p}}{2}\frac{d^2u(p)}{dp^2} + M_{\delta p}\frac{du(p)}{dp} = 0$$

with boundary conditions

(3.33) $$u(0) = 0,\ u(1) = 1.$$

Equation (3.29) may readily be extended to multivariate cases: consider $n$ independent loci each with a pair of alleles, a normal and a mutant allele. We will denote by $p^{(i)}$ the initial frequency of the mutant gene at the $i$th locus ($i = 1, 2, \cdots, n$). Let $u(p^{(1)}, p^{(2)}, \cdots, p^{(n)}; t)$ be the probability that all the $n$ mutant genes become fixed in the population by the $t$th generation, given that their frequencies are $p^{(1)}, p^{(2)}, \cdots, p^{(n)}$ at $t = 0$. Then $u(p^{(1)}, \cdots, p^{(n)}; t)$ satisfies

(3.34) $$\frac{\partial u}{\partial t} = \frac{1}{2}\sum_{i=1}^{n} V_{\delta p^{(i)}}\frac{\partial^2 u}{\partial^2 p^{(i)}} + \sum_{i>j} W_{\delta p^{(i)}\delta p^{(j)}}\frac{\partial^2 u}{\partial p^{(i)}\partial p^{(j)}} + \sum_{i=1}^{n} M_{\delta p^{(i)}}\frac{\partial u}{\partial p^{(i)}}.$$

In what follows, we will apply the method of partial differential equations to solve concrete problems arising in the theory of population genetics.

## II. Random Drift in the Narrow Sense

### 4. Random drift in a small finite population

We will start our discussion from the simplest situation where mutation, migration and selection are absent, but the gene frequency fluctuates from generation to generation because of the random sampling of gametes in a finite population. The process of change in gene frequency in this simplified form has attracted considerable attention among evolutionary geneticists, and various names have been given to it. Fisher (1922) called it the "Hagedoorn effect". Since Wright's work (Wright, 1931), the term *drift* has become quite popular among biologists, and terms such as the Wright drift or the Sewall Wright effect have been coined. However, in the mathematical theory of Brownian motion, the term drift originally connotes directional movement of the particle; therefore, to use this term to denote the random process in our context, the adjective *random* should be attached to it.

Let us consider an isolated population of $N$ breeding diploid individuals.

Let $A_1$ and $A_2$ be a pair of alleles with respective frequencies $x$ and $1-x$. We assume that mating is at random and that the mode of reproduction is such that $N$ male and $N$ female gametes are drawn as a random sample from the population to form the next generation. The mean and variance in the change of gene frequency $x$ per generation are: $M_{\delta x}=0$ and $V_{\delta x}=x(1-x)/(2N)$, the latter being the binomial variance corresponding to $2N$ genes. If mating is not random, or the distribution of the number of offspring does not follow a Poisson distribution, the effective number $N_e$ may be substituted for the actual number $N$ (cf. Kimura and Crow, 1963).

Substituting the above expressions for $M_{\delta x}$ and $V_{\delta x}$ into (3.10), we obtain the partial differential equation

$$\frac{\partial \phi}{\partial t} = \frac{1}{4N}\frac{\partial^2}{\partial x^2}\{x(1-x)\phi\}, \quad (0<x<1), \tag{4.1}$$

where $\phi = \phi(p,x;t)$ is the probability density that the gene frequency becomes $x$ in the $t$th generation, given that it is $p$ at $t=0$, i.e.

$$\phi(p,x;0) = \delta(x-p), \tag{4.2}$$

in which $\delta(\cdot)$ represents the Dirac delta function.

To solve (4.1) we try a solution of the form

$$\phi = TX,$$

where $T$ is a function of $t$ only and $X$ is a function of $x$ only. Substituting this into (4.1) and dividing both sides of the equation by $TX$, we have

$$\frac{1}{T}\frac{\partial T}{\partial t} = \frac{1}{4NX}\frac{\partial^2}{\partial x^2}\{x(1-x)X\}.$$

By assumption, $T$ is a function of $t$ only and hence the left-hand side of the above equation depends only on $t$, while $X$ is a function of $x$ only, and hence the right-hand side of the equation depends only on $x$. It follows then that both sides of the equation must equal a constant which we shall designate by $-\lambda$. Thus the above equation can be separated into two ordinary differential equations

$$\frac{dT}{dt} = -\lambda T \tag{4.3}$$

and

$$x(1-x)\frac{d^2X}{dx^2} + 2(1-2x)\frac{dX}{dx} - (2-4N\lambda)X = 0. \tag{4.4}$$

From the first equation (4.3) we have

$$T \propto e^{-\lambda t}$$

The second equation (4.4) is the hypergeometric equation

$$x(1-x)X'' + [\gamma - (\alpha+\beta+1)x]X' - \alpha\beta X = 0 \tag{4.5}$$

in which $\gamma = 2$, $\alpha + \beta = 3$ and $\alpha\beta = 2 - 4N\lambda$.

Thus we have

$$\alpha = \frac{3+\sqrt{1+16N\lambda}}{2} \quad \text{and} \quad \beta = \frac{3-\sqrt{1+16N\lambda}}{2}.$$

Though we cannot impose arbitrary conditions at the boundaries, we require a solution which is finite at the singular points ($x = 0$ and 1).

Among the two independent solutions of (4.5), only one, i.e. $F(\alpha,\beta,\gamma,x)$ is finite at $x = 0$ in this case. In order to find the condition which makes $F(\alpha,\beta,2,x)$ finite at the other singular point ($x = 1$), we make use of the following relation:

$$F(\alpha,\beta,2,x) = \frac{\Gamma(2)\Gamma(2-\alpha-\beta)}{\Gamma(2-\alpha)\Gamma(2-\beta)}F(\alpha,\beta,-1+\alpha+\beta,1-x)\,. \tag{4.6}$$

$$+\ \frac{\Gamma(2)\Gamma(\alpha+\beta-2)}{\Gamma(\alpha)\Gamma(\beta)}(1-x)^{2-\alpha-\beta}F(2-\alpha,2-\beta,3-\alpha-\beta,1-x)$$

If we note that $\alpha + \beta = 3$, we see that in order that $\lim_{x\to 1} F(\alpha,\beta,2,x)$ be finite, $2-\alpha$ must be a negative integer and $\beta$ must be 0 or a negative integer. Thus the only possible values of $\lambda$ are represented by

$$\lambda_i = i(i+1)/(4N),$$

where the $i$'s are positive integers ($i = 1,2,3,\cdots$). Corresponding to this eigenvalue $\lambda_i$, we have $\alpha_i = 2 + i$ and $\beta_i = 1 - i$. Thus we can write

$$X = F(2+i,1-i,2,x)$$

except that it may be multiplied by a constant. Here it may be convenient to use the Gegenbauer polynomial defined by

$$T_{i-1}^{1}(z) = \frac{i(i+1)}{2}F\left(i+2,1-i,2;\frac{1-z}{2}\right) \tag{4.7}$$

so that we can put

$$X = T_{i-1}^{1}(z)$$

where $z = 1 - 2x$. The properties of the polynomial have been thoroughly studied (see for example, Morse and Feshbach, 1953, pp. 782–783).

The complete solution of (4.1) may then be written in the form

$$\phi(p,x;t) = \sum_{i=1}^{\infty} C_i T_{i-1}^{1}(z)e^{-i(i+1)t/(4N)}. \tag{4.8}$$

The coefficients $C_i$ can be determined by applying the initial condition that the population starts from the gene frequency $p$, namely from (4.2), that

$$\delta(x-p) = \sum_{i=1}^{\infty} C_i T_{i-1}^1(z).$$

Multiplying by $(1-z^2)T_{i-1}^1(z)$ on both sides of the above equation, and using the orthogonality property

$$\int_{-1}^{1} (1-z^2)T_m^1(z)T_{i-1}^1(z)dz = \delta_{m,i-1}\frac{2(i+1)i}{(2i+1)},$$

where $m$ in the Kronecker $\delta_{m,i-1}$ represents zero or a positive integer, we obtain

$$2\{1-(1-2p)^2\}T_{i-1}^1(1-2p) = C_i\frac{2(i+1)i}{(2i+1)}$$

or

$$C_i = 4p(1-p)\frac{(2i+1)}{i(i+1)}T_{i-1}^1(1-2p).$$

Therefore, the required solution of (4.1) is

$$\phi(p,x;t) = \sum_{i=1}^{\infty}\frac{(2i+1)(1-r^2)}{i(i+1)}T_{i-1}^1(r)\,T_{i-1}^1(z)\,e^{-i(i+1)t/(4N)}, \tag{4.9}$$

where

$r=1-2p$ and $T_0^1(r)=1$, $T_1^1(r)=3r$, $T_2^1(r)=\frac{3}{2}(5r^2-1)$, $T_3^1(r)=\frac{5}{2}(7r^3-3r)$, etc.

In terms of the hypergeometric function $F(\cdot,\cdot,\cdot,\cdot)$, equation (4.9) may be expressed in the form

$$\phi(p,x;t) = \sum_{i=1}^{\infty} p(1-p)\,i(i+1)(2i+1)F(1-i,i+2,2,p)$$

$$\cdot F(1-i,i+2,2,x)e^{-i(i+1)t/(4N)} \tag{4.10}$$

$$= 6p(1-p)e^{-t/(2N)} + 30p(1-p)(1-2p)(1-2x)e^{-3t/(2N)} + \cdots.$$

For $t>0$, the series is uniformly convergent in $x$ and $p$. This may be easily seen if we note that the exponential term approaches zero rapidly.

Based on this solution, the process of change in the probability distribution of gene frequency when the population starts from $p=0.5$ and $0.1$ is illustrated in Figures 2a and 2b. In these figures, the abscissa represents the gene frequency $x$ and the ordinate the probability density $\phi$. In discussing such a distribution, it is often convenient to adopt the "frequency interpretation" of probability, regarding the distribution curve as representing relative frequencies of various gene frequency classes in the infinite collection of populations having the same size and subjected to the same conditions. The area under each curve represents

Figure 2a

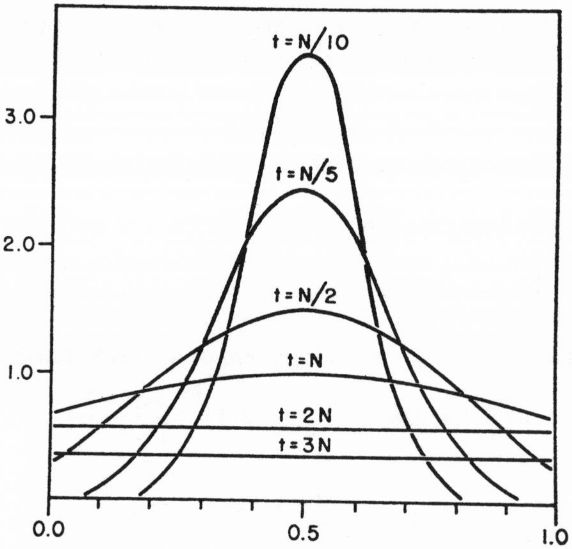


Figure 2b

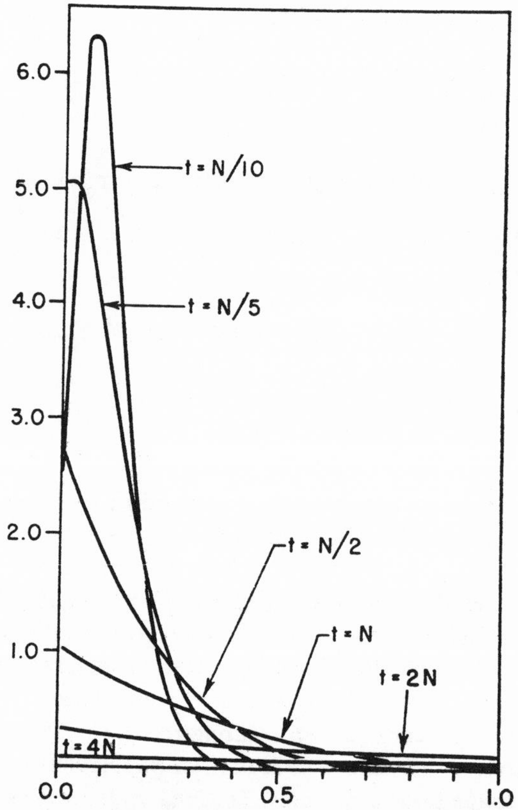


The process of change in the probability distribution of gene frequency, due to random sampling of gametes in reproduction. It is assumed that the population starts from the gene frequency 0.5 in Figure 2a, and 0.1 in Figure 2b. *t* indicates time (in generations), and *N* the effective size of the population. The abcissa is gene frequency, the ordinate is the probability density. (From Kimura, 1955a)

the probability that $A_1$ and $A_2$ coexist in the population. It may be seen from the figures that this probability gradually decreases with time. In other words, the frequency of unfixed classes decreases with increasing numbers of generations. This is because if, by random change, the gene frequency becomes either $x=1$ (fixation of $A_1$ or loss of $A_2$ from the population) or $x=0$ (loss of $A_1$), it cannot return to intermediate values because of our assumption that no mutation occurs. Namely, genes go irreversibly into fixation (or loss). In the language of probability theory, boundaries at $x=0$ and $x=1$ act as absorbing barriers. From (4.10) it can be seen that the probability distribution finally becomes flat and decreases its height at the rate of $1/(2N)$ per generation. This is known as the state of steady decay, and mathematically $1/(2N)$ corresponds to the smallest eigenvalue of the partial differential equation involved. This rate of decay is the most important single quantity used to describe the process of random drift in the narrow sense, and it was first determined by Wright (1931). Thus we have

$$\phi(p,x;t) \sim 6p(1-p)e^{-t/(2N)}, \quad (t \to \infty). \tag{4.11}$$

The number of generations after which this asymptotic formula becomes useful depends on the initial frequency $p$. For example, with $p=0.5$, it will be seen from Figure 2a that the distribution curve becomes almost flat after $2N$ generations, and the genes are still unfixed in about 50 per cent of the cases. On the other hand, with $p=0.1$ (see Figure 2b) it takes $4N$ or $5N$ generations before the distribution curves become practically flat. By that time, however, the genes are fixed in more than 90 per cent of the cases and the asymptotic formula (4.11) may not be as useful as in the case of $p=0.5$.

The probability that $A_1$ and $A_2$ coexist in the population of the $t$th generation is given by

$$\Omega_t = \int_0^1 \phi(p,x;t)dx = \sum_{j=0}^{\infty} [P_{2j}(r) - P_{2j+2}(r)]e^{-(2j+1)(2j+2)t/(4N)}, \tag{4.12}$$

where $r = 1-2p$ and $P(\cdot)$ represents the Legendre polynomials $P_0(r)=1$, $P_1(r)=r$, $P_2(r)=\frac{1}{2}(3r^2-1)$, $P_3(r)=\frac{1}{2}(5r^3-3r)$, etc. Thus we have the asymptotic formula

$$\Omega_t \sim 6p(1-p)e^{-t/(2N)}, \quad (t \to \infty). \tag{4.13}$$

The frequency of heterozygotes or the probability that an individual in a population is heterozygous can also be calculated by using (4.10) to obtain

$$H_t = \int_0^1 2x(1-x)\phi(p,x;t)dx = 2p(1-p)e^{-t/(2N)}. \tag{4.14}$$

This shows that heterozygosity decreases exactly at the rate of $1/(2N)$ per generation. Actually, this holds also for multiallelic cases and is independent of the number of alleles involved.

At this point a few remarks are in order. First, it should be noted that homozygosity or heterozygosity of an individual within a population is a distinct concept from the genetic homogeneity or heterogeneity of the population itself. Wright used the term homallelic or heterallelic; a population is homallelic if it contains only one kind of allele, and is heterallelic if it contains two or more. Secondly, the probability of heterozygosity decreases at the rate of exactly $1/(2N)$ per generation under random mating as shown by (4.14), while as shown in (4.12) the probability of coexistence of both alleles within a population, though continuously diminishing in each generation, does not generally decrease at a constant rate even for a population of constant size $N$. Its rate of decrease approaches the final value of $1/(2N)$ only asymptotically.

The above treatments do not directly give the probability of absorption, namely the probability of reaching fixation by a given generation $t$, starting from an intermediate gene frequency $p$. This may be obtained by the use of the backward equation (3.28) assuming $M_{\delta p}=0$ and $V_{\delta p}=p(1-p)/(2N)$. It turns out to be

$$(4.15)\quad u(p,t) = p + \sum_{i=1}^{\infty}(2i+1)p(1-p)(-1)^i F(1-i,i+2,2,p)e^{-i(i+1)t/(4N)} \quad ,$$

where this was first obtained from the study of the moments of the distribution (Kimura, 1955 a). In the present case, $u(p,t)$ is equivalent to $f(1,t)$. In terms of Legendre polynomials, (4.15) can be expressed also as

$$(4.16)\qquad f(1,t) = p + \sum_{i=1}^{\infty}\frac{(-1)^i}{2}\{P_{i-1}(r)-P_{i+1}(r)\}e^{-i(i+1)t/(4N)},$$

where $r=1-2p$. Using relation (3.19), it can also be obtained by integrating $\phi(p,1,\tau)/(4N)$ with respect to $\tau$ from $\tau=0$ to $\tau=1$. The probability $f(0,t)$ of $A_1$ being lost or $A_2$ being fixed by the $t$th generation is obtained simply by replacing $p$ with $1-p$ and $r$ with $-r$ in the above expressions. It is then possible to show that

$$f(1,t)+\Omega_t+f(0,t) = 1\,.$$

So far, we have assumed that the population contains a pair of alleles at the start. If a population contains more than two alleles, the problem becomes much more difficult. For example, suppose that the population contains three alleles $A_1$, $A_2$ and $A_3$ with respective frequencies of $x_1$, $x_2$ and $1-x_1-x_2$. Then, the probability density $\phi(p_1,p_2;x_1,x_2,t)$ that the frequencies of $A_1$ and $A_2$ become respectively $x_1$ and $x_2$ at time $t$, given that they are $p_1$ and $p_2$ at $t=0$, satisfies the following equation (cf. equation 3.22)

$$(4.17)\ \frac{\partial\phi}{\partial t}=\frac{1}{4N}\frac{\partial^2}{\partial x_1^2}\{x_1(1-x_1)\phi\}-\frac{1}{2N}\frac{\partial^2}{\partial x_1\partial x_2}\{x_1x_2\phi\}+\frac{1}{4N}\frac{\partial^2}{\partial x_2^2}\{x_2(1-x_2)\phi\},$$

where $0<x_1<x_1+x_2<1$.
The solution of this equation with the initial condition

$$\phi(p_1,p_2;x_1,x_2;0) = \delta(x_1-p_1)\delta(x_2-p_2)$$

is (cf. Kimura, 1956 a)

$$\phi(p_1,p_2;x_1,x_2;t)=\sum_{n=0}^{\infty}\sum_{j=0}^{\infty}C(n,j)(1-x_3)^n\cdot T_n^1\left(\frac{x_1-x_2}{1-x_3}\right)$$

$$(4.18)\qquad \cdot J_j(2n+5,2n+4,1-x_3)\exp\left\{-\frac{(j+n+2)(j+n+3)}{4N}t\right\}$$

$$=5!\,p_1p_2p_3e^{-3t/(2N)}+\frac{7!}{2!}p_1p_2p_3\left\{\sum_{i=1}^{3}\left(p_i-\frac{1}{3}\right)x_i\right\}e^{-6t/(2N)}+\cdots,$$

where

$$C(n,j)=\frac{4\cdot(j+2n+3)!(j+2n+4)!(2j+2n+5)}{j!(j+1)!(n+1)(n+2)\cdot(2n+2)!(2n+3)!}$$

$$\cdot\, p_1p_2p_3(1-p_3)^n\cdot T_n^1\left(\frac{p_1-p_2}{1-p_3}\right)J_j(2n+5,2n+4,1-p_3).$$

Here $x_3=1-x_1-x_2$, $p_3=1-p_1-p_2$ and, $T_n^1(\cdot)$ and $J_j(\cdot,\cdot,\cdot)$ denote respectively the Gegenbauer and Jacobi polynomials. The latter is expressed in terms of the hypergeometric function as follows:

$$J_n(a,c,\rho) = F(a+n,-n,c,\rho).$$

In particular $J_0(a,c,\rho)=1$, $J_1(5,4,\rho) = 1-\frac{3}{2}\rho$, etc.

For the general case of an arbitrary number of alleles, the exact solution has not been obtained. Nevertheless, the asymptotic behavior of the processes has been successfully analysed and we have the following result (cf. Kimura, 1955 b).

If we start from a population which contains $n$ alleles, say $A_1,A_2,\cdots,A_n$ with frequencies $p_1,p_2,\cdots,p_n$ respectively ($\sum_1^n p_i=1$), the probability density that it contains $k$ of them, say $A_1,A_2,\cdots,A_k$ with respective frequencies $x_1,x_2,\cdots,x_k$ ($\sum_1^k x_i=1$) in the $t$th generation is given asymptotically by

$$(4.19)\qquad \phi_{1,2,\cdots,k}(x_1,x_2,\cdots,x_{k-1};t)\sim(2k-1)!\left(\prod_{j=1}^{k}p_j\right)e^{-k(k-1)t/(4N)},$$

where $k\leq n$. The validity of this formula depends on the assumption that the population size $N$ is sufficiently large as compared with $n$, the number of alleles in question.

The above result indicates that as the number of coexisting alleles increases, the rate at which a particular state is eliminated by random drift increases rapidly. In this sense, random drift may be effective in keeping down the number of co-existing alleles in the population.

**5. An approximate treatment by the angular transformation**

It has been noted by Fisher (1922) that if we transform the gene frequency from $x$ to $\theta$ by $\cos\theta = 1 - 2x$, the sampling variance becomes independent of gene frequency. Here $\theta$ changes from 0 to $\pi$ as $x$ changes from 0 to 1. The rate of change of $\theta$ per generation, i.e. $\delta\theta$, is related to $\delta x$ as follows

$$\delta\theta = [x(1-x)]^{-1/2}\delta x - \frac{1}{4}(1-2x)[x(1-x)]^{-3/2}(\delta x)^2 + \cdots . \tag{5.1}$$

Thus, taking $M_{\delta x} = 0$ and $V_{\delta x} = x(1-x)/(2N)$, and neglecting the higher order terms, we get

$$M_{\delta\theta} = -\cot\theta/(4N), \quad V_{\delta\theta} = 1/(2N). \tag{5.2}$$

If we start from a fixed gene frequency, the variance in $\theta$ after $t$ generations may be given approximately by $V_\theta(t) = t/(2N)$, if $t$ is much smaller than $N$. It should be noted here, however, that the expected value of $\delta\theta$ is not strictly zero, i.e. the expression $M_{\delta\theta} = 0$ which might be obtained by neglecting the second and following terms on the right-hand side of (5.1) is incorrect. This may not produce any trouble in the treatment of variance for a short period, but will cause a serious error in the gene frequency distribution after a large number of generations. Let $\psi(\theta, t)$ be the probability density of $\theta$ at the $t$th generation. We obtain

$$\frac{\partial\psi}{\partial t} = \frac{1}{4N}\frac{\partial^2\psi}{\partial\theta^2} + \frac{1}{4N}\frac{\partial}{\partial\theta}(\psi\cos\theta), \tag{5.3}$$

which is the Kolmogorov forward equation in terms of $\theta$ for this case (cf. 3.10 and 5.2). The above equation was given by Fisher (1930) as the correct equation to replace the erroneous one which he had given earlier (Fisher, 1922), i.e.

$$\frac{\partial\psi}{\partial t} = \frac{1}{4N}\frac{\partial^2\psi}{\partial\theta^2}. \tag{5.4}$$

The latter equation was obtained by taking $M_{\delta\theta} = 0$ and gave the incorrect value of $1/(4N)$ as the rate of steady decay, rather than the correct value of $1/(2N)$ obtained by Wright (1931). The fact that the sampling variance becomes constant by the angular transformation is nevertheless convenient for treating data on random drift over a relatively short period (cf. Bodmer, 1960).

One of the most interesting applications of this type of transformation was given by Cavalli and Conterio (1960), who analysed the distribution of blood group genes in the Parma River Valley. Their method is based on the concept of "distance" as suggested by Fisher. Consider a locus with multiple alleles

$A_1, A_2, \cdots, A_n$. In order to characterize the genetic constitution of a population, we use $n$-dimensional Cartesian coordinates with each axis representing the square root of one of the allelic frequencies. A population which contains these alleles with respective frequencies of $x_1, x_2, \cdots, x_n$ may be located on a hypersphere with radius 1. Figure 3 illustrates the case with three alleles. Let $p_1, p_2, \cdots, p_n$ be the corresponding allelic frequencies in some other population. Then the coefficient of genetic distance $\theta$ between these two populations may be defined by

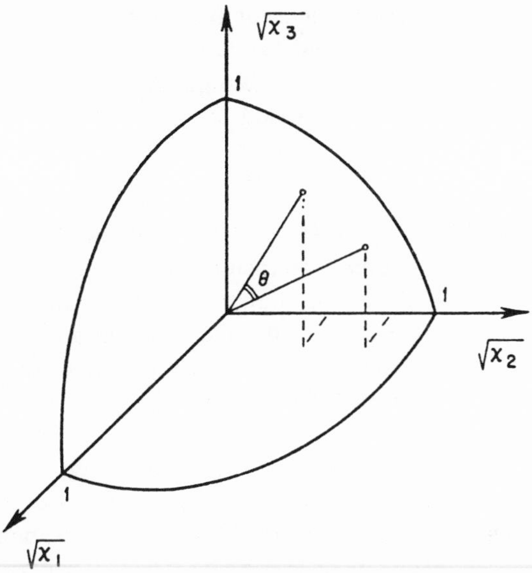


Figure 3
This illustrates the concept of the genetic distance for $n = 3$. (Redrawn from Cavalli and Conterio, 1960, with a slight modification)

$$\cos\theta = \sum_{i=1}^{n} \sqrt{x_i}\sqrt{p_i}. \tag{5.5}$$

Geometrically, $\theta$ is the angle made by two vectors $(\sqrt{x_1}, \sqrt{x_2}, \cdots, \sqrt{x_n})$ and $(\sqrt{p_1}, \sqrt{p_2}, \cdots, \sqrt{p_n})$ both radiating from the origin (see Figure 3 where $n=3$). If we take $x_1, x_2, \cdots, x_n$ as the allelic frequencies of a local population which is undergoing random drift, and take $p_1, p_2, \cdots, p_n$ as a set of reference points, such as the initial frequencies or average frequencies over infinitely many local populations, then it can be shown that the variance in the change of $\theta$ per generation is given approximately by

$$V_{\delta\theta} = \frac{1}{8N} = \frac{1}{4\,(\text{Number of genes})}, \tag{5.6}$$

where $N$ is the effective size of the population. As in the angular transformation $\theta = \cos^{-1}(1-2x)$, the expected value of $\delta\theta$ in this case is not zero, but this may not cause any serious error in treating the variance of $\theta$, as long as the number of generations involved is much smaller than $N$. Furthermore, if $\theta_a$ is the genetic distance of a population from the general population with respect to the first locus consisting of alleles $A_1, A_2, \cdots, A_n$ and if $\theta_b$ is the corresponding distance with respect to the second locus consisting of alleles $B_1, B_2, \cdots, B_m$, then the distance $\theta_{ab}$ with respect to these two loci combined is given by $\cos\theta_{ab} = \cos\theta_a \cos\theta_b$.

Using these relations, Cavalli and Conterio studied the regression of $\theta$ on population density, village size and "dimensionality". For details, their original paper (Cavalli and Conterio, 1960) should be referred to.

## III. Random Drift in the Wide Sense

### 6. A population under linear pressure and random sampling of gametes

Under the term linear pressure, we include the pressures of gene mutations and of migration. Usually the rate of mutation is so low that although supplying the raw material for evolution, it can hardly determine the course of change in gene frequency. On the other hand, migration between sub-populations may be of considerable significance in determining the gene frequency, as will be found in Wright's theory.

Consider a random mating population of effective size $N$ in which the frequencies of a pair of alleles $A_1$ and $A_2$ are $x$ and $1-x$ respectively. Let us suppose that this population exchanges individuals with a random sample taken from the total species at the rate of $m$ per generation. Then the mean and variance of the rate of change in $x$ are given by

$$M_{\delta x} = m(\bar{x} - x), \quad V_{\delta x} = x(1-x)/(2N), \tag{6.1}$$

where $\bar{x}$ is the frequency of $A_1$ in the immigrants. If mutation rates are not negligible, $m$ may be replaced by $m + \mu + v$ and $m\bar{x}$ by $m\bar{x} + v$, where $\mu$ and $v$ are respectively the mutation rates of $A_1$ to and from its allele $A_2$. Though the pressure of selection is intrinsically non-linear, in certain cases, like that of selection acting at the neighborhood of the equilibrium gene frequency, it may be treated as if it were linear with good approximation. However, the range of applicability is quite restricted. The solution of (3.10) when $M_{\delta x}$ and $V_{\delta x}$ are given by (6.1) was obtained by the author through the study of the moments of the distribution (Crow and Kimura, 1956), and it was found that it agrees with the "fundamental solution with flux zero boundary condition" derived by Goldberg (1950). The solution is given by

$$\phi(p,x;t) = \sum_{i=0}^{\infty} X_i(x)\exp\left\{-i\left(m+\frac{i-1}{4N}\right)t\right\} \tag{6.2}$$

where

$$X_i(x) = x^{B-1}(1-x)^{(A-B)-1}F(A+i-1,\ -i,\ A-B,\ 1-x)$$
$$\cdot F(A+i-1,-i,A-B,1-p)\ \frac{\Gamma(A-B+i)\Gamma(A+2i)\Gamma(A+i-1)}{i!\Gamma^2(A-B)\Gamma(B+i)\Gamma(A+2i-1)}$$

in which $A = 4Nm$ and $B = 4Nm\bar{x}$.

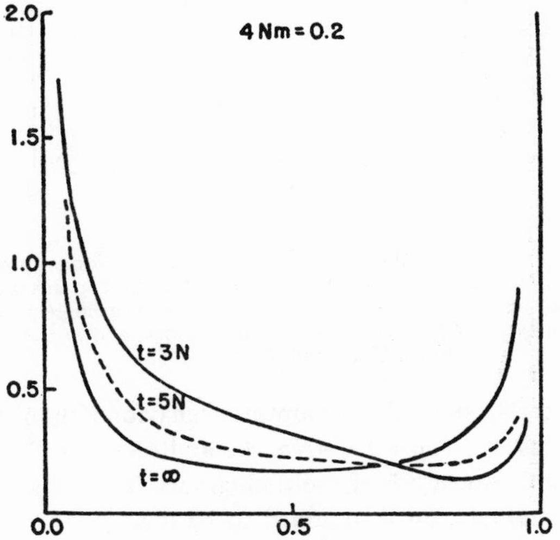


Figure 4a

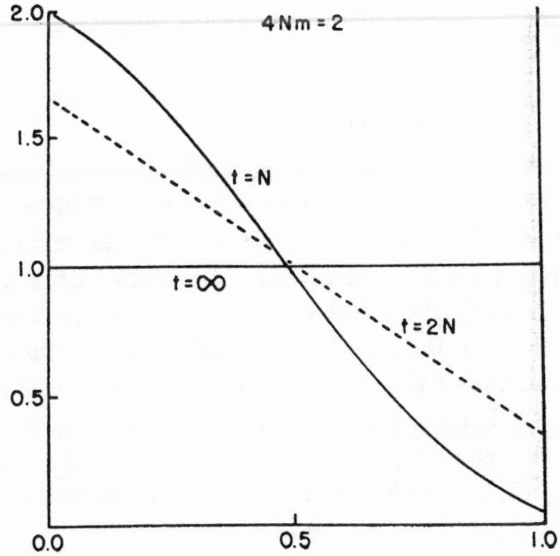


Figure 4b

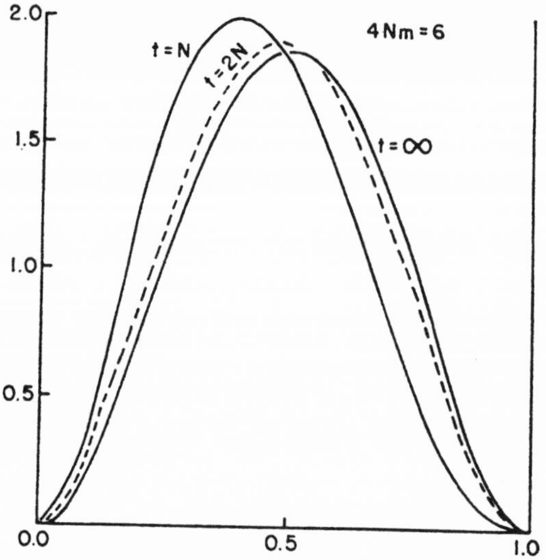


Figure 4c

Asymptotic behavior of the distribution curve for a finite population with migration or other linear pressure. In all three drawings, the gene frequency of the immigrants is assumed to be 0.5 and the initial frequency in the population 0.2. The abscissa is the gene frequency $x$, the ordinate is the probability density $\phi$. $N$ represent population number, and $m$ the rate of migration. (From Crow and Kimura, 1956)

Figures 4a, 4b and 4c show the asymptotic behavior of the distribution curve for three different cases: $4Nm = 0.2$, $4Nm = 2$ and $4Nm = 6$. In all three cases illustrated, the gene frequency $\bar{x}$ of the immigrants is 0.5 and the initial gene frequency $p$ of the population is assumed to be 0.2.

As $t \to \infty$, our formula (6.2) converges to Wright's well known formula for the steady state gene frequency distribution with migration

$$\phi(x) = \frac{\Gamma(4Nm)}{\Gamma(4Nm\bar{x})\,\Gamma(4Nm(1-\bar{x}))}\; x^{4Nm\bar{x}}(1-x)^{4Nm(1-x)-1} \; .$$

## 7. Change of gene frequency under selection and random sampling of gametes

7.1. *Genic selection (Case of no dominance).* Since selection, either natural or artificial, is always at work in the process of evolution or breeding, it is extremely important to study the effect of selection under random sampling of gametes. We will start from the simplest case of genic selection and consider random mating population of size $N$, in which $A_1$ and $A_2$ occur with respective frequencies $x$ and $1-x$. Let $s$ be the selective advantage of $A_1$ over $A_2$ such that the average rate of change in $x$ per generation is $M_{\delta x} = sx(1-x)$. We take, as before $V_{\delta x} = x(1-x)/(2N)$, which is the variance due to random sampling of gametes. With these expressions, (3.10) becomes

$$\frac{\partial\phi}{\partial t} = \frac{1}{4N}\frac{\partial^2}{\partial x^2}\{x(1-x)\phi\} - s\frac{\partial}{\partial x}\{x(1-x)\phi\}. \tag{7.1}$$

The process of change in gene frequency is analogous to the one studied in Section 4, but here selection is superimposed. The boundaries $x=0$ and $x=1$ act as absorbing barriers and the probability that a population contains both alleles $A_1$ and $A_2$ gradually decreases with time. Finally, it decreases at a constant rate which is given by the smallest eigenvalue ($\lambda_0$) of the above equation. In this state of steady decay, the distribution curve retains constancy of form, but its height decreases at the rate of $\lambda_0$ per generation. Probably, the smallest eigenvalue $\lambda_0$ is the most important single quantity in the representation of this stochastic process. The above equation has been used to analyse the gene frequency change in a very small experimental population of *Drosophila melanogaster* (Wright and Kerr, 1954). In this paper, Wright devised an ingenious method for analysing the process of steady decay; the complete solution of the above equation has been obtained by the present author (Kimura, 1955c and Crow and Kimura, 1956). In order to solve the equation (7.1), let us put

$$\phi \propto e^{2cx}V(x)e^{-\lambda t}$$

where $c = Ns$ and $V(x)$ is a function of $x$ only. If we substitute this in (7.1), we have

$$x(1-x)\frac{d^2V}{dx^2} + 2(1-2x)\frac{dV}{dx} - \{2+4c^2x(1-x)-4N\lambda\}V = 0. \tag{7.2}$$

Then by the substitution

$$x = (1-z)/2,$$

the above equation (7.2) becomes

$$(1-z^2)\frac{d^2V}{dz^2} - 4z\frac{dV}{dz} + \{(4N\lambda-2-c^2)+c^2z^2\}V = 0, \tag{7.3}$$

where $z = 1-2x$ $(-1<z<1)$. This type of differential equation is known as the oblate spheroidal equation. We want here the solutions which are finite at the singularities, $z=\pm 1$, and reduce to the Gegenbauer polynomials if there is no selection ($Ns = c = 0$). Such a solution has been studied by Stratton and others (1941) and is expressed in the form

$$V_{1k}^{(1)}(z) = \sum_{n=0,1}{}' f_n^k T_n^1(z), \tag{7.4}$$

where $k = 0,1,2,\cdots$ ($k$ here corresponds to $l$ in the notations of Stratton *et al.*). In the above expression, $f_n^k$'s are constants and $T_n^1(z)$'s are the Gegenbauer polynomials defined by (4.7). The primed summation is over even values of $n$ if $k$ is even, odd values of $n$ if $k$ is odd.

The desired solution of (7.2) is given by summing the $V_{1k}^{(1)}(z)$ for all possible values of $k$, after having multiplied through by $e^{2cx-\lambda_k t}$, where $\lambda_k$ is the $k$th eigenvalue; then

$$\phi(p,x;t) = \sum_{k=0}^{\infty} C_k e^{-\lambda_k t+2cx} V_{1k}^{(1)}(z). \tag{7.5}$$

The coefficient $C_k$ can be determined by the initial condition

$$\phi(p,x;0) = \delta(x-p),$$

using the orthogonal relation

$$\int_{-1}^{1} (1-z^2) V_{1k}^{(1)}(z) V_{1l}^{(1)}(z) dz = \delta_{kl} \sum_{n=0,1}' (f_n^k)^2 \frac{(n+2)!}{n!(2n+3)}.$$

Thus we have

$$C_k = \frac{(1-r^2)\, e^{-c(1-r)}\, V_{1k}^{(1)}(r)}{\sum_{n=0,1}' \frac{(n+1)(n+2)}{(2n+3)} (f_n^k)^2}, \tag{7.6}$$

where $r = 1-2p$, $c = Ns$ and the primed summation is over even values of $n$ if $k$ is even, over odd values of $n$ if $k$ is odd. The solution (7.5) with coefficients defined by (7.6) gives the probability distribution of the gene frequency among unfixed classes.

As $t$ increases, the exponential terms in (7.5) decrease in absolute value very rapidly, and for large $t$ only the first few terms are important. The numerical values of the first few eigenvalues $\lambda_0$, $\lambda_1$ and $\lambda_2$ can be obtained from the tables of the separation constants $(B_{1,k})$ in the book by Stratton *et al.* (1941), by using the relation

$$4N\lambda_k = c^2 - B_{1,k}.$$

Among them, the smallest eigenvalue $\lambda_0$ gives the final rate of decay and has special importance. For small values of $c, \lambda_0$ can be expanded into a power series in $c$,

$$2N\lambda_0 = 1 + \frac{2}{5}c^2 - \frac{2}{5^3 \cdot 7}c^4 - \frac{2^2}{3 \cdot 5^5 \cdot 7}c^6 - \frac{2 \cdot 31}{5^6 \cdot 7^3 \cdot 11}c^8 - \cdots. \tag{7.7}$$

In the new table of spheroidal wave functions by Stratton *et al.* (1956), "$t$" is tabulated for $c$ (denoted by $g$ in the table) up to $c = 8.0$ (pp. 506–508), from which values of $2N\lambda_0$ may be obtained by the relation

$$2N\lambda_0 = 1 + \left(\frac{2}{5} - \frac{t}{2}\right)c^2.$$

In Figure 5, the relation between $2N\lambda_0$ and $Ns$ is plotted for $Ns$ from 0 to 8.0.

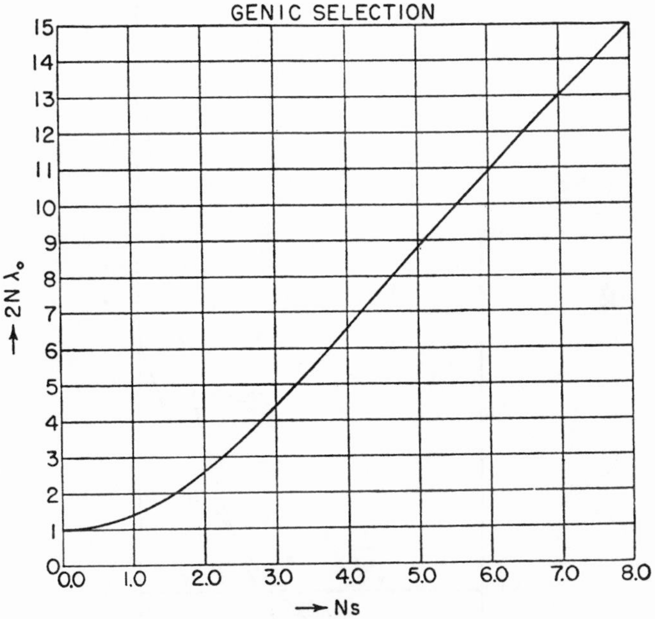


Figure 5

Relation between the rate of steady decay and intensity of selection as illustrated in terms of the relation between $2N\lambda_0$ and $Ns$, where $N$ is the effective size of the population, $\lambda_0$ is the rate of steady decay and $s$ is the selection coefficient relating to genic selection between a pair of alleles. (From Kimura, 1955c)

The eigenfunctions $V_{1k}^{(1)}(z)$ corresponding to the $\lambda_k$'s are given by (7.4). The coefficients $f_n^k$ corresponding to the first three eigenvalues are found in the table of Stratton *et al.* For $c = 0$, all the formulae given above reduce to the ones for the case of pure random drift studied in Section 4.

The first eigenfunction $V_{10}^{(1)}(z)$ which corresponds to $\lambda_0$ is of special significance, since it gives the frequency distribution of unfixed classes in the state of steady decay, when it is multiplied by $e^{c(1-z)}$. It is expressed by

(7.8) $$V_{10}^{(1)}(z) = f_0^0 T_0^1(z) + f_2^0 T_2^1(z) + f_4^0 T_4^1(z) + \cdots .$$

Figure 6 illustrates the shape of the distribution curve in the state of steady decay for three different cases; $Ns = 0$, $Ns = 1.0$ and $Ns = 1.7$. The area under each curve is adjusted so that it is unity.

The frequencies of both terminal classes can be obtained by using the relations (3.19) and (3.18):

(7.9) $$f(1;t) = e^{2c} \sum_{k=0}^{\infty} \frac{C_k}{4N\lambda_k}(1 - e^{-\lambda_k t} V)_{1k}^{(1)}(-1)$$

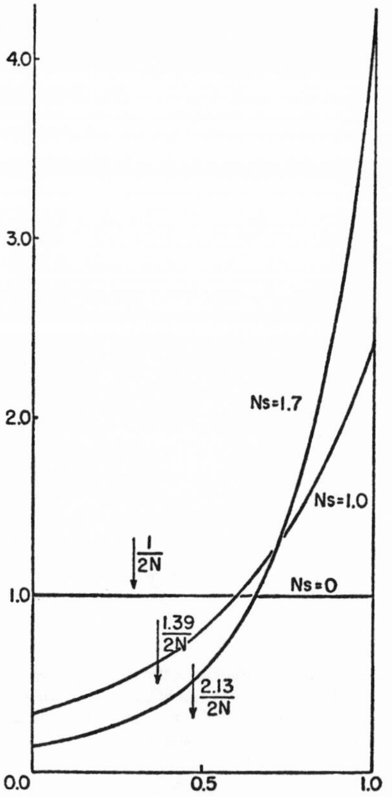


**Figure 6**

**Frequency distribution of unfixed classes at the stage of steady decay is illustrated for three values of *Ns*. The area under each curve is adjusted so that it is unity. Numbers beside the arrows indicate rates of steady decay. *N* is the effective size of population, and *s* is the selection coefficient. (From Kimura, 1955c)**

and

(7.10) $$f(0;t) = \sum_{k=0}^{\infty} \frac{C_k}{4N\lambda_k}(1-e^{-\lambda_k t})\, V_{1k}^{(1)}(1)$$

where

$$V_{1k}^{(1)}(-1) = \frac{1}{2} \sum_{n=0,1}{}' (-1)^n (n+1)(n+2) f_n^k$$

and

$$V_{1k}^{(1)}(1) = \frac{1}{2} \sum_{n=0,1}{}' (n+1)(n+2) f_n^k .$$

So far we have considered genic selection in which the gene is additive (i.e. no dominance) with respect to fitness. The case of complete dominance is more difficult to treat, but the process of steady decay has been worked out for weak selection (Kimura, 1957). Also, for the more general case of an arbitrary degree of dominance, the smallest eigenvalue can be given as a power series for weak selection. Namely, if the selective advantages of $A_1A_1$ and $A_1A_2$ over $A_2A_2$ are $s$ and $sh$ respectively, such that $M_{\delta x} = s[h + (1-2h)x]x(1-x)$, then

$$2N\lambda_0 = 1 + K_1c + K_2c^2 + K_3c^3 + \cdots, \tag{7.11}$$

where

$$K_1 = -\frac{1}{5}D,\quad K_2 = \frac{1}{2\cdot 5} + \frac{2^2\cdot 3}{5^3\cdot 7}D^2,\quad K_3 = \frac{1}{2\cdot 5^3\cdot 7}D - \frac{2^2}{5^5\cdot 7}D^3, \text{ etc.},$$

in which $c = Ns$ and $D = 2h - 1$. It may be noted that for the case of no dominance ($D = 0$), the above series (7.11) agrees with (7.7) provided that $2s$ is used instead of $s$ to express the selective advantage of the homozygous mutants.

7.2. *Case of overdominance.* It has been known since the early work of Fisher (1922) that, in an infinite population, heterozygote superiority in fitness for a pair of alleles leads to a stable polymorphism. Furthermore, a considerable number of claims have been made in recent years stating in effect that overdominance is the major factor for maintaining genetic variability in natural populations. Therefore, investigation of the overdominant case assuming a finite population number will be of interest. Let us assume a pair of overdominant alleles $A_1$ and $A_2$ and designate by $s_1$ and $s_2$ (both positive) the selection coefficients against the homozygotes $A_1A_1$ and $A_2A_2$ respectively, such that the average rate of change in the frequency of $A_1$ is $M_{\delta x} = [s_2 - (s_1 + s_2)x]x(1-x)$. The variance of $\delta x$ is again given by $V_{\delta x} = x(1-x)/(2N)$.

The partial differential equation corresponding to (3.10) with the $M_{\delta x}$ and $V_{\delta x}$ given above is

$$\frac{\partial\phi}{\partial t} = \frac{1}{4N}\frac{\partial^2}{\partial x^2}\{x(1-x)\phi\} - \frac{\partial}{\partial x}\{(s_2 - (s_1+s_2)x)x(1-x)\phi\}, \tag{7.12}$$

where $x$ is the frequency of $A_1$.

Let $\bar{s}$ be the average of the two selection coefficients, i.e., $\bar{s} = (s_1 + s_2)/2$, and let $\hat{x}$ be the equilibrium frequency of $A_1$ that may be expected in an infinitely large population, i.e. $\hat{x} = s_2/(s_1 + s_2)$, then the above equation may be expressed in the form,

$$4N\frac{\partial\phi}{\partial t} = \frac{\partial^2}{\partial x^2}\{x(1-x)\phi\} - 4N\bar{s}\frac{\partial}{\partial x}\{2(\hat{x} - x)x(1-x)\phi\}, \tag{7.13}$$

or denoting $2\hat{x} - 1 = \hat{z}$,

$$4N\frac{\partial\phi}{\partial t} = \frac{\partial^2}{\partial x^2}\{x(1-x)\phi\} + 4N\bar{s}\frac{\partial}{\partial x}\{(2x - 1 - \hat{z})x(1-x)\phi\}. \tag{7.14}$$

The smallest eigenvalue $\lambda_0$ of the above equation has been worked out by Miller (1962).

Without loss in generality, we can take $\hat{x} \geqq 0.5$ or $\hat{z} \geqq 0$. For a large value of $c \equiv N\bar{s} = N(s_1 + s_2)/2$ and for the range $1 > \hat{z} \geqq 0$, Miller obtained the asymptotic expansion

$$(7.15) \qquad 2N\lambda_0 = \frac{c}{T(x,\hat{z})}\left[\frac{(1-\hat{z})e^{-c(1-z)^2}}{S\{c(1-\hat{z})^2\}} + \frac{(1+\hat{z})e^{-c(1+\hat{z})^2}}{S\{c(1+\hat{z})^2\}}\right],$$

where

$$S(X) = 1 + \frac{1}{2}\frac{1}{X} + \frac{1\cdot 3}{2^2}\frac{1}{X^2} + \frac{1\cdot 3\cdot 5}{2^3}\frac{1}{X^3} + \cdots,$$

$$T(c,\hat{z}) = \sum_{i=0}^{\infty} \frac{C_{2i}\Gamma(i+\frac{1}{2})}{c^{i+\frac{1}{2}}} = \frac{C_0}{\sqrt{c}}\sqrt{\pi} + \cdots$$

in which the $C_i$'s are given by the recurrence relation

$$(1-\hat{z}^2)C_{i+1} = 2\hat{z}C_i + C_{i-1},$$

starting from

$$C_0 = 1/(1-\hat{z}^2) \text{ and } C_1 = 2\hat{z}/(1-\hat{z}^2)^2.$$

In particular, when $\hat{z} = 0$, that is when $s_1 = s_2$, (7.15) reduces to

$$2N\lambda_0 = \frac{2c^{3/2}e^{-c}}{\sqrt{\pi}\{S(c)\}^2}$$

Also, when $2N(s_1 + s_2)(1-\hat{x})^2$ is large, (7.15) may be replaced by

$$(7.16) \qquad 2N\lambda_0 = \sqrt{\frac{N^3(s_1+s_2)^3}{2\pi}}\cdot 4\hat{x}^2(1-\hat{x})^2\left\{\frac{e^{-4c(1-\hat{x})^2}}{\hat{x}} + \frac{e^{-4c\hat{x}^2}}{(1-\hat{x})}\right\}.$$

Miller has also obtained $\lambda_0$ for various values of $c(\geqq 0)$ up to $c = 12$ by numerical analysis.

It may be noted here that for small values of $N\bar{s}$, the eigenvalue may be calculated from the power series (7.11) by putting $c = N(s_2 - s_1)$ and $cD = N(s_1 + s_2) = 2N\bar{s}$ in it. Figure 7 is constructed on the basis of his numerical results, giving $2N\lambda_0$ as a function of $c$ for the cases of $\hat{x} = 0.5$, 0.7, 0.8 and 0.9. One of the most remarkable features disclosed in the figure seems to be that if $s_1$ and $s_2$ differ to such an extent that the equilibrium frequency $\hat{x}$ is higher than 0.8 (or, because of symmetry, less than 0.2), overdominance tends to *accelerate* fixation as compared with the neutral case, rather than retard fixation. This was first pointed out by Robertson (1962) who presented this fact in the form of a graph shown in Figure 8, where the term retardation factor is defined as the reciprocal of $2N\lambda_0$. According to him, selection for a heterozygote is a factor retarding fixation only if the equilibrium frequency lies within the range of

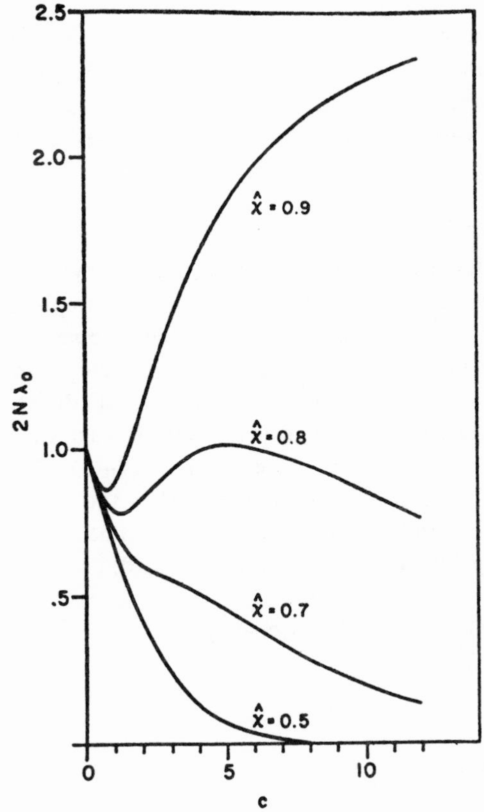


Figure 7
Relation between $2N\lambda_0$ and $c(=N(s_1+s_2)/2)$ is plotted for various equilibrium frequencies $\hat{x}$, when there is overdominance between a pair of alleles.

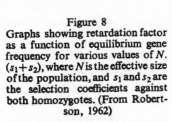

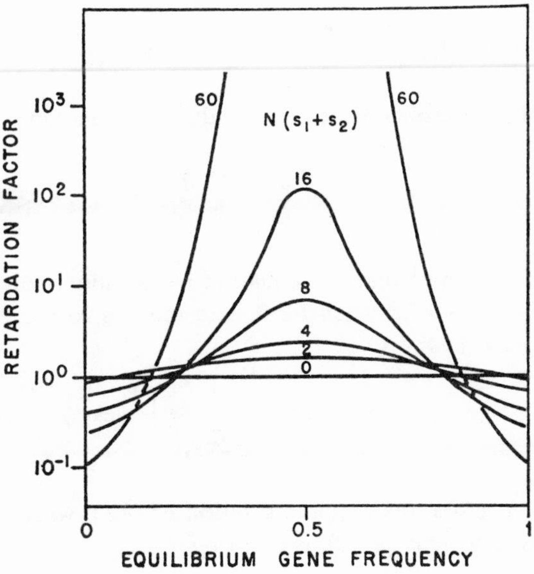


Figure 8
Graphs showing retardation factor as a function of equilibrium gene frequency for various values of $N.(s_1+s_2)$, where $N$ is the effective size of the population, and $s_1$ and $s_2$ are the selection coefficients against both homozygotes. (From Robertson, 1962)

0.2–0.8. For equilibrium gene frequencies outside this range, there is a range of values of $N(s_1+s_2)$ for which heterozygote advantage accelerates fixation, and the more extreme the equilibrium frequency the wider the range. However, for all values of $\hat{x}$ except 0 or 1, an increase in the values of $N(s_1+s_2)$ eventually leads to retardation.

**8. Randon fluctuation of selection intensities**

Among the factors which cause random fluctuation in gene frequencies, random fluctuation of selection intensities may often be as important as random sampling of gametes. To single out this factor, we will here assume that the population is infinitely large so that the effect of random sampling may be neglected. Also we will consider the simplest case of genic selection in which a pair of alleles, $A_1$ and $A_2$ are involved. Let $s$ be the selective advantage of $A_1$ over $A_2$ such that the rate of change in the frequency of $A_1$ for a fixed value of $s$ is $sx(1-x)$. Let us assume also that $A_1$ is selectively neutral on the average so that the mean value of $s$ over a long period is zero, and its variance $V_s$ is constant. Then $M_{\delta x}=0$, $V_{\delta x}=V_s x^2(1-x)^2$ and the partial differential equation (3.10) becomes

$$\frac{\partial\phi}{\partial t}=\frac{V_s}{2}\frac{\partial^2}{\partial x^2}\{x^2(1-x)^2\phi\} \qquad (0<x<1). \tag{8.1}$$

To solve this equation, let us put

$$u=\tfrac{1}{2}e^{V_s t/8}x^{3/2}(1-x)^{3/2}\phi$$

and

$$\xi=\log\frac{x}{1-x}.$$

Then we obtain the heat conduction equation

$$\frac{\partial u}{\partial t}=\frac{V_s}{2}\frac{\partial^2 u}{\partial\xi^2} \qquad (-\infty<\xi<\infty).$$

It is known that this equation has a unique solution which is continuous over $-\infty$ to $+\infty$ when $t\geqq 0$, and reduces to $u(\xi,0)$ when $t=0$. The solution is given by

$$u(\xi,t)=\frac{1}{\sqrt{2\pi V_s t}}\int_{-\infty}^{\infty}\exp\left\{-\frac{(\xi-\eta)^2}{2V_s t}\right\}u(\eta,0)d\eta.$$

Therefore, the required solution of (8.1), when the initial distribution of gene frequency is $\phi(x,0)$, is given by

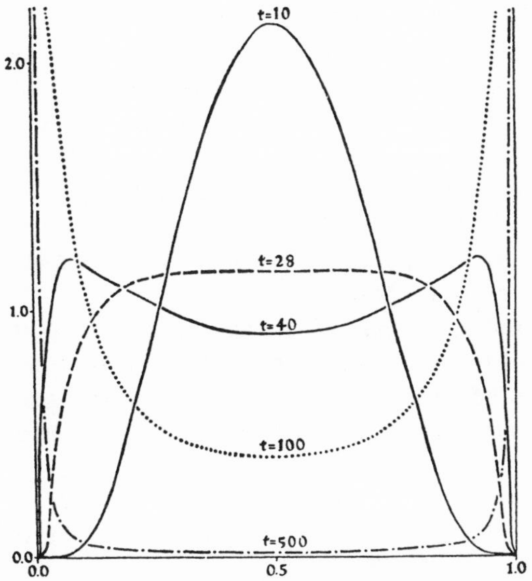


Figure 9

The process of change in the gene frequency distribution under random fluctuation of selection intensities. In this illustration it is assumed that the gene is selectively neutral when averaged over a very long period, that there is no dominance, and $p = 0.5$, $V = 0.0483$. The abscissa is the gene frequency $x$, the ordinate the probability density $\phi$. (From Kimura, 1954)

$$(8.2)\quad \phi(x,t) = \frac{1}{\sqrt{2\pi V_s t}} \cdot \frac{e^{-V_s t/8}}{\{x(1-x)\}^{3/2}} \cdot \int_0^1 \exp\left\{-\frac{\left[\log\frac{x(1-y)}{(1-x)y}\right]^2}{2V_s t}\right\} \cdot \sqrt{y(1-y)}\,\phi(y,0)dy.$$

If the initial condition is not a continuous distribution, but a fixed gene frequency $p$, the above formula reduces to

$$(8.3)\quad \phi(p,x;t) = \frac{1}{\sqrt{2\pi V_s t}} \exp\left\{-\frac{V_s}{8}t - \frac{\left[\log\frac{x(1-p)}{(1-x)p}\right]^2}{2V_s t}\right\} \frac{\sqrt{p(1-p)}}{(\sqrt{x(1-x)})^3}.$$

The process of change in the distribution curve with time is illustrated in Figure 9 assuming $p = 0.5$ and $V_s = 0.0483$. As will be seen in the figure the distribution curve is unimodal for a considerable number of generations (in the case illustrated, 27 generations), after which it becomes bimodal. In the 100th generation, gene frequencies in our example giving maximum probability (peaks in the curve) are approximately 0.0007 and 0.9993. As time goes on the distribution curve becomes more and more $U$-shaped, though it is not a true $U$-shaped curve, since its value at either terminal is always 0. This means that as time elapses the gene frequency shifts towards either terminal of the distribution ($x = 0$ or 1) indefinitely and accumulates in the neighborhood just short of fixation or loss, but never becomes fixed or lost completely (at least theoretically). To distinguish this from the fixation or loss in the case of random drift in small populations, the terms *quasi-fixation* and *quasi-loss* were proposed (Kimura, 1954).

If the genes are not neutral on the average, then $M_{\delta x} = 0$ should be replaced with $M_{\delta x} = \bar{s}x(1-x)$ in the partial differential equation, where $\bar{s}$ is the long term average of $s$. Unfortunately, the solution of the corresponding partial differential equation has not so far been found.

However, the following approximate treatment may be helpful in obtaining a rough picture of the process involved. If we transform the gene frequency $x$ into its logit $\xi$ by the relation $\xi = \log[x/(1-x)]$, $\xi$ changes continuously from $-\infty$ to $+\infty$ as $x$ changes from 0 to 1. For a small change of $\xi$, we have

$$\delta\xi = [x(1-x)]^{-1}\delta x + (2x-1)[2x^2(1-x)^2]^{-1}(\delta x)^2 + \cdots. \tag{8.4}$$

Neglecting terms of higher order than the first, and noting $M_{\delta x} = \bar{s}x(1-x)$ and $V_{\delta x} = V_s x^2(1-x)^2$, we obtain

$$M_{\delta\xi} = \bar{s}, \quad V_{\delta\xi} = V_s. \tag{8.5}$$

These expressions indicate that on the logit scale the mean and variance of the gene frequency distribution increase approximately linearly with time. Namely the probability distribution of $\xi$ in the $t$th generation is given by the normal distribution with mean $\xi_0 + \bar{s}t$ and variance $V_s t$, where $\xi_0$ is the logit of $p$, i.e. $\xi_0 = \log[p/(1-p)]$. Figure 10 illustrates the process of change obtained by this method for the case of a pair of alleles with $\bar{s} = 0.1$ and $V_s = 0.0025$ ($\sigma_s = 0.05$).

Actually, the case of genic selection in an infinite population with random fluctuation in selection intensities can most easily be treated by the following discrete model. Consider the multiple alleles, $A_1, A_2, \cdots, A_n$ at a locus and let $w_i$ be the fitness of $A_i$ measured in selective values. If $x_i(t)$ is the frequency of $A_i$ in the $t$th generation, then

$$x_i(t+1) = w_i x_i(t)/\bar{w}, \tag{8.6}$$

where $\bar{w}$ is the average selective value of the population in the $t$th generation,

$$\bar{w} = \sum_{i=1}^{n} w_i x_i(t).$$

Here $t$ takes on discrete values, 0, 1, 2, etc.

From (8.6), it follows that, for any pair of alleles, say for $A_i$ and $A_j$, we have

$$\log[x_i(t)/x_j(t)] = \log(w_i/w_j) + \log[x_i(t-1)/x_j(t-1)],$$

so that

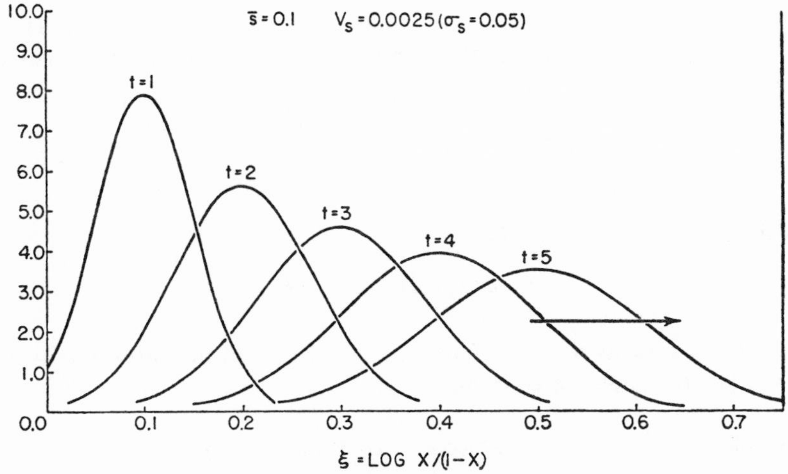


Figure 10
The process of change in gene frequency distribution due to genic selection with random fluctuation of selection intensity. Gene frequency is measured on a logit scale $\xi$, and $\bar{s}$ and $V_s$ are respectively the mean and variance of the selection intensity. In this figure the initial gene frequency is 0 on the logit scale, i.e. $p = 0.5$, and $t$ stands for time (in generations). (From Kimura, 1955 c)

$$z_{i/j}(t) = \sum_{\tau=0}^{t-1} s_{i/j}(\tau) + z_{i/j}(0) \qquad (t \geqq 1), \tag{8.7}$$

where $z_{i/j}(t)$ is the logarithmic gene ratio at the $t$th generation, i.e.

$$z_{i/j}(t) = \log[x_i(t)/x_j(t)],$$

and $s_{i/j}(\tau)$ is the value of $\log(w_i/w_j)$ in the $\tau$th generation. Thus, if $s_{i/j}(\tau)$ is distributed normally with mean $\bar{s}_{i/j}$ and variance $\sigma^2_{i/j}$, then $z_{i/j}(t)$ is distributed normally with mean $z_{i/j}(0) + \bar{s}_{i/j}t$ and variance $\sigma^2_{i/j}t$. Furthermore, even if $s_{i/j}$ is not distributed normally, the distribution of $z_{i/j}(t)$ for a large value of $t$ will approach the normal distribution by the central limit theorem. Since (8.7) holds for any pair of $i$ and $j$, and since

$$\sum_{i=1}^{n} x_i(t) = 1$$

in each generation, the joint distribution of gene frequencies for an arbitrary generation can easily be worked out assuming normality of $z_{i/j}(t)$.

Besides the effect of random fluctuation of selection intensity discussed in this section, that of random fluctuation in migration rate has also been studied. If $m$ is the migration rate which fluctuates randomly from generation to generation with mean $\bar{m}$ and variance $V_m$, then we have

$$\frac{\partial \phi}{\partial t} = \frac{V_m}{2}\frac{\partial^2}{\partial x^2}\{(x-\xi)^2\phi\} + \bar{m}\frac{\partial}{\partial x}\{(x-\xi)\phi\},$$

where $\xi$ is the frequency of $A_1$ in the migrants. Assuming that the initial gene frequency $p$ is higher than that of the migrants, the solution of this equation is

$$\phi(p,x;t) = \frac{1}{(x-\xi)\sqrt{\pi V_m t/2}} \exp\left\{-\frac{\left[\log\frac{x-\xi}{p-\xi}+\left(\frac{V_m}{2}+\bar{m}\right)t\right]^2}{2V_m t}\right\} \quad (x \geqq \xi).$$

It is then possible to show that

$$\lim_{t\to\infty} \phi(p,x;t) = \delta(x-\xi),$$

where $\delta(\cdot)$ represents the delta function. For details of this problem, Crow and Kimura (1956) should be consulted.

## IV. Genetic Equilibrium; Stationary Distributions and Gene Fixation

### 9. Gene frequency distribution at equilibrium

9.1. *Stationary distribution.* Random fluctuation of gene frequencies due to the factors so far discussed leads to fixation or loss of genes in a population. On the other hand, reversible mutation or migration tends to restore the intermediate gene frequencies. These two groups of factors will cause a non-trivial stable distribution of gene frequencies at equilibrium.

In the previous sections we have been mainly concerned with $\phi(p,x;t)$, the transition probability density for the gene frequency to become $x$ at the $t$th generation given that it is $p$ at $t=0$. Now, if we designate by $\phi(x)$ the probability density of such an equilibrium distribution, the existence of a stable distribution $\phi(x)$ means that

$$\lim_{t\to\infty} \phi(p,x;t) = \phi(x),$$

which is independent of the initial frequency $p$. Detailed knowledge of such a distribution is essential to understand the genetic structure of natural populations. For a single locus with a pair of alleles $A_1$ and $A_2$, with respective frequencies $x$ and $1-x$, a quite general formula has been obtained by Wright (1938a, 1949). His formula may be expressed as follows:

$$\phi(x) = \frac{C}{V_{\delta x}} \exp\left\{2\int \frac{M_{\delta x}}{V_{\delta x}} dx\right\}, \tag{9.1}$$

where $M_{\delta x}$ and $V_{\delta x}$ are the mean and the variance of $\delta x$. Here $\phi(x)$ stands for the probability density that the gene frequency is $x$ in the population. We can interpret $\phi(x)dx$ as the relative frequency or proportion of populations with gene frequency within the range $x \sim x+dx$ among a hypothetical aggregate consisting of the infinite number of populations satisfying the same conditions.

In the above formula $C$ is a constant which is usually adjusted in such a way that

$$\int_0^1 \phi(x)dx = 1. \tag{9.2}$$

It is remarkable that Wright (1938 a) first derived formula (9.1) from the simple consideration that at equilibrium the mean and variance of the frequency distribution are unchanged in successive generations.

From our standpoint it is more natural to derive the formula from (3.11) by imposing condition (3.20), namely that the probability flux must be zero at every point in the open interval (0, 1).

This leads to

$$\frac{1}{2}\frac{d}{dx}(V_{\delta x}\phi) - M_{\delta x}\phi = 0,$$

which, upon integration, gives (9.1). One of the tacit assumptions involved here is that the process is time homogeneous, so that $M_{\delta x}$ and $V_{\delta x}$ are independent of $t$.

As an example, let us consider the case discussed in Section 6, where

$$M_{\delta x} = m(\bar{x} - x)$$

and

$$V_{\delta x} = x(1-x)/(2N).$$

Substituting these in (9.1), we obtain

$$\phi(x) = Cx^{4Nm\bar{x}-1}(1-x)^{4Nm(1-\bar{x})-1}.$$

The coefficient $C$ as determined from (9.2) is

$$C = \frac{1}{B(4Nm\bar{x}, 4Nm(1-\bar{x}))} = \frac{\Gamma(4Nm)}{\Gamma(4Nm\bar{x})\Gamma(4Nm(1-\bar{x}))}.$$

Thus, the equilibrium distribution is given by

$$\phi(x) = \frac{\Gamma(4Nm)}{\Gamma(4Nm\bar{x})\Gamma(4Nm(1-\bar{x}))} x^{4Nm\bar{x}-1}(1-x)^{4Nm(1-\bar{x})-1}. \tag{9.3}$$

It is important to note that the above equation agrees with (6.3), which is obtained from (6.2) by taking the limit as $t \to \infty$.

For the application of (9.1) to various other cases, the reader may refer to a series of papers by Wright, especially Wright (1938 a, b, 1939, 1948), and also to Kimura (1955 c).

More recently, an interesting model for a haploid organism with overlapping generations was used by Moran (1958 c) to derive an exact distribution under reversible mutation.

So far, we have considered gene frequency distributions containing only one random variable $x$. Here, Wright's formula (9.1) is fundamental, and it has sufficient generality to be useful for most purposes. However, in order to study the effect of gene interaction on the frequency distribution involving multiple alleles or multiple loci, formulae for the distribution of more than one random variable are required. Unfortunately, no formulae of comparable !generality to (9.1) have been obtained for such cases, but Wright (1937) has obtained an important formula which deals with epistasis under the assumption of random mating and constant fitness of individual genotypes. For the case of two loci each with a pair of alleles, $A_1$ and $A_2$ at the first locus and $B_1$ and $B_2$ at the second, his formula can be expressed as follows:

$$\phi(x,y) = C\bar{w}^{2N}x^{4N\nu_1-1}(1-x)^{4N\mu_1-1}\cdot y^{4N\nu_2-1}(1-y)^{4N\mu_2-1}, \tag{9.4}$$

where $x$ and $y$ are respectively the frequencies of $A_1$ and $B_1$, and $\bar{w}$ is the average fitness of a population measured in selective values with respect to these two loci. In the above formula, $\mu_1$ is the mutation rate from $A_1$ to $A_2$ and $\nu_1$ is the rate in the reverse direction, while $\mu_2$ and $\nu_2$ are corresponding values for $B_1$ and $B_2$. If the fitness is measured in Malthusian parameters (cf. Fisher, 1958, Kimura, 1958), $\bar{w}$ in the above formula should be replaced by $e^{\bar{a}}$, where $\bar{a}$ is the average fitness measured in Malthusian parameters.

Wright (1937) derived the above equation through an ingenious but intuitive argument. From our standpoint, however, it is more natural, as in the case of a single variable, to derive the equation from a consideration of the probability flux. Actually, this enables us to view the problem in a much wider perspective.

Consider a region in two-dimensional Cartesian co-ordinates with $0 \leqq x \leqq 1$ and $0 \leqq y \leqq 1$. The probability flux which passes through the point $(x, y)$ along the $x$-axis is

$$P(x\,|\,y;t) = M_{\delta x}\phi(x,y) - \frac{1}{2}\frac{\partial}{\partial x}\{V_{\delta x}\phi(x,y)\} - \frac{1}{2}\frac{\partial}{\partial y}\{W_{\delta x\delta y}\phi(x,y)\}.$$

Similarly, the flux which passes through the same point along the y axis is

$$Q(y\,|\,x;t) = M_{\delta y}\phi(x,y) - \frac{1}{2}\frac{\partial}{\partial y}\{V_{\delta y}\phi(x,y)\} - \frac{1}{2}\frac{\partial}{\partial x}\{W_{\delta x\delta y}\phi(x,y)\}.$$

At equilibrium, when the flux is zero at every point,

$$P(x\,|\,y;t) = Q(y\,|\,x;t) = 0,$$

and we have

$$\begin{bmatrix} V_{\delta x} & W_{\delta x\delta y} \\ W_{\delta x\delta y} & V_{\delta y} \end{bmatrix} \begin{bmatrix} \dfrac{\partial\psi}{\partial x} \\ \dfrac{\partial\psi}{\partial y} \end{bmatrix} = \begin{bmatrix} 2M_{\delta x} - \dfrac{\partial V_{\delta x}}{\partial x} - \dfrac{\partial W_{\delta x\delta y}}{\partial y} \\ 2M_{\delta y} - \dfrac{\partial W_{\delta x\delta y}}{\partial x} - \dfrac{\partial V_{\delta y}}{\partial y} \end{bmatrix} \tag{9.5}$$

where $\psi = \log\phi(x,y)$. Thus, if the above simultaneous equation in $\partial\psi/\partial x$ and $\partial\psi/\partial y$ has a unique solution $(\psi_x, \psi_y)$ and if

$$\psi_x dx + \psi_y dy$$

is the total differential, which we will denote by $d\psi$, then the simultaneous distribution at equilibrium may be given by

$$\phi(x, y) = Ce^{\psi} \tag{9.6}$$

where the constant $C$ is determined by the condition

$$\int_0^1 \int_0^1 \phi(x, y)\, dx dy = 1. \tag{9.7}$$

In the special case in which random sampling of gametes is the sole factor for producing random fluctuation in gene frequencies,

$$V_{\delta x} = \frac{x(1-x)}{2N}, \quad V_{\delta y} = \frac{y(1-y)}{2N} \quad \text{and} \quad W_{\delta x \delta y} = 0. \tag{9.8}$$

Under the assumption of random mating and constant (but not necessarily equal) fitness of individual genotypes, the mean rates of change in gene frequencies are expressed by

$$\begin{cases} M_{\delta x} = \frac{1}{2}x(1-x)\frac{\partial \bar{a}}{\partial x} - \mu_1 x + \nu_1(1-x) \\ M_{\delta y} = \frac{1}{2}y(1-y)\frac{\partial \bar{a}}{\partial y} - \mu_2 y + \nu_2(1-y) \end{cases} \tag{9.9}$$

where $\bar{a} = \log \bar{w}$. Substituting (9.8) and (9.9) in (9.5), we get

$$\begin{cases} \psi_x = 2N\frac{\partial \bar{a}}{\partial x} - 4N\left(\frac{\mu_1}{1-x} - \frac{\nu_1}{x}\right) - \frac{d}{dx}\log[x(1-x)], \\ \psi_y = 2N\frac{\partial \bar{a}}{\partial y} - 4N\left(\frac{\mu_2}{1-y} - \frac{\nu_2}{y}\right) - \frac{d}{dy}\log[y(1-y)], \end{cases}$$

and since $\partial\psi_x/\partial y = \partial\psi_y/\partial x = 2N\partial^2\bar{a}/\partial x\partial y$, it is evident that $d\psi$ exists. Indeed it is given by

$$d\psi = d\{2N\bar{a} + (4N\mu_1 - 1)\log(1-x) + (4N\nu_1 - 1)\log x + (4N\mu_2 - 1)\log(1-y) + (4N\nu_2 - 1)\log y\}.$$

Thus, from (9.6), we obtain

$$\phi(x, y) = Ce^{2N\bar{a}}x^{4N\nu_1 - 1}(1-x)^{4N\mu_1 - 1}y^{4N\nu_2 - 1}(1-y)^{4N\mu_2 - 1}, \tag{9.10}$$

where $\bar{a} = \log \bar{w}$. This completes the derivation of (9.4).

9.2. *Distribution under irreversible mutation.* Our formula (3.11) may also be used to obtain the frequency distribution under steady flux. In this case we assume that the steady state is reached with respect to the distribution of intermediate gene frequencies $(0 < x < 1)$, but that there is a constant flow of probability from one terminal class to the other. Such an assumption may be justified if the loss of probability by the donor terminal class is negligible, as in the case of a deleterious mutation steadily reaching fixation in a finite population at an exceedingly low rate under the pressure of irreversible mutation against the strong action of selection.

The steady flux solution may be obtained from (3.11) by applying condition (3.21) as follows. Let $D$ be the probability flux, then

$$D = -\frac{1}{2}\frac{d}{dx}\{V_{\delta x}\phi(x)\} + M_{\delta x}\phi(x)$$

or

$$\frac{1}{2}\frac{d}{dx}\{V_{\delta x}\phi(x)\} - M_{\delta x}\phi(x) + D = 0.$$

The solution of this equation, i. e.

(9.11)
$$\phi(x) = \left\{C - 2D\int G(x)dx\right\}/\{V_{\delta x}G(x)\},$$

where

(9.12)
$$G(x) = \exp\left\{-\int\frac{2M_{\delta x}}{V_{\delta x}}dx\right\}$$

gives the steady flux distribution, in which $C$ is a constant. The formula (9.11) above was first obtained by Wright (1945).

In what follows I shall discuss the application of this formula to a more concrete genetical problem, and will also present some extensions of Wright's results on irreversible mutation.

Let us suppose that mutation is irreversible and occurs at a constant rate only in the direction $A_2 \to A_1$. In a finite population, $A_2$ will eventually be lost from the population even if $A_1$ is disadvantageous, because random drift may carry $A_1$ into fixation, and once this occurs $A_2$ is permanently lost from the population.

Let $x$ be the frequency of $A_1$ and suppose that mutation is occurring from the class $x = 0$ (i.e. from $A_2$) at an exceedingly minute rate $v$, with irreversible fixation in the class $x = 1$. Following Wright (1942), we assume two conditions,

(9.13)
$$f\left(\frac{1}{2N}\right) = 4Nv$$

and

(9.14)
$$f\left(1 - \frac{1}{2N}\right) = 0.$$

Both of these are approximations. The first condition means that in the neighborhood of $x = 0$, mutation and random extinction balance each other so that the number of new mutations $2Nv$ is half the frequency of the subterminal class (see equation 3.18). The second condition means that the height of the distribution curve is so low in the neighborhood of $x = 1$ as compared with the neighborhood of $x = 0$, that the frequency of the subterminal class with $x=1-1/(2N)$ may be set equal to zero. Here we assume that the majority of populations contain only $A_2$, i.e. $f(0) = 1$ approximately.

The selective advantages of $A_1A_1$ and $A_1A_2$ over $A_2A_2$ may be designated by $s$ and $sh$, both of which may be negative if $A_1$ is unconditionally deleterious. The mean rate of change in $x$ per generation by selection may be given by

$$M_{\delta x} = s\{h + (1-2h)x\}x(1-x).$$

It is more convenient, however, for the subsequent treatment to express $M_{\delta x}$ in the form

$$M_{\delta x} = (s_1 + s_2 x)x(1-x),$$

where $s_1 = sh$ and $s_2 = s(1 - 2h)$. If we combine this with

$$V_{\delta x} = x(1 - x)/2N$$

and substitute them in (9.11), we obtain

$$\phi(x) = \frac{2Ne^{4Ns_1x+2Ns_2x^2}}{x(1-x)}\left\{C - 2D\int_0^x e^{-4Ns_1\xi-2Ns_2\xi^2}\,d\xi\right\}. \tag{9.15}$$

The relative frequency, $f(x)$, of the class with gene frequency $x$ (discrete) will be given by $\phi(x)/(2N)$ for $x$ between $1/(2N)$ and $1-1/(2N)$. In the above formula, $C$ and $D$ are constants to be determined by the two conditions (9.13) and (9.14). In the following calculation, we will assume that $N$ becomes infinitely large and $|s_1|$ and $|s_2|$ infinitely small, while $2Ns_1$ and $2Ns_2$ remain finite.

First, from condition (9.13), we get, neglecting higher order terms,

$$\phi\left(\frac{1}{2N}\right)\frac{1}{2N} = 2N\left(C - \frac{2D}{2N}\right) = 4Nv. \tag{9.16}$$

Secondly, from (9.14),

$$\phi\left(1 - \frac{1}{2N}\right)\frac{1}{2N} = 2Ne^{4Ns_1+2Ns_2}\left\{C - 2D\int_0^1 e^{-4Ns_1\xi-2Ns_2\xi^2}d\xi\right\} = 0. \tag{9.17}$$

From the assumption that $|s_1|$ and $|s_2|$ are very small, it turns out that $D/N$ may be neglected as compared with $C$ in (9.16). Then, we obtain

$$C = 2v, \tag{9.18}$$

$$2D = 2v\Big/\int_0^1 e^{-4Ns_1\xi-2Ns_2\xi^2}\,d\xi \tag{9.19}$$

and

(9.20) $$f(x) = \frac{2ve^{4Ns_1x+2Ns_2x^2}}{x(1-x)} \cdot \frac{\int_x^1 e^{-4Ns_1\xi-2Ns_2\xi^2}d\xi}{\int_0^1 e^{-4Ns_1\xi-2Ns_2\xi^2}d\xi}.$$

For the special case of $s_2 = 0$ (no dominance), the above formula reduces to

(9.21) $$f(x) = \frac{2v}{x(1-x)} \cdot \frac{1-e^{-4Ns_1(1-x)}}{1-e^{-4Ns_1}}.$$

Fisher (1930) gave the frequency distribution of mutant genes when there is a supply of one mutation in each generation. His formula is

(9.22) $$df = \frac{2}{1-e^{-4an}}\{1-e^{-4an/(1+e^z)}\}\,dz,$$

where $df$ is the frequency element, i.e. $f(x)$ in our notation, $a$ is the selective advantage (our $s_1$) assumed to be very small, $n$ is the number of breeding individuals in a population (our $N$) and $z$ is the logit of the mutant gene frequency $x$. It is not difficult to show that (9.21) agrees with Fisher's formula (9.22) if we note that $n = N, a = s_1, x = \log_e\{x/(1-x)\}, dz = dx/\{x(1-x)\} = 1/\{2Nx(1-x)\}$ and $2Nv = 1$.

The net probability flux $D$ will give us the probability of ultimate fixation of an individual mutant gene if we divide $D$ by the number of mutations per generation, i,e, $2Nv$.

Thus we obtain, from (9.19),

$$u = \left(2N \int_0^1 e^{-4Ns_1x-2Ns_2x^2}dx\right)^{-1},$$

or putting $s_1 = sh$ and $s_2 = s(1-2h)$,

(9.23) $$u = \left[2N\int_0^1 e^{-4Nshx-2Ns(1-2h)x^2}dx\right]^{-1}.$$

A more general expression for the distribution under irreversible mutation may be obtained directly from (9.11) by imposing conditions similar to (9.13) and (9.14). In the following treatment, we will take a general form of selection in $M_{\delta x}$ so that selection coefficients may be gene-frequency dependent. Also, $V_{\delta x}$ may include the effect of random fluctuation in selection intensities. Since the rate of change in gene frequency by selection contains the factor $x(1-x)$, $V_{\delta x}$ may be expressed in the form

$$V_{\delta x} = x(1-x)/(2N_e) + x^2(1-x)^2 P(x),$$

where $P(x)$ is a poynomial in $x$ and $N_e$ is the effective size of the population. $N_e$ may differ from the actual number $N$ of the individuals.

First, consider the exchange of class frequencies in the neighborhood of $x = 0$. The flux due to the new production of the mutant genes from the terminal class ($x = 0$) is $2Nv$, while the flux towards the opposite direction due to the loss of the mutant genes is

$$\frac{1}{2}\left[\frac{d}{dx}(V_{\delta x}\phi(x))\right]_{x=1/(2N)} = \frac{1}{4N_e}\phi\left(\frac{1}{2N}\right),$$

where the higher order terms are neglected. Equating these two opposite fluxes,

$$f\left(\frac{1}{2N}\right) = 4N_e v. \tag{9.24}$$

Note that the above relation reduces to (9.13) if $N_e = N$. When applied to (9.11) it leads to

$$C = 2v \tag{9.25}$$

if higher order terms are neglected.

Next, we assume as before the condition (9.14). It leads to

$$C = 2D\int_0^1 G(x)dx,$$

or

$$D = v\bigg/\int_0^1 G(x)dx. \tag{9.26}$$

Substituting (9.25) and (9.26) in (9.11), we obtain a general formula for the gene frequency distribution under irreversible mutation

$$\phi(x) = \frac{2v}{V_{\delta x}G(x)} \cdot \frac{\int_x^1 G(x)dx}{\int_0^1 G(x)dx}, \tag{9.27}$$

where

$$G(x) = \exp\left\{-\int \frac{2M_{\delta x}}{V_{\delta x}}dx\right\}.$$

Note here that in calculating $M_{\delta x}$ only the effect of selection is assumed. The effect of mutation should not be included in this term for the present calculation.

The probability of fixation of an individual mutant gene may be obtained from $D$ by dividing by $2Nv$, so that

(9.28) $$u = \left[2N \int_0^1 G(x)dx\right]^{-1}.$$

As pointed out by Wright (1938a), the present treatment should have a bearing on the possible evolutionary modification caused by mutation pressure, the eye degeneration and loss of pigment of cave animals being especially suited examples.

## 10. Probability of fixation of mutant genes in a population

10.1. *Introductory remarks.* In the study of evolutionary genetics, it is important to know the probability of ultimate success (i.e. fixation) of mutant genes, because fixation of an advantageous gene is the key factor in the evolution of the species. Pioneering work has been done by Fisher (1922, 1930) and Haldane (1927) who obtained the approximate (but sufficiently accurate) probability of fixation of an individual mutant gene for the case of genic selection (i.e. no dominance). They made use of the method which is now standard in the treatment of branching processes. Recently, Moran (1961) was able to construct a rigorous theory for the probability of survival of a mutant gene in a finite population of a haploid organism, where complication by dominance is not involved.

Results equivalent to those of Fisher and Haldane have been obtained by Wright (1931) from the study of the frequency distribution under irreversible mutation. Also the probability for a recessive gene was estimated by Haldane (1927) and Wright (1942). Later, a more general result was obtained by the present author (Kimura, 1957) based on a diffusion model which covers any degree of dominance. The probability of eventual fixation $u(p)$ was expressed in terms of the initial frequency $p$, selection coefficients and the effective population number. This function was used by Robertson (1960) in his theory of selection limits in plants and animal breeding. A still more general, but quite simple expression for $u(p)$ was obtained by the author in terms of the initial frequency, and the mean and variance of the rate of change of gene frequency per generation (Kimura, 1962). It was applied to solve problems where there is a random fluctuation in selection intensity. These results by the author have been obtained by using the method of the Kolmogorov backward equation.

The method is quite far-reaching, and even enables us to obtain the probability of joint fixation of mutant genes at multiple loci under the assumption of random mating and constant (but not necessarily equal) selective values of individual genotypes. In the present article, the result for multiple loci will be presented for the first time. It enables us to study the effect of epistasis on the fixation of genes in a finite population.

10.2 *Single locus.* It was stated in Section 3 that if $u(p,t)$ is the probability of a mutant gene's reaching fixation by the $t$th generation, given that its frequency is $p$ at $t=0$, then $u(p,t)$ satisfies equation (3.29). It was also indicated that the required probability would be obtained by solving this equation with boundary conditions (3.30). In the simplest case of random drift in a finite population of size $N$ with no mutation and selection, the solution of the equation was given by (4.15), but in a more general case the exact solution is rather difficult to obtain.

Now let us consider the ultimate probability of fixation defined by

$$u(p) = \lim_{t\to\infty} u(p,t).$$

From the standpoint of long-term evolution, this may be the most important quantity relating to the fixation of mutant genes. Since $\partial u(p)/\partial t = 0$ for this quantity, equation (3.29) reduces to the ordinary differential equation

$$\frac{V_{\delta p}}{2}\frac{d^2u(p)}{dp^2} + M_{\delta p}\frac{du(p)}{dp} = 0 \tag{10.2}$$

with boundary conditions

$$u(0)=0, \quad u(1)=1. \tag{10.3}$$

Fortunately, the pertinent solution of this equation can easily be found and is expressed as follows (Kimura, 1962),

$$u(p) = \int_0^p G(x)dx \Big/ \int_0^1 G(x)dx, \tag{10.4}$$

where

$$G(x) = \exp\left\{-\int \frac{2M_{\delta x}}{V_{\delta x}}\,dx\right\},$$

in which $M_{\delta x}$ and $V_{\delta x}$ are the mean and variance of the change in gene frequency $x$ per generation.

The above formula for $u(p)$ is the steady state solution of the Kolmogorov backward equation, and is the counterpart of Wright's formula for $\phi(x)$ (cf. 9.1), which is the steady state solution of the Kolmogorov forward equation. Both formulae have a pleasing simplicity, and are yet of sufficient generality to cover the cases of sexually reproducing haploid, diploid and polyploid organisms, as well as asexually reproducing plants.

The probability of fixation of individual mutant genes in a population of $N$ diploid individuals may then be obtained by taking $p = 1/(2N)$, so that

$$u = u\left(\frac{1}{2N}\right). \tag{10.5}$$

However, caution is necessary when applying the above method to a dioecious population where the number of males, $N^*$, can be different from that of females,

$N^{**}$. In such a case, either the initial condition $p = 1/(4N^*)$ or $p = 1/(4N^{**})$ should be used depending on whether the mutant gene occurred in a male or in a female, as was pointed out by Moran (1961) and Watterson (1962).

In what follows I will discuss a few simple cases, assuming a population of sexually reproducing diploid individuals. We will denote by $A_1$ the mutant gene whose initial frequency is $p$.

The simplest case is that of genic selection, in which $A_1$ has a constant selective advantage $s$ over its alleles in a population of effective size $N_e$. In this case $M_{\delta x} = sx(1-x)$, $V_{\delta x} = x(1-x)/(2N_e)$ so that $2M_{\delta x}/V_{\delta x} = 4N_e s$, $G(x) = e^{-4N_e sx}$ and we obtain from (10.4)

$$u(p) = (1 - e^{-4N_e sp})/(1 - e^{-4N_e s}). \tag{10.6}$$

For $|2N_e s| < \pi$, the right-hand side of the above equation may be expanded in terms of $4N_e s$ as follows:

$$\begin{aligned} u(p) &= \sum_{i=1}^{\infty} \frac{\phi_i(p)(-1)^{i-1}}{i!}(4N_e s)^{i-1} \\ &= p + 2N_e sp(1-p) + \frac{(2N_e s)^2}{3} p(p-1)(2p-1) + \cdots, \end{aligned} \tag{10.7}$$

where the $\phi_i(p)$'s are Bernoulli polynomials. Thus for a small value of $2N_e s$, $u(p) - p$ is $2N_e$ times $sp(1-p)$. In other words, the total advance is $2N_e$ times the change in the first generation, as was pointed out by Robertson (1960).

If the effective size of the population is equal to the actual size, $N$ may be substituted for $N_e$. Assuming the sex ratio is unity, the probability of fixation of an individual mutant gene is obtained from formula (10.6) by putting $p = 1/(2N)$. If $|s|$ is small, we obtain

$$u = (2s)/(1 - e^{-4Ns}) \tag{10.8}$$

as a good approximation. This agrees with the result obtained by Fisher (1930) and by Wright (1931) using different methods. For a positive $s$ and very large $N$, we obtain the well known result that the probability of ultimate survival of an advantageous mutant gene is approximately twice the selection coefficient (Haldane, 1927). If $N_e$ differs from $N$, this value should be modified by a factor of $N_e/N$ so that

$$u = 2s(N_e/N). \tag{10.9}$$

According to Crow (1954) and also Crow and Morton (1955), estimated values of $N_e/N$ for a few cases are: *Drosophila* 0.48 ~ 0.9, *Lymnaea* 0.75, Man 0.69 ~ 0.95.

The above results were obtained on the basis of the method of diffusion approximation, and it is desirable to check some of them by a rigorous treatment. This was done by Moran (1961) who used a population model consisting of haploid individuals, with offspring number following a negative binomial distribution. He assumed that in each generation, the population consists of exactly $M$ individuals of which a random number $k$ are of one type (say $A_1$) and $M-k$ of the other type (say $A_2$). Suppose that the generating function for the probability distribution of the number of offspring is

$$P_2(z) = \left(\frac{a}{1-bz}\right)^{a/b}$$

for gene $A_2$ and

$$P_1(z) = \left(\frac{a}{1-bz}\right)^{\lambda a/b}$$

for gene $A_1$, where $b = 1 - a$ and $\lambda = 1 + s$ with $s = O(M^{-1})$. The mean and variance of the distribution generated by $P_2(z)$ are 1 and $a^{-1}$ respectively. The mean of the distribution generated by $P_1(z)$ is $\lambda = 1 + s$, so that $s$ represents the selective advantage of $A_1$ over $A_2$. It was then shown by Moran that the probability of ultimate fixation of $A_1$, say $P_1$, satisfies the relation

$$\text{(10.10)} \qquad \frac{1-e^{-\theta_0 k_0}}{1-e^{-\theta_0 M}} \leqq P_1 \leqq \frac{1-e^{-\theta_1 k_0}}{1-e^{-\theta_1 M}},$$

where $k_0$ is the initial number of $A_1$-genes, and that for large $M$ both $\theta_0$ and $\theta_1$ become asymptotically equal to $2sv^{-1}$, where $v = a^{-1}$ is the variance generated by $P_2(z)$. If we denote by $X$ the proportion of $A_1$ genes in the haploid population ($X = k/M$) and if $A_1$ were selectively neutral, the variance of $X$ per generation would be $vX(1-X)/(M-1)$. The effective size of the population, say $M_e$ may then be defined by equating this variance with the binomial variance $X(1-X)/M_e$, so that asymptotically $v^{-1} = M_e M^{-1}$. Since both $\theta_0$ and $\theta_1$ become asymptotically equal to $2sv^{-1}$ or $2sM_e M^{-1}$ and since Moran's $M$ and $k_0$ correspond respectively to our $2N$ and $2Np$, (10.10) shows that (10.6) is asymptotically correct.

Next, we will consider a more general case of zygotic selection. Let $s$ and $sh$ be respectively the selective advantages of the mutant homozygote ($A_1A_1$) and the heterozygote ($A_1A_2$) such that

$$M_{\delta x} = sx(1-x)[h + (1-2h)x].$$

For a random mating population of effective size $N_e$, we have

$$V_{\delta x} = x(1-x)/(2N_e)$$

and therefore

$$\text{(10.11)} \qquad G(x) = \exp\{-4N_e shx - 2N_e s(1-2h)x^2\}.$$

The ultimate probability of fixation in this case is obtained from (10.4) with $G(x)$ given by (10.11). The probability of fixation of an individual mutant gene may then be obtained by putting $p = 1/(2N)$ in $u(p)$. Assuming that $|s|$ and $|sh|$ are small, we obtain

$$u = \left[2N \int_0^1 e^{-4N_e shx - 2N_e s(1-2h)x^2} dx\right]^{-1}. \tag{10.12}$$

This agrees with (9.23) which was obtained in the previous section by a different method.

If the mutant gene is completely recessive ($h = 0$), but advantageous in the homozygous condition ($s > 0$) and if $N_e s$ is large, we have approximately

$$u = \frac{\sqrt{2N_e s}}{N\sqrt{\pi}}. \tag{10.13}$$

For an idealized situation in which $N_e = N$, the above formula reduces to

$$u = \sqrt{\frac{2s}{\pi N}} \approx 1.128 \sqrt{\frac{s}{2N}} \tag{10.14}$$

(Kimura, 1957). It is interesting to note that the value given in (10.14) above lies between $\sqrt{s/N}$, the value obtained by Haldane (1927) treating this as a branching process, and $\sqrt{s/(2N)}$, obtained by Wright (1942) using his method of integral equations. Wright's numerical approximation, $1.1\sqrt{s/(2N)}$, is very close to (10.14).

Finally I will discuss briefly the effect of random fluctuation of selection intensity on the fixation of mutant genes.

The simplest situation is again genic selection in which the mutant gene $A_1$ has a selective advantage $s$ over its alleles, but $s$ is now a random variable with mean $\bar{s}$ and variance $V_s$. Thus

$$\begin{cases} M_{\delta x} = \bar{s}x(1-x) \\ V_{\delta x} = V_s x^2(1-x)^2 + x(1-x)/(2N_e) \end{cases} \tag{10.15}$$

and $G(x)$ in (10.4) is given by

$$G(x) = \left(\frac{\lambda_1 - x}{x - \lambda_2}\right)^{k/\sqrt{1+4r}}$$

where

$$k = 2\bar{s}/V_s$$

and

$$\begin{cases} \lambda_1 = \frac{1}{2}(1 + \sqrt{1+4r}) \\ \lambda_2 = \frac{1}{2}(1 - \sqrt{1+4r}) \end{cases}$$

in which $r = 1/(2N_e V_s)$.

A few special cases have been studied by the author (Kimura, 1962), but the results obtained are not very satisfactory and further study is required.

10.3. *Multiple loci.* Let us consider $n$ independent loci each with a pair of alleles, conveniently called the normal and the mutant genes, and denote by $p^{(i)}$ the initial frequency of the mutant gene in the $i$th locus $(i = 1,2,\cdots,n)$. Let $u(p^{(1)},p^{(2)},\cdots,p^{(n)};t)$ be the probability that all the $n$ mutant genes become fixed in the population by the $t$th generation, given that their initial frequencies at $t=0$ are $p^{(1)},p^{(2)},\cdots,p^{(n)}$ respectively. As in the single locus case, we will be mainly concerned with the ultimate probability of joint fixation defined by

$$u(p^{(1)},p^{(2)},\cdots,p^{(n)}) = \lim_{t\to\infty} u(p^{(1)},p^{(2)},\cdots,p^{(n)};t). \tag{10.16}$$

Under the assumption of random mating and constant fitness of individual genotypes, it is possible to show that

$$u(p^{(1)},p^{(2)},\cdots,p^{(n)}) = \frac{\int_0^{p^{(1)}}\int_0^{p^{(2)}}\cdots\int_0^{p^{(n)}} e^{-2N\bar{a}}\,dx^{(1)}dx^{(2)}\ldots dx^{(n)}}{\int_0^1\int_0^1\cdots\int_0^1 e^{-2N\bar{a}}\,dx^{(1)}dx^{(2)}\ldots dx^{(n)}}, \tag{10.17}$$

where $\bar{a} = \bar{a}(x^{(1)},x^{(2)},\cdots,x^{(n)})$ is the average fitness of the population measured in Malthusian parameters, and is a function of the gene frequencies $(x^{(1)},\cdots,x^{(n)})$, in which $x^{(i)}\,(i = 1,\cdots,n)$ stands for the frequency of the mutant gene at the $i$th locus in the population. If selective values are used to measure the fitness of individual genotypes (discrete generation time model), $\log\bar{w}$ may be substituted for $\bar{a}$ in the above formula, where $w$ is the relative selective value in the sense used by Wright; $w$ coincides with the relative viability if no fertility differences are involved. The above formula (10.17) can be obtained as the steady state solution of the Kolmogorov backward equation for $n$ variables, assuming random mating, and constant but not necessarily equal fitness of individual genotypes. The formula enables us to study the effect of epistasis in fitness on the chance of joint fixation of mutant genes. To make our discussion simpler, let us choose a population of sexually reproducing haploid organisms and con-

TABLE 1

| Genotype | Fitness | Frequency |
|---|---|---|
| $A_1B_1$ | $s_1 + s_2 + \varepsilon$ | $xy$ |
| $A_1B_2$ | $s_1$ | $x(1-y)$ |
| $A_2B_1$ | $s_2$ | $(1-x)y$ |
| $A_2B_2$ | 0 | $(1-x)(1-y)$ |

sider epistasis involving two loci ($n = 2$) as in Table 1. In this table, $A_1$ and $B_1$ denote respectively the mutant genes at the first and second loci, each with respective frequencies $x$ and $y$. Here, fitness is measured in Malthusian parameters, taking the fitness of $A_2B_2$ as the standard. Thus $\varepsilon$ represents the amount of epistasis in fitness between the mutant genes. If the discrete generation time model is used, the value 1 may be added to each entry of the second column to represent fitness in selective values (=relative viability if no fertility differences are involved). However, as long as the selection coefficients $s_1$, $s_2$ and $\varepsilon$ are small, there will be no practical difference in the results, whichever model is used to measure fitness.

If $p$ and $q$ are the respective initial frequencies of $A_1$ and $B_1$, then from (10.17), we have

$$(10.18) \qquad u(p,q) = \int_0^p \int_0^q e^{-2N\bar{a}(x,y)} dxdy \Big/ \int_0^1 \int_0^1 e^{-2N\bar{a}(x,y)} dxdy$$

where

$$\bar{a}(x,y) = (s_1 + s_2 + \varepsilon)xy + s_1x(1-y) + s_2(1-x)y$$
$$= s_1x + s_2y + \varepsilon xy .$$

Writing $Ns_1 = S_1$, $Ns_2 = S_2$, $N\varepsilon = I$ and assuming that the initial frequencies, $p$ and $q$, are both very low so that

$$|\bar{a}(p,q)| \ll 1,$$

then we have approximately,

$$(10.19) \qquad u(p,q) = pqJ^{-1},$$

where

$$(10.20) \qquad J = \int_0^1 \int_0^1 e^{-2S_1x - 2S_2y - 2Ixy}\, dxdy .$$

In the simplest case where $S_1 = S_2 = 0$ and $I > 0$, namely in the case in which mutant genes are neutral separately but advantageous when combined, the above integral reduces to

$$(10.21) \qquad J = \frac{1}{2I}[\log_e(2I) + \gamma + E_i(2I)],$$

where $\gamma$ is Euler's constant (0.5772...) and $E_i(\cdot)$ denotes the exponential integral defined by

$$(10.22) \qquad E_i(z) = \int_z^\infty \frac{e^{-t}}{t} dt,$$

for which fairly extensive tabulation is available. As an example, let us take $N = 500$ and $\varepsilon = 0.01$. Then $2I = 10$, and $E_i(10) \approx e^{-10}/10$ is entirely negligible as compared with $\log_e 10 \approx 2.30$ and $\gamma \approx 0.58$, giving $J \approx 0.29$ or $J^{-1} \approx 3.5$. Thus, we obtain

$$u(p,q) \approx 3.5\,pq.$$

Since the probability of joint fixation of the two mutant genes at two independent loci (if they were completely neutral) is equal to $pq$, the above result shows that with 1% epistasis, in a population of $N = 500$, this probability is increased by a factor of about 3.5.

Let us now consider the biologically more interesting case in which each mutant gene is deleterious separately ($s_1 < 0$, $s_2 < 0$), but becomes advantageous when combined ($s_1 + s_2 + \varepsilon > 0$).

Writing $S_1 = -S_1'$ and $S_2 = -S_2'$ so that $S_1' > 0$ and $S_2' > 0$, (10.20) can be expressed as follows:

$$(10.23)\quad J = \frac{e^{2S_1'S_2'/I}}{2I}\left[E_i\left(\frac{2S_1'S_2'}{I}\right) + E_i\left(2I\left(1 - \frac{S_1'}{I}\right)\left(1 - \frac{S_2'}{I}\right)\right) + \bar{E}_i\left(2S_1'\left(1 - \frac{S_2'}{I}\right)\right) + \bar{E}_i\left(2S_2'\left(1 - \frac{S_1'}{I}\right)\right)\right],$$

where $E_i(\cdot)$ is the exponential integral defined in (10.22), while $\bar{E}_i(\cdot)$ is another exponential integral defined by

$$(10.24)\qquad \bar{E}_i(z) = \text{P.V.}\int_{\infty}^{-z} \frac{e^{-t}}{t}\,dt \quad (\text{P.V} = \text{principal value}).$$

For example, in a population of $N = 10^3$, if $s_1 = s_2 = -0.01$ (single mutant, 1% disadvantage) and $\varepsilon = 0.07$ (double mutant, 5% advantage), $S_1' = S_2' = 10$, $I = 70$ and $J$ turns out to be about $4.85 \times 10^8$. Thus, the probability of fixation is approximately

$$J^{-1} \approx 2.5 \times 10^{-6}$$

times the corresponding value for neutral genes. This may mean that, in this specific example, the probability is too low for this pair of mutant genes to be of much use in evolution.

A more detailed account of this subject together with the derivation of (10.17) will be published elsewhere.

## References


[1] BODMER, W. F. (1960) Discrete stochastic processes in population genetics. *J.R. Statist. Soc.* B **22**, 218–244.

[2] CAVALLI-SFORZA, L.L. AND CONTERIO, F. (1960) Analisi della fluttuazione di frequenze geniche nella popolazione della Val Parma. *Atti* A.G.I. **5**, 333–344.

[3] CROW, J.F. (1954) Breeding structure of populations II. Effective population number. *Statistics and Mathematics in Biology*. Kempthorne *et al.* (ed). Iowa State College Press, Ames, Iowa.

[4] CROW, J.F. AND MORTON, N. (1955) Measurement of gene frequency drift in small populations. *Evolution* **9**, 202–214.

[5] Crow, J. F. and Kimura, M. (1956) Some genetic problems in natural populations. *Proc. Third Berkeley Symp. on Math. Statist. and Prob.* **4,** 1–22.

[6] Ewens, W. J. (1963) Numerical results and diffusion approximations in a genetic process. *Biometrika* **50,** 241–249.

[7] Ewens, W. J. (1964) The pseudo-transient distribution and its uses in genetics. *J. Appl. Prob.* **1,** 141–156.

[8] Feller, W. (1951) Diffusion processes in genetics. *Proc. Second Berkeley Symp. on Math. Statist. and Prob.* 227–246.

[9] Feller, W. (1952) The parabolic differential equations and the associated semigroup of transformations. *Ann. Math.* **55,** 468–519.

[10] Fisher, R.A. (1922) On the dominance ratio. *Proc. Roy. Soc. Edin.* **42,** 321–341.

[11] Fisher, R.A. (1930) The distribution of gene ratios for rare mutations. *Proc. Roy. Soc. Edin.* **50,** 205–220.

[12] Fisher, R.A. (1953) Population genetics. *Proc. Roy. Soc. London* B **141,** 510–523.

[13] Fisher, R.A. (1958) *The Genetical Theory of Natural Selection* (2nd revised ed.). Dover, New York.

[14] Fokker, A.D. (1914) Die mittlere Energie rotierender elektrischer Dipole im Strahlungsfeld. *Ann. d. Phys.* **43,** 810–820.

[15] Goldberg, S. (1950) *On a singular diffusion equation.* Ph. D. thesis (unpublished). Cornell University.

[16] Haldane, J.B.S. (1924) A mathematical theory of natural and artificial selection. Part I. *Trans. Camb. Phil. Soc.* **23,** 19–41.

[17] Haldane, J.B.S. (1927) A mathematical theory of natural and artificial selection. Part V: Selection and mutation. *Proc. Camb. Phil. Soc.* **23,** 838–844.

[18] Haldane, J.B.S. (1932) *The Causes of Evolution.* Harper and Brothers, New York.

[19] Haldane, J.B.S. (1949) Suggestions as to quantitative measurement of rates of evolution. *Evolution* **3,** 51–56.

[20] Karlin, S. and McGregor, J. (1962) On a genetics model of Moran. *Proc. Camb. Phil. Soc.* **58,** 299–311.

[21] Kimura, M. (1954) Process leading to quasi-fixation of genes in natural populations due to random fluctuation of selection intensities. *Genetics* **39,** 280–295.

[22] Kimura, M. (1955 a) Solution of a process of random genetic drift with a continuous model. *Proc. Nat. Acad. Sci.* **41,** 144–150.

[23] Kimura, M. (1955 b) Random genetic drift in multi-allelic locus. *Evolution* **9,** 419–435.

[24] Kimura, M. (1955 c) Stochastic processes and distribution of gene frequencies under natural selection. *Cold Spring Harbor Symp.* **20,** 33–53.

[25] Kimura, M. (1956 a) Random genetic drift in a tri-allelic locus: Exact solution with a continuous model. *Biometrics* **12,** 57–66.

[26] Kimura, M. (1956 b) *Stochastic processes in population genetics.* Ph. D. thesis (unpublished). Univ. of Wisconsin.

[27] Kimura, M. (1957) Some problems of stochastic processes in genetics. *Ann. Math. Statist.* **28,** 882–901.

[28] Kimura, M. (1958) On the change of population fitness by natural selection. *Heredity* **12,** 145–167.

[29] Kimura, M. (1962) On the probability of fixation of mutant genes in a population. *Genetics* **47,** 713–719.

[30] Kimura, M. and Crow, J.F. (1963) The measurement of effective population number. *Evolution* **17,** 279–288.

[31] Kolmogorov, A. (1931) Über die analytischen Methoden in der Wahrscheinlichkeitsrechnung. *Math. Ann.* **104,** 415–458.

[32] MALÉCOT, G. (1948) *Les Mathématiques de l'Hérédité*. Masson et Cie, Paris.

[33] MILLER, G.F. (1962) The evaluation of eigenvalues of a differential equation arising in a problem in genetics. *Proc. Camb. Phil. Soc.* **58**, 588–593.

[34] MORAN, P.A.P. (1958 a) A general theory of the distribution of gene frequencies. I. Overlapping generations. *Proc. Roy. Soc. London.* B **149**, 102–112.

[35] MORAN, P.A.P. (1958 b) A general theory of the distribution of gene frequencies.II. Nonoverlapping generations. *Proc. Roy. Soc. London* B **149**, 113–116.

[36] MORAN, P.A.P. (1958 c) Random processes in genetics. *Proc. Camb. Phil. Soc.* **54**, 60–71.

[37] MORAN, P.A.P. (1961) The survival of a mutant under general conditions. *Proc. Camb. Phil. Soc.* **57**, 304–314.

[38] MORAN, P.A.P. (1962) *The Statistical Processes of Evolutionary Theory*. Clarendon Press, Oxford.

[39] MORSE, P.M. AND FESHBACH, H. (1953) *Methods of Theoretical Physics*. McGraw-Hill, iNew York.

[40] PLANCK, M. (1917) Über einen Satz der statistischen Dynamik und seine Erweiterung n der Quantentheorie. *Sitz. der. preuss. Akad.* 324–341.

[41] ROBERTSON, A. (1960) A theory of limits in artificial selection. *Proc. Roy. Soc. London* B **153**, 234–249.

[42] ROBERTSON, A. (1962) Selection for heterozygotes in small populations. *Genetics* **47**, 1291–1300.

[43] STRATTON, J.A., MORSE, P.M., CHU, L.J., AND HUTNER, R.A. (1941) *Elliptic Cylinder and Spheroidal Wave Functions*. John Wiley, New York.

[44] STRATTON, J.A., MORSE, P.M., CHU, L.J., LITTLE, J.D.C., AND CORBATÓ, F.J. (1956) *Spheroidal Wave Functions*. Technology Press of M.I.T. & John Wiley, New York.

[45] WATTERSON, G.A. (1962) Some theoretical aspects of diffusion theory in population genetics. *Ann. Math. Stat.* **33**, 939–957.

[46] WATTERSON, G.A. (1964) The application of diffusion theory to two population genetic models of Moran. *J. Appl. Prob.* **1**, 233—246.

[47] WRIGHT, S. (1931) Evolution in Mendelian populations. *Genetics* **16**, 97–159.

[48] WRIGHT, S. (1937) The distribution of gene frequencies in populations. *Proc. Nat. Acad. Sci.* **23**, 307–320.

[49] WRIGHT, S. (1938 a) The distribution of gene frequencies under irreversible mutation. *Proc. Nat. Acad. Sci.* **24**, 253-259.

[50] WRIGHT, S. (1938 b) The distribution of gene frequencies in populations of polyploids. *Proc. Nat. Acad. Sci.* **24**, 372–377.

[51] WRIGHT, S. (1939) The distribution of self-sterility alleles in populations. *Genetics* **24**, 538–552.

[52] WRIGHT, S. (1942) Statistical genetics and evolution. *Bull. Amer. Math. Soc.* **48**, 223–246.

[53] WRIGHT, S. (1945) The differential equation of the distribution of gene frequencies. *Proc. Nat. Acad. Sci.* **31**, 382–389.

[54] WRIGHT, S. (1948) On the roles of directed and random changes in gene frequency in the genetics of populations. *Evolution* **2**, 279–294.

[55] WRIGHT, S. (1949) Adaptation and selection. In *Genetics, Paleontology, and Evolution*. Jepsen *et al.* (ed.), Princeton Univ. Press.

[56] WRIGHT, S. (1950) Population structure as a factor in evolution. In *Moderne Biologie, Festschrift für Hans Nachtsheim*, F. W. Peters, Berlin.

[57] WRIGHT, S. (1952) The theoretical variance within and among subdivisions of a population that is in a steady state. *Genetics* **37**, 312–321.

[58] WRIGHT, S. AND KERR, W.E. (1954) Experimental studies of the distribution of gene frequencies in very small populations of *Drosophila melanogaster*. II. Bar. *Evolution* **8**, 225–240.

# PART TWO

## Fluctuation of Selection Intensity

### Introduction

When there is stochastic (temporal) variation in fitness (Wright 1948), the variance of selection intensity may be as important as the mean. Kimura (no. 5) derived and solved a diffusion equation of gene frequency changes when selection intensity fluctuates randomly from generation to generation (see also Ohta 1972).

Without other evolutionary forces, complete loss or fixation of an allele does not occur in an infinite population, but rather the distribution eventually becomes concentrated near the boundaries (quasi-loss or quasi-fixation). While the diffusion coefficient of this equation of course includes the variance of selection intensity, the drift coefficient does not. This is inconsistent if we base the derivation on the conventional Wright-Fisher model of selection (Gillespie 1973; Nei and Yokoyama 1976; Avery 1977; Takahata 1981), for the variance term in the drift coefficient has to be incorporated unless the mean relative fitness is constant.

As a simple example (Crow 1972), suppose that the fitness $1 + s$ of an allele with frequency x fluctuates randomly with a geometric mean one; thus the locus is neutral over time. Then $\Delta x = sx(1 - x)/(1 + sx) \approx sx(1 - x)(1 - sx)$. If the mean of $1 + s = 1$, $E[\log(1 + s)] \approx E[s - s^2/2] = 0$. But, $E(s^2) = V_s$, so $E(\Delta x) = V_s x(1 - x)(1 - 2x)/2$. Thus there is a centripetal tendency, with the allele frequency tending to move toward 1/2.

When criticism of the neutral theory revolved around the rather low heterozygosity in natural populations, the role of random fluctuation of selection intensity regained attention (reviewed by Felsenstein 1976). Kimura's paper (no. 6) may not appear to be related to this problem because his whole treatment was deterministic. However, under population number regulation the mean population fitness equilibrates rapidly so that gene frequency change per generation is given by $\Delta x = sx(1 - x)$, or similar equations with good approximation, instead of $\Delta x = sx(1 - x)/(1 + sx)$. If we base the diffusion approximation on such an equation, the variance term in the drift coefficient can disappear. The effect of fluctuating selection intensity is then to reduce the extent of variation without increasing the allelic substitution rate. All other diffusion equations for random fluctuation of selection intensity have stabilizing factors in the drift coefficient, such as that above, which act in much the same way that overdominance selection does (Gillespie 1973, 1986). In these models, the extent of genetic variation and the substitution rate can both be greatly enhanced compared to those expected from Kimura's 1954 diffusion equation (Takahata and Kimura 1979; Takahata 1981). These properties render most fluctuating selection models unsatisfactory as a means of accounting for the rough constancy of molecular evolution and rather low heterozygosity.

# PROCESS LEADING TO QUASI-FIXATION OF GENES IN NATURAL POPULATIONS DUE TO RANDOM FLUCTUATION OF SELECTION INTENSITIES *[1]

MOTOO KIMURA

*Department of Animal Husbandry, Iowa State College, Ames, Iowa*

Received September 7, 1953

AMONG factors that may produce random fluctuation of gene frequencies in natural populations, random sampling of gametes and random fluctuation of selection intensities may be especially important in relation to evolution. On the process of "random drift" that will be realized in finite populations due to the random sampling of gametes in reproduction, not only precise mathematical studies have been carried out (WRIGHT 1931, 1945; FISHER 1930; HALDANE 1939; FELLER 1950) but also several model experiments have been undertaken (cf. HOUSE 1953). Since 1931, WRIGHT has repeatedly emphasized the evolutionary significance of random drift in a natural population which is subdivided into many partially isolated sub-groups. His theory is now accepted by many evolutionists such as HALDANE (1949), MULLER (1949), DOBZHANSKY (1951) and others.

On the other hand, no special attention seems to have been paid to the random fluctuation of gene frequencies due to the random fluctuation in the selection intensities until FISHER and FORD (1947) emphasized its prevalence in natural populations and challenged the theory of Wright by denying any significance of random drift due to small population number in evolution. This led to a polemic (cf. WRIGHT 1948; FISHER and FORD 1950; WRIGHT 1951). Experimental studies on natural populations have been carried out by the school of FISHER and FORD (e.g., SHEPPARD 1951; DOWDESWELL and FORD 1952) and by LAMOTTE (1952).

In spite of these, no mathematical investigations seem to have been worked up on the process of change due to the random fluctuation of selection intensities, except a short article reported by the present author (KIMURA 1952a), though WRIGHT (1948) gave a distribution of gene frequencies in steady state for a special case.

In his report, the present author proved, using a method of transformation and approximation, that the process can be regarded as a deformed Gaussian process. In the present paper, a pair of alleles lacking dominance will be assumed. The process of change of their frequencies when their selection coefficients fluctuate fortuitously from generation to generation around a mean value

* Part of the cost of the mathematical formulae has been paid by the GALTON AND MENDEL MEMORIAL FUND.

[1] Contribution No. 57 of the National Institute of Genetics, Mishima-shi, Japan.

GENETICS 39: 280 May 1954.

0 is simplest for mathematical treatment. Investigation of this process is a main subject of this paper. The process of change which will be found in terminal portions of the frequency distribution curve is especially important in this connection, so that a precise analysis of it will be undertaken. Through this analysis the reader will be led to new concepts of " quasi-fixation " and " quasi-loss " of an allele. Comparison of this process with that of random drift due to small population number is another important subject in the present report. Though there are many theoretical studies on the process of random drift, they are usually restricted to the state that will be realized after a sufficient number of generations. In that state the distribution curve assumes a fixed form and the probabilities of all heterallelic classes decrease at a constant rate of 1/2N, with fixation and loss of the gene occurring at the same rate.

Hence more extensive studies may be needed to make such a comparison. In the present paper an asymptotic solution for the process of random drift due to small population number will be presented for the first time.

### PROCESS OF CHANGE OF FREQUENCIES OF ALLELES WHICH ARE NEUTRAL ON THE AVERAGE AND LACKING DOMINANCE

Consider a very large randomly mating population and assume a pair of alleles *A* and *A'*. If x is the relative frequency of the gene A in the population and s is the selection coefficient of A, the rate of change of the gene frequency due to selection is approximately

$$\delta x = sx(1-x)$$

per generation, when s is small and there is no dominance. If there is random fluctuation in the selection intensity, s and therefore $\delta x$ are random variables, and a certain irregularity is expected in the process of change in gene frequency from generation to generation. When the rate of change is small, this process may be safely treated as a continuous Markov process.

If $\phi(x,t)dx$ is the probability that the gene frequency lies between x and $x+dx$ in the t-th generation, it can be proved that $\phi(x,t)$ satisfies the partial differential equation,

$$\frac{\partial\phi(x,t)}{\partial t} = \frac{\partial^2}{\partial x^2}\left[\frac{V_{\delta x}}{2}\phi(x,t)\right] - \frac{\partial}{\partial x}\left[M_{\delta x}\,\phi(x,t)\right], \tag{1}$$

where $M_{\delta x}$ and $V_{\delta x}$ represent respectively the mean and the variance of $\delta x$. This equation which is known by mathematicians as " Kolmogorov's forward differential equation " is usually called " Fokker-Planck equation " by physicists, though this type of equation was already used by LORD RAYLEIGH (cited from FUSHIMI 1941). However, we are indebted principally to SEWALL WRIGHT (1945) for the application of this equation to the problem of population genetics.

A meaning of this equation can easily be understood by noting that the left hand side of this equation represents the rate of change of the relative probability of any class per generation and this can be decomposed into two parts as represented by two terms in the right; namely the part due to the random fluctuation (first term) and the one due to the directed change (second term).

If the gene *A* is selectively neutral on the average such that the mean value of its selection coefficient over very long periods is zero,

$$M_{\delta x} = 0 \quad \text{and} \quad V_{\delta x} = V_s x^2(1-x)^2$$

where $V_s$ is the variance of s. In this case equation (1) is written in the form:

$$\frac{\partial \phi}{\partial t} = \frac{V_s}{2}\frac{\partial^2}{\partial x^2}\{x^2(1-x)^2\phi\}. \tag{2}$$

This is a partial differential equation with singularities at the boundaries, so that no arbitrary conditions can be imposed there. But as will be seen in the following operations a continuous stochastic process satisfying the equation (2) is uniquely determined if an initial condition $\phi(x,0)$ is given.

As was demonstrated in the previous report (KIMURA 1952a), if the gene frequency x is transformed into a variate $\xi$ by the relation:

$$\xi = \log\left(\frac{x}{1-x}\right),$$

$\xi$ changes continuously from $-\infty$ to $+\infty$ as x changes from 0 to 1 and the distribution of $\xi$ becomes approximately normal; that is, the process of change of $\xi$ is approximately represented by a Gaussian process.

To solve the equation (2), the same transformation turns out to be very useful: Putting

$$u = \frac{1}{2} e^{\frac{V_s}{8}t} x^{\frac{3}{2}}(1-x)^{\frac{3}{2}}\phi$$

and

$$\xi = \log\left(\frac{x}{1-x}\right),$$

we obtain the heat conduction equation,

$$\frac{\partial u}{\partial t} = \frac{V_s}{2}\frac{\partial^2 u}{\partial \xi^2}. \tag{3}$$

It is already established that this equation has an unique solution which is continuous over $-\infty$ to $+\infty$ when $t \geqq 0$ and which reduces to $u(\xi,0)$ when $t = 0$.

$$u(\xi,t) = \frac{1}{\sqrt{2\pi V_s t}}\int_{-\infty}^{\infty} e^{-\frac{(\xi-\eta)^2}{2V_s t}} u(\eta,0)\,d\eta.$$

Therefore, if the initial distribution of gene frequencies $\phi(x,0)$ is given, the unique solution which satisfies (2) and is continuous between 0 and 1 is

$$\phi(x,t) = \frac{1}{\sqrt{2\pi V_s t}} \frac{e^{-\frac{V_s}{8}t}}{[x(1-x)]^{3/2}} \int_0^1 e^{-\frac{\left[\log\frac{x(1-y)}{(1-x)y}\right]^2}{2V_s t}} \sqrt{y(1-y)}\,\phi(y,0)dy. \quad (4)$$

If the initial condition is not a continuous distribution $\phi(x,0)$, but is a given gene frequency $x_0$, the relative probability that the gene frequency in the t-th generation will be between x and x + dx is given by the formula:

$$\phi(x,t) = \frac{1}{\sqrt{2\pi V_s t}} \exp\left\{-\frac{V_s}{8}t - \frac{\left[\log\frac{x(1-x_0)}{(1-x)x_0}\right]^2}{2V_s t}\right\} \frac{[x_0(1-x_0)]^{1/2}}{[x(1-x)]^{3/2}}. \quad (5)$$

The process of change of the distribution curve with generations is illustrated in figure 1 assuming that the initial gene frequency in the population is 50%. In this figure the variance of selection coefficient is 0.0483. This is a value which WRIGHT (1948) obtained for the *medionigra* gene in an isolated colony of *Panaxia dominura* (FISHER and FORD 1947), assuming that observed variance of change in gene frequency per year were due wholly to fluctuations in selection. As will be seen in the figure the distribution curve is unimodal before the 28th generation after which it becomes bimodal. In the 100th generation gene frequencies that give maximum probability (corresponding to peaks) are approximately 0.0007 and 0.9993, where the height of the curve ($\phi_{max}$) is about 11.37. This is 28.7 times higher than the height at the valley (about 0.397) in the middle part of the distribution. So the distribution curve looks like an U-shaped curve. The more precise form of the terminal part of the distribution curve where the gene frequency is very small is illustrated in figure 2.

With passage of time, the distribution curve becomes nearly U-shaped. The process of change is rather rapid and in the 1000th generation, the peaks of the distribution curve become so high, the gene frequencies corresponding to them become so close to the two termini of the distribution and the valley becomes so deep that it is practically impossible to illustrate the distribution curve in figure 1. More generally, if the initial gene frequency is 50%, the distribution curve is unimodal if the number of generations is less than $4/(3V_s)$ but becomes bimodal if it exceeds this value.

The mean of the distribution is always

$$\int_0^1 x\phi(x,t)\,dx = x_0, \quad (6)$$

But the variance,

$$V_t = \int_0^1 (x-x_0)^2\phi(x,t)\,dx,$$

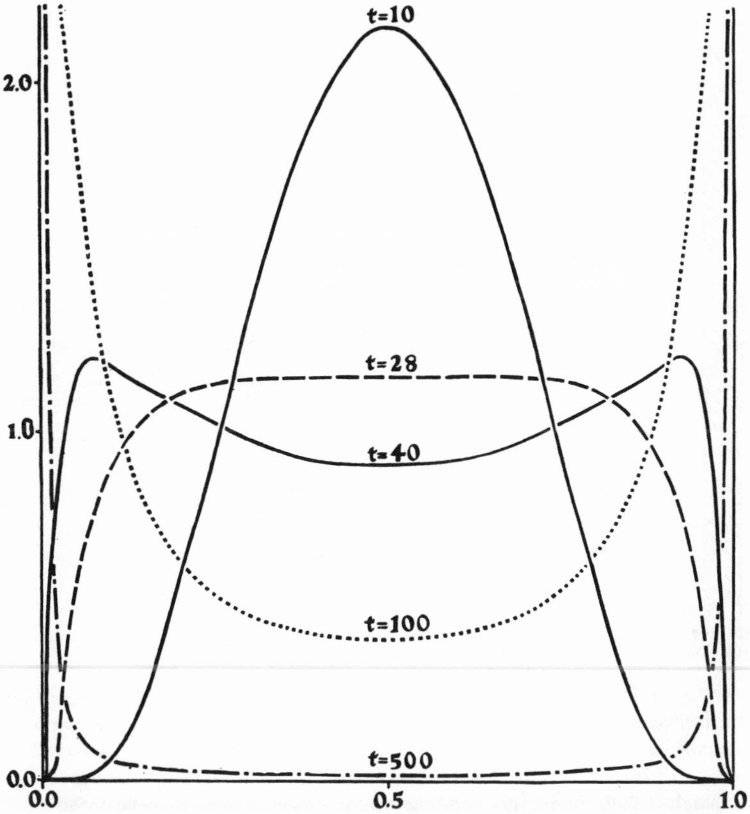


FIGURE 1.—A graph illustrating the process of the change in the distribution of gene frequencies with random fluctuation in the selection intensities. In this illusration it is assumed that the gene is selectively neutral when averaged over a very long period, that there is no dominance, that the initial gene frqunıcy of the population is 0.5 and that the variance of the selection coefficient is 0.0483. (Abscissa: gene frequency. Ordinate: relative probability.)

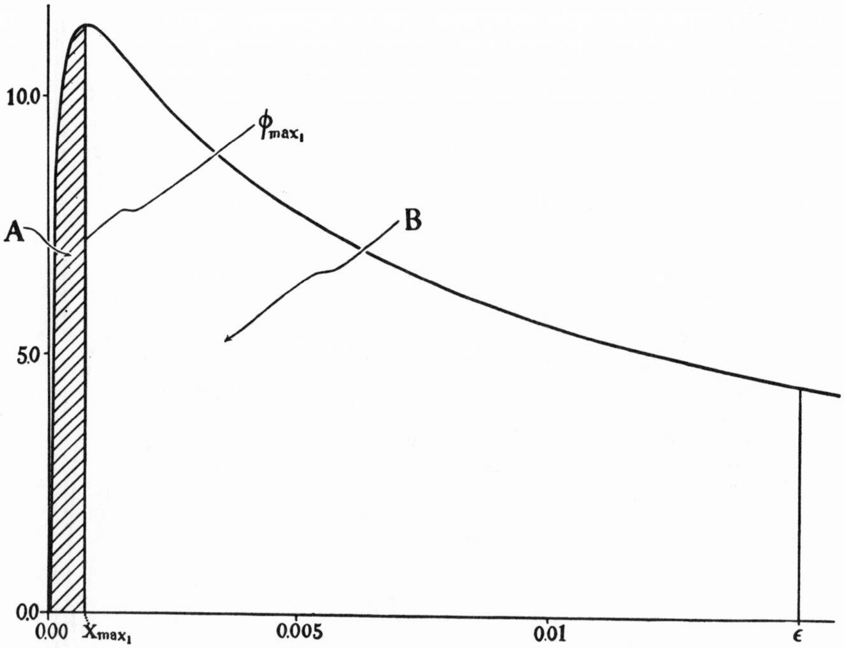


FIGURE 2.—A terminal portion of the distribution curve in the 100th generation where the gene frequency is very low. In this illustration it is assumed that the variance of the selection coefficient is 0.0483 and the initial gene frequency in the population is 0.5. The gene frequency that gives a maximum value in the distribution curve ($x_{max_1}$) is approximately 0.0007 and corresponding height ($\phi_{max_1}$) of the curve is about 11.37. A denotes the probability that the frequency of the gene is smaller than $x_{max_1}$. This is about 0.007 in this case. $\epsilon$ is an arbitrarily chosen gene frequency which is larger than this value. B stands for the probability that the gene frequency in a population is larger than $x_{max_1}$ but smaller than $\epsilon$. If we put $\epsilon$ as 0.015, B is approximately 0.098. B becomes ½ at the limit of $t \rightarrow \infty$. (Abscissa: gene frequency. Ordinate: relative probability.)

increases in successive generations and for large t it is represented asymptotically by the formula

$$V_t = x_0(1-x_0) - \sqrt{\frac{\pi x_0(1-x_0)}{2V_s t}}\, e^{-\frac{V_s t}{8}} + O\left(\frac{e^{-\frac{V_s t}{8}}}{t\sqrt{t}}\right). \tag{7}$$

Therefore its final rate of approach to the limiting value is very close to $V_s/8$ per generation.

CHANGE IN THE TERMINAL PARTS OF THE DISTRIBUTION AND THE PROCESS OF QUASI-FIXATION

As shown above, classes with the highest probability shift toward the terminals indefinitely with time so that the distribution curve appears to be U-shaped. But it is not a true U-shaped curve, since its value at either terminal is always 0. So it will be important to investigate how the distribution curve will continue to change after a sufficient number of generations, with special reference to its terminal parts.

First let us fix our attention to the terminal portion of the distribution where the frequency of the gene A is very low (see figure 2).

The gene frequency $x_{\max_1}$ that gives maximum relative probability $\phi_{\max_1}$ is asymptotically

$$x_{\max_1} \sim \frac{x_0}{1-x_0}\, e^{-\frac{3}{2}V_s t} \longrightarrow 0 \ (t \longrightarrow \infty). \tag{8}$$

While the corresponding relative probability is

$$\phi_{\max_1} \sim \frac{1}{\sqrt{2\pi V_s t}}\, \frac{(1-x_0)^2}{x_0}\, e^{V_s t} \longrightarrow \infty\, (t \longrightarrow \infty). \tag{9}$$

The gene frequency that gives the maximum value in the distribution curve will approach indefinitely to one terminal point (0), elevating indefinitely the corresponding height of the distribution curve.

Let A be the probability that the gene frequency in the population is lower than $x_{\max_1}$. To calculate this, we will start from a more general relation: The probability that the gene frequency in the population falls between two assigned values a and b is

$$\Pr\{a < x < b\} = \int_a^b \phi(x,t)dx$$

$$= \frac{1-x_0}{\sqrt{2\pi}} \int_{\alpha(a)}^{\alpha(b)} e^{-\frac{1}{2}\lambda^2} d\lambda + \frac{x_0}{\sqrt{2\pi}} \int_{\beta(a)}^{\beta(b)} e^{-\frac{1}{2}\lambda^2} d\lambda, \tag{10}$$

where

$$\alpha(\lambda) = \frac{\log\left(\frac{\lambda}{1-\lambda}\right) - \log\left(\frac{x_0}{1-x_0}\right)}{\sqrt{V_s t}} + \tfrac{1}{2}\sqrt{V_s t}$$

and

$$\beta(\lambda) = \frac{\log\left(\frac{\lambda}{1-\lambda}\right) - \log\left(\frac{x_0}{1-x_0}\right)}{\sqrt{V_s t}} - \tfrac{1}{2}\sqrt{V_s t}.$$

Using this relation it can be easily shown that

$$A = \Pr\{0 < x < x_{max_1}\}$$

$$\sim \frac{1-x_0}{\sqrt{2\pi}}\int_{\sqrt{V_s t}}^{\infty} e^{-\frac{1}{2}\lambda^2} d\lambda + \frac{x_0}{\sqrt{2\pi}}\int_{2\sqrt{V_s t}}^{\infty} e^{-\frac{1}{2}\lambda^2} d\lambda$$

$$= \frac{1-x_0}{\sqrt{2\pi V_s t}} e^{-\frac{1}{2}V_s t} + O\left(\frac{e^{-2V_s t}}{\sqrt{V_s t}}\right) \longrightarrow 0. \tag{11}$$

That is, this probability vanishes at the limit; $t \rightarrow \infty$.

On the other hand, Let B stand for the probability that the gene frequency in the population is larger than $x_{max_1}$ but smaller than $\epsilon$, where $\epsilon$ is an arbitrarily chosen gene frequency larger than $x_{max_1}$. Using the relation (10),

$$B = \int_{x_{max_1}}^{\epsilon} \phi(x,t)dx \sim \frac{1-x_0}{\sqrt{2\pi}}\int_{-\sqrt{V_s t}}^{\frac{\log\epsilon - \log\left(\frac{x_0}{1-x_0}\right)}{\sqrt{V_s t}} + \frac{1}{2}\sqrt{V_s t}} e^{-\frac{1}{2}\lambda^2} d\lambda$$

$$+ \frac{x_0}{\sqrt{2\pi}}\int_{-2\sqrt{V_s t}}^{\frac{\log\epsilon - \log\left(\frac{x_0}{1-x_0}\right)}{\sqrt{V_s t}} - \frac{1}{2}\sqrt{V_s t}} e^{-\frac{1}{2}\lambda^2} d\lambda.$$

Therefore, for any $\epsilon$, however small, B can be brought arbitrarily close to $1 - x_0$ by taking t sufficiently large such that

$$V_s t \gg -\log \epsilon.$$

This may be made more clear by the following relation:

$$B = \Pr\{x_{max_1} < x < \epsilon\}$$

$$\sim \frac{1-x_0}{\sqrt{2\pi}}\int_{-\sqrt{V_s t}}^{\frac{1}{2}\sqrt{V_s t}} e^{-\frac{1}{2}\lambda^2} d\lambda + \frac{x_0}{\sqrt{2\pi}}\int_{-2\sqrt{V_s t}}^{-\frac{1}{2}\sqrt{V_s t}} e^{-\frac{1}{2}\lambda^2} d\lambda$$

$$= (1 - x_0) - \frac{2(1 - 2x_0)}{\sqrt{2\pi V_s t}} e^{-\frac{V_s t}{8}} + O\left(\frac{e^{-\frac{V_s t}{2}}}{\sqrt{V_s t}}\right) \rightarrow 1 - x_0 \ (t \rightarrow \infty). \quad (12)$$

This shows that after a sufficient number of generations B approaches to $1 - x_0$ with the rate of $V_s/8$ per generation. Figure 2 illustrates the terminal portion of the distribution curve when $V_s = 0.0483$ and $x_0 = 0.5$.

Similar relations hold for the other terminal portion of the distribution where the frequency of the gene is very close to 1: If $x_{max_2}$ stands for the gene frequency giving the maximum value in the distribution curve and $\phi_{max_2}$ stands for the corresponding relative probability,

$$x_{max_2} \sim 1 - \frac{1 - x_0}{x_0} e^{-\frac{3}{2} V_s t} \rightarrow 1 \quad (13)$$

and

$$\phi_{max_2} \sim \frac{1}{\sqrt{2\pi V_s t}} \frac{x_0^2}{1 - x_0} e^{V_s t} \rightarrow \infty \quad (14)$$

as $t \rightarrow \infty$. The probability A′ that the frequency of the gene exceeds $x_{max_2}$ vanishes as t approaches infinity;

$$A' = \Pr\{x_{max_2} < x < 1\} \sim O\left(\frac{e^{-\frac{1}{2} V_s t}}{\sqrt{V_s t}}\right) \rightarrow 0. \quad (15)$$

On the other hand, even if $\epsilon' (> 0)$ is taken however small, the probability B′ that the gene frequency of the population will fall between $1 - \epsilon'$ and $x_{max_2}$ approaches to $x_0$ with the rate of $V_s/8$ per generation at the limit of $t \rightarrow \infty$;

$$B' = \Pr\{1 - \epsilon' < x < x_{max_2}\} \sim x_0 + O\left(\frac{e^{-\frac{V_s}{8} t}}{\sqrt{V_s t}}\right). \quad (16)$$

The gene frequency giving the minimum of this pseudo-U-shaped distribution curve ($x_{min}$) approaches ½ at $t \rightarrow \infty$ even if the initial gene frequency is not 50%:

$$x_{min} \sim \frac{1}{2} - \frac{\log \frac{x_0}{1 - x_0}}{\frac{3}{2} V_s t - 4}. \quad (17)$$

The corresponding relative probability $\phi_{min}$ vanishes at the limit:

$$\phi_{min} \sim \sqrt{\frac{x_0(1-x_0)}{2\pi V_s t}} \left\{ \frac{1}{4} - \left( \frac{\log \frac{x_0}{1-x_0}}{\frac{3}{2}V_s t - 4} \right)^2 \right\}^{-3/2}$$

$$\exp\left\{ - \frac{V_s t}{8} - \frac{9}{8}\left( \frac{\log \frac{x_0}{1-x_0}}{\frac{3}{2}V_s t - 4} \right)^2 V_s t \right\} \longrightarrow 0. \quad (18)$$

That is, the valley in the distribution curve deepens until the bottom reaches the abscissa.

As will be seen from the relation;

$$\lim_{\epsilon \to 0+} \int_{\epsilon}^{1-\epsilon} \phi(x,t) = 1, \quad (19)$$

the random fluctuation of selection intensities by itself cannot lead to the complete fixation or loss, in the strict sense, of the gene contrary to the case of random drift due to small population number. But as has been shown through (8) – (18), there is a strong tendency that the gene frequency will move toward either terminus with increasing time. In other words, after a sufficient number of generations almost all populations will be in such a situation that the gene is either almost fixed in the population or almost lost from it. To distinguish this from the fixation or loss in the case of small effective population number, the terms " quasi-fixation " and " quasi-loss " are proposed. As will be seen from (12) and (16), their rate can be taken as $V_s/8$.

In the long run, this process of quasi-fixation or -loss will be checked by the opposing mutation pressure.

In this state of statistical equilibrium, if the mutation rates of the gene *A* to and from its allele *A′* are u and v respectively, the frequency distribution of *A* in the population is given by the formula:

$$\phi(x) = C \frac{e^{-\frac{2}{V_s}\left(\frac{v}{x} + \frac{u}{1-x}\right)}}{x^2(1-x)^2} \left(\frac{x}{1-x}\right)^{-2\left(\frac{u-v}{V_s}\right)}, \quad (20)$$

where C is a constant chosen such that $\int_0^1 \phi(x)dx = 1$. Wright (1948) derived essentially the same formula assuming migration (p. 292).

COMPARISON WITH THE PROCESS OF RANDOM DRIFT DUE TO SMALL POPULATION NUMBER

In a finite population, owing to random sampling of gametes in reproduction, there occurs random fluctuation of the gene frequency from generation to generation. This process, as is well known, will finally lead to the complete fixation or loss of the gene if such factors as mutation, migration and selection are absent.

If N is the effective number of reproducing individuals in the population and p is the initial gene frequency the nth moment of distribution about zero in the t-th generation is given by the following formula if the order of t is not smaller than N (cf. KIMURA 1952b):

$$\mu_n'^{(t)} = p - 3pq\frac{n-1}{n+1}(1-\lambda_1)^t - 5pq(p-q)\frac{(n-2)(n-1)}{(n+1)(n+2)}(1-\lambda_2)^t$$
$$-7pq(-5pq+1)\frac{(n-3)(n-2)(n-1)}{(n+1)(n+2)(n+3)}(1-\lambda_3)^t$$
$$-9pq(14pq^2-7pq+p-q)\frac{(n-4)(n-3)(n-2)(n-1)}{(n+1)(n+2)(n+3)(n+4)}(1-\lambda_4)^t$$
$$+O\{(1-\lambda_5)^t\}, \quad (21)$$

where $q = 1 - p$. From this we can derive the probability that the gene will have become fixed in the population by the t-th generation:

$$f_t(1) = p - 3pq(1-\lambda_1)^t - 5pq(p-q)(1-\lambda_2)^t - 7pq(-5pq+1)(1-\lambda_3)^t$$
$$-9pq(14pq^2-7pq+p-q)(1-\lambda_4)^t + O\{(1-\lambda_5)^t\} \quad (22)$$

The corresponding probability of complete loss is:

$$f_t(0) = q - 3pq(1-\lambda_1)^t + 5pq(p-q)(1-\lambda_2)^t - 7pq(-5pq+1)(1-\lambda_3)^t$$
$$+9pq(14pq^2-7pq+p-q)(1-\lambda_4)^t + O\{(1-\lambda_5)^t\}(t\to\infty). \quad (23)$$

In these formulae

$$\lambda_1 = \frac{1}{2N}, \lambda_2 = \frac{3}{2N}, \lambda_3 = \frac{6}{2N}, \lambda_4 = \frac{10}{2N}, \lambda_5 = \frac{15}{2N}, \ldots.$$

In general $\lambda$'s are given by the formula;

$$\lambda_i = \frac{i(i+1)}{4N} \quad (i = 1, 2, \ldots).$$

The frequency of the gene in this case may take any one of a series of discontinuous values:

$$0, \frac{1}{2N}, \frac{2}{2N}, \ldots, 1-\frac{1}{2N}, 1.$$

Usually, however, the number of reproducing individuals (N) in a population is so large that practically the gene frequency (x) can be treated as a continuous variable with good approximation. Variance of the rate of change in gene frequency due to the random sampling of gametes is

$$V_{\delta x} = \frac{x(1-x)}{2N}.$$

Therefore if $\phi(x,t)$ is the relative probability that the frequency of the gene in the population will take any value between x and $x + dx (0 < x < 1)$ in the t-th generation, $\phi(x,t)$ satisfies the following partial differential equation:

$$\frac{\partial \phi}{\partial t} = \frac{1}{4N} \frac{\partial^2}{\partial x^2} \{x(1-x)\phi\}, \tag{24}$$

which is easily derivable from equation (1). To solve this, if we put

$$\phi \propto X_i(x) e^{-\lambda_i t} \qquad (i = 1, 2, 3, \ldots),$$

we obtain the ordinary differential equation;

$$x(1-x)\frac{d^2X_i}{dx^2} + (2-4x)\frac{dX_i}{dx} - (2-4N\lambda_i)X_i = 0,$$

where $\lambda_i$ corresponds to the eigen value of equation (24). Noting that $\lambda_i = i(i+1)/4N$, this becomes

$$x(1-x)\frac{d^2X_i}{dx^2} + (2-4x)\frac{dX_i}{dx} - (1-i)(i+2)X_i = 0. \tag{25}$$

This type of equation is known as Gauss's differential equation and (25) is satisfied by the following hypergeometric series:

$$
\begin{aligned}
X_i &= F(1-i,\ i+2,\ 2,\ x) \\
&= 1 + \frac{(1-i)(i+2)}{1\cdot 2}x + \frac{(1-i)(2-i)\cdot(i+2)(i+3)}{1\cdot 2\cdot 2\cdot 3}x^2 \\
&\quad + \frac{(1-i)(2-i)(3-i)\cdot(i+2)(i+3)(i+4)}{1\cdot 2\cdot 3\cdot 2\cdot 3\cdot 4}x^3 \\
&\quad + \frac{(1-i)(2-i)(3-i)(4-i)\cdot(i+2)(i+3)(i+4)(i+5)}{1\cdot 2\cdot 3\cdot 4\cdot 2\cdot 3\cdot 4\cdot 5}x^4 \\
&\quad + \ldots \quad (i = 1, 2, 3, \ldots)
\end{aligned}
$$

Therefore the asymptotic solution of (24) for large t is;

$$
\begin{aligned}
\phi(x,t) &= C_1 e^{-\lambda_1 t} + C_2(1-2x)e^{-\lambda_2 t} + C_3(1-5x+5x^2)e^{-\lambda_3 t} \\
&\quad + C_4(1-9x+21x^2-14x^3)e^{-\lambda_4 t} + O(e^{-\lambda_5 t}).
\end{aligned} \tag{26}
$$

To determine the constants $C_1$, $C_2$, $C_3$, . . . we can use the relation that the n-th moment of distribution obtained by this formula,

$$\int_0^1 x^n \phi(x,t)\,dx,$$

must be equal to

$$\mu'_n{}^{(t)} - 1^n \cdot f_t(1),$$

since the homallelic classes are excluded from the distribution curve to be given by (24). Thus we obtain the following values:

$C_1 = 6\,pq, C_2 = -30\,pq(p-q),\ C_3 = 84\,pq(-5\,pq+1),$

$C_4 = -180\,pq(14\,pq^2 - 7\,pq + p - q).$

How the distribution curve represented by (26) changes with generations is illustrated in figure 3, assuming that the initial gene frequency p is 0.1. As may be seen from this figure, the curve becomes gradually flat until finally every heterallelic class has equal probability and falls with the rate of 1/2N per generation. In this final stage, the fixation of the gene proceeds at the same rate, the correct value of which was first obtained by Wright to be 1/2N (see Wright 1931) by using a different method of calculation. This rate is usually known as the rate of fixation due to random sampling of gametes.

The probability $\Omega_t$ that the alleles A and A′ coexist in the population in the t-th generation can be obtained from (26):

$$\Omega_t = \int_0^1 \phi(x,t)dx$$

$$= 6pq\, e^{-\frac{1}{2N}t} + 14pq\,(-5pq+1)\, e^{-\frac{6}{2N}t} + O\left(e^{-\frac{15}{2N}t}\right)\ (t \rightarrow \infty). \quad (27)$$

Contrary to what was shown in (19) either complete fixation or loss of alleles is expected in this case and $\Omega_t$ vanishes at the limit of $t \rightarrow \infty$.

Variance of the distribution in the t-th generation is from (22), (23) and (26),

$$V_t = pq - pq\, e^{-\frac{1}{2N}t}, \quad (28)$$

namely the variance approaches its limiting value pq at the rate 1/2N per generation.

As has been demonstrated above, the process of change due to random fluctuation of selection intensity is quite different from that due to the random sampling of gametes. Therefore comparison of their effects must be made from various angles as Wright (1948) did in analyzing the data of *medionigra* gene in Panaxia.

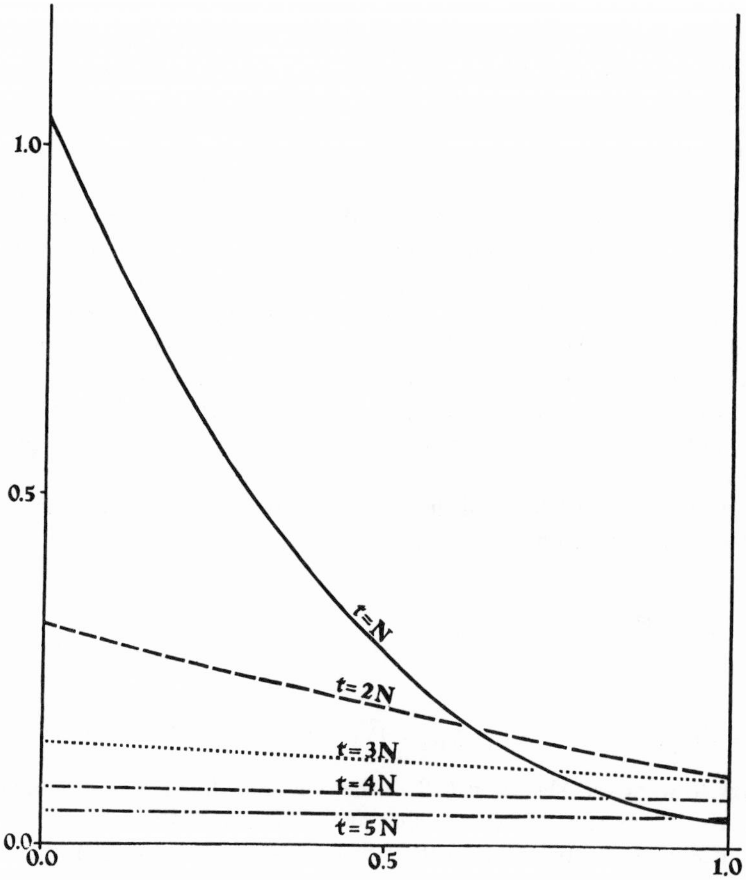


FIGURE 3.—Graphs showing the process of change of the frequency distribution curve due to the random drift in small populations. In this illustration it is assumed that the initial gene frequency in the population is 0.1. It will be seen that the distribution curve becomes more and more flat as the number of generations increases. (Abscissa: gene frequency in the population. Ordinate: relative probability. t: time in generation. N: effective population size.)

Thus, if we consider the process of change which will be realized after a sufficient number of generations, the rate of quasi-fixation, $V_s/8$, may be compared with the rate of fixation due to random sampling, $1/2N$, for the same purpose. Suppose $V_s$ is known to be 0.0483 as in figure 1, the equivalent N is calculated to be about 83 by using a equivalence relation;

$$\frac{1}{2N} \doteq \frac{V_s}{8},$$

though the applicability of this formula is rather restricted.

So far we have treated the two factors separately. But in nature not only these two factors but also systematic factors may work concurrently. The present writer (1951) reported briefly on the distribution of gene frequencies for such case. The more precise account will appear elsewhere, but the main conclusion derived from the analysis of the distribution curve is not difficult to present here: The effect produced by the random fluctuation in natural selection is relatively unimportant for small populations. But in large populations it has a remarkable effect in such ways that in the case of no dominance the distribution curve is modified markedly in the parts where the frequency of either allele is low and in the case of complete dominance in a part where the frequency of the recessive gene lies inside a certain range of higher frequencies. Also the product $NV_s$ is an important quantity. To estimate not only the distribution of population size (N) in nature but also the variability ($V_s$) of selection intensity for important loci may be an important task left for the future experiments.

ACKNOWLEDGMENT

The author is greatly indebted to MR. Y. KAWAGUCHI who made all the laborious computations from which the figures in this paper were constructed. Thanks are due to DR. SEWALL WRIGHT who read the manuscript and gave valuable criticisms and suggestions. The author wishes to express deep appreciation to DR. JAMES F. CROW without whose advice and help this paper could not have been published.

SUMMARY

When there are random fluctuations in selection intensity, the process of change in gene frequency in a population is represented by a stochastic process. In this paper an analysis of this process is presented for a gene lacking dominance and selectively neutral on the average. Especially interesting is the process of change that can be observed in the terminal portions of the distribution curve. Contrary to the case of random drift in small populations, if the population is very large, complete fixation or loss of an allele, in the strict sense, will not be realized. But there exists a strong tendency toward the state of almost fixation or almost loss. That is, if we allow a sufficient number of generations a situation will almost surely be realized in which the allele is either almost fixed in the population or almost lost from it. To distinguish this from the fixation or loss in the case of random drift in small populations, terms " quasi-fixation " and " quasi-loss " were proposed. Their rate per generation can be taken as $V_s/8$, where $V_s$ is the variance of the selection coefficient. Comparison of this process with that of random drift in small populations is another important subject in the present paper. In spite of many studies on the process of the drift very little is known about the process of the change before the fixation and loss of an allele proceeds at the constant rate of 1/2N. In this paper an asymptotic solution for this process is presented for the first time.

## REFERENCES

DOBZHANSKY, TH., 1951 Genetics and the origin of species. Columbia Univ. Press, New York.

DOWDESWELL, W. H., and E. B. FORD, 1952 The distribution of spot-number as an index of geographical variation in the butterfly *Maniola jurtia* L. (Lepidoptera: Satyridae). Heredity **6:** 99–109.

FELLER, W., 1950 Diffusion processes in genetics. Proceedings of Second Berkeley Symposium on Mathematical Statistics and Probability, Univ. of California Press, pp. 227–246.

FISHER, R. A., 1930 The genetical theory of natural selection. Oxford. Clarendon Press.

FISHER, R. A., and E. B. FORD, 1947 The spread of a gene in natural conditions in a colony of the moth *Panaxia dominula* L. Heredity **1:** 143–174.

1950 The " Sewall Wright Effect." Heredity **4:** 117–119.

FUSHIMI, K., 1941 Theories of probability and statistics (Japanese). Tokyo.

HALDANE, J. B. S., 1939 The equilibrium between mutation and random extinction. Ann. Eug. **9:** 400–405.

1949 Human evolution: Past and future. Genetics, paleontology and evolution, Princeton Univ. Press, pp. 405–418.

HOUSE, V. H., 1953 The use of the binomial expansion for a classroom demonstration of drift in small populations. Evolution **7:** 84–88.

KIMURA, M., 1951 Effect of random fluctuation of selective value on the distribution of gene frequencies in natural populations. Ann. Rep. Nat. Inst. Gen. **1:** 45–47.

1952a Process of irregular change of gene frequencies due to the random fluctuation of selection intensities. *Ibid.* **2:** 56–57.

1952b On the process of decay of variability due to random extinction of alleles. *Ibid.* **2:** 60–61.

LAMOTTE, M., 1952 Le rôle des fluctuations fortuites dans la diversité des populations naturelles de *Capaea nemoralis* (L.). Heredity **6:** 333–343.

MULLER, H. J., 1949 Redintegration of the symposium on genetics, paleontology, and evolution. Genetics, paleontology and evolution, Princeton Univ. Press, pp. 421–445.

SHEPPARD, P. M., 1951 Fluctuations in the selective value of certain phenotypes in the polymorphic land snail *Cepœa nemoralis* (L.). Heredity **5:** 125–134.

WRIGHT, S., 1931 Evolution in mendelian populations. Genetics **16:** 97–159.

1945 The differential equation of the distribution of gene frequencies. Proc. Nat Acad. Sci. **31:** 382–389.

1948 On the roles of directed and random changes in gene frequency in the genetics of populations. Evolution **2:** 279–294.

1951 Fisher and Ford on " The Sewall Wright Effect." Amer. Scientist **39:** 452–479.

# Change of gene frequencies by natural selection under population number regulation*

(population genetics/discrete deterministic treatment/quasi-equilibrium)

MOTOO KIMURA

National Institute of Genetics, Mishima, 411 Japan

*Contributed by Motoo Kimura, February 2, 1978*

**ABSTRACT** By incorporating a population number regulating mechanism into the formulation of genic selection involving a pair of alleles ($A_1$ and $A_2$) with respective frequencies $x$ and $1 - x$, it is shown that the change of $x$ in one generation is given by $\Delta x = sx(1-x)/\overline{W}$, in which $\overline{W}$ is the mean absolute selective value (in Wright's sense). It is also shown that, in the process in which advantageous allele (say $A_1$) increases from a low frequency to a high frequency, quasi-equilibrium is rapidly attained where $\Delta\overline{W} \approx 0$. In this state we have $\overline{W} \approx 1 + (s^2/\bar{c})x(1-x)$ in the case of logarithmic population number regulation, and $\overline{W} \approx 1 + s^2x(1-x)/(\bar{c}N)$ in the case of logistic regulation. In these expressions, $s$ is the selective advantage of $A_1$ over $A_2$, and $\bar{c}$ is a coefficient relating to the total population number regulation. It is pointed out that the approximation formula $\Delta x = sx(1-x)$ is valid under wider circumstances than usually suggested by the conventional treatment of genic selection.

It is customary in the mathematical treatment of population genetics to assume that relative fitnesses of various genotypes (or genes) are independent of the total population number. Actually, the ordinary formulae for the change of gene frequencies are those that can be obtained when we assume that there is no regulation of the total population number (see ref. 1, p. 26).

Such a treatment has the merit of simplifying the mathematics involved without too much affecting the essential nature of gene frequency change by natural selection.

On the other hand, it is clear that the number of offspring that survive to maturity depends much on the total population number (mortality being higher with more overcrowding), and this will affect the absolute fitness of individual genotypes. It is likely that relative fitnesses are also affected by the population number.

In the present paper, I shall attempt to clarify, by using a simple model that incorporates a mechanism of population number regulation, how the rate of change of gene frequency is affected by population regulation.

## Haploid population with two segregating alleles under logarithmic regulation of total number

Consider a population of haploid organisms consisting of two types of individuals (alleles) denoted by $A_1$ and $A_2$. We shall designate by $N$ the total population number (counted at the adult stage) and let $n_1$ and $n_2$ be the numbers of $A_1$ and $A_2$ individuals, respectively, so that $N = n_1 + n_2$.

We assume that generations are discrete, and in each generation, an $A_1$ individual produces on the average $R_1$ offspring of which $D_1$ die before maturity. In other words, $R_1 - D_1$ is the average number of offspring contributed to the next generation per $A_1$. Thus, the number of $A_1$ in the next generation is

$$n_1' = n_1(R_1 - D_1).$$

Let us assume that $D_1$ is expressed as

$$D_1 = c_1 \log_e N,$$

in which $c_1$ is a positive constant ($c_1 > 0$). This is a much weaker control of population number than the one usually considered—i.e., logistic control in which the term $D_1$ is of the form $c_1N$. Similarly, we assume that the number of offspring produced per $A_2$ individual is $R_2$, of which $c_2 \log_e N$ individuals die before maturity. Thus, the numbers of $A_1$ and $A_2$ in the next generations are

$$n_1' = n_1(R_1 - c_1 \log_e N) \text{ and}$$

$$n_2' = n_2(R_2 - c_2 \log_e N), \quad [1]$$

in which $N = n_1 + n_2$. If we let $W_1 = R_1 - c_1 \log_e N$ and $W_2 = R_2 - c_2 \log_e N$, then $W_1$ and $W_2$ are the absolute selective values of $A_1$ and $A_2$, respectively, in Wright's sense (ref. 2, p. 30). Under population number regulation, both $W_1$ and $W_2$ must be near unity, as will be shown below. The total population number in the next generation is

$$N' = N \cdot \overline{W}, \quad [2]$$

in which $\overline{W} = \overline{R} - \bar{c} \log_e N$ is the mean selective value of the population, and $\overline{R} = (n_1R_1 + n_2R_2)/N$ and $\bar{c} = (n_1c_1 + n_2c_2)/N$.

### Note on population number regulation

Before treating the process of change of relative frequencies of two types of individuals (alleles) by using Eq. 1, it might be convenient to investigate how the total population number is regulated in the present model when there is only one type of individual. Let $N_t$ be the population number in the $t$th ($t = 0, 1, 2, \ldots$) generation. In this case, Eq. 2 reduces to

$$N_{t+1} = N_t(R - c \log_e N_t), \quad [3]$$

in which $R$ and $c$ are positive constants. In the present treatment, we disregard random change. We also assume that the population is sufficiently large so that even if values of $N_t$ generated by Eq. 3 are not exactly equal to whole numbers, this does not cause any trouble for our purpose. The equilibrium in the population number is attained when

$$R - c \log_e N = 1, \quad [4]$$

and the corresponding equilibrium number is

$$\hat{N} = e^{(R-1)/c}. \quad [5]$$

The costs of publication of this article were defrayed in part by the payment of page charges. This article must therefore be hereby marked "*advertisement*" in accordance with 18 U. S. C. §1734 solely to indicate this fact.

* Contribution no. 1187 from the National Institute of Genetics, Mishima, Shizuoka-ken 411 Japan.

Let $\epsilon_t$ be a measure of departure of $N_t$ from its equilibrium value such that

$$N_t = \hat{N}(1 + \epsilon_t).$$

Then we have

$$\epsilon_{t+1} = (1 - c)\epsilon_t - \frac{c}{2}\epsilon_t^2 + \frac{c}{6}\epsilon_t^3 - \ldots, \quad [6]$$

in which we assume that $|\epsilon_t| \ll 1$ and $c > 0$.

If $c > 2$, we have $\epsilon_{t+1} \approx -(c-1)\epsilon_t$ for a small $|\epsilon_t|$, so that the equilibrium is unstable; any deviation from the equilibrium is enlarged in the next generation, with $N_t$ oscillating around $\hat{N}$ but departing from it farther and farther as time goes on. On the other hand, if $c < 2$, we have $\epsilon_{t+1} \approx (1-c)\epsilon_t$ when $c \neq 1$ and $\epsilon_{t+1} \approx -\epsilon_t^2/2$ when $c = 1$. In either case, any deviation from the equilibrium decreases in the next generation. So, the equilibrium is stable. For $c = 2$, we need more careful analysis. In this case we have $\epsilon_{t+1} \approx \epsilon_{t-1}[1 - (8/3)\epsilon_{t-1}^2]$, showing that $\epsilon_t \rightarrow 0$ as $t \rightarrow \infty$. However, the rate of approach to the equilibrium diminishes as $\epsilon_t$ becomes small. Thus, we conclude that regulation of the population number works effectively if $2 > c > 0$. Furthermore, we can show that $N_t$ converges to $\hat{N}$ if $N_0 < \exp\{(R + 1 - c)/c\}$.

## Change of gene frequencies

We first consider the case in which $A_1$ and $A_2$ differ in the numbers of offspring that they produce ($R_1 \neq R_2$) but are subject to the same amount of control from the population regulating factors ($c_1 = c_2 \equiv c$). We assume that $A_1$ has a constant selective advantage $s$ over $A_2$, so that $W_1 - W_2 = R_1 - R_2 = s > 0$.

If we denote by $x$ the frequency of $A_1$ in the population ($x = n_1/N$), then the change of $x$ per generation (i.e., $\Delta x = n_1'/N' - n_1/N$) is given by

$$\Delta x = sx(1 - x)/\overline{W}, \quad [7]$$

in which

$$s = R_1 - R_2 \text{ and } \overline{W} = xW_1 + (1 - x)W_2 = \overline{R} - c \log_e N.$$

Suppose that $A_1$ represents a mutant allele and that, because of its selective advantage, $A_1$ increases gradually from very low frequency to very high frequency. The equilibrium population number ($\hat{N}$) is $\exp\{(R_2 - 1)/c\}$ when $x = 0$ but becomes $\exp\{(R_2 + s - 1)/c\}$ when $x = 1$. So, we can assume a situation in which $N$ increases gradually as $x$ increases by natural selection. At every moment in this process, if regulation of population number were achieved instantaneously, $\overline{W}$ would be equal exactly to unity. Actually, however, there will always be a lag so that $\overline{W}$ departs from unity.

Consider the change of the mean selective value ($\overline{W} = \overline{R} - c \log_e N$) in one generation:

$$\Delta\overline{W} = R_1\Delta x + R_2\Delta(1 - x) - c\Delta \log_e N.$$

Because $\Delta \log_e N \approx \Delta N/N$, which is equal to $\overline{W} - 1$ from Eq. **2**, we have

$$\Delta\overline{W} \approx (R_1 - R_2)\Delta x - c(\overline{W} - 1).$$

This yields

$$\Delta\overline{W} = \frac{s^2x(1 - x)}{\overline{W}} - c(\overline{W} - 1) \quad [8]$$

if we apply Eq. 7. The first term in the right-hand side of Eq. 8 is equal to $\Delta\overline{R}$, the change of the intrinsic growth rate of the population, and this term corresponds to Fisher's (3) fundamental theorem of natural selection (formulated in terms of Wright's selective value). The second term in the right-hand side represents the effect of the population-regulating mechanism on the change of the mean selective value. From Eq. **8** we can see that, if $\overline{W}$ is less than or equal to unity, the right-hand side of the equation is positive and therefore $\overline{W}$ tends to increase. On the other hand, if $\overline{W}$ is much larger than unity, the right-hand side is negative and therefore it tends to decrease. If $s^2 \ll c$, we should expect that change of $\overline{W}$ occurs much more quickly than the change of $s^2x(1 - x)$ so that quasi-equilibrium is attained:

Table 1. Gene frequency change by natural selection under population number regulation

| Generation | $x$, % | $(\overline{W} - 1) \times 10^4$ | $\frac{(s^2/\overline{c})x}{(1-x)} \times 10^4$ |
|---|---|---|---|
| 0 | 10.00 | 50.31 | 1.50 |
| 1 | 10.44 | −22.73 | 1.55 |
| 6 | 12.97 | 2.59 | 1.88 |
| 10 | 15.36 | 2.16 | 2.16 |
| 20 | 22.89 | 2.88 | 2.94 |
| 40 | 44.44 | 4.09 | 4.11 |
| 60 | 68.56 | 3.63 | 3.59 |
| 80 | 85.73 | 2.08 | 2.03 |
| 100 | 94.34 | 0.91 | 0.88 |
| 120 | 97.89 | 0.35 | 0.34 |

$A_1$ has a higher intrinsic growth rate than $A_2$. Parameters involved are $R_1 = 16.05$ for $A_1$ and $R_2 = 16$ for $A_2$ ($s = 0.05$) and $c = 1.5$. The process of change of the frequency of $A_1$ ($x$) is listed at various generations, together with values of $\overline{W} - 1$, starting from $x = 0.1$ and $N = 22026$ at generation 0.

$$\Delta\overline{W} \approx 0. \quad [9]$$

In this quasi-equilibrium, we have

$$\overline{W} \approx 1 + \frac{s^2}{c}x(1 - x) \quad [10]$$

and the change of the frequency of $A_1$ per generation is given approximately by

$$\Delta x = \frac{sx(1 - x)}{1 + (s^2/c)x(1 - x)}. \quad [11]$$

Table 1 shows numerically the course of changes of $x$ and $\overline{W}$, assuming $s = R_1 - R_2 = 0.05$, $R_2 = 16$, and $c = 1.5$ and starting from $N = 22026$ and $x = 0.1$ (at generation = 0). Note that in this case $\hat{N} = 22026.47$ for $x = 0$ and $\hat{N} = 22773.06$ for $x = 1$. From the table, it will be seen that, in this example, the quasi-equilibrium is attained after several generations. Then Eq. **10** starts to give a very good approximation to the mean selective value of the population.

Next, let us consider the situation in which $A_1$ is advantageous over $A_2$ because of greater resistance against overcrowding. We assume that in Eqs. **1**, $R_1 = R_2 \equiv R$ but $c_1 < c_2$. The selective values of $A_1$ and $A_2$ are $W_1 = R - c_1 \log_e N$ and $W_2 = R - c_2 \log_e N$, and the change of the frequency of $A_1$ in one generation is

$$\Delta x = sx(1 - x)/\overline{W},$$

in which $s = (c_2 - c_1) \log_e N$ and $\overline{W} = R - \overline{c} \log_e N$ with $\overline{c} = c_1x + c_2(1 - x)$. Then, at quasi-equilibrium in which $\Delta\overline{W} \approx 0$, we have

$$\overline{W} \approx 1 + (s^2/\overline{c})x(1 - x). \quad [12]$$

Table 2. Different example of gene frequency change by natural selection under population number regulation

| Generation | $x$, % | $(\overline{W}-1)\times 10^4$ | $\dfrac{(s^2/\bar{c})x}{(1-x)\times 10^4}$ |
|---|---|---|---|
| 0 | 10.00 | 100.31 | 6.00 |
| 1 | 10.89 | −40.38 | 6.48 |
| 8 | 19.38 | 10.39 | 10.45 |
| 20 | 43.84 | 16.39 | 16.55 |
| 40 | 85.72 | 8.70 | 8.29 |
| 60 | 98.01 | 1.41 | 1.32 |
| 70 | 99.30 | 0.50 | 0.47 |

This example is similar to Table 1, but $A_1$ is advantageous over $A_2$ because of greater resistance against over-crowding. Parameters are $c_1 = 1.49$ for $A_1$, $c_2 = 1.5$ for $A_2$, and $R = 16$.

In Table 2, a numerical example is shown, assuming parameters $c_1 = 1.49$, $c_2 = 1.50$, and $R = 16$ and the initial conditions $x = 0.1$ and $N = 22026$. Again, agreement between the actual values of $(\overline{W} - 1)$ and its approximation $(s^2/\bar{c})x(1 - x)$ is excellent save for the first several generations. Note that, in this example, the equilibrium population numbers are roughly $\hat{N} = 22026$ at $x = 0$ and $\hat{N} = 23555$ and that the selective advantage of $A_1$ is roughly $s \approx 0.1$. The value of $\overline{W}$ is much nearer to unity than would be suggested by the conventional selection model with discrete generation for which $\overline{W} = 1 + sx$.

More generally, if $R_1 \neq R_2$ and $c_1 \neq c_2$, then letting $W_1 = R_1 - c_1 \log_e N$ and $W_2 = R_2 - c_2 \log_e N$ for $A_1$ and $A_2$, we have $\Delta x = sx(1 - x)/\overline{W}$ in which $s = (R_1 - R_2) - (c_1 - c_2) \log_e N$. Then we can show that, at quasi-equilibrium in which $\Delta\overline{W} = s^2x(1 - x)/\overline{W} - \bar{c}(\overline{W} - 1) \approx 0$, we have $\overline{W} = 1 + (s^2/\bar{c})x(1 - x)$ with good approximation.

## Discussion

We have considered a simple situation in which a pair of alleles $A_1$ and $A_2$ are segregating within a large population, and, assuming no dominance in fitness, we investigated how the changes of gene frequencies are affected by a population number regulating mechanism. The scheme of selection assumed is the one commonly called "genic selection."

We have shown that, in the process in which the advantageous allele ($A_1$) increases from a very low frequency to a very high frequency, quasi-equilibrium is rapidly attained in which $\Delta\overline{W} \approx 0$. In this quasi-equilibrium state, the mean absolute selective value (in Wright's sense) is very near to unity. For example, if $A_1$ is advantageous over $A_2$ because of higher intrinsic reproductive rate ($s = R_1 - R_2 > 0$), then the change of the frequency of $A_1$ per generation is given by $\Delta x = x(1 - x)/\overline{W}$ in which $\overline{W} = 1 + (s^2/\bar{c})x(1 - x)$ with good approximation under quasi-equilibrium. This should be compared with the corresponding expression in the model of genic selection commonly used in population genetics (ref. 2, p. 30). In the conventional model, relative selective values $1 + s$ and 1 are assigned respectively to $A_1$ and $A_2$. This yields $\Delta x = sx(1 - x)/\bar{w}$, in which $\bar{w} = 1 + sx$. It will be seen that, as compared with $\bar{w}$ in the conventional model, $\overline{W}$ in the present formulation is much nearer to unity. For example, if $s = 0.05$ and $x = 0.857$, we have $\overline{W} = 1.0002$ at quasi-equilibrium (see Table 1) but $\bar{w} = 1.0428$ in the conventional model.

It has long been regarded that the formula

$$\Delta x = sx(1 - x) \qquad [13]$$

[and also its continuous-time counterpart $dx/dt = sx(1 - x)$] is valid as an approximation only when the selection coefficient $s$ is very small so that $(1 + sx)$ in the denominator of the conventional formula can be neglected. The present treatment shows, however, that Eq. **13** is valid under wider circumstances.

Although we have considered only genic selection in the present paper, we can extend the treatment to cover zygotic selection. For example, let us assume that $A_1$ is completely dominant over $A_2$ in fitness, and that we can assign selective value $W_1 = R_1 - c \log_e N$ to the dominant ($A_1A_1$ and $A_1A_2$) and $W_2 = R_2 - c \log_e N$ to the recessive ($A_2A_2$). Then we can show that, under random mating, the change of the frequency of $A_1$ in one generation is given by

$$\Delta x = sx(1 - x)^2/\overline{W}$$

in which $s = R_1 - R_2$. Furthermore, we can show that under quasi-equilibrium ($\Delta\overline{W} \approx 0$) we have

$$\overline{W} \approx 1 + \frac{2s^2}{c}x(1 - x)^3$$

which can be much nearer to unity than the conventional expression $\bar{w} = 1 + sx(2 - x)$ (ref. 2, p. 33).

Finally, I would like to point out that the concept of quasi-equilibrium $\Delta\overline{W} \approx 0$ can be applied to other types of population number regulation. For example, consider the logistic control of the total population number for which we have

$$N_{t+1} = N_t(R - cN_t). \qquad [14]$$

This may be compared with Eq. 3 in the above treatment. In the present case we can show that the equilibrium population number $\hat{N} = (R - 1)/c$ is locally stable if $3 > R > 1$. The equations giving the course of change in the numbers of $A_1$ and $A_2$ may be expressed as follows:

$$\left.\begin{aligned} n_1' &= n_1(R_1 - c_1N) \\ n_2' &= n_2(R_2 - c_2N) \end{aligned}\right\} \qquad [15]$$

in which $N = n_1 + n_2$. It can then be shown that, when gene frequencies change by natural selection, quasi-equilibrium will be attained rapidly so that

$$\Delta\overline{W} = s^2x(1 - x)/\overline{W} - \bar{c}N)\overline{W} - 1) \approx 0,$$

in which $s = R_1 - R_2 - (c_1 - c_2)N$ and $\overline{W} = x(R_1 - c_1N) + (1 - x)(R_2 - c_2N)$. Then, the change of the frequency of $A_1$ in one generation is given with good approximation by

$$\Delta x = \frac{sx(1 - x)}{1 + s^2(\bar{c}N)^{-1}x(1 - x)}. \qquad [16]$$

Again, the denominator can be much nearer to unity than the conventional expression $(1 + sx)$. This may be illustrated by the following example. Suppose that parameters are $R_1 = 2.6$, $R_2 = 2.5$, and $c_1 = c_2 = 0.0001$. If we start ($t = 0$) from $N = 15000$ and $x = 0.1$, then at $t = 23$ we have $x = 0.51209$, $N = 15495.9$, and $\overline{W} = 1.00161321$ (exact). The approximation $\overline{W} = 1 + s^2(\bar{c}N)^{-1}x(1 - x)$ gives $\overline{W} = 1.00161238$ which is very near to the true value. This is much smaller than $1 + sx = 1.05120904$ (note that $s = 0.1$ and $\bar{c} = 0.0001$ in this example).

It is clear that we can extend the above treatments to cover a wider situation. Let us suppose that the process of change of the total population number from one generation to the next is given by

$$N_{t+1} = N_t[R - f(N_t)], \qquad [17]$$

in which $R$ is a positive constant and $f(N)$ is a non-negative monotone increasing function of $N$. The equilibrium population number (denoted by $\hat{N}$) may be obtained by solving $R - f(N)$

$-1 = 0$ for $N$. This yields $\hat{N} = f^{-1}(R - 1)$. The equilibrium $N = \hat{N}$ is locally stable if

$$-1 < 1 - \hat{N}f'(\hat{N}) < 1,$$

in which $f'(\hat{N})$ stands for the first derivative of $f(N)$ for $N = \hat{N}$ [if $1 - \hat{N}f'(\hat{N}) = 0$, we must have $2 + \hat{N}^2 f''(\hat{N}) > 0$ for the equilibrium to be stable]. We now assume that the numbers of $A_1$ and $A_2$ in the next generation are given by

$$\left.\begin{aligned} n_1' &= n_1[R_1 - f_1(N)] \\ n_2' &= n_2[R_2 - f_2(N)] \end{aligned}\right\} \quad [18]$$

in which $n_1$ and $n_2$ are the numbers of $A_1$ and $A_2$ in the present generation ($N = n_1 + n_2$). Let $\overline{R} = xR_1 + (1 - x)R_2$, $\overline{f} = xf_1(N) + (1 - x)f_2(N)$ and $\overline{W} = \overline{R} - \overline{f}$; then using the approximation $\Delta\overline{f} = (\partial\overline{f}/\partial N)\Delta N$, we have

$$\Delta\overline{W} = s\Delta x - \frac{\partial\overline{f}}{\partial N}\Delta N$$

$$= s\Delta x - \frac{\partial\overline{f}}{\partial N}\cdot N(\overline{W} - 1),$$

in which $s = R_1 - R_2 - [f_1(N) - f_2(N)]$ is the selection coefficient of $A_1$ over $A_2$, and $x = n_1/N$ stands for the frequency of $A_1$. If the allele $A_1$ increases gradually by natural selection ($s > 0$), the quasi-equilibrium will soon be attained so that

$$\Delta\overline{W} = \frac{s^2x(1 - x)}{\overline{W}} - \frac{\partial\overline{f}}{\partial N}\cdot N(\overline{W} - 1) \approx 0. \quad [19]$$

Under this condition, the change of $x$ in one generation is given by

$$\Delta x = sx(1 - x)/\overline{W}$$

with

$$\overline{W} \approx 1 + \frac{s^2x(1 - x)}{(\partial\overline{f}/\partial N)N}. \quad [20]$$

This includes the logarithmic and logistic controls as special cases. For the logistic case, $f_1(N) = c_1N$ and $f_2(N) = c_2N$, so that we have $s = R_1 - R_2 - (c_1 - c_2)N$ and $\overline{f} = xc_1N + (1 - x)c_2N$. Thus, $\partial\overline{f}/\partial N = xc_1 + (1 - x)c_2 = \overline{c}$. On the other hand, for the logarithmic case, $f_1(N) = c_1 \log_e N$ and $f_2(N) = c_2 \log_e N$, so that we have $s = R_1 - R_2 - (c_1 - c_2)\log_e N$ and $\overline{f} = xc_1 \log_e N + (1 - x)c_2 \log_e N$. Thus, $\partial\overline{f}/\partial N = \overline{c}/N$, and Eq. **20** reduces to $\overline{W} = 1 + (s^2/\overline{c})x(1 - x)$, as shown in the previous section.

I would like to thank Drs. Masatoshi Nei and Takeo Maruyama for stimulating discussions. Thanks are also due to Dr. J. F. Crow for reading the manuscript and giving valuable comments as well as correcting the English.

1. Crow, J. F. & Kimura, M. (1970) *An Introduction to Population Genetics Theory* (Harper and Row, New York).
2. Wright, S. (1969) *Evolution and the Genetics of Populations, Vol. 2. The Theory of Gene Frequencies* (The University of Chicago Press, Chicago, IL).
3. Fisher, R. A. (1930) *The Genetical Theory of Natural Selection* (Clarendon Press, Oxford).

PART THREE

# Population Structure

## Introduction

Wright (1931) pioneered the study of geographically structured populations with his island model. He assumed an infinite number of finite panmictic subpopulations, with migrants coming from any one of them with equal probability, regardless of distance. While still at the Genetics Institute in Japan, Kimura presented a model (no. 7) in which the migrants came from adjacent subpopulations. He also permitted occasional long-range migration, so Wright's island model is included as a special case.

Unknown to Kimura, Malécot, writing in obscure French journals, had been studying very similar models (see Nagylaki 1989a). Both he and Kimura provided for short- and long-range migration in a specified manner.

To study the correlation of genetic variation with geographic distance or isolation by distance (Wright 1943), the topology or distance between subpopulations must be taken into account, and the stepping-stone model, together with the continuous version (Malécot 1948, 1955, 1959; Felsenstein 1975), has become widely used for this purpose (see Nagylaki 1974a and Slatikin 1985 for reviews). A thorough mathematical analysis of the stepping-stone model under neutrality was made by Kimura and Weiss (no. 8) and Weiss and Kimura (1965). The extent of local differentiation is sensitively dependent on the number of dimensions (Malécot 1955; Cavalli-Sforza and Conterio 1960). Kimura and Weiss showed that the decay of genetic correlation with distance is slowest in the one-dimensional case and fastest in the three-dimensional case, as did Malécot (Nagylaki 1989a). (For a correction of errors in Weiss and Kimura 1965, see Nagylaki 1974b.)

Originally the stepping-stone model assumed infinite arrays of subpopulations, but it has been modified to apply also to a finite number of subpopulations by imposing some restrictions on the boundaries (Malécot 1951; Maruyama 1970; Fleming and Su 1974). An obvious difference between the original and modified models lies in the total number of individuals. The expected homozygosity in the entire population as well as the coefficient of kinship depends not only on the extent of gene flow and the size of each subpopulation but also on the number of subpopulations. High homozygosity and geographic uniformity of genetic polymorphisms, which Bulmer (1973) regarded as inconsistent with the neutral theory, actually become consistent if one assumes the stepping-stone model with a finite number of subpopulations and therefore a finite number of total individuals (Maruyama and Kimura 1974). Nagylaki (1983) and Crow and Aoki (1984) showed that, when migration is frequent, the finite island and stepping-stone models predict rather similar extents of genetic variability within and between subpopulations.

It has repeatedly been asked why genetic variability at a locus in diverse organisms is limited (e.g., Lewontin 1974): The heterozygosity, $H$, is less than 0.3 (usually 5–10%) and there is a positive correlation between $H$ and population size (Nei 1975; Nei and Graur 1984). If the neutral theory is correct, this means that the global effective population size ($N_e$) must be much smaller than $10^6$ or the reciprocal of mutation rate per locus per generation. Of various possibilities for the small value of $N_e$, local extinction and recolonization of subpopulations was emphasized by Wright (1940) and studied quantitatively by Slatkin (1977). The global effective size can be infinitely large with restricted gene flow, but it decreases to the number of breeding individuals in the entire population as gene flow becomes high (Nei and Takahata 1993). When local extinction and recolonization occur, $N_e$ may further decrease to the number of individuals within a subpopulation. Maruyama and Kimura (no. 9) showed that when this is frequent, $N_e$ is much smaller than the actual number of individuals in the whole population and there is little local differentiation. They applied their theoretical results to the genetic variation observed in *E. coli* and *D. willistoni* populations whose sizes were said to be enormous (e.g., Nei 1975).

## 2. *"Stepping-Stone" Model of Population.*

(Report by Motoo KIMURA)

It is recognized by many recent investigators that population structure is one of the most important factors in speciation. For theoretical studies on the problem of speciation, models of population structure are used. Wright has designed two different models, namely, the island model and the model of continuous distribution (cf. Ann. Eugen. 15, 1951). In the present the report writer proposes a new model of population structure which may be called the "stepping-stone" model.

In this model, a whole population is subdivided into many local subgroups within each of which mating occurs at random and exchange of individuals between the groups is allowed to occur only between adjacent ones. This may represent an intermediate situation between those of the two contrasting models stated above. Fig. 1 and Fig. 2. show respectively cases of linear and area distribution. Contrary to the case of the island model, there exists a high correlation between adjacent subgroups in the stepping-stone model, so that a considerable amount of random differentiation of gene frequencies may be expected among the subgroups even if the rate of migration is appreciably high. The following is a result obtained by the study on the linear distribution.

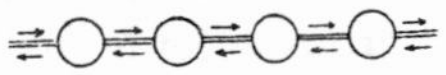

Fig. 1

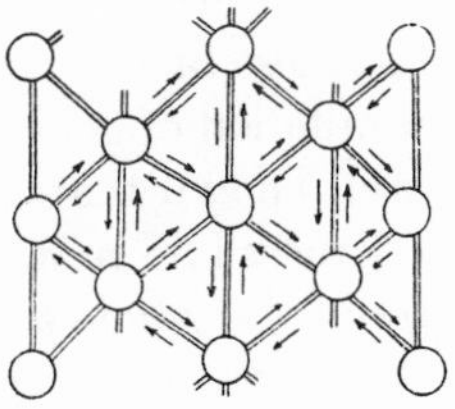

Fig. 2

If the number of breeding individuals in each subgroup is $N$ and the net rate of exchange of individuals between two adjacent subgroups is $\varepsilon$, then the distribution curve of gene frequency ($X$) among the subgroups will be given by

$$\phi(X)=CX^{8N\varepsilon(1-r)\bar{X}-1}(1-X)^{8N\varepsilon(1-r)(1-\bar{X})-1},$$

where $\bar{X}$ is a mean gene frequency in the whole population, and $C$ is a constant chosen such that $\int_0^1 \phi(X)dX=1$. In this formula $r$ represents a correlation coefficient between the gene frequencies of two adjacent subgroups and is a positive root of the quadratic equation;

$$\varepsilon r^2+(2-3\varepsilon)r+(4\varepsilon-2)=0.$$

Therefore if $\varepsilon$ is small, $r$ is approximately equal to $1-\varepsilon$.

In this case the variance of gene frequencies among subgroups becomes

$$V=\bar{X}(1-\bar{X})/[1+2N\varepsilon^2(4+\varepsilon)].$$

If $8N\varepsilon^2(1-\bar{X})$ and $8N\varepsilon^2\bar{X}$ are smaller than 1, a considerable amount of local differentiation due to random fixation of alleles will be expected. Since $2N\varepsilon$ is the number of immigrants coming from adjacent subgroups per generation, if $\bar{X}=0.5$, such conspicuous differentiation may be expected unless this number does not exceed $\sqrt{N}$. For example, if the size of the subgroup is 10,000, such a situation will be realized unless the number of the migrants is less than 100. On the other hand, in the island model the number of immigrants should be less than $\frac{1}{4}$, for the corresponding amount of differentiation to be realized.

In the stepping-stone model, if the long range dispersal of gametes is taken into consideration, it comprises the island model as a special case. Thus under certain circumstances the stepping-stone model may reflect the structure of a natural population more fully than the island model or the model of continuous distribution.

# THE STEPPING STONE MODEL OF POPULATION STRUCTURE AND THE DECREASE OF GENETIC CORRELATION WITH DISTANCE[1]

MOTOO KIMURA[2] and GEORGE H. WEISS[3]

*c/o Department of Genetics, L.I.G.B. Pavia Section, University of Pavia, Italy, and Institute of Fluid Dynamics and Applied Mathematics, University of Maryland, College Park*

WHEN a species occupies a very large territory, local differentiation is usually noticeable in the form of geographical races. Each race may in turn consist of numerous colonies which are differentiated to a less noticeable extent. The underlying differentiation in genetic constitution may reflect the local differences of selective pattern or may be the results of chance occurrence of different mutant genes, but these factors cannot act effectively unless some sort of isolation ensures the accumulation of genetic differences.

It is well known that existence of geographical barriers greatly favors the formation of races and new species. However, even if such barriers do not exist, the large size of the whole area as compared with the migration distance of an individual may prevent the species from forming a single panmictic unit, and this will produce a sort of isolation which WRIGHT called "isolation by distance" (WRIGHT 1943). He proposed a model of population structure in which a population is distributed uniformly over a large territory, but the parents of any given individual are drawn from a small surrounding region. He studied, by his method of path coefficients, the pattern of change in the inbreeding coefficient of subgroups relative to a larger population in which they are contained (WRIGHT 1940, 1943, 1946, 1951). The problem of local differentiation may also be studied in terms of change in correlation with distance as considered by MALÉCOT (1948, 1955, 1959); individuals living nearby tend to be more alike than those living far apart. In the mathematical theory of population genetics, the problem of local differentiation of gene frequencies in a structured population is one of the most intricate, and so far the main results are due to these two authors.

In natural populations, individuals often are distributed more or less discontinuously to form numerous colonies, and individuals may be exchanged between adjacent or nearby colonies. To analyze such a situation, one of us proposed a model which he termed "stepping stone model" of population structure (KIMURA

[1] Sponsored in part by the Mathematics Research Center, United States Army, Madison, Wisconsin under Contract No. DA-11-022-ORD-2059. This paper constitutes paper No. 480 of the National Institute of Genetics, Mishima-shi, Japan.

[2] On leave from the National Institute of Genetics, Mishima-shi, Japan.

[3] Present address: The Rockefeller Institute, New York 21, N.Y.

Genetics **49**: 561–576 April 1964.

1953). The purpose of the present paper is to present a solution of this model and to discuss its biological implications.

## ONE DIMENSIONAL CASE

Let us consider an infinite array of colonies with their position represented by integers on a line (Figure 1). The simplest situation for this one dimensional stepping stone model is that in each generation an individual can migrate at most "one step" in either direction between colonies. In other words, exchange of individuals is restricted to be between adjacent colonies.

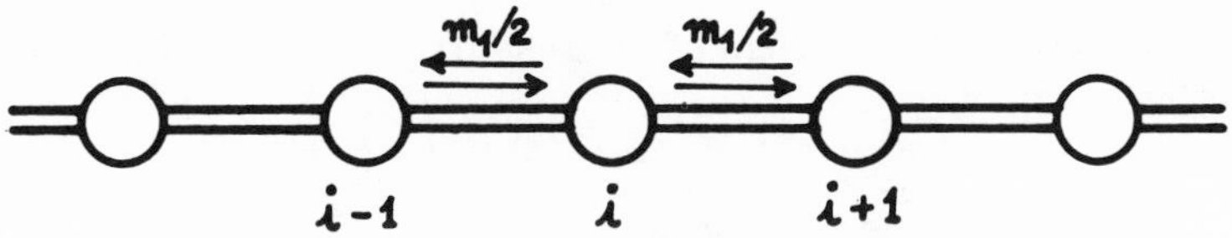


FIGURE 1.—One dimensional stepping stone model.

Consider a single locus with a pair of alleles $A$ and $A'$. The frequency of $A$ in each colony may change from generation to generation and to simplify the treatment we will assume discrete generation time. Also we assume that the gene frequency changes systematically by linear evolutionary pressures (mutation, migration but not selection, which will be considered later) and fortuitously by random sampling of gametes (small population number). If we denote by $p_i$ the (relative) frequency of $A$ in the $i$th colony in the present generation, then its value in the next generation may be given by

$$p'_i = (1 - m_1 - m_\infty)\, p_i + \frac{m}{2}\,(p_{i-1} + p_{i+1}) + m_\infty \bar{p} + \xi_i \tag{1.1}$$

In the above expression, $m_1$ stands for the rate of migration per generation to neighboring colonies such that $m_1/2$ is the proportion of individuals exchanged each generation between a pair of adjacent colonies. Also $m_\infty$ stands for the rate of long range dispersal per generation, namely the rate by which a colony exchanges individuals in each generation with a random sample taken from the entire population in which the frequency of $A$ is $\bar{p}$. The effect of the long range dispersal is formally equivalent to mutation; if there is mutation between $A$ and $A'$ in addition to long range dispersal of gametes, then $m_\infty$ should be replaced by $\mu + v + m_\infty$ and $m_\infty\bar{p}$ by $v + m_\infty\bar{p}$, where $\mu$ is the mutation rate from $A$ to $A'$ and $v$ is the mutation rate in the reverse direction. Furthermore, $\xi_i$ stands for the change in $p_i$ due to random sampling of gametes in reproduction, namely, due to the relatively small number of gametes being randomly chosen to form the next generation out of the very large number of gametes produced by the parents. Thus, if $N_e$ is the effective size (cf. WRIGHT 1940; KIMURA and CROW 1963) of the colony, $\xi_i$ follows the binomial distribution with mean and variance given by

$$E_\delta(\xi_i) = 0 \text{ and } E_\delta(\xi_i^2) = \frac{p_i(1 - p_i)}{2N_e}$$

where $E_\delta$ stands for an operator of taking expectation with respect to this random change.

Let us denote by $\tilde{p}_i$ the deviation of gene frequency from its mean, i.e.

$$\tilde{p}_i = p_i - \bar{p},$$

then (1.1) becomes

$$\tilde{p}'_i = \alpha\tilde{p}_i + \beta(\tilde{p}_{i-1} + \tilde{p}_{i+1}) + \xi_i, \tag{1.2}$$

where $\alpha = 1 - m_1 - m_\infty$ and $\beta = m_1/2$.

We will denote by $V_p$ the variance in the probability distribution of the gene frequency among colonies,

$$V_p = E_\phi(\tilde{p}_i^2),$$

and by $r_k$ the correlation coefficient of the gene frequencies between two colonies which are $k$ steps apart,

$$r_k = E_\phi(\tilde{p}_i\tilde{p}_{i+k})/V_p,$$

where $E_\phi$ stands for an operator for taking expectation with respect to gene frequency distribution among colonies.

In order, to obtain the formula for the variance, we square both sides of (1.2) and take expectations. Noting that the product terms between $\tilde{p}$'s and $\xi_i$ have expectation zero and that

$$E(\xi_i^2) = E_\phi\{E_\delta(\xi_i^2)\} = E_\phi\left\{\frac{p_i(1-p_i)}{2N_e}\right\}$$

$$= \{\bar{p} - (V_p + \bar{p}^2)\}/(2N_e),$$

we obtain

$$V_p' = \alpha^2 V_p + 4\alpha\beta V_p r_1 + 2\beta^2 V_p(1 + r_2) - \frac{V_p}{2N_e} + \frac{\bar{p}(1-\bar{p})}{2N_e}$$

where the prime indicates that it is the value in the next generation. At equilibrium in which $V'_p = V_p$, the above reduces to

$$V_p\left\{1 - \alpha^2 - 4\alpha\beta r_1 - 2\beta^2(1 + r_2) + \frac{1}{2N_e}\right\} = \frac{\bar{p}(1-\bar{p})}{2N_e} \tag{1.3}$$

In order to obtain the formula for the correlation coefficients, we will consider the expectation of product $\tilde{p}'_i\,\tilde{p}'_{i+k}$ $(k \neq 0)$. Noting that terms like $\tilde{p}_{i+k}\xi_i$ as well as $\xi_i\xi_{i+k}$ have zero expectations, we obtain

$$r'_k = \alpha^2 r_k + 2\alpha\beta(r_{k+1} + r_{k-1}) + \beta^2(r_{k+2} + 2r_k + r_{k-2})$$

At equilibrium in which $r'_k = r_k$, this reduces to

$$(\alpha^2 + 2\beta^2 - 1)\,r_k + 2\alpha\beta(r_{k+1} + r_{k-1}) + \beta^2(r_{k+2} + r_{k-2}) = 0. \qquad (k \neq 0) \tag{1.4}$$

Equation (1.4) holds for $k \geqq 1$. However, for $k = 1$, $r_{-1}$ should be replaced by $r_1$ to give

$$(\alpha^2 + 2\beta^2 - 1)\,r_1 + 2\alpha\beta(r_2 + 1) + \beta^2(r_3 + r_1) = 0. \tag{1.5}$$

In order to solve (1.4), let $r_k = \lambda^k$ and substitute in (1.4). This leads to a 4th order equation in $\lambda$ with the following four roots;

$$(1.6)\quad \begin{cases} \lambda_1 = \dfrac{1}{2\beta}\left\{(1-\alpha)+\sqrt{(1-\alpha)^2-(2\beta)^2}\right\} \\ \lambda_2 = \dfrac{1}{2\beta}\left\{(1-\alpha)-\sqrt{(1-\alpha)^2-(2\beta)^2}\right\} \\ \lambda_3 = \dfrac{-1}{2\beta}\left\{(1+\alpha)+\sqrt{(1+\alpha)^2-(2\beta)^2}\right\} \\ \lambda_4 = \dfrac{-1}{2\beta}\left\{(1+\alpha)-\sqrt{(1+\alpha)^2-(2\beta)^2}\right\} \end{cases}$$

where $\lambda_1 > 1$, $1 > \lambda_2 > 0$, $\lambda_3 < -1$, and $-1 < \lambda_4 < 0$.

The required solution should then be expressed in the form

$$(1.7)\qquad r_k = \sum_{i=1}^{4} C_i \lambda_i^k$$

where the $C_i$'s are constants, which may be determined through the following consideration: First, in order that $r_k$ vanishes at $k=\infty$, we must have $C_1 = C_3 = 0$, since both $\lambda_1$ and $\lambda_3$ are larger than unity in absolute value. Secondedly, $r_o = 1$ by definition and also $r_k$ has to satisfy relation (1.5). From these requirements, $C_2$ and $C_4$ are determined;

$$(1.8)\qquad C_2 = \frac{R_1}{R_1+R_2} \text{ and } C_4 = \frac{R_2}{R_1+R_2},$$

where

$$R_1 = \sqrt{(1+\alpha)^2-(2\beta)^2}$$

and

$$R_2 = \sqrt{(1-\alpha)^2-(2\beta)^2},$$

in which $\alpha = 1 - m_1 - m_\infty$ and $2\beta = m_1$.

Therefore, writing $r(k)$ instead of $r_k$, we obtain

$$(1.9)\qquad r(k) = C_2\lambda_2^k + C_4\lambda_4^k$$

as the correlation of gene frequencies between two colonies which are $k$ steps apart, with $\lambda_2$ and $\lambda_4$ given by (1.6) and $C_2$ and $C_4$ given by (1.8).

Substituting the values of $r_1$ and $r_2$ into (1.3), we get

$$(1.10)\qquad V_p = \frac{\bar{p}(1-\bar{p})}{1+2N_eC_0}$$

where $C_0 = 2R_1R_2/(R_1+R_2)$.

In the special case $m_1 = 0$, the above formula reduces to

$$(1.11)\qquad V_p = \frac{\bar{p}(1-\bar{p})}{1+2N_e(2m_\infty - m^2_\infty)}.$$

This case should correspond to WRIGHTS "island model" (WRIGHT 1943) and indeed (1.11) agrees with his formula except for the negligible term $m^2_\infty$ in the denominator.

For our purpose, however, a really interesting case is one in which $m_1 \gg m_\infty$. In this case, we have approximately

$$R_1 = 2\sqrt{1 - m_1}, \qquad R_2 = \sqrt{2m_1 m_\infty}$$

and (1.10) reduces to

$$V_p = \frac{\bar{p}(1-\bar{p})}{1 + 4N_e\sqrt{2m_1m_\infty}} \tag{1.12}$$

Furthermore, (1.9) may be approximated by

$$r(k) = e^{-\sqrt{\frac{2m_\infty}{m_1}}\,k} \tag{1.13}$$

with sufficient accuracy.

## GENERAL TREATMENT INCLUDING TWO AND THREE DIMENSIONS

Since the detailed account of the general treatment will be published elsewhere (Weiss and Kimura 1964), only the main results will be presented here.

In the one dimensional case, the correlation of gene frequencies between colonies which are $k$ steps apart may be given by

$$r(k) = \frac{C_0}{2\pi}\int_0^{2\pi}\frac{\cos k\theta\, d\theta}{1 - H^2(\cos\theta)} \tag{2.1}$$

where

$$C_0^{-1} = \frac{1}{2\pi}\int_0^{2\pi}\frac{d\theta}{1 - H^2(\cos\theta)} \tag{2.2}$$

and

$$H(\cos\theta) = 1 - m_\infty - m_1(1-\cos\theta). \tag{2.3}$$

Then the variance of gene frequencies between colonies is expressed in the form:

$$V_p = \frac{\bar{p}(1-\bar{p})}{1+2N_eC_0}. \tag{2.4}$$

It is possible to show that the above results agree with those given in the previous section; formula (2.1) and (2.4) respectively reduce to (1.9) and (1.10).

For numerical calculations, the following expressions turn out to be useful:

$$r(k) = \frac{A_1(k) + A_2(k)}{A_1(0) + A_2(0)} \tag{2.5}$$

$$C_0^{-1} = A_1(0) + A_2(0), \tag{2.6}$$

where

$$A_1(k) = \frac{1}{2}\cdot\frac{1}{2\pi}\int_0^{2\pi}\frac{\cos k\theta\ d\theta}{1 - H(\cos\theta)} = \frac{1}{4\pi}\int_0^{2\pi}\frac{\cos k\theta\ d\theta}{m_\infty + m_1(1-\cos\theta)} \tag{2.7}$$

and

$$A_2(k) = \frac{1}{2}\cdot\frac{1}{2\pi}\int_0^{2\pi}\frac{\cos k\theta\ d\theta}{1 + H(\cos\theta)} = \frac{1}{4\pi}\int_0^{2\pi}\frac{\cos k\theta\ d\theta}{2 - m_\infty - m_1(1-\cos\theta)} \tag{2.8}$$

For small values of $m_1$ and $m_\infty$ which we are interested in, the contribution of $A_2(k)$ is negligible in comparison with that of $A_1(k)$.

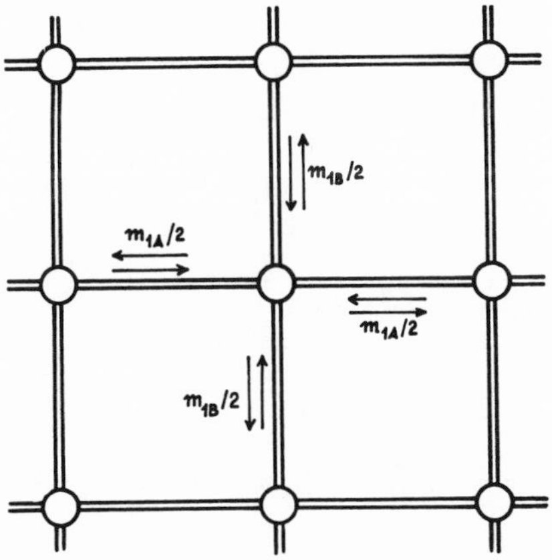


FIGURE 2.—Two dimensional stepping stone model.

In the two dimensional case (Figure 2), we assume that the entire population consists of a rectangular array of colonies, each of which occupies a point denoted by a pair of integers $(k_1,k_2)$. We will also assume that in each generation, an individual colony exchanges migrants with four surrounding colonies, but the effective population number in each colony remains the same $(N_e)$. The rate of migration may be different in $X$ and $Y$ directions: let $m_{1A}$ be the rate (taking one generation as a unit) of migration along the $X$-axis or horizontal direction, such that $m_{1A}/2$ is the proportion of individuals exchanged between a pair of adjacent colonies in this direction. Similarly let $m_{1B}$ be the rate of migration per generation along the $Y$-axis. The proportion of individuals which migrates to four neighboring colonies per generation is $m_1 = m_{1A} + m_{1B}$.

We will denote by $m_\infty$ the rate of long range dispersal as defined in the one dimensional case.

At equilibrium, the correlation of gene frequencies between colonies which are $k_1$ steps apart in the $X$ direction and $k_2$ steps apart in the $Y$ direction may be written

$$r(k_1, k_2) = \frac{A_1(k_1, k_2) + A_2(k_1, k_2)}{A_1(0,0) + A_2(0,0)} \tag{2.9}$$

where

$$A_1(k_1, k_2) = \frac{1}{2} \cdot \frac{1}{(2\pi)^2} \int_0^{2\pi} \int_0^{2\pi} \frac{\cos k_1\,\theta_1 \, \cos k_2\,\theta_2 \; d\,\theta_1 \; d\,\theta_2}{m_\infty + m_{1A}(1-\cos\theta_1) + m_{1B}(1-\cos\theta_2)} \tag{2.10}$$

and

$$A_2(k_1, k_2) = \frac{1}{2} \cdot \frac{1}{(2\pi)^2} \int_0^{2\pi} \int_0^{2\pi} \frac{\cos k_1\,\theta_1 \, \cos k_2\,\theta_2 \; d\,\theta_1 \; d\,\theta_2}{2 - m_\infty - m_{1A}(1-\cos\theta_1) - m_{1B}(1-\cos\theta_2)} \tag{2.11}$$

The variance of gene frequencies between colonies is given by

$$V_p = \frac{\bar{p}(1-\bar{p})}{1 + 2N_e C_0} \tag{2.12}$$

where $N_e$ is the effective population number of a colony and

(2.13) $$C_0^{-1} = A_1(0,0) + A_2(0,0).$$

The last formula can be expressed in terms of a complete elliptic integral:

(2.14) $$C_0^{-1} = \frac{1}{2\pi\sqrt{m_{1A}\, m_{1B}}} \left\{ \frac{1}{M_1} K\left(\frac{1}{M_1}\right) + \frac{1}{M_2} K\left(\frac{1}{M_2}\right) \right\}$$

where

$$M_1 = \sqrt{\left(1 + \frac{m_\infty}{2m_{1A}}\right)\left(1 + \frac{m_\infty}{2m_{1B}}\right)},$$

$$M_2 = \sqrt{\left(1 - \frac{2-m_\infty}{2m_{1A}}\right)\left(1 - \frac{2-m_\infty}{2m_{1B}}\right)}$$

and $K\ (\cdot)$ stands for the complete elliptic integral of the first kind defined by

(2.15) $$K(k) = \int_0^{\pi/2} \frac{d\theta}{\sqrt{1 - k^2 \sin^2\theta}}.$$

The following approximations are useful to evaluate $C_0^{-1}$ when $m_\infty$ is much smaller than $m_{1A}$ and $m_{1B}$, and also $m_{1A}$ and $m_{1B}$ themselves are small:

(2.16) $$K(1-\varepsilon) \approx \log_e \frac{4}{\sqrt{2\varepsilon}} \qquad (0 < \varepsilon \ll 1)$$

(2.17) $$K(\varepsilon) \approx \frac{\pi}{2}\left(1 + \frac{\varepsilon^2}{4}\right)$$

In order to obtain an accurate figure for the correlation from (2.9) for given values of $k_1$ and $k_2$, numerical integration has to be employed to evaluate (2.10) and (2.11). However, if the distance between colonies is large, simpler expressions are available. Namely, if we put

(2.18) $$\zeta = \left(\frac{k_1^2}{m_{1A}} + \frac{k_2^2}{m_{1B}}\right)^{1/2}$$

then, we obtain as an approximation for large $\zeta$

(2.19) $$A_1(k_1, k_2) = \frac{1}{2\pi\sqrt{m_{1A}\, m_{1B}}} K_0(\sqrt{2m_\infty}\, \zeta)$$

and

(2.20) $$A_2(k_1, k_2) = \frac{(-1)^{k_1+k_2}}{2\pi\sqrt{m_{1A}\, m_{1B}}} K_0(\sqrt{2(2-m_\infty-2\,(m_{1A}+m_{1B}))}\,\zeta)$$

where $K_0(\cdot)$ is the modified Bessel function of zeroth order.

If $m_\infty$ is much smaller than $m_{1A}$ and $m_{1B}$, which are themselves small as compared with unity, $A_2(k_1, k_2)$ is neglible as compared with $A_1(k_1, k_2)$. Then we get

(2.21) $$r(k_1, k_2) = \frac{C_0}{2\pi\sqrt{m_{1A}\, m_{1B}}} K_0(\sqrt{2m_\infty}\, \zeta)$$

Furthermore, if the rate of migration is equal in $X$ and $Y$ directions, so that $m_{1A} = m_{1B} = m_1/2$, the above formula reduces to

(2.22) $$r\ (\rho) = \frac{C_0}{\pi m_1} K_0\left(\sqrt{\frac{4m_\infty}{m_1}}\, \rho\right),$$

where $\rho = \sqrt{k_1^2 + k_2^2}$ is the distance between two colonies, and $m_1 = m_{1A} + m_{1B}$. Since asymptotically

$$K_0(z) = \sqrt{\frac{\pi}{2z}}\, e^{-z}, \qquad (z \to \infty),$$

the correlation at long distance is proportional to

$$e^{-\sqrt{\frac{4m_\infty}{m_1}}\,\rho} \Big/ \sqrt{\rho}\,,$$

which shows that the correlation falls off more quickly than in the one dimensional case.

The variance in gene frequency between colonies is

$$V_p = \frac{\bar{p}\,(1-\bar{p})}{1 + 2N_e C_0} \tag{2.23}$$

where $C_0$ is given by (2.14).

We will now turn to the three dimensional case. Let us assume that the cubic array of colonies extends to infinity in all directions. In the present case, each colony has six adjacent colonies to exchange individuals in each generation. The rates of migration in three perpendicular directions may be different and we will denote by $m_{1A}$, $m_{1B}$ and $m_{1C}$ the respective rates per generation in the $X$, $Y$ and $Z$ directions. Thus the amount of individuals exchanged per generation between two adjacent colonies parallel to the $X$ direction is $m_{1A}/2$, between those parallel to the $Y$ direction is $m_{1B}/2$, etc. As before $m_\infty$ stands for the rate of long range dispersal per generation. The position of each colony may be designated by a triplet of integers $(k_1, k_2, k_3)$, and we assume that each colony has the effective population number of $N_e$, which is constant in each generation.

The correlation of gene frequencies at equilibrium between two colonies which are respectively $k_1$, $k_2$ and $k_3$ steps apart in the $X$, $Y$ and $Z$ directions is

$$r\,(k_1, k_2, k_3) = \frac{A_1\,(k_1, k_2, k_3) + A_2\,(k_1, k_2, k_3)}{A_1\,(0, 0, 0,) + A_2\,(0, 0, 0)} \tag{2.24}$$

where

$$A_1(k_1,k_2,k_3) = \frac{1}{2\pi^3}\int_0^\pi\!\int_0^\pi\!\int_0^\pi \frac{\cos k_1\theta_1 \cos k_2\theta_2 \cos k_3\theta_3 d\theta_1 d\theta_2 d\theta_3}{m_\infty + m_{1A}(1-\cos\theta_1) + m_{1B}(1-\cos\theta_2) - m_{1C}(1-\cos\theta_3)} \tag{2.25}$$

and

(2.26) $A_2(k_1,k_2,k_3) =$

$$\frac{1}{2\pi^3}\int_0^\pi\!\int_0^\pi\!\int_0^\pi \frac{\cos k_1\theta_1 \cos k_2\theta_2 \cos k_3\theta_3\, d\theta_1\, d\theta_2\, d\theta_3}{2 - m_\infty - m_{1A}(1-\cos\theta_1) - m_{1B}(1-\cos\theta_2) - m_{1C}(1-\cos\theta_3)}$$

Generally, numerical integration has to be employed to evaluate these integrals. Fortunately, however, tables are available for the important case of $m_{1A} = m_{1B}$. This represents an isotropic migration in a plane but a different amount of migration in the third dimension. In this case, the integrals can be expressed in terms of the Green's functions for monatomic simple cubic lattices which are defined by

$$I(a,b,c;\alpha;\beta) = \frac{1}{\pi^3}\int_0^\pi\!\int_0^\pi\!\int_0^\pi \frac{\cos ax \cos bx \cos cz\, dx\, dy\, dz}{(2+\alpha)\beta - \cos x - \cos y - \alpha \cos z} \tag{2.27}$$

This function is extensively tabulated by MARADUDIN et al. (1960). In terms of this function, $A$'s are expressed as follows:

(2.28) $$A_1(k_1,k_2,k_3) = \frac{1}{2m_{1A}} I\left(k_1,k_2,k_3; \frac{m_{1C}}{m_{1A}}; 1 + \frac{m_\infty}{2m_{1A}+m_{1C}}\right)$$

(2.29) $$A_2(k_1,k_2,k_3) = \frac{(-1)^{k_1+k_2+k_3}}{2m_{1A}} I\left(k_1,k_2,k_3; \frac{m_{1C}}{m_{1A}}; \frac{2-m_\infty}{2m_{1A}+m_{1C}} - 1\right)$$

The variance of gene frequencies between colonies is given by

(2.30) $$V_p = \frac{\bar{p}(1-\bar{p})}{1+2N_e C_0}$$

where

(2.31) $$C_0^{-1} = A_1(0,0,0) + A_2(0,0,0).$$

If

$$R^2 = k_1^2 + k_2^2 + \frac{m_{1A}}{m_{1C}} k_3^2$$

is much larger than unity, a simple approximation formula is available for the correlation:

(2.32) $$r(k_1,k_2,k_3) = \frac{C_0 e^{-\sqrt{\frac{2m_\infty}{m_{1A}}} R}}{4\pi \sqrt{m_{1A} m_{1C}}\, R}.$$

In addition, if $m_{1A} = m_{1C}$ i.e. migration is completely isotropic and if $m_\infty$ is much smaller than $m_{1A}$ and also $m_{1A}$ is at most of the order of a few percent, we have approximately

$$C_0 = 4m_{1A}.$$

In this case, the correlation of gene frequencies between two colonies which are distance $\rho$ apart may be given by the following simple approximation formula

(2.33) $$r(\rho) = \pi^{-1} e^{-\sqrt{\frac{6m_\infty}{m_1}} \rho} / \rho$$

where

$$\rho = (k_1^2 + k_2^2 + k_3^2)^{1/2}$$

and

$$m_1 = m_{1A} + m_{1B} + m_{1C}$$

Thus, at long distance, the correlation falls off more quickly in the three dimensional case than in the two dimensional one, which in turn falls off more rapidly than the one dimensional case.

In Figure 3, the relation between the correlation coefficient $r(\rho)$ and distance $\rho$ is plotted for the one, two and three dimensional cases, taking $m_\infty = 4 \times 10^{-5}$, $m_1 = 0.1$ and assuming complete isotropic migration, namely, $m_{1A} = m_{1B} = 0.05$ for two dimensions and $m_{1A} = m_{1B} = m_{1C} = 0.0333$ for three dimensions. It can be seen that a distinct difference exists between the three cases.

## DISCUSSION

(i) *More general forms of migration:* In the foregoing treatments, we have as-

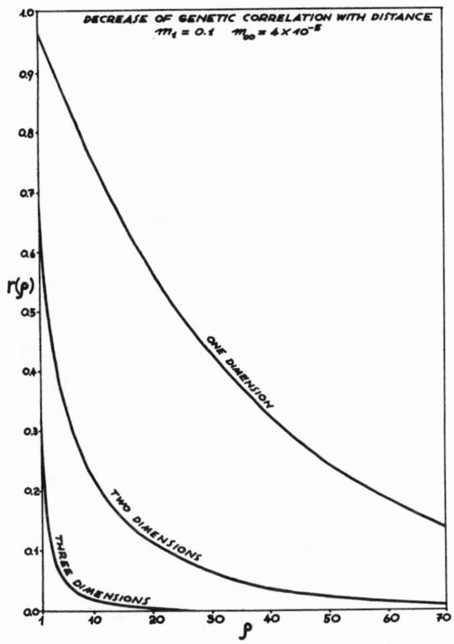


FIGURE 3.—Graphs showing the decrease of genetic correlation with distance for one, two and three dimensions taking $m_\infty = 4 \times 10^{-5}$, $m_1 = 0.1$ and assuming complete isotropic migration. Abscissa: distance between two colonies ($\rho$). Ordinate: correlation coefficient of gene frequencies between colonies, $r(\rho)$.

sumed that the migration is restricted to one step per generation. However, the results can be extended to the situation where there is more than one step migration per generation. For most purposes this can be done simply by substituting for $m_1$ the variance of migration distance per generation. For example, in the one dimensional case, if $m_j/2$ is the proportion of individuals exchanged per generation between two colonies which are $j$ steps apart, then

$$\sigma_m^2 = \sum_{j=1}^{\infty} j^2 m_j$$

should be replaced for $m_1$ in (1.13), to give

$$r(k) = e^{-\sqrt{\frac{2m_\infty}{\sigma_m}}\, k} \tag{3.1}$$

In the two dimensional case, let $m_{ij}/2$ be the proportion of individuals exchanged in each generation between two colonies, one at $(0,0)$ and the other at $(i,j)$. If we define by

$$\sigma_{mA}^2 = \sum_{i=0}^{\infty} \sum_{j=0}^{\infty} i^2 m_{ij}$$

and

$$\sigma_{mB}^2 = \sum_{i=0}^{\infty} \sum_{j=0}^{\infty} j^2 m_{ij},$$

the two variances, then the variance $\sigma^2_{mA}$ should be used instead of $m_{1A}$, and similarly the variance $\sigma^2_{mB}$ should be used instead of $m_{1B}$, to calculate the correlation coefficient and the variance.

More detailed discussion together with the mathematical justification will be given in WEISS and KIMURA (1964).

(ii) *Relationship between the stepping stone model and the "continuum model":* The latter model may be obtained as a limiting case of the former by letting the actual distance between the adjacent subgroups approach zero. The various results obtained for the stepping-stone model may easily be adapted to the continuum case by substituting $\sigma^2$ for $m_1$ and $(N_e/N)\delta$ for $N_e$, where $\sigma^2$ is the variance of the migration distance of individual per generation, $N_e/N$ is the ratio between the effective number and the actual number of individuals per subgroup and $\delta$ is the density of individuals or the number of inhabitants per unit size of the habitat (e.g. per unit area in the case of two dimensional habitat).

We may note here that the correlation of gene frequencies does not depend on the number of individuals or density, while the variance is clearly dependent on it. The latter, namely $V_p$, in the stepping stone model may be translated into the inbreeding coefficient of an individual in the continuum by the relation:

(3.2)
$$f_0 = \frac{V_p}{\bar{p}\,(1-\bar{p})}$$

This relation is derived easily by considering the expected frequency of heterozygotes, i.e.

$$E\,\{2p(1-p)\} = 2\bar{p}(1-\bar{p})(1-f_0)$$

which yields

$$2\,\{\bar{p} - (V_p + \bar{p}^2)\} = 2\bar{p}(1-\bar{p})(1-f_0)$$

or

$$V_p = \bar{p}(1-\bar{p})f_0$$

(iii) *Effect of selection:* In our original treatment, only mutation and migration are assumed as factors which cause systematic change in gene frequency. However, we can incorporate selection to the extent that its effect on the change of gene frequency can be expressed linearly. There are two important cases where this can be done as a good approximation.

Firstly, in a polymorphic locus, if deviation of gene frequency $p_i$ from its equilibrium frequency $\hat{p}$ is small, the amount of change in gene frequency per generation may be expressed in the form

(3.3)
$$-K(p_i - \hat{p})$$

where $K$ depends on the selection intensity and $\hat{p}$ may be substituted by $\bar{p}$. For example, if $A$ and $A'$ are heterotic and if $s_1$, and $s_2$ are respectively the selection coefficients against both homozygotes $AA$ and $A'A'$ as compared with heterozygote $AA'$, then we may take $K = s_1s_2/(s_1 + s_2)$ and $\hat{p} = s_2/(s_1 + s_2)$. Therefore expression (1.1) is unchanged, except that $m_\infty$ now include $K$. In such circumstance, the mutation rates may be negligible as compared with $K$ and if $m_\infty$ is also negligible, then we may take $m_\infty = K$ in all the subsequent formulae.

Secondly, if gene $A$ is unconditionally deleterious, kept in low frequency by the balance between recurrent mutation and selective elimination, then the selection coefficient against gene $A$ may be included in $m_\infty$. With no long range dispersal we may put $m_\infty = s$ and $m_\infty \bar{p} = v$ since $\mu + v$ may be neglected as compared with $s$. Here $s$ is equal to the reciprocal of the average number of generations through which a gene $A$ persists in a population. The incidence of mutant homozygotes among the offspring of marriages within a colony is $\Sigma(\bar{p}^2 + V_p)$, while that of marriages between two neighboring colonies is $\Sigma(\bar{p}^2 + V_p r(1))$ where the summation is over all relevant loci.

(iv) *Probability distribution of gene frequencies and the amount of random local differentiation:* In WRIGHT's island model, immigrants represent a random sample from the entire population and the probability distribution of gene frequency among colonies is given by

$$\phi(p) = \frac{\Gamma(4Nm)}{\Gamma(4Nm\bar{p})\,\Gamma(4Nm(1-\bar{p}))}\, p^{4Nm\bar{p}-1}\,(1-p)^{4Nm(1-\bar{p})-1} \tag{3.4}$$

where $m$ corresponds to our $m_\infty$. WRIGHT (1940) states that "if there is a correlation, $r$, between immigrants and receiving group, the $m$ of the formula must be replaced by $m(1-r)$ if $m$ is to continue to be the actual amount of replacement by immigration." Thus in our stepping stone model, we may take

$$m = m_\infty + m_1(1-r(1))$$

to obtain the approximate gene frequency distribution from (3.4). This means that in the typical case of $m_\infty \ll m_1$, $1-r(1)$ is an important quantity relating to the amount of local differentiation of gene frequencies due to random sampling of gametes. If $4N_e m_1(1-r(1))$ is less than $1/\bar{p}$ and $1/(1-\bar{p})$, the curve giving gene frequency distribution is $U$-shaped and strong local differentiation will be expected. More generally, the variance in gene frequency distribution may be given approximately by

$$V_p = \frac{\bar{p}(1-\bar{p})}{1 + 4Nm_1(1-r(1))} \tag{3.5}$$

if $m_\infty \ll m_1 \ll 1$.

In the one dimensional case, the exact value of $1 - r(1)$ may be obtained from

$$1 - r(1) = \frac{1}{4\alpha\beta}\left\{ R_1R_2 - 1 + (1-m_\infty)^2 \right\}. \tag{3.6}$$

If $m_\infty$ is much smaller than $m_1$, which in turn is much smaller than unity,

$$1 - r\,(1) = \sqrt{\frac{2m_\infty}{m_1}} \tag{3.7}$$

approximately. In the two dimensional case, assuming isotropy ($m_{1A} = m_{1B} = m_1/2$) the exact expression for $1-r(1)$ is

$$1-r(1,0) = \frac{1}{m_1}\left\{ \frac{(2-m_\infty)\,A_2\,(0,0) - m_\infty\,A_1\,(0,0)}{A_1\,(0,0) + A_2\,(0,0)} \right\} \tag{3.8}$$

If $m_\infty \ll m_1 \ll 1$, we have roughly

$$1 - r(1{,}0) = \frac{\pi}{2}\left(\log_e \frac{4}{\sqrt{\frac{2m_\infty}{m_1}}}\right)^{-1} \tag{3.9}$$

Again if we assume isotropy, the corresponding expression for the three dimensional case is

$$1-r(1{,}0{,}0) = \frac{1}{m_1}\left\{ \frac{(2-m_\infty)\, A_2\,(0{,}0{,}0)-m_\infty\, A_1\,(0{,}0{,}0)}{A_1\,(0{,}0{,}0) + A_2\,(0{,}0{,}0)} \right\} \tag{3.10}$$

where $m_1 = 3m_{1A} = 3m_{1B} = 3m_{1C}$. If in addition $m_\infty \ll m_1 \ll 1$, this quantity becomes almost independent of $m_\infty$ and $m_1$, and we have roughly

$$1 - r\,(1{,}0{,}0) = 2/3. \tag{3.11}$$

These results clearly show that the tendency toward local differentiation is very much dependent on the number of dimensions.

In the one dimensional case, $4N_e m_1(1-r(1))$ is approximately $4N_e\sqrt{2m_1 m_\infty}$ and it can easily be less than unity if $m_\infty$ is very small: with $m_1 = 0.1$ and $m_\infty = 2\times10^{-5}$, a considerable local differentiation will be expected if $N_e$ is less than 100. On the other hand, in two dimensions, a tendency toward local differentiation due to random sampling will generally be rather weak. For example, with $m_1 = 0.1$ and $m_\infty = 2\times10^{-5}$, $4N_e m_1(1-r(1{,}0))$ becomes less than unity only when $N_e$ is less then about nine. In the three dimensional case, the tendency will be still weaker, since $4N_e m_1(1-r(1))$ is approximately $8N_e m_1/3$. It is remarkable that this quantity is now independent of $m_\infty$. This means that however small the value of $m_\infty$ is, random differentiation is impossible in three dimension unless $m_1$ is very small.

WRIGHT (1940, 1943, 1946) studied the same type of problem for the continuum using an entirely different approach, and arrived at a similar conclusion with respect to one and two dimensions.

(v) *Decrease of genetic correlation with distance:* The fact that tendency toward random differentiation depends very much on the number of dimensions is also reflected in the way in which the correlation coefficient falls off with distance, especially when the distance is large ($\rho \to \infty$): In the one dimensional case, the correlation falls off exponentially with distance

$$r(\rho) \propto e^{-\sqrt{\frac{2m_\infty}{m_1}}\rho} \quad ; \tag{3.12}$$

in two dimensions, it falls off more rapidly, namely

$$r(\rho) \propto e^{-\sqrt{\frac{4m_\infty}{m_1}}\rho}\Big/\sqrt{\rho} \quad ; \tag{3.13}$$

and in three dimensions, it falls off still more rapidly

$$r(\rho) \propto e^{-\sqrt{\frac{6m_\infty}{m_1}}\rho} \Big/ \rho \tag{3.14}$$

MALÉCOT (1955) obtained a result, based on his elegant method of using an integro-differential equation for a continuum, that the coefficient of relationship as well as the correlation coefficient between gene frequencies decrease approximately in exponential form with distance independent of dimension. It is probable that our present results are more accurate than his result.

(vi) *Concept of dimension:* The one dimensional model can represent a population of organisms living along a river, coastal line or mountain ridge. The two dimensional model can represent a population on a plane and cover the most important cases in nature. CAVALLI-SFORZA and CONTERIO (1960) introduced "coefficient of dimensionality" to measure the pattern of geographical distribution of villages. It varies from 1 to 2. The intermediate case may be represented by the two dimensional stepping stone model with different migration rates in the $X$ and $Y$ directions. The three dimensional model can represent a population in an oceanic habitat and the treatment assuming equal rate of migration in horizontal ($X$ and $Y$) directions but different rate in a vertical ($Z$) direction will be useful.

The three dimensional model can also represent a population of organisms living on a plane, but there is a third dimension such as the social rank in which "migration" is restricted to the neighboring classes, if distance is well defined in this third dimension. The results presented in this paper can readily be extended to cover the cases of four and higher dimensions, but they do not appear to have an important application to natural populations.

(vii) *Variance of gene frequency within a restricted region:* In the present paper, $V_p$ represents the theoretical variance of gene frequencies between colonies in an infinitely large distribution range. On the other hand, actual observations cover only a restricted area. Therefore, in applying the present theory to actual data, it is necessary to derive, for each dimension, a formula for the variance of gene frequency within a restricted region. Here we will consider the one and two dimensional cases.

In the one dimensional case, the variance of gene frequency between colonies within a group of $n$ consecutive colonies may be given by

$$V_w(n) = V_p \left\{ 1 - \frac{1}{n} - \frac{2}{n^2} \sum_{i=1}^{n} (n-i)\, r(i) \right\}. \tag{3.15}$$

In the two dimensional case, if we consider a squared region of length $n_1$ in $X$ directions and length $n_2$ in $Y$ directions where $n_1 n_2$ colonies are contained, then the variance of gene frequency between colonies within such region is

$$V_w(n_1,n_2) = V_p \left[ 1 - \frac{1}{n_1 n_2} - \frac{2}{(n_1 n_2)^2} \left\{ \sum_{i=1}^{n_1} (n_1 - i) n_2 r(i,0) \right.\right. \tag{3.16}$$

$$+ \sum_{j=1}^{n_2} n_1 (n_2 - j)\, r(0,j)$$

$$\left.\left. + 2 \sum_{i=1}^{n_1} \sum_{j=1}^{n_2} (n_1 - i)(n_2 - j)\, r(i,j) \right\} \right] .$$

The main part of this work was done while the authors were visiting the University of Wisconsin and they express their thanks to PROFESSORS R. LANGER and J. F. CROW for kindly providing facilities and giving support during the course of this work. One of the authors (M.K.) also expresses his thanks to PROFESSOR L. L. CAVALLI-SFORZA for stimulting discussions and for help in the preparation of the manuscript. He also expresses his thanks to U. S. Atomic Energy Commission for partial support.

## SUMMARY

If the distance of individual migration is much smaller as compared with the entire distribution range of the species, the random local differentiation in gene frequency will be expected as shown by WRIGHT in his studies on "isolation by distance". In the present paper, the stepping stone model is used to study this phenomenon. The model assumes that the entire population is subdivided into colonies and the migration of individuals in each generation is restricted to nearby colonies.

The solution of this model is presented for one, two and three dimensional cases, with special reference to the correlation coefficient of gene frequencies between colonies. Also, the variance of gene frequencies between colonies is given for the three cases.

It has been shown that the decrease of genetic correlation with distance depends very much on the number of dimensions: In one dimension, the correlation decreases approximately exponentially with distance:

$$r(\rho) \propto e^{-A\rho}$$

where $\rho$ is the distance and $A$ is a constant which is equal to $\sqrt{2m_\infty/m_1}$. While in two dimensions, it falls off more rapidly and if the migration is isotropic in $X$ and $Y$ directions, we obtain, for a large value of $\rho$, the relation

$$r(\rho) \propto e^{-B\rho}/\sqrt{\rho},$$

where $B$ is a constant which is equal to $\sqrt{4m_\infty/m_1}$. In three dimensions, it falls off still more rapidly and asymptotically, we obtain

$$r(\rho) \propto e^{-C\rho}/\rho$$

where $C = \sqrt{6m_\infty/m_1}$.

The quantity $1-r(1)$, where $r(1)$ is the correlation coefficient between two adjacent colonies, is also pertinent in discussing the tendency toward random local differentiation. It has been shown that the relation of this quantity with mutation and migration rates depends very much on the number of dimensions. This, together with the above results on the decrease of correlation at a large distance, clearly indicates that the tendency toward random local differentiation is very much dependent on the number of dimensions; it is strongest in one dimension and becomes weaker as the number of dimension increases.

More general forms of migration and also some effects of selection are discussed.

## LITERATURE CITED

Cavalli-Sforza, L. L., and F. Conterio, 1960 Analisi della fluttuazione di frequenze geniche nella popolazione della Val Parma. Atti Ass. Genet. Ital. Pavia **5**: 333–343.

Kimura, M., 1953 "Stepping Stone" model of population. Ann. Rept. Nat. Inst. Genetics, Japan **3**: 62–63.

Kimura, M., and J. F. Crow, 1963 The measurement of effective population number. Evolution **17**: 279–288.

Malécot, G., 1948 *Les mathématiques de l'hérédité.* Masson, Paris. —— 1955 Remarks on decrease of relationship with distance, following paper by M. Kimura. Cold Spring Harbor Symp. Quant. Biol. **20**: 52–53. —— 1959 Le modèles stochastiques en génétique de population. Publ. Inst. Statist. Univ. Paris **8**: 173–210.

Maradudin, A. A., E. W. Montroll, G. H. Weiss, R. Herman, and H. W. Milnes, 1960 Green's functions for monatomic simple cubic lattices. Mémoires, Académie Royal de Belgique, Classe des Sciences **14**, Fascicule 7.

Weiss, G. and M. Kimura, 1964 A mathematical analysis of the stepping stone model of genetic correlation (in preparation).

Wright, S., 1940 Breeding structure of populations in relation to speciation. Am. Naturalist **74**: 232–248. —— 1943 Isolation by distance. Genetics **28**: 114–138. —— 1946 Isolation by distance under diverse systems of mating. Genetics **31**: 39–59. —— 1951 The genetical structure of populations. Ann. Eugenics **15**: 323–354.

# Genetic variability and effective population size when local extinction and recolonization of subpopulations are frequent

(population genetics/protein polymorphism/periodic selection/neutral mutation-random drift hypothesis)

TAKEO MARUYAMA AND MOTOO KIMURA

National Institute of Genetics, Mishima, 411 Japan

*Contributed by Motoo Kimura, July 17, 1980*

**ABSTRACT** If a population (species) consists of $n$ haploid lines (subpopulations) which reproduce asexually and each of which is subject to random extinction and subsequent replacement, it is shown that, at equilibrium in which mutational production of new alleles and their random extinction balance each other, the genetic diversity (1 minus the sum of squares of allelic frequencies) is given by $2N_e v/(1 + 2N_e v)$, where

$$N_e = \tilde{N} + n/(2\lambda) + n\tilde{N}v/\lambda,$$

in which $\tilde{N}$ is the harmonic mean of the population size per line, $n$ is the number of lines (assumed to be large), $\lambda$ is the rate of line extinction, and $v$ is the mutation rate (assuming the infinite neutral allele model). In a diploid population (species) consisting of $n$ colonies, if migration takes place between colonies at the rate $m$ (the island model) in addition to extinction and recolonization of colonies, it is shown that effective population size is

$$N_e = \tilde{N} + n/[4(v + \lambda + m)] + n\tilde{N}(v + m)/(v + \lambda + m).$$

If the rate of colony extinction ($\lambda$) is much larger than the migration rate of individuals, the effective population size is greatly reduced compared with the case in which no colony extinctions occur (in which case $N_e = n\tilde{N}$). The stepping-stone type of recolonization scheme is also considered. Bearing of these results on the interpretation of the level of genetic variability at the enzyme level observed in natural populations is discussed from the standpoint of the neutral mutation-random drift hypothesis.

The concept of effective population size, introduced by Wright (1), has played a fundamental role in treating the process of random gene frequency drift in finite populations. Useful formulae have been derived by him and others (2–6) to compute the effective sizes for various situations such as unequal numbers of males and females, different parents contributing widely different numbers of young, the population size fluctuating from time to time, and overlapping generations (see refs. 7–9 for reviews).

It is known that most species in nature have subdivided population structure, and extinction and recolonization of local populations may occur rather frequently in some groups of organisms such as insects (see refs. 10 and 11). This will greatly reduce the effective population size of the species.

Wright (12) pointed out that if local populations are liable to frequent extinction with restoration from the progeny of a few stray immigrants, the species may pass repeatedly through extremely reduced state of effective population size even though the species include at all times "countless millions of individuals in its range as a whole." He suggested that mutations such as reciprocal translocations that are very strongly selected against until half-fixed may require some such mechanism to become established.

A pioneering study of the effect of local extinction and recolonization of subpopulations on genetic variability of the species was made by Slatkin (11) using two models termed "propagule pool" and "migration pool." By pursuing the same problem further, we obtained some results which we report in this paper. Our model is similar to Slatkin's propagule pool model, but we use both "island model" (1) and "stepping-stone model" (13) of population structure. We shall also discuss some implication of the results for our interpretation of the amount of genetic variability observed in natural populations in the light of the neutral mutation-random drift hypothesis [the neutral theory, for short (14, 15)].

## Random replacement of haploid lines (island model)

Let us consider a population (species) of haploid organisms consisting of a finite number, $n$, of subpopulations which we refer to as lines. All the results in this section, however, hold equally well if the term "line" is replaced by "local colony" or "deme." Let us assume that each line is subject to extinction with rate $\lambda$ and that whenever a line is extinct it is immediately replaced by a line derived from individuals chosen from a single line in the population. In this section, we assume that, whenever extinction occurs, every existing line has an equal chance of becoming a donor, so that the geographical distance is irrelevant in the replacement (island model). We present the treatment using the stepping-stone model (where donors are chosen from neighboring colonies) in the next section. Throughout this paper, we assume the infinite allele model of Kimura and Crow (16)—that is, we assume that whenever a mutation occurs at a locus it represents a new, not a preexisting, allele.

Let $P_t$ be the probability that two randomly chosen homologous genes from a single line at time $t$ are identical by descent and therefore identical in allelic states. Similarly, let $Q_t$ be the probability of two randomly chosen genes, one each from two different lines at time $t$, being identical. Consider the species as a whole and ask what is the probability that two randomly chosen lines have descended from a single line a short time ($\Delta t$) ago. Since the probability of extinction of a line during $\Delta t$ is $\lambda\Delta t$, and since there are $n$ lines as a whole, the probability of two randomly chosen lines at time $t$ having descended from a single line $\Delta t$ ago is $2\lambda\Delta t/(n - 1)$. The factor 2 in the numerator comes from the consideration that, when two lines (tentatively called the first and the second lines) happen to have descended from a single colony $\Delta t$ ago, there are two possibilities—that is, either the first line is the donor (case $A$ in Fig. 1) or the first line is the recipient (case $B$ in Fig. 1). The probability that two lines at time $t + \Delta t$ have descended from two lines which are separate at time $t$ is therefore $1 - 2\lambda\Delta t/(n - 1)$.

The publication costs of this article were defrayed in part by page charge payment. This article must therefore be hereby marked "*advertisement*" in accordance with 18 U. S. C. §1734 solely to indicate this fact.

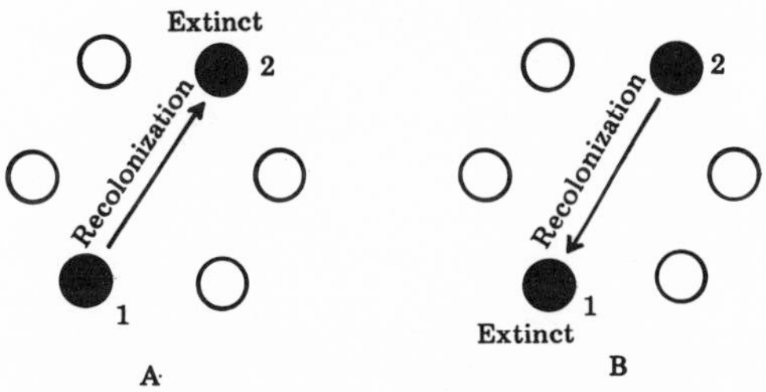


FIG. 1. Diagram illustrating extinction and recolonization for the case $n = 6$.

Let $v$ be the mutation rate per haploid individual per generation, then we have

$$Q_{t+\Delta t} = (1 - v\Delta t)^2 \left\{\left(1 - \frac{2\lambda\Delta t}{n-1}\right) Q_t + \frac{2\lambda\Delta t}{n-1} P_t + o(\Delta t)\right\} \quad [1]$$

where $o(\Delta t)$ stands for terms of higher order than $\Delta t$. Note that $(1 - v\Delta t)^2$, or $(1 - 2v\Delta t)$ if we neglect $o(\Delta t)$, represents the probability that no mutations occur during the time interval $\Delta t$ in two lines. An important assumption made in deriving Eq. **1** is that there is no exchange of individuals between lines. In other words, there is no exchange of genetic material (recombination) between asexual lines. If "lines" represent local colonies, within which random matings occur, this means no migration between colonies.

At equilibrium in which the identity coefficients remain constant with time so that $Q_{t+\Delta t} = Q_t$ and $P_{t+\Delta t} = P_t$, writing the equilibrium values of $P_t$ and $Q_t$ as $P$ and $Q$, we get from Eq. **1**,

$$-\left(2v + \frac{2\lambda}{n-1}\right) Q + \frac{2\lambda}{n-1} P = 0, \quad [2]$$

where we neglect terms of higher order than $\Delta t$. Thus, we obtain

$$Q/P = \frac{1}{1 + (n-1)v/\lambda}. \quad [3]$$

An analogous formula giving the relationship between two identity coefficients—that is, identity coefficients of gene members between and within chromosomes—was obtained by Ohta (17) in her theoretical study on genetic variation in a multigene family.

In order to derive a formula for $P_t$, let $N_t$ (>0) be the effective population size of a single line at time $t$. Consider random sampling of individuals within the line, still assuming no immigration of individuals from other lines. Then, we have

$$P_{t+\Delta t} = (1 - 2v\Delta t)\left\{\left(1 - \frac{\Delta t}{N_t}\right) P_t + \frac{\Delta t}{N_t}\right\}. \quad [4]$$

This is a continuous generation analogue of a more familiar expression

$$P_{t+1} = (1 - 2v)\left\{\left(1 - \frac{1}{N_t}\right) P_t + \frac{1}{N_t}\right\}, \quad [5]$$

which holds when generations are discrete [i.e., $t = 0, 1, 2, \ldots$ (see ref. 7, page 323)]. From Eq. **4**, neglecting terms involving $(\Delta t)^2$ and letting $\Delta P_t = P_{t+\Delta t} - P_t$, and substituting $dP_t/dt$ for $\Delta P_t/\Delta t$, we obtain

$$\frac{dP_t}{dt} = -\left(2v + \frac{1}{N_t}\right) P_t + \frac{1}{N_t}. \quad [6]$$

Let $h_t = 1 - P_t$; then $h_t$ gives the probability that two randomly chosen homologous genes within a line are distinct in allelic states. From Eq. **6**, we have

$$\frac{dh_t}{dt} = 2v - \left(2v + \frac{1}{N_t}\right) h_t. \quad [7]$$

This can be integrated to give

$$h_t = h_0 \exp\left\{-2vt - \int_0^t \frac{d\xi}{N_\xi}\right\} + 2v \int_0^t \exp\left\{-2v(t-\theta) - \int_\theta^t \frac{d\xi}{N_\xi}\right\} d\theta. \quad [8]$$

Also, in Eq. **6**, if we take expectations over all generations and if we assume (as an approximation) that the harmonic mean of $N_t$ and the arithmetic mean of $P_t$ are independent, we get

$$-(2v + 1/\tilde{N})P + 1/\tilde{N} = 0,$$

where $\tilde{N}$ is the harmonic mean of $N_t$ and $P$ is the average probability at equilibrium of two randomly chosen genes within a line being identical. This leads to

$$P = \frac{1}{1 + 2\tilde{N}v}. \quad [9]$$

Substituting this in Eq. **3**, we get

$$Q = \frac{1}{(1 + 2\tilde{N}v)[1 + (n-1)v/\lambda]}. \quad [10]$$

Let $\overline{H}_0$ be the probability of two homologous genes randomly extracted from the whole population being identical. Then,

$$\overline{H}_0 = (1 - 1/n)Q + (1/n)P \quad [11]$$

so that we obtain

$$\overline{H}_0 = \frac{1 + (n-1)k/n}{1 + (n-1)k} \cdot \frac{1}{1 + 2\tilde{N}v}, \quad [12]$$

where $k = v/\lambda$. The effective number of alleles, $n_e$, as defined by Kimura and Crow (16) is $1/\overline{H}_0$. If $n$ is large but $k$ is much smaller than unity, we have $n_e \approx (1 + nk)(1 + 2\tilde{N}v)$. This may be compared with the situation in which all the individuals in the population are panmictic rather than grouped into isolated lines which are subject to extinction and replacement. In such a situation $n_e = 1 + 2n\tilde{N}v$.

More generally, if $N_e$ is the effective size of a haploid population, then $n_e = 1 + 2N_e v$. Thus, equating this with $1/\overline{H}_0$ where $\overline{H}_0$ is given by Eq. **12**, and assuming $n \gg 1$ and $k \ll 1$, we obtain $1 + 2N_e v = (1 + nk)(1 + 2\tilde{N}v)$ or

$$N_e = \tilde{N} + n/(2\lambda) + n\tilde{N}v/\lambda. \quad [13]$$

Similarly, for a diploid population consisting of a large number, $n$, of local colonies each with the effective size $\tilde{N}$, if exchange of individuals between colonies is extremely rare and if extinction and recolonization occur frequently, $n_e = (1 + nk)(1 + 4\tilde{N}v)$. Equating this with $1 + 4N_e v$, we obtain

$$N_e = \tilde{N} + n/(4\lambda) + n\tilde{N}v/\lambda. \quad [14]$$

Note that this is much smaller than the effective size of a corresponding population which is completely panmictic, in which case we have $N_e = n\tilde{N}$.

**Stepping-stone model with local extinction and recolonization**

We consider a one-dimensional stepping-stone model with $n$ subpopulations arranged on a circle. Because of geographical structure, we refer to subpopulations as "colonies," although the term "lines" could still be used for them. We assume that every existing colony is subject to extinction with rate $\lambda$ per generation and that, whenever extinction of a colony occurs, it is replaced by individuals from either one of two adjacent colonies.

Let $g_{k,t}$ be the probability that two different colonies which are $k$ steps apart do not share a common ancestor during the past $t$ generations—that is, they are kept separate at least for the last $t$ generations. In what follows, although we treat $t$ as a continuous variable, it is convenient to measure time required for one generation as the unit. Then

$$g_{k,t+\Delta t} = (1 - 2\lambda\Delta t)g_{k,t} + \lambda\Delta t(g_{k-1,t} + g_{k+1,t}) + o(\Delta t). \quad [15]$$

Neglecting the terms $o(\Delta t)$ and substituting $dg_{k,t}/dt$ for $\Delta g_{k,t}/\Delta t$ where $\Delta g_{k,t} = g_{k,t+\Delta t} - g_{k,t}$, we obtain the following set of differential equations:

$$\frac{dg_{k,t}}{dt} = \lambda(g_{k-1,t} - 2g_{k,t} + g_{k+1,t}), \quad [16]$$

where $k = 1, 2, \ldots, n-1$, and

$$g_{0,t} = g_{n,t} = 0.$$

The solution of Eq. **16** which satisfies the initial condition $g_{k,0} = 1$ for $k = 1, 2, \ldots, n-1$ is as follows:

$$g_{k,t} = \frac{2}{n}\sum_{i=1}^{n-1} c_i e^{-2\lambda(1-\cos\frac{\pi i}{n})t} \sin\frac{k\pi i}{n}, \quad [17]$$

where

$$c_i = \sum_{k=1}^{n-1} \sin\frac{k\pi i}{n}.$$

Because $1 - g_{k,t}$ is the probability of two colonies that are $k$ steps apart sharing a common ancestral colony some time during the past $t$ generations, $d(1 - g_{k,t})/dt$ or $-dg_{k,t}/dt$ represents the probability density that these two colonies are derived from a common colony $t$ generations back. Thus, the probability $Q_k$ of two genes chosen from colonies which are geographically $k$ steps apart being identical is given by

$$Q_k = -P\int_0^\infty e^{-2vt}\frac{dg_{k,t}}{dt}dt,$$

where $e^{-2vt}$ represents the probability of two gene lineages remaining identical in allelic states after they diverged $t$ generations ago. Noting Eq. **17**, we obtain

$$Q_k = \frac{2P}{n}\sum_{i=1}^{n-1}\frac{c_i \sin(k\pi i/n)}{1 + v/\{\lambda[1 - \cos(\pi i/n)]\}}. \quad [18]$$

Then, the average probability of identity for two homologous genes each chosen randomly from two different colonies is

$$Q = \frac{2P}{n(n-1)}\sum_{i=1}^{n-1}\frac{c_i^2}{1 + v/\{\lambda[1 - \cos(\pi i/n)]\}}. \quad [19]$$

This corresponds to Eq. 3 in the island model.

## Effects of migration

In the preceding analyses, we assumed that there is no migration (equal exchange of individuals) among different subpopulations. Such an assumption is realistic if local extinction and recolonization occur much more frequently than exchange of individuals between existing subpopulations. It is obviously desirable to extend these models to allow migration.

We consider an island model with the entire population consisting of $n$ haploid colonies and assume that migrations among different colonies occur at a constant rate. Thus, every colony receives a fraction $m\Delta t$ of individuals from the entire population during a short time interval $\Delta t$. Using the same notations as before, we obtain

$$P_{t+\Delta t} = (1 - v\Delta t)^2\left\{(1 - m\Delta t)^2\left[P_t + \frac{1-P_t}{N_t}\Delta t\right] + 2m\Delta t Q_t\right\} + o(\Delta t) \quad [20]$$

and

$$Q_{t+\Delta t} = (1 - v\Delta t)^2(1 - \lambda\Delta t)^2(1 - m\Delta t)^2 Q_t + 2(\lambda + m)\Delta t\left(\frac{1}{n}P_t + \frac{n-1}{n}Q_t\right) + o(\Delta t), \quad [21]$$

where $P_t$ and $Q_t$ are the identity probabilities within and between lines, and $v$, $m$, $\lambda$, $N_t$, and $n$ are, respectively, the mutation rate, the migration rate, the colony extinction rate, the colony size, and the number of colonies.

From these equations we get

$$dP_t/dt = -2(v + m)P_t + (1 - P_t)/N_t + 2mQ_t \quad [22]$$

and

$$dQ_t/dt = -2(v + \lambda + m)Q_t + 2(\lambda + m)\{P_t + (n-1)Q_t\}/n. \quad [23]$$

At equilibrium in which $dP_t/dt = 0$ and $dQ_t/dt = 0$, we get

$$P = 1/[1 + 2\tilde{N}v + 2n\tilde{N}mv/(nv + \lambda + m)] \quad [24]$$

and

$$Q = 1/\{(1 + 2\tilde{N}v)[1 + nv/(\lambda + m)] + 2n\tilde{N}mv/(\lambda + m)\} \quad [25]$$

so that

$$Q/P = 1/[1 + nv/(\lambda + m)]. \quad [26]$$

For $m = 0$ (no migration), Eqs. **24** and **25** reduce, respectively, to Eqs. **9** and **10** except that $n$ in Eq. **25** has to be replaced by $n - 1$. This discrepancy arises because we ignored the difference between $n$ and $n - 1$ in deriving equations in this section.

The probability of identity of two randomly chosen homologous genes from the whole population is $\bar{H}_0 = [P + (n-1)\cdot Q]/n$, and if we equate $1/\bar{H}_0$ with $1 + 2N_e v$, we get

$$N_e = \tilde{N} + n/[2(v + \lambda + m)] + n\tilde{N}(v + m)/(v + \lambda + m). \quad [27]$$

For a diploid population, the corresponding formula is

$$N_e = \tilde{N} + n/[4(v + \lambda + m)] + n\tilde{N}(v + m)/(v + \lambda + m). \quad [28]$$

## Monte Carlo experiments

To test the validity of the above analyses, simulation experiments were carried out. An outline of the experiments is as follows. At the beginning, 10 lines each with 20 haploid individuals are assumed; and at the beginning, all the individuals have identical alleles. The first operation in each generation is to test if extinctions occur to the lines. For each line, a random number that is distributed uniformly in the interval (0, 1) is drawn, and this is compared with a given $\lambda$. If the drawn random number happens to be less than $\lambda$, that particular line is terminated and it is replaced by a certain number ($N_0$) of individuals randomly drawn from a single line which is chosen according to the model. For the island model, the donor line is chosen randomly from the remaining $n - 1$ lines. For the stepping-stone model the donor line is chosen from one of the two neighboring lines with equal probability. However, if the drawn random number is greater than λ, that line remains unchanged. The second operation in the simulation is the production of mutation. For each gene a random number is drawn and it is compared with a given value of $v$. If the random number happens to be less than $v$, mutation to a new allele occurs. Otherwise, the gene remains unchanged. The third operation is to increase the line size. Every line which is less than 20 individuals is increased by 1. The last operation is random sampling of gametes to form individuals in the next generation within each line.

These four operations constitute one generation for each line, and this cycle is repeated for a large number of times. The first $2/v$ generations were discarded to eliminate initial effects, and then data for $Q$, $P$, and $Q/P$ were taken at an interval of 10 generations. In Table 1, the averages of these quantities over 30 observations are presented, and they are compared with corresponding theoretical values. In these simulations rather high mutation and extinction rates and small population sizes were assumed to save computing time.

## Discussion

From the above analyses, it is clear that, when local extinction and recolonization occur frequently, not only the effective size of the total population (species) is much reduced but also divergence of subpopulations is largely prevented. As an example, let us consider a haploid population (species) consisting of 100,000 lines ($n = 10^5$). Let us assume that each line starts from a single individual and, although it may grow into a line (subpopulation) comprising an immense number of individuals, it then becomes extinct on the average in 1000 generations ($\lambda = 10^{-3}$) with the result that its harmonic size is only $\tilde{N} = 100$. Let us also assume that mutation rate per generation is $v = 10^{-8}$. Then, from Eq. 13, we get the effective size of the population of about 50 million—i.e., $N_e \approx 5 \times 10^7$.

The sum of squares of allelic frequencies for the total population is $\overline{H}_0 \approx 0.5$. This means that the genetic diversity (1 minus sum of squares of allelic frequencies) is only 50%. Yet, at any moment, this species may comprise an immense number, say $10^{20}$, of individuals. From Eqs. 9 and 10, we have $P \approx 1$, $Q \approx 0.5$, which means that genetic variability is almost entirely due to line differences. Such a situation most likely may be met by lower organisms which reproduce almost exclusively by asexual means and which can increase rapidly in number when conditions become favorable but then easily become extinct when conditions become unfavorable.

A remarkable example of this type of population structure is represented by *Escherichia coli* as revealed by recent studies by Levin (B. R. Levin, personal communication) and Selander and Levin (18). According to Levin (B. R. Levin, personal communication), "periodic selection" (i.e., appearance of a clone with a high fitness followed by its rapid expansion) occurs frequently, but gene exchange between clones through plasmid and phage-mediated mechanisms is extremely rare; its rate appears to be lower than the mutation rate. This is consistent with the finding by Selander and Levin (18), who surveyed electrophoretic variation at 20 enzyme loci in 109 clones of *E. coli* from natural populations, that the number of distinctive *E. coli* genotypes is rather limited; electrophoretically identical clones were obtained from unassociated hosts. They obtained an estimate of mean genetic diversity of 0.4718.

The genetic structure of *E. coli* population can most easily be understood by regarding it as a collection of asexual lines. Random sampling of such lines occurs frequently through periodic selection. We can think of each line being derived from a single individual (bacterium) in which a mutation that endows higher competitive ability happens to occur followed by rapid expansion of its progeny by asexual means to form countless individuals. Eventually such a line may become extinct to be replaced by a new "periodic" line. We may regard intestines of mammals as a sort of chemostat in which periodic selection goes on continuously. Thus, the effective size of *E. coli* is not really very large, contrary to the claim of Milkman (19) who considers that his observations on allozyme variation in *E. coli* (20, 21) are inconsistent with the neutral theory. It is now clear that Milkman's criticism against the neutral theory is not warranted, as pointed out by Levin (personal communication). In this respect, Nei (22) had a remarkable insight when he suggested in 1976 that the effective size of *E. coli* in the long evolutionary history must be much smaller than $10^{10}$, contrary to Milkman's (19) claim that *E. coli* has been at a population size well over $10^{10}$ for at least $4 \times 10^{10}$ generations. "This is because an *E. coli* colony rapidly grows under certain circumstances, while in other circumstances it easily becomes extinct" (22).

A similar criticism of the neutral theory of protein polymorphisms was made by Ayala *et al.* (23) on the ground that, in the neotropical fruit fly *Drosophila willistoni* which has immense population size (which they think has at least the ef-

Table 1. Comparisons of Monte Carlo simulation results (Sim.) with the theoretical expectations (Theo.)

| | | | $P$ | | $Q/P$ | |
|---|---|---|---|---|---|---|
| $v$ | λ | $\tilde{N}$ | Sim. | Theo. | Sim. | Theo. |
| | | | Island model | | | |
| 0.05 | 0.1 | 10.4 | 0.599 | 0.510 | 0.154 | 0.182 |
| 0.02 | 0.1 | 10.4 | 0.759 | 0.706 | 0.295 | 0.357 |
| 0.01 | 0.1 | 10.4 | 0.872 | 0.828 | 0.507 | 0.526 |
| 0.001 | 0.1 | 10.4 | 0.987 | 0.980 | 0.853 | 0.917 |
| 0.002 | 0.05 | 13.2 | 0.958 | 0.951 | 0.735 | 0.735 |
| 0.005 | 0.05 | 12.6 | 0.907 | 0.888 | 0.566 | 0.526 |
| | | | Stepping-stone model | | | |
| 0.002 | 0.05 | 13.4 | 0.951 | 0.949 | 0.683 | 0.593 |
| 0.005 | 0.05 | 13.4 | 0.889 | 0.883 | 0.316 | 0.386 |
| 0.01 | 0.01 | — | — | — | 0.077 | 0.081 |
| 0.01 | 0.1 | — | — | — | 0.415 | 0.386 |

In these simulation experiments listed, the initial line size after extinction ($N_0$) is assumed to be 2. Letters $v$, λ, and $\tilde{N}$ denote mutation rate, extinction rate, and harmonic mean of the population size per line, respectively. Ten lines ($n = 10$) each with the maximum of 20 haploid individuals are assumed.

fective size $N_e = 10^9$, with geographical distribution encompassing several million square kilometers), the observed heterozygosity is roughly 18%. They pointed out that, even if we assume a very low neutral mutation rate such as $v = 10^{-7}$, we still have $4N_ev = 400$ and, from the neutral theory, the predicted heterozygosity should practically be 100%, contrary to their observation. In this case, however, if we assume that the rate of local extinction ($\lambda$) of subpopulations is much higher than that of equal exchange of individuals ($m$) between subpopulations, the effective size of the species $N_e$ is much smaller than $n\tilde{N}$, the product of the number of colonies ($n$) and the effective size of individual colony ($\tilde{N}$). In fact, if $m/\lambda$ is small, $N_e \approx n\tilde{N}(m/\lambda)$ from Eq. **28**. In *D. willistoni*, it is possible that the effective size is 2 orders of magnitude less than what Ayala *et al.* claim and that their criticism of the neutral theory is unwarranted.

It is known that the average heterozygosity among loci per individual in diverse species, including those with apparently immense population sizes, is mostly restricted to the range 0–20% and seldom exceeds 30% (24). This observation has been used repeatedly as evidence against the neutral theory (see ref. 25). It is likely that local extinction and recolonization of subpopulations occur commonly in many species having very large apparent population sizes, and effective population sizes are therefore greatly reduced. In conjunction with the model of effectively neutral mutations (26) in which selective constraint is incorporated, such a difficulty of the neutral theory seems to be resolved.

We thank Drs. M. Slatkin and R. Lande for carefully reading the first draft and for making many useful suggestions to improve the presentation. Thanks are also due to Drs. B. R. Levin and R. K. Selander who kindly showed us their important papers before publication, which stimulated us to write this paper. Contribution no. 1327 from the National Institute of Genetics.

1. Wright, S. (1931) *Genetics* **16**, 97–159.
2. Wright, S. (1938) *Science* **87**, 430–431.
3. Kimura, M. & Crow, J. F. (1963) *Evolution* **17**, 279–288.
4. Nei, M. & Murata, M. (1966) *Genet. Res. Camb.* **8**, 257–260.
5. Hill, W. G. (1972) *Theor. Popul. Biol.* **3**, 278–289.
6. Hill, W. G. (1979) *Genetics* **92**, 317–322.
7. Crow, J. F. & Kimura, M. (1970) *An Introduction to Population Genetics Theory* (Harper & Row, New York), reprint by Burgess, Minneapolis, MN.
8. Kimura, M. & Ohta, T. (1971) *Theoretical Aspects of Population Genetics* (Princeton Univ. Press, Princeton, NJ).
9. Wright, S. (1969) *The Theory of Gene Frequencies. Evolution and the Genetics of Populations* (Univ. of Chicago Press, Chicago), Vol. 2.
10. Lande, R. (1979) *Evolution* **33**, 234–251.
11. Slatkin, M. (1977) *Theor. Popul. Biol.* **12**, 253–262.
12. Wright, S. (1940) *Am. Nat.* **74**, 232–248.
13. Kimura, M. & Weiss, G. H. (1964) *Genetics* **49**, 561–576.
14. Kimura, M. (1968) *Nature (London)* **217**, 624–626.
15. Kimura, M. (1968) *Genet. Res. Camb.* **11**, 247–269.
16. Kimura, M. & Crow, J. F. (1964) *Genetics* **49**, 725–738.
17. Ohta, T. (1978) *Genet. Res. Camb.* **31**, 13–28.
18. Selander, R. K. & Levin, B. R. (1980) *Science*, in press.
19. Milkman, R. (1976) *Trend Biochem. Sci.* **1**, N152–N154.
20. Milkman, R. (1973) *Science* **182**, 1024–1026.
21. Milkman, R. (1975) in *Isozymes IV*, ed. Markert, C. L. (Academic, New York), pp. 273–285.
22. Nei, M. (1976) *Trend Biochem. Sci.* **1**, N247–N248.
23. Ayala, F. J., Powell, J. R., Tracey, M. L., Mourão, C. A. & Pérez-Salas, S. (1972) *Genetics* **70**, 113–139.
24. Fuerst, P. A., Chakraborty, R. & Nei, M. (1977) *Genetics* **86**, 455–483.
25. Lewontin, R. C. (1974) *The Genetic Basis of Evolutionary Change* (Columbia Univ. Press, New York).
26. Kimura, M. (1979) *Proc. Natl. Acad. Sci. USA* **76**, 3440–3444.

PART FOUR

# Linkage and Recombination

## Introduction

Kimura (no. 10) first deduced the differential equations for the gametic frequencies at two diallelic loci and used them to confirm quantitatively Fisher's (1930) claim that epistatic interaction will always tend to diminish recombination and therefore tighter linkage is favored by natural selection (Bodmer and Felsenstein 1967; chapter 5 in Crow and Kimura 1970; chapter 6 in Kimura and Ohta 1971; chapter 8 in Nagylaki 1977 and references therein).

Kimura (no. 11) coined the term *quasi-linkage equilibrium*. Under loose linkage and relatively weak epistatic interaction, it represents a state in which the ratio of coupling and repulsion chromosome phases is kept approximately constant. This state is attained in a very short time, generally less than ten or twenty generations, while the equilibrium values of allele frequencies are attained in a much longer time. Under quasi-linkage equilibrium, the mean fitness change of the population is approximately equal to the genic variance, and Fisher's fundamental theorem of natural selection is usually a good approximation. This is because the contribution of linkage disequilibrium almost exactly cancels the epistatic variance.

It is interesting to ask to what extent genetic variation at a neutral locus is generated through linkage disequilibrium with selected loci. When selected loci are overdominant, the effect of linkage disequilibrium on a neutral locus is called associated or associative overdominance (Frydenberg 1963), whereas when selected loci are subjected to purifying selection by favorable alleles, the effect is called hitch-hiking (Maynard Smith and Haigh 1974). Ohta and Kimura (no. 12) studied associative overdominance, assuming that allele frequencies at selected loci are kept constant but neutral alleles experience random drift. There is little effect of associative overdominance unless the linkage is very tight. A similar conclusion is obtained when two loci are kept polymorphic by strong overdominance selection. This holds, however, only when the two loci are nearly independent and do not generate linkage disequilibrium. Recent DNA sequence data have revived the problem concerning associative overdominance and hitch-hiking (Kaplan, Hudson, and Langley 1989; Langley 1990; Takahata 1990; Hedrick, Whittam, and Parham 1991). For DNA regions in which recombination is rare, nonrandom association of polymorphic sites must have significant effects on genetic variation in populations. The hitch-hiking effect may be regarded as one of the reasons why the extent of heterozygosity is rather low in various organisms (Begun and Aquadro 1992). In *E. coli* populations, hitch-hiking of the whole genome is termed *periodic selection* (appearance of a clone with a high fitness followed by its rapid expansion) by Atwood, Schneider, and Ryan (1951) (see also Novick and Szilard 1950; Selander and Levin 1980).

Ohta and Kimura (no. 13) extended the previous work of Kimura (no. 38), incorporating linkage between two neutral sites, and concluded that marked linkage disequilibrium can arise due to random drift and tight linkage.

# A MODEL OF A GENETIC SYSTEM WHICH LEADS TO CLOSER LINKAGE BY NATURAL SELECTION [1]

MOTOO KIMURA

*Dept. of Genetics, University of Wisconsin*

Received February 13, 1956

## INTRODUCTION

It is generally assumed that linkage intensity may be modified by natural selection. Although there have been several discussions of the subject (e.g. Darlington, 1939; Fisher, 1930), no detailed mathematical analysis has been given. In the present paper a specific model in which selection leads to closer linkage will be examined.

In his 1930 book (p. 103) Fisher states, without going into detail, that "the presence of pairs of factors in the same chromosome, the selective advantage of each of which reverses that of the other, will always tend to diminish recombination, and therefore to increase the intensity of linkage in the chromosomes of that species." However, he does not discuss the ways in which the two loci are maintained in a polymorphic state, but gives a consequence (selection for decreased recombination) of such a state if it exists. Haldane (1931) has pointed out that the reversal of selective value of alleles at one locus according to which allele is present at another is not sufficient to maintain a stable polymorphism.

A specific model, in which a stable polymorphism is maintained, was proposed by Dr. P. M. Sheppard at the 1955 Cold Spring Harbor Symposium (XX: Population Genetics). Assume a locus with a pair of alleles, say $A_1$ and $A_2$, kept in balanced polymorphism by heterozygote superiority in fitness. Another pair of alleles, $B_1$ and $B_2$, are at another locus on the same chromosome, and interact with the genes in the first locus in such a way that $A_1$ is advantageous in combination with $B_1$ but is disadvantageous in combination with $B_2$, while the situation is reversed for the gene $A_2$. Then the second locus will remain polymorphic if linkage between the two loci is sufficiently close.

Under this model, close linkage would maintain a larger fraction of the fitter genotypes than loose linkage and a crossover reducing mechanism will be favored by selection.

It turns out that a full treatment of the Sheppard problem in general form is quite difficult. In the first part of this paper, therefore, I shall elaborate one specific model of the symmetrical type suggested by Sheppard to show that the polymorphism can be maintained and the type of selection on linkage intensity postulated by Fisher be effective. In the second part I shall consider more general models to point out that the main conclusion derived from the symmetrical model is still valid and applicable to such models.

## STABILITY CONDITION AND SHIFT OF EQUILIBRIUM IN THE SYMMETRICAL MODEL

The fitness of various genotypes are given in table 1 in terms of Malthusian parameters (see also Appendix I). We assume the s and t are non-negative and constant. Throughout this paper a large randomly mating population is assumed. Applying the rules for testing

[1] Paper No. 612 from the Department of Genetics, University of Wisconsin. Also Contribution No. 147 of the National Institute of Genetics, Mishima-shi, Japan. The work has been supported in part by grants from the Research Committee of the Graduate School with funds provided by the Wisconsin Alumni Research Foundation.

EVOLUTION 10: 278–287. September, 1956. 278

Table 1

| | $A_1A_1$ | $A_1A_2$ | $A_2A_2$ |
|---|---|---|---|
| $B_1B_1$ | s | t | −s |
| $B_1B_2$ | 0 | t | 0 |
| $B_2B_2$ | −s | t | s |

the stability of a selective polymorphism (Kimura, 1956), it can be shown that the proposed system can not be kept in non-trivial stable equilibrium if there is no linkage.

With complete linkage it is easily shown that the 4 types of chromosomes can not all be maintained in balanced polymorphism, though all 4 genes may be maintained; i.e., the equilibrium population may contain two chromosome types $A_1B_1$ and $A_2B_2$ or $A_1B_2$ and $A_2B_1$. The last case, however, is essentially unstable since mutation or rare recombination will break the equilibrium immediately. So we will not pay attention to this case hereafter.

Let x, y, z and u $(x + y + z + u = 1)$ be the relative frequencies of the 4 types of chromosomes $A_1B_1$, $A_1B_2$, $A_2B_1$ and $A_2B_2$ respectively in the population. It is convenient to write the fitness of various genotypes in the form of table 2.

Under a continuous model of generation time, as explained in Appendix I, the rate of change in the chromosome frequencies may be expressed by the set of differential equations:

$$\left.\begin{aligned} \dot{x} &= x(a_{x\cdot} - \bar{a}) - \rho \\ \dot{y} &= y(a_{y\cdot} - \bar{a}) + \rho \\ \dot{z} &= z(a_{z\cdot} - \bar{a}) + \rho \\ \dot{u} &= u(a_{u\cdot} - \bar{a}) - \rho \end{aligned}\right\} \quad (1)$$

where $a_{x\cdot} = sx + tz + tu$, $a_{y\cdot} = -sy + tz + tu$, $a_{z\cdot} = tx + ty - sz$ and $a_{u\cdot} = tx + ty + su$ in this specific case of table 2. The mean fitness of the population is

$$\bar{a} = a_{x\cdot}x + a_{y\cdot}y + a_{z\cdot}z + a_{u\cdot}u,$$

while $\rho$ represents the contribution by crossing over in double heterozygotes.

$$\rho = w\alpha(xu - yz),$$

where $\alpha$ is the recombination fraction between the two loci and w is the reproductive rate of the double heterozygotes defined in such a way that wdt is the probability that an $A_1A_2B_1B_2$ individual survives and reproduces in an infinitesimal time interval dt (see Appendix I).

To simplify the argument the equations (1) will be written in the following form:

$$\left.\begin{aligned} \mathring{x} &= \tilde{a}_x - \rho/x \\ \mathring{y} &= \tilde{a}_y + \rho/y \\ \mathring{z} &= \tilde{a}_z + \rho/z \\ \mathring{u} &= \tilde{a}_u - \rho/u \end{aligned}\right\} \quad (2)$$

where $\mathring{x}$, $\mathring{y}$, $\mathring{z}$ and $\mathring{u}$ are the rates of change of the logarithmic chromosome frequencies, i.e. $\mathring{x} = d \log x/dt$, etc. In these formulae $\tilde{a}_x$, $\tilde{a}_y$, $\tilde{a}_z$ and $\tilde{a}_u$ denote the mean fitness of the 4 types of chromosomes measured from the average fitness of the population and will be called the average excess in fitness of these chromosome types, i.e. $\tilde{a}_x = a_{x\cdot} - \bar{a}$, etc.

For a non-trivial equilibrium, the condition $\mathring{x} = \mathring{y} = \mathring{z} = \mathring{u} = 0$ must be satisfied, that is to say,

$$\begin{aligned} \tilde{a}_x - \rho/x &= 0 \\ \tilde{a}_y + \rho/y &= 0 \\ \tilde{a}_z + \rho/z &= 0 \\ \tilde{a}_u - \rho/u &= 0. \end{aligned}$$

Table 2

| | x $A_1B_1$ | y $A_1B_2$ | z $A_2B_1$ | u $A_2B_2$ |
|---|---|---|---|---|
| x $A_1B_1$ | s | 0 | t | t |
| y $A_1B_2$ | 0 | −s | t | t |
| z $A_2B_1$ | t | t | −s | 0 |
| u $A_2B_2$ | t | t | 0 | s |

By virtue of symmetry, the equilibrium chromosome frequencies are expressed in a simple form:

$$\left.\begin{aligned} \hat{x} = \hat{u} &= \tfrac{1}{2}\left(\tfrac{1}{2} - \beta + \sqrt{\tfrac{1}{4} + \beta^2}\right) \\ \hat{y} = \hat{z} &= \tfrac{1}{2}\left(\tfrac{1}{2} + \beta - \sqrt{\tfrac{1}{4} + \beta^2}\right) \end{aligned}\right\} \quad (3)$$

where $\beta = w\alpha/s$. At this point, the gene frequency in each locus is 50 percent. That this is the most natural gene frequency to be chosen among models in which equilibrium gene frequencies are preassigned is apparent from the very nature of the present problem, since natural selection acts somewhat symmetrically on the genetic system. The mean fitness of the population at equilibrium, which we shall designate by $\hat{a}$, is given by

$$\hat{a} = (s + t)/2 - 2sy. \qquad (4)$$

One of the crucial parts of our problem is to determine the stability condition of this equilibrium.

Since rigorous mathematical arguments are relegated to Appendices II and III, I will here sketch rather intuitively one of the methods.

Consider a set of equilibrium chromosome frequencies $(\hat{x}, \hat{y}, \hat{z}, \hat{u})$ as a point in a 4 dimensional frequency space. Suppose we make a small displacement denoted by a vector $V_1 = (\delta\hat{x}, \delta\hat{y}, \delta\hat{z}, \delta\hat{u})$ from the equilibrium point. Then the selection pressure will tend to move the displaced chromosome frequencies in some direction. This tendency will be measured by a vector $V_2$ whose components are $(\delta\mathring{x})_e$, $(\delta\mathring{y})_e$, $(\delta\mathring{z})_e$, $(\delta\mathring{u})_e$, in which the subscript e means the value at the equilibrium. It must be noted that these elements of $V_2$ are equivalent to $[\delta(\dot{x}/x)]_e$, $[\delta(\dot{y}/y)]_e$, etc., namely, each component is weighted inversely by the corresponding gene frequency.

The necessary and sufficient condition for the stability of the equilibrium is then that the vector have a tendency to go back to the original position. More precisely, the angle made by the above two vectors must be more than 90°, or equivalently their inner product $V_1 \cdot V_2$ must always be negative:

$$V_1 \cdot V_2 < 0. \qquad (5)$$

As shown in Appendix II, this leads to the condition

$$s^2 + 4\beta st < t^2, \qquad (6)$$

where s, t and $\beta$ are all non negative. The above argument, convincing as it may seem, may fail to deal with the case

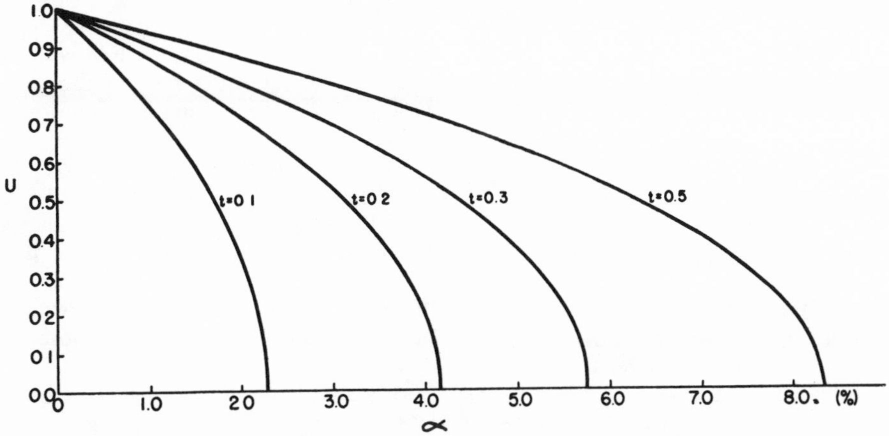


**FIG. 1.** **The upper limit, U, of the ratio s/t as a function of the recombination fraction, α, for selected values of t. Points to the left and below each curve lead to stable polymorphism for that particular value of t.**

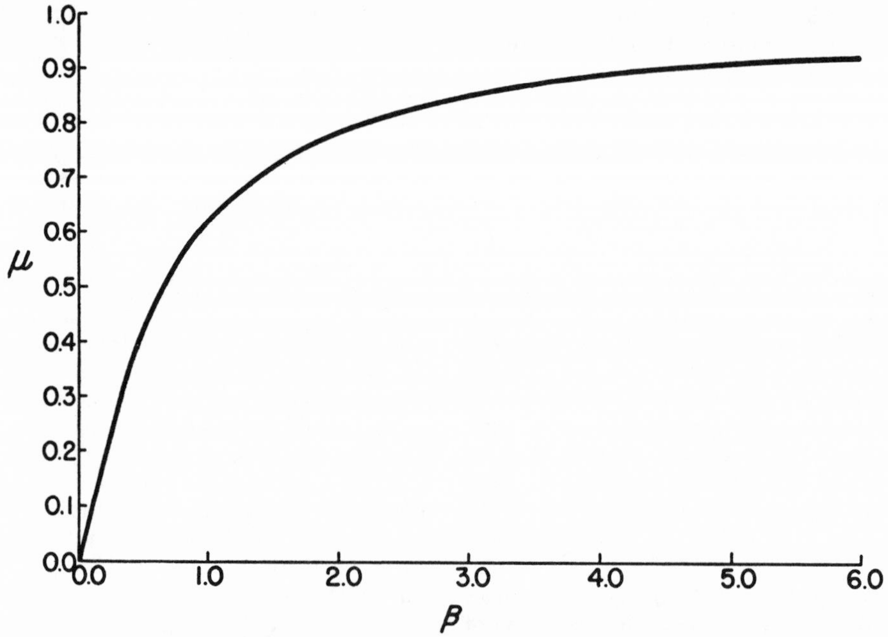


FIG. 2. The equilibrium ratio, $\mu$, of chromosomes in repulsion phase to those in coupling as a function of $\beta = w\alpha/s$. $\alpha$ = recombination fraction, s = selection intensity as defined in table 1, w is a measure of the reproductive rate of the double heterozygote.

in which the perturbed chromosome frequencies go back to the equilibrium point through a cyclically winding path. It turns out, as shown in Appendix III, that no such troublesome situation is involved in the present model and treatment by a more general method leads to exactly the same conclusion.

If we take $1 + t$ as w, the necessary and sufficient condition for stability may be written in the following form:

$$0 < s < tU, \qquad (7)$$

where

$$U = \sqrt{1 - 4(i + t)\alpha/t}.$$

From (7) we obtain the following conclusions: (i) For a given average selective advantage t of the heterozygote, the recombination fraction $\alpha$ must be smaller than $t/(4 + 4t)$, which never exceeds 1/4. For example, if $t = 0.1$, the second locus can remain polymorphic only if there is less than 2.3% recombination with the first, or, if $t = .5$, less than 8.3%. (ii) Even within these limits there are restrictions on s if a stable polymorphism is to be maintained; the more distant the loci, the greater the restriction. Figure 1 shows this restriction in terms of the allowable upper limit (U) for the ratio s/t. For example with $t = 0.1$, s must be less than one half t, if the recombination fraction is 1.7%.

In the equilibrium population the repulsion phase ($A_2B_1$ and $A_1B_2$) is usually much less frequent than the coupling phase ($A_1B_1$ and $A_2B_2$). The ratio between the two ($\mu = \hat{y}/\hat{x}$) is determined by $\beta = w\alpha/s$ as shown in figure 2. If the linkage is very close, we have approximately $\mu = \beta$.

In general, the closer the linkage the less frequent the repulsion phase.

Suppose now that a crossover reducing mechanism, has occurred in a chromosome in coupling phase. It is easily seen that such chromosomes will gradually increase in frequency and eventually replace the original chromosome. For example, consider the case in which a $B_1A_1$ chromosome arose from a $A_1B_1$ chromosome by inversion. If we denote the frequency of $B_1A_1$ by x′, then

$$\dot{x}' = x'\tilde{a}_x. \qquad (8)$$

As compared with the original $A_1B_1$ chromosome, it has no loss by crossing-over. At the present equilibrium point, the rate of increase in the logarithmic frequency is

$$(\dot{x}')_e = (\tilde{a}_x)_e = s\hat{y},$$

showing that the selective advantage is of the order of the recombination fraction which might be quite high (as compared with the rate of occurrence of the cross-over changing mechanism).

Through this process the population is led to the chromosome constitution consisting of 50% $B_1A_1$ and 50% $A_2B_2$. The mean fitness of the population is now $(s + t)/2$, with the gain of $2s\hat{y}$ as shown from (4). It is interesting to note that the resulting supergenes are less overdominant.

## More General Models

The main conclusions reached by the above model are by no means restricted by the assumption of the exact symmetry. For example we may assume the more general model given in table 3. If $s_1$, $s_2$, $s_3$ and $s_4$ are all positive and of approximately the same order of magnitude, while $t_1$ and $t_2$ are larger than any of them, the equilibrium population will contain predominantly coupling chromosomes $A_1B_1$ and $A_2B_2$ if linkage between the two loci is sufficiently close. Though the exact allowable upper limit of $\alpha$ is very difficult to set, it is not difficult to show if $\alpha$ is sufficiently small as compared with selective advantage or disadvantage, the equilibrium is stable.

Table 3

| | $A_1A_1$ | $A_1A_2$ | $A_2A_2$ |
|---|---|---|---|
| $B_1B_1$ | $s_1$ | $t_1$ | $-s_2$ |
| $B_1B_2$ | $\frac{s_1 - s_3}{2}$ | $\frac{t_1 + t_2}{2}$ | $\frac{s_4 - s_2}{2}$ |
| $B_2B_2$ | $-s_3$ | $t_2$ | $s_4$ |

At this point frequencies of the repulsion chromosomes are proportional to the recombination fraction. Suppose an inversion has occurred in a coupling phase say $A_1B_1$. It is possible to show that the new chromosome with inversion $B_1A_1$, other things being equal, has selective advantage of order of the recombination fraction $\alpha$, which may be much higher than the rate of occurrence of the chromosomal rearrangement. Thus the new chromosome $B_1A_1$ will replace the original $A_1B_1$ chromosome and by so doing, the mean fitness of the population increases.

Table 4

| | | Pink | | Yellow |
|---|---|---|---|---|
| | | $A_1A_1$ | $A_1A_2$ | $A_2A_2$ |
| Unbanded | $B_1B_1$ | $s_1$ | $s_1 + t$ | $-s_2$ |
| | $B_1B_2$ | $s_1$ | $s_1 + t$ | $-s_2$ |
| Banded | $B_2B_2$ | $-s_3$ | $-s_3 + t$ | $s_4$ |

The more realistic model suggested to me by Dr. Sheppard is given in table 4. In a species of land snail, pink, unbanded ($A_1-$, $B_1-$) or yellow, banded ($A_2A_2B_2B_2$) individuals are both more advantageous than yellow, unbanded ($A_2A_2B_1-$) or pink, banded ($A_1 - B_2B_2$) individuals. Furthermore it is assumed that an individual heterozygous for $A_1$ and $A_2$ has additional advantage due to overdominance. Exactly the same type of argument is applicable to this model though to set an exact allowable upper bound to $\alpha$ is again quite difficult due to the asymmetry involved.

## Concluding Remarks

The models given here assume that the crossover reducing influence is inherited with the chromosome influenced (as, for example, an inversion).

The effect of a crossover modifier not completely linked to the coupling chromosome is rather complicated, but would be selected with much less efficiency. These are, of course, only a few of many models that could be devised. The essential point seems to be that, if one locus already contains a mechanism for maintaining a stable polymorphism, a second linked locus may be incorporated into the system. Such a system gives a selective advantage to close linkage and thus provides a means for evolutionary adjustment of recombination values.

## Summary

As has been pointed out by various authors, frequency of recombination within a chromosome is subject to selective adjustment. In this sense, the history of the control of recombination may be regarded as a part of the history of the evolution of genetic system.

In 1930 Fisher said that with two pairs of linked loci such that $A_1$ is favored in the presence of $B_1$ while $A_2$ is favored with $B_2$ selection will act in such a way as to decrease the amount of recombination between the two loci. He did not, however, specify how a stable polymorphism of this type might be maintained. In this paper a model suggested by Sheppard is examined. It is shown that, if the first locus is maintained in balanced polymorphism by heterozygote superiority in fitness, the second locus will remain polymorphic if linkage is sufficiently close. Under this condition an inversion (or other crossover reducing mechanism) will be favored by selection and, other things being equal, will eventually be established in the population.

## Appendix I

*Derivation of the set of equations (1) based on the model of continuous generation time.*

Consider a very large randomly mating population. We assume that the population is continuous in time, that is, reproduction and death of the members of this population go on uniformly throughout the year. To facilitate concise mathematical argument we shall choose slightly different notations from the ones given in section 2. First we define the two quantities denoted by $b_{ijkl}$ and $d_{ijkl}$ in such way that

$b_{ijkl}\Delta t$: Probability that a randomly chosen individual of genotype $A_iA_jB_kB_l$ survives and reproduces in a small time interval $\Delta t$. (i, j, k, l = 1, 2)

$d_{ijkl}\Delta t$: Probability that a randomly chosen individual of genotype $A_iA_jB_kB_l$ dies in the interval $\Delta t$ without producing offspring in that interval.

Then the fitness of $A_iA_jB_kB_l$ is defined by

$$a_{ijkl} = b_{ijkl} - d_{ijkl}.$$
$$(a_{ijkl} = a_{jikl} = a_{jilk})$$

Any quantities of order $(\Delta t)^2$ will be neglected. The fitness so defined gives the geometric rate of increase or decrease of the number of individuals with genotype $A_iA_jB_kB_l$ and following Fisher (1930) will be called the fitness of the genotype measured in Malthusian parameters.

If we designate by $n_{ik}$ the number of $A_iB_k$ chromosomes in the population and by N the total number of chromosomes $(N = \sum_{ik} n_{ik})$, the frequency of the $A_iB_k$ is given by

$$x_{ik} = n_{ik}/N. \qquad (\sum_{ik} x_{ik} = 1)$$

Let us first consider the increase in the number of the chromosome $A_1B_1$ in the small time interval $\Delta t$. It is not difficult to see that

$$\begin{aligned}\Delta n_{11} = {} & b_{1111}\Delta t N x^2_{11} + b_{1112}\Delta t N x_{11}x_{12} \\ & + b_{1211}\Delta t N x_{11}x_{21} \\ & + b_{1212}\Delta t N x_{11}x_{22}(1 - \alpha) \\ & + b_{1212}\Delta t N x_{21}x_{12}\alpha \\ & - d_{1111}\Delta t N x^2_{11} \\ & - d_{1112}\Delta t N x_{11}x_{12} \\ & - d_{1211}\Delta t N x_{11}x_{21} \\ & - d_{1212}\Delta t N x_{11}x_{22},\end{aligned}$$

where $\alpha$ is the recombination fraction. Divide both sides by $N\Delta t$ and note that $a = b - d$, then

$$\frac{1}{N}\frac{\Delta n_{11}}{\Delta t} = x_{11}a_{1\cdot 1\cdot} - b_{1212}\alpha D$$

where $a_{1\cdot 1\cdot} = a_{1111}x_{11} + a_{1112}x_{12} + a_{1211}x_{21} + a_{1212}x_{22}$ and $D = x_{11}x_{22} - x_{12}x_{21}$.

Similarly we can derive

$$\frac{1}{N}\frac{\Delta n_{12}}{\Delta t} = x_{12}a_{1\cdot 2\cdot} + b_{1212}\alpha D$$

$$\frac{1}{N}\frac{\Delta n_{21}}{\Delta t} = x_{21}a_{2\cdot 1\cdot} + b_{1212}\alpha D$$

$$\frac{1}{N}\frac{\Delta n_{22}}{\Delta t} = x_{22}a_{2\cdot 2\cdot} - b_{1212}\alpha D$$

Therefore, if we note that $N = n_{11} + n_{12} + n_{21} + n_{22}$,

$$\frac{1}{N}\frac{\Delta N}{\Delta t} = \bar{a},$$

where $\bar{a} = x_{11}a_{1\cdot 1\cdot} + x_{12}a_{1\cdot 2\cdot} + x_{21}a_{2\cdot 1\cdot} + x_{22}a_{2\cdot 2\cdot}$. Thus

$$\begin{aligned}\frac{\Delta x_{11}}{\Delta t} &= \frac{\Delta}{\Delta t}\left(\frac{n_{11}}{N}\right) = \frac{\dfrac{\Delta n_{11}}{\Delta t}N - n_{11}\dfrac{\Delta N}{\Delta t}}{N^2} \\ &= \frac{1}{N}\frac{\Delta n_{11}}{\Delta t} - \frac{n_{11}}{N}\left(\frac{1}{N}\frac{\Delta N}{\Delta t}\right) \\ &= x_{11}a_{1\cdot 1\cdot} - \alpha b_{1212}D - x_{11}\bar{a}.\end{aligned}$$

Taking the limit $\Delta t \to 0$ and writing $dx_{11}/dt$ with the Newtonian symbol $\dot{x}_{11}$, we have

$$\dot{x}_{11} = x_{11}(a_{1\cdot 1\cdot} - \bar{a}) - \rho,$$

where $\rho = \alpha b_{1212}D$. Similarly we have

$$\dot{x}_{12} = x_{12}(a_{1\cdot 2\cdot} - \bar{a}) + \rho$$

$$\dot{x}_{21} = x_{21}(a_{2\cdot 1\cdot} - \bar{a}) + \rho$$

$$\dot{x}_{22} = x_{22}(a_{2\cdot 2\cdot} - \bar{a}) - \rho.$$

If we replace $x_{11}$, $x_{12}$, $x_{21}$, and $x_{22}$ respectively by new symbols x, y, z, and u, $a_{1\cdot 1\cdot}$, $a_{1\cdot 2\cdot}$, $a_{2\cdot 1\cdot}$, and $a_{2\cdot 2\cdot}$ by $a_{x\cdot}$, $a_{y\cdot}$, $a_{z\cdot}$ and $a_{u\cdot}$, and finally $b_{1212}$ by w, we have the equations (1) in general form.

## Appendix II

*Derivation of condition (6) in section 2.*

Given

$$V_1\cdot V_2 = (\delta\hat{x})(\delta\dot{x})_e + (\delta\hat{y})(\delta\dot{y})_e + (\delta\hat{z})(\delta\dot{z})_e + (\delta\hat{u})(\delta\dot{u})_e < 0,$$

we want to derive the condition (6) in section 2, i.e. $s^2 + 4\beta st < t^2$. If we use the relations:

$$(\delta\dot{x})_e = \left(\frac{\partial\dot{x}}{\partial x}\right)_e\delta\hat{x} + \left(\frac{\partial\dot{x}}{\partial y}\right)_e\delta\hat{y} + \left(\frac{\partial\dot{x}}{\partial z}\right)_e\delta\hat{z} + \left(\frac{\partial\dot{x}}{\partial u}\right)_e\delta\hat{u},$$

$$(\delta\dot{y})_e = \left(\frac{\partial\dot{y}}{\partial x}\right)_e\delta\hat{x} + \left(\frac{\partial\dot{y}}{\partial y}\right)_e\delta\hat{y} + \left(\frac{\partial\dot{y}}{\partial z}\right)_e\delta\hat{z} + \left(\frac{\partial\dot{y}}{\partial u}\right)_e\delta\hat{u}, \quad \text{etc.,}$$

where

$$\left(\frac{\partial\dot{x}}{\partial x}\right)_e = 2s\hat{y} - t - w\alpha\mu^2, \qquad \left(\frac{\partial\dot{y}}{\partial x}\right)_e = -\, 2s\hat{x} - t + w\alpha\mu^{-1}$$

$$\left(\frac{\partial\dot{x}}{\partial y}\right)_e = 2s\hat{y} - t + w\alpha, \qquad \left(\frac{\partial\dot{y}}{\partial y}\right)_e = -\, 2s\hat{x} - t - w\alpha\mu^{-2}$$

$$\left(\frac{\partial\dot{x}}{\partial z}\right)_e = 2s\hat{y} + w\alpha, \qquad \left(\frac{\partial\dot{y}}{\partial z}\right)_e = 2s\hat{y} - w\alpha$$

$$\left(\frac{\partial\dot{x}}{\partial u}\right)_e = -\, 2s\hat{x} - w\alpha, \qquad \left(\frac{\partial\dot{y}}{\partial u}\right)_e = -\, 2s\hat{x} + c\mu^{-1}, \qquad \text{etc.,}$$

in which $\mu = \hat{y}/\hat{x}$, we can express $V_2$ in terms of $\delta\hat{x}$, $\delta\hat{y}$, $\delta\hat{z}$ and $\delta\hat{u}$. Thus $V_1 \cdot V_2$ is expressed as a quadratic form in $\delta\hat{x}$, $\delta\hat{y}$, etc. However the 4 chromosome frequencies are not independent and $\delta\hat{u} = -\delta\hat{x} - \delta\hat{y} - \delta\hat{z}$. Substituting this in the quadratic form we obtain a new quadratic form in three variables $\delta\hat{x}$, $\delta\hat{y}$ and $\delta\hat{z}$, the matrix of which, after dividing by $s^3$, is

$$T = \begin{bmatrix} 2(1-\lambda+\beta-\beta\mu^2) & 1-2\lambda+\beta-\beta\mu^2 & 1+\beta-\beta\mu^2 \\ 1-2\lambda+\beta-\beta\mu^2 & -2\lambda-\beta(\mu^{-2}+\mu^{-1}+\mu+\mu^2) & 1-\beta(\mu^{-1}+1+\mu+\mu^2) \\ 1+\beta-\beta\mu^2 & 1-\beta(\mu^{-1}+1+\mu+\mu^2) & -\beta(\mu^{-2}+\mu^{-1}+\mu+\mu^2) \end{bmatrix}$$

where $\lambda = t/s$ and $\beta = w\alpha/s$. The condition $V_1 \cdot V_2 < 0$ is equivalent to the condition that a quadratic form associated with T is negative definite. From this we obtain the following three conditions with respect to $\lambda$:

$$\text{(i) } \lambda > 1 + \beta - \beta\mu^2$$

$$\text{(ii) } \lambda > \frac{(1 + \beta - \beta\mu^2)\{1 + \beta(2\mu^{-2} + 2\mu^{-1} + 1 + 2\mu + \mu^2)\}}{2\beta(\mu^{-2} + \mu^{-1} + \mu + \mu^2)}$$

$$\text{(iii) } \lambda > \frac{(1 + \beta - \beta\mu^2)(1 - \beta + \beta\mu^{-2})}{\beta(\mu^{-2} - 2 + \mu^2)}$$

The last condition turns out to be the strongest and if we substitute

$$\mu = \hat{y}/\hat{x} = \frac{1 + 2\beta - \sqrt{1 + 4\beta^2}}{1 - 2\beta + \sqrt{1 + 4\beta^2}}$$

in (iii), we obtain

$$s < (\sqrt{1 + 4\beta^2} - 2\beta)t;$$

which is equivalent to

$$s^2 + 4\beta st < t^2$$

if $s > 0$ and $t > 0$. If we note $\beta = w\alpha/s$, this may also be written in the form $s^2 + 4w\alpha t < t^2$.

## Appendix III

*General treatment of the stability condition with the demonstration that no spirally winding paths are involved.*

We start from the set of differential equation (1) given in section 2. Let $\delta x$, $\delta y$, $\delta z$ and $\delta u$ be the small deviations of chromosome frequencies x, y, z and u from their respective equilibrium frequencies $\hat{x}$, $\hat{y}$, $\hat{z}$ and $\hat{u}$. Then we have

$$\begin{bmatrix} \widehat{\delta x} \\ \widehat{\delta y} \\ \widehat{\delta z} \\ \widehat{\delta u} \end{bmatrix} = \begin{bmatrix} s\hat{y} & \frac{c}{2} & t\hat{x} + \frac{c}{2} & (t - s)\hat{x} \\ 0 & -s\hat{x} - \frac{c}{2} & (t + s)\hat{y} - \frac{c}{2} & t\hat{y} \\ t\hat{y} & (t + s)\hat{y} - \frac{c}{2} & -s\hat{x} - \frac{c}{2} & 0 \\ (t - s)x & t\hat{x} + \frac{c}{2} & \frac{c}{2} & s\hat{y} \end{bmatrix} \begin{bmatrix} \delta x \\ \delta y \\ \delta z \\ \delta u \end{bmatrix}$$

where $c = \alpha w$ and $\dot{\delta \hat{x}}$ means $(d(\delta x)/dt)_e$ etc. Since the 4 chromosome frequencies are not independent, we substitute

$$\delta u = -\delta x - \delta y - \delta z$$

in the above equation to get

$$\begin{bmatrix} \dot{\delta \hat{x}} \\ \dot{\delta \hat{y}} \\ \dot{\delta \hat{z}} \end{bmatrix} = \begin{bmatrix} \frac{s}{2} - t\hat{x} & (s-t)\hat{x} + \frac{c}{2} & s\hat{x} + \frac{c}{2} \\ -t\hat{y} & -s\hat{x} - t\hat{y} - \frac{c}{2} & s\hat{y} - \frac{c}{2} \\ t\hat{y} & (s+t)\hat{y} - \frac{c}{2} & -s\hat{x} - \frac{c}{2} \end{bmatrix} \begin{bmatrix} \delta \hat{x} \\ \delta \hat{y} \\ \delta \hat{z} \end{bmatrix}$$

For the derivation of the above equations, the following relations are found to be very useful:

$$2s\hat{x}\hat{y} = c(\hat{x} - \hat{y}), \quad 2s\hat{x}^2 = s\hat{x} - c(\hat{x} - \hat{y}),$$
$$2s\hat{y}^2 = s\hat{y} - c(\hat{x} - \hat{y}) \text{ and } \hat{x} + \hat{y} = 1/2.$$

The necessary and sufficient condition that the equilibrium is stable is that the real parts of all the characteristic roots of the matrix in the above equation be negative (see Bellman, 1953, p. 25).

After tedious but straightforward calculation we get the characteristic equation in the following form:

$$\lambda^3 + \frac{t+v}{2}\lambda^2 + \left\{\frac{(v-c)t}{2} - \frac{s^2}{4}\right\}\lambda - \frac{v}{8}(s^2 - tv + 2ct) = 0, \quad (\#1),$$

where $v = \sqrt{s^2 + 4c^2}$. Here we note that s, t, c and v are all non negative quantities. Let $\lambda = X/2$, then (#1) becomes

$$X^3 + (t+v)X^2 + \{2(v-c)t - s^2\}X - v(s^2 - tv + 2ct) = 0. \quad (\#2)$$

First we try to show that all the roots of (#2) and therefore all the roots of (#1) are real. If we put $X = Y - (t+v)/3$, we obtain the cubic equation of the form

$$Y^3 + pY + q = 0. \quad (\#3)$$

Now we proceed to show that the discriminant of (#3)

$$\Delta = q^2/4 + p^3/27$$

is negative. Calculations involved are again tedious but straightforward. They lead to

$$\frac{108\Delta}{c^2} = -16s^2t^2 + 112c^2t^2 + 64cs^2t - 16ct^3 - 4t^4 - 128c^3t - 64c^2s^2 + v(t-2c)^2 16t.$$

The last term is non negative and if we apply the inequality

$$v = \sqrt{s^2 + 4c^2} \leqq s + 2c,$$

we get

$$108\Delta \leqq -4c^2(t-2c)^2(t-2s)^2 \leqq 0.$$

Thus

$$\Delta \leqq 0$$

and all the roots of (#3) and therefore all the roots of (#1) are real. This is enough to show that there are no spirally winding paths to the equilibrium gene frequency.

Furthermore, once we know that no imaginary roots are involved, the necessary and sufficient condition that the real parts of all the roots of (#1) be negative can be worked out very simply. Namely we require that coefficients of $x^2$, $x^1$, and $x^0$ in (#2) are all positive.

$$t + v > 0$$
$$2(v-c)t - s^2 > 0$$
$$-v(s^2 - tv + 2ct) > 0.$$

It may be immediately seen that the last inequality is the strongest one and if we note $v = \sqrt{s^2 + 4c^2}$, we get

$$t^2 > s^2 + 4ct$$

which is exactly the same condition obtained in Appendix II since $c = \alpha w$.

## Acknowledgment

The author expresses his thanks to Dr. P. M. Sheppard for suggesting the problem and for stimulating discussions. Thanks are also due to Dr. deLeeuw for his helpful suggestions in completing the mathematical treatment given in Appendix III. Finally he expresses his indebtedness to Dr. J. F. Crow who helped him in completing the manuscript.

## References

Bellman, R. 1953. Stability Theory of Differential Equations. McGraw-Hill, New York.

Darlington, C. D. 1939. The Evolution of Genetic Systems. Cambridge University Press.

Fisher, R. A. 1930. The Genetical Theory of Natural Selection. Oxford, Clarendon Press.

Haldane, J. B. S. 1931. A mathematical theory of natural selection. Part VIII. Metastable population. Proc. Camb. Phil. Soc., **27**: 137–142.

Kimura, M. 1956. Rules for testing stability of selective polymorphism under simple situation. Proc. Nat. Acad. Sci. (in press).

# ATTAINMENT OF QUASI LINKAGE EQUILIBRIUM WHEN GENE FREQUENCIES ARE CHANGING BY NATURAL SELECTION

MOTOO KIMURA

*National Institute of Genetics, Mishima, Japan*

Received May 27, 1965

RECENTLY, a number of papers on population genetics have been published treating the effect of linkage and epistasis on selection (LEWONTIN and KOJIMA 1960; BODMER and PARSONS 1962; PARSONS 1963; LEWONTIN 1964a,b; NEI 1964a,b; WRIGHT 1965; FELSENSTEIN 1965 and others). These papers mainly treat particular problems and little is yet known about the general principles governing linked gene systems in evolution. The problem of finding such principles might appear to be difficult, since the recent paper by MORAN (1964) seems to show that WRIGHT's conception of an "adaptive topography" is untenable, at least as it stands.

The main purpose of the present paper is to show that linked gene systems own a remarkable property of rapidly settling to a state which I would like to call quasi linkage equilibrium. This state is attained if gene frequencies are changing under loose linkage and relatively weak epistatic interactions. On the other hand, linkage disequilibrium may be built up indefinitely when linkage is tight, epistatic interactions are relatively strong and gene frequencies are changing toward fixation.

I would like to show further that for a genetic system evolving under quasi linkage equilibrium, both WRIGHT's conception of an "adaptive topography" and FISHER's fundamental theorem of natural selection indeed hold.

Throughout this paper I will consider a very large random mating population and assume that the fitnesses of individual genotypes are constant, though not necessarily the same. Furthermore, I will restrict my consideration to the case of two linked loci, each with a pair of alleles, $A_1$ and $A_2$ in the first locus and $B_1$ and $B_2$ in the second, leaving more complex cases to future investigations.

## *Haploid population*

Let us consider a population of a haploid organism and designate by $x$, $y$, $z$, and $u$ the respective frequencies of four genotypes, $A_1B_1$, $A_2B_1$, $A_1B_2$ and $A_2B_2$, in the population before selection. Their respective frequencies after the selection will be denoted by capital letters, $X$, $Y$, $Z$, and $U$. We will assume that meiosis follows immediately after fertilization, to form the next generation.

Contribution No. 575 from the National Institute of Genetics, Mishima-shi, Shizuoka-ken, Japan. Aided in part by a Grant-in-Aid from the Ministry of Education, Japan, and also, by a Grant from Toyo Rayon Foundation.

Genetics **52**: 875–890 November 1965.

If $w_o$, $w_a$, $w_b$ and $w_{ab}$ are respectively the fitnesses of these genotypes measured in selective values (= viabilities if there are no fertility differences), then

$$\left.\begin{aligned} X &= xw_o/\bar{w} \\ Y &= yw_a/\bar{w} \\ Z &= zw_b/\bar{w} \\ U &= uw_{ab}/\bar{w} \end{aligned}\right\} \text{after selection,} \tag{1}$$

and

$$\left.\begin{aligned} x' &= X - cD \\ y' &= Y + cD \\ z' &= Z + cD \\ u' &= U - cD \end{aligned}\right\} \text{after recombination,} \tag{2}$$

where primed letters indicate values in the next generation. In the above formulae, $\bar{w}$ is the mean selective value of the population,

$$\bar{w} = w_o x + w_a y + w_b z + w_{ab} u, \tag{3}$$

$D$ is half the difference of the frequencies of the coupling and repulsion double heterozygotes after fertilization,

$$D = XU - YZ\,, \tag{4}$$

and $c$ is the recombination fraction between the two loci.

From (1) and (2), we have

$$\left.\begin{aligned} \Delta x &= \frac{x(w_o - \bar{w})}{\bar{w}} - cD \\ \Delta y &= \frac{y(w_a - \bar{w})}{\bar{w}} + cD \\ \Delta z &= \frac{z(w_b - \bar{w})}{\bar{w}} + cD \\ \Delta u &= \frac{u(w_{ab} - \bar{w})}{\bar{w}} - cD \end{aligned}\right. \tag{5}$$

where $\Delta$ is the finite difference operator with respect to time, measured by one generation as unit, so that $\Delta x = x' - x$, $\Delta y = y' - y$ etc.

Now, let us put

$$r = \frac{xu}{yz}\,, \tag{6}$$

$$R = \frac{XU}{YZ} = \left(\frac{w_o\, w_{ab}}{w_a\, w_b}\right) r \tag{7}$$

and consider the change of $\log r$ in one generation, assuming that changes in gene frequencies are going on.

Since

$$\Delta\log r = \Delta\log x - \Delta\log y - \Delta\log z + \Delta\log u\,,$$

we have, neglecting higher order terms,

$$\Delta\log r = \frac{\Delta x}{x} - \frac{\Delta y}{y} - \frac{\Delta z}{z} + \frac{\Delta u}{u}\,. \tag{8}$$

Then, using (5), we obtain

$$\bar{w}\Delta log\, r = (w_o - w_a - w_b + w_{ab}) \quad - \quad c\bar{w}D(\frac{1}{x}+\frac{1}{y}+\frac{1}{z}+\frac{1}{u}) \qquad (9)$$

$$= (w_o - w_a - w_b + w_{ab}) \quad - \quad c\bar{w}YZ(R-1)(\frac{1}{x}+\frac{1}{y}+\frac{1}{z}+\frac{1}{u})$$

or, noting (7), we have

$$\bar{w}\Delta log\, R = \varepsilon - cH(R-1)\ , \qquad (10)$$

where

$$\varepsilon = w_o - w_a - w_b + w_{ab} \qquad (11)$$

and

$$H = \bar{w}YZ\ (\frac{1}{x}+\frac{1}{y}+\frac{1}{z}+\frac{1}{u}). \qquad (12)$$

Let us consider the process of change in $R$ using equation 10. Without losing generality we can assume that $\varepsilon$ in the equation is non-negative, because if $\varepsilon$ is negative we may redefine $R^{-1}$ as $R$ and $-\varepsilon$ as $\varepsilon$ to carry out the same argument. Suppose that $R$ is less than unity, then the right side of (10) is positive and therefore $log\, R$ will increase. When $R$ reaches unity, $log\, R$ will increase roughly at the rate of $\varepsilon$ per generation if selection is mild. Thereafter, the rate of increase in $log\, R$ will diminish, since the second term on the right side of (10) starts to produce negative contribution to $\Delta log\, R$. However, if the epistatic effect $\varepsilon$ is very much larger than the recombination fraction $c$, the right side of (10) may remain positive and $R$ will increase indefinitely. On the other hand, if $\varepsilon$ is much smaller than $c$, term $cH(R\text{-}1)$ will approach $\varepsilon$ quickly, with the result that an equilibrium state will soon be reached, where

$$\Delta log\, R = 0. \qquad (13)$$

This state is stable, because if $R$ starts from a value which is larger than its equilibrium value ($\hat{R}$), it will decrease until $R=\hat{R}$. Deviation of $R$ from $\hat{R}$ will be reduced roughly by fraction $c$ in each generation.

It should be noted here that (13) is an approximation, since $H$ in (10) is not strictly a constant. However, in such a state, the change in $R$ may be so slow that we may treat $R$ as practically constant even if gene frequencies are changing. I would like to call such state, quasi linkage equilibrium. Furthermore, if selection coefficients are small, the equilibrium value of $R$ is expected to be near unity, with the result that $H$ is roughly unity and we have

$$\varepsilon - c(R-1) = 0$$

or

$$\hat{R} = 1 + \varepsilon/c \qquad (14)$$

approximately.

Though the above arguments are based on the assumption of nonnegative $\varepsilon$, equation 14 holds also for negative $\varepsilon$ as long as $|\varepsilon| \ll c$.

In order to check the validity of the above arguments, an extensive numerical study using a high-speed computer has been carried out. Table 1 shows an example with $w_o = 1.0$, $w_a = w_b = 0.98$, $w_{ab} = 1.06$ and $c = 0.5$. Namely, the two

TABLE 1

*A numerical example showing some properties of the quasi linkage equilibrium. Selective values,* $w_o = 1.00$, $w_a = w_b = 0.98$, $w_{ab} = 1.06$; *recombination fraction,* $c = 0.5$

| Generation | Percent chromosome frequency before selection | | | Linkage disequilibrium | | Change in population fitness | Additive and epistatic variances | |
|---|---|---|---|---|---|---|---|---|
| $t$ | $x \times 10^2$ | $y(=z) \times 10^2$ | $u \times 10^2$ | $R$ | $D \times 10^2$ | $\Delta\bar{w} \times 10^4$ | $V_A \times 10^4$ | $V_{EP} \times 10^4$ |
| 0 | 49.0000 | 21.0000 | 9.0000 | 1.10371 | 0.4419 | 2.63 | 0.42 | 4.41 |
| 5 | 47.8124 | 21.0096 | 10.1683 | 1.21565 | 0.9184 | 0.71 | 0.63 | 4.66 |
| 10 | 45.9617 | 21.4762 | 11.0860 | 1.21930 | 0.9752 | 0.81 | 0.80 | 4.88 |
| 20 | 41.5594 | 22.496 | 13.4485 | 1.21895 | 1.0662 | 1.30 | 1.29 | 5.34 |
| 40 | 29.5776 | 24.2446 | 21.9332 | 1.21811 | 1.2228 | 3.57 | 3.54 | 6.18 |
| 80 | 3.1992 | 14.0789 | 68.6430 | 1.22280 | 0.3955 | 10.77 | 11.07 | 2.13 |
| 100 | 0.2827 | 4.7859 | 90.1454 | 1.22814 | 0.0453 | 5.11 | 5.36 | 0.25 |
| 200 | 0.0000 | 0.0022 | 99.9955 | 1.23141 | 0.0000 | 0.00 | 0.00 | 0.00 |

loci are independent and either $A_2$ or $B_2$ alone decreases fitness by 2% but the two together increase fitness by 6%. The initial frequencies of $A_2$ and $B_2$ are each assumed to be 30% and also the initial frequencies of the four chromosomes are assumed to be in "linkage equilibrium" so that $x = 49\%$, $y = z = 21\%$ and $u = 9\%$ (i.e. $r = 1.0$ in the 0th generation). It may be seen from the table that quasi linkage equilibrium is reached after a few generations of random mating and then $R$ changes extremely slowly until it reaches the limiting value of 1.23141 . . . . The approximate value of $\hat{R}$ given by (14) is 1.20 because $\varepsilon = 1.0 - 0.98 - 0.98 + 1.06 = 0.1$ and $c = 0.5$. On the other hand, $D$ changes extensively throughout the process of selection.

Table 1 also shows another important property of quasi linkage equilibrium. Namely, the rate of change of population fitness ($\Delta\bar{w}$) is equal to the additive genetic variance ($V_A$), that is to say, the additive component of the total genotypic variance. This is again an approximation, but, as seen from the table, the agreement between these two quantities is close enough to rule out the existence of any appreciable contribution from epistatic variance ($V_{EP}$), except for the 0th generation for which $\Delta\bar{w} = V_A + \frac{1}{2} V_{EP}$ holds. This last relationship, however, is the result of artificially imposing "linkage equilibrium" ($r = 1.0$) where such an equilibrium can not be realized if the change in gene frequencies is kept going on due to natural selection.

The important property,

$$\Delta\bar{w} = V_A \tag{15}$$

at the state of quasi linkage equilibrium may be derived as follows: Since $\Delta \log r = \Delta \log R = 0$ in this state, we have, from equation 9,

$$\varepsilon - c\bar{w}DI = 0, \tag{16}$$

where

$$I = x^{-1} + y^{-1} + z^{-1} + u^{-1}.$$

Thus

$$\Delta\bar{w} = w_o\Delta x + w_a\Delta y + w_b\Delta z + w_{ab}\Delta u$$

$$= \frac{1}{\bar{w}}(w_o^2\, x + w_a^2\, y + w_b^2\, z + w_{ab}^2\, u - \bar{w}^2) - cD\varepsilon \qquad \text{(using (5))}$$

$$=(V_T - c\bar{w}D\varepsilon)/\bar{w}$$

$$=(V_T - \frac{\varepsilon^2}{I})/\bar{w} \qquad \text{(using (16))}$$

$$=(V_T - V_{EP})/\bar{w}$$

and therefore

$$\Delta\bar{w} = V_A/\bar{w}, \qquad (17)$$

where $V_T$ is the total genotypic variance and $V_{EP}$ is the epistatic variance. The additive genetic variance $V_A = V_T - V_{EP}$ is the variance due to the additive effects of the genes (see Appendix I). Since we assume small selection coefficients, we may put $\bar{w} \approx 1$ and (15) follows from (17). It is interesting to note in the above derivation that the amount of linkage disequilibrium is such that the term $-c\bar{w}D\varepsilon$ exactly cancels out the epistatic component of variance in fitness. This means that however small the epistatic effect may be, the assumption of "linkage equilibrium" is inappropriate here.

In the above example, free recombination ($c=0.5$) was assumed, but for closer linkage, the approach to quasi linkage equilibrium may be slower. Table 2 gives an example with the recombination fraction, $c=0.2$ and selective values, $w_o=1.00$, $w_a=w_b=0.99$, $w_{ab}=1.02$, i.e., $A_2$ or $B_2$ singly reduces fitness by 1% but in combination they increase fitness by 2%. It is assumed that initial frequencies of $A_2$ and $B_2$ are respectively 0.25 and 0.20, and that the four chromosome types are in linkage equilibrium so that $x=0.60$, $y=0.20$, $z=0.15$ and $u=0.05$. The table shows that at generation 40, $\Delta\bar{w}$ and $V_A$ agree with the error of about 4%, while $V_{EP}$ is more than 100 times as large as $\Delta\bar{w}$. The error is reduced to about 2% in generation 200.

## *Diploid population*

The essential part of the foregoing argument can be extended to cover the diploid population, as I would like to show in this section.

Let us denote by $X_1$, $X_2$, $X_3$ and $X_4$ the frequencies of four chromosome types $A_1B_1$, $A_2B_1$, $A_1B_2$ and $A_2B_2$ immediately after fertilization. It may be convenient

TABLE 2

*An example similar to that of Table 1, but with closer linkage. Selective values,* $w_o = 1.00$, $w_a = w_b = 0.99$, $w_{ab} = 1.02$; *recombination fraction,* $c = 0.2$

| Generation $t$ | Percent chromosome frequency before selection $x\times10^2$ | $y\times10^2$ | $z\times10^2$ | $u\times10^2$ | Linkage disequilibrium $R$ | Change in population fitness $\Delta\bar{w}\times10^7$ | Additive and epistatic variances $V_A\times10^7$ | $V_{EP}\times10^7$ |
|---|---|---|---|---|---|---|---|---|
| 0 | 60.0000 | 20.0000 | 15.0000 | 5.0000 | 1.04 | 392.6 | 7.500 | 480.0 |
| 5 | 60.4542 | 19.5306 | 14.7174 | 5.2970 | 1.16 | 138.1 | 5.302 | 493.1 |
| 10 | 60.6582 | 19.2918 | 14.6605 | 5.3895 | 1.20 | 50.59 | 4.629 | 496.8 |
| 20 | 60.8672 | 19.0073 | 14.7211 | 5.4044 | 1.22 | 9.388 | 4.042 | 496.9 |
| 40 | 61.1902 | 18.5743 | 14.9044 | 5.3312 | 1.23 | 3.576 | 3.441 | 492.5 |
| 80 | 61.9243 | 17.8173 | 15.1281 | 5.1303 | 1.23 | 3.353 | 3.153 | 480.1 |
| 160 | 64.0225 | 16.4390 | 14.9974 | 4.5410 | 1.23 | 5.811 | 5.678 | 440.4 |
| 200 | 65.5113 | 15.7009 | 14.6467 | 4.1411 | 1.23 | 9.015 | 8.816 | 411.6 |

TABLE 3

*Designation of chromosome frequencies and selective values of genotypes*

| | | $X_1$ $A_1B_1$ | $X_2$ $A_2B_1$ | $X_3$ $A_1B_2$ | $X_4$ $A_2B_2$ |
|---|---|---|---|---|---|
| $X_1$ | $A_1B_1$ | $w_{11}$ | $w_{12}$ | $w_{13}$ | $w_{14}$ |
| $X_2$ | $A_2B_1$ | $w_{21}$ | $w_{22}$ | $w_{23}$ | $w_{24}$ |
| $X_3$ | $A_1B_2$ | $w_{31}$ | $w_{32}$ | $w_{33}$ | $w_{34}$ |
| $X_4$ | $A_2B_2$ | $w_{41}$ | $w_{42}$ | $w_{43}$ | $w_{44}$ |

to give these chromosomes the numbers 1, 2, 3 and 4 so that the frequency of chromosome $i$ is $X_i$ ($i$=1,2,3,4). We will designate by $w_{ij}$ the selective values of the genotype formed by the union of chromosomes $i$ and $j$ as shown in Table 3 ($i, j$=1, . . . , 4). As before, the recombination fraction between the two loci will be denoted by $c$. Then the amount of change in one generation of these chromosome frequencies may be given by the following set of equations.

$$\Delta X_1 = \frac{X_1(w_{1.} - \bar{w}) - cD_W}{\bar{w}}$$

$$\Delta X_2 = \frac{X_2(w_{2.} - \bar{w}) + cD_W}{\bar{w}}$$

$$\Delta X_3 = \frac{X_3(w_{3.} - \bar{w}) + cD_W}{\bar{w}} \qquad (18)$$

$$\Delta X_4 = \frac{X_4(w_{4.} - \bar{w}) - cD_W}{\bar{w}}$$

where $w_{i.}$ is the average selective value of chromosome $i$, ($i$=1, 2, 3, 4), that is

$$w_{i.} = \sum_{j=1}^{4} w_{ij} X_j, \qquad (19)$$

$\bar{w}$ is the average selective value of the population

$$\bar{w} = \sum_{i=1}^{4} w_{i.} X_i = \sum_{i,j} w_{ij} X_i X_j \qquad (20)$$

and

$$D_W = w_{14} X_1 X_4 - w_{23} X_2 X_3. \qquad (21)$$

The above set of equations (18) is slightly more general than the one given by Kimura (1956) but equivalent to that given by Bodmer and Parsons (1962). Since equations 18 are a natural extension of equations 5, we again let

$$R = \frac{X_1 X_4}{X_2 X_3} \qquad (22)$$

and consider the rate of change of *log R* per generation as in the case of the haploid population. This leads to

$$\bar{w}\Delta log\, R = \bar{\varepsilon} - cD_W(X_1^{-1} + X_2^{-1} + X_3^{-1} + X_4^{-1}) \qquad (23)$$

or

$$\bar{w}\Delta log\, R = \bar{\varepsilon} - c(w_{14} R - w_{23})(\frac{X_1 + X_4}{R} + X_2 + X_3), \qquad (23')$$

TABLE 4

*A numerical example showing the attainment of quasi linkage equilibrium in a diploid population (completely recessive mutations). Selective values,* $w_{11}=w_{12}=w_{21}=w_{13}=w_{31}=w_{14}=w_{41}=w_{23}=w_{32}=1.00$, $w_{22}=w_{24}=w_{42}=0.99$, $w_{33}=w_{34}=w_{43}=0.985$, $w_{44}=1.02$; *Recombination fraction,* $c=0.5$

| Generation $t$ | Percent chromosome frequencies $X_1\times10^2$ | $X_2\times10^2$ | $X_3\times10^2$ | $X_4\times10^2$ | Linkage disequilibrium $R$ | Change in fitness $\Delta\bar{w}\times10^5$ | Additive and epistatic chromosomal variances $V_{AC}\times10^5$ | $V_{EPC}\times10^5$ |
|---|---|---|---|---|---|---|---|---|
| 0 | 20.0000 | 20.0000 | 30.0000 | 30.0000 | 1.000000 | 2.926 | 0.726178 | 2.18700 |
| 10 | 20.1063 | 20.4041 | 29.1205 | 30.3691 | 1.02766 | 0.661 | 0.658748 | 2.24996 |
| 50 | 19.6403 | 22.3886 | 26.6847 | 31.2864 | 1.02852 | 0.455 | 0.452730 | 2.40233 |
| 100 | 18.5716 | 24.2562 | 24.3791 | 32.7932 | 1.02990 | 0.370 | 0.365753 | 2.61448 |
| 200 | 14.6145 | 25.9359 | 20.9553 | 38.4943 | 1.03511 | 0.933 | 0.914888 | 3.32151 |
| 300 | 7.2680 | 22.8986 | 16.2207 | 53.6127 | 1.04908 | 5.7138 | 5.64348 | 4.45057 |
| 400 | 0.5350 | 8.4571 | 5.0317 | 85.9762 | 1.08091 | 15.29 | 15.4166 | 1.36214 |
| 500 | 0.0018 | 0.6407 | 0.2544 | 99.1031 | 1.09538 | 1.68 | 1.71846 | 0.00709 |
| 1500 | 0.0000 | 0.0000 | 0.0000 | 100.0000 | 1.09642 | 0.00 | 0.00000 | 0.00000 |

where

$$\bar{\varepsilon} = w_{1.} - w_{2.} - w_{3.} + w_{4.} \tag{24}$$

is the epistatic effect involving four chromosome types.

Let us consider a situation in which gene frequencies are changing slowly under loose linkage and weak selection. We may see from equation 23′ that starting from an arbitrary positive value, $R$ will be adjusted quickly to attain the quasi linkage equilibrium, where

$$\Delta \log R = 0 \tag{25}$$

or

$$R = \text{constant}. \tag{26}$$

Either of these is an approximation but seems to be good enough for practical purposes, though the accuracy might be less as compared with the haploid case, if $\bar{\varepsilon}$ does not behave nicely. A considerable number of numerical examples have been studied by computers and the results have shown that the assumption of quasi linkage equilibrium is satisfactory. Table 4 shows an example in which the two loci are independent ($c=0.5$) and both mutant alleles $A_2$ and $B_2$ are recessive. It is assumed that $A_2$, when homozygous, decreases fitness by 1%, and similarly, $B_2$ when homozygous, decreases fitness by 1.5%, while $A_2$ and $B_2$ in combination as double mutants increase fitness by 2%. In the 0th generation, frequencies of $A_2$ and $B_2$ are respectively 50% and 60%.
Also "linkage equilibrium" ($R=1$) is assumed between the two loci in the 0th generation so that $X_1=0.2$, $X_2=0.2$, $X_3=0.3$, $X_4=0.3$ at the start. The computation was carried out until generation 1,500. The quasi linkage equilibrium is attained in a few generations and then $R$ changes extremely slowly, as shown in the table. The table also reveals an important property of the quasi linkage equilibrium, i.e.

$$\Delta\bar{w} = V_{AC}, \tag{27}$$

where $V_{AC}$ is the additive chromosomal variance as defined in Appendix II. The above relationship can be derived as follows: From equation 20, we have, neglecting higher order terms,

$$\Delta\bar{w} = 2 \sum_{i=1}^{4} w_{i.} \Delta X_i. \tag{28}$$

Substituting (18) for $\Delta X_i$, we get

$$\Delta\bar{w} = \frac{2}{\bar{w}} [ \sum_{i=1}^{4} (w_{i.} - \bar{w})^2 X_i - cD_{W\bar{\varepsilon}} ]. \tag{29}$$

Now, from the assumption of quasi linkage equilibrium,

$$\bar{w}\Delta log\, R = \bar{\varepsilon} - cD_W(X_1^{-1} + X_2^{-1} + X_3^{-1} + X_4^{-1}) = 0,$$

(cf. equation 23)

and therefore

$$cD_W = \bar{\varepsilon}/J, \tag{30}$$

where $J = X_1^{-1} + X_2^{-1} + X_3^{-1} + X_4^{-1}$. From Appendix II,

$$V_{TC} = 2 \sum_{i=1}^{4} (w_{i.} - \bar{w})^2 X_i, \tag{31}$$

$$V_{EPC} = 2\bar{\varepsilon}^2/J, \tag{32}$$

and

$$V_{AC} = V_{TC} - V_{EPC}. \tag{33}$$

Thus, substituting (30) in (29), and noting (31), (32) and (33), we get

$$\begin{aligned} \Delta\bar{w} &= \frac{2}{\bar{w}} [ \sum_{i=1}^{4} (w_{i,} - \bar{w})^2 X_i - \bar{\varepsilon}^2/J] \\ &= (V_{TC} - V_{EPC})/\bar{w} \\ &= V_{AC}/\bar{w}, \end{aligned} \tag{34}$$

or assuming that selection coefficients are small so that $\bar{w}$ is approximately unity, we obtain $\Delta\bar{w} = V_{AC}$, as was to be shown. I would like to point out here that the good agreement found in Table 4 at the 0th generation between $\Delta\bar{w}$ and the total chromosomal variance ($V_{AC} + V_{EPC}$) comes from the artificial assumption of "linkage equilibrium" ($R = 1$). When gene frequencies are changing under natural selection, no "linkage equilibrium" could be attained if epistatic effects in fitness are involved. Kojima and Kelleher (1961) studied the rate of change of mean fitness of a population in a similar situation to that studied in the present paper. They argued that the correction term (corresponding to our $-2cD_{W\bar{\varepsilon}}$ in equation 29) may make positive or negative contribution to the rate, with the result that when the number of loci involved gets large, correction factors would tend to cancel each other and the rate of change in fitness would become twice the marginal variance (corresponding to our $V_{TC}$). However, under quasi linkage equilibrium $-2cD_{W\bar{\varepsilon}}$ is just enough for cancelling out the epistatic component of $V_{TC}$. More generally, when gene frequencies are changing under natural selection $D_W$ and $\bar{\varepsilon}$ would tend to have the same sign and $-cD_{W\bar{\varepsilon}}$ to make a negative contribution to the rate of increase in fitness as suggested by the recent work of Felsenstein (1965).

The attainment of quasi linkage equilibrium is not restricted to the case in which the gene frequencies are changing toward fixation. It can also be attained

TABLE 5

*A numerical example showing the attainment of quasi linkage equilibrium when overdominance is involved. Selective values,* $w_{11}=w_{22}=w_{33}=w_{44}=w_{23}=w_{32}=1.00$, $w_{12}=w_{21}=w_{13}=w_{31}=w_{24}=w_{42}=w_{34}=w_{43}=1.01$, $w_{14}=w_{41}=1.05$; *Recombination fraction,* $c=0.5$. *Chromosome frequencies at the 0 generation,* $X_1=0.6$, $X_2=X_3=0.0$, $X_4=0.4$

| Generation $t$ | Percent chromosome frequencies $X_1\times10^2$ | $X_2\times10^2=$ $X_3\times10^2$ | $X_4\times10^2$ | Linkage disequilibrium $R$ | Change in fitness $\Delta\bar{w}\times10^5$ | Additive and epistatic chromosomal variances $V_{AC}\times10^5$ | $V_{EPC}\times10^5$ |
|---|---|---|---|---|---|---|---|
| 1 | 47.4609 | 12.3047 | 27.9297 | 8.75510 | —319.76 | 2.24232 | 6.91295 |
| 5 | 35.9730 | 23.4166 | 17.1937 | 1.12797 | —17.30 | 0.599233 | 7.48018 |
| 10 | 34.9579 | 24.1481 | 16.7460 | 1.00390 | —0.04 | 0.502889 | 7.40367 |
| 20 | 34.3155 | 24.2625 | 17.1595 | 1.00029 | 0.43 | 0.445136 | 7.44652 |
| 30 | 33.7330 | 24.3458 | 17.5753 | 1.00025 | 0.39 | 0.395017 | 7.48770 |
| 40 | 33.1869 | 24.4200 | 17.9731 | 1.00022 | 0.34 | 0.352348 | 7.52441 |
| 50 | 32.6752 | 24.4860 | 18.3529 | 1.00020 | 0.30 | 0.313724 | 7.55707 |
| 60 | 32.1956 | 24.5446 | 18.7151 | 1.00018 | 0.27 | 0.272736 | 7.58614 |
| 70 | 31.7464 | 24.5967 | 19.0602 | 1.00016 | 0.24 | 0.252879 | 7.61195 |
| 80 | 31.3255 | 24.6429 | 19.3886 | 1.00014 | 0.22 | 0.221015 | 7.63488 |
| 90 | 30.9312 | 24.6840 | 19.7009 | 1.00012 | 0.19 | 0.197671 | 7.65524 |
| 99 | 30.5978 | 24.7169 | 19.9685 | 1.00011 | 0.18 | 0.175360 | 7.67159 |

when gene frequencies are changing toward intermediate equilibrium values, as shown in Table 5. In this example, overdominance is assumed.
A single heterozygote has 1% selective advantage; the double heterozygote has 5% advantage in the coupling phase, but none in the repulsion phase. The initial population (0th generation) is assumed to contain only two types of chromosomes, $A_1B_1$ and $A_2B_2$ with respective frequencies of 60% and 40% so that $R=\infty$ and $\bar{w}=1.024$. Also free recombination ($c=0.5$) is assumed. The table shows the process of change in the genetic constitution of the population for the succeeding 99 generations. Since the population starts from excess of the coupling phase and since only this phase is assumed to have selective advantage in double heterozygotes, the population fitness $\bar{w}$ decreases rather rapidly for the first few generations as shown in the table. However, the decrease stops at the 10th generation and then $\bar{w}$ starts to increase. By the 20th generation, quasi linkage equilibrium seems to have been established, as suggested by the good agreement between $\Delta\bar{w}$ and $V_{AC}$. Actually, change of $R$ is very slow from the 20th generation onward as the table shows.

It is important to note that the concept of quasi linkage equilibrium does not apply unless gene frequencies are changing. Change of chromosome frequencies alone is not enough. To show this, an example is given in Table 6, in which selective values and recombination fraction are exactly the same as in the above example (Table 5), but initial chromosome frequencies are $X_1=X_4=0.5$, $X_2=X_3=0.0$. In this example, there is no change in gene frequencies because they are equal to the equilibrium value of 0.5 from the start. Chromosome frequencies alone are changing and as the frequency of repulsion phase increases, the mean

TABLE 6

*An example in which the mean fitness of population ($\bar{w}$) is always decreasing*

| Generation $t$ | Percent chromosome frequencies $X_1\times10^2=X_4\times10^2$ | $X_2\times10^2=X_3\times10^2$ | Linkage disequilibrium $R$ | Change in fitness $\Delta\bar{w}\times10^5$ |
|---|---|---|---|---|
| 0 | 50.0000 | 0.0000 | $\infty$ | —735.51 |
| 1 | 37.1951 | 12.8049 | 8.43764 | —332.72 |
| 2 | 30.9918 | 19.0082 | 2.65836 | —157.35 |
| 4 | 26.4582 | 23.5418 | 1.26312 | —37.21 |
| 6 | 25.3561 | 24.6439 | 1.05864 | —9.03 |
| 8 | 25.0870 | 24.9130 | 1.01402 | —2.20 |
| 10 | 25.0213 | 24.9787 | 1.00341 | —0.54 |
| 12 | 25.0052 | 24.9948 | 1.00083 | —0.13 |
| 14 | 25.0013 | 24.9987 | 1.00020 | —0.03 |
| 16 | 25.0003 | 24.9997 | 1.00005 | —0.01 |
| 18 | 25.0001 | 24.9999 | 1.00001 | —0.00 |
| 20 | 25.0000 | 25.0000 | 1.00000 | —0.00 |

There is no change in gene frequencies, and therefore the concept of quasi linkage equilibrium is irrelevant in this case. Selective values and the recombination fractions are exactly the same as in Table 5, but the initial chromosome frequencies are $X_1=X_4=0.5$ and $X_2=X_3=0.0$.

fitness decreases until all the chromosome frequencies are in equilibrium. On the other hand, if we start from excess of the repulsion phase such as $X_1 = X_4 = 0.0$ and $X_2 = X_3 = 0.5$, the frequency of the coupling phase, and therefore the mean fitness of the population, would increase from generation to generation. Again, the concept of quasi linkage equilibrium is irrelevant because of no change in gene frequencies throughout the process.

Finally, I would like to show an example in which the recombination fraction is much smaller than the epistatic interaction and linkage disequilibrium is built

TABLE 7

*An example in which linkage disequilibrium is built up indefinitely. Selective values,* $w_{11}=w_{12}=w_{21}=w_{13}=w_{31}=w_{14}=w_{41}=w_{23}=w_{32}=1.0$, $w_{22}=w_{33}=w_{24}=w_{42}=w_{34}=w_{43}=0.95$, $w_{44}=1.10$; *Recombination fraction,* $c=0.01$

| Generation $t$ | Percent chromosome frequencies $X_1\times10^2$ | $X_2\times10^2=X_3\times10^2$ | $X_4\times10^2$ | Linkage disequilibrium $R$ | Change in fitness $\Delta\bar{w}\times10^5$ | Additive, epistatic and total chromosomal variances $V_{AC}\times10^5$ | $V_{EPC}\times10^5$ | $V_{TC}\times10^5$ |
|---|---|---|---|---|---|---|---|---|
| 0 | 25.000 | 25.000 | 25.000 | 1.00 | 31.84 | 0.002 | 31.25 | 31.25 |
| 10 | 27.776 | 21.805 | 28.614 | 1.67 | 38.762 | 1.456 | 40.26 | 41.72 |
| 20 | 29.365 | 18.526 | 33.583 | 2.87 | 53.561 | 8.790 | 52.52 | 61.31 |
| 40 | 26.299 | 11.441 | 50.819 | $1.02\times10^1$ | 144.62 | 79.88 | 88.86 | 168.73 |
| 80 | 2.474 | 0.702 | 96.122 | $4.83\times10^2$ | 82.82 | 76.65 | 22.64 | 99.29 |
| 100 | 0.325 | 0.086 | 99.503 | $4.38\times10^3$ | 11.59 | 11.01 | 3.00 | 14.01 |
| 120 | 0.040 | 0.010 | 99.940 | $3.81\times10^4$ | 1.41 | 1.35 | 0.36 | 1.71 |
| 140 | 0.005 | 0.001 | 99.993 | $3.22\times10^5$ | 0.17 | 0.17 | 0.04 | 0.21 |
| 200 | 0.000 | 0.000 | 100.000 | $1.82\times10^8$ | 0.00 | 0.00 | 0.00 | 0.00 |
| 400 | 0.000 | 0.000 | 100.000 | $2.59\times10^{17}$ | 0.00 | 0.00 | 0.00 | 0.00 |
| 450 | 0.000 | 0.000 | 100.000 | $5.03\times10^{19}$ | 0.00 | 0.00 | 0.00 | 0.00 |

up indefinitely as time goes on so that quasi linkage equilibrium is not attained. This is shown in Table 7. In this case, both mutant genes $A_2$ and $B_2$ are recessive and each in single homozygote reduces fitness by 5% but in combination they increase fitness by 10% in double mutant homozygotes. A tight linkage with recombination fraction $c = 0.01$ is assumed. The value of $R$ which is unity at generation 0 reaches $5.03 \times 10^{19}$ at generation 450, and it will continue to increase indefinitely. Note here that $D$ tends to zero as $R$ tends to infinity. Throughout the process, $\Delta\bar{w}$ is roughly equal to the total chromosomal variance ($V_{TC}$) and neither $V_{AC}$ nor $V_{EPC}$ alone fails to give good approximation to $\Delta\bar{w}$. This is understandable, since the two loci may behave as if they were a single locus under tight linkage and relatively strong epistatic interaction.

## DISCUSSION

I hope that the above demonstrations are enough to show that *when gene frequencies are changing slowly* under loose linkage and relatively weak epistatic interaction, the state is quickly realized in which chromosome frequencies are changing in such a way that $R =$ constant, where $R$ is the ratio between the frequencies of coupling and repulsion phases, i.e., $R = (X_1X_4)/(X_2X_3)$. The state is termed quasi linkage equilibrium and though its formulation is an approximation, it seems to be good enough for practical purposes. It may be most useful in the treatment of cases in which two loci are segregating independently and selection coefficients are at most of the order of a few percents. The term "quasi linkage equilibrium" should not be confused with "quasi gene frequency equilibrium" used by BODMER and PARSONS (1962) to denote an equilibrium for which $D$ is small. Furthermore, it may be noted that a small $D$ is not necessarily equivalent to a small $(R - 1)$, as shown by the example in Table 7. Generally, $R$ is more sensitive to linkage disequilibrium than $D$ and less dependent on gene frequencies.

The assumption of quasi linkage equilibrium leads to some important conclusions. I have already shown one of them that is related to the rate of change in the mean fitness of the population. Namely,

$$\Delta\bar{w} = V_{AC}/\bar{w}, \tag{35}$$

as shown in (34). In this formula $V_{AC}$ is the additive chromosomal variance. For weak selection, $\bar{w}$ is near unity and we have $\Delta\bar{w} = V_{AC}$ as a good approximation. It might be thought here that the additive chromosomal variance contains, besides the genic (or additive genetic) variance $V_g$, some epistatic components of variance in fitness. However, as shown in Appendix III, $V_{AC}$ is exactly equal to $V_g$ under random matings, so that we have

$$\Delta\bar{w} = V_g/\bar{w}, \tag{36}$$

even if linkage disequilibrium is present.

For overlapping generations, fitness may be measured in Malthusian parameters and the corresponding formulation should be

$$d\bar{a}/dt = v_g, \tag{37}$$

where $\bar{a}$ is the mean fitness of the population and $v_g$ is the genic variance in fitness. This means that FISHER's fundamental theorem of natural selection (cf. FISHER 1958) holds under quasi linkage equilibrium.

In one of my previous papers treating the change of population fitness by natural selection (KIMURA 1958), I attempted to resolve $d\bar{a}/dt$ into three components, one of which was $v_g$. I also suggested that the case of random mating with linkage may be approached by considering "additive chromosomal variance." At that time the remarkable property of quasi linkage equilibrium was not noticed and I could only carry out a formal analysis of the problem assuming general conditions.

Another important conclusion that follows from the assumption of quasi linkage equilibrium is that the direction of change in gene frequencies by natural selection is such that the mean fitness of the population is increased. This is derived from (35) noting that $V_{AC}$ is nonnegative and therefore

$$\Delta\bar{w} \geqq 0. \tag{38}$$

This result is significant, since it suggests that the stable equilibrium corresponds to the local maximum of $\bar{w}$ with respect to the change in gene frequencies. To be sure, one can find easily an example in which this inequality does not hold, when some unnatural value is artificially imposed upon $R$, as shown by the example in Table 5. In such an example, $R$ will change rapidly in a few generations, but as long as gene frequencies are changing, the quasi linkage equilibrium will soon be attained where (38) holds. Nature is simpler than some artificial examples suggest. MORAN (1964) gave an example in which $\bar{w}$ increases steadily from generation to generation until an equilibrium is reached if one starts from one set of chromosome frequencies, but decreases steadily toward the same equilibrium, if one starts from another set of chromosome frequencies. It turns out that in his example, only $R$ is changing and there is no change in gene frequencies. So the situation is exactly the same as explained by using the example of Table 6.

From his example, MORAN argues that WRIGHT's conception of an "adaptive topography" is not correct since populations do not in general tend to maximize their mean fitness if the latter is dependent on more than one locus. He also tries to show that stationary populations do not in general correspond to stationary values of $\bar{w}$ and that it is unlikely that equations giving equilibrium frequencies can be derived from any principle which maximizes a function of gametic frequencies.

I would like to assert that once the concept of quasi linkage equilibrium is introduced, the classical picture of adaptive topography remains to be useful for studying the change of gene frequencies in a population. For example, equilibrium chromosome frequencies can be derived by considering the stationary points of $\bar{w}$ and the stability of an equilibrium can be tested by seeing if $\bar{w}$ has a local maximum at the equilibrium point, *provided that the side condition of* R = *constant is imposed*, in addition to the ordinary condition that chromosome frequencies add up to unity.

However, a more exact specification of conditions under which quasi-linkage equilibrium holds in such a situation has to be worked out in future.

I would like to thank DR. M. NEI for stimulating discussions during preparation of the manuscript.

## SUMMARY

In a large random mating population, if gene frequencies are changing by natural selection, under loose linkage and relatively weak epistatic interactions, a state is quickly realized in which chromosome frequencies change in such a way that $R$ is kept constant, where $R$ is the ratio between the frequencies of coupling and repulsion phases. Such a state was termed *quasi linkage equilibrium*, and it was shown that several important conclusions follow from its formulation, namely: (1) The rate of change in the mean fitness of a population is equal to the additive genetic variance in fitness. So. FISHER's fundamental theorem holds. (2) The direction of change in gene frequencies by natural selection is such that the mean fitness is increased. (3) The stable equilibrium of gene frequencies corresponds to the local maximum of the mean fitness and WRIGHT's classical picture of "adaptive topography" continues to be useful if the concept of quasi linkage equilibrium is taken into account.

In order to corroborate the above points, an extensive numerical study was carried out with the help of high speed computers and some of the results are presented.

APPENDIX I. *Analysis of variances in fitness with two segregating loci each with a pair of alleles in a haploid population*

Let $\alpha$ be the average effect of substituting $A_2$ for $A_1$ and let $\beta$ be that of substituting $B_2$ for $B_1$. If we designate fitnesses and frequencies of four genotypes as in Table A.1, additive values $w$, $\alpha$ and $\beta$ may be obtained by minimizing

$$Q=(w_o-w)^2x+(w_a-w-\alpha)^2y+(w_b-w-\beta)^2z+(w_{ab}-w-\alpha-\beta)^2u.$$

Thus, from $\partial Q/\partial w=\partial Q/\partial\alpha=\partial Q/\partial\beta=0$, we obtain

$$(w_o-w)x=-(w_a-w-\alpha)y=-(w_b-w-\beta)z=(w_{ab}-w-\alpha-\beta)u\equiv K,$$

where

$$K=\varepsilon/I$$

in which $\varepsilon=w_o-w_a-w_b+w_{ab}$ and $I=x^{-1}+y^{-1}+z^{-1}+u^{-1}$. The epistatic variance ($V_{EP}$) is the sum of squares of deviations from additive approximation, i.e. the minimum value of $Q$:

$$V_{EP}=K^2x^{-1}+K^2y^{-1}+K^2z^{-1}+K^2u^{-1}=K^2I=\varepsilon^2/I.$$

The additive genetic variance ($V_A$) is the variance due to the additive values and can be shown to be equal to $V_T-V_{EP}$, where

$$V_T=w_o{}^2x+w_a{}^2y+w_b{}^2z+w_{ab}{}^2u-\bar{w}^2$$

is the total genotypic variance.

TABLE A.1

| Genotype | Fitness | Additive value | Frequency |
|---|---|---|---|
| $A_1B_1$ | $w_o$ | $w$ | $x$ |
| $A_2B_1$ | $w_a$ | $w+\alpha$ | $y$ |
| $A_1B_2$ | $w_b$ | $w+\beta$ | $z$ |
| $A_2B_2$ | $w_{ab}$ | $w+\alpha+\beta$ | $u$ |

TABLE A.2

| Chromosome | (i) | Average value | Additive value | Frequency |
|---|---|---|---|---|
| $A_1B_1$ | (1) | $w_{1.}$ | $W$ | $X_1$ |
| $A_2B_1$ | (2) | $w_{2.}$ | $W+A$ | $X_2$ |
| $A_1B_2$ | (3) | $w_{3.}$ | $W+B$ | $X_3$ |
| $A_2B_2$ | (4) | $w_{4.}$ | $W+A+B$ | $X_4$ |

APPENDIX II. *Additive and epistatic chromosomal variances in a population of a diploid organism.*

Let $w_{i.}$ be the average selective value of chromosome $i$ as defined in (19), that is

$$w_{i.} = \sum_{j=1}^{4} w_{ij} X_j \qquad (i=1, \ldots, 4)$$

The total chromosomal variance, $V_{TC}$, may be defined as twice the variance due to $w_{i.}$'s, because each individual has two homologous chromosomes (considering only autosomes);

$$V_{TC} = 2 \sum_{i=1}^{4} (w_{i.} - \bar{w})^2 X_i.$$

In order to extract the additive component from this, let $W$, $A$, and $B$ be additive values as shown in Table A.2, and choose these parameters so that

$$Q = X_1(w_{1.}-W)^2 + X_2(w_{2.}-W-A)^2 + X_3(w_{3.}-W-B)^2 + X_4(w_{4.}-W-A-B)^2$$

is minimized. From $\partial Q/\partial W = \partial Q/\partial A = \partial Q/\partial B = 0$, we obtain

$$(w_{1.}-W)X_1 = -(w_{2.}-W-A)X_2 = -(w_{3.}-W-B)X_3 = (w_{4.}-W-A-B)X_4 \equiv K,$$

where

$$K = \bar{\varepsilon}/J$$

in which $\bar{\varepsilon} = w_{1.} - w_{2.} - w_{3.} + w_{4.}$ and $J = X_1^{-1} + X_2^{-1} + X_3^{-1} + X_4^{-1}$. Then it can be shown that the minimum value of $Q$ is

$$Q_m = \bar{\varepsilon}^2/J.$$

It can also be shown that, if we define the additive chromosomal variance, $V_{AC}$, as twice the variance due to the additive values, and the epistatic chromosomal variance as twice the minimum value of $Q$, then

$$V_{AC} = V_{TC} - V_{EPC},$$

where

$$V_{EPC} = 2\bar{\varepsilon}^2/J.$$

Parameters $A$ and $B$ may be obtained by solving the following set of equations:

$$p_1 p_2 A + DB = C_p$$

$$\text{(A II.1)}$$

$$DA + q_1 q_2 B = C_q,$$

where $p_1$ and $p_2$ are frequencies of $A_1$ and $A_2$ in the first locus, $q_1$ and $q_2$ are those of $B_1$ and $B_2$ in the second locus and

$$C_p = X_2(w_{2.} - \bar{w}) + X_4(w_{4.} - \bar{w})$$
$$C_q = X_3(w_{3.} - \bar{w}) + X_4(w_{4.} - \bar{w})$$

In terms of $A$ and $B$, the additive chromosomal variance is

$$V_{AC} = 2A^2 p_1 p_2 + 4ABD + 2B^2 q_1 q_2, \qquad \text{(A II.2)}$$

where

$$D = X_1 X_4 - X_2 X_3.$$

APPENDIX III. *Proof that under random mating the additive component of the total chromosomal variance is equal to the genic or additive genetic variance.*

TABLE A.3

| Genotype | Fitness | Additive value | Frequency |
|---|---|---|---|
| $A_1B_1/A_1B_1$ | $w_{11}$ | $\omega$ | $X_1^2$ |
| $A_1B_1/A_2B_1$ | $w_{12}$ or $w_{21}$ | $\omega+\alpha$ | $2X_1X_2$ |
| $A_1B_1/A_1B_2$ | $w_{13}$ or $w_{31}$ | $\omega+\beta$ | $2X_1X_3$ |
| $A_1B_1/A_2B_2$ | $w_{14}$ or $w_{41}$ | $\omega+\alpha+\beta$ | $2X_1X_4$ |
| $A_2B_1/A_2B_1$ | $w_{22}$ | $\omega+2\alpha$ | $X_2^2$ |
| $A_2B_1/A_1B_2$ | $w_{23}$ or $w_{32}$ | $\omega+\alpha+\beta$ | $2X_2X_3$ |
| $A_2B_1/A_2B_2$ | $w_{24}$ or $w_{42}$ | $\omega+2\alpha+\beta$ | $2X_2X_4$ |
| $A_1B_2/A_1B_2$ | $w_{33}$ | $\omega+2\beta$ | $X_3^2$ |
| $A_1B_2/A_2B_2$ | $w_{34}$ or $w_{43}$ | $\omega+\alpha+2\beta$ | $2X_3X_4$ |
| $A_2B_2/A_2B_2$ | $w_{44}$ | $\omega+2\alpha+2\beta$ | $X_4^2$ |

Let $\alpha$ be the additive effect of substituting $A_2$ for $A_1$ and let $\beta$ be that of substituting $B_2$ for $B_1$ as shown in Table A.3. Parameters, $\omega$, $\alpha$ and $\beta$ are determined in such a way that the sum of squares of deviation from additive expectation is minimized. This leads to the following equations for $\alpha$ and $\beta$:

$$p_1p_2\alpha + D\beta = C_p$$

$$D\alpha + q_1q_2\beta = C_q, \tag{A III.1}$$

where $p_1=X_1+X_3$, $p_2=X_2+X_4$, $q_1=X_1+X_2$, $q_2=X_3+X_4$, $D=X_1X_4-X_2X_3$, and

$$C_p = X_2(w_{2.}-\bar{w}) + X_4(w_{4.}-\bar{w})$$
$$C_q = X_3(w_{3.}-\bar{w}) + X_4(w_{4.}-\bar{w}).$$

The genic or additive genetic variance, denoted by $V_g$, is the sum of squares due to additive values, $\alpha$ and $\beta$. It can be shown that

$$V_g = 2\alpha^2p_1p_2 + 4\alpha\beta D + 2\beta^2q_1q_2 \tag{A III. 2}$$

Comparison of A( III.1) with (A II.1) shows that $A=\alpha$ and $B=\beta$. Therefore, it is clear from the comparison of (A III.2) with (A II.2) that $V_{AC}=V_g$.

LITERATURE CITED


Bodmer, W. F., and P. A. Parsons, 1962 Linkage and recombination in evolution. Advan. Genet. **11**: 1–100.

Felsenstein, J., 1965 The effect of linkage on directional selection. Genetics **52**: 349–363.

Fisher, R. A., 1958 *The Genetical Theory of Natural Selection*. 2nd edition. Dover, New York.

Kimura, M., 1956 A model of a genetic system which leads to closer linkage by natural selection. Evolution **10**: 278–287. —— 1958 On the change of population fitness by natural selection. Heredity **22**: 145–167.

Kojima, K., and T. Kelleher, 1961 Changes of mean fitness in random mating populations when epistasis and linkage are present. Genetics **46**: 527–540.

Lewontin, R. C., 1964a The interaction of selection and linkage. I. General considerations; heterotic models. Genetics **49**: 49–67. —— 1964b The interaction of selection and linkage. II. Optimum models. Genetics **50**: 757–782.

Lewontin, R. C., and K. Kojima, 1960 The evolutionary dynamics of complex polymorphisms. Evolution **14**: 458–472.

MORAN, P. A. P., 1964 On the nonexistence of adaptive topographies. Ann. Human Genet. **27:** 383–393.

NEI, M., 1964a Effects of linkage and epistasis on the equilibrium frequencies of lethal genes. I. Linkage equilibrium. Japan. J. Genet. **39:** 1–6. —— 1964b Effects of linkage and epistasis on the equilibrium frequencies of lethal genes. II. Numerical solutions. Japan. J. Genet. **39:** 7–25.

PARSONS, P. A. P., 1963 Polymorphism and the balanced polygenic combination. Evolution **17:** 564–574.

WRIGHT, S., 1965 Factor interaction and linkage in evolution. Proc. Roy. Soc. London B **162:** 80–104.

# Development of associative overdominance through linkage disequilibrium in finite populations*

By TOMOKO OHTA AND MOTOO KIMURA

*National Institute of Genetics, Mishima, Japan*

(*Received* 18 *February* 1970)

SUMMARY

Associative overdominance arises at an intrinsically neutral locus through its non-random association with overdominant loci. In finite populations, even if fitness is additive between loci, non-random association will be created by random genetic drift.

The magnitude of such associative overdominance is roughly proportional to the sum of $\sigma_d^2$'s between the neutral and the surrounding overdominant loci, where $\sigma_d^2$ is the squared standard linkage deviation, defined between any two loci by the relation

$$\sigma_d^2 = E(D^2)/E\{p(1-p)q(1-q)\},$$

in which $p$ and $1-p$ are frequencies of alleles $A_1$ and $A_2$ in the first locus, $q$ and $1-q$ are frequencies of alleles $B_1$ and $B_2$ in the second locus, and $D$ is the coefficient of linkage disequilibrium. A theory was developed based on diffusion models which enables us to obtain formulae for $\sigma_d^2$ under various conditions, and Monte Carlo experiments were performed to check the validity of those formulae.

It was shown that if $A_1$ and $A_2$ are strongly overdominant while $B_1$ and $B_2$ are selectively neutral, we have approximately

$$\sigma_d^2 = 1/(4N_e c),$$

provided that $4N_e c \gg 1$, where $N_e$ is the effective population size and $c$ is the recombination fraction between the two loci. This approximation formula is also valid between two strongly overdominant as well as weakly overdominant loci, if $4N_e c \gg 1$.

The significance of associative overdominance for the maintenance of genetic variability in natural populations was discussed, and it was shown that $N_e s'$, that is, the product between effective population size and the coefficient of associative overdominance, remains constant with varying $N_e$, if the total segregational (overdominant) load is kept constant.

The amount of linkage disequilibrium expected due to random drift in experimental populations was also discussed, and it was shown that $\sigma_d^2 = 1/(n-1)$ in the first generation, if it is produced by extracting $n$ chromosomes from a large parental population in which $D = 0$.

* Contribution No. 757 from the National Institute of Genetics, Mishima, Shizuoka-ken, 411, Japan. Aided in part by a Grant-in-Aid from the Ministry of Education, Japan.

## 1. INTRODUCTION

It has been pointed out by several authors that linkage disequilibrium may create an apparent overdominance at intrinsically non-overdominant loci. In other words, non-random association with overdominant or ordinarily dominant loci may result in an apparent heterozygote advantage (Comstock & Robinson, 1952; Frydenberg, 1963; Chigusa & Mukai, 1964; Maruyama & Kimura, 1968). However, the underlying mechanism for such apparent overdominance has never been clarified.

Recently, Sved (1968); Ohta & Kimura (1969*b*) presented theoretical treatments of this problem by considering non-random association between neutral and overdominant loci due to random drift in finite populations. They showed that the degree of associative overdominance depends on the square of the coefficient of linkage disequilibrium, $D^2$, which in turn depends on the effective population size and the recombination fraction. Sved used a model in which all gene frequencies are assumed to be held at 50 % by strong overdominance while mutation is so rare as to be negligible. In considering natural populations, however, it may be more appropriate to assume a steady state in which random drift, recurrent mutation and natural selection balance each other.

Ohta & Kimura (1969*b*) developed a more general theory based on diffusion models to obtain the expected value of $D^2$ at steady state determined by random drift and mutation. They also showed that associative overdominance may appear at an intrinsically neutral locus when it is associated with overdominant or ordinarily dominant loci. Their treatment is valid under linear evolutionary pressures, such as mutation and migration, but, to extrapolate this to include selection, even if the selective change of gene frequencies may be linearized without serious error, should need justification. So, in the present paper, we treat a situation in which a neutral locus is linked with a strongly overdominant locus.

Thus, the present paper is an extension and elaboration of our previous work (Ohta & Kimura, 1969*b*), with special reference to the development of associative overdominance due to linkage disequilibrium. We will first present a theoretical treatment based on diffusion models and then demonstrate its validity using Monte Carlo methods.

Also, the bearing of associative overdominance on the maintenance of genetic variability in natural populations will be discussed.

## 2. ASSOCIATIVE OVERDOMINANCE

Let us consider two linked loci and assume that a pair of alleles, $A_1$ and $A_2$, are segregating (with respective frequencies $p$ and $1-p$) in the first locus, and the other pair, $B_1$ and $B_2$ (with frequencies $q$ and $1-q$) in the second locus. No selection is assumed at the $B$ locus and overdominance or ordinary dominance is assumed at the $A$ locus. Let the relative fitnesses of $A_1A_1$, $A_1A_2$ and $A_2A_2$ be respectively $1-s$, $1-hs$ and 1, and let $p_1$ and $p_2$ be the relative frequencies of $A_1$

among $B_1$-and $B_2$-carrying chromosomes. Then the mean fitnesses of $B_1B_1$, $B_1B_2$ and $B_2B_2$, for a given set of values of $p_1$ and $p_2$, are

$$\left.\begin{aligned} W_{B_1B_1} &= 1-2hsp_1-s(1-2h)p_1^2,\\ W_{B_1B_2} &= 1-hs(p_1+p_2)-s(1-2h)p_1p_2,\\ W_{B_2B_2} &= 1-2hsp_2-s(1-2h)p_2^2.\end{aligned}\right\} \tag{1}$$

In order to evaluate their expected values, let $p_1 = p+b_1$ and $p_2 = p-b_2$. If we denote by $g_1, g_2, g_3$ and $g_4$ the relative frequencies of the four types of chromosomes, $A_1B_1$, $A_1B_2$, $A_2B_1$ and $A_2B_2$, then

$$g_1 = q(p+b_1),\quad g_2 = (1-q)(p-b_2),\quad g_3 = q(1-p-b_1),\quad g_4 = (1-q)(1-p+b_2).$$

Therefore, we have $b_1 = D/q$ and $b_2 = D/(1-q)$, where $D = g_1g_4-g_2g_3$, $p = g_1+g_2$ and $q = g_1+g_3$. By substituting these relations in the formulae (1), we get the expected amount of associative overdominance at the $B$ locus:

$$\left.\begin{aligned} E\{W_{B_1B_2}-W_{B_1B_1}\} &= E\left\{(hs+s(1-2h)p)\frac{D}{q(1-q)}\right.\\ &\quad \left.+s(1-2h)\frac{D^2}{q^2(1-q)}\right\} = s(1-2h)E\left\{\frac{D^2}{q^2(1-q)}\right\},\\ E\{W_{B_1B_2}-W_{B_2B_2}\} &= E\left\{-(hs+s(1-2h)p)\frac{D}{q(1-q)}\right.\\ &\quad \left.+s(1-2h)\frac{D^2}{q(1-q)^2}\right\} = s(1-2h)E\left\{\frac{D^2}{q(1-q)^2}\right\}.\end{aligned}\right\} \tag{2}$$

Here, $E$ stands for the operator of taking expectations and we assume that $E(D) = 0$. The quantity, $D^2/q^2(1-q)$ or $D^2/q(1-q)^2$ may be considered as a measure of association in gene frequencies between neutral and overdominant loci, through which apparent overdominance is created at the neutral locus. Now, as reported earlier (Ohta & Kimura, 1969*b*), the squared correlation coefficient ($r^2$) of gene frequencies between two loci is approximately equal to the squared standard linkage deviation, i.e. $\sigma_d^2 = E(D^2)/E\{pq(1-p)(1-q)\}$. In the present paper, we are mainly interested in cases in which $q$ takes an intermediate value rather than considering the expected value. Hence we replace the above expressions by,

$$\left.\begin{aligned} E\{W_{B_1B_2}-W_{B_1B_1}\} &= s(1-2h)\sigma_d^2\left\{\frac{p(1-p)}{q}\right\},\\ E\{W_{B_1B_2}-W_{B_2B_2}\} &= s(1-2h)\sigma_d^2\left\{\frac{p(1-p)}{1-q}\right\}.\end{aligned}\right\} \tag{3}$$

For the special case of symmetric overdominance (with fitnesses of $A_1A_1$, $A_1A_2$ and $A_2A_2$ of $1-s$, 1 and $1-s$), it can easily be shown that the corresponding expressions are,

$$E\{W_{B_1B_2}-W_{B_1B_1}\} \approx \frac{s}{2}\frac{\sigma_d^2}{(q)} \quad\text{and}\quad E\{W_{B_1B_2}-W_{B_2B_2}\} \approx \frac{s}{2}\frac{\sigma_d^2}{(1-q)}. \tag{4}$$

The validity of these approximations was checked by Monte Carlo experiments, as will be shown later. The above expressions are clearly positive, and also expressions (2) and (3) are positive unless $h \geqslant \frac{1}{2}$. We will show in the following sections that $E(D) = 0$ at steady state, unless epistatic interaction is very strong or the recurrent mutations are of special type creating linkage disequilibrium. For example, if the two loci are multiplicatively overdominant, as shown by Bodmer & Felsenstein (1967), stable linkage disequilibrium will be established by selection only when the recombination fraction between them is less than $s^2/4$. Also, we can show that $E(D)$ is not zero at equilibrium, if the direction of mutation at one locus depends on the kind of alleles at another locus.

When $B$ locus is selectively neutral and $A$ locus is overdominant or ordinarily dominant, we need to estimate associative overdominance at $B$ locus. The gene frequency at $B$ locus may often deviate from its equilibrium value. So, in considering associative overdominance we might substitute a possible or observational value of $q$.

Comstock & Robinson (1952); Chigusa & Mukai (1964) reported the possibility of apparent overdominance for the explanation of their data. Their models are somewhat different from ours in that they assumed selection in all loci. For simplicity's sake, let us assume complete dominance at both $A$ and $B$ loci ($h = 0$). Then, if there is enough negative linkage disequilibrium, the excess of repulsion double heterozygotes will result in an apparent heterosis. This type of pseudo-overdominance may be responsible for many transient polymorphisms in experimental populations as well as for hybrid vigour in many crop plants including the maize. Frydenberg's (1963) interpretation of his experimental result is more similar to our present model. He concluded that the overdominance observed at his marker locus was, at least partly, due to its association with the inversion chromosome, and he termed this phenomenon associative overdominance, although he did not make any quantitative treatment of his model.

We will now proceed to present out basic theory based on diffusion models.

## 3. BASIC THEORY

The main aim of this section is to derive formulae for $\sigma_d^2$, that is, the square of the standard linkage deviation at steady state determined by mutation, selection and random drift, assuming a neutral and an overdominant locus. At the overdominant locus, it is assumed that the selection is so strong that gene frequencies are kept practically constant. If the selection is not strong (with $N_e s$ less than unity), one may use the result already obtained in our previous paper (Ohta & Kimura, 1969*b*). Namely, for a symmetric overdominance at one locus and with symmetric mutation rates at both loci, the square of the standard linkage deviation is

$$\sigma_d^2 = \frac{1}{3+4N_e(c+k_m)-4/(5+2N_e(c+2k_m)+N_e s)}, \tag{5}$$

where $c$ is the recombination fraction between the two loci, $s$ is the heterozygote advantage over both homozygotes and $k_m$ is the sum of the mutation rates. Therefore, $\sigma_d^2 \approx 1/(4N_e c)$ if $4N_e c$ is large. In the following treatments, we will show that even with a very strong overdominance at one locus the relation between $\sigma_d^2$ and $N_e c$ does not much differ from this.

As shown by Ohta & Kimura (1969*b*), if $f$ is a polynomial of random variables describing the stationary distribution, then we have

$$E\{L_B(f)\} = 0, \tag{6}$$

where $L_B$ denotes the differential operator such that if there are $n$ independent random variables, $x_1, x_2, \ldots, x_n$, the equation becomes

$$E\left\{\frac{1}{2}\sum_{i=1}^{n} V_{\delta x_i}\frac{\partial^2 f}{\partial x_i^2} + \sum_{i>j}^{n} W_{\delta x_i \delta x_j}\frac{\partial^2 f}{\partial x_i \partial x_j} + \sum_{i=1}^{n} M_{\delta x_i}\frac{\partial f}{\partial x_i}\right\} = 0, \tag{7}$$

where $M_{\delta x_i}$ and $V_{\delta x_i}$ are the mean and the variance of $\delta x_i$ and $W_{\delta x_i \delta x_j}$ is the covariance between $\delta x_i$ and $\delta x_j$ per unit time (generation).

Equation (6) enables us to calculate the moments of the frequency distribution. Let us apply this equation to the treatment of the present problem. We will assume that overdominance at the $A$ locus is so strong that the frequency $p$ of allele $A_1$ is constant which we denote by $\hat{p}$. At $B$ locus, we assume that a pair of alleles $B_1$ and $B_2$ are selectively neutral, and we denote their frequencies respectively by $q$ and $1-q$. Furthermore, let $q_1$ and $q_2$ be respectively the frequencies of $B_1$ among $A_1$- and $A_2$-carrying chromosomes. Both $q_1$ and $q_2$ are random variables. We will denote by $u$ the mutation rate from $B_1$ to $B_2$, and by $v$ the rate in the reverse direction. Let $N_e$ be the 'variance' effective size of the population and $c$ be the recombination fraction between $A$ and $B$ loci. Then the following equation can be obtained at steady state for $q_1$ and $q_2$ by taking account of mutation, recombination and random sampling of gametes.

$$E\left\{\frac{q_1(1-q_1)}{4N_e\hat{p}}\frac{\partial^2 f}{\partial q_1^2} + [(1-\hat{p})cq_2 + v - ((1-\hat{p})c+u+v)q_1]\frac{\partial f}{\partial q_1}\right.$$
$$\left. + \frac{q_2(1-q_2)}{4N_e(1-\hat{p})}\frac{\partial^2 f}{\partial q_2^2} + [c\hat{p}q_1 + v - (c\hat{p}+u+v)q_2]\frac{\partial f}{\partial q_2}\right\} = 0. \tag{8}$$

We now transform the set of independent random variables $q_1$ and $q_2$ into that of $q$ and $D$ using the relations,

$$q = \hat{p}q_1 + (1-\hat{p})q_2,$$

and

$$D = \hat{p}(1-\hat{p})(q_1 - q_2).$$

Then, equation (8) becomes,

$$E\left\{\frac{1}{4}\left[q(1-q) - \frac{D^2}{\hat{p}(1-\hat{p})}\right]\frac{\partial^2 f}{\partial q^2} + \frac{1}{2}\left[(1-2q)D + \frac{2\hat{p}-1}{\hat{p}(1-\hat{p})}D^2\right]\frac{\partial^2 f}{\partial q \partial D}\right.$$
$$+ \frac{1}{4}\left[\hat{p}(1-\hat{p})q(1-q) + (1-2\hat{p})(1-2q)D - \frac{1-3\hat{p}(1-\hat{p})}{\hat{p}(1-\hat{p})}D^2\right]\frac{\partial^2 f}{\partial D^2}$$
$$\left. + N_e[v-(u+v)q]\frac{\partial f}{\partial q} - N_e(c+u+v)D\frac{\partial f}{\partial D}\right\} = 0. \tag{9}$$

We should note here that the same result can be obtained by computing directly the means, the variances and the covariance of changes in $q$ and $D$ per generation. Using this equation, we will derive the moments of the distributions of $q$ and $D$ and therefore $\sigma_d^2$ at steady state.

Let $f = D$ in (9), then we get $E(D) = 0$. Similarly, we get $E(q) = v/(u+v)$ by putting $f = q$ in (9). Next, if we substitute three functions, $D^2$, $q^2$ and $qD$ for $f$, we get the following simultaneous equations for $E(D^2)$, $E(q^2)$ and $E(qD)$.

$$\left.\begin{aligned} &E\left\{\hat{p}(1-\hat{p})q(1-q)+(1-2\hat{p})(1-2q)D - \frac{1-3\hat{p}(1-\hat{p})}{\hat{p}(1-\hat{p})}D^2 - 4N_e(c+u+v)D^2\right\} = 0, \\ &E\left\{q(1-q)-\frac{D^2}{\hat{p}(1-\hat{p})}+4N_e[vq-(u+v)q^2]\right\} = 0, \\ &E\left\{(1-2q)D+\frac{2\hat{p}-1}{\hat{p}(1-\hat{p})}D^2+2N_e[vD-(u+v)qD]-2N_e(c+u+v)qD\right\} = 0. \end{aligned}\right\} \quad (10)$$

Solving these equations, we obtain the following formula for

$$\sigma_d^2 \equiv E(D^2)/\hat{p}(1-\hat{p})E\{q(1-q)\}.$$

$$\sigma_d^2 = \frac{1}{1+4N_e(c+k'_m)+\dfrac{(1-2\hat{p})^2}{\hat{p}(1-\hat{p})}\dfrac{N_e(c+2k'_m)}{1+N_e(c+2k'_m)}}. \quad (11)$$

In this equation, $k'_m = u+v$, $c$ is the recombination fraction between the two loci, and $\hat{p}$ is the frequency of the overdominant allele $A_1$ supposed to be kept constant in a population of effective size $N_e$. For a special case of symmetrical overdominance at $A$ locus, $\hat{p} = 1/2$ and the last term in the denominator vanishes giving

$$\sigma_d^2 = \frac{1}{1+4N_e(c+k'_m)}. \quad (12)$$

If we compare this formula with the corresponding formula, equation (5) obtained assuming weak overdominance, we note that for a large value of $N_e c$, they become practically the same. We may also note that the total mutation rate, $k'_m$ in this formula is the sum only for the $B$ locus, whereas $k_m$ in formula (5) is the sum for both $A$ and $B$ loci, since the effect of mutation is neglected at the overdominant locus in the present treatment.

When $N_e c$ is small, and especially at the limit of $N_e(c+k_m) \to 0$, these two formulae give somewhat different values. Namely, at this limit, $\sigma_d^2$ in formula (11) or (12) approaches 1, whereas $\sigma_d^2$ in formula (5) approaches a value between $1/2{\cdot}2 \sim 1/3$. Also, for such an extremely tight linkage, and for an intermediate intensity of selection such as $N_e s = 2$, the exact evaluation of $\sigma_d^2$ appears to be very difficult.

Considering all these results, we may conclude that if $N_e c$ is much larger than unity, we have

$$\sigma_d^2 \approx 1/(4N_e c), \tag{13}$$

with good approximation. This simple approximation formula should have a wide applicability, because in most of the natural populations and for most of the linked loci, $4N_e c \gg 1$ is expected.

It is interesting to note that this approximation formula is also valid for the case of steady decay (cf. Ohta & Kimura, 1969*a*).

## 4. MONTE CARLO EXPERIMENTS

Using the IBM 360 computer, Monte Carlo experiments were performed simulating a two-locus genetic system. A simple scheme following Ohta (1968) was used for the experiments, that is, selection and recombination were carried out deterministically and sampling and mutation were performed by generating uniform pseudo-random numbers $X(0{\cdot}0 \sim 1{\cdot}0)$ using subroutine RANDU in FORTRAN IV. Each generation consists of mutation, selection, recombination and sampling. The initial frequencies of four gamete types were read into the computer and the simulation experiments were continued up to 200 generations so that the results represent the equilibrium state.

Let us assign the numbers 1, 2, 3 and 4 to four gamete types, $A_1B_1$, $A_1B_2$, $A_2B_1$ and $A_2B_2$. We will denote by $g_i$ the frequency of gamete $i$ and by $z_{ij}$ the frequency of the zygote formed by the union of gametes $i$ and $j$ ($i, j = 1, 2, 3, 4$).

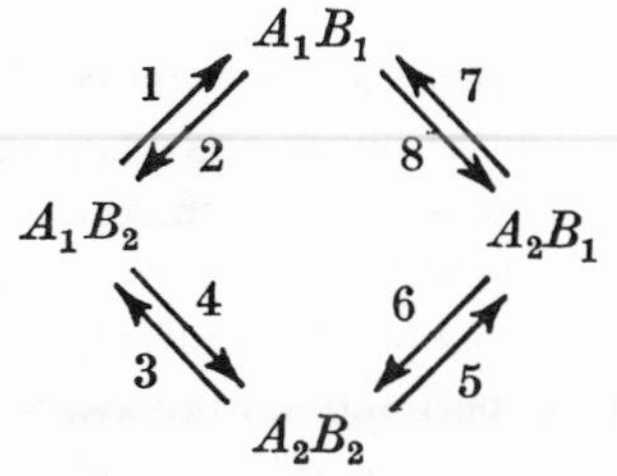


Also we will number eight directions of mutations according to the above diagram. The process of mutation is as follows: We generate a sequence of eight random numbers. Then, one mutation of type 1 is induced among the gametes with $A_1B_2$ if the first random number is less than $m_1 g_2$. Similarly, one mutation of type 2 is induced if the second random number is less than $m_2 g_1$ and so on. Here, $m_1 \sim m_8$ are constants representing the mutation rates. Next, selection was exerted on zygotes using the equations,

$$z'_{ij} = z_{ij} + \Delta z_{ij},$$

and

$$\Delta z_{ij} = \frac{z_{ij}(w_{ij} - \overline{w})}{\overline{w}},$$

where $w_{ij}$ is the fitness (in selective values) of individuals with genotype $ij$ and $\overline{w}$ is the average selective value of individuals in the population. The sampling of zygotes was made by generating pseudo-random numbers $N$ times each generation. Finally, the recombination was carried out deterministically and the frequencies of four gamete types to form the next generation were determined.

Table 1. *Results of Monte Carlo experiments to check the formula* (12)

(Each experimental value is the average of 1200 generations starting with gametic frequencies of 1/4 for all four types. Throughout the experiments the effective population number ($N_e$) was assumed to be 50 and $k$, the sum of mutation rates was $k = 0{\cdot}01$ so that $N_e k = 0{\cdot}5$.

| | | Monte Carlo | | | |
|---|---|---|---|---|---|
| | | $N_e s = 4$ | | $N_e s = 20$ | |
| $c$ | Theoretical $\sigma_d^2$ | $\sigma_d^2$ | $r^2$ | $\sigma_d^2$ | $r^2$ |
| 0 | 0·333 | 0·362 | 0·257 | 0·258 | 0·134 |
| 0·005 | 0·250 | 0·183 | 0·123 | 0·205 | 0·173 |
| 0·01 | 0·200 | 0·128 | 0·080 | 0·168 | 0·133 |
| 0·02 | 0·143 | 0·113 | 0·074 | 0·134 | 0·108 |
| 0·03 | 0·111 | 0·076 | 0·048 | 0·082 | 0·065 |
| 0·04 | 0·091 | 0·100 | 0·077 | 0·080 | 0·060 |
| 0·05 | 0·077 | 0·074 | 0·060 | 0·073 | 0·054 |
| 0·06 | 0·067 | 0·074 | 0·048 | 0·064 | 0·043 |
| 0·07 | 0·059 | 0·051 | 0·040 | 0·041 | 0·037 |
| 0·08 | 0·053 | 0·033 | 0·025 | 0·042 | 0·037 |
| 0·09 | 0·048 | 0·042 | 0·028 | 0·041 | 0·035 |
| 0·1 | 0·043 | 0·038 | 0·035 | 0·033 | 0·031 |

Both cases of large and intermediate selection coefficients were tried. The main purpose of the experiments was to check the validity of formula (12). The symmetric overdominance ($s_1 = s_2 \equiv s$) was assumed at $A$ locus and no selection was assumed at $B$ locus. In one set of experiments, we assumed $N_e s = 4$ and in another, $N_e s = 20$. The population size was 50 and mutation rates were equal in all directions with $N_e k_m = 0{\cdot}5$. The experiments were carried out for various levels of recombination ranging from $c = 0$ to $c = 0{\cdot}1$. Each experiment consisting of 1200 generations started with gene frequencies of 1/2 at both loci and without linkage disequilibrium. In Table 1 theoretical and experimental values of $\sigma_d^2$ are presented together with corresponding values of $r^2$ obtained from the experiments. In computing $\sigma_d^2$ from the experiments, we took the ratio between the mean of $D^2$ and that of $pq(1-p)(1-q)$ each averaged over all 1200 generations. On the other hand, $r^2$ was obtained by taking the average of the ratios of these two statistics over 1200 generations.

In order to show the level of accuracy of formulae (3) and (4), we have produced Table 2 in which $\langle D^2/q^2(1-q)\rangle$ and $\langle D^2/q(1-q)^2\rangle$ are compared respectively with $\sigma_d^2\langle p(1-p)\rangle/\langle q\rangle$ and $\sigma_d^2\langle p(1-p)\rangle/\langle 1-q\rangle$, where $\langle\,\rangle$ denotes the average obtained from the experiments. Also, experimental values are used for $\sigma_d^2$. In the present case $k_m$ (sum of mutation rates at $A$ and $B$ loci) is substituted for $k_m'$ (sum of mutation rates at only $B$ locus) in the formula, since mutation is not negligible at the overdominant locus in our experiments. As seen from the tables, the agreement between the theoretical predictions and the experimental results is satisfactory. However, the approximation formula (12) seems to overestimate slightly the true value. The reason for this appears to be that the gene frequency at the overdominant locus is not strictly fixed but slightly fluctuating.

Table 2. *Experimental check on the approximation involved in formula* (3)

(The data are derived from the same experiments which were performed to construct Table 1. In the table, the symbol $\langle\ \rangle$ denotes the average obtained from the experiments. For details, see text.)

| $N_e s$ | $c$ | $\left\langle \frac{D^2}{q^2(1-q)} \right\rangle$ | $\sigma_d^2 \frac{\langle p(1-p)\rangle}{\langle q\rangle}$ | $\left\langle \frac{D^2}{q(1-q)^2} \right\rangle$ | $\sigma_d^2 \frac{\langle p(1-p)\rangle}{\langle 1-q\rangle}$ |
|---|---|---|---|---|---|
| 4 | 0·0 | 0·139 | 0·125 | 0·178 | 0·224 |
| | 0·005 | 0·099 | 0·096 | 0·076 | 0·069 |
| | 0·01 | 0·090 | 0·079 | 0·059 | 0·043 |
| | 0·02 | 0·058 | 0·041 | 0·088 | 0·065 |
| | 0·03 | 0·026 | 0·019 | 0·075 | 0·053 |
| | 0·04 | 0·058 | 0·044 | 0·056 | 0·051 |
| | 0·05 | 0·056 | 0·040 | 0·043 | 0·030 |
| | 0·06 | 0·061 | 0·046 | 0·026 | 0·023 |
| | 0·07 | 0·013 | 0·014 | 0·067 | 0·055 |
| | 0·08 | 0·047 | 0·021 | 0·024 | 0·011 |
| | 0·09 | 0·045 | 0·024 | 0·026 | 0·016 |
| | 0·1 | 0·035 | 0·017 | 0·020 | 0·014 |
| 20 | 0·0 | 0·058 | 0·077 | 0·165 | 0·341 |
| | 0·005 | 0·101 | 0·117 | 0·098 | 0·090 |
| | 0·01 | 0·073 | 0·063 | 0·116 | 0·114 |
| | 0·02 | 0·090 | 0·088 | 0·055 | 0·051 |
| | 0·03 | 0·072 | 0·064 | 0·029 | 0·029 |
| | 0·04 | 0·081 | 0·065 | 0·032 | 0·028 |
| | 0·05 | 0·047 | 0·033 | 0·058 | 0·038 |
| | 0·06 | 0·040 | 0·026 | 0·063 | 0·039 |
| | 0·07 | 0·059 | 0·032 | 0·019 | 0·014 |
| | 0·08 | 0·019 | 0·016 | 0·049 | 0·034 |
| | 0·09 | 0·031 | 0·021 | 0·038 | 0·019 |
| | 0·1 | 0·014 | 0·012 | 0·039 | 0·024 |

## 5. DISCUSSION

The main aim of the present paper is to estimate, using our formulae for the square of the standard linkage deviation ($\sigma_d^2$), the approximate magnitude of associative overdominance created by linkage disequilibrium. Certainly, it is based on several approximations, but as demonstrated by the Monte Carlo experiments, these formulae give us a sufficiently accurate estimate for $\sigma_d^2$. One problem in our procedure of estimating associative overdominance is that we substitute the ratio of the expectations for the expectation of the ratio, cf. formulae (2) and (3). To see the magnitude of errors involved, we compared $\sigma_d^2 = E(D^2)/E\{p(1-p)q(1-q)\}$

which is the ratio of the expectations, with $r^2 = E\{D^2/pq(1-p)(1-q)\}$ which is the expectation of the ratio. The latter is the square of the correlation of $p$ and $q$ in the usual sense. In Table 1, values of $\sigma_d^2$ and $r^2$ are compared. The values of $r^2$ are slightly smaller than the corresponding values of $\sigma_d^2$ mainly due to deviation of gene frequency ($q$) from 1/2 at the neutral locus. The reason for $r^2 < \sigma_d^2$ may most easily be understood by considering the situation in which $q$ happens to become extremely small. Then $r^2$ must approach 0, but $\sigma_d^2$ will not be much influenced by such a deviation. Thus, excluding such extreme situations, we may conclude that, for large $N_e c$, the approximation $\sigma_d^2 \approx 1/\{4N_e(c+k_m)\}$ is valid and useful to evaluate associative overdominance.

It is quite interesting that essentially the same formula for $\sigma_d^2$ holds also at the transient state in which the variability is steadily decaying, as long as $N_e c$ is large. Hill & Robertson (1968) found that the square of the correlation of gene frequencies ($r^2$) at steady decay becomes approximately $1/(4N_e c)$ when $N_e c$ is large. Ohta & Kimura (1969*a*) also showed that under such a situation $\sigma_d^2$ becomes approximately $1/(4N_e c)$. On the other hand, when $N_e c$ is small, there are some differences in $\sigma_d^2$ between the stationary state and the steadily decaying state. Also, the effects of selection and mutation become important.

Let us investigate how much associative overdominance will be developed if a neutral locus is linked with a number of overdominant loci on the same chromosome. Our model is as follows. We assume that there are $n_1$ overdominant loci on the left and $n_2$ overdominant loci on the right of the neutral locus in such a way that the recombination fraction between the neutral locus and the $i$th overdominant locus, either on the left or on the right, is $ic_0$. Let $s$ be the selection coefficient against either homozygote in each overdominant locus. We then ask how much associative overdominance will develop at the neutral locus ($B$), where alleles $B_1$ and $B_2$ are segregating with respective frequencies of $q$ and $1-q$. We assume that linkage among overdominant loci is loose in comparison with the selection intensity, so that these overdominant loci do not constitute 'super genes' and that they are more or less randomly combined. However, a small amount of linkage disequilibrium will be created among them by random genetic drift. If we assume strict additivity in fitness among the overdominant loci, i.e. additive overdominance, and if $\sigma_d^2$ between the neutral and each overdominant locus is adequately given by our formula (12), the associative overdominance at the neutral locus simply becomes the sum of the effects of all linked overdominant loci.

For a multiplicative overdominance, we may take logarithms of the individual fitnesses in the following formulation. Let us suppose that $q$ happens to take an intermediate value not very far from 1/2. That is, the frequency of $B_1$ or $B_2$ has increased in the population by random frequency drift. Then, from equation (3), if we take $q = 1/2$, the coefficient of associative overdominance $s'$ becomes approximately,

$$s' = E\{W_{B_1B_2} - W_{B_1B_1}\} = E\{W_{B_1B_2} - W_{B_2B_2}\} = s\left\{\sum_{i=1}^{n_1} \sigma_d^2(i) + \sum_{i=1}^{n_2} \sigma_d^2(i)\right\},$$

where $\sigma_d^2\,(i)$ is the standard linkage deviation between the neutral locus and the $i$th overdominant locus, either on the right or on the left. Then, using

$$\sigma_d^2(i) \approx 1/(4N_e c_0 i),$$

we have
$$s' = \frac{s}{4N_e c_0}\{2\gamma + \log n_1 + \log n_2\}, \tag{14}$$

where $\gamma = 0{\cdot}577$ is Euler's constant. For example, if the effective population size is 1000 and if there are 100 overdominant loci on the chromosome such that $n_1 = n_2 = 50$ covering roughly the map length of 100 units ($c_0 = 0{\cdot}01$), the neutral locus being located just in the middle, $s'$ becomes about $0{\cdot}22s$. If the population is ten times as large, $s'$ is $0{\cdot}022s$. An important point to note here is that $N_e s'$ remains constant with varying $N_e$. If we change the number of overdominant loci such that the total effect on the given segment does not change (for example, $2n$ loci each with homozygous disadvantage $s/2$ instead of $n$ loci each with $s$), then the amount of associative overdominance $s'$ changes relatively little (for example, by the factor $\log 2n/\log n$). For a large value of $N_e$ such as $10^5$ or $10^6$, the value of $s'$ may be quite small, but it does retard fixation at the neutral locus, since $N_e s'$ remains constant.

So far, we have investigated the amount of pseudo-overdominance developed at a neutral locus through its non-random association with overdominant loci in a finite population.

In discussing the effect of such associative overdominance on the amount of heterozygosity at the neutral locus, we must be careful not to overlook the effect of mutation at that locus. Namely, at the neutral locus $B$, if we disregard the effect of associative overdominance, it can be shown that the expected amount of heterozygosity is

$$H_B = E\{2q(1-q)\} = \frac{8N_e u\bar{q}}{1+4N_e(u+v)}, \tag{15}$$

where $\bar{q} = v/(u+v)$. On the other hand, if we include the effect of associative overdominance from the $A$ locus, it can be shown using equations (10) that the expected heterozygosity is

$$H_{B(A)} = \frac{8N_e u\bar{q}}{1+4N_e(u+v)-\sigma_d^2} = \frac{H_B}{1-\sigma_d^2/\{1+N_e(u+v)\}}, \tag{16}$$

where $\sigma_d^2$ is given by equation (11).

If $4N_e(u+v)$ is much larger than unity, $H_B$ approaches $2\bar{q}(1-\bar{q})$ by mutation, and there may be little room left for pseudo-overdominance to enhance heterozygosity. However, if $4N_e(u+v)$ is small, we have

$$H_{B(A)} \approx (1+\sigma_d^2)H_B,$$

provided that $\sigma_d^2 \approx 1/(4N_e c)$ is small. Thus, the heterozygosity at the $B$ locus is enhanced by the fraction $1/(4N_e c)$ by associative overdominance caused by $A$ locus.

When a large number of overdominant loci are linked to the neutral locus, their effect on enhancing heterozygosity at the neutral locus would probably be proportional to $\exp(\Sigma\sigma_d^2)$, as long as each $\sigma_d^2$'s is small. A more detailed study on this subject will be left to our future reports. However, we should mention here that when both $A$ and $B$ loci are kept polymorphic by strong overdominance, linkage disequilibrium is created between these two overdominant loci just as between a neutral and an overdominant loci. Namely, the approximation equation (13) also holds for them if $c$ is the recombination fraction between them. This can be shown as follows. Let $\hat{p}$ and $\hat{q}$ be respectively the frequencies of $A_1$ and $B_1$ in these two loci, and suppose that both $\hat{p}$ and $\hat{q}$ are kept constant by strong overdominance. At the neighbourhood of $D = 0$, the sampling variance of $D$ is approximately

$$V_{\delta D} = \hat{p}(1-\hat{p})\hat{q}(1-\hat{q})/(2N_e),$$

because the problem is analogous to that of sampling in $2 \times 2$ contingency table in statistics, and we can show that

$$\chi_1^2 = 2N_e D^2/[\hat{p}(1-\hat{p})\hat{q}(1-\hat{q})],$$

follows approximately Chi-square distribution with one degree of freedom, so that $E(\chi_1^2) = 1$. Thus, noting that $M_{\delta D} = -cD$, by recombination, the equation corresponding to (7) becomes

$$E\left\{\frac{1}{4N_e}\hat{p}(1-\hat{p})\,\hat{q}(1-\hat{q})\frac{\partial^2 f}{\partial D^2} - cD\frac{\partial f}{\partial D}\right\} = 0.$$

By setting successively $f = D$ and $f = D^2$ in this equation, we obtain $E(D) = 0$ and $E(D^2) = 4N_e c\hat{p}(1-\hat{p})\hat{q}(1-\hat{q})$. Therefore,

$$\sigma_d^2 = E(D^2)/[\hat{p}(1-\hat{p})\hat{q}(1-\hat{q})] = 1/(4N_e c),$$

as was to be shown. This formula should be valid when $\sigma_d^2$ is small.

Finally we intend to discuss problems of linkage disequilibrium in experimental populations. There are many reports on the experimental measure of fitness values with respect to isozyme alleles or other marker genes. Very often, however, the results merely reflect the effects of a group of surrounding genes and hence the effect of individual alleles on fitness is very difficult to measure. In experimental populations, initial linkage disequilibrium may be produced by sampling a relatively small number of chromosomes from a large parental population. As shown by Hill & Robertson (1968), if $n$ chromosomes are sampled to form the first generation of experimental populations, the values of $X = E\{pq(1-p)(1-q)\}$, $Y = E\{D(1-2p)(1-2q)\}$ and $Z = \{D^2\}$ in the first generation is given by

$$\begin{bmatrix} X_1 \\ Y_1 \\ Z_1 \end{bmatrix} = \begin{bmatrix} \left(1-\frac{1}{n}\right)^2 & \frac{1}{n}\left(1-\frac{1}{n}\right)^2(1-c) & \frac{2}{n^2}\left(1-\frac{1}{n}\right)(1-c)^2 \\ 0 & \left(1-\frac{1}{n}\right)\left(1-\frac{2}{n}\right)^2(1-c) & \frac{4}{n}\left(1-\frac{1}{n}\right)\left(1-\frac{2}{n}\right)(1-c)^2 \\ \frac{1}{n}\left(1-\frac{1}{n}\right) & \frac{1}{n}\left(1-\frac{1}{n}\right)^2(1-c) & \left(1-\frac{1}{n}\right)\left[\frac{1}{n^2}+\left(1-\frac{1}{n}\right)^2\right](1-c)^2 \end{bmatrix} \begin{bmatrix} X_0 \\ Y_0 \\ Z_0 \end{bmatrix}, \tag{17}$$

where $X_0$, $Y_0$ and $Z_0$ are the corresponding values in the parental population.

Assuming that the linkage disequilibrium in the parental population is negligible so that approximately $Y_0 = Z_0 = 0$, then the squared standard linkage deviation in the first generation of the experimental populations is

$$\sigma^2_{d,1} = \frac{E(Z_1)}{E(X_1)} = \frac{1}{n-1}. \tag{18}$$

Usually, the chromosomes thus sampled are rapidly multiplied from $n$ to $n'$ in the succeeding generations and then they are used for measuring the fitness of the marker gene. The following is a very rough estimate of linkage disequilibrium in such experimental populations. Applying the results of Ohta & Kimura (1969*a*), $\sigma^2_d$ in the $t$-th generation may be given as follows,

$$\sigma^2_{d,t} = \frac{\sum_{i=1}^{3} C_{J_i}\left[\frac{X_1}{2(1+\lambda_i)} + \frac{1}{4}(3+4R+2\lambda_i)Y_1 + Z_1\right]e^{2\lambda_i t/n'}}{\sum_{i=1}^{3} C_{H_i}\left[\frac{X_1}{2(1+\lambda_i)} + \frac{1}{4}(3+4R+2\lambda_i)Y_1 + Z_1\right]e^{2\lambda_i t/n'}}, \tag{19}$$

where $\lambda_i$'s are the first three eigenvalues of the Kolmogorov forward equation involved and $C_{H_i}$'s and $C_{J_i}$'s are the functions of $\lambda_i$'s. These parameters are given in Ohta & Kimura (1969*a*). The value of $\sigma^2_{d,t}$ rapidly approaches to $1/2n'c$ for sufficiently large value of $n'c$. By using the formula (3), one can estimate the average degree of associative overdominance. Of course this method gives the overall average due to non-epistatic genes.

If there are strong epistatic effects between loci so that super-genes are formed, $E(D)$ is not zero and the result becomes much more complex.

## REFERENCES


Bodmer, W. F. & Felsenstein, J. (1967). Linkage and selection: Theoretical analysis of the deterministic two locus random mating model. *Genetics* **57**, 237–265.

Chigusa, S. & Mukai, T. (1964). Linkage disequilibrium and heterosis in experimental populations of *Drosophila melanogaster* with particular reference to the *sepia* gene. *Japanese Journal of Genetics* **39**, 289–305.

Comstock, R. E. & Robinson, H. F. (1952). Estimation of average dominance of genes. In *Heterosis*, pp. 494–516. Ames: Iowa State College Press.

Frydenberg, O. (1963). Population studies of a lethal mutant in *Drosophila melanogaster*. I. Behaviour in populations with discrete generations. *Hereditas* **50**, 89–116.

Hill, W. G. & Robertson, A. (1968). Linkage disequilibrium in finite populations. *Theoretical and Applied Genetics* **38**, 226–231.

Maruyama, T. & Kimura, M. (1968). Development of temporary overdominance associated with neutral alleles. *Proceedings of XII International Congress of Genetics, Tokyo*, vol. 1, p. 229.

Ohta, T. (1968). Effect of initial linkage disequilibrium and epistasis on fixation probability in a small population, with two segregating loci. *Theoretical and Applied Genetics* **38**, 243–248.

Ohta, T. & Kimura, M. (1969*a*). Linkage disequilibrium due to random genetic drift. *Genetical Research* **13**, 47–55.

Ohta, T. & Kimura, M. (1969*b*). Linkage disequilibrium at steady state determined by random genetic drift and recurrent mutation. *Genetics* **63**, 229–238.

Sved, J. A. (1968). The stability of linked systems of loci with a small population size. *Genetics* **59**, 543–563.

# LINKAGE DISEQUILIBRIUM BETWEEN TWO SEGREGATING NUCLEOTIDE SITES UNDER THE STEADY FLUX OF MUTATIONS IN A FINITE POPULATION[1]

TOMOKO OHTA AND MOTOO KIMURA

*National Institute of Genetics, Mishima, Japan*

Received July 31, 1970

IN our previous reports (OHTA and KIMURA 1969a, b, 1970), we have studied linkage disequilibrium, that is, nonrandom association of genes between loci, caused by random frequency drift in finite populations. We have considered the situation in which a stationary distribution is reached under recurrent mutation or overdominance. We have also considered the case in which genetic variability decays each generation due to random sampling of gametes.

The present paper is an extension of the work of KIMURA (1969a) who studied the number of heterozygous nucleotide sites under the steady flux of molecular mutations in a finite population. Here, we intend to study the amount of linkage disequilibrium between two segregating sites using the same model; it assumes that the total number of nucleotide sites making up the genome is so large and the mutation rate per site is so low that whenever a mutant appears, it represents a mutation at a previously homoallelic site, that is, a site in which no mutants are currently segregating in the population.

## BASIC THEORY

Consider a random mating diploid population of actual size $N$ and effective size $N_e$. We assume that each generation mutations occur in $v_m$ sites distributed throughout the population.

Since each mutant becomes fixed in the population or lost from it within a finite length of time, if mutations continue to occur at a constant rate over many generations, a steady state will be reached with respect to the frequency distribution of mutants among different sites, provided that we restrict our consideration to only those sites in which mutant forms are segregating.

In the following treatment, we shall consider two segregating nucleotide sites with the recombination fraction $c$ between them. Let $X_1$ be the frequency of chromosomes having no mutants at both sites, $X_2$ and $X_3$ be the frequencies of chromosomes having a mutant at the first and the second sites, respectively, and $X_4$ be the frequency of chromosomes having mutants at both sites. Then, $x=X_2+X_4$ is the frequency of the mutant at the first site, $y=X_3+X_4$ is that at the second site, and $D=X_1X_4-X_2X_3$ is the index of linkage disequilibrium.

[1] Contribution No. 782 from the National Institute of Genetics, Mishima, Shizuoka-ken 411 Japan. Aided in part by a Grant-in-Aid from the Ministry of Education, Japan.

Genetics **68**: 571–580 August, 1971.

Let $\Phi(x,y,D)$ be the steady flux distribution involving two nucleotide sites such that $\Phi(x,y,D)\mathrm{d}x\mathrm{d}y\mathrm{d}D$ is the expected number of the pairs of sites having mutant frequencies and the disequilibrium index within the intervals $(x,x+\mathrm{d}x)$, $(y,y+\mathrm{d}y)$, and $(D,D+\mathrm{d}D)$. This includes all pairs whose distance apart corresponds to a recombination fraction $c$. We assume that mutants are simultaneously segregating at both sites $(0<x<1, 0<y<1)$.

As shown in the APPENDIX, if $f(x,y,D)$ is a function (polynomial) of $x,y$, and $D$, we have, with a steady flux of mutations,

$$E\{L(f)\} + \Delta_{mut}E(f) = 0. \qquad (1)$$

In this equation $L$ is the differential operator for the diffusion process involving the two sites, and if we assume that the molecular mutants are selectively neutral,

$$L = \frac{x(1-x)}{4N_e}\frac{\partial^2}{\partial x^2} + \frac{y(1-y)}{4N_e}\frac{\partial^2}{\partial y^2} + \frac{D}{2N_e}\frac{\partial^2}{\partial x \partial y} + \frac{D(1-2x)}{2N_e}\frac{\partial^2}{\partial x \partial D}$$
$$+ \frac{D(1-2y)}{2N_e}\frac{\partial^2}{\partial y \partial D} + \frac{1}{4N_e}[xy(1-x)(1-y)+D(1-2x)(1-2y)-D^2]\frac{\partial^2}{\partial D^2}$$
$$-\left(\frac{1}{2N_e}+c\right)D\frac{\partial}{\partial D}\;. \qquad (2)$$

This operator is equivalent to $L_B$ in formula (12) of OHTA and KIMURA (1969b) except that $L$ here does not contain terms involving the effect of mutation. In the present model, considering nucleotide sites rather than conventional genetic loci, mutation is essentially irreversible and the effect of mutation is represented by the term $\Delta_{mut}E(f)$ in equation (1). Specifically, $\Delta_{mut}E(f)$ represents the contribution made each generation by new mutations to $E(f)$, where $E$ is the operator for taking the expectation with respect to the steady flux distribution, that is,

$$E(f) = \iiint f(x,y,D)\Phi(x,y,D)\mathrm{d}x\mathrm{d}y\mathrm{d}D,$$

in which the integral is over $0<x<1$, $0<y<1$, $-\tfrac{1}{4}\leqq D\leqq+\tfrac{1}{4}$.

In choosing $f$ we must keep in mind that equation (1) is valid only for $f$ such that $f(x,y,D)\Phi(x,y,D)$ vanishes at $x=0$, $x=1$, $y=0$, and $y=1$; in other words, $f\Phi$ must be zero on the periphery of the square $0\leqq x\leqq 1$, $0\leqq y\leqq 1$. Note that on the periphery $D$ is also zero.

First, let $f=xy(1-x)(1-y)$ in equation (1), then we have

$$E\{L(f)\} = \frac{1}{2N_e}(-2X+Y)\;, \qquad (3)$$

where $X=E\{xy(1-x)(1-y)\}$ and $Y=E\{D(1-2x)(1-2y)\}$. To determine $\Delta_{mut}E(f)$ in equation (1), let $v_s$ be the number of pairs of nucleotide sites that start segregating simultaneously in the entire population each generation, considering only those pairs of sites that are separated by a distance corresponding to a recombination fraction $c$. We assume that simultaneous segregation always starts from the situation in which one of the sites is already segregating while a new mutant is just added to the other site. Thus for $f=x(1-x)y(1-y)$, we have

$$\Delta_{mut}E(f) = v_s\overline{x(1-x)}p(1-p),$$

where $\overline{x(1-x)}$ is the average value of $x(1-x)$ among segregating sites, and $p$ is the frequency of mutants at the time of occurrence.

If we denote by $I_1(p)$ the total number of segregating sites in the population and by $E\{x(1-x)\}$ the sum of $x(1-x)$ over all these sites, then

$$\overline{x(1-x)} = E\{x(1-x)\}/I_1(p),$$

and as shown by KIMURA (1969a),

$$E\{x(1-x)\} = 2N_e v_m p(1-p) \tag{4}$$

and

$$I_1(p) = -4N_e v_m\{p \log_e p + (1-p) \log_e(1-p)\}.$$

An independent derivation of (4) is given in the APPENDIX (see formula A7).

We note here that we may put $p=1/(2N)$ in the above expressions because, in our model, the mutation rate per site is so low that each mutant is likely to be represented only once at the moment of appearance. Thus, if we write

$$\Delta_{mut}E(f) = K$$

for $f=x(1-x)y(1-y)$, we have

$$K = v_s\overline{x(1-x)}/(2N) = v_s/[4N(\log_e 2N+1)] \tag{5}$$

approximately.

Next, letting $f=D(1-2x)(1-2y)$ in (1), we obtain

$$E\{L(f)\} = -\frac{1}{2N_e}\{(5+2N_e c)Y - 4Z\},$$

where $Z = E\{D^2\}$. In determining $\Delta_{mut}E(f)$ for this case, we note that when a mutant is just introduced into the second site, the mutant appears either on a chromosome carrying a mutant in the first site in which case $X_4 = 1/2N$, $X_3 = 0$, $X_2 = x - 1/(2N)$, $X_1 = 1 - x$, or on a chromosome carrying a nonmutant in the first site in which case $X_4 = 0$, $X_3 = 1/2N$, $X_2 = x$, $X_1 = 1 - x - 1/2N$, where $x$ is the frequency of the mutant in the first site.

The frequencies of these alternative events are $x$ and $1 - x$, respectively. Therefore,

$$\Delta_{mut}E(f) \propto E\Big[x(\frac{1-x}{2N})(1-2x)(1-\frac{2}{2N}) + (1-x)(-\frac{x}{2N})(1-2x)(1-\frac{2}{2N})\Big] = 0.$$

Finally, let $f = D^2$ in (1), then we get

$$E\{L(f)\} = \frac{1}{2N_e}\{X + Y - (3+4N_e c)Z\},$$

and

$$\Delta_{mut}E(f) = v_s\overline{x(1-x)}/(2N)^2 \approx K/(2N),$$

because for a particular value of $x$, the contribution of a new mutant to $E(D^2)$ is

$$v_s\Big[x(\frac{1-x}{2N})^2 + (1-x)(\frac{-x}{2N})^2\Big] = v_s x(1-x)/(2N)^2.$$

Therefore, we have a set of equations for $X$, $Y$, and $Z$.

$$\begin{aligned} &-2X+Y=-2N_eK \\ &-(5+2N_ec)Y+4Z=0 \qquad (6) \\ &X+Y-(3+4N_ec)Z=-2N_eK/(2N) \end{aligned}$$

Solving this we find

$$\begin{aligned} &X=E\{x(1-x)y(1-y)\}=N_eK(11+26R+8R^2+2/N)/(9+26R+8R^2) \\ &Y=E\{D(1-2x)(1-2y)\}=4N_eK(1+1/N)/(9+26R+8R^2) \qquad (7) \\ &Z=E\{D^2\}=N_eK(1+1/N)(5+2R)/(9+26R+8R^2), \end{aligned}$$

where $R=N_ec$.

To express the degree of linkage disequilibrium between the two nucleotide sites, we use the quantity,

$$\sigma_d^2=\frac{Z}{X}=\frac{E\{D^2\}}{E\{x(1-x)y(1-y)\}}, \qquad (8)$$

where $\sigma_d$ is the *standard linkage deviation* (OHTA and KIMURA 1969b). The quantity $D^2/[x(1-x)y(1-y)]$ has been used by statisticians as a measure of degree of association and is called the *mean square contingency coefficient* (see KENDALL 1948). Our $\sigma_d^2$ is closely related, being the ratio of the expected value of the numerator to that of the denominator. Unless the mutant frequencies at one or both sites take the extreme values near 0 or 1, $\sigma_d^2$ is approximately equal to the expected value of the mean square contingency coefficient.

From (7) we obtain

$$\sigma_d^2=\frac{(5+2R)(1+1/N)}{11+26R+8R^2+2/N}\approx\frac{5+2R}{11+26R+8R^2} \qquad (9)$$

where $R=N_ec$.

Thus if $N_ec$ is large, $\sigma_d^2$ is approximately equal to $1/(4N_ec)$, while if $N_ec$ is much smaller than unity, $\sigma_d^2$ is approximately 5/11.

Note that $E(D)=0$, namely, $D$ has the mean zero, as may be seen by setting $f=D$ and $\Delta_{mut}E(f)=0$ in (1). However, for any finite population, $D$ is likely to deviate from this theoretical mean, and $\sigma_d$ is a more appropriate measure of the amount of linkage disequilibrium.

## MONTE CARLO EXPERIMENTS

In order to check the validity of these theoretical predictions, several Monte Carlo experiments were performed. It is desirable to simulate as closely as possible a natural population of organisms having a very large number of nucleotide sites in its genome with a steady flux of mutations occurring over many generations. However, we used a simpler model having only two sites, corresponding to the two sites treated by the method of this paper. Each site is supplied with a new mutant as soon as it becomes homoallelic. In this model we cannot control mutation rate, although for the present purpose, the simulation may be used for checking formula (9) giving the squared standard linkage deviation. Also, it may be used to compare $\sigma_d$ with the contingency coefficient between the two nucleotide sites.

TABLE 1

| $N_ec$ | Theoretical value (formula 9) | $\sigma_d^2$ Monte Carlo results | | |
|---|---|---|---|---|
| | | $N_e$=100 | $N_e$=200 | Mean |
| 0.0 | 0.4545 | 0.7380 | 0.2933 | 0.5157 |
| 0.1 | 0.3824 | 0.7056 | 0.6888 | 0.6972 |
| 0.2 | 0.3273 | 0.4571 | 0.4637 | 0.3922 |
| 0.3 | 0.2872 | 0.3903 | 0.2930 | 0.3414 |
| 0.4 | 0.2557 | 0.4526 | 0.2989 | 0.3758 |
| 0.5 | 0.2308 | 0.1890 | 0.2532 | 0.2211 |
| 0.6 | 0.2102 | 0.2008 | 0.1197 | 0.1603 |
| 0.7 | 0.1932 | 0.1823 | 0.1348 | 0.1586 |
| 0.8 | 0.1788 | 0.1700 | 0.2606 | 0.2153 |
| 0.9 | 0.1663 | 0.1273 | 0.0053 | 0.0663 |
| 1.0 | 0.1555 | 0.1334 | 0.2554 | 0.1944 |
| 2.0 | 0.0947 | 0.1208 | 0.0858 | 0.1033 |
| 3.0 | 0.0683 | 0.0622 | 0.0768 | 0.0695 |
| 4.0 | 0.0535 | 0.0391 | 0.0380 | 0.0386 |
| 5.0 | 0.0440 | 0.0444 | 0.0506 | 0.0475 |
| 6.0 | 0.0374 | 0.0319 | 0.0310 | 0.0315 |
| 7.0 | 0.0325 | 0.0313 | 0.0273 | 0.0293 |
| 8.0 | 0.0287 | 0.0263 | 0.0231 | 0.0247 |
| 9.0 | 0.0258 | 0.0238 | 0.0211 | 0.0225 |
| 10.0 | 0.0233 | 0.0197 | 0.0192 | 0.0195 |

Results of Monte Carlo experiments performed to check equation (9) on the squared standard linkage deviation between two segregating sites under the steady flux of molecular mutations in a finite population. Each experimental value is the average over 10,000 generations. $N_e$ stands for the effective population number and $c$ the recombination fraction between the two segregating nucleotide sites.

The procedure of the experiment was as follows. In the first generation, each site contains one mutant; crossing over and zygote formation are performed deterministically; sampling of $N$ zygotes for the parents of the next generation are carried out as follows using pseudorandom numbers with uniform distribution in the interval [0,1] (RAND 20 in TOSBAC 3400). Let $f_i$ be the frequency of the $i$th genotype ($i = 1,2, \ldots, 10$). Then we pick out an individual of the $i$th genotype if a random number happens to lie between $\sum_{j=0}^{i-1} f_j$ and $\sum_{j=0}^{i} f_j$, where we set $f_0 = 0$. The procedure used here is essentially the same as the one used by OHTA (1968) except for mutation production. A new mutation is supplied whenever a locus becomes homoallelic, regardless of whether the previous mutant was lost or fixed. In Table 1, values of $\sigma_d^2$ obtained from the experiment are compared with the corresponding theoretical values derived from equation (9). The experiments were performed assuming two levels of the effective population number, $N_e = 100$ and 200. The recombination fraction ranges from 0 to $N_ec = 10$. The values listed are the averages over 10,000 generations.

As seen from the table, agreement between experimental and theoretical results appears to be satisfactory.

## DISCUSSION

In discussing the biological implication of the above results, we must keep in mind that we are here concerned with nonrandom association of mutants between two nucleotide sites, both of which are *simultaneously segregating* in the population. This differs from the nonrandom association reported by JOSSE, KAISER and KORNBERG (1961) for base composition of adjacent nucleotide sites (nearest neighbor). Their analysis involves the overall base composition, most of which is from nonsegregating sites. Furthermore, our analysis considers nonrandom associations between nucleotide pairs, considered as individual pairs, but with an average value $D$ not differing from zero, whereas they found significant deviations from $D = 0$.

The most likely explanation of the nearest-neighbor association is that the mutation rate at a nucleotide is somehow influenced by the nucleotide at the adjacent site. The alternative, that the nonrandomness is due to selection favoring different amino acids is more difficult to understand in view of the fact that the same paired nucleotide sequences occur in the codons of many amino acids and many different amino acids are used in each polypeptide. If a majority of nucleotide substitutions in evolution are the result of chance fixation of molecular mutants through random frequency drift as suggested by KIMURA (1968, 1969b), KING and JUKES (1969), and CROW (1969), nonrandomness of base arrangements between adjacent sites is still less likely to be caused by selection, increasing the strength of the evidence for neighbor-influenced mutation rates.

Returning to the dynamic aspect, we note from equation (9) that marked linkage disequilibrium will arise between the segregating sites if $N_ec$, the product of the effective population number and the recombination fraction, is less than unity. For example, if $N_ec = 0.5$, we have $\sigma_d^2 = 3/13$ or $\sigma_d \approx 0.48$. Since $\sigma_d$ is roughly equal to the correlation coefficient, this means that roughly 50% correlation exists between the frequencies of mutants at both sites.

According to NEI (1968), who listed the number of nucleotide pairs per unit map length for various organisms ranging from viruses to mammals, the recombination fraction between neighboring sites is about $4 \times 10^{-8}$ for Drosophila and $4 \times 10^{-9}$ for the mouse. These values are about the same order of magnitude as the mutation rate per nucleotide (cf. KIMURA 1968a,b).

Since the effective number (but not the actual number) of many species may not reach $10^6$ and may be much less, it is expected that $N_ec$ is usually much smaller than unity for two segregating nucleotide sites within a cistron. This means that strong linkage disequilibrium is expected to be very common between segregating sites within a cistron. Although four "alleles" are theoretically possible by segregation in a single nucleotide site, it is much more likely that when three or more alleles are maintained in the population, two or more sites are involved. In such a case, polymorphism involving exactly three alleles necessarily means linkage disequilibrium within a cistron.

The fact that $\sigma_d^2$ approaches 5/11 or roughly 1/2 rather than 1 at the limit of $N_ec = 0$ may be understood by noting that under a steady flux of mutations, the mutant frequencies at two segregating sites are usually different, and $\sigma_d^2 = 1$ cannot be attained even when there is no recombination.

Another interesting point related to our present result is that $\sigma_d^2$ becomes approximately $1/(4N_ec)$ if $4N_ec$ is much larger than unity.

$$\sigma_d^2 \approx 1/(4N_ec) \tag{10}$$

For two segregating sites that are 0.1 map unit or more apart, this should give a good approximation for most natural populations. In this case, the contingency coefficient of mutant frequencies between the two sites is roughly equal to $1/\sqrt{4N_ec}$.

We have already shown in our previous papers that this approximation holds for the case of steady decay (OHTA and KIMURA 1969a). It also holds for the stationary state attained under recurrent mutation or overdominance (OHTA and KIMURA 1969b, 1970).

Thus the approximation formula (10) seems to be applicable quite generally to two segregating sites as long as $4N_ec$ is large.

We would like to thank Dr. J. F. CROW for reading the manuscript and making valuable suggestions.

## SUMMARY

Linkage disequilibrium or nonrandom association of mutant forms between two segregating nucleotide sites in a finite population was studied using diffusion models, assuming that the number of nucleotide sites making up the genome is so large while the mutation rate per site is so low that whenever a mutant appears, it represents a mutation at a homoallelic site, i.e., a site in which no mutant forms are currently segregating in a population.—It was shown that under steady flux of molecular mutations in a finite population, if we measure the amount of linkage disequilibrium between two segregating sites by

$$\sigma_d^2 = E\{D^2\}/E\{x(1-x)y(1-y)\},$$

where $D$ is the ordinary coefficient of linkage disequilibrium, and $x$ and $y$ are the frequencies of mutants at the two sites, then we have

$$\sigma_d^2 \approx (5+2R)/(11+26R+8R^2),$$

where $R = N_ec$ in which $N_e$ is the effective size of the population and $c$ is the recombination fraction between the two sites.—It was pointed out that if multiple alleles in a random mating population are generated through segregation at two or more nucleotide sites within a cistron, a strong linkage disequilibrium is usually expected between those sites, even in the absence of selection.

## LITERATURE CITED

CROW, J. F., 1969 Molecular genetics and population genetics. Proc. 12th Intern. Congr. Genet. **3**: 105–113.

JOSSE, J., A. D. KAISER and A. KORNBERG, 1961 Enzymatic synthesis of deoxyribonucleic acid. VIII: Frequencies of nearest-neighbor base sequences in deoxyribonucleic acid. J. Biol. Chem. **236**: 864–875.

KENDALL, M. G., 1948 *The Advanced Theory of Statistics*, Vol. 1. Charles Griffin & Co. Ltd., London.

KIMURA, M., 1968a Evolutionary rate at the molecular level. Nature **217**: 624–626. ——, 1968b Genetic variability maintained in a finite population due to mutational production of neutral and nearly neutral isoalleles. Genet. Res. **11**: 247–269. ——, 1969a The number of heterozygous nucleotide sites maintained in a finite population due to steady flux of mutations. Genetics **61**: 893–903. ——, 1969b The rate of molecular evolution considered from the standpoint of population genetics. Proc. Natl. Acad. Sci. U.S. **63**: 1181–1188.

KING, J. L. and T. H. JUKES, 1969 Non-Darwinian evolution: Random fixation of selectively neutral mutations. Science **164**: 788–798.

NEI, M., 1968 Evolutionary change of linkage intensity. Nature **218**: 1160–1161.

OHTA, T., 1968 Effect of initial linkage disequilibrium and epistasis on fixation probability in a small population, with two segregating loci. Theoret. Appl. Genet. **38**: 243–248.

OHTA, T. and M. KIMURA, 1969a Linkage disequilibrium due to random genetic drift. Genet. Res. **13**: 47–55. ——, 1969b Linkage disequilibrium at steady state determined by random genetic drift and recurrent mutation. Genetics **63**: 229–238. ——, 1970 Development of associative overdominance through linkage disequilibrium in finite populations. Genet. Res. **16**: 165–177.

## APPENDIX

*Basic equations for deriving the moments of the steady flux distribution. I. The moment equations for nonequilibrium population:*

We will first consider the single-variable case. Let $x_t$ be the frequency of a mutant form at a given site at time $t$ (conveniently measured with one generation as the unit length of time), and let $\delta x_t$ be the amount of change in $x_t$ during a short time interval between $t$ and $t+\delta t$ so that

$$x_{t+\delta t} = x_t + \delta x_t. \tag{A1}$$

Let $f(x)$ be a polynomial of $x$ and consider the expected value of this function with respect to the frequency distribution at time $t+\delta t$, i.e.,

$$E\{f(x_{t+\delta t})\}.$$

Substituting (A1) in this expression, we have

$$E\{f(x_{t+\delta t})\} = E_{\phi} E_{\delta}\{f(x_t+\delta x_t)\}, \tag{A2}$$

where $E_\delta$ is the operator of taking the expectation with respect to the change $\delta x$, and $E_\phi$ is that of taking the expectation with respect to the frequency distribution at time $t$.

Expanding the right-hand side of (A2) in terms of $\delta x_t$ and neglecting the higher-order terms containing $(\delta x_t)^3$ etc., we get

$$E\{f(x_{t+\delta t})\} = E_{\phi}\,\{f(x_t)+E_{\delta}(\delta x_t)f'(x_t)+\frac{E_{\delta}(\delta x_t)^2}{2!}f''(x_t)\},$$

or

$$\frac{E\{f(x_{t+\delta t})\}-E\{f(x_t)\}}{\delta t}=E\left\{\frac{E_\delta(\delta x_t)}{\delta t}f'(x_t)+½\frac{E_\delta(\delta x_t)^2}{\delta t}f''(x_t)\right\},$$

where $E$ now denotes $E_\phi$. At the limit of $\delta t \to 0$, we obtain

$$\frac{d}{dt}E\{f(x)\}=E\left\{\frac{V_{\delta x}}{2}f''(x)+M_{\delta x}f'(x)\right\}, \qquad (A3)$$

where $M_{\delta x}$ and $V_{\delta x}$ are the mean and the variance, respectively, of the rate of change in mutant frequency per generation and are approximations to $\lim_{\delta t\to 0}\{E_\delta(\delta x_t)/\delta t\}$ and $\lim_{\delta t\to 0}\{E_\delta(\delta x_t)^2/\delta t\}$. For the stationary distribution, the left-hand side of equation (A3) vanishes and it reduces to the equation (A2′) of OHTA and KIMURA (1969b).

Extension of equation (A3) to the multivariable case is immediate. For $n$ random variables $x_1, x_2, \ldots, x_n$, let $f \equiv f(x_1, x_2, \ldots, x_n)$. Then we have

$$\frac{d}{dt}E(f)=E\{L(f)\}, \qquad (A4)$$

where $L$ is the differential operator

$$L=½\sum_{i=1}^{n}V_{\delta x_i}\frac{\partial^2}{\partial x_i^2}+\sum_{i>j}W_{\delta x_i\,\delta x_i}\frac{\partial^2}{\partial x_i\partial x_j}+½\sum_{i=1}^{n}M_{\delta x_i}\frac{\partial}{\partial x_i} \qquad (A5)$$

in which $M$, $V$, and $W$ designate, respectively, the mean, the variance, and the covariance in the rate of change in the random variables that appear as subscripts.

II. *Steady flux case:* In the single-variable case, we assume that mutational input occurs at $x=p$ and output due to extinction or fixation occurs either at $x=0$ or $x=1$. At the state of steady flux, input and output balance each other and a steady state is reached with respect to the probability distribution, $\Phi(x)$, of mutants among segregating sites. Thus, writing $\Delta_{mut}E(f)$ for the input by mutation with respect to $E(f)$, we have

$$\Delta_{mut}E(f)+\frac{dE(f)}{dt}=0 \qquad (A6)$$

or

$$E\left\{\frac{V_{\delta x}}{2}f''(x)+M_{\delta x}f'(x)\right\}+\Delta_{mut}E(f)=0. \qquad (A6')$$

It is important to note here that equation (A6) or (A6′) is valid only for $f(x)$ which vanishes both at $x=0$ and $x=1$, because the steady state distribution $\Phi(x)$ refers only to unfixed classes $(1/2N \leqq x \leqq 1-1/2N)$, and new fixations that occur each generation at the terminal classes $x=0$ and $x=1$ should not be included in $E(f)$.

As an example of application of equation (A6′), consider the case of selectively neutral mutations. We assume that each generation mutation occurs in the entire population at $v_m$ sites and that the effective population size is $N_e$. If we take $f(x)=2x(1-x)$, $H\equiv E(f)$ represents the number of heterozygous nucleotide sites per individual as considered by KIMURA (1969a). Since for this case $M_{\delta x}=0$, $V_{\delta x}=x(1-x)/(2N_e)$, and $\Delta_{mut}E(f)=2v_m p(1-p)$; noting $f'(x)=2(1-2x)$ and $f''(x)=-4$, we obtain, from equation (A6′),

$$-\frac{1}{2N_e}E\{2x(1-x)\}+2v_m p(1-p)=0$$

or

$$H=E\{2x(1-x)\}=4N_e v_m p(1-p). \tag{A7}$$

This agrees with the result obtained by KIMURA (1969a) using a different method.

The above treatment may readily be extended to the multivariate case, and we obtain

$$E\{L(f)\}+\Delta_{mut}E(f)=0, \tag{A8}$$

where $L$ is the differential operator given by (A5).

## PART FIVE

# Evolutionary Advantages of Sexual Reproduction

## Introduction

The evolutionary advantages of sexual reproduction have been extensively discussed since Fisher (1930) and Muller (1914, 1932). If the main difference between sexual and asexual reproduction is in the presence or absence of the Mendelian segregation process occurring more or less independently at loci on a chromosome, the question is equivalent to asking whether or not there are evolutionary advantages of recombination. If offspring only duplicate the parental genotype, it is easy to see that the rate of accumulating beneficial mutations is slower than otherwise, while the rate of accumulating harmful mutations is faster than otherwise. Crow and Kimura (no. 14), following Muller's approach, first formulated this problem quantitatively. For subsequent, more quantitative, studies mainly based on computer simulations, see Felsenstein (1974); Felsenstein and Yokoyama (1976); Maynard Smith (1978, 1989); Haigh (1978); Takahata (1982); and Pamilo, Nei, and Li (1987).

Crow and Kimura (no. 14) further argued that diploidy provides an immediate advantage over haploidy by covering recessive deleterious alleles. This advantage is only transient, however, since eventually the diploid load will be twice as large; but the diploid state may have become fixed before this time. Kondrashov and Crow (1991) showed that with truncation-like selection there is a window of dominance values, around .25, that permits a permanent advantage of diploidy.

Muller's argument, as developed by Crow and Kimura (no. 14), is given less weight now than in the past. Most discussions now center on the advantages of sexuality in keeping up with a changing environment and with reducing the mutation load (for reviews, see Crow 1989a, 1992; Michod and Levin 1989).

Vol. XCIX, No. 909 The American Naturalist November–December, 1965

# EVOLUTION IN SEXUAL AND ASEXUAL POPULATIONS*

JAMES F. CROW AND MOTOO KIMURA

University of Wisconsin, Madison, Wisconsin, and National Institute of Genetics, Mishima-shi, Shizuoka-ken, Japan

It has often been said that sexual reproduction is advantageous because of the enormous number of genotypes that can be produced by a recombination of a relatively small number of genes (for example, *Issues in Evolution*, p. 114–115). The number of potential combinations is indeed great, but the number produced in any single generation is limited by the population size, and gene combinations are broken up by recombination just as effectively as they are produced by it. Furthermore, for a given amount of variability, the efficiency of selection is greater in an asexual population than in one with free recombination since the rate is measured by the total genotypic variance rather than by just the additive component thereof.

On the other hand, unless new mutations occur, an asexual population has a selection limit determined by the best existing genotype, whereas directional selection in a sexual population can progress far beyond the initial extreme, as has been demonstrated by selection experiments. The purpose of this article is to compare sexual and asexual systems as to the rate at which favorable gene combinations can be incorporated into the population, considering the effect of gene interaction, mutation rate, population size, and magnitude of gene effect. Most of the material is not new, but the various ideas have not been brought together in this context and we have introduced some refinements.

## HISTORICAL

The question was first discussed from the viewpoint in which we are here interested by Fisher (1930) and Muller (1932). We shall follow mainly the argument given by Muller.

In an asexual population, two beneficial mutants can be incorporated into the population only if the second occurs in a descendant of the individual in which the first occurred. On the other hand, in a sexual population the various mutants can get into the same individual by recombination. Only if the mutation rate were so low or the population so small that each mutant became established before another favorable mutant occurred would the two systems be equivalent.

The situation is illustrated in figure 1, adapted from Muller's original drawings. The three mutants, A, B, and C are all beneficial. In the asexual

*Paper number 987 from the Division of Genetics, University of Wisconsin; also contribution number 534 from the National Institute of Genetics, Mishima-shi, Japan. This work was aided by grants from the National Institutes of Health (RG-8217 and RG-7666).

population when all three arise at approximately the same time only one can persist. In figure 1, A is better adapted than B or C (or perhaps luckier in happening to occur in an individual that for other reasons was more fit than those in which B and C arose) so that A is eventually incorporated. B is incorporated only after it occurs in an individual that already carries A, and C only in an individual that already carries both A and B. In the sexual system, on the other hand, all three mutants are incorporated approximately as fast as any one of them is in the asexual system.

The lower part of the figure shows a small population. Here the favorable mutants are so infrequent that one has time to be incorporated before another occurs. Thus there is no advantage to the sexual system.

In general, several favorable mutants arising at the same time can all be incorporated in a sexual system whereas only one can be without recombination. There is, of course, a high probability of random loss of even a favorable mutation in the first few generations after its occurrence. This problem has been solved by Fisher (1930), but since the result is essentially the same in an asexual and sexual system it is irrelevant to the present discussion.

Muller's verbal argument was made quantitative in later papers (1958, 1964). We have improved Muller's calculations slightly by taking into consideration the decelerating rate of increase in the frequency of a favorable mutant as it becomes common.

## MATHEMATICAL FORMULATION

Consider first a population without recombination. We ignore the large majority of mutants that are unfavorable, for these are eliminated and thus are not incorporated into the population. Our interest is only in mutants that are favorable.

Let $N$ = the population number
- $U$ = the total rate of occurrence per individual per generation of favorable mutations at all loci
- $g$ = the average number of generations between the occurrence of a favorable mutation and the occurrence of another favorable mutation in a descendant of the first
- $x = 1/U$ = the number of individuals such that on the average one favorable mutation will occur
- $s$ = the average selective advantage of a favorable mutant

Thus, $g$ is the number of generations required for the cumulative number of descendants of a mutant to equal $x$, this being a number of such size that on the average one mutant will have occurred. Letting $p_i$ be the proportion of individuals carrying the mutant gene in the $i^{\text{th}}$ generation, $g$ is given by

$$x = Np_1 + Np_2 + \dots + Np_g \tag{1}$$

the summation being continued until there have been enough generations, $g$, to make the total number of mutant individuals equal to $x$. We assume that

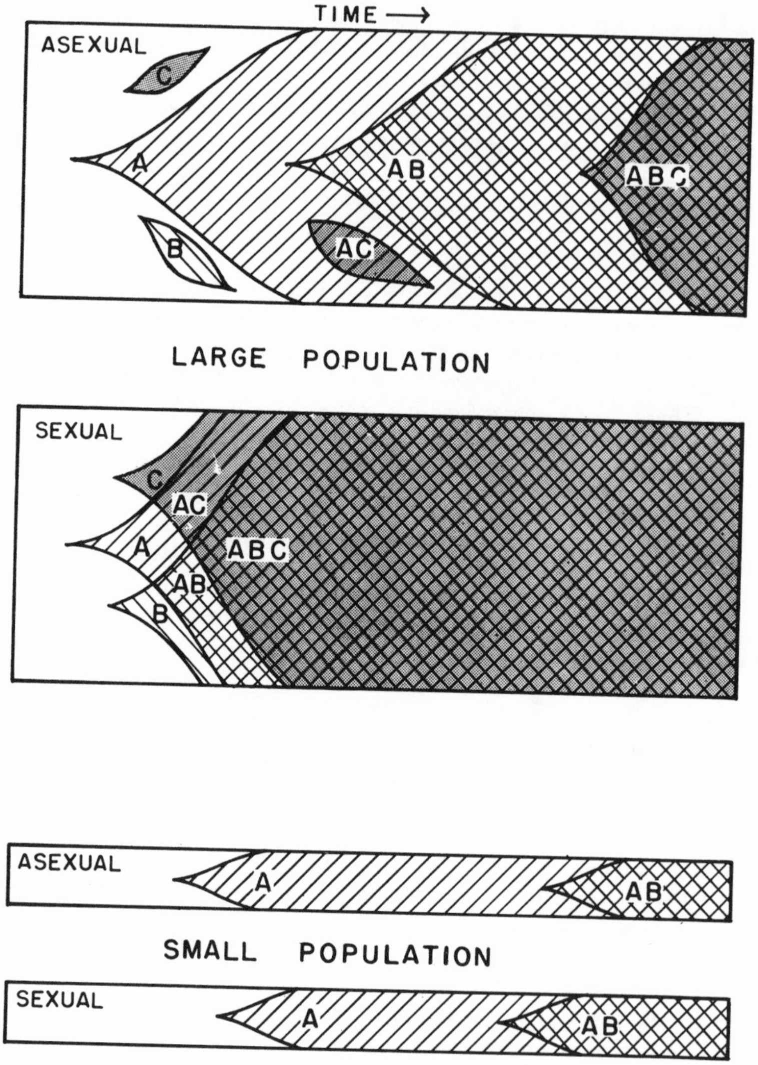


FIGURE 1. Evolution in sexual and asexual populations. The hatched and shaded areas show the increased number of mutant individuals following the occurrence of a favorable mutation. The abscissa is time. Modified from Muller (1932).

$s$ is small, and therefore that $p$ changes very slowly so that it is appropriate to replace addition by integration. This leads to

$$x = \int_0^g Np\, dt \qquad (2)$$

In the absence of recombination, $p$ follows the logistic curve

$$p = \frac{p_0}{p_0 + (1 - p_0)e^{-st}} \quad (3)$$

where $p_0$ is the initial proportion of mutants, as first shown by Haldane (1924). If we start with a single mutant, $p_0 = 1/N$, and (3) becomes

$$p = \frac{1}{1 + (N - 1)e^{-st}} \quad (4)$$

Substituting this for $p$ in (2) and integrating, we obtain

$$x = \frac{N}{s} \ln \frac{N - 1 + e^{sg}}{N}$$

or, rewriting and recalling that $x = 1/U$,

$$g = \frac{1}{s} \ln [N(e^{s/UN} - 1) + 1] \quad (5)$$

In an asexual population one new mutant that will eventually be incorporated into the population arises every $g$ generations. On the other hand, if reproduction is sexual, all the mutants that occur during this interval can eventually be incorporated. The number of mutants that arise per generation is $NU$, or in $g$ generations, $NUg$. Thus the ratio of incorporated mutations in a sexual population to that in an asexual population is $NUg:1$, or

$$\frac{NU}{s} \ln [N(e^{s/UN} - 1) + 1] \quad (6)$$

The favorable genes need not be mutants that are occurring for the first time. They may, for example, be genes brought in by immigrants or previously harmful mutants that have become beneficial because of a changed environment and already exist in the population at low frequencies.

Some numerical results are shown in table 1. For example, if the selective advantage of a favorable mutation is 0.01 and the total rate of occur-

TABLE 1

The relative rate of incorporation of new mutations into the population with and without recombination

| $U/s$ | $N$: $10^3$ | $10^4$ | $10^5$ | $10^6$ | $10^7$ | $10^8$ | $10^9$ |
|---|---|---|---|---|---|---|---|
| $10^{-7}$ | 1.0007 | 1.01 | 1.12 | 2.38 | 16.7 | 162 | $1.6 \times 10^3$ |
| $10^{-6}$ | 1.007 | 1.09 | 2.15 | 14.4 | 139 | $1.4 \times 10^3$ | $1.4 \times 10^4$ |
| $10^{-5}$ | 1.07 | 1.92 | 12.1 | 116 | $1.2 \times 10^3$ | $1.2 \times 10^4$ | $1.2 \times 10^5$ |
| $10^{-4}$ | 1.69 | 9.75 | 92.6 | 922 | $9.2 \times 10^3$ | $9.2 \times 10^4$ | $9.2 \times 10^5$ |
| $10^{-3}$ | 7.50 | 69.6 | 691 | $6.9 \times 10^3$ | $6.9 \times 10^4$ | $6.9 \times 10^5$ | $6.9 \times 10^6$ |
| $10^{-2}$ | 46.7 | 462 | $4.6 \times 10^3$ | $4.6 \times 10^4$ | $4.6 \times 10^5$ | $4.6 \times 10^6$ | $4.6 \times 10^7$ |
| $10^{-1}$ | 240 | $2.4 \times 10^3$ | $2.4 \times 10^4$ | $2.4 \times 10^5$ | $2.4 \times 10^6$ | $2.4 \times 10^7$ | $2.4 \times 10^8$ |

The ratio of the two rates is given in the body of the table. $N$ is the population number, $U$ is the total rate of occurrence of all favorable mutants, and $s$ is the average selective advantage of such mutants.

rence of such mutations is $10^{-8}$, the ratio $U/s$ is $10^{-6}$. As can be seen from the table, the advantage of recombination is negligible in a population of $10^3$, but the rate ratio is 2.15 in a population of $10^5$ and 1380 in a population of $10^8$. If the selective advantage is smaller, the advantage of a sexual system is greater. Likewise, with a higher mutation rate the advantage is greater.

We have discussed mutants that were beneficial at the time of their first occurrence. Similar considerations are involved when a previously deleterious mutant type is rendered beneficial, such as by a change in the environment.

We do not intend to imply that the very high values in the lower right part of the table are realistic. Doubtless other factors become limiting. But the table does show the general trend and emphasizes the enormous advantage of an evolutionary system with recombination.

It is likely that the probability of a mutant being favorable is greater when the effect of the mutant is small. Thus, with small $s$, $U$ tends to become larger. Since increasing $U$ and decreasing $s$ both have the effect of enhancing the advantage of recombination, the more that evolution proceeds by small micromutational steps, the greater the advantage of sexuality.

It is interesting that $U$ and $s$ enter the formula always in the form $U/s$, and never separately. This exact reciprocal dependence is understandable; for with slow selection the number of generations required for a given gene frequency change is inversely proportional to $s$ (Haldane, 1924). Thus a reduction in $s$ means that proportionately more mutations will occur during the time that one is being incorporated.

Table 1 also shows that the advantage of recombination increases with an increase in the population size. In fact, with large populations the advantage is nearly proportional to the population number.

To summarize: The advantage of a reproductive system that permits free recombination is greatest for the incorporation of mutant genes with individually small effects, occurring at relatively high rates, and in a large population.

## THE EFFECT OF GENE INTERACTION

So far we have been concerned only with mutant genes that are beneficial. We have also assumed that the combination of two mutant genes is more beneficial than either by itself; otherwise there would be no advantage of incorporating the second one.

The situation is quite different with some kinds of gene interaction. Where two or more mutants are individually harmful, but beneficial in combination, sexual reproduction may actually be disadvantageous.

The essential situation is clear with a haploid model, so we shall consider this simpler case. Suppose that the existing wild type in the population is genotype *ab*. The mutant types *Ab* and *aB* have fitnesses that, relative to *ab*, are reduced by the proportions $s_1$ and $s_2$. On the other hand, we assume that the double mutant has an enhanced fitness, greater than *ab* by a proportion $t$. The quantities $s_1$, $s_2$, and $t$ are all taken to be positive.

Both single mutant types, *Ab* and *aB*, will be found in low frequency in the population, their exact numbers being determined by the ratio of their rate of occurrence by mutation to their rate of elimination by selection. The double mutant, *AB*, will occasionally arise, but infrequently.

However, once such a double mutant does arise, its fate will be quite different in a population with and without recombination. Ignoring the question of chance elimination during the early generations (which, as we said earlier, is not significantly different in the two kinds of populations), an *AB* double mutant in an asexual population will increase and eventually be incorporated at a rate determined by the value of $t$.

On the other hand, in a sexual population, an *AB* individual will ordinarily mate with an *ab* genotype, in which case the progeny will consist of all four genotypes in proportions depending on the amount of recombination. Only if the fitness of the *AB* type is great enough to compensate for the loss of *AB* types through recombination will this genotype increase. The relationships can be set forth as follows:

| Genotype | *ab* | *Ab* | *aB* | *AB* |
|---|---|---|---|---|
| Relative fitness, $w$ | 1 | $1 - s_1$ | $1 - s_2$ | $1 + t$ |
| Frequency | $(1 - x)(1 - y)$ | $x(1 - y)$ | $(1 - x)y$ | $xy$ |

$$\bar{w} = 1 - s_1 x(1 - y) - s_2(1 - x)y + txy$$

$$\frac{\partial w}{\partial x} = y(s_1 + s_2 + t) - s_1$$

$$\frac{\partial w}{\partial y} = x(s_1 + s_2 + t) - s_2$$

These relationships assume that the two loci change independently under the action of natural selection, which is not strictly true unless $(1 + t) = (1 - s_1)(1 - s_2)$; but the formulae are approximately correct for unlinked loci when $s_1$, $s_2$, and $t$ are small.

Gene *A* will increase when $\partial \bar{w}/\partial x$ is positive and decrease when this is negative. Therefore there is an unstable equilibrium at $y = s_1/(s_1 + s_2 + t)$. Below this value of $y$, $x$ will decrease; above this value, $x$ will increase. The situation is symmetrical for $x$ and $y$ by interchanging $s_1$ and $s_2$. Thus there is no way for the frequency of the *AB* type to increase unless it somehow gets past the equilibrium point. This problem was discussed extensively by Haldane (1931).

The formulae are identical in a diploid population with complete dominance, on replacement of $x$ by $p^2$ and $y$ by $q^2$, where $p$ and $q$ are the frequencies of the recessive alleles at the two loci (see Wright, 1959, p. 442).

The situation is the familiar bottleneck frequently discussed by Wright (1931 and later). In his metaphor, the population is at one adaptive peak composed mainly of *ab* genotypes and there is no way for it to go to the higher peak composed mainly of *AB* genotypes without passing through a valley where *Ab* and *aB* types predominate.

There are several ways in which a sexual population might conceivably solve this problem. Some populations have several generations of asexual reproduction intervening between sexual generations. Another possibility would be strong assortative mating among the *AB* types; but the *a priori* probability of the genes that gave the increased fitness also producing the right type of mating behavior seems small indeed. Another possibility is random drift across the adaptive valley because of variable conditions or small effective population number, but this, as Wright (1931 and later) has emphasized, would lead to a considerable lowering of fitness (see also Kimura, Maruyama, and Crow, 1963). Furthermore, Kimura (unpublished) has shown that the probability of joint fixation of two genes such as are being discussed here is very small, even in small populations. For example, in a population of effective size $N = 1000$, the single mutants with 1 per cent selective disadvantage and the double mutant with 5 per cent advantage, the probability of joint fixation is about $2.5\ p_0 q_0 \times 10^{-6}$, where $p_0$ and $q_0$ are the initial frequencies of the single mutants. The corresponding probability for completely neutral genes is $p_0 q_0$. Note that $p_0\, q_0$ is a very small quantity. For individually deleterious but collectively advantageous mutant genes to have a reasonably high probability of joint fixation, the population must be so small that the inbreeding effect causes a serious effect on the viability.

In general, sexual reproduction can be a distinct disadvantage if evolution progresses mainly by putting together groups of individually deleterious, but collectively beneficial mutations. It seems to us that if this type of gene action were the limiting factor in evolution at the time sexual reproduction first evolved, sexual recombination might never have been "invented."

## THE EFFECT OF LINKAGE

Two closely linked genes in a sexual organism can be quite similar to genes in an asexual organism as far as their relations to each other are concerned, for they may stay together for a great length of time. If $r$ is the recombination frequency between two linked genes, they will stay together $1/r$ generations on the average before being separated by crossing over. This can easily be seen by noting that the probability that they will remain together $g$ generations and separate in the next is $(1 - r)^g r$. Then the average number of generations during which they remain together is

$$\bar{g} = r + 2(1 - r)r + 3(1 - r)^2 r + 4(1 - r)^3 r + \dots$$
$$= r(1 + 2x + 3x^2 + 4x^3 + \dots$$

where $x = 1 - r$. But $1 + 2x + 3x^2 + \dots$ is the derivative of $1 + x + x^2 + x^3 + \dots = 1/(1 - x)$. Therefore

$$\bar{g} = r \frac{d}{dx}\left(\frac{1}{1 - x}\right) = \frac{1}{r}$$

Thus, two genes linked together with a recombination value of 0.1 per cent would remain linked on the average for 1000 generations before separating.

Consider again the earlier model where the four haploid genotypes, *ab*, *Ab*, *aB*, and *AB* have fitnesses in the ratio $1 : 1 - s_1 : 1 - s_2 : 1 + t$, and assume that the amount of recombination between the loci is $r$. If the rare *AB* individual mates with an *ab* type, which will usually be the case, the proportion of *AB* progeny will be reduced by a fraction $r$ because of recombination. However, the *AB* type will increase from these matings if the extra fitness of the *AB* type is enough to more than compensate for this; that is, if $(1 + t)(1 - r) > 1$, or $t > r/(1 - r)$.

The conditions for increase of *AB* genotypes in general are a little less stringent because some *AB* matings are with *AB*, *aB*, or *Ab* types and in these there is no effect of crossing over. Furthermore, *AB* types are being added by recombination from $Ab \times aB$ matings. Finally, $\overline{w}$, the average fitness is not 1, but slightly less. However, these do not change the direction of the inequality, so we can still say that a sufficient condition for the double mutant type to increase is $t > r/(1 - r)$. This is also the condition for increase in diploids, where $t$ is now the advantage of the double heterozygote over the prevailing type (Bodmer and Parsons, 1962, p. 73).

In Wright's metaphor, the effect of linkage is to raise the valley between the two adaptive peaks and with extremely close linkage to provide a direct bridge.

For closely linked genes where $r$ is small, the *AB* type will increase and ultimately become fixed if $t > r$. Thus, the closer the linkage, the greater the tendency to build up coadapted complexes—provided, of course, that such closely linked, mutually beneficial mutants occur. The extreme example is the high degree of functional interdependence within a cistron.

## COADAPTATION

We have seen that asexual organisms are in a better position than sexual species to build up coadapted complexes, except under conditions of close linkage. In an asexual population the mutants accumulate in a certain sequence; first we have mutant *A*, then *AB*, then *ABC*, and so on. In this case the effect of *B* in the absence of *A* is irrelevant; it may be beneficial or harmful, or simply be a modifier that is neutral in the absence of *A*. Of all the mutants that arise in the species after *A* has been incorporated, the one that is most likely to persist is the one that in the presence of *A* gives the greatest fitness. Therefore, there will be a tendency for combinations to be mutually coadapted, and these genes may be less beneficial or even harmful in other combinations (Cavalli and Maccacaro, 1952). That is to say, they may well be what Mayr has called "narrow specialists."

In a sexual population, on the other hand, genes *A* and *B* are likely to be incorporated only if they are beneficial both individually and in combination. The type of gene that is most efficiently selected in a sexual population is one that is beneficial in combination with a large number of genes. We can only guess about the *a priori* distribution of gene interactions; but it is clear that in a population with free recombination the "good mixers" (that is, those having a large additive component) will be most efficiently selected.

The best opportunity to test these possibilities would be populations exposed to an entirely new environment. Drug resistance in bacteria and insecticide resistance in insects offer such a possibility. Chloramphenicol resistance in *Escherichia coli* is polygenic and has been analyzed by Cavalli and Maccacaro (1952). During the selection for resistance the reproduction was asexual, though recombination was used later for analyzing the genetic basis of the resistance. Recombinants between resistant and susceptible strains were skewed in the direction of greater susceptibility, as were crosses between different resistant strains. The results, therefore, suggest considerable coadaptation with complementary action of the genes accumulated during the selection process.

DDT resistance in Drosophila is also polygenic and has been analyzed genetically (Crow, 1957; King and Somme, 1958). Analysis of variance of the contribution of various chromosomes to the resistance showed an almost complete additivity, as would be expected in a sexual species according to the view we have been discussing. Thus, at least in these two cases, there is good agreement with what our theoretical speculations would predict.

We should emphasize that the genetic variability in sexual populations that have had a long history of selection for the traits under consideration may not have a large additive component. The genes that act additively may already have been incorporated into the population so that those that remain in unfixed condition are the ones that are not responsive to selection; that is, they are genes with complex interactions. Thus it is not surprising if, in a stable natural environment or in an artificial population where selection has been practiced for a long time, the nonadditive components of variance predominate.

## HAPLOIDY VERSUS DIPLOIDY

The evolutionary advantages of recombination can be obtained in haploid as well as diploid species. Yet diploidy is the rule in a great many complex organisms and there must have been a regular trend of evolution from haploidy to diploidy.

At first glance it would appear that there is an obvious advantage of diploidy in that dominant alleles from one haploid set can prevent the deleterious effects of harmful recessive alleles in the other. However, when equilibrium is reached the situation is roughly the same in a diploid as in a haploid. In a haploid species the mutation load will equal the total mutation rate when equilibrium is reached. With diploidy the load will be somewhere between this value and twice this value, depending on the level of dominance. With any substantial heterozygous effect of deleterious recessive genes, the mutation load is nearly twice the mutation rate (Haldane, 1937; Kimura, 1961). So the effect of diploidy is generally to double the mutation load by doubling the number of genes. From this standpoint diploidy certainly offers no advantages, only disadvantages.

However, when the population has reached equilibrium as a haploid, a change to diploidy offers an immediate advantage (Muller, 1932). To be sure, when the population reaches a new diploid equilibrium the advantage is lost; but by then there is no turning back, for a return to haploidy would greatly increase the load by uncovering deleterious recessives. Thus, it is easy to see how diploidy might evolve from haploidy, even if the population did not gain any permanent benefit therefrom.

On the other hand, there are some other possible advantages of diploidy, of which we shall mention two. One that has frequently been suggested is the possibility of overdominance. To the extent that the heterozygote is fitter than either homozygote at some loci there is an advantage of diploidy, provided the average fitness of the diploid population is enough greater than the haploid to compensate for the greater mutation load.

A second possible advantage of diploidy is the protection it affords against the effects of somatic mutation, a possibility that also occurred independently to Muller. The zygote in a diploid species or the post-meiotic cell from which the organism develops in a haploid species might have approximately the same fitness at equilibrium, but the effects of somatic mutation would be quite different. If the soma were large and complicated, as in higher plants and especially animals, a diploid soma might provide a significant protection against the effects of recessive mutations in critical organs.

## THE EVOLUTION OF SEXUALITY

The development of sexual reproduction confers no immediate advantage on the individual in which this occurs. In fact, the result is far more likely to be deleterious. The benefit is only to the descendants, perhaps quite remote, and to the population as a whole. Thus, it seems likely that the selective mechanism by which recombination was established was intergroup selection. Fisher (1930) goes so far as to suggest that sexuality may be the only character that evolved for species rather than for individual advantage.

On the other hand, despite the great evolutionary advantages of sexual reproduction, there are immediate advantages in a return to asexual reproduction. An advantageous type whose recombinant progeny were disadvantageous would have an advantage for its immediate descendants by developing an asexual mode of reproduction, other things being equal. In diploids there is the additional advantage of fixing heterotic combinations.

This all accords with the conventional belief that sexuality developed very early in the evolution of living forms a' is therefore found in all major groups; but that numerous independent retrogressions to vegetative reproduction continue to occur, conferring an immediate advantage but a long time evolutionary disadvantage.

SUMMARY

In an asexual population two favorable mutants can be incorporated into the population only if one occurs in a descendant of the individual in which the other occurred. In a sexual population both mutants can be incorporated through recombination. A mathematical formulation is given of the relative rates of incorporation of the new mutations with and without recombination. Recombination is of the greatest advantage when the double mutant is more advantageous than either single mutant, when the mutant effects are small, when mutations occur with high frequency, and when the population is large.

On the other hand, for the incorporation of individually deleterious but collectively beneficial mutations, recombination can be disadvantageous. Close linkage has effects similar to those of asexual reproduction. Experimental data on DDT resistance in Drosophila and chloramphenicol resistance in bacteria are cited showing greater development of coadaptation in an asexual system.

The evolution of diploidy from haploidy confers an immediate reduction in the mutation load by concealment of deleterious recessives, but this advantage is lost once a new equilibrium is reached. Thus the development of diploidy may be because of an immediate advantage rather than because of any permanent benefit. On the other hand, there are other possible advantages of diploidy, such as heterosis and protection from somatic mutations.

LITERATURE CITED

Bodmer, W. F., and P. A. Parsons, 1962, Linkage and recombination in evolution. Adv. in Genet. 11: 1–100.

Cavalli, L. L., and G. A. Maccacaro, 1952, Polygenic inheritance of drug-resistance in the Bacterium *Escherichia coli.* Heredity 6: 311–331.

Crow, J. F., 1957, Genetics of insecticide resistance to chemicals. Ann. Rev. Entomol. 2: 227–246.

Fisher, R. A., 1930, The Genetical Theory of Natural Selection. The Clarendon Press, Oxford. 2nd ed., 1958. Dover Pub., New York.

Haldane, J. B. S., 1924, The mathematical theory of natural and artificial selection. Part I. Trans. Cambridge Phil. Soc. 23: 19–41.

1931, A mathematical theory of natural selection. Part VIII. Metastable populations. Proc. Cambridge Phil. Soc. 27: 137–142.

1937, The effect of variation on fitness. Amer. Natur. 71: 337–349.

Kimura, M., 1961, Some calculations on the mutational load. Jap. J. Genet. 36 (Suppl.): 179–190.

Kimura, M., T. Maruyama, and J. F. Crow, 1963, The mutation load in small populations. Genetics 48: 1303–1312.

King, J. C., and L. Somme, 1958, Chromosomal analysis of the genetic factors for resistance to DDT in two resistant lines of *Drosophila melanogaster.* Genetics 43: 577–593.

Muller, H. J., 1932, Some genetic aspects of sex. Amer. Natur. 8: 118–138.

1958, Evolution by mutation. Bull. Amer. Math. Soc. 64: 137–160.

1964, The relation of recombination to mutational advance. Mutation Res. 1: 2–9.

Tax, Sol, and C. Callender (eds.), 1959, Issues in Evolution. Vol. 3 of Evolution after Darwin. University of Chicago Press.

Wright, S., 1931, Evolution in mendelian populations. Genetics 16: 97–159.

1959, Physiological genetics, ecology of populations, and natural selection. Perspect. Biol. Med. 3: 107–151.

# PART SIX

# Natural Selection

## Introduction

Fisher's (1930) Fundamental Theorem of Natural Selection states that the rate of increase of fitness of a population is equal to the additive genetic variance. He provided extensions to include the effect of crowding and deterioration of the environment (Frank and Slatkin 1992). Fisher's writing is elegant, but obscure, and many writers have attempted to explain what Fisher "really meant" (Edwards 1971; Price 1972; Nagylaki 1989b, 1991; Ewens 1989; Crow 1990a). Kimura (no. 15) provided an explicit formulation in which the terms are explicitly derived and identified. The number of alleles and loci is arbitrary, as is the mating system, and the fitness coefficients need not be constant. Unfortunately, however, only the single-locus results in Kimura no. 15 are correct (Akin 1979; Nagylaki 1989b, 1991). Although Fisher agreed to publish Kimura's paper, he did not regard it as a generalization of his theorem, but rather a more explicit specification.

Section 4 of the paper exemplifies the generality, and some pathological, counterintuitive cases are impressively demonstrated. The genic variance is consumed quite slowly by natural selection; the population is not likely to exhaust its genic variance unless it is small, the number of loci (involved in a quantitative trait) is small, or selection is very intense (chapter 5 in Crow and Kimura 1970; chapter 12 in Falconer 1960). Although this holds for short-term results, as in breeding experiments, the maintenance mechanisms of the heritable variation that is observed for most characters in natural populations are still poorly understood. For recent work on this problem, see Lande (1975, 1977); Hill (1982, 1990); Barton and Turelli (1987, 1989); Weir et al. (1988); and Tachida and Cockerham (1990).

Kimura's view (no. 16) was that the genetic information (or improbability) of organisms was gained by allelic substitutions owing to natural selection of random mutations. He was a strong neo-Darwinist at that time and related the information content to the substitutional load (Kimura 1960). He calculated the total amount of genetic information

(*G*) for the period of 500 million years since the Cambrian epoch. Although necessarily crude because of a number of uncertain assumptions, the estimate was 1% of the maximum that could be stored in the human genome. This was based on the horotelic (standard) rate of substitution (1/300 per generation) of Haldane (1957). Because of the small value of *G*, Kimura suggested that the genome may be highly redundant (Schmalhausen 1958) and that parts of a gene rather than the whole are duplicated. We may compute *G* from the nucleotide substitution rate which has become available since 1960. If the genome experiences one nucleotide substitution every other year (no. 48), *G* becomes 5% of the maximum, while for a more recent estimate of nucleotide substitution ($10^{-9}$ per site per year; Li, Luo, and Wu 1985; Satta et al. 1993), it becomes 20%. These amounts of genetic information gained may be consistent with the development of complex organisms during the last 500 million years, but the fact is that a large part of the genome in higher organisms is non-coding, and the genome is redundant not only in subunits of genes but also in functional units, as in multigene families (Watson et al. 1987).

# ON THE CHANGE OF POPULATION FITNESS BY NATURAL SELECTION * †

MOTOO KIMURA
*National Institute of Genetics, Mishima-shi, Japan*

Received 26.v.57

## 1. INTRODUCTION

In *The Origin of Species*, Darwin (1859) put forward the view that natural selection, by preserving and accumulating small inherited modifications advantageous to each organism, is continuously effecting the improvement of the species in relation to its organic and inorganic environment. At that time the mechanism of particulate inheritance was unknown and he could not reject the theory of blending inheritance nor even the inheritance of acquired characters. Nevertheless it is remarkable that he did arrive at the right conclusion on the role of natural selection in evolution.

Later, based on the theory of Mendelian inheritance, Fisher (1930) formulated his " fundamental theorem of natural selection " by equating the rate of increase of population fitness with the genetic variance in fitness. Since the variance is a non-negative quantity, fitness in this formulation always increases with time. This view seems to be compatible with the general picture of evolution which has taken place in the past and may be considered as a quantitative description of an essential feature of evolution. In fact, Fisher (1930) claims that this law should hold the same position among the biological sciences as the second law of thermodynamics in physical sciences. However, if one considers actual cases in which competition with other species or deterioration of environment are involved, additional terms may be necessary and Fisher (1930) derived an equation for the rate of increase of the Malthusian parameter of the population in which the effect of overpopulation (C), genetic variance (W) and the effect of deterioration of environment (D) are included.

The validity of the " fundamental theorem " for the case of variable selection coefficients has been questioned by Wright (1949) who later derived an additional term for such a case (Wright, 1950). Attempts to obtain more general expressions for the rate of increase of population fitness have been made by Crow and Kimura (1956), Wright (1955) and Crow (1955).

The present paper deals with four problems : (1) I shall give a general equation for the role of additive, dominance, and epistatic components of fitness in determining the rate of change in population fitness. No restriction is placed on the mating system. A few of the many interesting special cases are discussed. (2) The process of

* Contribution No. 240 from the National Institute of Genetics, Mishima-shi, Japan.

† Dedicated to Dr Taku Komai on the occasion of his seventieth birthday anniversary as a token of gratitude.

Reprinted from *Heredity* 12:145–67 by permission of the publisher, Blackwell Scientific Publications, Oxford, England.

natural selection will be treated from a different view, and a new formulation given which may be called a " maximum principle in the genetic theory of natural selection ". (3) A short section is devoted to the change in population variability with selection. (4) Finally these results will be applied to artificial selection, with special reference to the roles of dominance and epistatic components when inbreeding and linkage are involved.

## 2. MEASUREMENT OF FITNESS WITH TWO MODELS OF GENERATION TIME

Before going into the detailed argument on the main subject, it might be preferable to explain the mathematical models used and to define " fitness " for each model.

### 2.1. *A model with a discrete time parameter*

In this model, all the members of the population at a given moment belong to the same generation. After each breeding cycle the population is completely replaced by its progeny to form the next generation. Thus the generation time, $t$, takes only integral values ; 0, 1, 2, etc. The fitness of each genotype is measured by the average number of offspring per individual. The letter $w$ will be used to designate fitness in this sense and following Wright (1942 and later) will be called the selective value of the genotype. The population is stationary in size if the mean fitness ($\bar{w}$) is unity. The selection coefficient of the genotype $A_iA_j$ denoted by $s_{ij}$ will be defined by

$$s_{ij} = w_{ij} - 1,$$

where $w_{ij}$ is the selective value of $A_iA_j$.

Quite often, an arbitrary value of unity is assigned to one of the genotypes, but this is not meaningful unless the aim of study is restricted to the change of gene frequencies.

Consider an arbitrary number of alleles at one locus and let $x_i$ be the relative frequency of the $i^{\text{th}}$ allele $A_i$. Under this model, the rate of change of the gene frequency per generation, $\Delta x_i$, is given by

$$\Delta x_i = \frac{x_i(w_{i\cdot} - \bar{w})}{\bar{w}}, \tag{2.1.1}$$

where $w_{i\cdot} = \sum_j w_{ij}P_{ij}/x_i$ and $\bar{w} = \sum_{ij} w_{ij}P_{ij} = \sum_i w_{i\cdot}x_i$. $P$ denotes the zygotic frequency such that $P_{ii}$ is the frequency of the homozygote $A_iA_i$ while $2P_{ij}(i \neq j)$ is the frequency of the heterozygote $A_iA_j(P_{ij} = P_{ji})$. The definition of $P$ is complicated for the multi-locus case, but the difficulty may be avoided by saying that $P_{ij}$ (for arbitrary $i$ and $j$) stands for the frequency of the *ordered genotype* $A_iA_j$, i.e. for mathematical convenience we distinguish between $A_iA_j$ and $A_jA_i$.

### 2.2. *A model with a continuous time parameter*

In this model a population may contain individuals of different ages as in human populations. For a rigorous treatment of such a problem integral equations are needed (Haldane, 1927).

An appropriate scale for measuring fitness is the Malthusian parameter, $m$ (Fisher, 1930), which measures the rate of population increase or decrease. It is defined as the real root of the equation ;

$$\int_0^\infty e^{-mx} l_x b_x dx = 1,$$

where $l_x$ is the proportion of the population which survive to age $x$, and $b_x$ is the rate of reproduction at that age. To obtain the fitness, $m_{ij}$, of the genotype $A_iA_j$, the above equation may be applied to a group of individuals of genotype $A_iA_j$, though perhaps differing at other loci. If there are differences in the vital statistics of the two sexes, " we may, certainly in mankind, and with a strong presumption in all other species, effectively average out the sex differences " (Fisher, 1953).

In the present paper, I will designate the fitness measured in Malthusian parameters by the letter $a$ such that $a_{ij}$ is the fitness of $A_iA_j$. Let $n_{ij}$ be the number of individuals with the ordered genotype $A_iA_j$ in the population. Since $a_{ij}$ represents the net reproductive rate of the genotype $A_iA_j$ such that the contribution to the rate of increase of the population number $N$ from $A_iA_j$ individuals is $n_{ij}a_{ij}$, then if $N_i$ is the number of $A_i$ genes such that

$$x_i = N_i/N,$$

we have

$$\frac{dN_i}{dt} = \sum_j n_{ij}a_{ij} = N\sum_j P_{ij}a_{ij}. \tag{2.2.1}$$

Therefore

$$\frac{dN}{dt} = N \sum_{ij} P_{ij}a_{ij}$$

and we obtain

$$\frac{dx_i}{dt} = \frac{d}{dt}\left(\frac{N_i}{N}\right) = \frac{1}{N}\frac{dN_i}{dt} - \frac{N_i}{N^2}\frac{dN}{dt}$$

$$= \sum P_{ij}a_{ij} - x_i \sum_{ij} P_{ij}a_{ij}$$

or more simply

$$\frac{dx_i}{dt} = x_i(a_{i\cdot} - \bar{a}), \tag{2.2.2}$$

where $a_{i\cdot} = \sum_i a_{ij}P_{ij}/x_i$ and $\bar{a} = \sum_{ij} a_{ij}P_{ij}$. The quantity $a_{i\cdot} - \bar{a}$ will be called the average excess of the gene $A_i$ and will be designated by $\tilde{a}_{i\cdot}$. Throughout this paper a bar will be used to designate the population mean while a tilde will designate the deviation from the mean. Thus

(2.2.2) is equivalent to saying that the logarithmic rate of increase of gene frequency is equal to the average excess of the gene in fitness, *i.e.*

$$\mathring{x}_i = \tilde{a}_{i\cdot}, \tag{2.2.3}$$

in which

$$\mathring{x}_i = \frac{d}{dt}\log x_i.$$

### 2.3. *Interrelation between the two models*

The above two models are both abstractions from reality and each has its advantages and disadvantages. For example, the discrete model might be suitable for an annual plant population while the continuous model would be preferable for the treatment of a human population. Though the scales of fitness for the two models are different, there are certain correspondences between them. Roughly speaking, $m_{ij}$ or $a_{ij}$ is equivalent to $\log w_{ij}$. A formal passage from (2.1.1) to (2.2.2) is attained as follows. Write $w_{ij} = 1 + s_{ij}\Delta t$ and substitute in (2.1.1) to give

$$\frac{\Delta x_i}{\Delta t} = \frac{x_i(s_{i\cdot} - \bar{s})}{1 + \bar{s}\Delta t}.$$

By letting $\Delta t \to 0$, we have (2.2.2) except that $s_{ij}$ appears instead of $a_{ij}$. This suggests a formal correspondence between $s_{ij}$ in the discrete model and $a_{ij}$ in the continuous model.

Mathematically, the discrete model requires the calculus of finite differences, while the continuous model permits application of the familiar differential calculus which often leads to a more elegant formulation. In the present paper I shall mainly employ the continuous model for this reason.

## 3. RATE OF CHANGE IN POPULATION FITNESS

We start by considering a single locus with an arbitrary number, say $n$, of alleles. No restriction will be imposed on the mating system. The fitness $a_{ij}$ may be expressed in the following form,

$$a_{ij} = \bar{a} + \alpha_i + \alpha_j + d_{ij},$$

where $\bar{a}$ is the mean fitness of the population, $\alpha_i$ and $\alpha$ are respectively additive effects of the genes $A_i$ and $A_j$ on fitness and $d_{ij}$ is the dominance deviation from linearity. Following Fisher (1930, 1941) we shall call $\alpha_i$ the average effect of the gene $A_i$. The $\alpha_i$'s are determined by minimising

$$\sum_{ij} d_{ij}^2 P_{ij}.$$

By writing $\tilde{a}_{ij}$ for $a_{ij} - \bar{a}$, the quantity to be minimised is

$$Q = \sum_{ij} P_{ij}(\tilde{a}_{ij} - \alpha_i - \alpha_j)^2.$$

Since $-\frac{1}{2}\frac{\partial Q}{\partial \alpha} = 0$ leads to

$$\sum_j P_{kj}(\tilde{a}_{kj}-\alpha_k-\alpha_j) + \sum_i P_{ik}(\tilde{a}_{ik}-\alpha_i-\alpha_k) = 0,$$

we have the normal equations for $\alpha_k$'s ;

$$(3.1) \qquad 2\alpha_k x_k + 2\sum_i P_{ki}\alpha_i = 2\sum_i P_{ki}\tilde{a}_{ki}. \quad (k = 1, \ldots, n)$$

If we adopt the notation $\tilde{a}_{k\cdot}$ for $\sum_i P_{ki}\tilde{a}_{ki}/x_k$, the right side of (3.1) is written as $2x_k\tilde{a}_{k\cdot}$ and the sum of squares due to the average effects is given by

$$(3.2) \qquad V_g = 2\sum_k \alpha_k x_k \tilde{a}_{k\cdot}\,.$$

If we note (2.2.2) this may also be written as

$$(3.2)' \qquad V_g = 2\sum_i \alpha_i \frac{dx_i}{dt}.$$

It may be convenient to note for later discussion that under random mating $\alpha_i$ is equal to $\tilde{a}_{i\cdot}$ and

$$(3.2)'' \qquad V_g = 2\sum_i x_i \tilde{a}_{i\cdot}^2\,.$$

For the derivation of normal equations (3.1) and the extraction of the sum of squares due to $\alpha_i$'s, the reader may refer to Kempthorne (1957).

We will call $V_g$ the *genic* variance in fitness ; it has also been called the additive genetic variance in fitness. Fisher (1930) called the same quantity the genetic variance.

Now we introduce a new quantity $\theta_{ij}$ defined by

$$(3.3) \qquad \theta_{ij} = \frac{P_{ij}}{x_i x_j}. \quad (\theta_{ij} = \theta_{ji}).$$

$\theta_{ij}$ may be called the coefficient of departure from random combination of alleles and is unity if mating is random. Substituting $P_{ki} = x_k x_i \theta_{ki}$ in (3.1) and dividing both sides of the equation by $2x_k$, we have

$$\alpha_k + \sum_i x_i\theta_{ki}\alpha_i = \tilde{a}_{k\cdot} = \sum_i x_i\theta_{ki}\tilde{a}_{ki},$$

or

$$(3.4) \qquad \sum_j x_j\theta_{ij}\tilde{a}_{ij} = \alpha_i + \sum_j \alpha_j\theta_{ij}x_j.$$

Let us now consider the rate of increase in population fitness under natural selection. Since

$$\bar{a} = \sum_{ij} a_{ij}P_{ij},$$

we have

$$\frac{d\bar{a}}{dt} = \sum_{ij}\frac{da_{ij}}{dt}P_{ij} + \sum_{ij} a_{ij}\frac{dP_{ij}}{dt},$$

or more simply, employing the Newtonian method of denoting a time derivative by a superior dot,

$$(3.5) \qquad \dot{\bar{a}} = \sum_{ij}\dot{a}_{ij}P_{ij} + \sum_{ij} a_{ij}\dot{P}_{ij}.$$

The first term on the right side is the population average of the rate of change of individual fitness and may be denoted by

(3.6) $$\overline{\frac{da}{dt}} \text{ or } \bar{\dot{a}}.$$

The second term on the right is

$$\sum_{ij} a_{ij}\dot{P}_{ij} = \sum_{ij} (\bar{a}+\tilde{a}_{ij})\dot{P}_{ij} = \bar{a} \sum_{ij} \dot{P}_{ij} + \sum_{ij} \tilde{a}_{ij}\dot{P}_{ij},$$

and since $\sum_{ij} P_{ij} = 1$ or $\sum_{ij} \dot{P}_{ij} = 0$, it reduces to

$$\sum_{ij} \tilde{a}_{ij}\dot{P}_{ij},$$

which upon substitution of $P_{ij} = x_i x_j \theta_{ij}$ becomes

(3.7) $$\sum_{ij} \tilde{a}_{ij}\dot{x}_i x_j \theta_{ij} + \sum_{ij} \tilde{a}_{ij} x_i \dot{x}_j \theta_{ij} + \sum_{ij} \tilde{a}_{ij} x_i x_j \dot{\theta}_{ij}.$$

If we substitute (3.4) in the first term of the above expression, we have

(3.8) $$\sum_i \dot{x}_i(\alpha_i + \sum_j \alpha_j \theta_{ij} x_j) = \sum_i \alpha_i \dot{x}_i + \sum_{ij} \alpha_j \dot{x}_i x_j \theta_{ij}.$$

It will be immediately seen from (3.2)′ that the first term of the last expression is

$$\tfrac{1}{2} V_g.$$

By definition $\sum_i P_{ij} = x_j$ or $\sum_i x_i x_j \theta_{ij} = x_j$, and we get $\sum_i x_i \theta_{ij} = 1$ or differentiating both sides we get $\sum_i \dot{x}_i \theta_{ij} + \sum_i x_i \dot{\theta}_{ij} = 0$. Thus the second term of (3.8) reduces to

$$\sum_j \alpha_j x_j(-\sum_i x_i \dot{\theta}_{ij}) = -\sum_{ij} \alpha_j x_i x_j \dot{\theta}_{ij}.$$

Similarly the second term of (3.7) becomes

$$\tfrac{1}{2} V_g - \sum_{ij} \alpha_i x_i x_j \dot{\theta}_{ij}.$$

Therefore (3.7) can be written in the form

$$V_g + \sum_{ij} (\tilde{a}_{ij} - \alpha_i - \alpha_j) x_i x_j \dot{\theta}_{ij},$$

or

$$V_g + \sum_{ij} d_{ij} P_{ij} (\dot{\theta}_{ij}/\theta_{ij}).$$

Thus the required expression for the rate of change of population fitness by selection is

(3.9) $$\frac{d\bar{a}}{dt} = V_g + \overline{\frac{da}{dt}} + \sum_{ij} P_{ij} d_{ij} \frac{d}{dt} \log \theta_{ij}$$

or in more simple notation

(3.10) $$\dot{\bar{a}} = V_g + \bar{\dot{a}} + \overline{d \cdot \dot{\theta}},$$

in which the open circle over $\theta$ denotes the time derivative of log $\theta$ as in (2.2.3). Also it is convenient to note that

$$\sum_{ij} P_{ij}d_{ij} \frac{d}{dt} \log (P_{ij}/x_i x_j) = \sum_{ij} P_{ij}d_{ij} \frac{d}{dt} \log P_{ij} = \sum_{ij} d_{ij} \frac{d}{dt} P_{ij}.$$

Thus, as shown in (3.9) or (3.10), the rate of increase of the population fitness is decomposed into three components : (i) Genic variance in fitness. (ii) Average increase in the fitness of individual genotypes. This term is zero if genotypic fitnesses remain constant. (iii) A contribution due to dominance deviations and mating system. This term vanishes under random mating ($\theta_{ij} = 1$), or under any system of mating if there is no dominance. It is interesting to note the simple relation by which dominance components appear in the expression. Each component is weighted by the rate of change of the logarithm of the corresponding coefficient of departure from random combination of alleles. This would suggest a similar relation for epistatic components when two or more loci are involved.

Without going into the details of derivation, the result for the two loci may be summarised as follows. We decompose the fitness of the genotype $A_iA_jB_kB_l$ in the form

$$a_{ijkl} = \bar{a} + A_{ij} + B_{kl} + \epsilon_{ijkl},$$

where $A_{ij}$ is the additive effect of the first locus ($a$), $B_{kl}$ that of the second locus ($b$) and $\epsilon_{ijkl}$ is the epistatic deviation. Let

$$A_{ij} = \alpha_i + \alpha_j + d_{ij}^{(a)}$$

and

$$B_{kl} = \beta_k + \beta_l + d_{kl}^{(b)}$$

in which $\alpha$ and $\beta$ stand for the additive effects of the alleles for the first and second loci respectively and $d^{(a)}$ and $d^{(b)}$ are dominance deviations for the two loci. Let $P_{ijkl}$ be the frequency of the ordered genotype $A_iA_jB_kB_l$ and let $P_{ij}^{(a)}$ be the frequency of the ordered genotype $A_iA_j$ in which no attention is paid to the second locus, i.e. $P_{ij}^{(a)} = \sum_{kl} P_{ijkl}$, and $P_{kl}^{(b)}$ be the similar frequency of the ordered genotype $B_kB_l$ at the second locus. If we designate the frequencies of genes $A_i$ and $B_k$ by $x_i$ and $y_k$ respectively, we can define coefficients of departure from random combination of alleles at the two loci by

$$\theta_{ij}^{(a)} = \frac{P_{ij}^{(a)}}{x_i x_j} \text{ and } \theta_{kl}^{(b)} = \frac{P_{kl}^{(b)}}{y_k y_l},$$

respectively. Similarly we introduce the coefficient of departure involving two loci by

$$\theta_{ijkl} = \frac{P_{ijkl}}{P_{ij}^{(a)} P_{kl}^{(b)}}.$$

Then

$$\frac{d\bar{a}}{dt} = V_g + \overline{\frac{da}{dt}} + \sum_{ij} P_{ij}^{(a)} d_{ij}^{(a)} \frac{d}{dt} \log \theta_{ij}^{(a)}$$

$$+ \sum_{kl} P_{kl}^{(b)} d_{kl}^{(b)} \frac{d}{dt} \log \theta_{kl}^{(b)} + \sum_{ijkl} P_{ijkl} \epsilon_{ijkl} \frac{d}{dt} \log \theta_{ijkl},$$

or more simply

$$(3.11) \qquad \dot{\bar{a}} = V_g + \bar{\dot{a}} + \overline{\mathring{\theta}_{ij}^{(a)} d_{ij}^{(a)}} + \overline{\mathring{\theta}_{kl}^{(b)} d_{kl}^{(b)}} + \overline{\mathring{\theta}_{ijkl} \epsilon_{ijkl}}.$$

The epistatic effect may further be subdivided into additive $\times$additive, additive $\times$dominance and dominance $\times$dominance effects. To simplify the following formulation we adopt a rule that the letters $i$, $j$, $k$ and $l$ refer respectively to the left gene in the first locus, the right gene in the first locus, the left gene in the second locus and the right gene in the second locus in the ordered genotype $A_iA_jB_kB_l$. With this abbreviation, not only subscripts $a$ and $b$ are unnecessary but also simple expressions like $P_{ik}$, $P_{ijk}$, . . ., etc. may be used for the frequencies of ordered genotypes $A_i$-$B_k$-, $A_iA_jB_k$-, etc.

Let

$$\begin{aligned}\epsilon_{ijkl} &= (\alpha\beta)_{ik} + (\alpha\beta)_{il} + (\alpha d)_{ikl} \\ &+ (\alpha\beta)_{jk} + (\alpha\beta)_{jl} + (\alpha d)_{jkl} \\ &+ (d\beta)_{ijk} + (d\beta)_{ijl} + (dd)_{ijkl}.\end{aligned}$$

Corresponding to each of the dominance and epistatic components, coefficients of departure from random combination are defined in the obvious way :

$$\theta_{ij} = P_{ij}/x_i x_j,\ \theta_{ik} = P_{ik}/x_i y_k,\ \ldots,$$
$$\theta_{ijk} = P_{ijk}/P_{ij} y_k,\ \theta_{ijkl} = P_{ijkl}/P_{ij} P_{kl}.$$

Then

$$(3.12) \qquad \begin{aligned}\dot{\bar{a}} &= V_g + \bar{\dot{a}} + \overline{\mathring{\theta}_{ij} d_{ij}} + \overline{\mathring{\theta}_{kl} d_{kl}} \\ &+ \overline{\mathring{\theta}_{ik} (\alpha\beta)_{ik}} + \ldots + \overline{\mathring{\theta}_{ijkl} (dd)_{ijkl}}.\end{aligned}$$

The law of formulation of each term is self-evident so no further explanation is necessary.

Thus for an arbitrary number of loci each with an arbitrary number of alleles, we can write the rate of increase in population fitness as follows :

$$(3.13) \qquad \frac{d\bar{a}}{dt} = V_g + \overline{\frac{da}{dt}} + \overline{\Sigma\, \epsilon \frac{d}{dt} \log \theta},$$

in which $\epsilon$ stands for an interaction effect (dominance or epistatic) and $\theta$ is the corresponding coefficient of departure from random combination. The summation extends over all possible intra- and inter-locus interactions.

## 4. SOME SPECIAL SITUATIONS

4.1. For constant genotypic fitnesses, $\bar{\dot{a}} = 0$ and (3.10) becomes

$$\dot{\bar{a}} = V_g + \overline{d \cdot \mathring{\theta}}.$$

If there exists a constant $\lambda_{ij}$ such that

$$\lambda_{ij} = P_{ij}^2 / P_{ii} P_{jj}$$

for all $i$ and $j$, then $\dot{\bar{a}} = V_g$. This is easily shown as follows:

Since $P_{ij} = x_i x_j \theta_{ij}$, we have $\lambda_{ij}\theta_{ii}\theta_{jj} = \theta_{ij}^2$ or

$$\log \lambda_{ij} + \log \theta_{ii} + \log \theta_{jj} = 2 \log \theta_{ij}.$$

Differentiating both sides with respect to time,

$$\mathring{\theta}_{ii} + \mathring{\theta}_{jj} = 2\mathring{\theta}_{ij}.$$

Thus
$$\overline{d \cdot \mathring{\theta}} = \sum_{ij} d_{ij}\mathring{\theta}_{ij}P_{ij} = \tfrac{1}{2}\sum_{ij} d_{ij}(\mathring{\theta}_{ii} + \mathring{\theta}_{jj})P_{ij}$$

$$= \tfrac{1}{2}\sum_i \mathring{\theta}_{ii} \sum_j d_{ij}P_{ij} + \tfrac{1}{2}\sum_j \mathring{\theta}_{jj} \sum_i d_{ij}P_{ij} = 0,$$

because
$$\sum_j d_{ij}P_{ij} = \sum_i d_{ij}P_{ij} = 0.$$

The equivalent result was obtained by Kempthorne (1957) by using matrix algebra. However, Fisher (1941) had already demonstrated such a result for the two allelic case. Extension to the multi-locus case is immediately apparent from (3.13).

In the case of inbreeding in which zygotic frequencies are expressed by gene frequencies and the inbreeding coefficient $F$, $P_{ii} = Fx_i + (1 - F)x_i^2$ for homozygotes and $P_{ij} = (1 - F)x_i x_j$ for heterozygotes. Therefore $\theta_{ii} = \frac{1}{x_i} F + (1 - F)$ and $\theta_{ij} = (1 - F)$ $(i \neq j)$.

We will consider this case later in more detail.

4.2 A quantity analogous to the potential function in a conservative physical system may exist in some special cases. Wright (1955) considered such a quantity and termed it the internal selective value. I shall use a similar quantity which may be called the internal fitness ($I$) in the following way: We say that the internal fitness exists if

$$2 \sum_i \alpha_i dx_i \qquad (\sum_i x_i = 1)$$

is an exact differential with respect to gene frequencies $(x_1, x_2, \ldots, x_n)$. In that case there exists a function $I$ such that the above expression is equated with $dI$. Therefore

$$\frac{dI}{dt} = 2 \sum_i \alpha_i \frac{dx_i}{dt} = V_g,$$

*i.e.* the rate of increase of internal fitness is equal to the genic variance at that moment. Thus if the internal fitness is equal to the mean fitness of the population $\bar{a}$ except for an arbitrary constant, the fundamental theorem is satisfied. However, the reverse is not necessarily true and the derivation of Fisher's fundamental theorem does not presuppose or imply the existence of any such potential function.

4.3. When competition between genotypes is involved we sometimes encounter a rather peculiar situation. As an example let us consider three alleles $A_1$, $A_2$, and $A_3$ with frequencies $x_1$, $x_2$, and $x_3$. For simplicity we assume no dominance or more conveniently a haplont situation. Suppose an individual with gene $A_1$ has its fitness decreased by the amount $c$ when it is surrounded by individuals with $A_2$, while $A_2$ if surrounded by the former $A_1$ gains fitness by the same amount

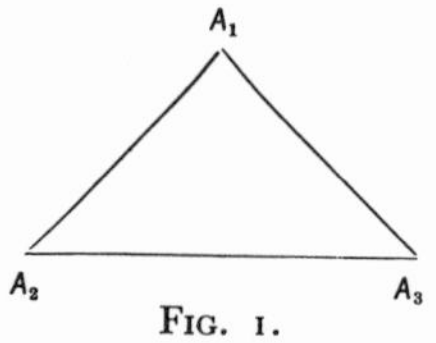


FIG. 1.

$c$. Similarly $A_2$ when surrounded by $A_3$ individuals loses fitness by $c$, while the reverse situation would add fitness $c$ to an $A_3$ individual, and so on cyclically (fig. 1). Then the law of change of gene frequencies will be given by

$$(4.3.1)\qquad \begin{cases} \dfrac{dx_1}{dt} = cx_1(x_3-x_2) \\ \dfrac{dx_2}{dt} = cx_2(x_1-x_3) \\ \dfrac{dx_3}{dt} = cx_3(x_2-x_1). \end{cases}$$

The gene frequencies will move perpetually on a closed path * which is the intersection of a plane

$$x_1+x_2+x_3 = 1$$

and a surface

$$x_1x_2x_3 = C_0,$$

where $C_0$ is a constant determined by the initial conditions. In the present case $\alpha_1 = \bar{a}_{1\cdot} = c(x_3-x_2)$, $\alpha_2 = \bar{a}_{2\cdot} = c(x_1-x_3)$, $\alpha_3 = \bar{a}_{3\cdot} = c(x_2-x_1)$, and $\bar{a} \equiv 0$. This shows that, while gene frequencies change perpetually, no change in population fitness occurs:

$$\frac{d\bar{a}}{dt} = 0.$$

* With a discrete model, we have a set of finite difference equations $\Delta x_1 = cx_1(x_3-x_2)$, etc. instead of (4.3.1) and the path is not quite closed. In the neighbourhood of (1/3, 1/3, 1/3), the solution contains a factor $(1+c^2/3)^{t/2}$ showing that the path is an outwardly proceeding spiral. The author owes to Professor S. Wright and Mr Y. Hiraizumi for calling his attention to this fact.

Nevertheless the population at a given moment may contain plenty of genic variance, for

$$V_g = c^2\{x_1(x_3-x_2)^2+x_2(x_1-x_3)^2+x_3(x_2-x_1)^2\}.$$

The reason for this seeming contradiction is that the second term in our formula (3.9) always cancels out the first term, namely

$$\overline{\frac{da}{dt}} = -V_g,$$

so that

$$\frac{d\bar{a}}{dt} = V_g + \overline{\frac{da}{dt}}$$

still holds.

4.4 In our definition of fitness in (2.2.1), sexual difference in fitness for each genotype, if it exists, is in a sense averaged out in the expression. This is reasonable since under a complicated mating system it would not be practicable to analyse contributions due to different sexes. However, under simple circumstances this may be done to give interesting results. Let $a_{ij}$ be the fitness of an $A_iA_j$ female in producing offspring and let $c_{ij}$ be the competitive ability of the $A_iA_j$ male in fertilising females. We assume random mating and no dominance. Then we have

$$\frac{d\bar{a}}{dt} = \sigma_{aa}+\sigma_{ac}, \tag{4.4.1}$$

where $\sigma_{aa}$ is half the genic variance,

$$\sigma_{aa} = \sum_i x_i(a_{i\cdot}-\bar{a})^2$$

and $\sigma_{ac}$ is the covariance between average excess (or effect in this case) of fitness in females and that of competitive ability in males.

$$\sigma_{ac} = \sum_i x_i(a_{i\cdot}-\bar{a})(c_{i\cdot}-\bar{c}),$$

in which $\bar{c}$ may be chosen to be zero. Let $b$ be the regression coefficient of $c_{i\cdot}$ on $a_{i\cdot}$, i.e.

$$c_{i\cdot} = \bar{c}+b(a_{i\cdot}-\bar{a}),$$

where $b = \sigma_{ac}/\sigma_{aa}$. Then if $b$ is less than $-1$,

$$\sigma_{ac}<-\sigma_{aa},$$

and therefore

$$\frac{d\bar{a}}{dt} = \sigma_{aa}+\sigma_{ac}<0,$$

leading to the deterioration of the population. Yet the population may contain genic variance through this process.

$$V_g = 2\sigma_{aa}>0.$$

In fact, Huxley (1942) put forward an explanation for the extinction of species due to fixation of genes which enable males to be more successful in competition but make the species less fit to the external environment.

Fisher (1941) considered a hypothetical gene which enables the possessor to be self-fertilised but which has no effect on the seed productivity or cross-fertilising ability of the pollen. This is essentially equivalent to the case in which $\sigma_{aa} = 0$ in the above formulation. For, in this case $\sigma_{aa} = \sigma_{ac} = 0$ and

$$\frac{d\bar{a}}{dt} = 0,$$

which agrees well with what is expected from the fundamental theorem since

$$V_g = 2\sigma_{aa} = 0.$$

4.5. In a natural population there always exists an element of stochastic indeterminacy in the process of change in fitness. We consider here a simple single locus case with random mating.

Let $\delta x_i$ be the rate of change in the frequency of $A_i$ per unit time interval which may be taken as one generation, we can write

$$\delta x_i = x_i(a_{i.} - \bar{a}) + \delta' x_i,$$

where $\delta' x_i$ is the change due to random sampling of gametes. Then the rate of change in population fitness is

$$\delta \bar{a} = \sum_{ij} \delta a_{ij} x_i x_j + V_g + 2 \sum_i a_{i.} \delta' x_i.$$

Here $a_{ij}$ is a random variable which may fluctuate from generation to generation. $\delta' x_i$'s are variables from multinomial distribution such that their mean, variance and covariance are given respectively by

$$M(\delta' x_i) = 0, \; V(\delta' x_i) = x_i(1 - x_i)/2N$$

and

$$\mathrm{Cov}(\delta' x_i \delta' x_j) = -x_i x_j / 2N \text{ for } i \neq j,$$

where $N$ is the number of reproducing individuals in the population. Then the mean and variance of $\delta \bar{a}$ are

(4.5.1) $$M(\delta \bar{a}) = \overline{M(\delta a)} + V_g,$$

and

(4.5.2) $$V(\delta \bar{a}) = \overline{\sigma_{\delta a \delta a'}} + V_g/2N,$$

where

$$\overline{M(\delta a)} = \sum_{ij} M(\delta a_{ij}) x_i x_j$$

and

$$\overline{\sigma_{\delta a \delta a'}} = \sum_{iji'j'} \mathrm{Cov}(\delta a_{ij} \delta a_{i'j'}) x_i x_j x_{i'} x_{j'}.$$

The last term $V_g/(2N)$ corresponds to Fisher's error variance for his fundamental theorem (Fisher, 1930).

## 5. A MAXIMUM PRINCIPLE IN THE GENETICAL THEORY OF NATURAL SELECTION

So far we have considered how natural selection increases population fitness, but have paid little attention to how natural selection changes gene frequencies.

Let us start our consideration from a simple case with random mating and constant fitness of individual genotypes. Consider a locus with $n$ alleles, say $A_1, A_2, \ldots, A_n$ with respective frequencies $x_1, x_2, \ldots, x_n$. First we will establish the following theorem : *For a given short time interval $\delta t$, natural selection causes gene frequency changes $\delta x_1, \delta x_2, \ldots, \delta x_n$, in such a way that the increase of population fitness, $\delta \bar{a}$, shall be maximum under the restriction*

$$(5.1) \qquad \sum_{i=1}^{n} \frac{(\delta x_i)^2}{x_i} = \tfrac{1}{2} V_g(\delta t)^2.$$

To prove this theorem, we put $\delta\bar{a} = \eta$, $\delta x_i = \xi_i$ and $\frac{1}{2}V_g(\delta t)^2 = c$. Since

$$\bar{a} = \sum_{ij} a_{ij} x_i x_j,$$

we have, for any small set of gene frequency changes $(\xi_1, \xi_2, \ldots, \xi_n)$,

$$\delta\bar{a} = 2 \sum_{i=1}^{n} a_{i.}\delta x_i$$

or

$$(5.2) \qquad \eta = 2 \sum_{i=1}^{n} a_{i.}\xi_i.$$

The restriction (5.1) is now written as

$$(5.1)' \qquad \sum_{1}^{n} \frac{\xi_i^2}{x_i} = c.$$

In addition, since the sum of all the gene frequencies is unity, we inevitably have the obvious restriction :

$$(5.3) \qquad \sum_{1}^{n} \xi_i = 0.$$

We are going to show that among all possible sets of gene frequency changes which satisfy (5.1)′ and (5.3), the one which is caused by natural selection is the set which makes $\eta$ maximum.

For this purpose we introduce the Lagrange multipliers $\lambda$ and $\mu$ and consider a function

$$(5.4) \qquad \psi = \eta - \lambda\left(\sum_{1}^{n} \frac{\xi_i^2}{x_i} - c\right) - \mu\left(\sum_{1}^{n} \xi_i\right).$$

A set of $\xi_i$'s which make $\psi$ maximum may be obtained from

$$\frac{\partial\psi}{\partial\xi_i} = 2a_{i\cdot} - 2\lambda\frac{\xi_i}{x_i} - \mu = 0,$$

or

$$2a_{i\cdot}x_i - 2\lambda\xi_i - \mu x_i = 0. \tag{5.5}$$

Summing up both sides of (5.5) from $i = 1$ to $n$, and noting (5.3), we have

$$2\bar{a} = \mu.$$

Therefore (5.5) yields

$$\xi_i = (a_{i\cdot} - \bar{a})x_i/\lambda. \tag{5.6}$$

By substituting this back into (5.1)′ we have

$$\sum_i x_i(a_{i\cdot} - \bar{a})^2/\lambda^2 = c.$$

The denominator in the left side of the above equation is $\frac{1}{2}V_g$ from (3.2)″, while $c = \frac{1}{2}V_g(\delta t)^2$, so that

$$1/\lambda^2 = (\delta t)^2 \text{ or } 1/\lambda = \delta t,$$

because negative $\lambda$ is inadequate. Thus (5.6) is

$$\xi_i = x_i(a_{i\cdot} - \bar{a})\delta t$$

or

$$\frac{\delta x_i}{\delta t} = x_i(a_{i\cdot} - \bar{a}), \quad (i = 1, \ldots, n)$$

which is equivalent to (2.2.2) and gives the law of change in gene frequencies due to natural selection.

Geometrical consideration may help in understanding the above theorem (fig. 2). We consider an $(n+1)$ dimensional space in which there are $n$ coordinates for gene frequencies and one coordinate for population fitness. A population with a given set of gene frequencies is expressed as a point $P$ in this space. A set of displacements $\xi_i$'s together with a corresponding change in population fitness $\eta$ define a vector radiating from $P$. Projection of this vector on the $n$ dimensional subspace of gene frequencies is inevitably on the hyper plane given by (5.3). Then the theorem states that among all such vectors whose projections on such $n$ dimensional subspace are on the hyper-ellipsoid given by (5.1)′, the vector corresponding to the change by natural selection is the one which gives maximum length of projection on the $\bar{a}$ axis.

The theorem can easily be extended to any number of loci. For example for two loci, we designate the displacement of gene frequencies in the second loci having $m$ alleles by $\zeta_1, \zeta_2, \ldots, \zeta_m$, then (5.1)′ should be replaced by

$$\sum_1^n \frac{\xi_i^2}{x_i} + \sum_1^m \frac{\zeta_j^2}{y_j} = \frac{1}{2}V_g(\delta t)^2 \tag{5.6}$$

and (5.3) by

$$(5.7) \qquad \sum_{1}^{n} \xi_i = 0, \sum_{1}^{m} \zeta_j = 0$$

respectively.

The theorem may be further extended to cover a quite general situation in which the above restrictions of random mating and constant fitness of genotypes are removed. In what follows we will

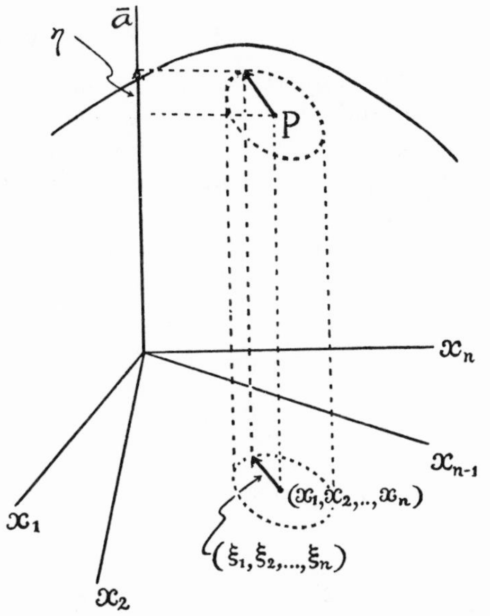


FIG. 2.

consider only the case of a single segregating locus, but the argument may easily be extended to any number of loci.

Let $$A = (\tilde{a}_{1.}, \tilde{a}_{2.}, \ldots, \tilde{a}_{n.})$$

be a vector whose components are average excesses of genes and let

$$D = (\xi_1, \xi_2, \ldots, \xi_n)$$

be a displacement vector in the set of gene frequencies. Then natural selection acts in such a way that the product

$$AD' = \sum_{1}^{n} \tilde{a}_{i.} \xi_i$$

shall be maximum under the restriction

$$\sum_{1}^{n} \xi_i^2 / x_i = \sigma_a^2 (\delta t)^2,$$

where $\sigma_a^2$ is the variance of average excess in fitness per ploid, *i.e.*,

$$\sigma_a^2 = \sum_{1}^{n} x_i (a_{i.} - \bar{a})^2.$$

In this form the theorem is applicable not only to diploid but also to $2K$ ploid in general in which $K = \frac{1}{2}, 1, 2, 3, \ldots$, etc.

The above theorem is somewhat analogous to the principle of least action in physics and may be called the maximum principle in the genetical theory of natural selection. Whereas from the least action principle the laws of motion may be derived, we may from this theorem derive the rules of change in gene frequencies.

## 6. RATE OF CHANGE OF POPULATION VARIABILITY

Considering the great importance of the genic variance in evolution, it is appropriate to study how quickly the genic variance will be consumed by natural selection for the improvement of the species. Though the following result is quite restricted, it may help our understanding of the problem.

Consider a random mating population and assume constant fitness of genotypes. In this case the average effect is equal to the average excess and we have

$$V_g = 2 \sum_{i=1}^{n} x_i(a_{i.} - \bar{a})^2.$$

Taking the time derivative we have

$$\dot{V}_g = 2 \sum_{1}^{n} \dot{x}_i(a_{i.} - \bar{a})^2 + 4 \sum_{1}^{n} x_i(a_{i.} - \bar{a})(\dot{a}_{i.} - \dot{\bar{a}})$$

or

$$\dot{V}_g = 2 \sum_{1}^{n} x_i(a_{i.} - \bar{a})^3 + 4 \sum_{ij} \dot{x}_i\dot{x}_j a_{ij}.$$

The first term in the right side is the third moment in average effect and will be designated by $\mu_{3(g)}$. The second term reduces to

$$4 \sum_{ij} \dot{x}_i\dot{x}_j d_{ij} \text{ or } 4\overline{\alpha_i\alpha_j d_{ij}}$$

so that

$$\frac{dV_g}{dt} = \mu_{3(g)} + 4\overline{\alpha_i\alpha_j d_{ij}}. \tag{6.1}$$

With no dominance the second term vanishes and we have

$$\frac{dV_g}{dt} = \mu_{3(g)}.$$

If the distribution of fitness among individuals in a population is near normal $\mu_{3(g)}$ will be very small as compared with $V_g$ itself. This result seems to indicate that the additive genetic variance is consumed quite slowly by natural selection. This conclusion may not be much influenced in most cases by the presence of dominance, since the second term in (6.1) will usually be a small quantity.

## 7. APPLICATION TO ARTIFICIAL SELECTION

Here we consider only mass selection with respect to a quantitative character governed by a large number of factors.

Let $P_{ijkl\ldots}$ be the frequency of the ordered genotype $A_iA_jB_kB_l\ldots$, whose phenotypic value or yield is $Y_{ijkl\ldots}$. Applying (3.13), the increase of population mean per generation becomes

$$\text{(7.1)} \qquad \Delta\bar{Y} = \sum_{ijkl\ldots} (\Delta Y_{ijkl\ldots}) P_{ijkl\ldots} + 2\left(\sum_i \alpha_i \Delta x_i + \sum_k \beta_k \Delta y_k + \ldots\right) + \sum \overline{\epsilon \Delta \log \theta},$$

where $\alpha$, $\beta$, ..., are average effects of the genes on yield at the 1st, 2nd, etc., locus, and $\epsilon$ denotes an interaction effect on yield.

The first term in the right side of (7.1) gives the increase of yield by improvement of environment.

In order to handle the remaining terms in the right side of (7.1), we have to establish the relation between the yield ($Y_{ijkl\ldots}$) and the fitness or selective value ($w_{ijkl\ldots}$) of each genotype. For this we will employ the method used by Crow (1955), and based on an earlier finding of Haldane (1932).

Let $\sigma^2$ be the total phenotypic variance. Consider various phases in any locus, say the first, and designate the mean genotypic value of individuals with the phase $A_iA_j$ by

$$Y_{ij} = \sum_{kl\ldots} P_{ijkl\ldots} Y_{ijkl\ldots} \Big/ \sum_{kl\ldots} P_{ijkl\ldots}.$$

We assume that the means of these phases have very small variance around $\bar{Y}$ compared with $\sigma^2$, while individuals with a given phase, say $A_iA_j$, distribute normally with variance $\sigma'^2$ around the mean $Y_{ij}$.

Let $\tilde{c}$ be the point of truncation measured from the population mean $\bar{Y}$, then the selective value of the phase $A_iA_j$ is

$$\text{(7.2)} \qquad w_{ij} = \frac{1}{\sqrt{2\pi}\sigma'} \int_{\tilde{c}}^{\infty} e^{-\frac{(y-\tilde{Y}_{ij})^2}{2\sigma'^2}} dy,$$

where $\tilde{Y}_{ij} = Y_{ij} - \bar{Y}$. Now we assume that $(\tilde{Y}_{ij}/\sigma')^2$ is a negligible quantity. Thus we have approximately

$$w_j = S + \frac{\tilde{Y}_{ij}}{\sigma^2} IS \propto 1 + \frac{\tilde{Y}_{ij}}{\sigma^2} I,$$

where

$$S = \frac{1}{\sqrt{2\pi}\sigma} \int_{\tilde{c}}^{\infty} e^{-\frac{y^2}{2\sigma^2}} dy$$

or the proportion saved, and

$$I = \frac{1}{\sqrt{2\pi}\sigma} \int_{\tilde{c}}^{\infty} y e^{-\frac{y^2}{2\sigma^2}} dy / S$$

or the selection differential,

The average excess in fitness of $A_i$ is

$$\tilde{w}_i = \frac{\tilde{Y}_{i.}}{\sigma^2} I.$$

Thus the yield is transformed into fitness by multiplying by the factor $I/\sigma^2$. Therefore

$$\Delta x_i = x_i \tilde{Y}_{i.} \frac{I}{\sigma^2}, \tag{7.3}$$

where $\tilde{Y}_{i.}$ is the average excess of the gene $A_i$ on yield. Substituting this in the second term of (7.1) we have

$$2 \sum_i \alpha_i \Delta x_i = 2 \frac{I}{\sigma^2} \sum_i \tilde{Y}_{i.} \alpha_i x_i = \frac{I}{\sigma^2} \sigma^2_{g(1)},$$

where $\sigma^2_{g(1)}$ is the additive or genic variance in yield due to the first locus. Thus if we designate by $\sigma^2_g$ the total genic variance due to all loci, we have

$$2(\sum_i \alpha_i \Delta x_i + \ldots) = \frac{I}{\sigma^2} \sigma^2_g = h^2 I, \tag{7.4}$$

where $h^2$ is the heritability in the narrow sense.

Next we will consider the effect of gene interactions collectively expressed in the last term of (7.1). The effect of dominance in the first locus appears as

$$\sum_{ij} P_{ij} d_{ij} \Delta \log \theta_{ij},$$

or

$$\sum_{ij} x_i x_j d_{ij} \Delta \theta_{ij}.$$

The most important case of non-random mating is inbreeding in which

$$\theta_{ii} = F/x_i + 1 - F \text{ and } \theta_{ij} = 1 - F \text{ for } i \neq j.$$

For this case, we have

$$\sum_{ij} x_i x_j d_{ij} \Delta \theta_{ij} = \frac{\Delta F}{1 - F} \sum_i d_{ii} x_i - F \sum_i d_{ii} \Delta x_i, \tag{7.5}$$

where $\Delta x_i$ is given by (7.3). The total contribution by dominance is the sum of such terms for all loci.

Next we will consider terms due to epistasis. As pointed out by Fisher (1918), epistasis involving more than two loci may be rather unimportant and we will consider only dual epistacy. Let $J_{ab}$ be the epistatic contribution due to interaction between the first and the second loci, then

$$J_{ab} = \sum \overline{\epsilon_{ijkl} \Delta \log \theta_{ijkl}} = \sum_{ijkl} \epsilon_{ijkl} \Delta P_{ijkl},$$

where $\epsilon_{ijkl}$ stands for the epistatic effect on yield rather than on fitness. Perhaps the simplest but most important case in which $J_{ab}$

is not zero is the case in which under random mating the two loci on the same chromosome have not reached linkage equilibrium. Let $X_{ik}$ be the frequency of the chromosome with $A_iB_k$. Then the frequency of the ordered genotype $A_iA_jB_kB_l$ is

$$P_{ijkl} = \tfrac{1}{2}(X_{ik}X_{jl} + X_{il}X_{jk}).$$

Thus

(7.6) $$\mathcal{J}_{ab} = 2 \sum_{ijkl} \epsilon_{ijkl} \Delta X_{ik} X_{jl} = 2 \sum_{ik} \epsilon_{i\cdot k\cdot} \Delta X_{ik},$$

where

$$\epsilon_{i\cdot k\cdot} = \sum_{jl} \epsilon_{ijkl} X_{jl}, \quad (\epsilon_{ijkl} = \epsilon_{ijlk} = \epsilon_{jikl} = \epsilon_{jilk}).$$

However

(7.7) $$\Delta X_{ik} = X_{ik}\tilde{w}_{i\cdot k\cdot} - c \sum_{jl} \begin{vmatrix} X_{ik} & X_{il} \\ X_{jk} & X_{jl} \end{vmatrix} w_{ijkl},$$

in which

$$\tilde{w}_{i\cdot k\cdot} = \sum_{jl} w_{ijkl} X_{jl} - \bar{w}$$

and $w_{ijkl}$ is the selective value of $A_iA_jB_kB_l$ while $c$ is the recombination fraction between the two loci. For the derivation of (7.7), the reader may refer to my previous paper (Kimura, 1956). Substituting (7.7) in (7.6) noting that fitness may be obtained from yield by multiplying $I/\sigma^2$.

(7.8) $$\mathcal{J}_{ab} = \frac{I}{\sigma^2} \left\{ E - 2c \sum_{ijkl} \epsilon_{i\cdot k\cdot} \begin{vmatrix} X_{ik} & X_{il} \\ X_{jk} & X_{jl} \end{vmatrix} Y_{ijkl} \right\},$$

where

$$E = 2 \sum_{ik} \epsilon_{i\cdot k\cdot} \tilde{Y}_{i\cdot k\cdot} X_{ik}.$$

The total contribution may be obtained by adding up terms like $\mathcal{J}_{ab}$ for all relevant pairs of loci. The case of random mating with linkage may also be approached by considering additive chromosomal variance rather than additive genic variance.

## 8. CONCLUDING REMARKS

Any discussion on the change of population fitness must start from the definition of the term fitness. Though there may be diverse ways of defining fitness, the term, to be useful for population genetics, must be related to organism's ability to leave descendants. If it is defined in terms of Malthusian parameters expressing the geometric rate of population increase, the rate of increase of population fitness, as shown from (3.9) through (3.13), is expressed as the sum of three terms : A term due to genic variance, one due to change in environment, and one due to the joint effect of genetic interaction (dominance and epistasis) and change in mating system :

(8.1) $$\frac{d\bar{a}}{dt} = V_g + \overline{\dot{a}} + \Sigma \overline{\epsilon \dot{\theta}}.$$

From the standpoint of the genetical theory of evolution, the first term $V_g$ is by far the most important, since it represents the part of increase in fitness due to the improvement of the *genic* constitution of the species. The effects due to the remaining terms are generally speaking unpredictable from the *genic* constitution of the species and may conveniently be treated as environmental factors. However, if we come down to a specific situation, we may handle these terms to obtain more information as shown in section 4.

In this section we will consider the effect of overpopulation on fitness. Since a given niche can support only a finite number of individuals at a given moment, the population mean of the Malthusian parameters, averaged over the long period that the species has lived, must be very near to zero. Whenever the genic constitution of the species is improved by natural selection, the increase in the Malthusian parameters will soon be cancelled out by the increase of population number. Though the equation (8.1) is general enough to include the above process as a special aspect, it may be interesting to separate it for special consideration.

Let $M$ be the Malthusian parameter of an actual population and let $N$ be the population number at a given moment, then we have

$$\frac{1}{N}\frac{dN}{dt} = M \tag{8.2}$$

by definition. Let $a_{ijkl...}$ be the Malthusian parameter of the genotype $A_iA_jB_kB_{l...}$ which would be realised if there were no effect of overcrowding, then $\bar{a}$ is the intrinsic rate of population increase and

$$\frac{1}{N}\frac{dN}{dt} = \bar{a}-f(N), \tag{8.3}$$

where $f(N)$ is a function representing the effect of population number or, when the population is confined to a restricted area, the effect of density. In other words $f(N)$ represents the resistance against population increase due to overcrowding.

From (8.2) and (8.3)

$$M = \bar{a}-f(N)$$

or taking time derivatives of both sides,

$$\frac{dM}{dt} = \frac{d\bar{a}}{dt} - \frac{df(N)}{dt}. \tag{8.4}$$

Since the second term in the right side may be written as

$$\frac{df(N)}{dt} = \frac{df(N)}{dN}\frac{dN}{dt} = MN\frac{df(N)}{dN},$$

(8.4) becomes

$$\frac{dM}{dt} + MN\frac{df(N)}{dN} = \frac{d\bar{a}}{dt}, \tag{8.5}$$

where the right side is given by (8.1). Since we are considering the deleterious effect of *over*population, we may assume that $f(N)$ is a monotone non-decreasing function and therefore $f'(N)$ is non-negative. However, it should be kept in mind that in nature extreme low population density is also disadvantageous for population growth (*cf.* Haldane, 1953).

Returning to (8.3), we will consider a few special cases. For the logistic law of growth

$$f(N) = KN,$$

where $K$ is a constant, and therefore $f'(N) = K$. Thus the second term in the left side of (8.5) becomes $KMN$.

For weaker resistance, we may take

$$f(N) = K \log N.$$

In this case $f'(N) = K/N$ and we can get rid of a troublesome factor $N$ out of the equation (8.5) :

$$\frac{dM}{dt} + \frac{M}{C} = V_g + \bar{\dot{a}} + \overline{\Sigma \epsilon \mathring{\theta}}, \tag{8.6}$$

where $C = 1/K$. (8.6) may be considered to be equivalent to Fisher's formulation :

$$\frac{dM}{dt} + \frac{M}{C} = W - D, \tag{8.7}$$

in which " $W$ is the rate of actual increase in fitness determined by natural selection, and $D$ is the rate of loss due to the deterioration of the environment " (Fisher, 1930). $W$ is equal to our $V_g$ and $-D$ corresponds to

$$\bar{\dot{a}} + \overline{\Sigma \epsilon \mathring{\theta}}.$$

Thus Fisher's " environment " implies not only physical and biological environment in the usual sense, but also genetical environment the effect of which is represented by

$$\overline{\Sigma \epsilon \mathring{\theta}},$$

in our formulation. Considering the great diversity of ways in which non-genic (non-genetic in Fisher's sense) factors come into play, to treat such factors collectively as an environment is quite adequate. However, it is possible, as shown in the present paper, to represent the contribution by such factors in terms of gene interaction and change in mating system in a rather simple way when these are known.

## 9. SUMMARY

1. When fitness ($a$) of individual genotypes is measured in Malthusian parameters, the rate of increase of population fitness due to natural selection is expressed as a sum of the three terms : (i) a term due to *genic* (or additive genetic) variance in fitness ($V_g$), (ii) one due

to change in environment, and (iii) one due to joint effect of genetic interaction (dominance and epistasis) and change in mating system. Thus we have

$$\frac{d\bar{a}}{dt} = V_g + \overline{\frac{da}{dt}} + \overline{\Sigma\epsilon \frac{d}{dt} \log \theta},$$

in which $\epsilon$ stands for an interaction effect in fitness and $\theta$ is the corresponding coefficient of departure from random combination. The summation extends over all possible interactions. The above formulation establishes the roles of additive, dominance and epistatic components on the rate of increase of population fitness. It is the reformulation of the results published by Fisher in 1930 under the title *The Fundamental Theorem of Natural Selection*, and in no way differs from it in substance. $\theta$ corresponds with $\lambda$ of Fisher's (1941) treatment.

2. An attempt to formulate the law of change in gene frequencies due to natural selection has led to a theorem which, for a simple case of random mating with constant fitness of individual genotypes, may be expressed in the following form : For a given short time interval $\delta t$, the natural selection causes gene frequency changes, $\delta x_1, \delta x_2, \ldots, \delta x_n$, in such a way that the increase of population fitness, $\delta\bar{a}$, shall be maximum under the restriction of

$$\sum_{i=1}^{n} \frac{(\delta x_i)^2}{x_i} = \tfrac{1}{2} V_g (\delta t)^2.$$

This theorem is extended to cover a quite general situation. It is remarkable that the set of differential equations describing the law of change in gene frequencies by natural selection can be derived from the theorem. It may be called " a maximum principle in the genetical theory of natural selection " analogous to the principle of least action in physics.

3. The rate of change of the genic variability in fitness has been studied for a simple situation. The result seems to indicate that the genic variance is consumed rather slowly by natural selection.

4 The rate of increase of yield by artificial selection has been given with special reference to the roles of dominance and epistatic components when inbreeding and linkage are involved.

5. The present formulation on the change of population fitness is compared with the results by Fisher (1930) and the effect of population number on fitness has been discussed.

*Acknowledgment.*—The author expresses his appreciation to Dr James F. Crow for reading the manuscript and making valuable suggestions.

## 10. REFERENCES

CROW, J. F. 1955. General theory of population genetics : synthesis. *Cold Spring Harbor Symp. Quant. Biol.*, *20*, 54-59.

CROW, J. F., AND KIMURA, M. 1956. Some genetic problems in natural populations. *Proc. Third Berkeley Symp. Math. Stat. Prob.*, Vol. IV, pp. 1-22.

DARWIN, CHARLES. 1859. *The Origin of Species.* First Ed. London, John Murray.

FISHER, R. A. 1918. The correlation between relatives on the supposition of Mendelian inheritance. *Trans. Roy. Soc. Edin.*, *52*, 399-433.

FISHER, R. A. 1930. *The Genetical Theory of Natural Selection.* Oxford, The Clarendon Press.

FISHER, R. A. 1941. Average excess and average effect of a gene substitution. *Ann. Eugen.*, *11*, 53-63.

FISHER, R. A. 1953. Population genetics. *Proc. Roy. Soc.* B, *141*, 510-523.

HALDANE, J. B. S. 1927. A mathematical theory of natural and artificial selection. IV. *Proc. Cam. Phil. Soc.*, *23*, 607-615.

HALDANE, J. B. S. 1932. *The Causes of Evolution.* New York, Harper and Brothers.

HALDANE, J. B. S. 1953. Animal populations and their regulation. *New Biology*, *15*, 9-24.

HUXLEY, J. 1942. *Evolution, the Modern Synthesis.* London, Allen and Unwin.

KEMPTHORNE, O. 1957. *An Introduction to Genetic Statistics.* New York, John Wiley.

KIMURA, M. 1956. A model of a genetic system which leads to closer linkage by natural selection. *Evolution*, *10*, 278-287.

WRIGHT, S. 1942. Statistical genetics and evolution. *Bull. Amer. Math. Soc.*, *48*, 223-246.

WRIGHT, S. 1949. Adaptation and selection. In G. L. Jepsen *et al.*, *Genetics, Paleontology and Evolution.* Princeton Univ. Press, pp. 365-389.

WRIGHT, S. 1950. Variable selective values of genotypes. Mimeographed notes on lectures at Edinburgh.

WRIGHT, S. 1955. Classification of the factors of evolution. *Cold Spring Harbor Symp. Quant. Biol.*, *20*, 16-24.

# Natural selection as the process of accumulating genetic information in adaptive evolution*

By MOTOO KIMURA

*National Institute of Genetics, Mishima, Japan*

(*Received* 3 *October* 1960)

## INTRODUCTION

Modern genetic studies have shown that the instructions for forming an organism are contained in the nucleus of the fertilized egg. In the language of information theory, we may say that in the process of development the genetic (hereditary) information of an organism is transformed into its phenotypic (organic) information. Thus, to account for the tremendous intricacy of organization in a higher animal, there must exist a sufficiently large amount of genetic information in the nucleus.

What is the origin of such genetic information? If the Lamarckian concept of the inheritance of acquired characters were accepted, one might be justified in saying that it was acquired from the environment. However, since both experimental evidence and logical deductions have entirely failed to corroborate such a concept, we must look for its source somewhere else.

We know that the organisms have evolved and through that process complicated organisms have descended from much simpler ones. This means that new genetic information was accumulated in the process of adaptive evolution, determined by natural selection acting on random mutations.

Consequently, natural selection is a mechanism by which new genetic information can be created. Indeed, this is the only mechanism known in natural science which can create it. There is a well-known statement by R. A. Fisher that 'natural selection is a mechanism for generating an exceedingly high degree of improbability', owing to which, as will be seen, the amount of genetic information can be measured. It may be pertinent to note here that the remarkable property of natural selection in realizing events which otherwise can occur only with infinitesimal probability was first clearly grasped by Muller (1929).

The purposes of the present paper are threefold. First, a method will be proposed by which the rate of accumulation of genetic information in the process of adaptive evolution may be measured. Secondly, for the first time, an approximate estimate of the actual amount of genetic information in higher animals will be derived which might have been accumulated since the beginning of the Cambrian epoch (500 million years), and thirdly, there is a discussion of problems involved in the storage and transformation of the genetic information thus acquired. There is a vast field

* Contribution No. 340 of the National Institute of Genetics, Mishima, Japan.

of fundamental importance which awaits the fruitful activities of statisticians and other applied mathematicians collaborating with biologists.

## THE CONCEPT OF A SUBSTITUTIONAL LOAD

A unit process in adaptive evolution is the replacement in a Mendelian population of one allele by another which is better fitted to a new environment. It was pointed out by Haldane (1957) that if this is carried out by premature death of less fit individuals, it may cost a number of deaths equal to about thirty times the population number. I proposed the term substitutional (or evolutional) load to express the decrease of population fitness (in the Darwinian sense) in the process of such a gene substitution (Kimura, 1960*a*, *b*).

Let us consider the simplest situation in which the population consists of haploid organisms, such as some fungi. In such a case, each gene exists in a single dose in a somatic cell. Let $x$ be the frequency (relative proportion) of a gene $A$ which is in the process of being substituted for its allele $A'$ because of its selective advantage over $A'$. Then the rate of change in gene frequency $x$ is given by

$$\frac{d}{dt}\log x - \frac{d}{dt}\log(1-x) = s,$$

or

$$\frac{dx}{dt} = sx(1-x),$$

where $s$ $(>0)$ is the selective advantage measured in Malthusian parameters (Fisher, 1930), i.e. in terms of its contribution to the geometric growth-rate of the population, and $t$ is the time.

Since the population at a given moment contains the unfit genotype $A'$ in the proportion of $1-x$, the total decrease in population fitness, also measured in Malthusian parameters, throughout the process of substitution is

$$L = \int_0^\infty s(1-x)\,dt = \int_p^1 s(1-x)\frac{dx}{sx(1-x)} = \int_p^1 \frac{dx}{x} = -\log_e p,$$

where $p$ is the initial value of $x$. Thus we have

$$L = -\log_e p. \tag{1}$$

This is the expected substitutional load for a haplont if the substitution proceeds at the rate of one gene per generation. The actual substitutional load may be obtained by summing the above quantity over all relevant gene loci, each weighted according to the rate of substitution per locus per generation.

$$L_e = \sum \epsilon L = -\sum \epsilon \log_e p, \tag{2}$$

where $\epsilon$ is the rate of substitution per locus.

The situation is much more complicated for higher animals and plants, in which each gene exists in double dose within a somatic cell (diploidy) and gene interaction within a locus (dominance) becomes important. It can be shown (cf. Kimura, 1960*b*) that, if the selective advantages of the genotypes $AA$ and $AA'$ over $A'A'$ are $s$ and $sh$ respectively, then the load produced by substituting $A$ for $A'$ is

$$L = -\frac{1}{h}\left\{\log_e p + (1-h)\log_e \frac{1-h}{h+(1-2h)p}\right\}, \qquad (s \geqq 0,\ 1 \geqq h \geqq 0) \qquad (3)$$

where $p$ is the initial frequency of $A$ in the population and $h$ represents the degree of dominance of $A$ over $A'$ in fitness. One salient feature of this result is that $L$ does not depend directly on $s$, the magnitude of selective advantage involved.

In Fig. 1(*a*) and (*b*), values of $L$ are plotted for various values of $h$ and $p$. It may

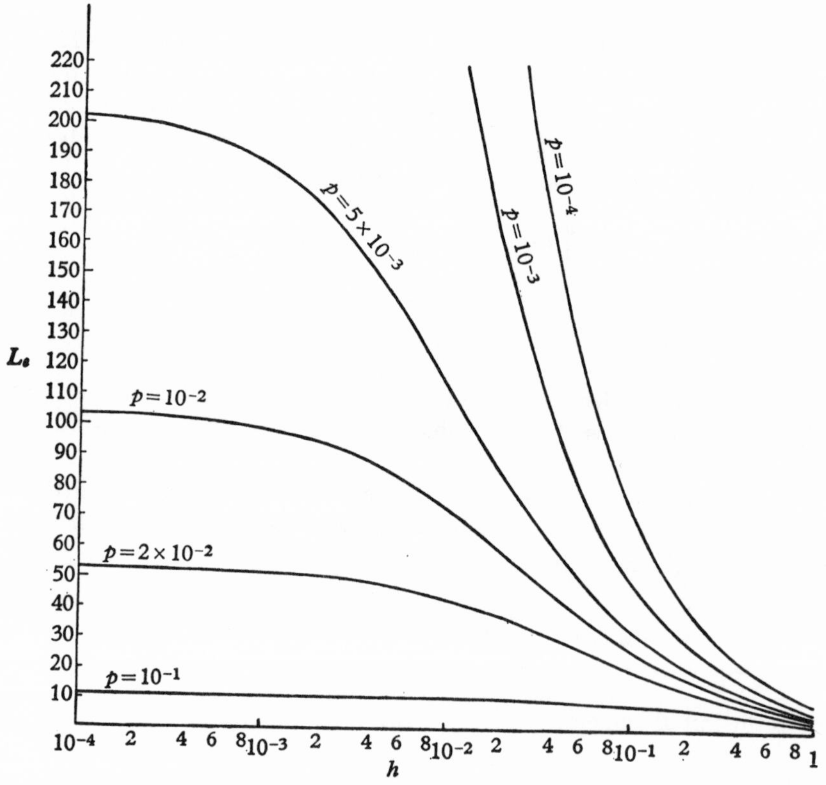


Fig. 1*a*.

be seen from this figure that $L$ increases as $p$ decreases, while it decreases as $h$ increases. In man, the typical frequency of 'recessive' deleterious genes is of the order of 1%, and if we assume that their dominance in fitness is about 2% as in recessive lethals of the fruit-fly *Drosophila melanogaster*, $L$ turns out to be about 59.

As in the case of the haplonts, the substitutional load is given by

$$L_e = \sum \epsilon L = -\sum \frac{\epsilon}{h}\left\{\log_e p + (1-h)\log_e \frac{1-h}{h+(1-2h)p}\right\}. \qquad (4)$$

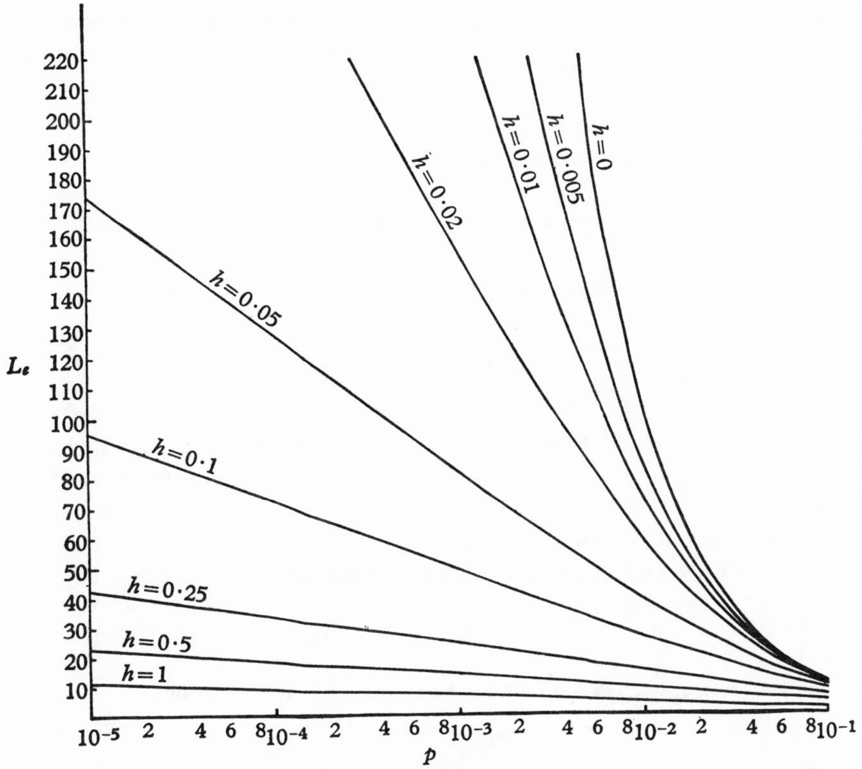


Fig. 1*b*.

Fig. 1 (*a*) and (*b*). Graphs showing the substitutional load $L$ as a function of initial gene frequency $p$ and degree of dominance $h$. $L$ is the decrease in fitness which is expected if the gene substitution proceeds at the rate of one gene per generation.

## THE SUBSTITUTIONAL LOAD AS A MEASURE OF GAIN IN GENETIC INFORMATION

I now propose to show that the rate of accumulation of genetic information denoted by $H$ is directly proportional to the substitutional load, namely,

$$H = \frac{L_e}{\log_e 2} \approx 1{\cdot}44 L_e \quad \text{bits per generation,} \tag{5}$$

where bit is a commonly used unit of the amount of information equivalent to the information content of choosing between a pair of alternatives, say 0 or 1, with equal probability (0·5). The above relation may be derived from two independent courses of reasoning:

(1) If those individuals which are to be eliminated by natural selection in the process of progressive evolution were kept alive and allowed to reproduce at the same rate as the favoured individuals, the population number would become, after $t$ generations,

$$e^{L_e t}$$

times its initial value. This means that natural selection allows an incident to occur with probability one, which, without selection, could occur only with a probability of

$$1/e^{L_e t} = e^{-L_e t}.$$

Thus information gained through $t$ generations amounts to

$$-\log_2 e^{-L_e t} = \frac{L_e t}{\log_e 2} \text{ bits},$$

and therefore information gained per generation is

$$H = \frac{L_e}{\log_e 2} = (1{\cdot}442\ldots) L_e \text{ bits},$$

as was to be shown.

(2) Consider a population of haploid organisms. Let $p$ be the initial frequency of an advantageous gene $A$. The probability that the gene $A$ is ultimately established in the population is 1 under natural selection, while it is only $p$ if natural selection were not working and the fixation of genes were left to the action of random genetic drift. Thus the amount of information corresponding to this gene substitution is

$$H = \log_2 \frac{1}{p} = -\log_2 p \text{ bits}.$$

On the other hand, we have shown that for one-gene substitution

$$L = -\log_e p. \qquad \text{(cf. (1))}$$

Therefore

$$H = -\log_2 e^{-L} = \frac{L}{\log_e 2}.$$

## ESTIMATION OF THE INFORMATION GAIN IN THE ACTUAL PROCESS OF EVOLUTION

As shown above, the gain of genetic information is directly proportional to the substitutional load, and the problem of estimating the former ($H$) is now reduced to that of estimating the latter ($L_e$). However, since evolution is usually an exceedingly slow process in comparison to our ordinary life-span, it may be very difficult to determine $L_e$ from direct observation. Haldane (1949) has shown, based on paleontological data of Simpson (1944), that the standard rate of evolution in morphological characters is of the order of one-tenth of a darwin, one darwin standing for a change by a factor of $e$ per million years.

In an attempt to derive theoretically some fundamental genetic parameters such as mutation rate ($\mu$) and degree of dominance ($h$) from the standard rate of evolution in the past, I proposed what I called the principle of minimum genetic load (Kimura, 1960*b*), a hypothesis that in the process of evolution the genetic parameters tend to be adjusted such that the total genetic load is minimized. In particular,

$$L_T = L_e + L_m$$

may be minimized in adaptive evolution. Here $L_m$ stands for the mutational load which arises through the elimination of deleterious genes produced by recurrent mutation (cf. Crow, 1958). Based on this principle, it was demonstrated that the spontaneous mutation rate per gamete, $\sum \mu$, and the harmonic mean of the degree of dominance of mutant genes in fitness, $\bar{h}$, can be derived from the rate of substitution of genes and the total amount of hidden deleterious effect, per gamete, of mutant genes:

$$\left.\begin{aligned} \Sigma\mu &= \frac{0{\cdot}3419E}{\bar{h}}(1+1{\cdot}720\bar{h}+\ldots), \\ \bar{h} &= 0{\cdot}6838\sqrt{\frac{E}{2D}}\left(1+1{\cdot}018\sqrt{\frac{E}{2D}}+\ldots\right), \end{aligned}\right\} \quad (6)$$

where $E$ is the rate of substitution of genes in horotelic evolution (standard rate evolution, cf. Simpson, 1944).

$$E = \sum \epsilon,$$

and $D$ is the total amount of genetic damage per gamete expressed in lethal equivalents (cf. Morton, Crow & Muller, 1956). If we take $E = \frac{1}{300}$, an approximate value suggested by Haldane (1957), and $D = 2$, the one obtained by Morton, Crow & Muller from the study of inbreeding in man, we get

$$\bar{h} = 0{\cdot}0203 \quad \text{and} \quad \sum \mu = 0{\cdot}0581,$$

both of which agree fairly well with the corresponding observed values in the fruit-fly *Drosophila* ($\bar{h}$ about 2%, $\sum \mu$ about 4%). This is remarkable since the calculation is based on a simplified assumption that evolution has proceeded at a constant rate over an indefinitely long time. The calculation also supplies, at the same time, the substitutional and mutational loads:

$$L_e = 0{\cdot}206, \qquad L_m = 0{\cdot}099.$$

I will take $L_e = 0{\cdot}2$ and $L_m = 0{\cdot}1$ for the present purpose.

Then, the rate of accumulation of genetic information becomes approximately

$$H = 0{\cdot}29 \text{ bits per generation,}$$

if we apply relation (5). Similarly, we may calculate the amount of information gained by eliminating deleterious mutant genes by using the relation $H = L_m/\log_e 2$. This is an amount which exactly cancels out the loss of information by mutation. Table 1 is a balance-sheet of genetic information in evolution.

Table 1. *Balance sheet of genetic information in the process of horotelic evolution (bits/generation)*

| | | |
|---|---|---|
| Gene substitution | +0·29 | |
| Appearance of deleterious mutant genes | −0·14 | 0·00 |
| Elimination of the deleterious mutant genes | +0·14 | |
| Total | +0·29 | |

We are now in a position to estimate the total amount of genetic information which has been accumulated since the beginning of the Cambrian epoch. Prior to this we know very little about the actual forms of life on the earth because of the scarcity of fossil records. Through the following epochs our knowledge of the major course of evolution is fairly good. Before we do this, it may be instructive to see how effectively a high level of improbability in genetic organization can be generated by natural selection. With $H = 0{\cdot}29$ bit per generation, the amount of genetic information accumulated over 1,000 generations is 290 bits. On the other hand, according to Eddington, the total number of electrons in the universe is $\frac{3}{2}.136.2^{256}$, or approximately $2{\cdot}36 \times 10^{79}$. Thus, the probability that a randomly chosen electron out of the universe happens to be the preassigned one is the reciprocal of this number and the corresponding measure of improbability is about 263 bits. This means that 1,000 generations of natural selection can achieve something more improbable than this. But, for the actual process of organic evolution, a duration of 1,000 generations is a very short time indeed.

We do not know how old life on earth is, though there are some reasons to believe that it has existed 2 billion years. We do know, however, that in the Cambrian epoch, which started about 500 million years ago, the earth was already inhabited by organisms such as jellyfish, annelid worms, trilobites, crustaceans, etc.

If we assume then that the genetic information has been gained at the rate of 0·29 bit per generation, the total amount of genetic information accumulated since the beginning of Cambrian epoch is

$$\frac{0{\cdot}29 \times 5 \times 10^8}{\bar{G}} = 1{\cdot}45 \times 10^8/\bar{G},$$

where $\bar{G}$ is the harmonic mean of the duration of one generation in years. Unfortunately, for organisms which do not exist except as fossils, it is impossible to measure the exact length of their generations. All we can do is to infer them from contemporary analogues. For various groups of animals, there is some tendency for smaller and less differentiated members to mature more quickly than the larger and well-differentiated ones. Now, in the history of evolution, it is known that it was always the former type of animal which has succeeded in leaving descendants. Furthermore, $\bar{G}$ is expected to be much smaller than the arithmetic average of the lengths of one generation. With no reliable estimates available at present, I assume, as a biologically reasonable guess, that $\bar{G}$ is of the order of one year.

We may conclude then, that the total amount of genetic information which has been accumulated since the beginning of the Cambrian epoch along the lineage leading to higher mammals may be of the order of one hundred million bits ($10^8$ bits).

Corresponding to this increase in genetic information, there has occurred a tremendous improvement of phenotypic organization which is implied in the term evolution in the usual sense of the word.

## STORAGE AND TRANSFORMATION OF GENETIC INFORMATION

Owing to the recent development of bacterial and viral genetics and also of DNA chemistry, it has become increasingly clear that DNA (deoxyribonucleic acid) molecules forming chromosomes are the carriers of genetic information. From this standpoint, a chromosome may be considered as a linear sequence of nucleotide pairs, of which four kinds are discriminated.

Muller (1958) estimated the total number of nucleotide pairs which may be present in the chromosome set of man as approximately $4 \times 10^9$, by dividing the total mass of DNA contained in a human sperm (*ca.* $4 \times 10^{-12}$ gr.) by the mass of one nucleotide pair (*ca.* $10^{-21}$ gr.). Thus, with four kinds of nucleotide pairs, the maximum amount of genetic information that may be stored in the haploid chromosome set of man amounts to

$$\log_2 4^{4\times 10^9} = 8 \times 10^9 \text{ bits},$$

and twice as much for the diploid set:

$$1{\cdot}6 \times 10^{10} \text{ bits}.$$

This is the maximum amount of genetic information that may possibly be stored in the nucleus of a fertilized human egg, if the four kinds of nucleotide pairs are equally efficient.

It is generally accepted that the information in DNA is transferred, via RNA (ribonucleic acid) molecules, to proteins, and if, as some workers in this field assume (cf. Crick *et al.*, 1957), a sequence of three nucleotides determines one of twenty possible amino acids, the above value should be reduced by a factor of

$$\log_2 20/\log_2 4^3 \approx 0{\cdot}72,$$

giving

$$1{\cdot}15 \times 10^{10} \text{ bits},$$

or roughly $10^{10}$ bits as the maximum amount of genetic information that might effectively be stored in the diploid chromosome set.

Here the chromosome set may be compared with an electronic computer. In IBM 650, for example, there are 2,000 memory locations and it can store 20,000 digits, or about $6{\cdot}64 \times 10^4$ bits of information. A more interesting object for comparison is the self-reproducing machine envisaged by von Neumann. According to Kemeny (1955), von Neumann's machine consists of a basic box of $80 \times 400$ squares plus a tail containing 150,000 squares. The basic box has a function analogous to the soma, while the tail contains the instructions of the machine and is analogous to the chromosome set. This tail consists of 150,000 cells which are in either an 'on' or 'off' state, and therefore it may store

$$1{\cdot}5 \times 10^5 \text{ bits}$$

of 'genetic information'.

These comparisons not only show the tremendous complexity of chromosome structure, but also reveal an indeed amazing efficacy of DNA codes—efficacy of such an extent, as pointed out originally by Muller (1935), that all of the chromosomes present initially in the fertilized eggs from which the present population of the world (some two thousands of millions) developed would occupy a volume about equal to that of an ordinary aspirin tablet.

Deciphering DNA codes, i.e. learning to read the genetic language is a very fascinating problem which was vigorously attacked for the first time only a few years ago (cf. Yčas, 1958) and, without a Rosetta stone, it would be solved only by statistical treatment, though no success seems to have been obtained so far.

We have estimated, in the previous section, that the total amount of genetic information which has been accumulated since the Cambrian epoch is of the order of $10^8$ bits. On the other hand, as shown above, the maximum amount of genetic information that might be stored in the diploid chromosome set of man may be of the order of at least $10^{10}$ bits. If the first estimate ($10^8$ bits) is correct, the difference between these two estimates must be real, even if we admit that our Cambrian ancestors had already accumulated a considerable amount of genetic information. If so, I believe that this difference can be interpreted in two ways, namely, either the amount of genetic information which has been accumulated is a small fraction of what can actually be stored in the chromosome set or, more probably, the DNA code itself is highly redundant. In a stimulating paper, to which more attention should be paid by Western geneticists, Schmalhausen (1958), a Russian geneticist, points out that higher reliability of transmitted information may be achieved by the repetition of information, such as repetition of equal genes (polygenes), 'repeats' of gene complexes and in particular diploidy or polyploidy. Furthermore, there may be repetition or a certain kind of redundancy of information within each gene or 'cistron'.

Recently, Sueoka (1960) has made an extensive survey on the guanine–cytosine (G–C) content of deoxyribonucleic acid (DNA) taken from various organisms ranging from bacteria to man. It has been found that for vertebrates the average content lies within the range of 40 to 44%. For various species of bacteria, it varies over a much wider range of 25 ~ 75%, indicating the marked divergence in phylogeny. On the other hand, the G–C content of DNA molecules within an organism has a rather narrow distribution with $2\sigma$ (twice the standard deviation) of some 6% or usually less around its specific mean value $\bar{p}$. In the case of native DNA taken from the calf thymus, the average G–C content, $\bar{p}$, is about 40% and its heterogeneity, $2\sigma$, is 9·6%, which has been the largest value ever observed.

It may not be difficult to see that if the arrangement of G–C pairs and A–T pairs were entirely at random, the proportion, $p$, of G–C pairs between molecules within an organism should be distributed unimodally with a binomial variance of

$$\sigma_p^2 = \frac{\bar{p}(1-\bar{p})}{b}, \tag{7}$$

where $b$ stands for the number of base pairs (sum of G–C and A–T pairs) composing a DNA molecule. In the actual case, the number of base pairs per molecule is not constant and we should use its harmonic mean for $b$. In the case of calf thymus DNA, the harmonic mean is roughly $10^4$ and in most cases $b$ should be at least several thousands. Thus, with $b = 10^4$, the expected heterogeneity ($2\sigma_p$) calculated for $\bar{p} = 0{\cdot}3 \sim 0{\cdot}7$ is

$$0{\cdot}009 < 2\sigma_p < 0{\cdot}01,$$

i.e. it lies between 0·9 and 1·0%. If $b$ is half as large, the heterogeneity will become about 1·4 times as large, and even if $b$ is $\frac{1}{4}$ as large, it will only double the above value. On the other hand, the observed heterogeneity is some 6% in vertebrates (Sueoka, 1960), much larger than that expected from the binomial distribution. Thus the ratio between the observed and the expected heterogeneity is roughly 6 in terms of standard deviation and 36 in terms of variance.

I assume that this discrepancy between the observed and the expected is due to repetition in the pattern of base arrangement within the DNA molecule. A simple analysis of our ordinary language will help us greatly to clarify this point. It is known that, in English sentences, the most frequent letter is 'e', and this is followed by 't' or 'a' in frequency. These three letters, together with a space between words, make up about 40% (analogous to G–C pairs in 'genetic language'). I extracted fifty lines, each with seventy letter positions, from a paper on genetics and calculated the mean frequency of 'e', 't', 'a' and space per line and the standard deviation of the frequency between different lines. As shown in the second row of Table 2, the ratio between the observed and the expected standard deviation is about 0·92.* Here the expected standard deviation is calculated from (7) using $\bar{p} = 0{\cdot}439$ and $b = 70$. Similar calculations were performed for lines taken from genetical papers written in French, German and Russian. The results are also listed in Table 2.

Table 2. *Mean and standard deviation of the relative frequency of the sum of 'e', 't', 'a' and space per line in samples of seventy letter positions. The figures in the last column denote the ratio between observed and expected standard deviations. For each language, fifty lines were extracted for calculation.*

| Language | Mean ($p$) | Standard deviation ($\sigma$) observed | Standard deviation ($\sigma_p$) expected | Ratio |
|---|---|---|---|---|
| English | 0·439 | 0·0547 | 0·0593 | 0·922 |
| French | 0·450 | 0·0539 | 0·0616 | 0·875 |
| German | 0·381 | 0·0544 | 0·0601 | 0·905 |
| Russian | 0·290 | 0·0464 | 0·0535 | 0·867 |

It may be seen from this table that the ratio between observed and expected heterogeneity in 'e, t, a, space' content per line is roughly 0·9, which is very different from what we have found in the G–C content per DNA molecule.† Suppose we duplicate each letter fifty times so that each line now consists of 50 × 70 or 3,500 letter positions. By this duplication, the observed heterogeneity does not change but the expected heterogeneity may be reduced to about 1/7·1 because $b$ should be taken as 3500. Then, the ratio between the observed and expected becomes about 6·4, which is similar to the ratio obtained for DNA.

Granting that there may be some thirty-six repetitions in the arrangement of

* This value is mainly due to the negative correlation between neighbouring letters. Giving value 1 for 'e', 't', 'a' and space and value 0 for the remaining letters, I obtained correlation coefficients of −0·20 between two adjacent letters, −0·02 between two neighbouring letters once removed, etc.

† See note at end of paper.

base pairs in DNA molecule, the next question is how such repetitions can be visualized. The analogy with our ordinary language may help us again.

Let us take, as an example, the following sentence:

IT IS SO.

From this sentence we can derive various forms of letter arrangements by duplicating each letter twice. Among them, the following three are especially significant, which I will tentatively call letter, word and sentence repetitions respectively:

(i) Letter repetition: IITT IISS SSOO
(ii) Word repetition: ITIT ISIS SOSO
(iii) Sentence repetition: IT IS SO IT IS SO

These three forms are, as such, indistinguishable with respect to the ratio between the observed and the expected 'heterogeneity'. However, by splitting each of these into pieces of certain length and studying their heterogeneity, we will find that they behave quite differently. Suppose we split each into two pieces of equal length. In the case of letter repetition, the observed variance will become double, because each piece contains only one half of independent letters of the original sentence. On the other hand, in the case of sentence repetition, the variance will remain the same because each piece contains the full sentence. The situation is intermediate in the case of word repetition and the variance becomes 5/3. Returning to the problem of heterogeneity in G–C content, Sueoka (1959) found that by splitting calf thymus DNA molecules by ultrasonic vibration into pieces of about one-tenth in size, the heterogeneity variance increases very little (order of 10% if any). Since we have already assumed that there may be thirty-six repetitions of letters in DNA, this result may be interpreted as showing that repetition in the DNA molecule must be more near to the type of sentence repetition than that of word repetition. Assuming $10^4$ base pairs per DNA, then each sentence consists of roughly 300 letters.

The above hypothesis may be tested by splitting DNA molecules into much smaller pieces consisting of less than 300 base pairs and by seeing if the observed and the expected heterogeneity agree with each other.

It should be noted here that the repetition may not be exact in the actual DNA molecule, rather at each repetition the 'word' may be slightly modified from one to the other, like variations in music.

At any rate, through the process of individual development (ontogeny), the genetic information is finally transformed into phenotypic information, with its various aspects in morphology, physiology and behaviour, admittedly a large amount of redundancy being involved among them. Then, how large is the phenotypic information of higher mammals, or specifically that of man? Perhaps the more pertinent question to ask here is how much more phenotypic information is contained in higher animals or man as compared to their Cambrian ancestors. In this sense, the information content should not be counted in terms of atomic or molecular configurations, but should be done in terms of the three-dimensional anatomical structure plus chemical data, as pointed out by Elsasser (1958). He suggests that, since the information content of human species pertaining to gross anatomy alone could hardly be diagrammed on a plane area of 1 $m^2$ in which the smallest unit of discrimination is 1 $mm^2$, and since gross anatomy can only be a moderate fraction of the information content of the organism, the information content of the human organism must be at least of the order of $10^7$ bits or, more probably, $10^8$ bits. Elsasser states that even a figure of $10^9$ bits would hardly appear fantastic. However, since the phenotypic information is transformed genetic information, the former cannot be larger than the latter, which we have estimated as being of the order of $10^8$ bits. The correspondence between the genetic and phenotypic information turns out to be quite close considering that, while new genetic information can only be gained through natural selection acting on genotypes, this action is mediated by the phenotypes which are determined by the genotypes. A more reliable estimate will be supplied in the future by anatomists or chemists who will have access to a proper statistical methodology.

In my opinion the creative role of natural selection, which is still not infrequently overlooked by evolutionists, may most convincingly be brought to light by calculating its power of accumulating genetic information and considering the phenotypic complexity as its product. Lerner (1959) states that the meaning of natural selection as a creative process may be well illustrated by quoting Michelangelo's concept of creation: 'The sculptor's hand can only break the spell to free the figures slumbering in the stone.' Indeed, any elaborate work of art must contain a large amount of information.

## SUMMARY

1. In the course of evolution, complicated organisms have descended from much simpler ones. Since the instructions to form an organism are contained in the nucleus of its fertilized egg, this means that the genetic constitution has become correspondingly more complex in evolution. If we express this complexity in terms of its improbability, defining the amount of genetic information as the negative logarithm of its probability of occurrence by chance, we may say that genetic information is increased in the course of progressive evolution, guided by natural selection of random mutations.

2. It was demonstrated that the rate of accumulation of genetic information in adaptive evolution is directly proportional to the substitutional load, i.e. the decrease of Darwinian fitness brought about by substituting for one gene its allelic form which is more fitted to a new environment. The rate of accumulation of genetic information is given by

$$H = \frac{L_e}{\log_e 2} \approx 1{\cdot}44 L_e \quad (\text{'bits'/generation}),$$

where $L_e$ is the substitutional load measured in 'Malthusian parameters'.

3. Using $L_e = 0{\cdot}199$, a value obtained from the application of the 'principle of minimum genetic load' (cf. Kimura, 1960*b*), we get

$$H = 0{\cdot}29 \text{ bit/generation.}$$

It was estimated that the total amount of genetic information accumulated since the beginning of the Cambrian epoch (500 million years) may be of the order of $10^8$ bits, if evolution has proceeded at the standard rate.

Since the genetic information is transformed into phenotypic information in ontogeny, this figure ($10^8$ bits) must represent the amount of information which corresponds to the improved organization of higher animals as compared to their ancestors 500 million years back.

4. Problems involved in storage and transformation of genetic information thus acquired were discussed and it was pointed out that the redundancy of information in the form of repetition in linear sequence of nucleotide pairs within a gene may play an important role in the storage of genetic information.

## REFERENCES

Crick, F. H. C., Griffith, J. S. & Orgel, L. E. (1957). Codes without commas. *Proc. nat. Acad. Sci., Wash.*, **43**, 416–421.

Crow, J. F. (1958). Some possibilities for measuring selection intensities in man. *Hum. Biol.* **30**, 1–13.

Elsasser, W. M. (1958). *The Physical Foundation of Biology.* London: Pergamon Press.

Fisher, R. A. (1930). *The Genetical Theory of Natural Selection.* Oxford: Clarendon Press.

Haldane, J. B. S. (1949). Suggestions as to quantitative measurement of rates of evolution. *Evolution,* **3**, 51–56.

Haldane, J. B. S. (1957). The cost of natural selection. *J. Genet.* **55**, 511–524.

Kemeny, J. G. (1955). Man viewed as a machine. *Sci. Amer.* **192**, 58–67.

Kimura, M. (1960*a*). Genetic load of a population and its significance in evolution. (Japanese with English summary.) *Jap. J. Genet.* **35**, 7–33.

Kimura, M. (1960*b*). Optimum mutation rate and degree of dominance as determined by the principle of minimum genetic load. *J. Genet.* **57**, 21–34.

Lerner, I. M. (1959). The concept of natural selection: A centennial view. *Proc. Amer. Phil. Soc.* **103**, 173–182.

Morton, N. E., Crow, J. F. & Muller, H. J. (1956). An estimate of the mutational damage in man from data on consanguineous marriages. *Proc. nat. Acad. Sci., Wash.*, **42**, 855–863.

Muller, H. J. (1929). The method of evolution. *Sci. Mon., N.Y.* **29**, 481–505.

Muller, H. J. (1935.) *Out of the Night.* New York: Vanguard Press.

Muller, H. J. (1958). Evolution by mutation. *Bull. Amer. math. Soc.* **64**, 137–160.

Schmalhausen, I. I. (1958). Control and regulation in evolution. (Russian with English summary.) *Bull. Soc. Nat. Moscow*, **63**, 93–121.

Simpson, G. G. (1944). *Tempo and Mode in Evolution.* New York: Columbia University Press.

Sueoka, N. (1959). A statistical analysis of deoxyribonucleic acid distribution in density gradient centrifugation. *Proc. nat. Acad. Sci., Wash.*, **45**, 1480–1490.

Sueoka, N. (1960). Some genetic and evolutionary considerations on the base composition of deoxyribonucleic acids. (In press.)

Yčas, M. (1958). The protein text. *Symposium on Information Theory in Biology*, pp. 70–102. London: Pergamon Press.

## NOTE ADDED IN PROOF

After this paper had been sent to press, I had the privilege of seeing a preprint of a paper by J. Josse, A. D. Kaiser and A. Kornberg, who successfully determined the nearest neighbour sequence of nucleotides in DNA taken from various organisms. From their Table VI, I calculated the correlation between two adjacent nucleotide pairs in calf thymus DNA, giving value 1 for a G–C pair and 0 for an

A–T pair. The correlation coefficient obtained was about −0·09, a value not drastically different from the one obtained for English sentences. Similar calculations for bacterial and bacteriophage DNA's (Tables VIII and IX) gave correlation coefficients of at most a few per cent (either positive or negative). Cf. Josse, Kaiser & Kornberg (1960), Enzymatic synthesis of deoxyribonucleic acid. VIII. Frequencies of nearest neighbor base sequences in deoxyribonucleic acid, *Jour. biol. Chem.* (in press).

PART SEVEN

# Meiotic Drive

## Introduction

Segregation Distorter (*SD*) in *Drosophila melanogaster* was first discovered by Hiraizumi in 1956 (Sandler and Hiraizumi 1960; Hiraizumi and Crow 1960), and *SD* and the *t*-haplotype in *Mus* are the two best known meiotic drive systems. *SD* violates the Mendelian segregation rule by producing an excess of *SD* gametes in heterozygous males (see Crow 1979, 1991 for reviews). Because of its transmission advantage, *SD* can spread over a population even though it is generally harmful, at least in homozygotes. The special supplement in *The American Naturalist* (Temin et al. 1991) shows that *SD* is now the best characterized meiotic drive system at the molecular level. Theoretical aspects of *SD* have also been worked out (e.g., Charlesworth and Hartl 1978; Hartl 1980).

Meiotic drive is not restricted to animals but exists in plants as well. One such example in plants is supernumerary (or accessory or **B**) chromosomes in the lily, *Lilium callosum*. Combining experiments and observations with population genetic theory, Kimura and Kayano showed (no. 17) that **B** chromosomes are predominantly deleterious (when there are more than two in a cell), and they are kept in the population by preferential segregation in embryo sac mother cells. Meiotic drive is a phenomenon that can be studied at the population, cellular, and molecular levels. It may therefore provide a way to understand the poorly known connection between phenotypic and genotypic changes (Wu and Hammer 1991).

# THE MAINTENANCE OF SUPERNUMERARY CHROMOSOMES IN WILD POPULATIONS OF LILIUM CALLOSUM BY PREFERENTIAL SEGREGATION[1]

MOTOO KIMURA AND HIROSHI KAYANO

*National Institute of Genetics and Department of Biology, Faculty of Science, Kyushu University*

Received August 25, 1961

A CONSIDERABLE number of plant species have been reported to contain supernumerary or accessory chromosomes (B chromosomes) in some members of their population, and the list of such species is steadily increasing. Recently, MÜNTZING and his collaborators (MÜNTZING 1954) have carried out extensive cytogenetical investigations of supernumerary chromosomes in Secale, Festuca, Centaurea, Poa and Anthoxanthum. One of the most interesting points involved is that in such a case as in rye, the total effect of B chromosomes is predominantly deleterious but at the same time there is a special mechanism which effects a numerical increase of the B chromosomes in the functional male gametes, thus leading to a balance between two opposing forces. A similar result has been reported by the junior author (KAYANO 1956b, 1957) in *Lilium callosum*, in which a supernumerary chromosome denoted by $f_1$ (meaning long telocentric fragment) has a tendency to increase in numbers due to preferential segregation in embryo-sac-mother cells. Thus by crossing a plant with one $f_1$ chromosome as a seed parent and one with no $f_1$ chromosome as a pollen parent, he obtained plants with one $f_1$ and no $f_1$ in an approximate ratio of 80:20. This agreed well with the cytological observation that in about 80 percent of the cases $f_1$ goes to the micropylar side to be included in the egg cell. Pollen fertility and seed setting have also been investigated in plants with various numbers of $f_1$ chromosomes. As will be shown in the next section, $f_1$ chromosomes depress the reproductive ability of the possessor rather severely so that plants with more than two $f_1$ chromosomes are almost completely sterile. In order to investigate the distribution of the supernumerary chromosome among individuals of wild populations, samples were collected from several localities and examined cytologically. The pattern of distribution was quite clear, and it was realized by the senior author that with observational data on pollen fertility and seed setting it might be possible to analyse the mechanism of distribution of the $f_1$ chromosome by developing a mathematical model. In the present paper, the junior author is responsible for experimental and observational data, while the senior author is responsible for mathematical and statistical analyses.

[1] Contribution No. 390 from the National Institute of Genetics, Mishima-shi, Japan. Aided in part by a Grant-in-Aid for Fundamental Scientific Research from the Ministry of Education, Japan, No. 0030.

Genetics 46: 1699–1712 December 1961.

## *Experimental and observational data*

Samples were collected from four colonies or populations denoted by I, II, III and IV.

Populations I, II and III are located on a calcareous terrain of Hirao-Dai near Kokura-shi, Fukuoka Prefecture and they are about two kilometers apart. Population IV was found on a slope in the suburbs of Karatsu-shi, Saga Prefecture and is quite far from the other three populations. In this area the soil is not calcareous. In all these places, *L. callosum* was found growing together with *Miscanthus sinensis*.

In Figures 1 a–d, the results of karyotype analysis are presented together with maps showing the spatial distribution of individuals within each population. It may be seen from these maps that the distribution of supernumerary chromosomes is quite striking, one type denoted by $f_l$ being found in a large majority of plants and yet its number per individual being restricted to one or two (for the description of the types of supernumerary chromosomes, see KAYANO 1956a). Frequency distributions of the number of $f_l$ chromosomes within the four samples are summarized in Table 1. Pollen and seed fertilities were investigated using individuals from population I and the results are summarized in Tables 2 and 3. These figures represent averages over two years in the case of pollen fertility (1953 and 1954) and over three years for seed fertility (1952, 1953 and 1954).

The characteristic mode of transmission of the $f_l$ chromosome has already been reported by the junior author (KAYANO 1956b), but it may be convenient to

TABLE 1

*Frequency distribution of the number of $f_l$ chromosomes*

| Populations sampled / No. of $f_l$ | I (percent) | II (percent) | III (percent) | IV (percent) | Sum (percent) |
|---|---|---|---|---|---|
| 0 | 17 (28.8) | 10 (18.5) | 18 (39.1) | 29 (46.0) | 74 (33.3) |
| 1 | 30* (50.9) | 33‡ (61.1) | 26 (56.5) | 31 (49.2) | 120 (54.1) |
| 2 | 12† (20.3) | 11 (20.4) | 2 ( 4.3) | 3 ( 4.8) | 28 (12.6) |
| Total no. of individuals observed | 59 (100) | 54 (100) | 46 (100) | 63 (100) | 222 (100) |

* Contains two individuals having $1f_l + 1f_s$, and one individual having $1f_l + 1F_l$. ($f_s$ means a short supernumerary fragment).
† Contains one individual with $2f_l + 1f_s$.
‡ Contains one individual with $1f_l + 3f_s$.

TABLE 2

*Pollen fertility of individuals with various numbers of $f_l$ chromosomes*

| No. of $f_l$ | No. of plants investigated | No. of fertile pollen grains | No. of sterile pollen grains | Total no. counted | Fertility (percent) |
|---|---|---|---|---|---|
| 0 | 24 | 17,621 | 879 | 18,500 | 95.25 |
| 1 | 38 | 26,739 | 3,761 | 30,500 | 87.67 |
| 2 | 21 | 8,521 | 7,479 | 16,000 | 53.26 |

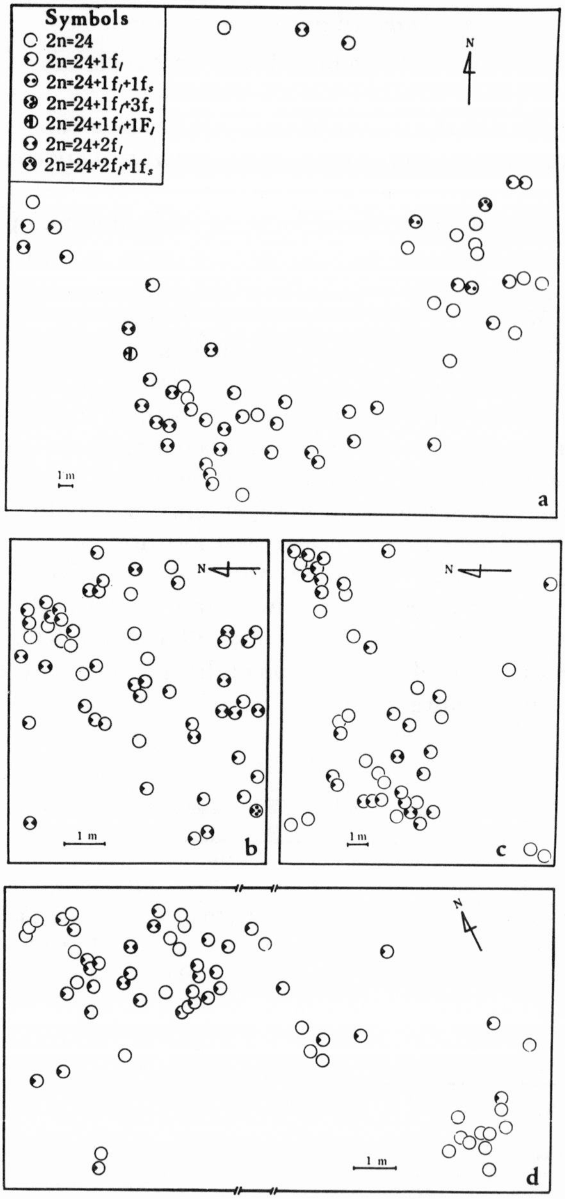


FIGURE 1a–d.—Spatial distribution of individuals with various number of $f_1$ chromosomes within four colonies of *Lilium callosum*. a, b and c (populations I, II and III) were located on a calcareous terrain of Hirao-Dai near Kokura-shi, Fukuoka Prefecture; these are about two kilometers apart. d (population IV) was found on a slope in the suburbs of Karatsu-shi, Saga Prefecture; it is quite far apart from the others.

reproduce the experimental results in the form of Table 4, in which the figures given in parentheses are the approximate values used as bases for calculation.

A crossing experiment conducted in 1958 revealed that this lily was self-incompatible. The results of this experiment are listed in Table 5. Thus *L. callosum* may be regarded as an outbreeding species. Furthermore, it reproduces, at least in part, asexually through development of new bulbs.

## *Mathematical and statistical analyses*

1. *Mathematical model:* In order to analyse the present case, we must take into account two new features which do not appear in the ordinary mathematical treatment in population genetics. Namely, (a) both sexual and asexual reproduction have to be considered and (b) the mode of segregation of the $f_l$ chromosome does not follow the simple rules characteristic of ordinary diploid loci. With these two points in mind, we will formulate the problem as follows, in which an abbreviated symbol f will often be used to represent the $f_l$ chromosome.

TABLE 3

*Seed fertility of individuals with various numbers of $f_l$ chromosomes*

| No. of $f_l$ | No. of individuals investigated | No. of fertile seeds | No. of sterile seeds | Total no. of seeds counted | Fertility (percent) |
|---|---|---|---|---|---|
| 0 | 15 | 4,662 | 2,907 | 7,569 | 61.59 |
| 1 | 25 | 9,102 | 8,305 | 17,407 | 52.29 |
| 2 | 12 | 3,228 | 5,859 | 9,087 | 35.52 |

TABLE 4

*Mode of transmission of $f_l$ chromosome in artificial crosses*

| No. of $f_l$ in the offspring / No. of $f_l$ in the parents ♀ × ♂ | 0 | 1 | 2 | 3 | No. of offspring observed |
|---|---|---|---|---|---|
| 0 × 1 | 45(50) | 54(50) | . . | . . | 99* |
| 1 × 0 | 16(20) | 83(80) | 1 | . . | 100 |
| 1 × 1 | 8(10) | 53(50) | 38(40) | 1 | 100 |
| 0 × 2 | 7 | 88 | 5 | . . | 100 |
| 2 × 0 | 5 | 76 | 18 | 1 | 100 |

* One individual having chromosome $2n=24+1F_1$ is excluded.

TABLE 5

*Comparison of the effects of outcrossing and selfing on the development of capsules*

| | No. of flowers pollinated | No. of capsules developed | No. of capsules failed to develop |
|---|---|---|---|
| Outcrossing | | | |
| Between individuals with no $f_l$ | 11 | 10 | 1 |
| Between individuals with 1 $f_l$ | 21 | 19 | 2 |
| Between individuals with different karyotypes | 17 | 17 | 0 |
| Total | 49 | 46 | 3 |
| Selfing | 20 | (2)* | 18 |

* Development was incomplete and did not contain fertile seeds.

Let $\Phi_i$ be the relative frequency of mature individuals with $i$ f-chromosomes ($i = 0, 1, 2, \ldots$) in the population, and let $w_i^*$ and $w_i^{**}$ be the relative fertilities or reproductive values of individuals with $i$ f's as pollen and seed parent respectively. We will designate by $g_{j/i}^*$ the relative proportion of pollen with $j$ f-chromosomes among fertile pollen produced by individuals having $i$ f-chromosomes, and by $g_{j/i}^{**}$ the proportion of fertile egg cells with $j$ f-chromosomes produced by the same individual. Then we can write the male gametic array giving the frequency distribution of the number of f chromosomes in the male gametes which contribute to the next generation as

$$(1) \qquad \sum_j G_j^* f^j \equiv G_0^* + G_1^* f + G_2^* ff + \ldots + G_j^* f^j + \ldots,$$

in which

$$G_j^* = \frac{\sum_i g_{j/i}^* w_i^* \Phi_i}{\sum_i w_i^* \Phi_i} \qquad (j = 0, 1, 2, \ldots)$$

is the proportion of pollen with $j$ f's. Similarly the female gametic array may be written in the form:

$$(2) \qquad \sum_k G_k^{**} f^k \equiv G_0^{**} + G_1^{**} f + G_2^{**} ff + \ldots + G_k^{**} f^k + \ldots,$$

where

$$G_k^{**} = \frac{\sum_i g_{k/i}^{**} w_i^{**} \Phi_i}{\sum_i w_i^{**} \Phi_i} \qquad (k = 0, 1, 2, \ldots)$$

is the relative proportion of egg cells with $k$ f's in the totality of eggs which contribute to the next generation. If there were no asexual reproduction, the zygotic array immediately after fertilization could be generated by the product of male and female gametic arrays, i.e., by

$$\left(\sum_j G_j^* f^j\right)\left(\sum_k G_k^{**} f^k\right),$$

because self-sterility of the present material ensures outcrossing, making pollination approximately random. The above product may be rearranged to give

$$\sum_i \phi_i f^i \equiv \phi_0 + \phi_1 f + \phi_2 ff + \ldots + \phi_i f^i + \ldots, \quad (i = 0, 1, 2, \ldots)$$

where

$$(3) \qquad \phi_i = \sum_{j+k=i} G_j^* G_k^{**} \qquad (j, k = 0, 1, 2, \ldots)$$

is the relative frequency of zygotes with $i$ f's immediately after fertilization.

However, we cannot neglect asexual reproduction and it is necessary to consider the relative contributions made by sexual and asexual reproduction to the next generation. Let $M$ be the mean number of viable seeds per plant, and let $b_i$ be the probability of a randomly chosen viable seed with $i$ f's reaching maturity. Furthermore, we designate by $A_i$ the mean number of plants which are produced asexually from a plant with $i$ f-chromosomes and which reach maturity in the next generation.

With these definitions, the frequency of mature individuals containing $i$ f-chromosomes in the next generation may be given by

$$(4) \qquad \Phi_i' = \frac{Mb_i\phi_i + A_i\Phi_i}{\sum_j (Mb_j\phi_j + A_j\Phi_j)} \qquad (i = 0, 1, 2, \ldots).$$

Thus in order to calculate the expected frequency distribution at equilibrium, we have to know all the parameters involved in segregation and reproduction, namely, the $w$'s, $g_{j/i}$'s, $b$'s, $A$'s and $M$. If these are given, a set of equilibrium frequencies denoted by $\hat{\Phi}_0, \hat{\Phi}_1, \hat{\Phi}_2, \ldots$ may be worked out by using relation (4) and applying the trial and error method, starting from a set of arbitrary frequencies and repeating calculations until we get

$$\Phi_i' = \Phi_i \equiv \hat{\Phi}_i, \qquad (i = 0, 1, 2, \ldots)$$

with desired accuracy. The stability of equilibrium will usually become evident in the process of iteration.

The equation (4) may be simplified if we assume that $A_i$ is proportional to $b_i$, namely, genotypes with higher viability tend to have a higher rate of asexual reproduction. If we accept such an assumption, (4) is reduced to

$$(5) \qquad \Phi_i' = \frac{b_i(\phi_i + \theta\Phi_i)}{\sum_j b_j(\phi_j + \theta\Phi_j)}, \qquad (i = 0, 1, 2, \ldots)$$

where $\theta$ is the ratio of the relative contributions by asexual and sexual reproduction to form the next generation such that

$$(6) \qquad \frac{A_i}{Mb_i} = \theta$$

for all $i$, i.e., $i = 0, 1, 2, \ldots$. Then at equilibrium we have

$$(7) \qquad b_i = \frac{C\hat{\Phi}_i}{\hat{\phi}_i + \theta\hat{\Phi}_i}$$

where $\hat{\phi}_i$ is the equilibrium value of $\phi_i$ and $C$ is a positive undetermined constant and may be adjusted so that $b_0$ is unity.

2. *Application of the mathematical model to the actual data*: We will first try to see whether the observed distribution of the f chromosome in natural populations (Table 1) may be explained satisfactorily from the observed mode of its transmission and its effect on fertility as shown in Tables 2, 3 and 4.

If we assume that different genotypes have the same viability, namely

(8) $$b_0 = b_1 = b_2 = \ldots$$

and that reproduction is almost exclusively sexual, i.e.

(9) $$\theta = 0 \qquad (i = 0, 1, 2, \ldots)$$

then we can calculate the equilibrium distribution using (4) or (5), taking

$$w_0^* = 0.9525 \quad w_1^* = 0.8767 \quad w_2^* = 0.5326 \quad w_3^* = w_4^* = \ldots = 0$$

(10)
$$\begin{array}{llll} g^*_{0/0} = 1.00 & g^*_{0/1} = 0.50 & g^*_{0/2} = 0.07 & \ldots \\ g^*_{1/0} = 0 & g^*_{1/1} = 0.50 & g^*_{1/2} = 0.88 & \ldots \\ g^*_{2/0} = 0 & g^*_{2/1} = 0 & g^*_{2/2} = 0.05 & \ldots \end{array}$$

with respect to the pollen parent, and

$$w_0^{**} = 0.6159 \quad w_1^{**} = 0.5229 \quad w_2^{**} = 0.3552 \quad w_3^{**} = w_4^{**} = \ldots = 0$$

(11)
$$\begin{array}{llll} g^{**}_{0/0} = 1.00 & g^{**}_{0/1} = 0.20 & g^{**}_{0/2} = 0.05 & \ldots \\ g^{**}_{1/0} = 0 & g^{**}_{1/1} = 0.80 & g^{**}_{1/2} = 0.76 & \ldots \\ g^{**}_{2/0} = 0 & g^{**}_{2/1} = 0 & g^{**}_{2/2} = 0.18 & \ldots \\ g^{**}_{3/0} = 0 & g^{**}_{3/1} = 0 & g^{**}_{3/2} = 0.01 & \ldots \end{array}$$

with respect to the seed parent.

After a considerable number of iterations, the equilibrium distribution (in %) turned out to be as follows:

(12)
$$\begin{array}{llll} \hat{\Phi}_0 = 20.2 & \hat{\Phi}_1 = 48.4 & \hat{\Phi}_2 = 28.9 & \hat{\Phi}_3 = 2.4 \\ \hat{\Phi}_4 = 0.1 & \hat{\Phi}_5 = \hat{\Phi}_6 = \ldots = 0. & & \end{array}$$

This expected distribution may be compared with the observed distribution in population I, from which materials for the study of segregation and fertility were obtained. In Table 6, classes with the number of $f_l$ larger than two are

TABLE 6

*Observed and expected frequency distributions in population I, assuming equal viabilities of individuals with different numbers of $f_l$ chromosomes.* $\chi^2 = 5.2$, *df 3,* $0.1 < P < 0.2$

| No. of $f_l$ | Frequencies (numbers) Observed | Expected |
|---|---|---|
| 0 | 17 | 11.92 |
| 1 | 30 | 28.56 |
| 2 | 12 | 17.05 |
| 3 and over | 0 | 1.47 |
| Total | 59 | 59.00 |

grouped together. Application of the $\chi^2$ test gives the $\chi^2$ value of 5.2. With three degrees of freedom, $0.1 < P < 0.2$, and therefore the deviation of observed frequencies from the expected is not significant. However, this does not prove that different genotypes have the same viability. On the contrary, Table 6 suggests that individuals having two or more f chromosomes are less viable than those having one f chromosome or none. In fact, if we add all the observed frequencies from four populations and compare the result with the expected, the deviation is now highly significant ($\chi^2 = 46.4$, df 3, $P < 0.01$). Observed frequencies of individuals having two or more f's are definitely smaller than the corresponding expected frequencies.

However, it might not be desirable to add the observed frequencies from four populations, since the four samples from populations I, II, III and IV are heterogeneous ($\chi^2 = 19.3$, df 6, $P < 0.01$). The main source of heterogeneity comes from the fact that the two groups of samples I + II and III + IV have different distributions ($\chi^2 = 17.2$, df 2, $P < 0.01$), though within each group the difference between samples is not significant. If we test the observed frequency distribution of I + II against the expectation, the deviation of the observed from expected is significant ($\chi^2 = 7.84$, df 3, $P < 0.05$), again showing that those having two or more f's have lower viabilities than those with one or no f.

Therefore, we have to reject the null hypothesis that individuals with different numbers of f chromosomes have the same viability, i.e., the assumption (8) must be abandoned. It may be noted here that though the expected distribution given in (12) is obtained by taking $\theta = 0$, actually the distribution is independent of the value of $\theta$, unless $\theta = \infty$, i.e., the reproduction is entirely asexual.

Having rejected the above hypothesis, next step is evidently to estimate the relative viabilities of individuals with different numbers of f's. For this purpose we assume that the observed populations I, II, III and IV are in equilibrium. The procedure is as follows:

For the sample from population I, we have

$$\hat{\Phi}_0 = 17/59 = 0.2881, \quad \hat{\Phi}_1 = 30/59 = 0.5085$$
$$\hat{\Phi}_2 = 12/59 = 0.2034, \quad \hat{\Phi}_3 = \hat{\Phi}_4 = \ldots = 0.$$

Then using relations (2) and (3), we obtain

$$\hat{\phi}_0 = 0.2769, \quad \hat{\phi}_1 = 0.4908, \quad \hat{\phi}_2 = 0.2177, \quad \hat{\phi}_3 = 0.0139,$$
$$\hat{\phi}_4 = 0.0007, \quad \hat{\phi}_5 = \hat{\phi}_6 = \ldots = 0.$$

The relative viabilities are computed from formula (7) if we know the value of $\theta$, as exemplified in Table 7. Unfortunately no experimental or observational data are available which allow us to determine the value of $\theta$ with confidence. Indeed, we have never encountered a report that such a value was determined in any perennial plant species in its natural habitat. However, there are some reasons to believe that, in *L. callosum*, the average value of $\theta$ is of the order of a few percent. The senior author is informed by MR. M. SHIMIZU, a lily specialist in Japan, that most of the lily species, except for the triploid tiger lily, appear to propagate predominantly by seeds. He guessed that the relative contribution made by asexual reproduction might be around five percent in a wild state. It is interesting to find that this figure is substantiated by the following calculation based on our observational data.

From the inspection of Figure 1a, it becomes immediately evident that there is considerable similarity in chromosome constitution between neighboring individuals. By taking, rather arbitrarily, 40 random pairs of nearby individuals in Figure 1a (see Table 8) and calculating the intraclass correlation, we have $r = 0.347$. In order to get a more reliable estimate, the field observation must be specially designed to have a better sample of pairs.

Such a correlation is probably due to one of the following two causes: (a) The neighboring individuals are derived asexually from a same individual through division of bulbs, in which case the similarity is perfect, i.e., the correlation coefficient is unity. (b) The neighboring individuals are derived sexually but are close relatives, such as sibs or more likely half-sibs from the same seed parent.

If we assume, as a first approximation, that the adjacent individuals are either derived asexually from the same bulb or sexually from the same seed parent, the

TABLE 7

*Relative viabilities of individuals with various numbers of $f_l$'s. In this table $\theta$ represents the ratio between the contributions made by asexual and sexual reproduction to form the next generation*

| Mode of reproduction<br>No. of $f_l$ | $\theta = 0$ | $\theta = 0.1$ | $\theta = 1$ |
|---|---|---|---|
| 0 | 1.000 | 1.000 | 1.000 |
| 1 | 0.996 | 0.996 | 0.998 |
| 2 | 0.898 | 0.907 | 0.947 |
| 3 and over | 0.000 | 0.000 | 0.000 |

TABLE 8

*Intraclass correlation in the number of f chromosomes between two nearby individuals randomly sampled from population I in Figure 1a. In this table, the symbol* (i, j) *denotes the pair of individuals with* i *and* j *f-chromosomes*

| Pairs | No. observed | |
|---|---|---|
| (0,0) | 6 | |
| (0,1) | 7 | $r = 0.347$ |
| (0,2) | 3 | $z = 0.362$ |
| (1,1) | 13 | $\sigma_z = 0.161$ |
| (1,2) | 7 | |
| (2,2) | 4 | |
| | Total n = 40 | |

correlation coefficient in the number of the f chromosomes between the adjacent individuals may be given by

$$(13) \qquad r = \frac{\sigma_b^2}{\sigma_w^2 + \sigma_b^2},$$

where $\sigma_b^2$ is the variance between sibs (including the case of asexual reproduction which occurs at the proportion of $\theta/(1+\theta)$ of the total), i.e.

$$(14) \qquad \sigma_b^2 = \frac{\sum_i M_i^2 \Phi_i w_i^{**}}{\sum_i \Phi_i w_i^{**}} - \overline{M}^2 \quad (\overline{M}\text{: mean of } M_i\text{'s})$$

and $\sigma_w^2$ is the variance within sibs, i.e.

$$(15) \qquad \sigma_w^2 = \frac{\sum_i \sigma_i^2 \Phi_i w_i^{**}}{\sum_i \Phi_i w_i^{**}},$$

in which

$$M_i = \frac{\sum_{j+k=0}^{\infty} (j+k)\, g_{j/i}^{**} G_k^* b_{j+k} + \theta i b_i}{\sum_{j+k=0}^{\infty} g_{j/i}^{**} G_k^* b_{j+k} + \theta b_i} \qquad (i = 0, 1, 2, \ldots)$$

and

$$\sigma_i^2 = \frac{\sum_{j+k=0}^{\infty} (j+k)^2\, g_{j/i}^{**} G_k^* b_{j+k} + \theta i^2 b_i}{\sum_{j+k=0}^{\infty} g_{j/i}^{**} G_k^* b_{j+k} + \theta b_i} - M_i^2 \quad (i = 0, 1, 2, \ldots)$$

are the mean and the variance of the number of f chromosomes among sibs reaching maturity and which are derived from individuals having $i$ f-chromosomes.

Though the calculation involved is again laborious, it is possible to estimate $\theta$ using the above formulae: Using the values in (10) (11) and observed frequencies of

$$\Phi_0 = 0.2881, \quad \Phi_1 = 0.5085, \quad \Phi_2 = 0.2034, \quad \Phi_3 = \Phi_4 = \ldots = 0,$$

the calculation is repeated until we get $r = 0.347$ from (13), the values of $\theta$, and $b_1, b_2, \ldots$ ($b_0 \equiv 1$) being adjusted at every cycle of iteration. The results obtained were as follows:

$$\theta = 0.07, \quad b_0 = 1, \quad b_1 = 0.996, \quad b_2 = 0.905, \quad b_3 = b_4 \ldots = 0.$$

If we put all four samples (in Table 1) together, estimated values of $\theta$ and relative viabilities were approximately as follows:

$$\theta = 0.06, \quad b_0 = 1.000, \quad b_1 = 1.066, \quad b_2 = 0.691, \quad b_3 = b_4 = \ldots = 0.$$

In Table 9, frequency distributions, relative fertilities and viabilities are summarized.

TABLE 9

*Summary of frequency distributions, relative fertilities and viabilities. The estimated value of $\theta$ is approximately 0.06*

| No. of f chromosomes | 0 | 1 | 2 | 3 or more |
|---|---|---|---|---|
| Observed frequencies (sum of four samples) $\Phi$ | 0.333 | 0.541 | 0.126 | 0.000 |
| Expected frequencies at fertilization $\phi$ (estimated) | 0.323 | 0.489 | 0.180 | 0.008 |
| Pollen fertility (observed) $w^*$ | 1.000 | 0.920 | 0.559 | .... |
| Fertility as seed parents (observed) $w^{**}$ | 1.000 | 0.849 | 0.577 | .... |
| Relative viability (estimated) $b$ | 1.000 | 1.066 | 0.691 | 0.000 |

## DISCUSSION AND CONCLUSION

Although a considerable number of cases of supernumerary chromosomes have been reported in populations of various species, the present case is rather exceptional for its simplicity of frequency distribution and completeness of experimental and observational data which enable mathematical analysis based on a theoretical model.

The result of the present analysis has shown that individuals with f chromosomes are not only less fertile but also less viable than those with no f chromosomes. This deleterious effect of f chromosomes on viability is especially evident for the case of two or more f chromosomes: With two f's viability is about 70 percent and with more than two f's, it is practically zero. On the other hand for the case of one f chromosome, there is no definite indication of lowered viability. It is possible that even a slight increased viability may exist in plants with a single f, as suggested by the result of analysis on the four samples together. Furthermore, the difference in the details of frequency distribution between samples from different localities such as I + II *vs.* III + IV mentioned earlier, may very well be due to difference in ecologic conditions which influence viabilities of individuals with various numbers of f chromosomes. BOSEMARK (1956) reported an interesting observation that in *Festuca pratensis*, there was a marked correlation between frequency of accessory chromosomes and the clay content of the soil in Sweden.

Whatever modification is made in the selective pattern by local ecologic conditions, the conclusion is inevitable that in *Lilium callosum* the f chromosome is predominantly deleterious and it may only be kept in the population by preferential segregation in embryo sac mother cells.

Analogous situations occur with the *t* allele found by DUNN in the mouse (DUNN 1957) and SD factor discovered by HIRAIZUMI in *Drosophila melanogaster* (HIRAIZUMI, SANDLER and CROW 1960). In the former, the *t* allele (actually consisting of multiple alleles) is a recessive gene located on an autosome and is lethal (or causes complete sterility) when homozygous. Nevertheless, it is found with very high frequencies in wild populations of mice sampled in USA, due to excessive frequency of the *t* allele among the sperms produced by heterozygous males ($t/+$). If we designate by $k$ ($> 0.5$) the segregation proportion of $t$ among the sperms produced from $t/+$ males, and if we assume that the heterozygotes, either male or female, have completely normal reproductive ability and that $t/t$ is lethal, then it is possible to show that in equilibrium the proportion of $t$ allele among the pools of male and female gametes are

$$k-\sqrt{k(1-k)} \quad \text{and} \quad \frac{1}{2k}\left\{k-\sqrt{k(1-k)}\right\}$$

respectively. Thus the proportion of zygotes which fail to develop is

$$\frac{1}{2k}\left\{k-\sqrt{k(1-k)}\right\}^2$$

or

$$\frac{1}{2}\left\{1-2\sqrt{k(1-k)}\right\}$$

In other words, this is the proportion of individuals which have to be eliminated to maintain the deleterious allele in the population. Recently the senior author proposed the term distortional load (or load due to meiotic drive) to designate such a load (KIMURA 1960). In the present case, this is given by

$$L_{SD}=\frac{1}{2}\left\{1-2\sqrt{k(1-k)}\right\}$$

According to DUNN, segregation percentage of $t$ ranges from 89 percent to 99 percent ($k = 0.89$–$0.99$), so that $\hat{L}_{SD}$ lies between 0.19 and 0.40. This means that in wild populations of mice, even disregarding possible effects of overdominance of $t$ allele, the load due to maintaining $t$ amounts to about 20–40 percent.

A similar concept may be applied to the present case and the distortional load due to f chromosomes may be calculated as follows: From Table 9, the average loss of fertility for plants with one f chromosome is $1 - (0.920 + 0.849)/2$ or 0.115. Similarly for plants with two f's it is 0.432. Thus the component of load due to fertility differences is $0.541 \times 0.115 + 0.126 \times 0.432$ or 0.117, when the population reproduces sexually. The component of load due to viability differences is $0.180 \times 0.309 + 0.08 \times 1$ or 0.064 under sexual reproduction, and $0.126 \times 0.309$ or 0.039 under asexual reproduction (see Table 9). In both cases we assume that plants with one f chromosome have the same viability as those with no f's. Combining all these values and taking $\theta = 0.06$, the load becomes $(1 - 0.06)(0.117 + 0.064) + 0.06 \times 0.039$ or 0.1724. Thus in wild populations of *L. callosum,* the distortional load due to f chromosomes amounts to about 17 percent. If apparent increased viability of plants with a single f ($b_1 = 1.066$) is taken into consideration, this value may be increased by three percent.

Though the existence of the increased viability with a single B chromosome is not conclusive in the present case, there may be a considerable possibility for it in a population of organisms in which B chromosomes reach such a high frequency that the majority of individuals carry one or more of them. This is because, under such a circumstance, modifiers which improve viability in combination with B are more advantageous, other things being equal, than those which improve viability without B. If this modification goes on, fitness of individuals with one B will become higher than that with no B's. Through this process, the frequency of B in the population will be gradually enhanced and deleterious effect of B will be diminished. We may say that the B chromosome changes from "parasitic" to "symbiotic."

## SUMMARY

1. In *Lilium callosum,* a supernumerary chromosome (B chromosome) denoted by $f_1$ has a tendency to increase its number due to preferential segregation in embryo-sac mother cells in such a way that in plants with one $f_1$ chromosome, it moves to the micropylar side in 80 percent of the cases to be included in the egg cells. No such tendency was found in its transmission through the pollen. (KAYANO 1956b, 1957).

2. By crossing experiments involving plants with various numbers of $f_1$ chromosomes, the mode of transmission and its effect on pollen and seed fertility were investigated. It was found that $f_1$ chromosomes, when more than one is present, markedly reduced both pollen and seed fertility (Tables 2 and 3).

3. In order to study the frequency distribution of the number of $f_1$ chromosomes among individuals in natural populations of this lily, samples were collected from four localities in Kyushu and their chromosome constitutions were determined.

The result of this study showed that the frequencies of $f_1$ are uniformly high: Individuals with one $f_1$ are most frequent and they together with those having two $f_1$'s compose nearly 70 percent of a population on the average. On the other hand, no individuals were found with more than two $f_1$'s (Table 1, Figure 1 a–d).

4. A mathematical theory was developed to analyse the mechanism of distribution of $f_1$ chromosomes in the population. The theory contains two new features which do not appear in the ordinary mathematical treatment in population genetics, in that it takes into account of (a) the existence of both sexual and asexual reproduction and (b) the fact that the mode of segregation of the $f_1$ chromosome does not follow the simple rules characteristic of diploid loci.

5. The analysis has shown that individuals with $f_1$ chromosomes are not only less fertile but also less viable than those with no $f_1$ chromosomes. This deleterious effect of $f_1$ chromosome on viability is especially evident for the case of two or more $f_1$'s. Namely, with two $f_1$'s viability is about 70 percent and with more than two, it is practically zero. For the case of one $f_1$, there exists no definite indication of lowered viability and it is suspected that there is even a slight increase.

6. The concept of distortional load (or load due to meiotic drive) was applied to the maintenance of $f_1$ chromosomes in natural populations of *L. callosum* and it was estimated that this load amounts to about 17 percent. A mechanism was suggested by which the deleterious effect of $f_1$ chromosome may be reduced in the course of evolution.

ACKNOWLEDGMENTS

We would like to express our indebtedness to DR. T. HAGA under whose guidance the experimental and observational work by the junior author had been carried out. Thanks are also due to DR. J. F. CROW who kindly reviewed the draft and made valuable suggestions for improving the manuscript.

LITERATURE CITED

BOSEMARK, N. O., 1956 On accessory chromosomes in *Festuca pratensis*. III. Frequency and geographical distribution of plants with accessory chromosomes. Hereditas **42:** 189–210.

DUNN, L. C., 1957 Evidence of evolutionary forces leading to the spread of lethal genes in wild populations of house mice. Proc. Natl. Acad. Sci. U.S. **43:** 158–163.

HIRAIZUMI, Y., L. SANDLER, and J. F. CROW, 1960 Meiotic drive in natural populations of *Drosophila melanogaster*. III. Populational implications with special application to the segregation-distorter locus. Evolution. **14:** 433–444.

KAYANO, H., 1956a Cytogenetic studies in *Lilium callosum*. I. Three types of supernumerary chromosomes. Mem. Fac. Sci. Kyushu Univ. Ser. E **2:** 45–52.

1956b Cytogenetic studies in *Lilium callosum*. II. Preferential segregation of a supernumerary chromosome. Mem. Fac. Sci. Kyushu Univ. Ser. E **2:** 53–60.

1957 Cytogenetic studies in *Lilium callosum*. III. Preferential segregation of a supernumerary chromosome in EMCs. Proc. Japan Acad. **33:** 553–558.

KIMURA, M., 1960 Genetic load of a population and its significance in evolution. (Japanese with English summary.) Japan. J. Genet. **35:** 7–33.

MÜNTZING, A., 1954 Cytogenetics of accessory chromosomes (B chromosomes). (Suppl.) Caryologia **6:** 282–301.

PART EIGHT

# Genetic Load

## Introduction

In 1960 Kimura was awarded the annual prize from the Genetics Society of Japan for his studies of theoretical population genetics. In his award lecture, he talked about the genetic load and its significance in evolution. In the same year, he presented a mathematical treatment of an old idea that if the mutation rate is too high the species will be crushed under a heavy mutational load; if it is too low the species will not be able to cope with adverse environmental changes (no. 18). Thus between these conflicting demands there must be an optimum mutation rate which may correspond to the spontaneous mutation rate. By minimizing the sum of mutational and substitutional loads (the so-called principle of minimum genetic load), he derived a simple relationship between the mutation rate and the horotelic evolutionary rate by advantageous mutations with arbitrary degrees of dominance. Kimura later reexamined the same subject (no. 20) and argued that modifiers which enhance the mutation rate of other loci are selected against, whereas those which decrease the mutation rate are favored by natural selection. The present theory of modifiers is concerned not only with the mutation rate (Leigh 1970; Ishii et al. 1989), but also with other genetic parameters such as the recombination fraction (Nei 1969; Feldman 1972; Eshel and Feldman 1982; Feldman and Liberman 1986). Compared to the then-assumed mutation rate per gamete per generation ($M \approx 0.04$ in *Drosophila*), however, the principle of minimum genetic load led to a much smaller value of optimum mutation rate. The optimum rate equals the total rate of allelic substitutions per generation ($E$) that can be attained by positive selection. Being puzzled by this large discrepancy, and later finding a still larger value of $M(0.5)$ from amino acid sequence data (no. 48), he came to the conclusion that $E = M$ does not hold at the molecular level. This might have suggested to him that most substitutions are not driven by natural selection.

The Haldane-Muller principle (Haldane 1937) states that the mutational load (the reduction in mean population fitness) by deleterious mutations is twice as large as the mutation rate in a randomly mating diploid population with partial dominance and independence between loci. Some experimental data (Mukai 1965) suggested, however, that the deleterious effects reinforce each other as the number of mutations increases. Kimura and Maruyama (no. 19) formulated the mutational load under various forms of epistasis with and without recombination. Those results may be understood if one notices under what circumstances one genetic death can efficiently eliminate more or fewer deleterious mutations than under the circumstances that Haldane (1937) and Muller (1950) envisioned. Synergistic interaction (reinforcing type) is effective if a large number of deleterious genes are involved. Truncation or quasi-truncation selection (Crow and Kimura, no. 25) is especially effective in removing deleterious genes in clusters, whereas this cannot happen in asexual species. This is one argument for the prevalence of sexual reproduction (Kondrashov 1984).

# OPTIMUM MUTATION RATE AND DEGREE OF DOMINANCE AS DETERMINED BY THE PRINCIPLE OF MINIMUM GENETIC LOAD*

By MOTOO KIMURA

*National Institute of Genetics, Mishima-shi, Japan*

(Received Feb. 5, 1959)

## Introduction

The genetical properties of an organism are the products of its evolutionary history. Thus we should be able to deduce some of the main properties common to a wide range of organisms from the general picture of evolution in the past. For example, it is well known that the majority of gene mutations are harmful to the organism. Admitting that the directions of mutations are sufficiently diverse, this must follow from the fact that adaptive evolution has been achieved through natural selection acting on random mutations.

Again, it is commonly accepted that in most organisms, wild type genes tend to be dominant over mutant alleles. Whatever theory of dominance is invoked to explain this phenomenon, one has to assume that dominance of wild type genes was gradually strengthened by natural selection.

In the present paper, I will take up the problem of spontaneous mutation rate from such a point of view.

As noted above, the majority of gene mutations are harmful to an organism in environmental conditions to which it has become adapted. Thus at any given moment, natural selection appears largely directed toward the maintenance of *status quo* by eliminating extreme forms (Haldane 1954). This type of selection is often referred to as normalising or stabilising selection (Waddington, 1957), but Haldane (1958) prefers to call it centripetal selection. The higher the mutation rate the more the reproductive potential of a species will be impaired. Yet, without heritable variation, adaptive evolution by natural selection will be impossible. If gene mutation ceases to occur, the store of genetic variability of a species will soon be depleted; and when environmental conditions change, the species will no longer be able to readjust itself to the new environment. The result can only be extinction of the species.

These considerations inevitably suggest that there must be an optimum mutation rate for the survival of a species under a given rate of environmental change. If the mutation rate is too high the species will be crushed under a heavy mutational load; if it is too low the species will not be able to cope with adverse environmental changes.

* Contribution No. 276 of the National Institute of Genetics, Mishima-shi Japan. A part of this paper was orally presented to the 29th Meeting of the Genetics Society of Japan held in Sapporo, September 1957, under the title "The principle of minimum genetic load of a population".

## 22 *Mutation rate and dominance*

The organisms that have managed to survive up to the present must be such that have been able to adjust their mutation rate to the optimum level through inter-group as well as intra-group selection. What we call "the spontaneous mutation rate" is thus a product of natural selection acting on the hereditary variations, just as it acts on any other morphological or physiological character of an organism. Evidence for the genetic control of the mutation rate comes not only from the fact that different strains in the same species may have different mutation rates, but also from the fact that genes which increase the mutation rate of other genes have been found in wild populations (Ives 1950). Such an idea is by no means novel, and in her popular book, Genetics in the Atomic Age, Auerbach (1956) states: "Thus each species has to strike a balance between the short-term requirement for a low frequency of mutation and the long-term requirement for an ample store of mutant genes. A species in which mutations are too frequent will die out because too many of its individuals are weak, short-lived or sterile. A species in which mutations are too rare may do well for a time, but will not survive when altered conditions demand adaptations for which it does not possess the necessary genes.

The final balance between these conflicting requirements is the so-called "spontaneous mutation rate", that is, the average frequency with which genes of a given species mutate."

However, no quantitative formulation of the idea, as far as I know, has ever been worked out. In the present paper, I try to show that, by introducing what may be called the principle of minimum genetic load, a mathematical formulation of the problem can be achieved. Though this is the first attempt, I hope that it will throw a new light on quantitative statements about evolution of genetic systems.

### The principle of minimum genetic load

We owe the mathematical formulation underlying the concept of genetic load chiefly to Haldane (1937, 1957). Here and throughout this paper, I will use the term genetic load strictly in the sense defined by Crow (1958), namely as "the proportion by which the population fitness is decreased in comparison with an optimum genotype". In this paper, fitness will be measured in Malthusian parameters (cf. Kimura 1958). This is roughly equivalent to the natural logarithm of the average number of offspring. In what follows we will be concerned with three kinds of genetic loads: (*i*) *Mutational load.* The population fitness may be depressed by the elimination of recurrent harmful mutations. It has been shown by Haldane (1937) that the amount of depression is roughly equal to the sum of all mutation rates and is independent of selection coefficients. An equivalent result has been obtained independently by Muller (1950 and earlier writings) who employed the concept of "genetic death". (*ii*) *Segregational load.* The population fitness may be decreased if the heterozygote is superior in fitness to either homozygote, because the inferior homozygotes are produced in each generation by segregation. If $s_1$ and $s_2$ are the selection coefficients of the two homozygotes, the segregational load per locus is $s_1s_2/(s_1+s_2)$ (Haldane 1937; Crow 1958).

## MOTOO KIMURA 23

(*iii*) *Substitutional* (*Evolutional*) *load.* The process of substituting one allele by another through natural selection involves lowering of population fitness and thus creates a genetic load. I propose to call this the substitutional or evolutional load. Again, it has been shown by Haldane (1957) that the evolutional load per gene substituted is almost independent of the intensity of selection and rather depends on the initial gene frequency and the degree of dominance. The load at any moment also varies as the rate at which genes are being substituted.

Let us now consider a large random mating population placed in an environment which changes steadily at a very low constant rate such that in each given time interval mutant genes in a certain number of loci start to increase their frequencies because of the change in the environment which makes "wild type" alleles in such loci disadvantageous. On an average, that number must be a very small fraction if counted per generation. In the majority of loci, "mutant genes" are maintained at low frequency due to the balance between mutation and selection.

Let $\mathbf{A}_1$ and $\mathbf{A}_2$ be a pair of alleles in a locus in which the mutant gene $\mathbf{A}_2$ exists at a low frequency. We will designate the fitnesses of the three genotypes $\mathbf{A}_1\,\mathbf{A}_1$, $\mathbf{A}_1\,\mathbf{A}_2$ and $\mathbf{A}_2\,\mathbf{A}_2$ in terms of Malthusian parameters by 0, $-sh$ and $-s$ respectively as shown in the second line of Table 1.

If we designate the frequency of $\mathbf{A}_2$ by $x$, then the average fitness of the population may be given by

$$\bar{a} = -sx^2 - 2shx\,(1-x)$$

so that the rate of change in the frequency of $\mathbf{A}_2$ per generation by selection is

$$\frac{dx}{dt} = \frac{x(1-x)}{2}\,\frac{d\bar{a}}{dx} = -sx(1-x)\,\{h+(1-2h)x\}.$$

Table 1

| Fitnesses \ Genotypes and their frequencies | $\mathbf{A}_1\mathbf{A}_1$ $(1-x)^2$ | $\mathbf{A}_1\mathbf{A}_2$ $2x(1-x)$ | $\mathbf{A}_2\mathbf{A}_2$ $x^2$ |
|---|---|---|---|
| Before the change of environment | 0 | $-sh$ | $-s$ |
| After the change of environment | $-s'$ | $-s'(1-h)$ | 0 |

The rate of change of the frequency due to mutation is

$$\mu\,(1-x),$$

where $\mu$ is the mutation rate from $\mathbf{A}_1$ to $\mathbf{A}_2$.

Thus at equilibrium we should have

$$\mu(1-x) = sx(1-x)\,\{h+(1-2h)x\}$$

## 24 Mutation rate and dominance

or writing the equilibrium frequency of $\mathbf{A}_2$ as $x_0$, we obtain

$$\frac{\mu}{s}=x_0\{h+(1-2h)x_0\}. \tag{1}$$

Here we assume that $s>sh>0$.

Thus the mutational load denoted by $L_m$ is

$$L_m=\Sigma s x_0\{x_0+2h(1-x_0)\},$$

where the summation is over all relevant loci.

We suppose now that, because of environmental changes, the mutant gene $\mathbf{A}_2$ becomes advantageous over its "wild type" allele $\mathbf{A}_1$. The situation may be expressed by substituting $-s$ by $s'$ ($>0$). It may be natural to assume that $h$ remains constant, because the rate of modification of dominance may be exceedingly slow as compared with the rate by which $\mathbf{A}_2$ increases its frequency under a new environment, though $\mathbf{A}_2$ will ultimately be modified to become dominant over $\mathbf{A}_1$. The relative fitnesses of the three genotypes $\mathbf{A}_1\mathbf{A}_1$, $\mathbf{A}_1\mathbf{A}_2$ and $\mathbf{A}_2\mathbf{A}_2$ are now 0, $s'h$ and $s'$, or equivalently, $-s'$, $-s'$ (1-$h$) and 0, as shown in the third line of Table 1. The average fitness of the population is now

$$\bar{a}=-s'(1-h)2x(1-x)-s'(1-x)^2$$

and the rate of change in $x$ is

$$\frac{dx}{dt}=\frac{x(1-x)}{2}\frac{d\bar{a}}{dx}=s'x(1-x)\{h+(1-2h)x\}.$$

Thus the total amount of decrease in fitness throughout the process of substituting $\mathbf{A}_1$ by $\mathbf{A}_2$ is

$$\int_0^\infty|\bar{a}|\,dt=\int_0^\infty\{s'(1-h)2x(1-x)+s'(1-x)^2\}\,dt$$

$$=\int_{x_0}^1\frac{s'(1-h)2x(1-x)+s'(1-x)^2}{s'x(1-x)\ \{h+(1-2h)x\}}dx$$

$$=-\frac{1}{h}\left\{\log x_0+(1-h)\log\frac{1-h}{h+(1-2h)x_0}\right\} \tag{2}$$

Since the above quantity is the amount of decrease in fitness which would be realized if substitution by new alleles proceeded at the rate of one locus per generation, we will define the substitutional or evolutional load of the population by

$$L_e=-\Sigma\frac{\epsilon}{h}\left\{\log x_0+(1-h)\log\frac{1-h}{h+(1-2h)x_0}\right\}, \tag{3}$$

where $\epsilon$ is the rate of substitution *per locus* per generation, and summation is over all relevant loci.

We will designate the total number of substitution per generation by $E$, namely

$$E=\Sigma\epsilon.$$

The idea underlying the evolutional load was first formulated in Haldane's (1957) article entitled "The cost of natural selection" who anticipated the importance of such an idea "in all future discussions of evolution". The crucial point here is that this load does not depend on the intensity of selection, i.e. $s'$. It rather depends on $h$ and $x_0$.

I now assume that, in the course of evolution, $\mu$ and $h$ are adjusted by the accumulation of modifiers such that $\bar{a}$ will be maximized, or more strictly the total genetic load under consideration, i.e.,

$$L = L_m + L_e$$

will be minimized. More generally, I believe that in the course of evolution, important genetic parameters of each species tend to be adjusted in such direction that the total genetic load

$$L = L_m + L_s + L_e + \ . \ . \ .$$

will be minimized. This is my view on the course of evolution, and I would like to call it the principle of minimum genetic load. Here $L_s$ denotes the segregational load, and certainly there must be other types of genetic load yet to be defined.

The picture of evolution that I have in mind is that at any given moment the genetic composition of the species is almost in equilibrium, and this equilibrium state is gradually shifted by the secular change of environmental conditions.

Returning to our original problem, we require that

$$L = \Sigma s x_0 \{x_0 + 2h(1-x_0)\} - \Sigma \frac{\epsilon}{h} \left\{ \log x_0 + (1-h) \log \frac{1-h}{h + (1-2h)x_0} \right\} \tag{4}$$

shall be minimized with respect to $\mu$ and $h$ under the side condition given in (1).

The problem may be reduced to minimizing

$$\psi = \Sigma s x_0 \{x_0 + 2h\ (1-x_0)\} - \Sigma \frac{\epsilon}{h} \left\{ \log x_0 + (1-h) \log \frac{1-h}{h+(1-2h)x_0} \right\} + \Sigma\lambda \{\mu - s x_0 (h + (1-2h)\ x_0)\},$$

where $\lambda$ is the Lagrange multiplier and $\mu$, $h$ and $x_0$ may be treated as if they were all independent.

Since

$$\frac{\partial \psi}{\partial \mu} = \lambda = 0,$$

we have

$$\frac{\partial \psi}{\partial x_0} = 2s\{h + x_0(1-2h)\} - \frac{\epsilon}{h} \left\{ \frac{1}{x_0} - \frac{(1-h)\ (1-2h)}{h + (1-2h)x_0} \right\} = 0 \tag{5}$$

$$\frac{\partial \psi}{\partial h} = 2s x_0 (1-x_0) + \frac{\epsilon}{h} \left\{ \frac{1}{h} \log \frac{(1-h)x_0}{h+(1-2h)x_0} + 1 + \frac{(1-h)(1-2x_0)}{h+(1-2h)x_0} \right\} = 0 \tag{6}$$

26 *Mutation rate and dominance*

together with (1) or

$$\frac{\partial\psi}{\partial\lambda}=\mu-sx_0\{h+(1-2h)x_0\}=0. \tag{7}$$

For a given small value of $\epsilon/s$, values of $h$ and $x_0$ should be determined by the set of two equations, (5) and (6). However, because of the logarithmic as well as fractional terms involved, explicit analytical solutions seem to be very difficult to obtain. So numerical analysis was applied to work out relations between these three parameters, with the results shown in the following table (Table 2).

Table 2

| $x_0$ | $1\times10^{-3}$ | $5\times10^{-3}$ | $1\times10^{-2}$ | $2\times10^{-2}$ | $5\times10^{-2}$ | $1\times10^{-1}$ |
|---|---|---|---|---|---|---|
| $h$ | $2{\cdot}172\times10^{-3}$ | $1{\cdot}104\times10^{-2}$ | $2{\cdot}252\times10^{-2}$ | $4{\cdot}664\times10^{-2}$ | $1{\cdot}259\times10^{-1}$ | $2{\cdot}640\times10^{-1}$ |
| $\epsilon/s$ | $2{\cdot}004\times10^{-8}$ | $2{\cdot}525\times10^{-6}$ | $2{\cdot}037\times10^{-5}$ | $1{\cdot}648\times10^{-4}$ | $2{\cdot}570\times10^{-3}$ | $1{\cdot}849\times10^{-2}$ |

It may be seen from this table that the mutant gene frequency at equilibrium is roughly one half of the degree of dominance, i.e.

$$x_0\approx h/2. \tag{8}$$

This suggests that a more precise result may be obtained by applying the power series expansion

$$x_0=C_1h+C_2h^2+C_3h^3+\ \ldots \tag{9}$$

and determining the coefficients involved.

If we eliminate $\epsilon/s$ from (5) and (6), we have

$$\left(\frac{h+x_0-2hx_0}{h}\right)^2\left\{-\log(1-h)-h+\log\frac{h+x_0-2hx_0}{x_0}\right\}$$

$$=\frac{h+x_0-2hx_0}{h}(1-h)(1-2x_0)+(1-x_0)(1+x_0-2hx_0) \tag{10}$$

Now, substituting (9) in the above equation and comparing coefficients of $h^0$, $h^1$, etc. on both sides, we can successively determine the values of $C_1$, $C_2$, etc.

The first coefficient $C_1$ satisfies the following transcendental equation

$$(1+C_1)^2\log\left(1+\frac{1}{C_1}\right)=2+C_1, \tag{11}$$

from which we obtain

$$C_1=0{\cdot}4624\ \ldots$$

The second coefficient is given by

$$C_2=C_1(1-C_1)^{-1}(1-4C_1-C_1^2),$$

or approximately

$$C_2=-0{\cdot}9146,$$

and the third coefficient by

$$C_3=C_1(1-C_1)^{-3}(1-7C_1+22C_1^2-6C_1^3-3C_1^4+C_1^5)$$

which is approximately $+5{\cdot}232$.

A similar expression for the fourth coefficient turned out to be quite complicated, with numerical value of about—19·14.

The resulting series gives satisfactory approximation for a small value of $x_0$. For example with $h=1/50$, the error is less than 0·4% if we use the first two terms in the series. For a larger value of $x_0$, the approximation becomes less satisfactory, giving nearly 10% error when $x_0$ is 0·1, even if we include terms up to the fourth power of $h$ for the calculation. However, this will not give us much trouble, since, as will be seen from the discussion, we will be interested in the values of $x_0$ corresponding to about $h=1/50$.

From equation (5), we have

$$\frac{\epsilon}{s}=\frac{2x_0(h+x_0-2hx_0)^2}{1+x_0-2x_0h} \tag{12}$$

and applying (9) to the right side, we obtain

$$\frac{\epsilon}{s}=2C_1(1+C_1)^2\,h^3\left\{1+\frac{C_2(1+3C_1)-C_1^2(5+C_1)}{C_1(1+C_1)}h+\ .\ .\ .\right\}. \tag{13}$$

From equation (7), we have

$$\frac{\mu}{s}=x_0(h+x_0-2hx_0) \tag{14}$$

and combining this with (12) we get

$$\mu=\frac{\epsilon(1+x_0-2x_0h)}{2(h+x_0-2x_0h)} \tag{15}$$

or

$$\mu=\frac{-\epsilon}{2(1+C_1)h}\left\{1+(C_1-\frac{C_2-2C_1}{1+C_1})\ h+\ .\ .\ .\right\}. \tag{16}$$

Also from (7), we obtain

$$sx_0=\frac{\mu}{(1+C_1)h}\left\{1-\frac{C_2-2C_1}{1+C_1}h+\ .\ .\ .\right\} \tag{17}$$

Thus, mutational and evolutional loads are expressed as follows:

$$L_m=\frac{2+C_1}{1+C_1}\,\Sigma\mu\,\left\{1+\frac{2C_1-C_2}{(1+C_1)(2+C_1)}\,h+\ .\ .\ .\right\} \tag{18}$$

$$L_e=\Sigma\epsilon\left\{-\log x_0+\frac{1}{h}\log\left(\frac{1+C_1}{C_1}\right)-\log\left(\frac{1+C_1}{C_1}\right)+\left(\frac{C_2-2C_1}{1+C_1}-\frac{C_2-C_1}{C_1}\right)+...\right\} \tag{19}$$

28 *Mutation rate and dominance*

Though the above two relations (16) and (17) are derived for each locus, it may be more appropriate to sum each of these relations over all relevant loci to get

$$\Sigma\mu = \frac{E}{2(1+C_1)\bar{h}_\epsilon}\left\{1 + \left(C_1 - \frac{C_2-2C_1}{1+C_1}\right)\bar{h}_\epsilon + \dots\right\} \tag{20}$$

and,

$$\Sigma s x_0 = \frac{\Sigma\mu}{(1+C_1)\bar{h}_\mu}\left\{1 - \frac{C_2-2C_1}{1+C_1}\bar{h}_\mu + \dots\right\}, \tag{21}$$

where $E$ is the total number of substitutions per generation. In these formulae, $\bar{h}_\epsilon$ and $\bar{h}_\mu$ respectively denote the harmonic means of $h$ weighted according to the rate of substitution and the rate of mutation at each locus. The left side of the above equation (21) may be identified with the total number of recessive deleterious genes per gamete measured in lethal equivalents, or using the term of Morton, Crow and Muller (1956), with the total mutational damage per gamete which we will designate by $D$. The actual value of $D$ for an outbreeding population may be estimated through studies of inbreeding depression in fitness.

Combining the last two relations (20) and (21), we obtain

$$\Sigma\mu = \frac{E}{2(1+C_1)\bar{h}_\epsilon}\left\{1 + \left(C_1 - \frac{C_2-2C_1}{1+C_1}\right)\bar{h}_\epsilon + \dots\right\}$$

$$= D(1+C_1)\bar{h}_\mu\left\{1 + \frac{C_2-2C_1}{1+C_1}\bar{h}_\mu + \dots\right\}, \tag{22}$$

where

$$D = \Sigma\, s x_0 \tag{23}$$

It is remarkable that the total mutation rate and average degree of dominance are connected with the rate of substitution of alleles in horotelic evolution, i.e. standard rate evolution (cf. Simpson, 1944), by rather a simple relationship.

## Discussion

In order to examine the feasibility of such formulation, we must check our results using data from actual observation. Before doing this, however, we have to make the simplifying assumption that the unweighted harmonic mean may be substituted for the weighted harmonic means, $\bar{h}_\epsilon$ and $\bar{h}_\mu$. At least this may be accepted as a first approximation. If we designate the unweighted harmonic mean by $\bar{h}$, (22) may be substituted by the following pair of equations

$$\left.\begin{aligned} \Sigma\mu &= \frac{E}{2(1+C_1)\bar{h}}\left\{1 + \left(C_1 - \frac{C_2-2C_1}{1+C_1}\right)\bar{h} + \dots\right\} \\ \bar{h} &= \frac{1}{1+C_1}\sqrt{\frac{E}{2D}}\left\{1 + \left(\frac{C_1}{2} - \frac{C_2-2C_1}{1+C_1}\right)\frac{1}{1+C_1}\sqrt{\frac{E}{2D}} + \dots\right\} \end{aligned}\right\} \tag{24}$$

or using $C_1=0{\cdot}4624$ and $C_2=-0{\cdot}9146$,

$$\left.\begin{aligned}\Sigma\mu&=\frac{0{\cdot}3419E}{\bar{h}}(1+1{\cdot}720\bar{h}+\ldots)\\ \bar{h}&=0{\cdot}6838\sqrt{\frac{E}{2D}}\left(1+1{\cdot}018\sqrt{\frac{E}{2D}}+\ldots\right)\end{aligned}\right\}\qquad(25)$$

Now, the following data are available to us: As to the rate of substitution of alleles in horotelic evolution, Haldane (1957) suggests the figure of one per 300 generations, i.e. $E=1/300$, based on his well-considered conception of the cost of natural selection and the observed speed of evolution in phenotypic characters. As to the inbreeding depression in fitness, the recent study by Morton, Crow and Muller (1956) on human population gives $1{\cdot}5\sim2{\cdot}5$ as the possible values of $D$. So we will take $D=2$. The same authors calculated the harmonic mean of the degree of dominance using data from lethal genes in *Drosophila*; and it turns out to be about 2%, i.e. $h=1/50$. As to the spontaneous mutation rate, Muller (1956) estimated that one to two tenths of gametes in man contain a newly arisen mutation. He also estimated that about 4·3% of gametes in *Drosophila* contain a newly arisen detrimental mutation (Muller 1955). So $\Sigma\mu$ should be of the order of 10%.

We will first check the second formula in (25). Taking $E=1/300$ and $D=2$, we obtain

$$\bar{h}=0{\cdot}0203$$

which agrees remarkably well with the observed value of about 2%. One of the interesting features of this result is that it supplies a new "Theory of dominance", independent of the previous theories (Fisher 1930; Wright 1934; Haldane 1930; Muller 1932). Furthermore it has the merit of predicting the amount of "dominance" of recessive deleterious genes.

Next, we will check the first formula in (25). Using $E=1/300$ and $\bar{h}=0{\cdot}0203$, we obtain

$$\Sigma\mu=0{\cdot}0581$$

or about 5·8%. The agreement with the expected value of the order of 10% is not wholly unsatisfactory, because the formula is concerned with a typical mutation rate determined by a typical speed of evolution extending over enormous lengths of time. Using the above figures and applying (18) and (19), the mutational and the evolutional loads turn out to be about 9·9% and 20·6% respectively. It may be rather unexpected that the organism is paying twice as much for improvement as for prevention from deterioration, even under an extremely slow rate of evolution. Haldane suggested the value of about 5% for mutational load (Haldane 1937) and about 10% for evolutionary load (Haldane 1957). Thus our values for these loads are twice as high.

Another interesting parameter which it is worth while to estimate is the average magnitude of the selection coefficient $s$. It may be calculated using equation (13),

(14) or (23), but in any case we have to know the total number of loci involved. For our purpose, we adopt 10,000 as a possible figure to represent the total number of loci. This figure is known as a ratio of total detrimental to per-locus visible mutation rates in Drosophila (Muller 1956). First, we will use formula (13). With $h=0{\cdot}02$, the ratio $\epsilon/s$ becomes about $5{\cdot}6\times10^{-6}$. If we take the average value of $\epsilon$ per locus as $(1/300)\times10^{-4}$, the harmonic mean of $s$ becomes about 0·059 or 6%. However, it might be more appropriate to use the arithmetic mean of $h$ rather than the harmonic mean to calculate the right side of (13). For lethal genes in *Drosophila*, the arithmetic mean of $h$ is known to be about 0·04 (Morton, Crow and Muller 1956). Thus with $h=0{\cdot}04$, $\epsilon/s$ turns out to be about $4\times10^{-5}$, with the result that the harmonic mean of $s$ becomes about 0·8%. Secondly, we may use formula (14), in which we take $\mu=0{\cdot}0581\times10^{-4}$ or about $5{\cdot}8\times10^{-6}$ as the average mutation rate per locus. With $h=0{\cdot}02$, the harmonic mean of $s$ becomes about 0·02, while with $h=0{\cdot}04$ it becomes about 0·5%. Finally equation (23) may be used to get the arithmetic mean of $s$, assuming $D=2$. With $h=0{\cdot}02$, $x_0$ is approximately $9{\cdot}1\times10^{-3}$ or roughly 1%, so that $\Sigma x_0$ may be of the order of 100. So the arithmetic mean of $s$ turns out to be about 2%. With $h=0{\cdot}04$, $x_0$ is approximately 2%, and the arithmetic mean of $s$ becomes about 1%.

From the above results, we may conclude that the arithmetic mean of the selection coefficient $s$ is probably 1 to 2%, while the harmonic mean is at the most half as large.

We can now form some idea about the number of loci in which substitution of alleles is going on at a given moment. Since the rate of increase in the frequency of an advantageous mutant gene in such a locus is expressed by the equation

$$\frac{dx}{dt}=s'x(1-x)\,\{h+(1-2h)x\}\ ,$$

the number of generations required for the mutant gene to increase by selection pressure from $x_0$ to $1-x_0$ is given by

$$T=\frac{1}{s'h(1-h)}\left\{\log\frac{1-x_0}{x_0}-(1-2h)\log\frac{h+(1-2h)\,(1-x_0)}{h+(1-2h)x_0}\right\}\ ,$$

or roughly

$$T=\frac{1{\cdot}15}{s'h}, \tag{26}$$

when $h$ is small. If we assume that the harmonic mean of $s'$ is equal to that of $s$, the average value of the right side of (26) turns out to be about 5,000. The number of loci in which substitution of alleles is under way may be obtained from the ratio $T/E^{-1}$, which is about 17 if we assume $E=1/300$. Thus, at a given moment, somewhat over a dozen loci may be involved in the process of gene substitution.

There are several comments to be made on our results on mutation rate: First, it should be emphasized that we are not sharing the erroneous view, often held by some biologists, that the rate of evolution is directly determined by the mutation rate. On the contrary, we assume that the natural selection determines the rate of evolution, and that the mutation rate is adjusted in the most economical way for the perpetuation of the species. Here intergroup selection may play as important a role as intragroup selection. The adjustment may be achieved through the accumulation of modifiers, which, in itself, is a part of the process of evolution. Secondly, the present result shows, rather unexpectedly, that the mutation rate does not depend on the amount of reproductive waste. Namely, those organisms which deposit a tremendous number of eggs per individual like fruit flies may not necessarily have a higher spontaneous mutation rate than slowly breeding organisms like elephants, when counted per generation. Thirdly, the part of a spontaneous mutation rate which is due to natural radiation is rather unlikely to be easily modified. However, this will not cause us much trouble, since it is known in fruit flies that natural radiation accounts only for about one ten-thousandth (0·0001) of their spontaneous mutations, and in man the corresponding figure is estimated to be about 2% (cf. "The Hazards to Man of Nuclear and Allied Radiations" 1956).

So far we have considered the "strategy" of genes against nonrecurrent or secular environmental changes. What then, would be the strategy of genes against recurrent or cyclic environmental changes ? Here we will consider an extremely simplified situation in which two environments, denoted by $e_1$ and $e_2$, alternate each for an equal length of time. We assume that in $e_1$, genotype $\mathbf{A_1A_1}$ is more advantageous than $\mathbf{A_2A_2}$, while in $e_2$, it is less advantageous than $\mathbf{A_2A_2}$, as shown in Table 3, where $s>0$. Assume further,

Table 3

| Fitnesses \ Genotypes | $\mathbf{A_1A_1}$ | $\mathbf{A_1A_2}$ | $\mathbf{A_2A_2}$ |
|---|---|---|---|
| In environment I ($e_1$) | 0 | $-sh$ | $-s$ |
| In environment II ($e_2$) | $-s$ | $-sh$ | 0 |

for the purpose of simplification, that selection is mild and the gene frequency remains very close to 0·5 throughout the cycle. Then the average fitness of the population is approximately

$$-\frac{s(2h+1)}{4},$$

so the genetic load of a population in each environment with respect to this polymorphic state is

$$L_s = \max\ \{0,\ -sh\} + \frac{s(2h+1)}{4} \tag{27}$$

where max $\{0, -sh\}$ denotes the larger of the two values, 0 or $-sh$. The principle of the minimum genetic load may then be applied to find out what value of $h$ makes $L_S$ a minimum. The required value of $h$ may be obtained as follows: First assume that $h$ is non-negative. Then the right side of (27) becomes

$$\frac{s(2h+1)}{4},$$

which takes its minimum value of $s/4$ when $h=0$. Next assume that $h$ is non-positive so that $h' = -h$ is non-negative. Then the right side of (27) is

$$\frac{s(2h'+1)}{4},$$

which takes its minimum value of $s/4$ when $h'=0$. It is obvious then that $h=0$ is the required solution for minimizing $L_S$. Namely, the dominance should be modified in such a way that $\mathbf{A}_1$ is completely dominant over $\mathbf{A}_2$ in environment $e_1$, and $\mathbf{A}_2$ is completely dominant over $\mathbf{A}_1$ in $e_2$. To put it in another way, the more advantageous allele should become dominant over the less advantageous in each environment. If this is achieved through the accumulation of modifiers, the relative fitnesses of the three genotypes $\mathbf{A}_1\mathbf{A}_1$, $\mathbf{A}_1\mathbf{A}_2$ and $\mathbf{A}_2\mathbf{A}_2$ become respectively $-s/2$, 0 and $-s/2$ when averaged over one cycle of environmental changes. This means a development of overdominance in the averaged condition and the minimum value of $L_S$,

$$s/4$$

represents the segregational load. To be sure, this modification is achieved through the substitution of new alleles in other loci and it involves substitutional load in its process. A similar model of assuming alternative changes of selection intensity in successive generations was considered by Dempster (1955) to account for the maintenance of genetic variability. He rightly pointed out that "Variable selection pressures in successive generations operating in conjunction with an average degree of overdominance could result in variance that is largely additive *within* generations and yet would not provide the possibility of varietal improvement by selection."

In our simplified situation as given in Table 3 and assuming $h=0$, about two-thirds of the genotypic variance within each environment appears as additive variance, though the genotypic variance is entirely non-additive if averaged over two environments.

As to the physiological side of the problem, one might suspect a difficulty in attaining the ideal situation of $h=0$. For a decrease in the value of $h$ toward 0 in $e_1$ may act to increase the value of $h$ toward unity in $e_2$ and vice versa. This may most easily be revealed by substituting $s$ by $-s$ while keeping $h$ constant in the fitness array in $e_1$ to get the new array of 0, $sh$, $s$ or equivalently $-s$, $-s(1-h)$, 0 rather than $-s$, $-sh$, 0 which is assumed for $e_2$ (see Table 3). However, such a difficulty will be overcome in the long run.

The segregational load of $s/4$, which is the minimum value of $L_s$ with respect to $h$, may still be decidedly large. Certainly the next step should be to reduce the value of $s$ itself. This will be achieved again by the adjustment of genotypes which involves substitutional load. Physiologically, reduction of $s$ must be attained by the development of homeostatic devices in organisation. Indeed, physiological homeostasis must be the most efficient means for an organism to escape from the heavy load of segregation.

I should like to conclude this discussion by quoting from Eddington, who wrote in one of his books, "This is a new adventure, and I do not wish to insist on the accuracy or finality of the first attempt. I cannot see how there can be anything seriously wrong with it; but then one never does see these faults until some new circumstance arises or some ingenious person comes along to show us how blind we have been. But there are two kinds of scientific misadventure; we may start off on a false trail altogether, or we may make temporary blunders in following the true path. I am content if in this chapter I can justify my belief that at any rate we are not committing the first error".

## SUMMARY

It is demonstrated that by introducing what may be called the principle of minimum genetic load, the spontaneous mutation rate and the average degree of dominance of deleterious mutant genes may be derived theoretically from the total genetic damage and the rate of substitution of genes in horotelic evolution. The relations connecting these quantities may be expressed by a pair of equations :

$$\left.\begin{aligned} \Sigma\mu &= \frac{0{\cdot}3419E}{\bar{h}}\,(1+1{\cdot}720\bar{h}+\ldots) \\ \bar{h} &= 0{\cdot}6838\sqrt{\frac{E}{2D}}\left(1+1{\cdot}018\sqrt{\frac{E}{2D}}+\ldots\right) \end{aligned}\right\},$$

where $\Sigma\mu$ is the spontaneous mutation rate per gamete per generation, $h$ is the average degree of dominance in fitness of deleterious mutant genes, $D$ is the total mutational damage or approximately the rate of inbreeding depression in fitness per unit increase in the inbreeding coefficient and $E$ is the rate of substitution of genes in horotelic evolution. The above formulae are sufficiently simple to be checked by observational data now available. The present result offers, as a byproduct, a new theory of dominance which can account for the partial dominance in fitness of the normal alleles of "recessive" deleterious genes.

The implication of the principle of minimum genetic load for a cyclical change in environmental condition is also discussed.

The author expresses his thanks to Prof. J. B. S. Haldane who kindly suggested that it was better to solve equations (5) and (6) in terms of power series expansions rather than to rely on the rough approximation (8) which was used in the original manuscript. He also thanks Mr. T. Yasuda for calling his attention to some computational errors.

Thanks are also due to Dr. J. F. Crow for his helpful suggestions and constant encouragement in the course of this work.

# 34 *Mutation rate and dominance*

## REFERENCES


AUERBACH, C. (1956). Genetics in the Atomic Age. Oliver and Boyd, Edinburgh.

CROW, J. F. (1958). Some possibilities for measuring selection intensities in man. *Human Biology* **30:** 1-13.

DEMPSTER, E. R. (1955). Maintenance of genetic heterogeneity. *Cold Spring Harbor Symposia* Vol. **20:** 25-32.

FISHER, R. A. (1930). The Genetical Theory of Natural Selection. Oxford, Clarendon Press.

HALDANE, J. B. S. (1930). A note on Fisher's theory of the origin of dominance, and on a correlation between dominance and linkage. *Amer. Nat.*, **64:** 87-90.

HALDANE, J. B. S. (1937). The effect of variation on fitness. *Amer. Nat.*, **71:** 337-349.

HALDANE, J. B. S. (1954). The measurement of natural selection. *Caryologia 6, Vol. suppl.* pp. 480-487.

HALDANE, J. B. S. (1957). The cost of natural selection. *J. Genet.*, **55:** 511-524.

HALDANE, J. B. S. (1958). The theory of evolution, before and after Bateson. *J. Genet.*, **56:** 11-27.

IVES, P. T. (1950). The importance of mutation rate genes in evolution. *Evolution*, **4:** 236-252.

KIMURA, M. (1958). On the change of population fitness by natural selection. *Heredity*, **12:** 145-167.

MORTON, N. E., J. F. CROW AND H. J. MULLER (1956). An estimate of the mutational damage in man from data on consanguineous marriages. *Proc. Nat. Acad. Sci.*, **42:** 855-863.

MULLER, H. J. (1932). Further studies on the nature and causes of gene mutations. *Proc. 6th Int. Congr. Genet.* (Ithaca), **1:** 213-255.

MULLER, H. J. (1950). Our load of mutations. *Amer. Jour. Hum. Genet.* **2:** 111-176.

MULLER, H. J. (1955). How radiation changes the genetic constitution. *Bull. Atomic Scientists*, **11:** 329-339.

MULLER, H. J. (1956). Further studies bearing on the load of mutations in man. *Acta Genetica et Statistica Medica*, **6:** 157-168.

SIMPSON, G. G. (1944). Tempo and Mode in Evolution. Columbia University Press, New York.

The Hazards to Man of Nuclear and Allied Radiations (1956). Her Majesty's Stationery Office, London.

WADDINGTON, C. H. (1957). The Strategy of the Genes. George Allen and Unwin, London.

WRIGHT, S. (1934). Physiological and evolutionary theories of dominance. *Amer. Nat.*, **68:** 24-53.

# THE MUTATIONAL LOAD WITH EPISTATIC GENE INTERACTIONS IN FITNESS[1]

MOTOO KIMURA AND TAKEO MARUYAMA

*National Institute of Genetics, Mishima, Japan and University of Wisconsin, Madison*

Received July 14, 1966

THE mutational load is defined as the proportion by which the population fitness is decreased through the elimination of recurrent harmful mutations (CROW 1958). In a very large population, where the mutant genes are kept in low frequencies by the balance between mutation and selection, it represents the intensity of natural selection at the genotypic level.

The mutational load in a large population was first calculated by HALDANE (1937) without assuming an epistatic component in fitness. Later, a similar but more detailed calculation was carried out by KIMURA (1961). Also, the mutational load in a small population was studied by KIMURA, MARUYAMA and CROW (1963).

The purpose of the present paper is to investigate the effect of epistasis on the mutational load, using a model which assumes that the fitness is a function of the number of mutant genes in an individual. In particular, we will elaborate the case of quadratic interaction in fitness, namely, the deleterious effect of mutant genes to an individual is given by the quadratic expression of the number of mutant genes. This includes a case where the deleterious effect is proportional to the square of the number of mutant genes. Such a model may be realistic if the phenotypic suppression of mutational damage by developmental homeostasis breaks down rapidly as the number of mutant genes increases.

In what follows, we will assume a very large population and investigate first the case of free recombination among mutant genes. Then, in order to clarify the effect of restricted recombination, we will investigate a population of a hypothetical organism having only one pair of chromosomes within which no crossing over takes place. We will also study the mutational load under asexual reproduction. Finally, these results will be compared with other types of epistasis such as threshold character and "diminishing type" epistasis.

Throughout this paper, the senior author (M. K.) is responsible for the theoretical treatments, while the junior author (T. M.) is responsible for the numerical treatments based on a computer.

1. *Free recombination among mutant genes:* Let us consider a very large random mating population of diploid organism and assume that the fitness of an individual having $i$ mutant genes is given by

$$w_i = 1 - h_1 i - h_2 i^2 , \tag{1.1}$$

[1] Contribution No. 616 from the National Institute of Genetics, Mishima, Shizuoka-ken, Japan. Aided in part by a Grant-in-Aid from the Ministry of Education, Japan, by a Grant from Toyo Rayon Foundation, and by the U.S. Public Health Service (GM07666).

where $h_1$ and $h_2$ are non-negative constants. The right side of the above expression becomes negative for a large $i$. So, we assume that

$$w_i = 0 \quad \text{for} \quad i \geqq n , \tag{1.2}$$

where $n$ is the smallest number of mutant genes for which the right side of (1.1) becomes negative.

Let $f_i$ be the frequency of individuals having $i$ mutant genes before selection, then the average selection coefficient, $h$, against each mutant gene is

$$-h = \sum_i f_i(w_{i+1} - w_i) / \sum_i f_i w_i . \tag{1.3}$$

Under free recombination between mutant genes, $i$ may be distributed with a Poisson distribution,

$$f_i = e^{-\lambda} \frac{\lambda^i}{i!} \tag{1.4}$$

where $\lambda$ is the average number of mutant genes per individual before selection. With this distribution,

$$- \sum_i f_i(w_{i+1} - w_i) = (h_1+h_2)(1-\varepsilon_{n-1}) + 2h_2\lambda(1-\varepsilon_{n-2}) + (\varepsilon_{n-1} - \varepsilon_{n-2}) w_{n-1}$$

and

$$\sum_i f_i w_i = 1 - \varepsilon_n - (h_1+h_2)\lambda(1-\varepsilon_{n-1}) - h_2\lambda^2(1-\varepsilon_{n-2}), \tag{1.5}$$

where $\varepsilon_n$ represents the tail of the Poisson distribution such that

$$\varepsilon_n = \sum_{i=n}^{\infty} \frac{\lambda^i}{i!} e^{-\lambda} .$$

As a function of $n$ and $\lambda$, $\varepsilon_n$ may be evaluated by using "Tables of the Incomplete Γ-Function" by KARL PEARSON (1922), since

$$\varepsilon_n = \gamma(n, \lambda) / \Gamma(n),$$

where $\gamma(n, \lambda)$ is the incomplete gamma function of the first kind, i.e.

$$\gamma(n, \lambda) = \int_0^{\lambda} e^{-t} t^{n-1} dt$$

and $\Gamma(n)$ is the ordinary gamma function. In the present study it turns out that $\varepsilon$'s in (1.5) are generally very small and therefore may be neglected. For example, in the case of $M = 0.1$, $h_1 = 0$ and $h_2 = 0.0025$ (see Table 1), we have $n = 21$ and $\lambda \approx 5.76$, giving

$$\varepsilon_n \approx 9 \times 10^{-7}, \quad \varepsilon_{n-1} \approx 3 \times 10^{-6} \text{ and } \varepsilon_{n-2} \approx 10^{-5}.$$

This means that individuals having a large enough number of mutant genes, for which $1 - h_1 i - h_2 i^2$ becomes negative, are so rare that they may be neglected in the calculation.

Therefore,

$$h = \frac{(h_1+h_2) + 2h_2\lambda}{1 - (h_1+h_2)\lambda - h_2\lambda^2} \tag{1.6}$$

with good approximation.

Consider a particular locus and let $u$ be the mutation rate and $p$ be the frequency of the mutant gene. Assuming that the gene frequency is very low, its rate of change per generation may be given by

$$\frac{dp}{dt} = u - hp , \tag{1.7}$$

where $h$ is the average selection coefficient against the mutant gene. Multiplying both sides of (1.7) by 2 and summing up over all relevant loci, we have

$$\frac{d\lambda}{dt} = 2M - h\lambda \,, \tag{1.8}$$

where $M = \sum u$ is the number of new mutations produced per gamete per generation and $\lambda = \sum 2p$.

At equilibrium, where mutation and selection balance each other, we have $d\lambda/dt = 0$, or

$$2M = h\lambda \tag{1.9}$$

The above relations (1.7), (1.8) and (1.9) are approximations, but they are satisfactory as long as $h$ is much larger than $u$. Note also that (1.9) follows from

$$\frac{dp}{dt} = u(1-p) - hp(1-p)$$

which is a more exact expression than (1.7).

Thus, substituting (1.6) in (1.9), we get the following equation for $\lambda$:

$$2h_2(M+1)\lambda^2 + (2M+1)(h_1+h_2)\lambda - 2M = 0, \tag{1.10}$$

from which we obtain, as the relevant root,

$$\lambda = \frac{-(h_1+h_2)(2M+1) + \sqrt{(2M+1)^2(h_1+h_2)^2 + 16h_2M(M+1)}}{4h_2(M+1)} \,. \tag{1.11}$$

With this $\lambda$, the mutational load may be calculated either from

$$L = (h_1+h_2)\lambda + h_2\lambda^2 \,, \tag{1.12}$$

or, equivalently, from

$$L = \frac{2M + (h_1+h_2)\lambda}{2(M+1)} \,. \tag{1.13}$$

In Table 1, values of $\lambda$ and $L$ which were computed using (1.11) and (1.13)

TABLE 1

*Mutational load under a quadratic gene interaction in fitness with free recombination between mutant genes*

| Case | $M$ | $h_1$ | $h_2$ | $\lambda$ From equation (1.11) | $\lambda$ From numerical analysis | $L$ From equation (1.13) | $L$ From numerical analysis |
|---|---|---|---|---|---|---|---|
| 1 | 0.1000 | 0.00000 | 0.0025 | 5.76 | 5.68 | 0.0975 | 0.100 |
| 2 | 0.1000 | 0.00238 | 0.00238 | 5.66 | 5.57 | 0.103 | 0.106 |
| 3 | 0.1000 | 0.00025 | 0.00249 | 5.75 | 5.66 | 0.0981 | 0.101 |
| 4 | 0.1000 | 0.01666 | 0.00167 | 4.97 | 4.85 | 0.132 | 0.145 |
| 5 | 0.1414 | 0.00000 | 0.0025 | 6.76 | 6.64 | 0.131 | 0.136 |
| 6 | 0.1414 | 0.00238 | 0.00238 | 6.68 | 6.54 | 0.138 | 0.143 |
| 7 | 0.1414 | 0.00025 | 0.00249 | 6.75 | 6.62 | 0.132 | 0.137 |
| 8 | 0.1414 | 0.01666 | 0.00166 | 6.08 | 5.91 | 0.173 | 0.177 |

The average number of mutant genes per individual before selection ($\lambda$) and the mutational load ($L$) are listed for eight different combinations of $h_1$, $h_2$ and $M$, where $M$ is the mutation rate per gamete. In each case, values obtained from the theory are compared with those obtained by purely numerical treatments by a computer.

are listed for various combinations of $M$, $h_1$ and $h_2$. The table also contains, for comparison, the corresponding values of $\lambda$ and $L$ derived from purely numerical treatments with the help of a computer as detailed in Appendix I. Briefly, the set of zygotic frequencies, $\{f_i\}$, in one generation is transformed into that of the next under a given scheme of mutation, selection and recombination and this operation is repeated until an equilibrium is reached. Then $\lambda$ and $L$ are computed from the equilibrium distribution. The table shows good agreement between the results obtained from these two different methods. Also, it suggests an interesting fact that if $h_1$ is zero,

$$L \approx M, \tag{1.14}$$

namely, the mutational load is approximately equal to the mutation rate per gametes, rather than twice this value as expected from the HALDANE-MULLER principle (cf. CROW 1957). The relation (1.14) appears to hold as long as $|h_1|$ is smaller than $h_2$. Actually, if $\lambda >> 1$ and $|h_1| \leqq h_2$, the first order terms are less important than the second order terms in (1.10) and (1.12), thus giving $L = M/(M+1)$, or roughly $L = M$, if $M$ is small. On the other hand, if $h_2 = 0$ (no epistasis), we have $L = 2M/(1+2M)$ or roughly $L = 2M$ if $M$ is small. Namely the mutational load is roughly equal to the total mutation rate per zygote. For example, if $M = 0.05$ and no epistatic interaction in fitness, $L = 0.09$. For a larger value of $M$, the more accurate formula $L = 2M/(1+2M)$ is preferable. For example, if $M = 0.1$, we have $L = 0.167$.

2. *One pair of chromosomes with no crossing over:* In the above treatment, completely free recombination was assumed between mutant genes. In actual situations, however, slight restriction of recombination may occur and even if its effect on the load is negligible, it might be worthwhile to investigate the effect assuming an extreme situation. So, in this section, we will consider a random mating population of a hypothetical organism having only one pair of chromosomes within which no crossing over takes place.

We will denote by $g_i$ the frequency before selection of chromosomes having $i$ mutant genes. Under random mating the frequencies of individuals having various number of mutant genes may be obtained by expanding $(\sum_0^\infty g_i)^2$. As before the selective value of individuals is given by (1.1). Thus the relative selective value, $v_i$, of chromosomes having $i$ mutant genes is

$$v_i = \sum_{j=0}^{\infty} w_{i+j} g_j = 1 - h_1(i+\mu_1') - h_2(i^2 + 2i\mu_1' + \mu_2'), \tag{2.1}$$

where $\mu_1'$ and $\mu_2'$ are the first and the second moments of the distribution of the number of mutant genes in a chromosome, namely,

$$\mu_1' = \sum_i i g_i, \qquad \mu_2' = \sum_i i^2 g_i .$$

Let us consider the process by which the frequency distribution changes from one generation to the next: After selection the frequency of chromosomes having $i$ mutant genes changes from $g_i$ to $(g_i v_i)/\bar{v}$, where $\bar{v}$ is the mean selective value

$$\bar{v} = 1 - 2h_1\mu_1' - 2h_2(\mu_1'^2 + \mu_2'). \tag{2.2}$$

The selection is followed by mutation and we assume, as an approximation, that proportion $M$ of the chromosomes having $i$ mutant genes move to the class having

$i+1$ mutant genes. For a more exact treatment, we should assume that the number of new mutations follow Poisson distribution with mean $M$, but this makes the following treatment much more difficult. However, as will be shown later, the approximation is satisfactory as long as $M$ is small. Thus the frequency of chromosomes in the next generation having $i$ mutant genes is

$$g_0' = \frac{g_0 v_0}{\bar{v}} (1-M) \qquad \text{if } i=0$$

and (2.3)

$$g_i' = \frac{g_i v_i (1-M)}{\bar{v}} + \frac{g_{i-1} v_{i-1} M}{\bar{v}} \quad \text{if } i \geqq 1 .$$

At equilibrium where $g_i' = g_i$, we may drop the primes in the above set of equations. Let $\phi(\theta)$ be the moment generating function of the distribution defined by

$$\phi(\theta) = \sum_{i=0}^{\infty} \theta^i g_i \ . \tag{2.4}$$

Then, (2.3) at equilibrium gives the following second order differential equation for $\phi(\theta)$:

$$\phi''(\theta) + \frac{A}{\theta}\phi'(\theta) - \frac{B}{\theta(1-M+M\theta)}\phi(\theta) = 0, \tag{2.5}$$

where

$$A = (h_1/h_2)+1+2\mu_1' = S+1+2\mu_1'$$

and (2.6)

$$B = Mv_0/h_2 = Rv_0 \ ,$$

in which $S = h_1/h_2$, $R = M/h_2$ and

$$v_0 = 1 - h_1\mu_1' - h_2\mu_2'. \tag{2.7}$$

If the solution, $\phi(\theta)$, of the above differential equation is obtained, we may use it to calculate

$$\mu_1' = \left.\frac{\phi'(\theta)}{\phi(\theta)}\right|_{\theta=1} , \tag{2.8}$$

from which $\mu_1'$ may be obtained. For this purpose, we introduce an approximation and substitute (2.5) by

$$\phi''(\theta) + \frac{A}{\theta}\phi'(\theta) - \frac{B}{\theta}\phi(\theta) = 0, \tag{2.9}$$

namely, we omit the term $(1-M+M\theta)$ in (2.5). This should not cause any serious error as long as $M$ is small and $\theta$ is very near to unity. We note here that the above approximation equation (2.9) has the same form as the exact equation in the continuous treatment given in Appendix (II.4). Thus, the pertinent solution is

$$\phi(\theta) \propto \sum_{i=0}^{\infty} \frac{B^i\theta^i}{i!\Gamma(A+i)} = (B\theta)^{-(A-1)/2} I_{A-1} (2\sqrt{B\theta}), \tag{2.10}$$

where $I(\cdot)$ stands for the modified Bessel function.
Thus,

$$\mu_1' = \left.\frac{\phi'(\theta)}{\phi(\theta)}\right|_{\theta=1} = B^{1/2} \frac{I_A(2B^{1/2})}{I_{A-1}(2B^{1/2})}$$

or, writing $\alpha \equiv A-1 = S+2\mu_1'$ and dividing both side by $B^{1/2}$, we obtain

$$\frac{\alpha-S}{2B^{1/2}} = \frac{I_{\alpha+1}(2B^{1/2})}{I_\alpha(2B^{1/2})} \tag{2.11}$$

The average number of mutant genes per individual is

$$\lambda = 2\mu_1' \tag{2.12}$$

and this can be obtained by solving the above equation (2.11) for $\alpha$ through iteration by using tables of the modified Bessel functions (cf. SHIBAGAKI 1955). The iteration may be carried out as follows. First, take $v_0 = 1$ so that $B = R = M/h_2$ and solve (2.11) for $\alpha$, from which we get the first approximation of $\lambda(=\alpha-S)$ and $\mu_1'(=\lambda/2)$. Use this $\mu_1'$ to calculate $\mu_2'$ from

$$\mu_2' = \frac{R-S(1+M)\mu_1'-2\mu_1'^2}{1+M} \tag{2.13}$$

which is derived from the first equation in (2.3), namely from $\bar{v} = v_0(1-M)$. With these values of $\mu_1'$ and $\mu_2'$, the values of $v_0$ and therefore $B = Rv_0$ may be obtained, enabling us to start the second cycle of calculation, from which we get the better approximations of $\lambda$ and $\mu_1'$ by using equation (2.11). The process may be repeated until the desired accuracy is reached. Usually two cycles of iteration were sufficient for our purpose.

The mutational load is calculated from

$$L = 1 - \bar{w} = 2h_1\mu_1' + 2h_2(\mu_1' + \mu_2'). \tag{2.14}$$

In Table 2, values of $\lambda$ and $L$ thus obtained are listed for five different cases, together with corresponding values of $\lambda$ and $L$ obtained by the purely numerical treatment by a computer (see Appendix I). The table shows fairly good agreements between the values obtained by the two different methods. For all these five cases, the mutation rate per gamete ($M$) is 0.1. The load is roughly 0.13 for

TABLE 2

*The mutational load under no crossing over in an organism having one pair of chromosomes*

| | | | | $\lambda$ | | $L$ | |
|---|---|---|---|---|---|---|---|
| Case | $M$ | $h_1$ | $h_2$ | From theory, eq. (2.11) | From computer | From theory, eq. (2.14) | From computer |
| 1 | 0.1 | 0.00000 | 0.0100 | 3.31 | 3.25 | 0.137 | 0.130 |
| 2 | 0.1 | 0.00000 | 0.0025 | 6.92 | 6.76 | 0.133 | 0.127 |
| 3 | 0.1 | 0.00238 | 0.00238 | 6.78 | 6.60 | 0.137 | 0.131 |
| 4 | 0.1 | 0.00025 | 0.00249 | 6.90 | 6.74 | 0.133 | 0.127 |
| 5 | 0.1 | 0.01666 | 0.00167 | 5.75 | 5.57 | 0.159 | 0.152 |

The average number of mutant genes per individual before selection ($\lambda$) and the load ($L$) are listed for five different combinations of values in $h_1$ and $h_2$, assuming mutation rate $M=0.1$. In each case, values obtained from the theory are compared with those obtained by the purely numerical treatment by a computer.

the first four cases in which $h_1 \leqq h_2$. In the fifth case, $h_1 >> h_2$ and the load is larger, being roughly 0.15.

3. *Asexual reproduction:* In the previous section we have considered a random mating population of an organism having only one pair of chromosomes within which no crossing over takes place. In this case, recombination still occurs in the sense that different chromosomes recombine through fertilization. We now proceed to investigate a population of asexually reproducing oragnism, where no recombination takes place.

Let $f_i$ be the frequency before selection of individuals having $i$ mutant genes, whose fitness is $w_i$ and let $2M$ be the average number of mutant genes produced per individuals per generation.

If the number of new mutations follows Poisson distribution with mean $2M$, then the frequency of individuals having $i$ mutant genes in the next generation is

$$f_i' = \sum_{j=0}^{i} \frac{w_{i-j} f_{i-j}}{\bar{w}} \frac{(2M)^j}{j!} e^{-2M} \tag{3.1}$$

where

$$\bar{w} = \sum_{i=0}^{\infty} f_i w_i \, .$$

The above relation (3.1) is quite general and no restriction is given to the form of $w_i$'s.

In particular, the frequency of individuals having no mutant genes is

$$f_0' = \frac{w_0 f_0}{\bar{w}} e^{-2M} \, . \tag{3.2}$$

Thus, at equilibrium in which $f_0' = f_0$, we have

$$\bar{w} = w_0 e^{-2M} \, . \tag{3.3}$$

So, if we assume that mutations are deleterious and the individuals with no mutant genes have the highest fitness, then the mutational load is

$$L = \frac{w_0 - \bar{w}}{w_0} = 1 - e^{-2M} \tag{3.4}$$

For example, if $M = 0.1$ we have $L = 0.181$.
For a smaller $M$, we have

$$L = 2M$$

approximately. Thus the load is roughly equal to the total mutation rate per individual. The above treatment shows that under asexual reproduction, epistasis has no effect in reducing the mutational load.

## DISCUSSION

The model employed in the present paper assumes that the fitness is a function of the number of mutant genes contained in an individual. Such a model may be useful to investigate the following two situations. (1) Mutant genes in each locus are sufficiently rare so that only heterozygotes need to be considered. The selective elimination is mainly through their deleterious effect in heterozygous condition. (2) Mutant genes are mildly deleterious in homozygous as well as heterozygous states and they are semidominant in fitness, namely, for each mutant gene, the heterozygous condition is only half as deleterious as the homozygous one. This seems to be nearly the case in "viability polygenes" studied by MUKAI (1965a).

The foregoing treatments have shown that under random mating the mutational load is nearly half the total mutation rate per individual if the deleterious effect of the mutant genes to an individual is roughly proportional to the square of its number, unless the mean number of mutant genes per individual is very small. This result may be particularly pertinent for assessing the load due to the viability polygenes, whose total mutation rate is estimated to be at least 70% per individual (cf. MUKAI 1964). If the ordinary HALDANE-MULLER principle (cf. CROW 1957) is applied to such genes, the load becomes at least about 0.5, which may be too high even for Drosophila.

The term mutational load has still been used by some to mean undesirable genes or gene complexes produced by mutation, but we use this strictly in the sense defined by CROW (1958). Namely, the proportional decrease of population fitness through the elimination of recurrent harmful mutations. The elimination may either be carried out by premature death or by sterility of the carriers and it is convenient to call such elimination the genetic death (MULLER 1950). Then, if one mutant gene is eliminated through one genetic death, the proportion of genetic deaths within an equilibrium population should be equal to the mutation rate per individual. In this case, the intensity of natural selection as expressed by the fraction of genetic deaths is equal to the mutation rate per individual but independent of the selection coefficient against individual mutant genes. Such a principle does not hold under epistatic interaction in fitness, but it does suggest that if two mutant genes are eliminated on the average through one genetic death, the mutational load becomes only half as large as the above. The present quadratic model appears to correspond to this latter situation. More generally, if $m$ mutant genes are eliminated through one genetic death, the mutational load may become only $1/m$ as large. This might be expected if the deleterious effect

TABLE 3

*The average number* ($\lambda$) *of mutant genes per individual before selection and the mutational load* (L) *when the viability is a threshold character. Mutant genes act as lethal when* m *or more are present in one individual. The mutation rate per gamete is assumed to be 0.1*

| $m$ | $\lambda$ | $L$ |
|---|---|---|
| 1 | 0.200 | 0.181 |
| 2 | 0.558 | 0.108 |
| 3 | 1.00 | 0.0803 |
| 4 | 1.49 | 0.0654 |
| 5 | 2.03 | 0.0556 |
| 10 | 5.07 | 0.0342 |

is proportional to the $m$th power of the mutant genes and if the average number of mutant genes per individual is fairly large.

This also might be the case, if viability is a threshold character such that the mutant genes have no deleterious effect when its number is less than $m$, but produce lethal effect when $m$ or more are present in one individual. In this case, (1.3) gives

$$h = \frac{f_{m-1}}{1-\varepsilon_m} \,, \tag{4.1}$$

where

$$f_{m-1} = e^{-\lambda} \frac{\lambda^{m-1}}{(m-1)!}$$

and

$$\varepsilon_m = \sum_{i=m}^{\infty} \frac{\lambda^i}{i!} e^{-\lambda} \,.$$

The mean number of mutant genes ($\lambda$) may be obtained by solving (1.9), with $h$ given by (4.1), that is, by solving

$$2M = \frac{\lambda f_{m-1}}{1-\varepsilon_m} \,. \tag{4.2}$$

With this value of $\lambda$, the mutation load is

$$L = \varepsilon_m \,. \tag{4.3}$$

Table 3 lists $\lambda$ and $L$ for several values of $m$ when $M = 0.1$. The table shows that $L$ becomes progressively small as $m$ increases, even though the relation $L = 2M/m$ does not hold as might be expected. The above model is rather artificial in assuming that the character is solely determined genetically. Rather, it is likely that most threshold traits are strongly influenced by the environment. The following is a model suggested to us by DR. J. F. CROW: Suppose that, among individuals having no mutant genes, a character ($x$) is distributed normally with mean 0 and variance $\sigma^2$, while for those having $i$ mutant genes, $x$ is distributed normally with mean $-i\alpha\sigma$, ($\alpha>0$), and variance $\sigma^2$. The threshold is $c\sigma$ such that only those having $x$ value larger than $c\sigma$ survive. Then the fraction of survivors among those having $i$ mutant genes is

$$W_i = \frac{1}{\sqrt{2\pi}} \int_{c+i\alpha}^{\infty} e^{-x^2/2}\, dx \,.$$

Let $w_i$ be the relative fitness so that $w_i = W_i/W_0$. If $n$ loci are involved and if $p$ is the frequency of a mutant gene in each locus, then the frequency of individuals having $i$ mutant genes is

$$f_i = \binom{n}{i} (1-p)^{n-i} p^i, \qquad (i = 0, 1, 2, \dots).$$

Thus,

$$\bar{w} = (1-p)^n + nw_1(1-p)^{n-1}p + \dots \,,$$

$$-h = \{(w_1-1)(1-p)^n + (w_2-w_1)n(1-p)^n p + \ldots\}/\bar{w}\,.$$

If $p$ is small, the higher order terms in $p$ are negligible and we can compute the equilibrium frequency $p$ using the relation $u = hp$.

For example, assuming haploid organism with $n = 4$, $c = -1.5$, $\alpha = 0.05$, $u = 10^{-5}$, we get

$$p = 1.388 \times 10^{-3}$$

$$L = 3.98 \times 10^{-5},$$

showing that the load is very near to $4u$, as expected from the HALDANE-MULLER principle. Table 4 lists results of more exact calculations for several cases.

The good agreement of these examples with the HALDANE-MULLER principle is mainly due to the fact that the individuals having two or more mutant genes are so rare that there is almost no room for epistatic interaction. On the other hand, KING (1966) presented examples of threshold character in which there is a substantial departure from the HALDANE-MULLER principle. He assumed a very high threshold, large effect of a single gene, rather many loci, and an environmental component of variance of the same order of magnitude as the genetic component.

So far, we have considered a type of epistasis in which deleterious effect becomes disproportionately large as the number of mutant genes increases. Such an epistasis may be called the "reinforcing type." On the other hand, under some circumstances, deleterious effect per mutant gene might become progressively small as the number of mutant genes increases. Such an epistasis may be called the "diminishing type." The balance between mutation and selection in this type of epistasis will be somewhat delicate, because the selection against individual mutant genes becomes less intense as their frequencies increase. In

TABLE 4

*Mutational load in a haploid organism for a threshold trait determined by four segregating loci. The mutation rate per locus is $10^{-5}$*

| Threshold $c$ | Gene effect $\alpha$ | Gene frequency $p$ | Load | Fraction of survivors $W_0$ |
|---|---|---|---|---|
| −1.5 | 0.05 | $1.39 \times 10^{-3}$ | $3.999 \times 10^{-5}$ | 0.9332 |
| | 0.1 | $6.66 \times 10^{-4}$ | $3.999 \times 10^{-5}$ | |
| | 0.2 | $3.11 \times 10^{-4}$ | $3.999 \times 10^{-5}$ | |
| | 0.3 | $1.93 \times 10^{-4}$ | $4.000 \times 10^{-5}$ | |
| −2.0 | 0.05 | $3.486 \times 10^{-3}$ | $4.00 \times 10^{-5}$ | 0.9772 |
| | 0.10 | $1.65 \times 10^{-3}$ | $4.00 \times 10^{-5}$ | |
| | 0.20 | $7.45 \times 10^{-4}$ | $4.00 \times 10^{-5}$ | |
| | 0.30 | $4.45 \times 10^{-4}$ | $4.00 \times 10^{-5}$ | |
| −3.0 | 0.05 | $4.08 \times 10^{-2}$ | $3.96 \times 10^{-5}$ | 0.9982 |
| | 0.10 | $1.89 \times 10^{-2}$ | $3.96 \times 10^{-5}$ | |
| | 0.20 | $8.17 \times 10^{-3}$ | $3.96 \times 10^{-5}$ | |
| | 0.30 | $4.63 \times 10^{-3}$ | $3.97 \times 10^{-5}$ | |

order to treat this case more quantitatively, let us suppose that the fitness of an individual having $i$ mutant genes is given by

$$w_i = 1 - s\left(\frac{i}{i+1}\right), \tag{4.4}$$

where $0 \leqq s \leqq 1$. Note that $w_1 = 1 - s/2$ and $w_\infty = 1 - s$. Under random mating and free recombination among mutant genes, we obtain, by using (1.3) and (1.4), the following expression for the mean selection coefficient.

$$h = \frac{s}{\lambda} \cdot \frac{1-(1+\lambda)e^{-\lambda}}{\lambda - s(\lambda - 1 + e^{-\lambda})} \tag{4.5}$$

With this $h$, the value of $\lambda$ at equilibrium may be obtained from

$$\frac{d\lambda}{dt} = 2M - h\lambda = 0. \tag{4.6}$$

The equilibrium is stable, if

$$\frac{1}{d\lambda}\left(\frac{d\lambda}{dt}\right) < 0 \tag{4.7}$$

and unstable, if

$$\frac{1}{d\lambda}\left(\frac{d\lambda}{dt}\right) > 0\,. \tag{4.8}$$

Table 5 gives some numerical results obtained by assuming $M = 0.1$. For $s = 1.0$, the mean number of mutant genes per individual at equilibrium is 0.431 with load $L = 0.188$. The equilibrium is stable. For $s = 0.5$, there are two equilibrium values of $\lambda$, one ($\lambda = 1.66$) is stable and the other ($\lambda = 3.16$) is unstable. The mutational load corresponding to the stable equilibrium is $L = 0.256$ and this is definitely larger than the total mutation rate per individual, that is, $2M = 0.2$. For a slightly smaller value of $s$, that is for $s = 0.4871$, we have again two equilibrium values of $\lambda$; $\lambda = 2.00$ (stable) and $\lambda = 2.60$ (unstable). The load corresponding to the stable equilibrium is $L = 0.276$, which is still larger than in the case of $s = 0.5$. For values of $s$ less than about 0.4845, no equilibrium exists in $\lambda$, because $d\lambda/dt = 2M - h\lambda > 0$ for all values of $\lambda$ and $\lambda$ tends to increase indefinitely. This means that for $s < 0.4845$ the selection can not check the spread of mutant genes.

These results suggest that the diminishing type epistasis among mutant genes

TABLE 5

*Some numerical results for $\lambda$ (mean number of mutant genes per individual) and L (mutational load) in the case of "diminishing type" epistasis, in which the fitness is given by $w_i = 1 - si/(i+1)$. The mutation rate per gamete is assumed to be 0.1*

| $s$ | $\lambda$ Stable | $\lambda$ Unstable | $L$ |
|---|---|---|---|
| 1.00 | 0.431 | . . . | 0.188 |
| 0.50 | 1.66 | 3.16 | 0.256 |
| 0.4871 | 2.00 | 2.60 | 0.276 |
| <0.4845 | . . . | . . . | . . . . |

tend to create much larger genetic loads (both expressed and hidden) than in the case of no epistasis and probably this type of epistasis is unfavourable for the evolution of the species. On the other hand, the reinforcing type epistasis tend to reduce the load and appear to be favourable for evolution. Also, this type of epistasis must be more common because of the physiological reason, namely the developmental homeostasis breaks down rapidly as mutant genes accumulate. Therefore, as far as the effect of deleterious mutant genes on fitness is concerned the reinforcing type epistasis will be found more often than the diminishing type in nature.

Recent studies of MATSUDAIRA (1963), SPASSKY *et al.* (1965), MUKAI (1965b) and KITAGAWA (1966) seem to support such a view. In the last study, up to seven lethal genes were accumulated experimentally in one individual to see their effect on fitness in heterozygous condition. It was found that their deleterious effects tend to reinforce each other as their number increases.

We would like to express our thanks to DR. J. F. CROW for reading the manuscript and for giving valuable suggestions, especially those pertaining to the load for the threshold character.

## SUMMARY

The effect of epistasis on the mutational load was studied using a model which assumes that fitness is a function of the number of mutant genes in an individual. It was shown that if the deleterious effect of the mutant genes is proportional to the square of their number, the load under random mating becomes roughly half as large as in the case of no epistasis, provided that the average number of such genes per individual is fairly large. Under asexual reproduction, however, the epistasis has no effect in reducing the load. The situation is intermediate for a random mating population of a hypothetical organism having only one pair of chromosomes within which no crossing over takes place.—The mutational load is also reduced under random mating if the fitness is a threshold character such that the mutant genes produce no deleterious effect when their number is less than $m$ but act as lethals when $m$ or more of them are present in one individual.—Epistatic interaction in fitness among deleterious mutant genes is classified into two types, namely, the reinforcing type and the diminishing type. In the former, the deleterious effect becomes disproportionately large as their number in an individual increases. On the other hand, in the latter, the deleterious effect per mutant gene becomes smaller as their number increases.—Reasons are presented to believe that the reinforcing type of epistasis among deleterious mutant genes must be more common than the diminishing type in nature.

## APPENDIX I. NUMERICAL TREATMENT OF THE QUADRATIC GENE INTERACTION MODEL WITH A COMPUTER

The process of calculation consists of the following steps: (1) As a starting point, a population of individuals having no mutant genes is assumed so that $f_0 = 1.0$, $f_1 = f_2 = \ldots = 0$. When gametes are formed, segregation takes place before mutation, so that from this initial population, $g_0 = 1$, $g_1 = g_2 = \ldots = 0$. (2) When mutation occurs, mutant genes are produced with their frequencies given by the Poisson distribution with mean $M$, say with $M = 0.1$, such that from the above population the frequencies of gametes carrying 0,1,2, ... mutant genes are $e^{-0.1}$, $(0.1)e^{-0.1}$, $(0.1)^2e^{-0.1}/2!$, ... respectively. More generally, if $g_0$, $g_1$, $g_2$, ... are repsectively the frequencies of gametes carrying 0,1,2, ... mutant genes before the production of new mutations, the gametic frequencies after mutation ($g_0'$, $g_1'$, $g_2'$, ...) are given by

$$\begin{bmatrix} g_0' \\ g_1' \\ g_2' \\ \cdot \\ \cdot \\ \cdot \\ \cdot \end{bmatrix} = e^{-M} \begin{bmatrix} 1 & 0 & 0 & \cdots \\ M & 1 & 0 & \cdots \\ \frac{M^2}{2!} & M & 1 & \cdots \\ \cdot & \cdot & \cdot & \\ \cdot & \cdot & \cdot & \\ \cdot & \cdot & \cdot & \\ \cdot & \cdot & \cdot & \end{bmatrix} \begin{bmatrix} g_0 \\ g_1 \\ g_2 \\ \cdot \\ \cdot \\ \cdot \\ \cdot \end{bmatrix} \qquad \text{(I.1)}$$

(3) Zygotes are formed assuming random mating, such that $f_0=g_0'^2$, $f_1=2g_0'g_1'$, etc. (4) Selection is practiced with fitness $w_i=1-h_1 i-h_2i^2$ for a given set of values of $h_1$ and $h_2$, for example, $h_1=0.0$, and $h_2=0.0025$. This transforms $\{f_i\}$ into $\{f_iw_i/\bar{w}\}$. (5) Segregation takes place when the gametes are formed, but, the mode of segregation depends on whether crossing over occurs or not. In the case of free recombination between mutant genes, segregation follows binomial distribution in such a way that an individual having $i$ mutant genes produces gametes with various number of mutant genes following the expansion of $(\frac{1}{2} + \frac{1}{2})^i$. In the case of no crossing over, however, homologous chromosomes which formed an individual at fertilization again segregate intact. (6) Mutation follows segregation as described in (2).

The cycle of "zygote formation-selection-segregation-mutation in gametes-zygote formation" is repeated many times until equilibrium is reached with respect to the frequency distribution $\{f_i\}$, and this was carried out using the computer, CDC 1604. Usually 120 cycles (generations) of iteration seemed to be sufficient for our purpose, but sometimes the computations were carried out as many as 500 cycles. When equilibrium is reached, the difference in the average number of mutant genes before and after selection must be equal to twice the number of new mutation per gametes, that is $2M$, and this was checked in all cases.

APPENDIX II. ONE PAIR OF CHROMOSOMES WITH NO CROSSING OVER. TIME CONTINUOUS TREATMENT.

Consider a random mating population of an organism having only one pair of chromosomes within which no crossing over takes place. We will denote by $g_i$ the frequency of chromosomes having $i$ mutant genes. Let $M\Delta t$ be the probability that one mutation occurs in a chromosome during the short time interval of length $\Delta t$, i.e. $(t, t+\Delta t)$, and let $w_{i+j}=1-[h_1(i+j)+h_2(i+j)^2]\Delta t$ be the fitness of an individual having $(i+j)$ mutant genes, the fitness being measured during the same time interval, $(t, t+\Delta t)$.

Assuming that the combination of homologous chromosomes occur at random, the relative fitness of chromosomes having $i$ mutant genes is

$$v_i=1-[h_1(i+\mu_1')+h_2(i^2+2i\mu_1'+\mu_2')]\Delta t,$$

where $\mu_1'$ and $\mu_2'$ are respectively the first and second moment around 0 of the number of mutant genes in a chromosome. The amount of change in $g_i$ during the time interval $(t, t+\Delta t)$ is

$$\Delta g_0 = \frac{g_0v_0(1-M\Delta t)}{\bar{v}} - g_0 \qquad \text{for } i=0,$$

and (II.1)

$$\Delta g_i = \frac{g_iv_i(1-M\Delta t)}{\bar{v}} + \frac{g_{i-1}v_{i-1}M\Delta t}{\bar{v}} - g_i \qquad \text{for } i\geqq 1,$$

where

$$\bar{v}=1-2[h_1\mu_1'+h_2(\mu_1'^2+\mu_2')]\Delta t.$$

At the limit of $\Delta t \to 0$, the above set of equations reduces to

$$\frac{dg_0}{dt} = g_0\{h_1\mu_1' + h_2(\mu_2' + 2\mu_1'^2) - M\}$$

and

$$\frac{dg_i}{dt} = g_i\{h_1(\mu_1' - i) + h_2(\mu_2' + 2\mu_1'^2 - i^2 - 2i\mu_1') - M\} + Mg_{i-1} . \tag{II.2}$$

Let $\sum_{i=0}^{\infty} \theta^i g_i = \phi(\theta)$, $M/h_2 = R$ and $h_1/h_2 = S$, then the above set of equations yield

$$\frac{1}{h_2}\frac{\partial\phi(\theta)}{\partial t} = -\theta^2\frac{\partial^2\phi(\theta)}{\partial\theta^2} - (S+1+2\mu_1')\theta\frac{\partial\phi(\theta)}{\partial\theta} + R\theta\phi(\theta) \tag{II.3}$$

When the equilibrium is reached with respect to the frequency distribution $\{g_i\}$, $\partial\phi/\partial t = 0$ and we have the following ordinary differential equation,

$$\theta\phi''(\theta) + A\phi'(\theta) - R\phi(\theta) = 0, \tag{II.4}$$

where $A = S+1+2\mu_1'$.

The pertinent solution of the above equation is

$$\phi(\theta) = C\sum_{i=0}^{\infty}\frac{R^i\theta^i}{i!\,\Gamma(A+i)} = C(R\theta)^{(1-A)/2} I_{A-1}(2\sqrt{R\theta}), \tag{II.5}$$

where $C$ is a constant such that $\phi(1) = 1$, and $I(\cdot)$ stands for the modified Bessel function. Thus the equation for $\mu_1'$ is

$$\mu_1' = \left.\frac{\phi'(\theta)}{\phi(\theta)}\right|_{\theta=1} = R^{1/2}\frac{I_A(2R^{1/2})}{I_{A-1}(2R^{1/2})} \tag{II.6}$$

or

$$\frac{A\text{-}S\text{-}1}{2R^{1/2}} = \frac{I_A(2R^{1/2})}{I_{A-1}(2R^{1/2})}, \tag{II.7}$$

where $R = M/h_2$ and $S = h_1/h_2$. Equation (II.7) may be solved numerically for $A$ by using tables of the modified Bessel functions and the average number of mutant genes per individual is then obtained from

$$\lambda = 2\mu_1' = A\text{-}S\text{-}1.$$

The mutational load per short time interval $\Delta t$ is

$$L = \frac{1-\bar{w}}{\Delta t} = 2[h_1\mu_1' + h_2(\mu_1'^2 + \mu_2')] ,$$

which, at equilibrium, reduces to

$$L = 2M - \left(\frac{h_2}{2}\right)\lambda^2 .$$

## LITERATURE CITED


CROW, J. F., 1957 Possible consequences of an increased mutation rate. Eugenics Quarterly **4:** 67–80. —— 1958 Some possibilities for measuring selection intensities in man. Human Biology **30:** 1–13.

HALDANE, J. B. S., 1937 The effect of variation on fitness. Am. Naturalist **71:** 337–349.

KIMURA, M., 1961 Some calculations on the mutational load. Japan. J. Genet. **36** (Suppl.): 179–190.

KIMURA, M., T. MARUYAMA, and J. F. CROW, 1963 The mutation load in small populations. Genetics **48:** 1303–1312.

KING, J. L., 1966 The gene interaction component of the genetic load. Genetics **53:** 403–413.

KITAGAWA, O., 1966 Epistatic interaction in fitness between lethal genes in heterozygous condition. (in preparation).

MATSUDAIRA, Y., 1963 On the interaction between the chromosomes carrying the detrimental genes or gene blocks in viability. J. Radiation Research **4:** 111–119.

MUKAI, T., 1964 The genetic structure of natural populations of *Drosophila melanogaster*. I. Spontaneous mutation rate of polygenes controlling viability. Genetics **50:** 1–19. —— 1965a Probable factors suppressing the manifestation of overdominance in natural populations of *Drosophila melanogaster*. (Abstr.) Genetics **52:** 460–461. —— 1965b Synergistic interaction between spontaneous mutant polygenes controlling viability in *Drosophila melanogaster*. Ann. Rep. Nat. Inst. Genet. **15:** 28–29.

MULLER, H. J., 1950 Our load of mutations. Am. J. Human Genet. **2:** 111–176.

PEARSON, K., 1922 *Tables of the Incomplete Γ-Function*. Cambridge University Press.

SIBAGAKI, W., 1955 *0.01% Tables of Modified Bessel Functions*. Baifukan, Tokyo.

SPASSKY, B., TH. DOBZHANSKY, and W. W. ANDERSON, 1965 Genetics of natural populations. XXXVI. Epistatic interactions of the components of the genetic load in *Drosophila pseudoobscura*. Genetics. **52:** 653–664.

# On the evolutionary adjustment of spontaneous mutation rates*

MOTOO KIMURA

*National Institute of Genetics, Mishima, Japan*

(*Received* 10 *May* 1966)

## 1. INTRODUCTION

There seems to be enough evidence to show that mutation rates are under genetic control (UN report 1958, Crow 1959). Thus the mutation rate characteristic of each species must be a product of past evolution like any other morphological or physiological character. The purpose of the present paper is to consider evolutionary factors which influence mutation rates through natural selection and to discuss mechanisms by which spontaneous mutation rates are adjusted in the course of evolution.

## 2. SELECTION AGAINST HIGHER MUTATION RATES

Since mutation is a random process, the majority of mutant genes must be deleterious to any organism in any environment. In other words, the genetic blue print of an organism will more often be impaired than improved by randomly modifying its drawing. Thus, on the average, a higher mutation rate is deleterious in so far as short term effects on future generations are concerned. This means that a modifier which enhances the mutation rates of other genes will be selected against through intra-group selection.

To illustrate this point quantitatively, let us consider a very large, random mating population and denote by $B$ a modifier which enhances the mutation rate of another locus, say, a locus containing mostly the wild-type gene $A$. Let us assume that with a single dose of $B$, the mutation rate from $A$ to its allele $a$ is increased by $\Delta\mu$, and that mutant allele $a$ in single dose decreases the fitness of an individual by $sh$, where $s$ is the selection coefficient against the mutant homozygote and $h$ is the degree of dominance. We will calculate the selection coefficient, $k$, against modifier $B$ by calculating its total number of descendants, $T$, when single $B$ appears in a population where allele $b$ is predominant. We will assume that the mutation from $A$ to $a$ is induced by $B$ when gametes are formed from an individual

* Contribution No. 615 from the National Institute of Genetics, Mishima, Shizuoka-ken, Japan. Aided in part by a Grant-in-Aid from the ministry of Education, Japan, and also, by a Grant from Toyo Rayon Foundation.

containing $B$. The rationale of this method is that $T$ is given by

$$\begin{aligned} T &= 1+(1-k)+(1-k)^2+ \dots \\ &= 1/k \qquad \text{if } 0 < k \leqq 1 \end{aligned}$$

Let $T$ be the total number of descendants of a gene $B$ (including the parental gene itself) over all the subsequent generations when in the initial generation $A$ is on the same chromosome, and let $R$ be that of $B$ when, in the initial generation, $a$ is on the same chromosome. Disregarding the ground mutation rate $\mu$, we obtain

$$\left.\begin{aligned} T &= 1+(1-\Delta\mu)T+\Delta\mu . R \\ R &= (1-sh)\{1+r(1-\Delta\mu)T+[1-r(1-\Delta\mu)]\,R\} \end{aligned}\right\} \tag{1}$$

The first relation is derived by considering the immediate offspring of a chromosome of type $AB$: Among the offspring chromosomes containing $B$, proportion $1-\Delta\mu$ are

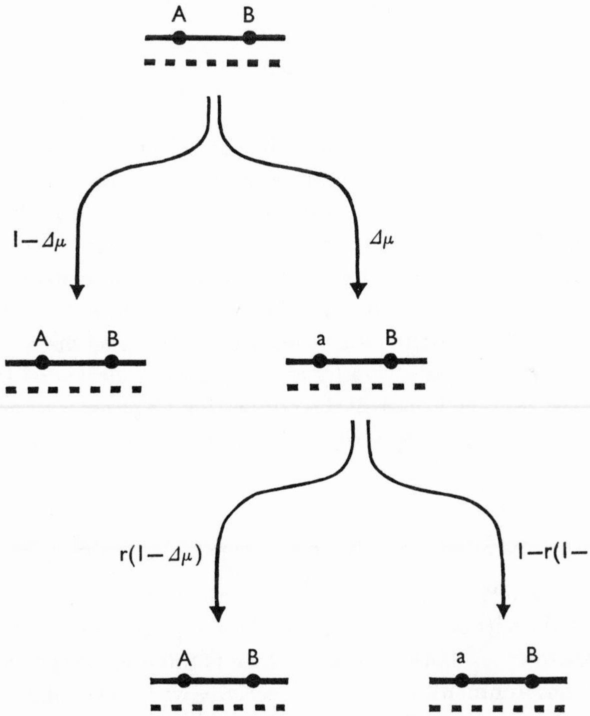


Fig. 1. An explanatory figure for the derivation of the set of equations (1). Broken lines represent chromosome $Ab$.

of type $AB$ while proportion $\Delta\mu$ are of $aB$ (see Fig. 1). The total number of descendants from each of these chromosomes are $T$ and $R$ respectively. Similarly, the second relation is derived by considering the offspring of a chromosome of type $aB$, whose fitness is lower by $sh$ as compared with $AB$: Among the offspring chromosomes containing $B$, proportion $r$ have lost $a$ due to crossing-over, while $1-r$ retain it.

However, the recombination fraction $r$ has to be multiplied by $(1-\Delta\mu)$ because of the effect of induced mutation. By solving the above set of equations (1), we obtain

$$T = \frac{1-(1-sh)(1-r)(1-\Delta\mu)}{sh\Delta\mu}$$

Since $T$ is equal to $1/k$, we can obtain the selection coefficient against $B$ as its reciprocal, i.e.

$$k = \frac{sh\Delta\mu}{1-(1-sh)(1-r)(1-\Delta\mu)} \qquad (2)$$

In the above derivation, the frequency of the mutant gene $a$ is assumed to be so low that no $aa$ individuals need to be considered. The above formula shows that $k$ becomes larger as $r$ becomes smaller. When $\mathrm{r}=0$, it reaches its maximum value and is roughly $\Delta\mu$ if $\Delta\mu$ is much smaller than $sh$. If $B$ enhances the mutation rates of many other loci, the actual disadvantage is the sum $(\Sigma)$ of the above expression (2) over all relevant loci. For $\mathrm{r}=0{\cdot}5$, we have

$$k = \sum \frac{2sh\Delta\mu}{1+sh+\Delta\mu(1-sh)} \qquad (3)$$

If both $\Delta\mu$ and $sh$ are much smaller than unity, this reduces to $k=\sum 2sh\Delta\mu$, which is equivalent to the result obtained by Crow (lecture note on population genetics). For example, if $\Delta\mu=10^{-5}$, $sh=0{\cdot}05$ and $10^4$ loci are modified, the selection coefficient is about 1%. So the selection against a modifier which increases the mutation rate of many other loci will be fairly strong. On the other hand, a modifier which decreases the mutation rate of other loci will be favoured.

A modifier which enhances the mutation rate of other loci may also be selected against through inter-group selection by creating extra mutational load, that is to say, by lowering the average fitness of the population. The amount of loss in fitness due to modifier $B$ is $\sum 2\Delta\mu . p$ approximately, where $p$ is the frequency of $B$ in the population.

### 3. SELECTION AGAINST LOWER MUTATION RATES

Evolution is ultimately dependent on the production of favourable mutations. A species can not do without mutations in the long run, even if the majority of mutations are deleterious. If the mutation rate is too low, the species will not be able to cope with environmental changes, especially with the evolutionary progress of competing species, and it will ultimately become extinct. Therefore, through inter-group selection, genotypes having mutation rates that are too low will be selected against in the course of evolution. This process can not proceed as rapidly as the one by which mutation rates are reduced through intra-group selection, yet when we consider the evolution of organisms over a very long period, it may be effective. The situation is analogous to the evolution of sexuality. As first pointed out by Muller (cf. Muller, 1932, 1958) and later somewhat elaborated by Crow and the present author (Crow and Kimura, 1965), the essential advantage of sexuality consists in its ability to combine *simultaneously* advantageous mutations in one individual through recombination and thereby to speed up the evolutionary progress tremendously. Sexuality is a character that could be interpreted as evolved for the specific rather than for the individual advantage (Fisher, 1930). Since the very advantage of sexuality presupposes the simultaneous occurrence of advantageous mutations within a population, the evolution of sexuality and the evolutionary adjustment for higher mutation rates should be closely related.

## 4. OPTIMUM MUTATION RATES

The above considerations may suggest that there is an optimum mutation rate for the survival of a species depending on how rapidly the environment changes. Those organisms which have managed to survive up to the present may be those which happened to have their mutation rate near the optimum level. In order to treat this idea quantitatively, the author proposed (Kimura, 1960) what he called the principle of minimum genetic load. This is a hypothesis that, in the course of evolution, important genetic parameters of each species tend to be adjusted in such a way that the total genetic load will be minimized.

More specifically, it was assumed that in the course of evolution, the mutation rate and degree of dominance of mutant genes are adjusted such that

$$L = L_m + L_e \tag{4}$$

will be minimized. Such adjustment may be achieved through intergroup selection.

In the above expression $L_m$ and $L_e$ respectively stand for the mutational and substitution load (cf. Kimura, 1960): The mutational load is the load or the decrease of population fitness due to the elimination of recurrent harmful mutations. It has been shown by Haldane (1937) and Muller (1950*a*) that this quantity does not depend on the selection coefficients against mutant genes, as long as they are not selected for in the heterozygote. It depends only on the mutation rate and the degree of dominance (for more detail see Kimura, 1961). The mutational load may be exprssed in the form

$$L_m = \sum 2\mu d_m \tag{5}$$

where $d_m$ is the sum total of the fraction of genetic elimination over all generations following the appearance of a single mutation (Muller's genetic death) and $\mu$ is the mutation rate per locus per generation. The summation is over all relevant loci. For a dominant mutation $d_m = 1$ and for a completely recessive mutation $d_m = 0{\cdot}5$. For a more general expression of $L$ see Kimura (1961). The substitutional load is the load due to substituting new alleles for old through natural selection in the course of adaptive evolution and it may be expressed in the following form

$$L_e = \sum \epsilon d_e \tag{6}$$

where $d_e$ is the sum of the fractions of genetic elimination over all generations in the process of substituting one allele for another in the population (Haldane's cost of natural selection) and $\epsilon$ is the rate per locus per generation of such substitution. $\epsilon$ is supposed to be a very small fraction. If $p_0$ is the initial frequency and $h$ is the degree of dominance of the mutant gene being substituted, we have (cf. Kimura, 1960),

$$d_e = -\frac{1}{h}\left\{\log p_0 + (1-h)\log\frac{1-h}{h+(1-2h)p_0}\right\} \tag{7}$$

This formula is an approximation which may be sufficient for weak selection. It shows that $d_e$ is independent of the selection coefficient $s$. For a semi-dominant favourable mutation ($h=0{\cdot}5$), $d_e = 2\log_e(1/p_0)$. For example, if $p_0 = 10^{-6}$ we have $d_e \approx 27{\cdot}6$. If $p_0$ is larger, say, $10^{-3}$, then $d_e \approx 13{\cdot}8$. In general, $d_e$ becomes larger as $p_0$ becomes smaller. Thus the principle of minimum genetic load as applied to the present situation can be stated roughly as follows: Too high a mutation rate is clearly disadvantageous in that the mutational load will be too high. On the other hand, too low a rate will maintain potentially useful variance at such a low level that, when the environment alters, the population has to do too much work (in terms of selective death) to bring such alleles to final fixation. Therefore there must be an optimum mutation rate for the survival of the species.

Using general expressions for $L_m$ and $L_e$, it was shown that if we chose $\mu$ and $h$ in such a way that $L$ is minimized we get

$$\mu = \frac{0{\cdot}3419E}{\bar{h}}(1+1{\cdot}720\bar{h}+\ldots)$$

$$\bar{h} = 0{\cdot}6838\sqrt{\left(\frac{E}{2D}\right)}\left[1+1{\cdot}018\sqrt{\left(\frac{E}{2D}\right)}+\ldots\right]$$

where $\mu$ is the spontaneous mutation rate per gamete per generation, $\bar{h}$ is the average degree of dominance in fitness of deleterious mutant genes, $D$ is the total mutational damage or approximately the rate of inbreeding depression in fitness per unit increase in the inbreeding coefficient and $E$ is the rate of substitution of genes in horotelic (i.e. standard rate) evolution. It was shown also that if we take $E = \sum\epsilon = 1/300$, a value suggested by Haldane (1957) and $D=2$, a value given by Morton, Crow and Muller (1956), we obtain about 0·02 for $\bar{h}$ and about 0·058 for $\sum\mu$. These values agree fairly well with the corresponding observed values in *Drosophila*. However, one of the assumptions in the treatment, i.e. that the evolutionally useful genes have at the start the same degree of dominance as the lethals and sub-vitals, may be implausible (Morton, 1965). Furthermore, from the study of Greenberg and Crow (1960), mildly detrimental mutants appear to have a higher degree of dominance than lethals and near lethals. Also, recent work of Mukai and Yamazaki (1964) suggests that polygenic mutations are nearly semi-dominant.

It is probable that the degree of dominance is modified solely by intra-group selection, as is generally assumed in the ordinary theories of dominance (cf. Fisher, 1930; Muller, 1950*b*).

In the present paper, we will investigate a hypothesis that only the mutation rate $\mu$ is adjusted so that the total genetic load $L$ is minimized. To simplify the mathematical treatment, we will assume that deleterious mutations have enough degree of dominance so that

$$d_m = 1$$

and also that advantageous mutations are nearly semi-dominant so that

$$d_e = 2\log_e\left(\frac{1}{p_0}\right) \tag{8}$$

The mutant genes that are used for substitution may not necessarily be advantageous at the start. More probably, most of them may be initially deleterious, but become advantageous after a change of environment. The mutation rate for such potentially useful genes may be a tiny fraction (say $c$) of the mutation rate ($\mu$) at each locus. Fisher (1930) argued that favourable mutations can scarcely be permitted to continue to occur for long, even at rate 1000-fold less than that of unfavourable mutations. This implies that in his opinion $c \leqslant 0{\cdot}001$. If $s_1$ is the selection coefficient against such a mutant gene before the change of environment, we may put

$$p_0 = c\mu/s_1$$

Here we assume that the change of environment takes place suddenly. Assuming that the great majority of mutations are deleterious in any environment, the total load (4) is approximately

$$L = \sum 2\mu - \sum 2\epsilon \log(c\mu/s_1) \tag{9}$$

In the following treatment for minimizing $L$, we will take $\epsilon$, $c$, and $s_1$ as constants. Especially, $\sum \epsilon$ ($\equiv E$) will be considered as given *a priori* to keep up with the evolutionary progress of competing species. This is consistent with the idea of treating the substitution of advantageous alleles as a 'load' in the evolution of species. Differentiating the above expression for $L$ with respect to $\mu$, we obtain

$$2 - 2\epsilon/\mu = 0$$

or

$$\mu = \epsilon,$$

and therefore

$$M = E, \tag{10}$$

where $M = \sum \mu$ and $E = \sum \epsilon$. Thus the mutation rate per gamete that minimizes the genetic load $L$ is equal to the rate of gene substitution per individual per generation. Note that the value of $c$ is irrelevant in arriving at this relation. According to Haldane (1957) a probable figure for $E$ is 1/300 when evolution proceeds at the standard rate. This gives the substitutional load of $0{\cdot}1$, or using his terminology, the intensity of selection $I = 0{\cdot}1$: He considered that a typical value of $p_0$ would be about $5 \times 10^{-5}$ for a partially or wholly dominant gene. With this value of $p_0$, $d_e$ is roughly 20, but to allow for occasionally high values, he took $d_e$ as 30. This leads to $I = 30/300 = 0{\cdot}1$. So, if we adopt his figure of $E = 1/300$, we get $\sum \mu = 1/300$ as an optimum mutation rate. This gives a mutational load of about $0{\cdot}67\%$ plus a substitutional load of about $11{\cdot}5\%$, if we take $c = 10^{-3}$, $s_1 = 0{\cdot}01$ and the total number of loci $N_L = 10^4$. On the other hand, the actual rate of mutation per gamete is at least a few per cent for higher organisms. So the theoretically derived value is less than 1/10 of the actual value.

What is the cause of this discrepancy if the present hypothesis is at all correct? It is unlikely that the substitutional load is much larger than 0·1 or 0·2.

One possibility is that the actual value of $E$ may be larger, yet the substitutional load remains essentially the same, for the following reason.

In the original derivation of equation (7), it was assumed that a change of environment takes place suddenly so that the selection coefficient also changes abruptly. For example, in the case of a semi-dominant mutation, the selection coefficient for the mutant gene changes from $-s_1$ to $+s_1$ in one generation. However, it may be more likely that the change of environment will take place only gradually so that the gene passes through the neutral stage. This enables the mutant gene to increase its frequency by mutation pressure and the substitutional load is thereby reduced (Haldane, personal communication). To treat such a situation mathematically, let us assume that $s_1$ is expressed by $kt$, where $k$ is a positive constant and $t$ is time measured in generations. The gene is neutral at $t = 0$. The initial frequency $p_0$ may then be calculated as follows: The expected number at $t = 0$ of the mutant gene which appeared $n$ generations earlier is

$$\prod_{t=-n}^{0} e^{kt} = \exp\left(k \sum_{t=-n}^{0} t\right) \approx e^{-kn^2/2}$$

Thus if one mutation appears in each generation, the expected number at $t = 0$ is

$$\sum_{n=-\infty}^{0} e^{-kn^2/2} \approx \int_{-\infty}^{0} e^{-kx^2/2}\, dx = \sqrt{\left(\frac{\pi}{2k}\right)}$$

It follows then that the frequency of the mutant gene at $t = 0$ is

$$p_0 = c\mu\sqrt{\left(\frac{\pi}{2k}\right)}$$

It can be shown that $d_e = 2\log_e(1/p_0)$ still holds in this case and therefore we have

$$L = \sum 2\mu - \sum 2\epsilon \log\left[c\mu\sqrt{\left(\frac{\pi}{2k}\right)}\right] \qquad (11)$$

Thus, if the selection coefficient changes from $-0{\cdot}01$ to 0 in 10,000 generations, $k$ is $10^{-6}$ and $p_0$ is roughly $1{\cdot}3 \times 10^3\, c\mu$. In this case, $d_e$ may become about 2/3 as large as in the case of $s_1 = 0{\cdot}01$ where the change of environment takes place in one generation. If in addition a substitutional load of about 0·25 is tolerated, allelic substitution may proceed at the rate of $E = 1/100$, but the theoretically optimum mutation rate $M = E = 0{\cdot}01$ still seems to be too low, where $M$ is the mutation rate per gamete.

To see how sharply the minimum value of $L$ is attained as a function of $M$, values of $L$ are plotted against values of $M$ in Fig. 2. The solid line shows the case where the change of environment takes place gradually as explained above ($k = 10^{-6}$) with substitution rate $E = 0{\cdot}01$. To construct the graph it was assumed that there are $10^4$ loci with equal mutation rate and $c = 10^{-3}$. The graph shows that even if $L$ is minimum at $M = 0{\cdot}01$, the curve is so flat around $M = 0{\cdot}01$ that $L$ changes only slightly with the change of $M$. Actually change of $M$ from 0·01 to 0·04, or from 0·01 to 0·0009, increases $L$ only by about 10% (or 0·03 in absolute magnitude).

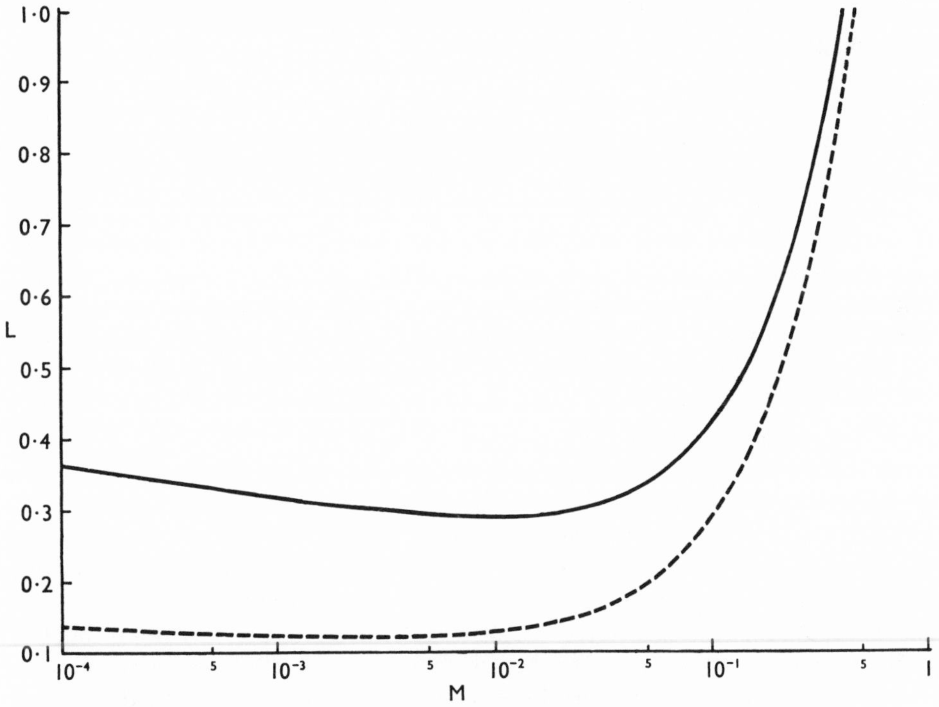


Fig. 2. The total load $L$ as a function of the mutation rate per gamete, $M$. The solid line illustrates a case in which the change of environment takes place very slowly, i.e. the selection coefficient of the potentially advantageous gene changes from $-0{\cdot}01$ to 0 in 10,000 generations. The rate of allelic substitution, $E$, is assumed to be 0·01 in this case. The dashed line illustrates a case in which the change of environment takes place abruptly, i.e. the selection coefficient changes from $-0{\cdot}01$ to $+0{\cdot}01$ in one generation. $E = 1/300$ is assumed.

In both cases, it is assumed also for the purpose of drawing the graph that there are $10^4$ loci with equal mutation rate and the proportion of potentially advantageous mutations among newly arisen mutations is $10^{-3}$.

Since $M = 0{\cdot}04$ is close to the actual mutation rate in *Drosophila*, we may regard the actual $M$ as being in the 'optimum region' even if it is not the strict optimum. In this case, the substitutional load becomes about 0·24. A substitutional load of this magnitude might appear to be somewhat large, but some of the advantageous mutant genes would increase the reproductive potential of the species, allowing the species to tolerate the slightly heavier load than in the case of detrimental mutations.

The situation is somewhat analogous to the overdominance load (i.e. Crow's segregation load) if the heterozygosity is associated with a higher reproductive potential.

The dashed line in Fig. 2 illustrates $L$ as a function of $M$ when the change of environment takes place suddenly and the total allelic substitution proceeds at the rate $E = 1/300$. As before $10^4$ loci with equal mutation rate and $c = 10^{-3}$ are assumed to construct the graph. Again, the curve is so flat near the optimum ($M = 1/300$) that a considerable latitude in $M$ is allowed with only slight increase in $L$: $L$ increases from 0·1215 to 0·1601 as $M$ changes from 0·003 to 0·03.

## 5. IS THE SPONTANEOUS MUTATION RATE THE MINIMUM THAT CAN BE EFFECTIVELY ATTAINED BY NATURAL SELECTION?

In our arguments leading to the hypothesis of optimum mutation rate, there is little doubt as to the existence of selection for lower mutation rates. Such selection must be immediately effective and must have been at work since the origin of life: In the earlier stages of evolution, those genes that can replicate with fewer errors must have a definite selective advantage over those which produce errors more often. In the later stages of evolution in which each individual contains a large number of genes, a more subtle type of selection must have become also important, i.e., the selection against modifiers which increase mutation rates at other loci. Compared with this the existence of selection against too low a mutation rate is less certain. Its effectiveness in the course of evolution can at least be doubted.

So let us consider the following hypothesis: In the course of evolution, natural selection has been at work only toward lowering mutation rates. However, mutation as the replication error of the genetic material cannot entirely be eliminated because of physical or physiological limitations. Random processes at the submolecular level may not be completely excluded because of physical principles, or an elaborate apparatus that must be developed for checking and eliminating errors in replication might be physiologically so costly relative to the gain thereby achieved that it did not pay in adaptive evolution. In either case, we will assume that the mutation rates of organisms that are now living are very near to the physical or physiological limit.

If the spontaneous mutation rate is at the physically attainable limit, it may be expressed as the probability of replication error per nucleotide pair per division. This probability may be calculated from

$$u = \frac{M}{N.n} \tag{12}$$

where $M$ is the total mutation rate per gamete per generation, $N$ is the number of nucleotide pairs in DNA contained in a gamete, and $n$ is the number of cell divisions along a germ line from the fertilized egg to a gamete. Furthermore, the value of $u$ must be constant for different organisms. No exact comparison of this value is possible at present for various organisms. For man, probable figures of $M$, $N$ and $n$

are roughly 0·2, $4 \times 10^9$ and 50, giving $u = 10^{-12}$ approximately. For *Drosophila*, $M$ may be about 1/4 as large as that of man, $N$ is about 1/20 as large and $n$ about $1/2 \sim 1/3$. So the probable value of $u$ is roughly $10^{-11}$. This means that values of $u$ for man and *Drosophila* are quite different and the validity of the hypothesis that the spontaneous mutation rate is the physically attainable limit is doubtful.

The difference, however, seem to be much smaller than the difference of mutation rates between these two species when measured in absolute time units. The other criterion that may be used to check this hypothesis is the nature of mutations. If the limit is physical, almost all the modifiers that appear by mutation should have the property of increasing the mutation rates of other loci.

On the other hand, if the limit is physiological, modifiers that decrease the present mutation rates, if found, should always be accompanied by some deleterious effect, enough to upset the advantage that comes from reducing the mutation rates of other loci. However, it may be difficult to test this hypothesis.

## 6. DISCUSSION

The idea that the spontaneous mutation rate is the optimum rate for the evolution of species is not new (cf. Auerbach, 1956). No quantitative formulations of the idea, however, have been published except for the author's paper on the subject (Kimura, 1960), in which it was assumed that the mutation rate and the degree of dominance are adjusted in the course of evolution in such a way that the sum of mutational and substitutional load is minimized.

In the present paper the theory is re-examined from the standpoint that only the mutation rate is adjusted to minimize the total load. It turns out that the optimum mutation rate is equal to the rate of allelic substitution, irrespective of how rapidly the environmental change takes place. We have then $M = E$, where $M$ is the total mutation rate per gamete and $E$ is the total rate of allelic substitution per individual per generation. Consideration of mutational and substitutional loads involved seems to suggest that the theoretical mutation rate thus derived is too low to explain the actual mutation rate. This is mainly due to the fact that occurrence of one detrimental mutation leads to one genetic death, while one substitution of an allele in the whole population leads to at least twenty times as many genetic deaths as the population number. So the allelic substitution per individual cannot proceed at a rate comparable to the mutation rate per gamete which is at least a few per cent for higher organisms.

We must conclude therefore that the actual mutation rate is above the optimum from the standpoint of minimizing the genetic load $L$. However, as shown in Fig. 2, considerable latitude is allowed for $M$ without much increasing the actual value of $L$.

On the other hand, Crow (1959) considered the possibility that the mutation rate would usually be below the optimum from the standpoint of long-term evolutionary progress, because selection to reduce the mutation load has an immediate beneficial effect. He was cautious, however, in not drawing a definite conclusion because the mutation-rate-adjusting gene may also have a direct effect on the organism such as an effect on size or metabolic rate.

In the present paper, we have also considered an alternative hypothesis that the spontaneous mutation rate is the minimum rate that can be attained by natural selection. Unfortunately, in such a problem it is not easy to set up alternative hypotheses and destroy one, due to paucity of reliable data and also the impossibility of direct experimental tests.

Thus the possibility cannot at present be excluded that the spontaneous mutation rate is near the minimum that may be attained under the present mode of organization of the genetic material, and at the same time is not very far from the optimum in the sense of minimizing the genetic load. The whole question of the evolutionary modification of the spontaneous mutation rate is quite puzzling and more evidence is needed to clarify the problem.

## SUMMARY

Evolutionary factors which tend to decrease the mutation rate through natural selection and those which tend to increase the mutation rate are discussed from the standpoint of population genetics. The author's theory of optimum mutation rate based on the principle of minimum genetic load is re-examined, assuming that mutation rate is adjusted in the course of evolution in such a way that the sum of mutational and substitutional load is minimized. Another hypothesis is also examined that only selection toward lowering the mutation rate is effective and the present mutation rate in each organism represents the physical or physiological limit that may be attained by natural selection.

The possibility cannot be excluded that the spontaneous mutation rate is near the minimum that may be attained under the present mode of organization of the genetic material, and at the same time is not very far from the optimum in the sense of minimizing the genetic load.

The author would like to express his thanks to Drs L. L. Cavalli-Sforza and Masatoshi Nei for stimulating discussions and valuable criticism during preparation of the manuscript. Thanks are also due to Drs James F. Crow and Alan Robertson for reading the manuscript and making valuable suggestions.

## REFERENCES

AUERBACH, C. (1956). *Genetics in the Atomic Age*. Edinburgh: Oliver and Boyd.

CROW, J. F. (1959). Ionizing radiation and evolution. Scient. Am. **201** (Sept. issue), 138–160.

CROW, J. F. (1960). Lecture notes on population genetics.

CROW, J. F. & KIMURA, M. (1965). Evloution in sexual and asexual populations. *Am. Nat.* **99,** 439–450.

FISHER, R. A. (1930). *The Genetical Theory of Natural Selection*. Oxford: Clarendon Press.

GREENBERG, R. & CROW, J. F. (1960). A comparison of the effect of lethal and detrimental chromosomes from *Drosophila* populations. *Genetics*, **45,** 1153–1168.

HALDANE, J. B. S. (1937). The effect of variation on fitness. *Am. Nat.* **71,** 337–349.

HALDANE, J. B. S. (1957). The cost of natural selection. *J. Genet.* **55,** 511–524.

HALDANE, J. B. S. (1960). Personal Communication.
KIMURA, M. (1960). Optimum mutation rate and degree of dominance as determined by the principle of minimum genetic load. *J. Genet.* **57,** 21–34.
KIMURA, M. (1961). Some calculations on the mutational load. *Jap. J. Genet.* **36** (Suppl.), 179–190.
MORTON, N. E., CROW, J. F. & MULLER, H. J. (1956). An estimate of the mutational damage in man from data on consanguineous marriages. *Proc. natn. Acad. Sci. U.S.A.* **42,** 855–863.
MORTON, N. E. (1965). Models and evidence in human population genetics. *Proc. XI Int. Congr. Genet.*, pp. 935–951.
MUKAI, T. & YAMAZAKI, T. (1964). Position effect of spontaneous mutant polygenes controlling viability in *Drosophila melanogaster*. *Proc. Japan Acad.* **40,** 840–845.
MULLER, H. J. (1932). Some genetic aspects of sex. *Am. Nat.* **66,** 118–138.
MULLER, H. J. (1950*a*). Our load of mutations. *Am. J. hum. Genet.* **2,** 111–176.
MULLER, H. J. (1950*b*). Evidence of the precision of genetic adaptation. *Harvey Lect.*, Series 18, pp. 165–229, Springfield: Charles C. Thomas.
MULLER, H. J. (1958). Evolution by mutation. *Bull. Am. math. Soc.* **64,** 137–160.
Report of the United Nations Scientific Committee on the Effects of Atomic Radiation, 1958, New York.

PART NINE

# Inbreeding Systems

## Introduction

Wright (1918) first used his path analysis to quantify each cause of the variability of quantitative characters by a method similar to that of partitioning the variance. Later he refined the method and defined the path coefficient as the standard deviation of the effect to be found when all causes are constant except the one in question (Wright 1920); it is a standardized partial regression coefficient. As described in Chapter 5 in Provine (1986), the method of path coefficients proved to have much wider application than was apparent in Wright's first use of it. One application is to compute effects of inbreeding in various systems of mating (Wright 1921, 1922; see also appendices in Wright 1950). Haldane and Waddington (1931), Cotterman (1940), Malécot (1948), and Falconer (1960) considered the effect of inbreeding on homozygosity due to the more readily understood concept of identity by descent.

Kimura (no. 21) developed a means of handling inbreeding systems in terms of probability. It is based on the concept of identity by descent and therefore is rather similar to Malécot's, but it is applied to identity of state, and Kimura derived known results in a simpler way. The method was applied to the so-called *circular mating:* Circular mating was an attempt to find the mating system that can conserve heterozygosity in the long run. By arranging males and females alternately in a circle and allowing mating only between the neighboring individuals, Kimura found that the effective number of individuals in this system is proportional to the square of the actual number. He might have anticipated that this system would become widely adopted in animal breeding practice, particularly for preserving endangered species in captivity (Yamada 1985). Kimura and Crow (no. 22) worked out the system of circular mating in addition to *circular pair mating* suggested by Haldane, *circular subpopulation mating,* and a mixture of circular and circular pair mating suggested by Yamada. Kimura and Crow distinguished two related, but distinct, effects of inbreeding (random drift and decrease of heterozygosity), and compared various mating systems together with the one for maximum avoidance of inbreeding (Wright 1921; see also Wright 1965). Among these systems, circular mating has the lowest rate of decrease of heterozygosity in the long run, though the rate is highest initially, and has the smallest eventual value of drift variance per generation. Boucher and Cotterman (1990) systematically examined regular systems of mating, finding that when the number of individuals is larger than four, there are many other systems that have slower rates of approach to genetic uniformity than circular mating.

These treatments were based on neutrality of genes and the concept of identity by descent. As remarked by Nagylaki (1989a), identity by descent is a fundamental concept that was discovered independently by Cotterman (1940) and Malécot (1941), and it lies at the heart of the powerful genealogical approach to population genetics.

# A PROBABILITY METHOD FOR TREATING INBREEDING SYSTEMS, ESPECIALLY WITH LINKED GENES[1]

MOTOO KIMURA[2]

*University of Wisconsin,*
*Madison, Wisconsin, U.S.A.*

## SUMMARY

A simple method was developed to handle inbreeding systems in terms of probability rather than in terms of correlation as in Wright's theory. As applied to the case of a single gene, the method is very similar to Malécot's approach which makes use of the concept of identity by descent. As applied to linked loci, it yields all the main results obtained by Wright [1933b] through the method of path coefficients. Though it should be equivalent to Wright's method, the present method may have an advantage in that it utilizes only simple probability calculations and no new concept or preliminary knowledge of correlation or path analysis is required.

In addition to the conventional inbreeding systems, the method was applied to cases of partial self-fertilization and a circular mating system.

The utility and limitation of the present method to handle more general cases have also been discussed.

## 1. INTRODUCTION

The usual method of enumerating all possible genotypes or mating types and writing their frequencies in successive generations becomes very unwieldy if the number of mating types is large. In their study of inbreeding and linkage, Haldane and Waddington [1931] had to consider 100 different mating types and a set of 22 linear equations to treat the case of brother-sister mating for 2 linked autosomal loci each with only 2 alleles. They also studied the case of parent-offspring mating and confirmed the earlier results of Robbins [1918] on self-fertilization, but they could not go beyond inbreeding systems involving two individuals per generation. For their purpose of merely finding out the final proportion of recombinant types, the method was too

[1]Sponsored by the Mathematics Research Center, United States Army, University of Wisconsin. Also aided by grants from the National Institute of Health (RG-8217 and RG-7666). A part of this paper was orally presented at the Biometric Society Meeting held at the University of North Carolina, Chapel Hill, on April 12, 1962.

[2]On leave from the National Institute of Genetics, Mishima-shi, Japan.

This paper also constitutes Contribution No. 428 from the National Institute of Genetics, Mishima-Shi, Japan.

cumbersome. They state that the case of double first cousin matings with autosomal linkage involves the consideration of 10,000 different mating types and other systems are still more complex.

They suggested that such problems might be solved by an extension of Wright's correlation method.

Two years later, the solution was supplied by Wright [1933b] who utilized his correlation method, i.e. the method of path analysis. This is one of the most remarkable demonstrations of the power of the path coefficient method. However, for the case of a single locus, it has been shown by Malécot [1948] that equivalent results can be obtained in terms of probability by introducing the concept of identity of genes by descent.

The main purpose of the present paper is to show that, even for linked loci, results equivalent to those obtained by Wright can be derived by rather simple probability calculations if certain crucial quantities are properly defined.

Before we consider linked genes, the method will be applied to the case of a single locus. This will reveal the similarities and differences of the present approach and that of Malécot.

## 2. SINGLE LOCUS

### *2.1. Autosomal gene.*

First assume a population of $N$ monoecious diploid individuals and consider an autosomal locus. Let $I_t$ be the probability that the two homologous chromosomes in an individual, at the $t$th generation ($t = 0, 1, 2, \cdots$), carry the same allele.

With an arbitrary number of alleles $A_1$, $A_2$, $\cdots$ etc. in a population, $I_t$ represents the probability that an individual is homozygous for one or another of these alleles. Similarly, let $J_t$ be the probability that two homologous chromosomes taken one from each of two different individuals in the $t$th generation carry the same allele.

Under random mating which includes self-fertilization in $1/N$ of the cases, two genes are either derived (1) from the same parental individual with probability $1/N$ or (2) from different individuals with probability $(1 - 1/N)$ (see Figure 1). In the former case they either come from the same gene or from homologous genes with equal probability of $\frac{1}{2}$. The probability $I_t$ that the two homologous chromosomes in an individual in the $t$th generation share the same allele may be obtained by the following consideration: (1) When the two genes are descended from the same gene in the previous generation the probability is 1 that they share the same allele, while the probability is by definition

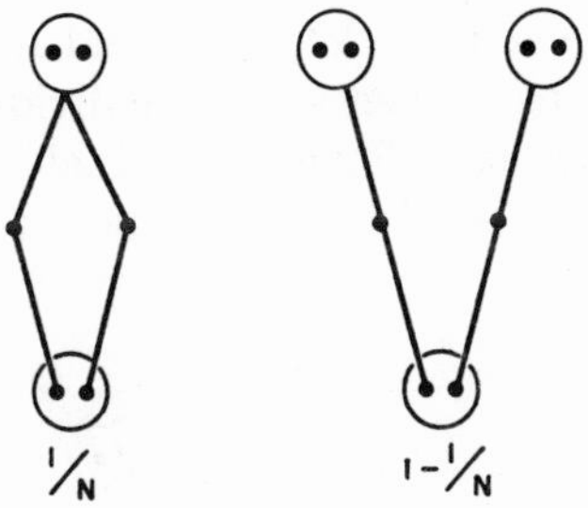


FIGURE 1.

$I_{t-1}$ when they come from homologous loci in the same individual; (2) When the two genes come from two different individuals, the probability is $J_{t-1}$ which is by definition the probability that homologous loci in two different individuals have the same allele in the $(t - 1)$th generation. Thus we have

$$I_t = \frac{1}{N}\left(\tfrac{1}{2} + \tfrac{1}{2}I_{t-1}\right) + \left(1 - \frac{1}{N}\right)J_{t-1} \tag{2.1.1}$$

$$J_t = \frac{1}{N}\left(\tfrac{1}{2} + \tfrac{1}{2}I_{t-1}\right) + \left(1 - \frac{1}{N}\right)J_{t-1}\,. \tag{2.1.2}$$

Note that with random mating $I_t = J_t$ for $t \geqq 1$, but it is not necessarily true that $I_0 = J_0$ unless the initial population is itself a product of random mating. For $t \geqq 2$ the above two equations (2.1.1) and (2.1.2) reduce to

$$I_t = \frac{1}{2N} + \left(1 - \frac{1}{2N}\right)I_{t-1}\,, \qquad (t \geqq 2). \tag{2.1.3}$$

The probability that the homologous loci of an individual carry the different alleles, namely the probability of heterozygosity, is given by

$$H_t = 1 - I_t\,,$$

so that

$$H_t = \left(1 - \frac{1}{2N}\right)H_{t-1} = \left(1 - \frac{1}{2N}\right)^{t-1} H_1 \qquad (t \geqq 1), \tag{2.1.4}$$

where

$$H_1 = 1 - I_1\,,$$

in which

$$I_1 = \frac{1}{2N}(1 + I_0) + \left(1 - \frac{1}{N}\right)J_0\,. \tag{2.1.5}$$

If the population consists of $N^*$ males and $N^{**}$ females mating at random, the two homologous genes in an individual either come from the same individual or from different individuals two generations back (cf. Figure 2). Since the probability of the two genes coming both

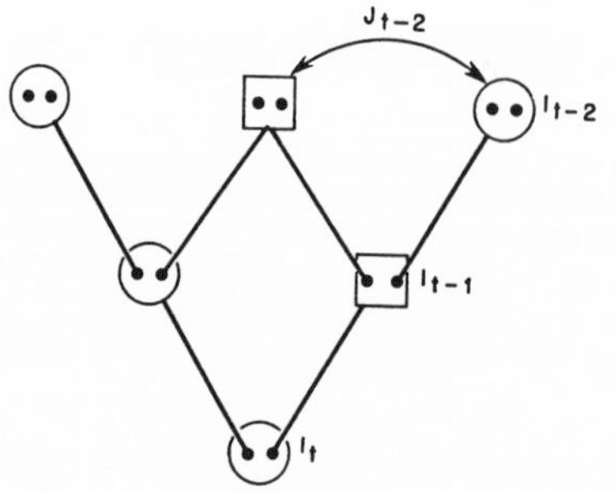


FIGURE 2.

from the male side is $\frac{1}{4}$, the probability that both come from the same male is $1/(4N^*)$. Similarly, the probability of both coming from the same female is $1/(4N^{**})$. Thus the probability of two homologous genes coming from the same individual two generations back is $1/(4N^*) + 1/(4N^{**})$ which we will denote by $1/N_e$, conforming with Wright's well-known formula,

$$\frac{1}{N_e} = \frac{1}{4}\left(\frac{1}{N^*} + \frac{1}{N^{**}}\right), \tag{2.1.6}$$

where $N_e$ is defined as the effective number of the population. When two genes come from the same grandparent, they either come from the same gene or from two homologous genes with equal probability of $\frac{1}{2}$. Thus we have

$$\begin{aligned} I_t &= \frac{1}{N_e}\left(\tfrac{1}{2} + \tfrac{1}{2}I_{t-2}\right) + \left(1 - \frac{1}{N_e}\right)J_{t-2} \\ I_{t-1} &= J_{t-2} \end{aligned} \tag{2.1.7}$$

or

$$I_t = \frac{1}{2N_e} + \left(1 - \frac{1}{N_e}\right)I_{t-1} + \frac{1}{2N_e} I_{t-2} . \tag{2.1.8}$$

The probability of heterozygosity is

$$H_t = 1 - I_t$$

and from (2.1.8) we obtain

$$H_t = \left(1 - \frac{1}{N_e}\right)H_{t-1} + \frac{1}{2N_e} H_{t-2} \, . \qquad (2.1.9)$$

For brother-sister mating, $N^* = N^{**} = 1$, $N_e = 2$ and we obtain the well known relation

$$H_t = \tfrac{1}{2}H_{t-1} + \tfrac{1}{4}H_{t-2} \, .$$

*2.2. Sex-linked gene.*

The situation is a little more complicated for a sex-linked gene. We will assume that the male is heterogametic so that $I_t$ is defined only for the female. Furthermore, the probability of the two homologous loci in two different individuals sharing the same allele will be defined separately for the three different situations, being denoted by $J_t^*$ when the two individuals are both males, by $J_t^{**}$ when both are females and by $M_t$ when one is a male and the other a female.

Under random mating, we have

$$J_t^* = \frac{1}{N^{**}}\left(\tfrac{1}{2} + \tfrac{1}{2}I_{t-1}\right) + \left(1 - \frac{1}{N^{**}}\right)J_{t-1}^{**} \qquad (2.2.1)$$

$$J_t^{**} = \frac{1}{4}\left[\frac{1}{N^*} + \left(1 - \frac{1}{N^*}\right)J_{t-1}^*\right] + \tfrac{1}{2}M_{t-1} + \tfrac{1}{4}J_t^* \qquad (2.2.2)$$

$$M_t = \tfrac{1}{2}M_{t-1} + \tfrac{1}{2}J_t^* \qquad (2.2.3)$$

$$I_t = M_{t-1} \, . \qquad (2.2.4)$$

The first relation (2.2.1) is based on the fact that males receive genes only from their mother and therefore the homologous genes in two males are both derived from the same female of the previous generation with probability $1/N^{**}$ and from different females with the probability of $1 - 1/N^{**}$. On the other hand, females receive genes equally from their father and mother (cf. Figure 3). Thus two homologous genes, taken one from each of two females, both come from their fathers with probability $\tfrac{1}{4}$, one from the father and the other from the mother with

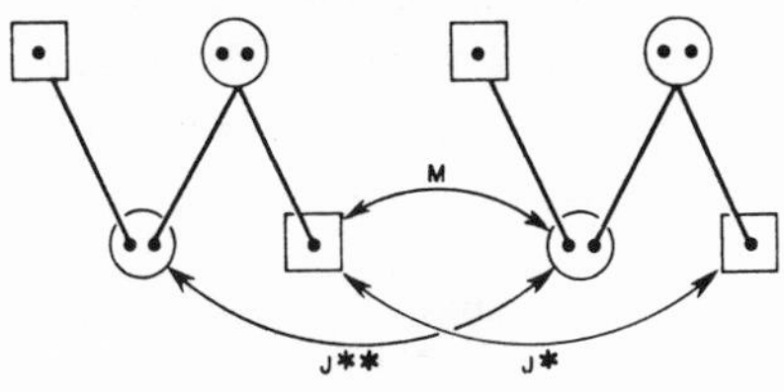


FIGURE 3.

the probability $\frac{1}{2}$ and both from mothers with probability $\frac{1}{4}$. In the first case they come from the same male of the previous generation with probability $1/N^*$ and different males with probability $1 - 1/N^*$. Therefore the first term in the right side of (2.2.2) follows naturally. The second term follows from the definition of $M_{t-1}$. The third term is based on the consideration that when the two homologous genes are both derived from females, their relation is exactly the same as two homologous genes in two males produced from the females. Note the subscript $t$ rather than $t - 1$. Formula (2.2.3) is derived by a similar consideration.

From the above four relations, we obtain the recurrence relation:

$$2I_t - \left(\frac{3}{2} - \frac{1}{2N^{**}}\right)I_{t-1} - \left(\frac{3}{4} - \frac{1}{2N^*} - \frac{1}{4N^{**}} + \frac{1}{2N^*N^{**}}\right)I_{t-2}$$
$$+ \frac{1}{4}\left(1 - \frac{1}{N^*}\right)\left(1 - \frac{1}{N^{**}}\right)I_{t-3} - \left(\frac{1}{2N^{**}} + \frac{1}{4N^*} - \frac{1}{4N^*N^{**}}\right) = 0. \quad (2.2.5)$$

Substituting $I_t = 1 - H_t$, we obtain

$$H_t - \frac{1}{4}\left(3 - \frac{1}{N^{**}}\right)H_{t-1} - \frac{1}{4}\left(\frac{3}{2} - \frac{1}{N^*} - \frac{1}{2N^{**}} + \frac{1}{N^*N^{**}}\right)H_{t-2}$$
$$+ \frac{1}{8}\left(1 - \frac{1}{N^*}\right)\left(1 - \frac{1}{N^{**}}\right)H_{t-3} = 0, \quad (2.2.6)$$

which agrees with Wright [1933a].

For the simplest case of brother-sister mating, $N^* = N^{**} = 1$, (2.2.6) reduces to

$$H_t - \tfrac{1}{2}H_{t-1} - \tfrac{1}{4}H_{t-2} = 0$$

which is the same as for an autosomal gene.

## 3. TWO LINKED LOCI

### *3.1. Autosomal genes.*

We will again start from the simplest case of $H$ monoecious individuals mating at random in which self-fertilization is included in the proportion of $1/N$. Let us consider two linked genes, the first consisting of alleles $A_1$, $A_2$, $\cdots$ etc. and the second consisting of alleles $B_1$, $B_2$, $\cdots$ etc.

Let

$C_t(A_iB_k) = C_t =$ the probability that a chromosome chosen at random carries alleles $A_i$ and $B_k$

$S_t(A_iB_k) = S_t =$ the probability that one chromosome chosen at random carries $A_i$ and the other chromosome in the same individual carries $B_k$

$T_t(A_iB_k) = T_t =$ the probability that a chromosome chosen at random carries $A_i$ and that another chromosome chosen from another individual carries $B_k$

then, if $r$ is the recombination fraction between the two genes

$$C_t = (1 - r)C_{t-1} + rS_{t-1} \tag{3.1.1}$$

$$S_t = \frac{1}{2N} C_{t-1} + \frac{1}{2N} S_{t-1} + \left(1 - \frac{1}{N}\right)T_{t-1} \tag{3.1.2}$$

$$T_t = \frac{1}{2N} C_{t-1} + \frac{1}{2N} S_{t-1} + \left(1 - \frac{1}{N}\right)T_{t-1}\,. \tag{3.1.3}$$

The first relation (3.1.1) is immediately derived from the condition that two loci on a chromosome are either from two loci on a chromosome of the previous generation without crossing-over, or one from each of the homologous chromosomes by crossing-over. The second and the third relation are derived through a similar argument. The first locus on one chromosome and the second locus on the other chromosome are either from the same individual in the previous generation with the probability $1/N$ or from different individuals with probability $1 - 1/N$. In the former case they are either derived from the same chromosome or from homologous chromosomes with equal probability of $\frac{1}{2}$. Note that $S_t = T_t$ for $t \geqq 1$ but not necessarily $S_0 = T_0$ unless the starting population itself is produced by random mating.

Thus, for $t \geqq 2$, the three relations are simplified to

$$\begin{aligned} C_t &= (1 - r)C_{t-1} + rS_{t-1} \\ S_t &= \frac{1}{2N} C_{t-1} + \left(1 - \frac{1}{2N}\right)S_{t-1}\,. \end{aligned} \tag{3.1.4}$$

It may be observed that if $C_t$ converges to $C_\infty$ at the limit of $t = \infty$, so does $S_t$ and $S_\infty = C_\infty$. This should be the case because at the limit the population becomes genetically homogeneous due to random fixation of alleles.

In order to find $C_\infty$, consider the linear combination

$$L_t = \alpha C_t + \beta S_t \tag{3.1.5}$$

and choose two constants $\alpha$ and $\beta$ such that

$$L_t = L_{t-1} \qquad (t \geqq 2).$$

A few algebraic steps show that the last relation is satisfied with

$$\alpha = 1/(2N), \qquad \beta = r. \tag{3.1.6}$$

With these values of $\alpha$ and $\beta$,

$$L_\infty = L_1$$

or

$$\left(\frac{1}{2N} + r\right)C_\infty = \frac{1}{2N} C_1 + rS_1$$

and we obtain

$$C_\infty = (C_1 + 2NrS_1)/(1 + 2Nr) \tag{3.1.7}$$

where

$$\begin{aligned} C_1 &= (1 - r)C_0 + rS_0 \\ S_1 &= \frac{1}{2N} C_0 + \frac{1}{2N} S_0 + \left(1 - \frac{1}{N}\right)T_0 \,. \end{aligned} \tag{3.1.8}$$

The solution of the set of equations (3.1.4) for an arbitrary $t$ ($t \geqq 1$) is

$$\begin{aligned} C_t &= C_\infty + (C_1 - C_\infty)\left(1 - r - \frac{1}{2N}\right)^{t-1} \\ S_t &= S_\infty + (S_1 - S_\infty)\left(1 - r - \frac{1}{2N}\right)^{t-1}, \end{aligned} \tag{3.1.9}$$

where $C_\infty = S_\infty$ is given by (3.1.7).

*Examples*

(1) Starting with a population 100% $AB/ab$, we obtain $C_0(AB) = \frac{1}{2}$, $S_0 = 0$, $T_0 = \frac{1}{4}$.

$$C_1 = (1 - r)/2, \qquad S_1 = 1/4$$

$$C_\infty(AB) = C_\infty(ab) = (1 - r + Nr)/2(1 + 2Nr)$$

$$C_\infty(Ab) = C_\infty(aB) = (r + Nr)/2(1 + 2Nr)$$

$$\%\ \text{recombinants} = (r + Nr)/(1 + 2Nr)$$

as first shown by Wright [1933b].

(2) Initial population 50% $AB/AB$, 50% $ab/ab$

$$C_0(AB) = 1/2, \qquad S_0 = 1/2, \qquad T_0 = (N - 2)/4(N - 1)$$

$$C_1 = 1/2, \qquad S_1 = 1/4$$

$$C_\infty(AB) = C_\infty(ab) = (1 + Nr)/2(1 + 2Nr)$$

$$C_\infty(Ab) = C_\infty(aB) = Nr/2(1 + 2Nr)$$

$$\%\ \text{recombinants} = Nr/(1 + 2Nr)$$

again agreeing with Wright [1933b].

When sexes are separate and the population consists of $N^*$ males and $N^{**}$ females, we have

$$C_t = (1 - r)C_{t-1} + rS_{t-1} \tag{3.1.10}$$

$$S_t = M_{t-1} \tag{3.1.11}$$

$$T_t = \frac{1}{N_e}\left(\tfrac{1}{2}C_{t-1} + \tfrac{1}{2}S_{t-1}\right) + \left(1 - \frac{1}{N_e}\right)T_{t-1} \tag{3.1.12}$$

$$M_t = T_t \qquad (t \geqq 1), \tag{3.1.13}$$

where $M_t$ is the probability that the first and the second loci taken from the two mating individuals in the $t$th generation carry alleles $A_i$ and $B_k$ respectively, while $T_t$ as before is the similar probability for two randomly chosen individuals. Furthermore $N_e$ in (3.1.12) represents the effective population number given in (2.1.6).

As compared with the monoecious case, a pair of the first and the second loci do not come from the same individual from the previous generation but do come from the same individual two generations back with the probability of $1/N_e$. For $t \geqq 1$, $M_t = T_t$ and therefore the two relations (3.1.11) and (3.1.13) may be replaced by

$$S_t = T_{t-1} \qquad (t \geqq 2). \tag{3.1.14}$$

Note that this is different from $S_t = T_t$ $(t \geqq 1)$ which we had for the monoecious case.

As $t \to \infty$, $C_t$, $S_t$ and $T_t$ all converge to the same limiting value $(C_\infty = S_\infty = T_\infty)$. In order to calculate this, we choose constants $\alpha$, $\beta$ and $\gamma$ in the expression

$$L_t = \alpha C_t + \beta S_t + \gamma T_t$$

such that

$$L_t = L_{t-1} .$$

This is satisfied with

$$\alpha = 1/2N_e , \qquad \beta = r/N_e , \qquad \gamma = r$$

and therefore we have

$$C_\infty = \left(\frac{1}{2N_e}C_1 + \frac{r}{N_e}S_1 + rT_1\right)\Big/\left(\frac{1}{2N_e} + \frac{r}{N_e} + r\right)$$

or

$$C_\infty = (C_1 + 2rS_1 + 2rN_eT_1)/(1 + 2r + 2rN_e) \tag{3.1.15}$$

where $C_1$, $S_1$ and $T_1$ are calculated from (3.1.10), (3.1.11) and (3.1.12) by setting $t = 1$. In the special case of $T_0 = M_0$, we may use the formula:

$$C_\infty = (C_0 + 2rS_0 + 2rN_eT_0)/(1 + 2r + 2rN_e). \qquad (3.1.16)$$

*Example*:

Starting with 100% $AB/ab$

$$C_0(AB) = 1/2, \qquad S_0 = 0, \qquad T_0 = M_0 = 1/4$$

$$C_\infty(AB) = C_\infty(ab) = (1 + rN_e)/2(1 + 2r + 2rN_e)$$

$$\%\text{ recombinants} = D_\infty = (2r + rN_e)/(1 + 2r + 2rN_e) \qquad (3.1.17)$$

which agrees with Wright [1933b]. If the recombination fraction is 0.5, the final proportion of recombinants is 50%. For the special case of brother-sister mating $N_e = 2$ and $D_\infty = 4r/(1 + 6r)$ as shown by Haldane and Waddington [1931].

*3.2. Sex-linked genes.*

Here the situation is more complicated, because the relation between the two loci may depend on the sexes of the individuals in which they happen to be contained. Let $C_t^*$ be the probability that a chromosome in a male of the $t$th generation carries alleles $A_i$ and $B_k$ and let $C_t^{**}$ be the similar probability for a female. The probability that the two loci from homologous chromosomes in an individual carry $A_i$ and $B_k$ is denoted by $S_t$ as in the autosomal case, but here it is defined only for females. Let $T_t^*$ be the probability that the two loci taken from different males in the $t$th generation are $A_i$ and $B_k$. A similar probability for two different females will be denoted by $T_t^{**}$, and for two different individuals, one being a male and the other female, will be denoted by $M_t$.

If we note that a chromosome in a male is derived only from his mother, while a chromosome in a female is either derived from her mother or father with equal probability, we obtain the following relations:

$$C_t^* = (1 - r)C_{t-1}^{**} + rS_{t-1} \qquad (3.2.1)$$

$$C_t^{**} = \tfrac{1}{2}C_{t-1}^* + \tfrac{1}{2}C_t^* \qquad (3.2.2)$$

$$S_t = M_{t-1} \qquad (3.2.3)$$

$$T_t^* = \frac{1}{N^{**}}\left(\tfrac{1}{2}C_{t-1}^{**} + \tfrac{1}{2}S_{t-1}\right) + \left(1 - \frac{1}{N^{**}}\right)T_{t-1}^{**} \qquad (3.2.4)$$

$$T_t^{**} = \frac{1}{4}\left[\frac{1}{N^*}C_{t-1}^* + \left(1 - \frac{1}{N^*}\right)T_{t-1}^*\right] + \tfrac{1}{2}M_{t-1} + \tfrac{1}{4}T_t^* \qquad (3.2.5)$$

$$M_t = \tfrac{1}{2}M_{t-1} + \tfrac{1}{2}T_t^*. \qquad (3.2.6)$$

The first relation (3.2.1) is derived by an argument similar to that which was used to derive (3.1.10). The second relation (3.2.2) is derived from the consideration that a chromosome in a female either comes from her father or mother; if it comes from the father, it is exactly the same as the paternal chromosome, while if it comes from the mother, the resulting chromosome is equivalent to a chromosome in a male of the contemporary generation. Note that the subscript of the last term in the right side is $t$ rather than $t - 1$. The other relation which might need explanation is (3.2.5). As to the two loci, each taken from different females, one of the following three may happen: (1) Both are derived from males in the previous generation. This occurs with probability $\frac{1}{4}$. When this happens, they are derived from the same male with probability $1/N^*$ and different males with probability $1 - 1/N^*$; (2) One is derived from a male and the other derived from a female. This occurs with the probability $\frac{1}{2}$. In this case the relation between the two loci is the same as between those in two mating individuals in the previous generation; (3) Both from females. This occurs with probability $\frac{1}{4}$ and the two loci have the same relation as those of two males in the contemporary generation.

At the limit of $t \to \infty$, all the probabilities, $C_t^*$, $C_t^{**}$, $S_t$, $T_t^*$, $T_t^{**}$, and $M_t$ converge to the same limiting value ($C_\infty$). In order to calculate this, we put

$$L_t = \alpha C_t^* + \beta C_t^{**} + \gamma S_t + \delta T_t^* + \epsilon T_t^{**} + \phi M_t$$

and choose constants $\alpha$, $\beta$, $\gamma$, $\delta$, $\epsilon$ and $\phi$ such that

$$L_t = L_{t-1} .$$

The last relation is satisfied by

$$\begin{aligned} \alpha &= \frac{N^*}{r} + \frac{N^{**} - 1}{2}\left(\frac{1}{r} + 1\right) \\ \beta &= \frac{2N^*}{r} + (N^{**} - 1)\left(\frac{1}{r} + 1\right) \\ \gamma &= 4N^* + N^{**} - 1 \\ \delta &= (N^* - 1)(N^{**} - 1) \\ \epsilon &= 4N^*(N^{**} - 1) \\ \phi &= 2(2N^*N^{**} + 2N^* + N^{**} - 1). \end{aligned} \tag{3.2.7}$$

The final proportion of $A_iB_k$ chromosomes is

$$C_\infty = \frac{\alpha C_0^* + \beta C_0^{**} + \gamma S_0 + \delta T_0^* + \epsilon T_0^{**} + \phi M_0}{\alpha + \beta + \gamma + \delta + \epsilon + \phi}. \tag{3.2.8}$$

In the special case of $N^{**} = 1$, these coefficients are $\alpha = N^*/r$, $\beta = 2N^*/r$, $\gamma = 4N^*$, $\delta = 0$, $\epsilon = 0$, $\phi = 8N^*$ or we may use $\alpha = 1/r$, $\beta = 2/r$, $\gamma = 4$, $\delta = \epsilon = 0$, $\phi = 8$.

*Example*:

Let $N^{**} = 1$, and suppose that the population starts from $AB/AB \times ab$. Let $C$, $S$, $T$, and $M$ denote the proportion of either $A$ and $B$ or $a$ and $b$. Since $C_0^* = C_0^{**} = 1$, $S_0 = 1$, $T_0^* = T_0^{**} = 1$ and $M_0 = 0$, we obtain

$$C_\infty = (3 + 4r)/(3 + 12r).$$

The final proportion of recombinants is

$$D_\infty = 1 - C_\infty = 8r/(3 + 12r)$$

which agrees with Wright [1933b].

## 4. DISCUSSION

The present method may be extended to handle directly the cases of three or more linked genes by defining a larger number of appropriate quantities, though the calculation involved might be very complicated. However, for the case of three linked genes, there is a short cut whereby the problem is solved by a simple application of the result already worked out for two linked genes. As pointed out by Haldane and Waddington [1931], if $f(r)$ is the proportion of recombinants for two linked genes with recombination fraction $r$, then the proportion of double recombinants for three linked genes $A$, $B$, $C$ (in this order) is given by

$$\tfrac{1}{2}[f(r_{AB}) + f(r_{BC}) - f(r_{AC})]. \tag{4.1}$$

Suppose, for example, a random mating population of $N$ monoecious individuals starts from the composition of 100% $ABC/abc$, then

$$f(r) = D_\infty = (1 + N)r/(1 + 2Nr)$$

and the final proportion of the double recombinants, $AbC$ and $aBc$ is given by

$$\frac{1 + N}{2}\left[\frac{r_{AB}}{1 + 2Nr_{AB}} + \frac{r_{BC}}{1 + 2Nr_{BC}} - \frac{r_{AC}}{1 + 2Nr_{AC}}\right].$$

The present method also applies to the situation in which inbreeding and random mating are mixed. As an example, consider a population of partially self-fertilizing plants, in which each individual is fertilized by its own pollen with probability $\mu$ and by a pollen randomly taken from an entire population with probability $1 - \mu$. If a gene consists of alleles $A_1$, $A_2$, $\cdots$ etc, we have

$$I_t = \frac{\mu}{2}(1 + I_{t-1}) + (1 - \mu) \sum_i p_i^2 , \tag{4.2}$$

where $p_i$ is the frequency of the $i$th allele in the population which we assume to be infinitely large. The above equation may readily be derived from the consideration that two homologous genes in an individual either come from the same parent with probability $\mu$ as the result of self-fertilization or from two unrelated parents with probability $1 - \mu$. In the former case, they either come from the same gene of that parent or from two homologous genes with equal probability. In the latter case, the result is the same with the case of random mating and the probability of the resulting two homologous chromosomes share the same particular allele, say $A_i$, is $p_i^2$. The proportion of homozygotes in the population at equilibrium is obtained from (4.2) by putting $I_t = I_{t-1} = I_\infty$. Thus we have

$$I_\infty = \frac{\mu + 2(1 - \mu) \sum_i p_i^2}{2 - \mu}. \tag{4.3}$$

The proportion of heterozygotes at equilibrium is

$$H_\infty = 1 - I_\infty = \frac{2(1 - \mu)}{2 - \mu} H(0), \tag{4.4}$$

where $H(0) = 1 - \sum_i p_i^2$ is the proportion of heterozygotes which is expected under exclusive random mating ($\mu = 0$).

For a pair of linked genes, we obtain, by similar argument

$$\begin{aligned} C_t &= (1 - r)C_{t-1} + rS_{t-1} \\ S_t &= \frac{\mu}{2}(C_{t-1} + S_{t-1}) + (1 - \mu)T_{t-1} \\ T_t &= T_{t-1} . \end{aligned} \tag{4.5}$$

If the frequencies of $A_i$ and $B_k$ are $p$ and $q$, then

$$T_t = pq \qquad (t \geqq 1).$$

At the limit of $t = \infty$, $C_\infty = S_\infty = T_\infty$ and we have

$$C_\infty = S_\infty = pq. \tag{4.6}$$

This shows that at the final equilibrium the alleles of the different loci on a chromosome combine randomly. Furthermore the coupling and repulsion phases become equally frequent. However, as shown by

Bennett and Binet [1956], and also by the present author Kimura [1958], there still exists a correlation between the first and second loci with respect to homozygosity. Our treatment here is not informative enough to bring out this fact. Actually it can not give full information to enable us to write down the frequencies of all the genotypes with respect to linked genes. In order to do this, a method has to be developed to obtain the probability that the first and the second loci are both homozygous, or equivalently the joint probability of identity by descent as considered by Haldane [1947] and Schnell [1961].

One of the most interesting applications of the present method of handling inbreeding systems is found in a mating system which the author would like to call circular mating. The simplest type of circular mating is illustrated in Figure 4, where males and females are arranged,

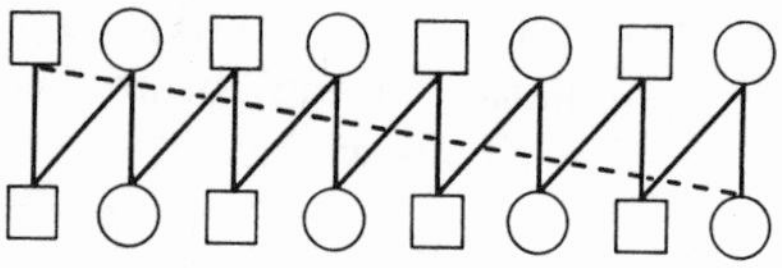

FIGURE 4.

so to speak, alternately in a circle and mating is made between the neighboring individuals.

Let the number of males or females each equal $n$ such that the total number $N$ is equal to $2n$. First consider the single gene case. Let $J_t(1)$ be the probability that two homologous gene in two adjacent individuals in the $t$th generation share the same allele. In general, let $J_t(k)$ be the similar probability for two homologous genes taken from two individuals which are $k$ steps ($k = 1, 2, \cdots 2n - 1$) removed from each other on the circle. Note that if $k>n$, then $J_t(k) = J_t(2n-k)$ because of the circular arrangement of individuals. We then obtain the following relations:

$$I_t = J_{t-1}(1) \tag{4.7}$$

$$J_t(1) = \tfrac{1}{4}[\tfrac{1}{2}(1 + I_{t-1}) + 2J_{t-1}(1) + J_{t-1}(2)] \tag{4.8}$$

$$J_t(k) = \tfrac{1}{4}[J_{t-1}(k-1) + 2J_{t-1}(k) + J_{t-1}(k+1)] \quad (k = 2, \cdots, n-1) \tag{4.9}$$

$$J_t(n) = \tfrac{1}{2}[J_{t-1}(n-1) + J_{t-1}(n)]. \tag{4.10}$$

The first relation (4.7) is simply an expression of the fact that each individual is produced from the mating of two adjacent individuals in the previous generation. Then, the two homologous genes taken one from each of the two adjacent individuals in the present generation, are derived from the same individual in the previous generation with probability $\frac{1}{4}$, from adjacent individuals with probability $\frac{1}{2}$ and from individuals two steps removed with probability $\frac{1}{4}$. When they are derived from the same individual, they are either derived from the same gene or from homologous genes with equal probability of $\frac{1}{2}$. Thus follows the relation (4.8). The relation (4.9) is obtained by similar arguments, and (4.10) follows from (4.9) by noting that $J_{t-1}(n+1) = J_{t-1}(n-1)$.

Let $H_t = 1 - I_t$, $K_t(k) = 1 - J_t(k)$, $k = 1, \cdots, n$ and writing the relation of these quantities between two consecutive generations using matrix notation, we have

$$\begin{pmatrix} H_t \\ K_t(1) \\ K_t(2) \\ \cdot \\ \cdot \\ \cdot \\ K_t(n-1) \\ K_t(n) \end{pmatrix} = \begin{bmatrix} 0 & 1 & 0 & \cdots & & & 0 \\ \frac{1}{8} & \frac{1}{2} & \frac{1}{4} & 0 & & & \vdots \\ 0 & \frac{1}{4} & \frac{1}{2} & \frac{1}{4} & 0 & & \vdots \\ \vdots & & \ddots & \ddots & \ddots & \ddots & \vdots \\ \vdots & & & \ddots & \ddots & \ddots & 0 \\ \vdots & & & & \ddots & \ddots & \\ \vdots & & & 0 & \frac{1}{4} & \frac{1}{2} & \frac{1}{4} \\ 0 & \cdots & & & 0 & \frac{1}{2} & \frac{1}{2} \end{bmatrix} \begin{pmatrix} H_{t-1} \\ K_{t-1}(1) \\ K_{t-1}(2) \\ \cdot \\ \cdot \\ \cdot \\ K_{t-1}(n-1) \\ K_{t-1}(n) \end{pmatrix} \tag{4.11}$$

We will denote the $(n+1) \times (n+1)$ matrix of the above transformation by $A$. All the characteristic roots of $A$ lie between 1 and $-1$ exclusive, and the proportion of heterozygotes decreases ultimately with the ratio $\lambda$ per generation, where $\lambda$ is the largest characteristic root. Letting

$$\epsilon = 1 - \lambda,$$

and expanding the characteristic equation of $A$ in terms of $\epsilon$, we have

$$\begin{aligned} | A - \lambda I | = -(-\tfrac{1}{4})^n[1 - 2n(n+2)\epsilon \\ + \tfrac{2}{3}n(n^3 + 4n^2 - n + 2)\epsilon^2 - \cdots]. \end{aligned} \tag{4.12}$$

Thus the rate of ultimate decrease of heterozygosis may be given approximately by

$$\epsilon = 1/2n(n+2), \tag{4.13}$$

which shows that the present mating system is equivalent to random mating with

$$N_e \approx n(n+2)$$

individuals (The exact treatment gives $\epsilon = \pi^2/[16(n + 1)^2]$ as a better approximation especially for a large value of $n$). The ratio between the effective and the actual number is asymptotically

$$\frac{N_e}{N} \sim \frac{2N}{\pi^2} \approx \frac{N}{5} \tag{4.14}$$

which shows that it becomes increasingly large as the number of individuals increases. Probably this is the most effective mating system from the standpoint of conserving heterozygosity in the long run. A fuller and more detailed account of circular mating systems, including practically more important cases of circular subpopulation mating, will be published elsewhere.

Let us now consider linked genes.

We will define by $T_t(k)$ the probability that two chromosomes, taken one from each of the two individuals which are $k$ steps ($k = 1, \cdots, n$) apart at the $t$th generation carry alleles $A_i$ and $B_k$, respectively. Then by a similar argument which we used to derive (4.7)–(4.10), we obtain

$$\begin{aligned}
C_t &= (1 - r)C_{t-1} + rS_{t-1} \\
S_t &= T_{t-1}(1) \\
T_t(1) &= \tfrac{1}{4}(\tfrac{1}{2}C_{t-1} + \tfrac{1}{2}S_{t-1}) + \tfrac{1}{2}T_{t-1}(1) + \tfrac{1}{4}T_{t-1}(2) \\
T_t(k) &= \tfrac{1}{4}T_{t-1}(k - 1) + \tfrac{1}{2}T_{t-1}(k) + \tfrac{1}{4}T_{t-1}(k - 1) \\
&\qquad (k = 2, \cdots, n - 1) \\
T_t(n) &= \tfrac{1}{2}T_{t-1}(n - 1) + \tfrac{1}{2}T_{t-1}(n)
\end{aligned} \tag{4.15}$$

At the limit of $t = \infty$, $C_t$, $S_t$ and all $T_t(k)$'s converge to the same value, $C_\infty$ and this can be obtained by using the linear combination of $C_t$, $S_t$ and $T_t(k)$'s,

$$L_t = \frac{1}{8r} C_t + \tfrac{1}{4}S_t + \sum_{k=1}^{n-1} T_t(k) + T_t(n),$$

which satisfies the relation

$$L_t = L_{t-1} .$$

Thus we obtain

$$C_\infty = \frac{\dfrac{1}{8r} C_0 + \tfrac{1}{4}S_0 + \sum_{1}^{n-1} T_0(k) + \tfrac{1}{2}T_0(n)}{\dfrac{1}{8r} + \tfrac{1}{4} + (n - 1) + \tfrac{1}{2}}. \tag{4.16}$$

*Example*:

Starting from 100% $AB/ab$, $C_0(AB) = \frac{1}{2}$, $S_0 = 0$, $T_0(k) = \frac{1}{4}$ $(k = 1, 2, \cdots, n)$ and we have

$$2C_\infty = (1 + 4nr - 2r)/(1 + 8nr - 2r)$$

and the proportion of recombinants

$$D_\infty = 1 - 2C_\infty = 4nr/(1 + 8nr - 2r).$$

With 50% recombination between the two genes, the final proportion of recombinants is $\frac{1}{2}$. As compared with the case of $n$ males and $n$ females mating at random, in which the proportion of final recombinants was shown to be

$$D_\infty\,(\text{random}) = (2r + 2nr)/(1 + 2r + 4nr) \qquad \text{(see 3.1.17)},$$

the present case gives a higher proportion of final recombinants, since

$$D_\infty\,(\text{circular}) - D_\infty\,(\text{random}) = \frac{4nr}{1 - 2r + 8nr} - \frac{2r + 2nr}{1 + 2r + 4nr}$$
$$= \frac{2r(n - 1)(1 - 2r)}{(1 - 2r + 8nr)(1 + 2r + 4nr)} \geqq 0.$$

This is expected since heterozygosity decreases more slowly in circular mating than in the random mating.

ACKNOWLEDGEMENT

The present paper is an outcome of stimulating discussions with Dr. J. F. Crow, to whom the author expresses his sincere thanks, both for these discussions and also for helping to improve the manuscript. Thanks are also due to Drs. Sewall Wright and C. C. Li who carefully read the manuscript and gave valuable criticisms.

REFERENCES

Bennett, J. H. and F. E. Binet [1956]. Association between mendelian factors with mixed selfing and random mating. *Heredity 10*, 51–5.

Haldane, J. B. S. and C. H. Waddington [1931]. Inbreeding and linkage. *Genetics 16*, 357–74.

Haldane, J. B. S. [1947]. The association of characters as a result of inbreeding and linkage. *Ann. Eugen. 15*, 15–23.

Kimura, M. [1958]. Zygotic frequencies in a partially self-fertilizing population. *Ann. Rep. National Institute of Genetics Japan No. 8*, 104–5.

Malécot, G. [1948]. *Le Mathématiques de L'Hérédité*. Masson et Cie, Paris.

Robbins, R. B. [1918]. Some applications of mathematics to breeding problems III. *Genetics 3*, 375–89.

Schnell, F. W. [1961]. Some general formulas of linkage effects in inbreeding. *Genetics 46*, 947–57.

Wright, S. [1933a]. Inbreeding and homozygosis. *Proc. Nat. Acad. Sci. 19*, 411–20.

Wright, S. [1933b]. Inbreeding and recombination. *Proc. Nat. Acad. Sci. 19*, 420–33.

# On the maximum avoidance of inbreeding*

By MOTOO KIMURA† AND JAMES F. CROW
*University of Wisconsin, Madison, Wisconsin, U.S.A.*

(*Received* 16 *April* 1963)

## 1. INTRODUCTION

In his classical 1921 paper Sewall Wright determined the decrease in heterozygosity in successive generations with various mating systems. Some systems, such as regular mating of double first cousins in a population of 4, quadruple second cousins in a population of 8, octuple third cousins in a population of 16, etc., have the property that the matings are between individuals that are least related. Wright designated such systems as having maximum avoidance of inbreeding.

It might be thought that such systems would lead to the minimum decrease of heterozygosity, but this is not generally true. Although the 'maximum avoidance' systems have the slowest initial rate of decrease of heterozygosity, there are other systems which have more remaining heterozygosity in later generations. The purpose of this article is to discuss some mating systems for which this is true.

Inbreeding has two related, but distinct effects. One is the decrease in average heterozygosity; the other is random drift in gene frequencies. A system that minimizes one of these processes is not necessarily minimum for the other. We shall consider first some systems that have very slow ultimate rates of decrease in heterozygosity, and compare these with those having maximum avoidance of inbreeding. Later, in section 6, we shall consider systems that minimize gene frequency drift.

Other things being comparable, either kind of inbreeding effect will be less when the number of progeny per parent is constant. We shall consider only mating systems in which this is true; we assume that the number of progeny per parent (or per pair) is exactly two each generation so that the population size remains constant.

Figures 1, 2, and 3 illustrate the three principal systems of mating to be compared for a population of size 8. Figure 1 shows maximum avoidance of inbreeding (quadruple second cousins), Figs. 2 and 3 show, respectively, the systems that we shall call circular mating and circular pair mating.

* Paper Number 916 from the Division of Genetics; also paper number 464 from the National Institute of Genetics, Mishima-shi, Japan. Supported in part by the National Institutes of Health (RG-8217). Sponsored by Mathematics Research Center, U.S. Army, Madison, Wisconsin (Contract No. DA-11-022-ORD-2059).

† On leave from the National Institute of Genetics, Mishima-shi, Japan.

## 2. CIRCULAR MATING

In the simplest system to be considered, $n$ males and $n$ females are arranged alternately so that each individual is mated with its neighbour. The last individual is mated with the first, so that the system is most easily visualized as circular. Under this system, each individual after the second generation is the product of a half-sib mating. Figure 2 illustrates continued circular mating for the case, $n = 4$.

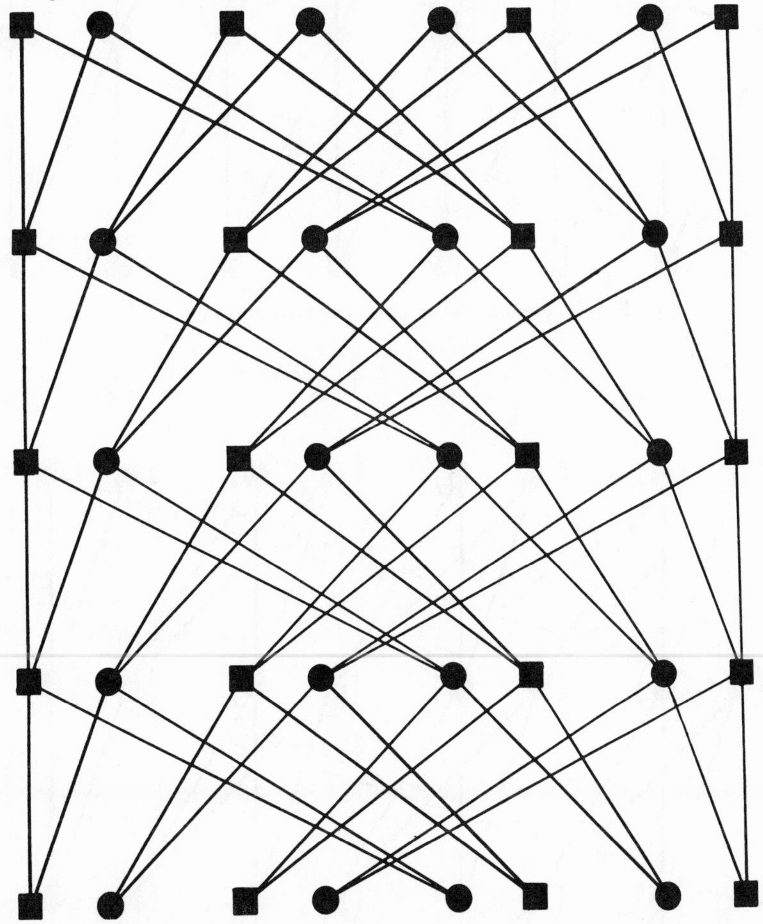

Fig. 1. Maximum avoidance of inbreeding in a population of constant size 8. This and the other pedigrees are to be read down; the oldest generation is at the top.

Let the total number of individuals in a generation be $N = 2n$. We designate by $I_t$ the probability that two homologous loci in an individual in generation $t (t = 0, 1, 2, \ldots)$ share the same allele; that is, that an individual is homozygous for any one of the possible alleles at this locus. Similarly, let $J_t(1)$ be the probability that two randomly chosen homologous genes in two adjacent individuals in the $t^{\text{th}}$ generation share the same allele. In general, let $J_t(k)$ be the similar probability

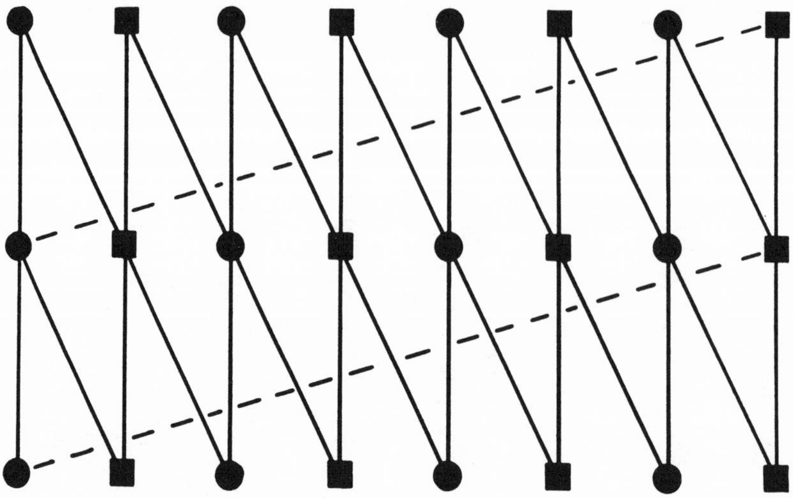

Fig. 2. Circular mating in a population of size 8.

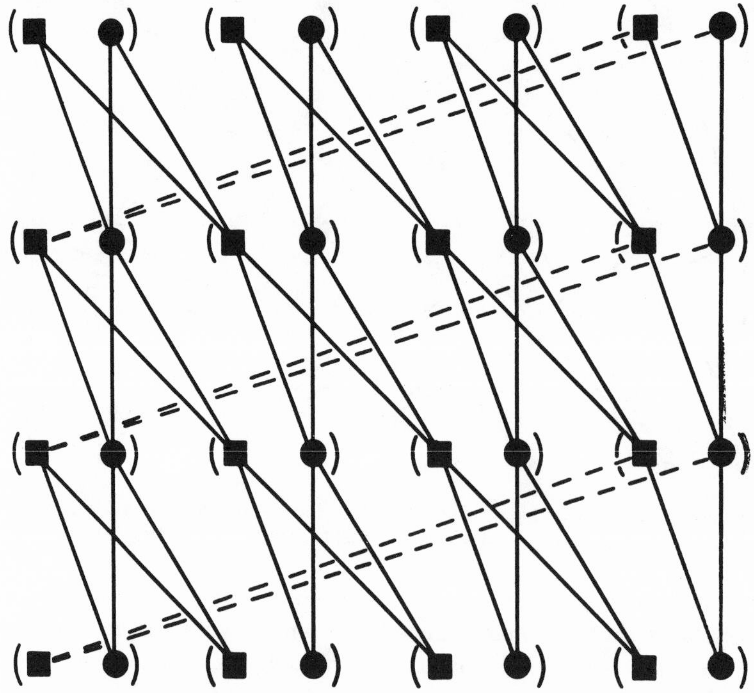

Fig. 3. Circular pair mating in a population of size 8 (4 pairs). Parentheses designate cages, and the animals are shown in the cages at the time of mating.

that two homologous genes in individuals $k$ steps removed are identical ($k = 1, 2, \ldots 2n-1$). Note that because of the circular arrangement $J_t(k) = J_t(2n-k)$. We then have:

$$I_t = J_{t-1}(1) \tag{2.1}$$

$$J_t(1) = \tfrac{1}{4}[\tfrac{1}{2}(1+I_{t-1})+2J_{t-1}(1)+J_{t-1}(2)] \tag{2.2}$$

$$J_t(2) = \tfrac{1}{4}[J_{t-1}(1)+2J_{t-1}(2)+J_{t-1}(3)] \tag{2.3}$$

$$J_t(k) = \tfrac{1}{4}[J_{t-1}(k-1)+2J_{t-1}(k)+J_{t-1}(k+1)] \quad (n-1 \geqq k \geqq 2) \tag{2.4}$$

$$J_t(n-1) = \tfrac{1}{4}[J_{t-1}(n-2)+2J_{t-1}(n-1)+J_{t-1}(n)] \tag{2.5}$$

$$J_t(n) = \tfrac{1}{2}[J_{t-1}(n-1)+J_t(n)] \tag{2.6}$$

The first relation (2.1) follows immediately from the fact that two homologous genes in an individual are derived from two adjacent individuals in the preceding generation. The second relation (2.2) is derived from the consideration that two homologous genes, taken one from each of the adjacent individuals, are derived from the same parent with probability 1/4, from two adjacent individuals with probability 1/2 and from two individuals two steps removed with probability 1/4. When they are derived from the same individual, they either come from the same gene or from two homologous genes with equal probability of 1/2. In the former case the probability is 1 that they share the same allele, while in the latter case, the probability is $I_{t-1}$ by definition. Relations (2.3) to (2.5) are derived from similar considerations and (2.6) follows from (2.4) by putting $k = n$ and noting that $J_t(n-1) = J_t(n+1)$.

The heterozygosity at time $t$ is proportional to $H_t = 1 - I_t$. From the above relations the proportion of heterozygosity can be worked out in successive generations. This was done by digital computer. The heterozygosity as a proportion of the initial value is shown for populations of size $N = 4$, 8, 16, and 32 in Table 1, along with the corresponding changes for maximum avoidance and for circular pair mating. In each case it will be seen that circular mating has a more rapid initial loss of heterozygosity, but that eventually the rate of loss is much less.

Relations (2.1) through (2.6) may easily be expressed in matrix form by letting $H_t = 1 - I_t$ and $K_t(k) = 1 - J_t(k)$. This is given by

$$\begin{bmatrix} H_t \\ K_t(1) \\ K_t(2) \\ \cdot \\ \cdot \\ K_t(n-2) \\ K_t(n-1) \\ K_t(n) \end{bmatrix} = \begin{bmatrix} 0 & 1 & 0 & 0 & 0 & . & . & 0 & 0 & 0 & 0 \\ \frac{1}{8} & \frac{1}{2} & \frac{1}{4} & 0 & 0 & . & . & 0 & 0 & 0 & 0 \\ 0 & \frac{1}{4} & \frac{1}{2} & \frac{1}{4} & 0 & . & . & 0 & 0 & 0 & 0 \\ . & . & . & . & . & . & . & . & . & . & . \\ . & . & . & . & . & . & . & . & . & . & . \\ 0 & 0 & 0 & 0 & 0 & . & . & \frac{1}{4} & \frac{1}{2} & \frac{1}{4} & 0 \\ 0 & 0 & 0 & 0 & 0 & . & . & 0 & \frac{1}{4} & \frac{1}{2} & \frac{1}{4} \\ 0 & 0 & 0 & 0 & 0 & . & . & 0 & 0 & \frac{1}{2} & \frac{1}{2} \end{bmatrix} \begin{bmatrix} H_{t-1} \\ K_{t-1}(1) \\ K_{t-1}(2) \\ \\ \\ K_{t-1}(n-2) \\ K_{t-1}(n-1) \\ K_{t-1}(n) \end{bmatrix} \tag{2.7}$$

**Table 1.** *Decrease in heterozygosity with maximum avoidance of inbreeding (M), circular pair mating (CP), and circular mating (C) in populations of size 4, 8, 16, and 32. The value given is* $H_t/H_0$, *where* $H_t$ *is the heterozygosity in generation* t. $1-\lambda$ *is the asymptotic rate of decrease in heterozygosity per generation*

| | $N=4$ | | $N=8$ | | | $N=16$ | | | $N=32$ | | |
|---|---|---|---|---|---|---|---|---|---|---|---|
| $t$ | M, CP | C | M | CP | C | M | CP | C | M | CP | C |
| 0 | 1·000 | 1·000 | 1·000 | 1·000 | 1·000 | 1·000 | 1·000 | 1·000 | 1·000 | 1·000 | 1·000 |
| 1 | 1·000 | 1·000 | 1·000 | 1·000 | 1·000 | 1·000 | 1·000 | 1·000 | 1·000 | 1·000 | 1·000 |
| 2 | 1·000 | 0·875 | 1·000 | 1·000 | 0·875 | 1·000 | 1·000 | 0·875 | 1·000 | 1·000 | 0·875 |
| 3 | 0·875 | 0·813 | 1·000 | 0·938 | 0·813 | 1·000 | 0·938 | 0·813 | 1·000 | 0·938 | 0·813 |
| 4 | 0·813 | 0·750 | 0·938 | 0·906 | 0·758 | 1·000 | 0·906 | 0·758 | 1·000 | 0·906 | 0·758 |
| 5 | 0·750 | 0·695 | 0·906 | 0·875 | 0·715 | 0·969 | 0·879 | 0·715 | 1·000 | 0·879 | 0·715 |
| 10 | 0·492 | 0·477 | 0·755 | 0·732 | 0·577 | 0·891 | 0·777 | 0·577 | 0·953 | 0·777 | 0·577 |
| 15 | 0·324 | 0·327 | 0·628 | 0·614 | 0·494 | 0·817 | 0·709 | 0·499 | 0·915 | 0·711 | 0·499 |
| 20 | 0·213 | 0·224 | 0·522 | 0·514 | 0·432 | 0·750 | 0·655 | 0·446 | 0·878 | 0·662 | 0·446 |
| 25 | 0·140 | 0·154 | 0·434 | 0·431 | 0·379 | 0·688 | 0·608 | 0·407 | 0·842 | 0·623 | 0·407 |
| 30 | 0·092 | 0·105 | 0·361 | 0·361 | 0·334 | 0·632 | 0·566 | 0·377 | 0·808 | 0·592 | 0·377 |
| 40 | 0·040 | 0·050 | 0·250 | 0·254 | 0·259 | 0·532 | 0·490 | 0·333 | 0·744 | 0·542 | 0·333 |
| 50 | 0·017 | 0·023 | 0·173 | 0·178 | 0·202 | 0·448 | 0·424 | 0·300 | 0·685 | 0·503 | 0·302 |
| 70 | 0·003 | 0·005 | 0·083 | 0·088 | 0·122 | 0·318 | 0·318 | 0·252 | 0·581 | 0·443 | 0·259 |
| 100 | 0·000 | 0·000 | 0·027 | 0·030 | 0·057 | 0·190 | 0·207 | 0·199 | 0·454 | 0·374 | 0·219 |
| 150 | | | 0·004 | 0·005 | 0·016 | 0·080 | 0·101 | 0·136 | 0·301 | 0·287 | 0·180 |
| 200 | | | 0·000 | 0·001 | 0·004 | 0·034 | 0·049 | 0·092 | 0·199 | 0·220 | 0·156 |
| 300 | | | | 0·000 | 0·000 | 0·006 | 0·012 | 0·043 | 0·087 | 0·130 | 0·123 |
| 400 | | | | | | 0·001 | 0·003 | 0·020 | 0·038 | 0·077 | 0·099 |
| 500 | | | | | | | 0·001 | 0·009 | 0·017 | 0·045 | 0·080 |
| $1-\lambda$ | 0·0804 | 0·0727 | 0·0362 | 0·0347 | 0·0249 | 0·0170 | 0·0142 | 0·0076 | 0·0082 | 0·0053 | 0·0021 |

We designate this $(n+1)\times(n+1)$ matrix by $A$. Also we designate $|A-\lambda I|$ by $F_n(\lambda)$, so that the characteristic equation of $A$ is written as

$$F_n(\lambda) = |A-\lambda I| = 0$$

The ultimate ratio by which the frequency of heterozygotes decreases from one generation to the next is given by the largest root of this equation. For $N = 2, 4, 6$ and 8, the characteristic equations and the corresponding dominant roots are as follows:

$$N = 2;\quad F_1(\lambda) = \lambda^2-\tfrac{1}{2}\lambda-\tfrac{1}{4} = 0$$
$$\lambda = 0{\cdot}8090$$

$$N = 4;\quad F_2(\lambda) = -\lambda^3+\lambda^2-\tfrac{1}{16} = 0$$
$$\lambda = 0{\cdot}9273$$

$$N = 6;\quad F_3(\lambda) = \lambda^4-\tfrac{3}{2}\lambda^3+\tfrac{7}{16}\lambda^2+\tfrac{3}{32}\lambda-\tfrac{1}{64} = 0$$
$$\lambda = 0{\cdot}9606$$

$$N = 8;\quad F_4(\lambda) = -\lambda^5+2\lambda^4-\tfrac{9}{8}\lambda^3+\tfrac{1}{16}\lambda^2+\tfrac{1}{16}\lambda-\tfrac{1}{256} = 0$$
$$\lambda = 0{\cdot}9751$$

The case with $N = 2$ is ordinary brother–sister mating with recurrence relation $H_t = \frac{1}{2}H_{t-1} + \frac{1}{4}H_{t-2}$ for heterozygote frequencies (Wright, 1921). To compute the dominant root for a larger $n$, a more general treatment of the characteristic equation is required and this can be done as follows.

By expanding the determinant $|A - \lambda I|$ with respect to the first two and the last two rows, we get

$$F_n(\lambda) = -\lambda(\tfrac{1}{2}-\lambda)(\tfrac{1}{4})^{n-1}\phi_{n-1}(2-4\lambda) + \left(\frac{\lambda}{4}-\frac{1}{16}\right)(\tfrac{1}{4})^{n-2}\phi_{n-2}(2-4\lambda) + (\tfrac{1}{8})^2(\tfrac{1}{4})^{n-3}\phi_{n-3}(2-4\lambda), \tag{2.8}$$

where $\phi_n(\cdot)$ is an $n \times n$ determinant known as Wolstenholme's determinant, defined by

$$\phi_n(x) = \begin{vmatrix} x & 1 & 0 & 0 & . & . & 0 & 0 & 0 \\ 1 & x & 1 & 0 & . & . & 0 & 0 & 0 \\ 0 & 1 & x & 1 & . & . & 0 & 0 & 0 \\ . & . & . & . & . & . & . & . & . \\ . & . & . & . & . & . & . & . & . \\ 0 & 0 & 0 & 0 & . & . & 0 & 1 & x \end{vmatrix} \tag{2.9}$$

(cf. Rutherford, 1952). One of the important properties of this determinant is that if $x = -2\cos\theta$, then

$$\phi_n(x) = (-1)^n \frac{\sin(n+1)\theta}{\sin\theta}. \tag{2.10}$$

Thus, if we put

$$2 - 4\lambda = -2\cos\theta$$

or

$$\lambda = \tfrac{1}{2}(1+\cos\theta) \tag{2.11}$$

then, (2.8) reduces to

$$F_n(\lambda) = (-\tfrac{1}{4})^n[\sin\theta \sin n\theta - \cos n\theta]. \tag{2.12}$$

This shows that the dominant root and, indeed, all the roots of the characteristic equation are given by (2.11) with $\theta$ satisfying the equation.

$$\sin\theta = \frac{\cos n\theta}{\sin n\theta}. \tag{2.13}$$

Though (2.12) was derived by assuming $|A - \lambda I|$ has at least 4 rows, it turns out that (2.13) is valid for all $n \geqq 1$.

For a large $n$, the following series expansion is useful to compute the dominant root:

$$\lambda = 1 - \frac{\pi^2}{16(n+1)^2} + \frac{\pi^4}{4!2^5(n+1)^4} + \cdots \tag{2.14}$$

Since $H_t$ is the probability that an individual in the $t^{\text{th}}$ generation is heterozygous with respect to the locus under consideration, $1-\lambda$ with $\lambda$ given by (2.14) is equal to the rate of decrease of heterozygosity at the state of steady decay. As $n$ becomes large, the higher terms of the series decrease in absolute value very rapidly, so that for $n > 4$, the first two terms of the series are sufficient to evaluate the dominant root for any practical purpose. Thus we obtain

$$\lambda \approx 1 - \frac{\pi^2}{4(N+2)^2} \quad (N \to \infty) \tag{2.15}$$

Since it is known that $\lambda = 1-(1/2N)$ for a randomly mating population of $N$ breeding individuals, the above result shows that under circular individual mating the decrease of heterozygosity is extremely slow on a long-term basis. However it may be many generations before the superiority over other mating systems is manifest—so long that much of the heterozygosity is already lost.

For a monoecious organism, circular mating can be carried out among odd numbers of individuals, but we will not consider such a case in this paper.

## 3. CIRCULAR PAIR MATING

Although circular mating is extremely effective in slowing down the progress toward homozygosity in the long run, it is not a convenient mating system in practice because each individual has to be mated twice. A rather similar system is what we have called circular pair mating. This is illustrated for a population of 8 in Fig. 3. If the animals are thought of as being in four cages, indicated by parentheses, then in each generation a male is mated with a female in the cage to his right. This system, being monogamous, should be very convenient for insects or litter-bearing

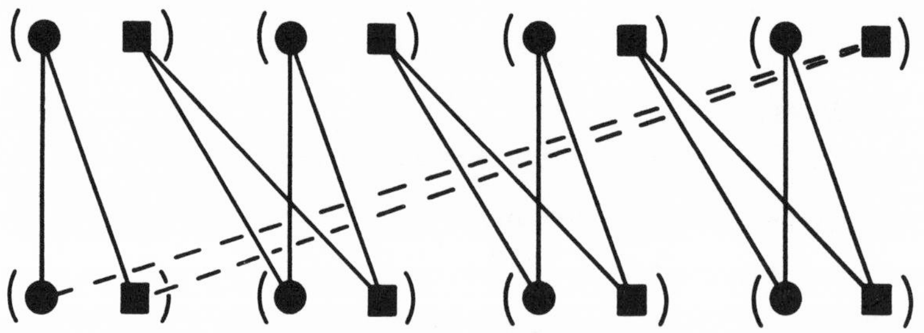

Fig. 4. Circular pair mating; same as Fig. 3 except that the animals are shown in cages at the time of birth.

mammals. It was suggested by J. B. S. Haldane in a personal communication to Professor Wright.

Figure 3 is drawn with the animals in their cages at the time of mating. For analytical purposes we have found it somewhat simpler to draw the pedigree with the animals in their positions at the time of birth, as shown in Fig. 4.

As before, let $I_t$ be the probability that an individual in generation $t$ is homozygous. Let $J_t(0)$ be the probability that the homologous genes taken one from each of the sibs born in the same cage in generation $t$ are the same allele. Similarly, we will denote by $J_t(k)$ the probability that two homologous genes taken one from each of two newly born individuals in cages $k$ steps apart are identical alleles.

In circular pair mating, the number of pairs may be either even or odd. In the former case we will denote the number of pairs by $2n$ and in the latter by $2n+1$.

First consider the even case ($2n$ pairs) where the total number of individuals is $N = 4n$. The recurrence relation connecting the $I$ and $J$'s in two consecutive generations are as follows:

$$I_t = J_{t-1}(1) \tag{3.1}$$

$$J_t(0) = \tfrac{1}{2}[\tfrac{1}{2}(I_{t-1}+1)+J_{t-1}(1)] \tag{3.2}$$

$$J_t(k) = \tfrac{1}{4}[J_{t-1}(k-1)+2J_{t-1}(k)+J_{t-1}(k+1)] \quad (1 \leqq k \leqq n-1) \tag{3.3}$$

$$J_t(n) = \tfrac{1}{2}[J_{t-1}(n-1)+J_{t-1}(n)] \tag{3.4}$$

These are readily derived by arguments very similar to the ones through which (2.1)–(2.6) were derived.

Let $H_t = 1-I_t$ and $K_t(k) = 1-J_t(k)$, $k = 0, 1, 2, \ldots, n$, then we have

$$\begin{bmatrix} H_t \\ K_t(0) \\ K_t(1) \\ K_t(2) \\ \cdot \\ \cdot \\ K_t(n-1) \\ K_t(n) \end{bmatrix} = \begin{bmatrix} 0 & 0 & 1 & 0 & 0 & . & . & 0 & 0 & 0 \\ \frac{1}{4} & 0 & \frac{1}{2} & 0 & 0 & . & . & 0 & 0 & 0 \\ 0 & \frac{1}{4} & \frac{1}{2} & \frac{1}{4} & 0 & . & . & 0 & 0 & 0 \\ 0 & 0 & \frac{1}{4} & \frac{1}{2} & \frac{1}{4} & . & . & 0 & 0 & 0 \\ . & . & . & . & . & . & . & . & . & . \\ . & . & . & . & . & . & . & . & . & . \\ 0 & 0 & 0 & 0 & 0 & . & . & \frac{1}{4} & \frac{1}{2} & \frac{1}{4} \\ 0 & 0 & 0 & 0 & 0 & . & . & 0 & \frac{1}{2} & \frac{1}{2} \end{bmatrix} \begin{bmatrix} H_{t-1} \\ K_{t-1}(0) \\ K_{t-1}(1) \\ K_{t-1}(2) \\ \cdot \\ \cdot \\ K_{t-1}(n-1) \\ K_{t-1}(n) \end{bmatrix} \tag{3.5}$$

The characteristic equation of the above $(n+2)\times(n+2)$ matrix can again be expressed in terms of Wolstenholme's determinant $\phi_n(\cdot)$ as follows:

$$\begin{aligned} F_n(\lambda) = {} & (\tfrac{1}{4})^n[4\lambda^2(\tfrac{1}{2}-\lambda)\phi_{n-1}(2-4\lambda)+4(\tfrac{1}{8}-\lambda^2)\phi_{n-2}(2-4\lambda)- \\ & -(\lambda+\tfrac{1}{2})\phi_{n-3}(2-4\lambda)] = 0. \end{aligned} \tag{3.6}$$

This leads to

$$\left.\begin{aligned} \lambda &= \tfrac{1}{2}(1+\cos\theta) \\ \sin\theta &= \frac{\cot n\theta}{2+\cos\theta} \end{aligned}\right\} \tag{3.7}$$

For a large $n$, we obtain

$$\lambda = 1-\frac{\pi^2}{16(n+3)^2} \tag{3.8}$$

or, putting $n = N/4$,

$$\lambda = 1-\frac{\pi^2}{(N+12)^2} \tag{3.9}$$

Numerical values are given in Table 1.

In the case of odd $(2n+1)$ pairs, the total number of individuals is $N = 4n+2$ and the equation corresponding to (3.5) becomes

$$\begin{bmatrix} H_t \\ K_t(0) \\ K_t(1) \\ \cdot \\ \cdot \\ K_t(n-1) \\ K_t(n) \end{bmatrix} = \begin{bmatrix} 0 & 0 & 1 & 0 & \cdot & \cdot & 0 & 0 & 0 \\ \frac{1}{4} & 0 & \frac{1}{2} & 0 & \cdot & \cdot & 0 & 0 & 0 \\ 0 & \frac{1}{4} & \frac{1}{2} & \frac{1}{4} & \cdot & \cdot & 0 & 0 & 0 \\ \cdot & \cdot & \cdot & \cdot & \cdot & \cdot & \cdot & \cdot & \cdot \\ \cdot & \cdot & \cdot & \cdot & \cdot & \cdot & \cdot & \cdot & \cdot \\ 0 & 0 & 0 & 0 & \cdot & \cdot & \frac{1}{4} & \frac{1}{2} & \frac{1}{4} \\ 0 & 0 & 0 & 0 & \cdot & \cdot & 0 & \frac{1}{4} & \frac{3}{4} \end{bmatrix} \begin{bmatrix} H_{t-1} \\ K_{t-1}(0) \\ K_{t-1}(1) \\ \cdot \\ \cdot \\ K_{t-1}(n-1) \\ K_{t-1}(n) \end{bmatrix} \qquad (3.10)$$

In this case, the equation giving $\lambda$ turns out to be as follows:

$$\left.\begin{aligned} \lambda &= \tfrac{1}{2}(1+\cos\ \theta) \\ \tan\frac{\theta}{2} &= \left(\frac{\cos^2\theta+\cos\theta-1}{\cos^2\theta+3\cos\theta+3}\right)\cot n\theta \end{aligned}\right\} \qquad (3.11)$$

For a large $n$, we obtain

$$\lambda = 1-\frac{\pi^2}{16}\frac{1}{(n+\frac{7}{2})^2}, \qquad (3.12)$$

or, putting $n = (N-2)/4$,

$$\lambda = 1-\frac{\pi^2}{(N+12)^2} \qquad (3.13)$$

which fortunately is the same as (3.9). Comparison of (3.13) or (3.9) with (2.15) shows that asymptotically circular pair mating needs twice as many individuals as the circular individual mating to attain the same rate of decrease in heterozygosity.

## 4. CIRCULAR SUBPOPULATION MATING

Under some circumstances it may be convenient to have several parents in a single cage. It is natural to extend the circular mating system to include this possibility. For example, each generation the male progeny could be transferred one cage to the right.

Let the number of cages be $2n$ (or $2n+1$ if the number is odd) and in each cage let $m^*$ be the number of males and $m^{**}$ be the number of females. Figure 5 illustrates an example where $2n = 4$, $m^* = 2$, and $m^{**} = 3$.

We assume that mating within a cage or subpopulation is random and that each parent of a given sex has the same expectation of progeny. As before, we denote by $I$ the probability that an individual in the $t^{\text{th}}$ generation is homozygous with respects to any of the alleles, $A_1$, $A_2$, etc. Let $J_t(k)$ be the probability that two homologous genes taken one from each of two newly born individuals coming from subpopulations $k$ steps apart are identical. In particular, $J_t(0)$ stands for the probability of two homologous genes taken one from each of two newly born individuals within a subpopulation are the same allele.

In the case of an even number of subpopulations, if $2n$ is the number of subpopulations such that the total number of breeding individuals is

$$N = 2n(m^* + m^{**}),$$

we have the following recurrence relations:

$$I_t = J_{t-1}(1) \tag{4.1}$$

$$J_t(0) = \frac{1}{2}\left\{\frac{1}{2}\left[\frac{1}{m^*}\left(\frac{1+I_{t-1}}{2}\right) + \left(1 - \frac{1}{m^*}\right)J_{t-1}(0)\right] + \right.$$
$$\left. + \frac{1}{2}\left[\frac{1}{m^{**}}\left(\frac{1+I_{t-1}}{2}\right) + \left(1 - \frac{1}{m^{**}}\right)J_{t-1}(0)\right]\right\} + \tfrac{1}{2}J_{t-1}(1) \tag{4.2}$$

$$J_t(k) = \tfrac{1}{4}[J_{t-1}(k-1) + 2J_{t-1}(k) + J_{t-1}(k+1)] \quad (1 \leqq k \leqq n-1) \tag{4.3}$$

$$J_t(n) = \tfrac{1}{2}[J_{t-1}(n-1) + J_{t-1}(n)] \tag{4.4}$$

These relations are similar to (3.1)–(3.4) and only (4.2) requires explanation: Two homologous genes, taken one from each of two individuals within a sub-

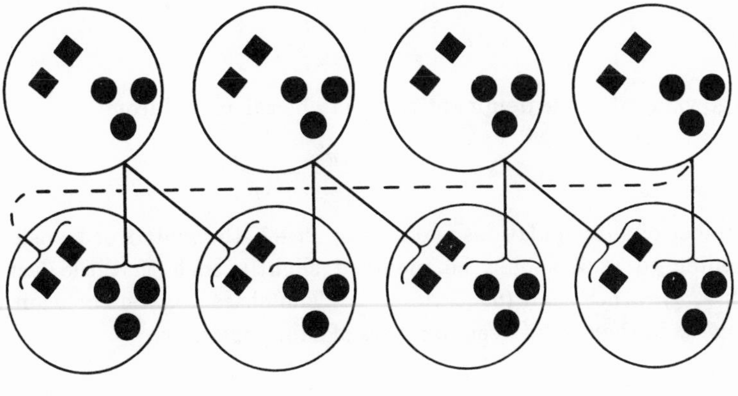

Fig. 5. Circular population mating for 4 populations, each with 2 males and 3 females.

population, are either derived from the same parental subpopulation or from two adjacent parental subpopulations with equal probability of 1/2. In the former case, they are either derived from a male, or a female, with equal probability of 1/2. If paternal, they are derived from the same male with probability $1/m^*$, and from different males with probability $1 - 1/m^*$. Similarly, if maternal, they are derived from the same female with probability $1/m^{**}$ and different females with probability $1 - 1/m^{**}$. Irrespective of whether they are derived from male or female, the probability of the two homologous genes sharing the same allele is $(1 + I_{t-1})/2$ if they are derived from the same individual and $J_{t-1}(0)$ if they are derived from different individuals originated from the same subpopulation. On the other hand, if they are derived from adjacent parental subpopulations, the probability that they share the same allele is $J_{t-1}(1)$.

Let $H_t = 1_t - I_t$ and $K_t(k) = 1 - J_t(k)$, $k = 0, 1, 2, \ldots, n$. Then

$$\begin{bmatrix} H_t \\ K_t(0) \\ K_t(1) \\ \cdot \\ \cdot \\ K_t(n-1) \\ K_t(n) \end{bmatrix} = \begin{bmatrix} 0 & 0 & 1 & 0 & . & . & 0 & 0 & 0 \\ \frac{1}{4m} & \left(\frac{1}{2}-\frac{1}{2m}\right) & \frac{1}{2} & 0 & . & . & 0 & 0 & 0 \\ 0 & \frac{1}{4} & \frac{1}{2} & \frac{1}{4} & . & . & 0 & 0 & 0 \\ \cdot & \cdot & \cdot & \cdot & \cdot & \cdot & \cdot & \cdot & \cdot \\ \cdot & \cdot & \cdot & \cdot & \cdot & \cdot & \cdot & \cdot & \cdot \\ 0 & 0 & 0 & 0 & . & . & \frac{1}{4} & \frac{1}{2} & \frac{1}{4} \\ 0 & 0 & 0 & 0 & . & . & 0 & \frac{1}{2} & \frac{1}{2} \end{bmatrix} \begin{bmatrix} H_{t-1} \\ K_{t-1}(0) \\ K_{t-1}(1) \\ \cdot \\ \cdot \\ K_{t-1}(n-1) \\ K_t(n) \end{bmatrix} \tag{4.5}$$

where $m$ is the harmonic mean of the number of males and females;

$$m \doteq 2\bigg/\left(\frac{1}{m^*}+\frac{1}{m^{**}}\right). \tag{4.6}$$

Performing the calculation as before, the characteristic equation giving $\lambda$ becomes

$$\left.\begin{aligned} \lambda &= \tfrac{1}{2}(1+\cos\theta) \\ \sin\theta &= \frac{\cot n\theta}{m(1+\cos\theta)+1} \end{aligned}\right\} \tag{4.7}$$

For a large value of $n$, the dominant root may be calculated from

$$\lambda = 1 - \frac{\pi^2}{16(n+2m+1)^2} \tag{4.8}$$

If the number of subpopulations is odd, say $2n+1$, the matrix of transformation corresponding to (4.5) should be modified slightly such that the last row is $(0, \ldots, 0, \frac{1}{4}, \frac{3}{4})$ rather than $(0, \ldots, 0, \frac{1}{2}, \frac{1}{2})$. The characteristic equation may be worked out as in the case of even numbers of subpopulations.

## 5. COMPARISON OF VARIOUS MATING SYSTEMS WITH RESPECT TO THE RATE OF DECREASE OF HETEROZYGOSITY

It has been shown by Wright (1931) that in a random-mating population of $N$ breeding individuals equally divided between males and females, the rate of decrease of heterozygosity is approximately $1/(2N)$ per generation, i.e.

$$1-\lambda \sim \frac{1}{2N}. \tag{5.1}$$

With $N$ breeding individuals, but under 'maximum avoidance' of consanguineous mating, the ultimate rate of decrease of heterozygosity is asymptotically $1/(4N)$ per generation (see Wright, 1951), i.e.

$$1-\lambda \sim \frac{1}{4N}. \tag{5.2}$$

From the common notion that the decrease of heterozygosity is directly related to the intensity of inbreeding, these results seem to show that $1/4N$ is the minimum rate of decrease of heterozygosity that can possibly be attained with $N$ breeding individuals per generation. Furthermore, Wright (1931) has shown that if each individual leaves exactly two offspring per generation, the effective size of population is approximately twice the actual size. In other words, we again have relation (5.2). Thus it is natural to infer that in addition to keeping the number of offspring equal between different individuals, if matings between close relatives are avoided, 'the inbreeding coefficient in any generation is slightly lower and is more uniform between the individuals in the generation than if matings between close relatives are allowed; but the rate of inbreeding is the same' (Falconer, 1960).

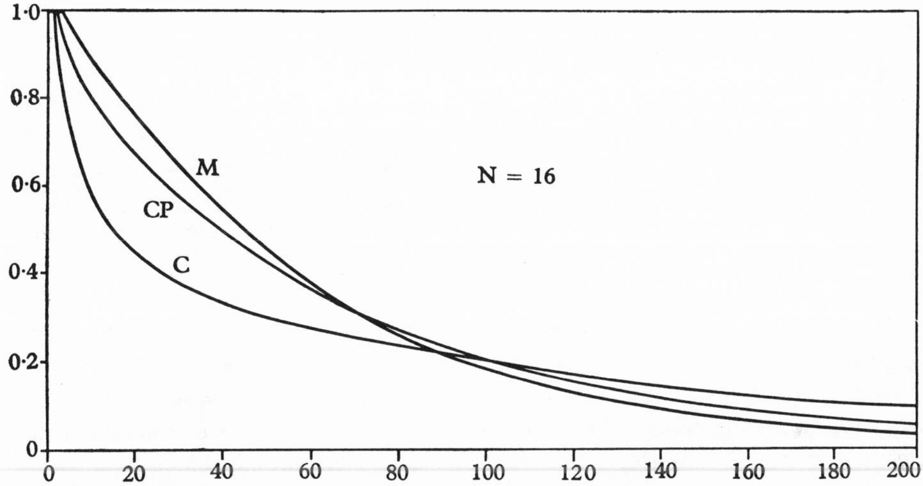


Fig. 6. Decrease of heterozygosity in a population of size 16 for circular mating (C), circular pair mating (CP), and maximum avoidance of inbreeding (M). Ordinate: heterozygosity as a fraction of initial heterozygosity; abscissa: time in generations.

Actually, an inquiry on this point from Dr Y. Yamada to one of us (M.K.) provided the stimulus to work on the problem. Thus the general result that, under circular mating, the ultimate rate of decrease in heterozygosity is proportional to $1/N^2$ rather than to $1/N$ was quite an unexpected one.

The processes of change in heterozygote frequencies under circular matings and the 'maximum avoidance of inbreeding' are illustrated in Figs. 6 and 7 for $N = 16$ and 32.

They show that the decrease of heterozygosity is always more rapid in the circular matings than in the maximum avoidance system during the earlier stages of inbreeding, after which the latter starts to lose heterozygosity more rapidly. The point at which this transition occurs depends on the population size as well as the type of circular mating.

The practical utility of circular mating systems for maintaining heterozygosity is limited by the fact that much of the heterozygosity is lost before the circular mating systems become advantageous. The point of transition is difficult to determine analytically, but some idea of the numerical values can be gotten from Figs. 6 and 7 and Table 1.

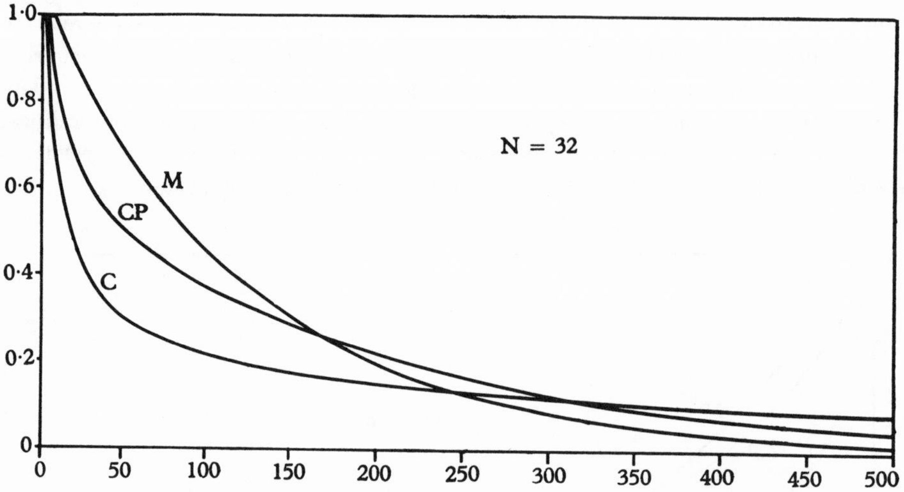


Fig. 7. Decrease of heterozygosity in a population of size 32.

Although the ultimate rate of decrease in heterozygosity, $1-\lambda$, may not be very important in judging the merit of various mating systems for practical use, it is of considerable theoretical interest. A problem that immediately comes up is whether there is any mating system that gives a smaller value of $1-\lambda$ than circular mating for the same population size. We have not found any that is applicable to populations that have separate sexes. For example, Fig. 8 shows a system suggested by Dr Y.

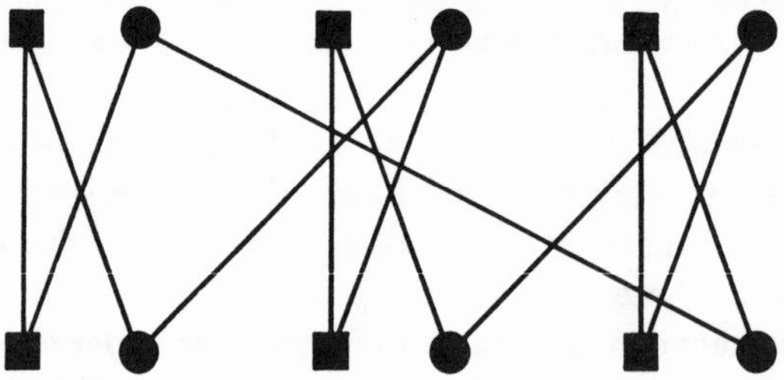

Fig. 8. A mating system that combines circular and circular pair systems.

Yamada that is a mixture of circular and circular pairs. For the case of $N = 6$, the equation determining $\lambda$ is

$$-(4\lambda)^6+3(4\lambda)^5+5(4\lambda)^4-(4\lambda)^3-23(4\lambda)^2-6(4\lambda)+12 = 0$$

from which we get $1-\lambda = 0{\cdot}0465$ as the ultimate rate of decrease in heterozygosity. Since the corresponding values for the circular individual and the circular pair matings are respectively $1-\lambda = 0{\cdot}0395$ and $0{\cdot}0494$, this system is intermediate between these two circular mating systems with respect to the ultimate rate of decrease in heterozygosity.

For a monoecious population in which self-fertilization is allowed, it is possible to produce an example which gives a smaller value of $1-\lambda$ than circular individual mating (for equal $N$). Figure 9 illustrates a mating scheme in which selfing and sib-mating are systematically mixed among three monoecious individuals compared

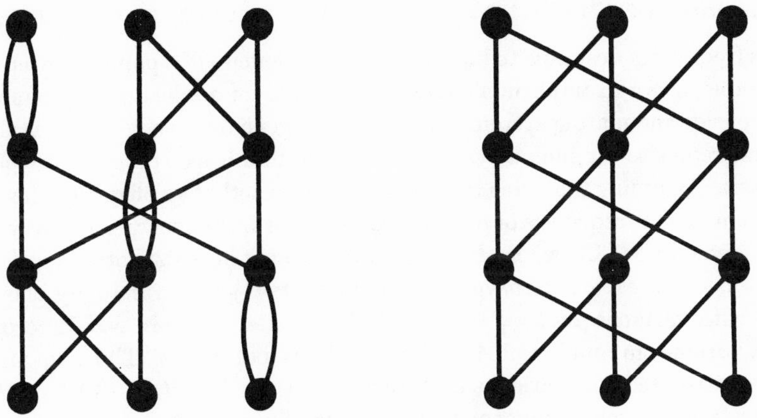

Fig. 9. Partial self-fertilization without subdivision into lines for $N = 3$ compared to circular mating.

with circular mating involving three monoecious individuals. In the former case, the equation determining $\lambda$ turns out to be

$$32\lambda^4 - 15\lambda^3 - 8\lambda^2 - 2\lambda - 1 = 0$$

from which we get $1-\lambda = 0{\cdot}1013$. In the latter case,

$$8\lambda^2 - 6\lambda - 1 = 0$$

and the ultimate rate of decrease in heterozygosity is $1-\lambda = 0{\cdot}1096$.

Thus, more intense inbreeding of individuals within the line produces a lower ultimate rate of decrease in heterozygosity, provided there is no permanent splitting of the population into isolated lines. An extreme example would be a system of self-fertilization or sib-mating for several generations, with random mating interspersed. By lengthening the interval between the random matings the ultimate rate of decrease in heterozygosity could be made as small as desired. Conversely, a system that avoids mating of relatives for as long as possible does so at the expense of a more rapid final approach to homozygosis.

The circular and circular pair systems have the same mating pattern each generation, as do the maximum avoidance systems. Among such systems it would appear that circular mating has the slowest ultimate rate of approach to homozygosity.

It is interesting to note that circular mating and circular pair mating for populations of infinite size have already been worked out, although they are usually not thought of in this sense. The half sib mating scheme of Wright (1921, p. 138) would be equivalent to circular mating if the right and left ends of the pedigree were joined. Likewise, Wright's first cousin mating scheme (p. 140) is equivalent to circular pair mating in a population of infinite size. As expected, the heterozygosity change in the first few generations of our systems agree with those of Wright.

## 6. MATING SYSTEMS THAT MINIMIZE RANDOM GENE FREQUENCY DRIFT

It is frequently desirable to keep the gene frequencies of a population as constant as possible, as for example in a control population for a selection experiment.

In considering mating systems that minimize gene frequency drift it is clear that the first requisite for such a system is that the number of progeny per parent be constant. Assuming that each individual leaves exactly two offspring, the variance in the change in frequency of a particular allele $A$ from generation $t$ to generation $t+1$ is given by $H_t/8N$, where $H_t$ is the proportion of heterozygotes in generation $t$. This is because each heterozygote contributes two genes, randomly chosen, and hence with variance $2 \times \frac{1}{2} \times \frac{1}{2} = \frac{1}{2}$. With $N$ parents there are $NH_t$ heterozygotes, with a variance in number of $A$ alleles contributed of $NH_t/2$. Thus, among the $2N$ genes contributed, the variance in $A$ allele frequency is $NH_t/2 \div 4N^2$, or $H_t/8N$.

Therefore, the gene frequency variance after $T$ generations is

$$V_T = \frac{1}{8N} \sum_{t=0}^{T-1} H_t$$

that is, the total random drift is proportional to the sum of the heterozygote frequencies in all previous generations. If there are $L$ isolated lines, each with $N$ individuals, the variance in average gene frequency becomes $1/L$ of the above value. Thus it is desirable to adopt a mating system in which heterozygosity decreases rapidly and the population breaks up into as many lines as possible.

For a randomly mating population of $N$ monoecious individuals and constant number of progeny per parent $H_t = \lambda H_{t-1}$ where $1-\lambda = 1/(4N-2)$ (see Kimura & Crow, 1963), so that

$$V_\infty = \frac{1}{8N} \sum_{t=0}^{\infty} H_0 \lambda^t = \frac{H_0}{8N(1-\lambda)} = \frac{H_0}{8N}(4N-2) \tag{6.2}$$

If, instead, the population is split into $N$ self-fertilizing lines, $\lambda = \frac{1}{2}$ in each line and

$$V_\infty = \frac{1}{N} \cdot \frac{H_0}{8(1-\lambda)} = \frac{2H_0}{8N} \tag{6.3}$$

thus by splitting into $N$ selfing lines the total accumulated variance is reduced to $1/(2N-1)$ of the value in a randomly mated population.

If there are separate sexes the change of heterozygosity is given by

$$H_t = (1-P_t) H_{t-1} + \frac{P_t}{2} H_{t-2}$$

where $P_t$ is the probability that two homologous genes in generation $t$ come from the same individual in generation $t-2$ (Wright, 1951; Kimura & Crow, 1963). If each parent leaves the same number of progeny (which implies equal numbers of male and female parents), $P_t = 1/(2N-2)$ and we have

$$H_t = \frac{2N-3}{2N-2} H_{t-1} + \frac{1}{4N-4} H_{t-2}$$

$$\sum_{t=0}^{\infty} H_t = H_0(4N-2)$$

$$V_\infty = \frac{H_0}{8N}(4N-2) \tag{6.4}$$

Note that (6.2) and (6.4) are the same; i.e. the situation is not changed by having seperate sexes.

If the population is split into $N/2$ sib-mated lines

$$V_\infty = \frac{6H_0}{8N}$$

and the drift variance is decreased to $3/(2N-1)$ of the value with random mating.

These examples clearly show that in order to keep the total drift to a minimum, the population should be split into as many lines as possible.

If the population is not split into lines the system that has the least total heterozygosity over the generations considered is preferred. From Figs. 6 and 7 it is clear that circular mating is the best of the systems considered, since for any particular generation the area under this curve to the left of the generation is least.

If we ask for the limit, each of these systems (or any other that does not lead to population subdivision) leads to the same result, for eventually one allele will become fixed. Therefore

$$V_\infty = p(1-p) \tag{6.5}$$

where $p$ is the initial frequency of the allele under consideration.

This can be verified by noting that, with random mating in a finite population,

$$H_0 = 2p(1-p)(1-\alpha)$$

$$\alpha = \frac{-1}{2N-1} \quad \text{(Kimura \& Crow, 1963)}$$

Substituting these into (6.2) or (6.4) leads to (6.5).

The relationship between minimum gene frequency drift and minimum ultimate decrease in heterozygosity can now be seen. Minimization of random drift is accomplished by intense inbreeding in the early generations and subdivision of the population into as many lines as possible; for example, self-fertilization if this is possible. By crossing such lines at random a population can be reconstituted and the final approach to homozygosity of such populations, successively reconstituted, will be minimal.

## 7. SUMMARY

Mating systems in which the least related individuals are mated have been designated by Wright as having maximum avoidance of inbreeding. For such systems the initial rate of decrease in heterozygosity is minimum. However, some other systems have a lower rate of decrease in later generations.

Circular mating, in which each individual is mated with the one to his right and to his left, leads to an asymptotic rate of decrease in heterozygosity of $1-\lambda \sim \pi^2/(2N+4)^2$ compared with $1/4N$ for maximum avoidance systems. Circular pair mating, in which for example each male progeny is moved one cage to the right, leads to $1-\lambda \sim \pi^2/(N+12)^2$. Other similar systems are discussed.

For minimum gene frequency drift, a mating system should have a constant number of progeny per parent and the population should be broken up as rapidly as possible into the maximum number of lines. The gene frequency variance at generation $T$ within a line is

$$V_T = \frac{1}{8N} \sum_{t=0}^{T-1} H_t$$

where $N$ is the number in the line and $H_t$ is the proportion of heterozygotes in generation $t$. Although the three mating systems, circular, circular pair, and maximum avoidance (and many others) have the same amount of random drift ultimately, at any generation circular mating has the smallest drift variance, $V_T$, and circular pair next smallest.

We are indebted to Mr Joseph Felsenstein for computer programming, and to Drs Hans Schneider, Gene Golub, and Joel L. Brenner for assisting one of us (M. K.) in obtaining eigenvalues. We should like also to thank Drs Sewall Wright and Alan Robertson for suggestions that helped us see the problem more clearly and for reading the manuscript.

## REFERENCES

Falconer, D. S. (1960). *Introduction to Quantitative Genetics.* ix + 365 pp. New York: The Ronald Press.

Kimura, M. & Crow, J. F. (1963). The measurement of effective population number. *Evolution* (in press).

Rutherford, D. E. (1952). Some continuant determinants arising in physics and chemistry. *Proc. roy. Soc. Edinb.* A, **63**, 232–241.

Wright, S. (1921). Systems of mating. *Genetics*, **6**, 111–178.

Wright, S. (1931). Evolution in mendelian populations. *Genetics*, **16**, 97–159.

Wright, S. (1951). The genetical structure of populations. *Ann. Eugen.* **15**, 323–354.

# PART TEN

# Evolution of Quantitative Characters

## Introduction

Kimura (no. 23), following Crow and Kimura (1964), derived a formula for the equilibrium genic variance if mutation and selection are in balance, the distributions of allelic effects at a locus are approximately Gaussian, loci are independent, and there are no pleiotropic effects of the loci. As noted by Lande (1975) and others, an implicit assumption is that the variance associated with new mutations at each locus is small, relative to the variance of the effects of the alleles already segregating. In the opposite situation, in which individual mutational effects are large, Turelli (1984) obtained a quite different formula, and Slatkin (1987) developed an approximation that bridges these extremes. Although mutation-selection balance is the most commonly invoked explanation for the maintenance of additive genetic variance, there are other suggestions (Turelli 1988; Barton and Turelli 1987, 1989; see also Bulmer 1980).

In evolutionary quantitative genetics, there are two complementary interests. One is to seek the rate of phenotypic evolution with specified intensities of selection on phenotypes, and the other is to consider selection at individual loci under a given intensity of phenotypic selection. The former approach was taken by Lande (1976), for example, to account for observed paleontological changes. The latter is the traditional approach in quantitative genetics, and this was greatly extended by Kimura and Crow (no. 24). To explain their extension, let $F(X)$ and $W(X)$ be the frequency and fitness functions of individuals with a quantitative character whose value is $X$. In order to derive the relationship between the average excess of a gene or the average effect of an allele involved in $X$, say $A$, and the selective advantage of a particular allele, say $s$, it is customarily assumed that $F(X)$, $W(X)$, or both are normally (Gaussian) distributed. For example, it is shown under the assumption of $F(X)$ being Gaussian that truncation selection with $W(X) = W_{max}$ for $X > C$ and $W(X) = 0$ for $X < C$($C$; truncation point) leads to

$$s = AW_{max}F(C)/\overline{W}$$

in which $\overline{W}$ is the mean fitness defined by $\int W(X)F(X)dX$. Kimura and Crow demonstrated that no such assumption on $F(X)$ is needed; the only requirements are that the gene effects on the character be additive between loci, the individual effects be small, and the character distribution be differentiable. But by assuming that $F(X)$ is Gaussian, they derived various interesting results. Truncation selection is the most efficient form of directional selection at individual loci. The efficiency depends not only on the magnitude of the phenotypic effect, but on the mean fitness (and in general the shape of the fitness function). Cockerham and Burrows (1980) used this general theoretical framework to study strategies for maximizing the truncation selection limit for a quantitative trait in a finite population.

Crow and Kimura (no. 25) quantified the efficiency of truncation selection using a modification, due to Milkman (1978), of the standard linear ranking of character values. The modified model was called a broken-line alternative. The aim of the paper is to reconcile the very high mutation rate in viability polygenes (Mukai 1964, 1990; Ohnishi 1977; Simmons and Crow 1977) with a small mutational load. Truncation selection as well as the broken-line alternative can eliminate deleterious mutants in groups even when their allelic effects are additive, so that Haldane-Muller's principle "one mutant—one genetic death" is bypassed. Hence, Crow and Kimura concluded that truncation selection, or something even very roughly approximating it, can be very effective in reducing the mutational load for mutants having minor deleterious effects on viability.

Rank order selection had earlier been suggested as a segregation load-reducing mechanism (King 1967; Milkman 1967; Sved, Reed, and Bodmer 1967). The question as to whether nature ranks and truncates, or does something approximating it, was also raised by Nei (1971, 1975). But not until the work of Milkman (1978) and Crow and Kimura (no. 25) was it realized that a very crude approximation to truncation is almost as effective.

Another fundamental question centers on the connection of neutral evolution at the molecular level with adaptive evolution at the phenotypic level. It is possible to extract the effect of natural selection acting on a quantitative character related to Darwinian fitness on allele dynamics at loci which determine the character. Assuming the most prevalent type of phenotypic selection (stabilizing selection) and using the general formula in Kimura and Crow (no. 24), Kimura demonstrated (no. 26) that allele dynamics become underdominant (Bulmer 1972). Since the selection coefficient is of the second order of allelic effect, stabilizing phenotypic selection leads to near neutrality at individual loci. This may substantiate the "very slightly deleterious" hypothesis, but an important difference from it is that the population fitness does not drift downward under stabilizing selection. More significant is the picture of evolution depicted in the paper. That is captured (no. 26) as "From time to time, the position of the optimum (of characters) shifts due to changes in environment and the species tracks such changes rapidly by altering its mean. But, most of the time, stabilizing selection predominates. Under this selection, neutral evolution (random fixation of alleles by sampling drift) occurs extensively, transforming all genes, including those of living fossils, profoundly at the molecular level." This line of quantitative pursuit is still premature, but the extensive neutral evolution under stabilizing selection forms one of the bases of Kimura's recent view of evolution (Kimura 1989, 1990, 1991a, 1991b).

# *A STOCHASTIC MODEL CONCERNING THE MAINTENANCE OF GENETIC VARIABILITY IN QUANTITATIVE CHARACTERS**

BY MOTOO KIMURA

NATIONAL INSTITUTE OF GENETICS, MISHIMA-SHI, SHIZUOKA-KEN, JAPAN

*Communicated by James F. Crow, July 15, 1965*

The mechanism by which genetic variability is maintained in natural populations for quantitative characters is not well understood. Many observations show that there is a considerable amount of genetic variability in a large population, and unless the character is closely correlated with fitness the optimum is usually near the mean, with fitness decreasing as the distance from the mean increases. Probably, Fisher[1] was the first to investigate a model in which the fitness was assumed to decrease in proportion to the squared deviation from the optimum. This model was also used by Haldane[2] and Wright.[3] Robertson[4] has shown that if genes are maintained by overdominance, but act additively with respect to a quantitative character, then the optimum is at the mean and the decrease of fitness is proportional to the squared deviation from the optimum.

In all these treatments the relation between the mutation rate and the amount of genetic variability maintained is either ambiguous or left out of consideration. The purpose of this paper is to propose a new model which enables one to make predictions about the relations between mutation rate, genotypic variance, and genetic load or amount of selective elimination involved in the maintenance of genetic variability.

*Assumptions and Mathematical Formulation.*—The basic assumptions are:

(1) At every locus involved with the quantitative character under discussion, mutation can produce an infinite sequence of alleles. Every mutation may produce a new allele different from the pre-existing ones.

(2) The effect of a new allele on the quantitative character is only slightly different from the parent allele from which it was derived by a single mutational step.

(3) The genes are additive with respect to their effect on the quantitative character.

(4) The optimum phenotype is fixed, and fitness decreases in proportion to the squared deviation from the optimum.

Consider a particular locus. We denote by $x$ the average effect of an allele on the quantitative character, taking the optimum as the origin. We assume that by mutation an allele having an average effect $x$ changes to another allele with an average effect $x + \xi$ with probability density given by $f(\xi)$.

If we denote by $p(x,t)$ the relative frequency of the alleles having an average effect $x$ in a large population at time $t$ (measured in generations), then the rate of change in $p$ per generation is given by the sum of the following two components:

(a) *Change due to mutation:* If $\mu$ is the mutation rate per gene per generation (assumed to be constant), the contribution to $\partial p/\partial t$ by mutation is

$$-\mu p(x,t) + \mu \int_{-\infty}^{\infty} p(x - \xi,t)f(\xi)d\xi. \tag{1}$$

The first term gives the rate of loss of alleles having an average effect $x$ by mutation to other alleles having a different effect. The second term gives the rate of gain from the mutation to alleles with average effect $x$ from other alleles.

(b) *Change due to selection:* We assume that the fitness of an individual with total genotypic value $Y$ with respect to the quantitative character is less on the average by an amount $KY^2$ in comparison with those having a total genotypic value of 0 (the optimum), fitness being measured in Malthusian parameters.

Consider an allele having an average effect $x$ and let $Y'$ be the value of the genetic background. Since the relative fitness of a genotype with genotypic value $x + Y'$ is $-K(x + Y')^2$, if $Y'$ is distributed with mean $M'$ and variance $V'$, then the relative fitness of an allele having average effect $x$ is $-K[(x + M')^2 + V']$. Therefore, the rate of change in $p$ by selection is

$$p(x,t)[-Kx'^2 + K\int_{-\infty}^{\infty} x'^2 p(x,t)dx], \tag{2}$$

where $K$ is a positive constant and $x' = x + M'$.

Thus, $\partial p(x,t)/\partial t$ is expressed as the sum of (1) and (2). We are mainly interested in the frequency distribution at equilibrium which is denoted by $p(x)$ and for which $\partial p/\partial t = 0$.

Let

$$\mu\int_{-\infty}^{\infty} \xi f(\xi)d\xi = -m, \tag{3}$$

$$\mu\int_{-\infty}^{\infty} \xi^2 f(\xi)d\xi = v \qquad (>0), \tag{4}$$

and assume that the terms,

$$\mu\int_{-\infty}^{\infty} \xi^n f(\xi)d\xi \qquad (n \geqq 3), \tag{5}$$

are negligible. Then the contribution from mutation, i.e., (1), may be reduced as follows:

$$-\mu p(x,t) + \mu\int_{-\infty}^{\infty} p(x - \xi,t)f(\xi)d\xi$$

$$= -\mu p + \mu\int_{-\infty}^{\infty}\left\{p - \xi\frac{\partial p}{\partial x} + \frac{\xi^2}{2}\frac{\partial^2 p}{\partial x^2} - \ldots\right\}f(\xi)d\xi$$

$$= m\frac{\partial p(x,t)}{\partial x} + \frac{v}{2}\frac{\partial^2 p(x,t)}{\partial x^2}.$$

Thus, at equilibrium when $\partial p/\partial t = 0$, we obtain the following ordinary differential equation:

$$\frac{v}{2}\frac{d^2p(x)}{dx^2} + m\frac{dp(x)}{dx} + K(k^2 - x'^2)p(x) = 0, \tag{6}$$

where

$$k^2 = \int_{-\infty}^{\infty} x'^2 p(x)dx. \tag{7}$$

The above equation may also be written in the form:

$$\frac{v}{2}\frac{d^2p}{dx'^2} + m\frac{dp}{dx'} + K(k^2 - x'^2)p = 0, \tag{6'}$$

where $x' = x + M'$.

Putting

$$\left.\begin{aligned} p &= e^{-\alpha x'}U(x') \\ x' &= \beta y \end{aligned}\right\} \tag{8}$$

in which $\alpha = m/v$ and $\beta^4 = v/8K$, equation (6′) reduces to

$$\frac{d^2U}{dy^2} + \left(A - \frac{y^2}{4}\right) U = 0, \tag{9}$$

where

$$A = \frac{k^2}{4\beta^2} - \alpha^2\beta^2. \tag{10}$$

Equation (9) is a Weber equation. The solution of this which is nonnegative and satisfies the conditions $U(-\infty) = U(\infty) = 0$ is uniquely determined, except for the scaling factor $C$ (>0), as

$$U(y) = Ce^{-y^2/4}. \tag{11}$$

Therefore,

$$p(x) = Ce^{-\alpha x' - x'^2/4\beta^2}. \tag{12}$$

Furthermore, it is known in the theory of the Weber equation that for the solution given by (11), the constant $A$ in (9) is equal to $^1/_2$. This leads to

$$k^2 = \beta^2(2 + 4\alpha^2\beta^2). \tag{13}$$

On the other hand, substituting (12) in the right side of (7), we obtain

$$k^2 = C\sqrt{4\pi}\beta^3(2 + 4\alpha^2\beta^2)e^{\alpha^2\beta^2}. \tag{14}$$

Comparing (13) and (14), we get

$$C = e^{-\alpha^2\beta^2}/\sqrt{4\pi\beta^2},$$

and (12) becomes

$$p(x) = \frac{1}{\sqrt{4\pi\beta^2}} e^{-(x'+2\alpha\beta^2)^2/4\beta^2}. \tag{15}$$

Thus, $x$ is normally distributed with mean

$$-2\alpha\beta^2 - M' = -\frac{m}{\sqrt{2vK}} - M', \tag{16}$$

and variance

$$2\beta^2 = \sqrt{\frac{v}{2K}}. \tag{17}$$

Since a diploid organism has two genes at each locus, the distribution of the genotypic value among individuals is given by the normal distribution with mean

$$-\frac{2m}{\sqrt{2vK}} - M'' \tag{18}$$

and variance

$$2\sqrt{\frac{v}{2K}}, \tag{19}$$

where $-m$ and $v$ are the amount of increase in mean and variance of the genotypic value per gene per generation. $M''$ is the mean of the genotypic value produced by the remaining loci.

Considering all relevant loci, the genotypic value is distributed normally with mean

$$M = -\Sigma_i \frac{2m_i}{\sqrt{2v_iK}} \tag{20}$$

and variance

$$V_G = \Sigma_i \sqrt{\frac{2v_i}{K}}, \tag{21}$$

where the subscript $i$ refers to the $i$th locus.

*The Genetic Load Associated with This Mechanism.*—Since the selective disadvantage of a genotype with genotypic value $Y$ is $KY^2$, the total genetic load for a diploid organism with respect to this quantitative character is

$$L = K\overline{Y^2} = K(V_G + M^2)$$
$$= \Sigma_i\sqrt{2Kv_i} + 2\left(\Sigma_i \frac{m_i}{\sqrt{v_i}}\right)^2, \tag{22}$$

where $m_i = -\mu\overline{\xi_i}$ and $v_i = \mu\overline{\xi_i^2}$.

In expression (22), the first term is the component caused by the increase in variance, while the second is caused by the mean being shifted away from the optimum. As might be expected, this is proportional to the mutation rate.

*Relations among Observable Quantities.*—If we denote by $r$ the ratio of the new genotypic variance produced by mutation in one generation to the total genotypic variance, then

$$r = \frac{\Sigma 2v_i}{\Sigma\sqrt{\frac{2v_i}{K}}}. \tag{23}$$

From (22) we let $L_v = \Sigma\sqrt{2Kv_i}$, this being the genetic load associated with increased variance. Substituting this into (23) leads to

$$r = \frac{L_v\Sigma v_i}{(\Sigma\sqrt{v_i})^2} \geqq \frac{L_v}{n},$$

where $n$ is the number of loci involved, so that

$$n \geqq L_v/r, \tag{24}$$

where the inequality changes to an equality when the $v_i$'s are all equal.

The following relation also follows from this model. From (21) and (22) we obtain

$$V_G = L_v/K.$$

Therefore,

$$\frac{V_G}{V_T} = \frac{L_v}{KV_T} = \frac{L_v}{-\log_e w(\sqrt{V_T})} \tag{25}$$

where $w(\sqrt{V_T})$ is the relative selective value of an individual whose phenotype deviates from the optimum by a standard deviation of total phenotypic value.

*Effect of Continued Truncation Selection.*—If the quantitative character is subject to continued truncation selection, the relative fitness for this selection of alleles having an average effect $x$ is $1 + sx$, where $s = z/b\sqrt{V_T}$. $V_T$ is the total phenotypic variance, $z$ is the ordinate at the truncation point, and $b$ is the proportion saved, the latter two quantities being defined for the normal distribution with mean zero and unit variance. The contribution of this selection to $\partial p/\partial t$ is $s(x - \bar{x})$, where

$$\bar{x} = \int_{-\infty}^{\infty} xp(x,t)dx.$$

Thus, the equation corresponding to (6) is now

$$\frac{v}{2}\frac{d^2p(x)}{dx^2} + m\frac{dp(x)}{dx} + [K(k^2 - x'^2) + s(x - \bar{x})]p(x) = 0. \tag{26}$$

It is possible to show then that $x$ is distributed normally with mean

$$\frac{s}{2K} - \frac{m}{\sqrt{2vK}}$$

and variance

$$\sqrt{\frac{v}{2K}}.$$

It follows that, under continued truncation selection, the mean of the genotypic value of an individual increases by an amount

$$\frac{zn}{Kb\sqrt{V_T}}$$

more than without the truncation selection, but the genotypic variance remains the same. Thus, the variance is more stable than the mean.

*Summary.*—A new model was proposed to explain the maintenance of genetic variability in quantitative characters. The model assumes that at every locus involved with the quantitative character, mutation can produce an infinite sequence of alleles and the effect of a new allele is only slightly different from the parental allele from which it was derived by a single mutational step. The new model is in sharp contrast with the conventional models in which mutation is assumed to occur only between a pair of alleles, say $A$ or $a$. Together with the additional assumptions that the genes are additive with respect to the quantitative character, that the optimum phenotype is fixed, and that fitness decreases in proportion to the squared deviation from the optimum, the properties of the model were worked out, enabling one to make predictions about the relation between mutation rates, genotypic variance, and mutational load.

The author would like to thank Dr. J. F. Crow for his stimulating discussions, for help in the completion of the manuscript, and for suggesting the idea of assuming small mutational steps in quantitative characters. Thanks are also due to Dr. Felix M. Arscot for his help in the solution of a differential equation.

* Contribution no. 535 from the National Institute of Genetics.

1 Fisher, R. A., *The Genetical Theory of Natural Selection* (Oxford: Clarendon Press, 1930).
2 Haldane, J. B. S., *The Causes of Evolution* (London: Longmans, 1932).
3 Wright, S., *J. Genet.*, **30**, 243–256 (1935).
4 Robertson, A., *J. Genet.*, **54**, 236–248 (1956).

# Effect of overall phenotypic selection on genetic change at individual loci*

(selection differential/truncation selection/rank-order selection/selection intensity/fitness potential)

MOTOO KIMURA† AND JAMES F. CROW‡

†National Institute of Genetics, Mishima, Shizuoka-ken 411, Japan; and ‡Genetics Laboratory, University of Wisconsin, Madison, Wisconsin 53706

*Contributed by James F. Crow, September 27, 1978*

**ABSTRACT** The selective advantage of an allele $G_i$ (relative to the mean of alleles at this locus) is given by

$$s_i = -A_i \int_{-\infty}^{\infty} W(X)F'(X)dX/\overline{W}$$
$$= A_i \int_{-\infty}^{\infty} W'(X)F(X)dX/\overline{W},$$

in which $A_i$ is the average excess of the allele on the character, $X$; $W(X)$ is the fitness function; $F(X)$ is the frequency function; $\overline{W}$ is the mean fitness; and the prime denotes differentiation. With truncation selection $s_i = A_iF(C)/\overline{w}$, in which $F(C)$ is the ordinate at the culling level and $\overline{w}$ is the proportion saved; this does not depend on any assumption about the distribution of $F(X)$. If the character is normally distributed, $s_i = A_iI/\sigma^2$, in which $I$ is the selection differential and $\sigma^2$ is the variance of the character distribution. Finally, if the logarithm of the fitness is proportional to the squared deviation from the optimum and the character is distributed normally, $s_i = A_iK(X_{op} - m)$, in which $X_{op}$ is the optimum value of the character, $m$ is the mean value, and $K$ is a constant determined by the variances of the fitness function and the frequency function. Truncation is the most efficient form of directional selection in the sense of producing the maximum gene frequency change for a given effect of the gene on the character, but fitness functions can depart considerably from sharp truncation without greatly reducing the efficiency.

Milkman (1) and Wills (2), following earlier work by King (3), Sved *et al.* (4), and Milkman (5), have argued that "soft selection" (6, 7) can account for a high degree of polymorphism without excessive selection or segregation load. This theory assumes a linear ranking of value for some character that, along with the environment, determines individual fitness. The mean fitness is determined by the total reproductive capacity and the carrying capacity of the environment. The severity of the environment may greatly alter individual and average fitnesses, but is assumed not to change the rank order. Milkman (1) has termed the character scale "fitness potential" and the kind of selection thus implied is designated by Wills (2) as "rank-order" selection.

In this paper we derive a general relation between the selection on a measured character and the selection coefficient of a gene at a locus. From this we easily find Milkman's (1) formula, but with less restrictive assumptions, and consider other specific frequency and fitness functions.

## The model

A general diagram is shown in Fig. 1. $X$ is a measured character, such as size or viability, a genotypic property, such as heterozygosity, or a more abstract quantity, such as fitness potential. Three fitness functions are illustrated. We shall use the word

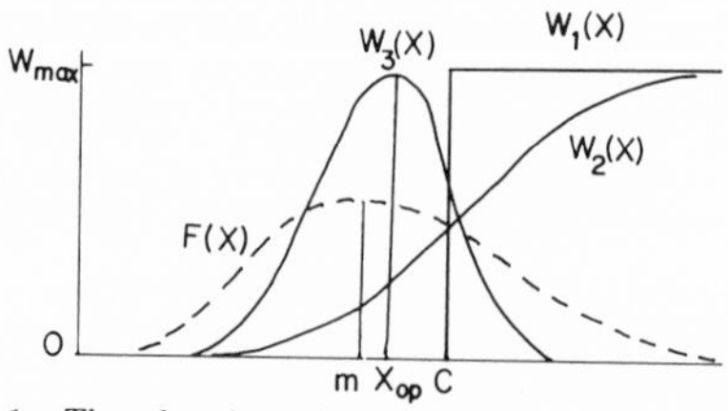


FIG. 1. Three functions relating fitness, $W(X)$, to the character value, $X$. The frequency of the character value is given by $F(X)$ and has mean $m$. $W_1(X)$ is the fitness function with truncation selection in which all individuals with $X$ above $C$ have fitness $W_{max}$, whereas all below this value have fitness 0. $W_2(X)$ shows a less extreme function of directional selection. $W_3(X)$ shows selection for an intermediate optimum, with $X_{op}$ designating the value of $X$ with optimum fitness.

"character" for the measured variable $X$. Later, we specifically consider viability, but the general discussion applies to any character that can be measured or ranked on a linear scale.

Let $F(X)$ and $W(X)$ be the frequency and fitness of individuals with character value $X$. $F(X)$ is the density function such that $F(X)dX$ is the fraction of individuals whose character measure is in the range $X$ to $X + dX$. The mean and variance of $X$ are given by

$$m = m_X = \int_{-\infty}^{\infty} XF(X)dX \qquad [1]$$

and

$$\sigma^2 = \sigma_X^2 = \int_{-\infty}^{\infty} (X-m)^2F(X)dX, \qquad [2]$$

and the mean fitness by

$$\overline{W} = \int_{-\infty}^{\infty} W(X)F(X)dX. \qquad [3]$$

Letting $P_{ij}$ be the proportion of the population that are of the ordered genotype $G_iG_j$ ($P_{ij} = P_{ji} = \frac{1}{2}$ the proportion of $G_iG_j$ heterozygotes) and $W_{ij}$ be the mean fitness of all individuals of genotype $G_iG_j$, the mean character value and fitness of the allele $G_i$ are

$$X_i = \sum_j P_{ij}X_{ij}/p_i \qquad [4]$$

and

$$W_i = \sum_j P_{ij}W_{ij}/p_i, \; p_i = \sum_j P_{ij}. \qquad [5]$$

The average amount by which individuals carrying allele $G_i$,

The publication costs of this article were defrayed in part by page charge payment. This article must therefore be hereby marked "*advertisement*" in accordance with 18 U. S. C. §1734 solely to indicate this fact.

* This is the first of two related papers. The second paper will appear in the January 1979 issue of the *Proceedings*.

weighted by the number of $G_i$ alleles carried, exceed the population average for the character is

$$A_i = X_i - m. \qquad [6]$$

Following Fisher (8), we call $A_i$ the average excess of the gene $G_i$. If the genotypes are in Hardy–Weinberg ratios, the average excess is the same as the average effect of the allele. The selective advantage of $G_i$ relative to the mean at this locus is

$$s_i = \frac{W_i - \overline{W}}{\overline{W}}. \qquad [7]$$

Note that $s_i$ is not the usual selection coefficient measured as a deviation from a reference genotype, but is measured as a standardized deviation from the population average.

Finally, the standard Wright equation for gene frequency change is

$$\Delta p_i = p_i \frac{W_i - \overline{W}}{\overline{W}} = p_i s_i, \qquad [8]$$

in which $p_i = \Sigma_j P_{ij}$. Thus, $s_i$ is a direct measure of the effect of selection in bringing about a change in the frequency of gene $G_i$.

We now seek a relationship between $s_i$ and $A_i$. Assume that the effect of an allele replacement at the $G$ locus is very small relative to the range of $X$ values. This may be because many loci contribute to the character, or because environmental effects are large, or both. The distribution of individuals carrying a $G_i$ allele will then be the same as the population distribution except for a small displacement, as illustrated in Fig. 2. Increasing the character value $X$ by an amount $A_i$ is equivalent to assigning a fitness of $W(X)$ to character value $X - A_i$. Alternatively, it is equivalent to giving the character $X$ a fitness of $W(X + A_i)$. These lead to two alternative equations for $W_i$,

$$W_i = \int_{-\infty}^{\infty} W(X)F(X - A_i)dX \qquad [9a]$$

$$W_i = \int_{-\infty}^{\infty} W(X + A_i)F(X)dX. \qquad [9b]$$

Note that dominance can be arbitrary. With a more elaborate model, epistasis and linkage disequilibrium could be included, but here we assume additivity between loci.

For small $A_i$, using the Taylor expansion

$$F(X - A_i) = F(X) - A_iF'(X) + (A_i^2/2)F''(X) - \ldots$$

$$W(X + A_i) = W(X) + A_iW'(X) + (A_i^2/2)W''(X) + \ldots$$

If terms in $A_i^2$ and higher order are ignored, we have two approximate expressions for the selection coefficient,

$$s_i = -\frac{A_i}{\overline{W}} \int_{-\infty}^{\infty} W(X)F'(X)dX \qquad [10a]$$

$$s_i = \frac{A_i}{\overline{W}} \int_{-\infty}^{\infty} W'(X)F(X)dX. \qquad [10b]$$

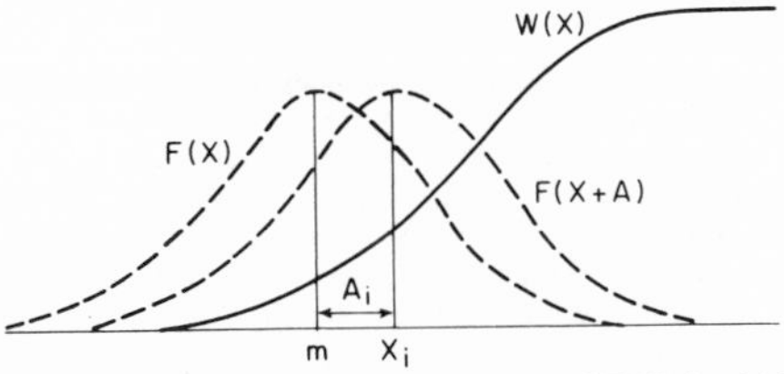


FIG. 2. Relationship between $F(X + A_i)$ and $F(X)$, in which $A_i$ is the average excess of the allele $G_i$ on the character scale $X$.

If $X$ is a quantitative character such that $F(X)$ is differentiable at all but a finite number of points, then Eq. **10a** is appropriate. There is then no restriction on the fitness function. It need be neither continuous nor nondecreasing; for example, the maximum fitness may be at an intermediate value of $X$, or the function could even be multimodal. On the other hand, Eq. **10b** is more useful when the phenotype is discontinuous. This might be true of a meristic trait, such as vertebra or bristle number, or of a rank-order character for which the ranking is by discontinuous classes. In this case, however, $W(X)$ must be differentiable.

### Truncation selection on a continuous character

For truncation selection, as is practiced by an animal or plant breeder who selects all individuals above a certain culling level, $C$, and rejects the rest, $W(X) = W_{max}$ for $X > C$ and $W(X) = 0$ for $X < C$. From Eq. **10a** we immediately obtain

$$s_i = -\frac{A_iW_{max}}{\overline{W}} \int_{C}^{\infty} F'(X)dX = A_i \frac{F(C)}{\overline{w}}, \qquad [11]$$

in which

$$\overline{w} = \frac{\overline{W}}{W_{max}}.$$

In words, the selection coefficient of a gene (or genotype) is its effect on the character (average excess) multiplied by the ratio of the ordinate at the truncation point to the proportion selected. This principle is well known (9–12), but is customarily derived by assuming that the character is normally distributed. No such assumption is needed; the only requirements are that the gene effects on the character be additive between loci, the individual effects be small, and the character distribution be differentiable.

### Phenotype normally distributed

Most quantitative characters are approximately normally distributed or may be transformed to be so. If normal distribution can be assumed, a further simplification occurs. Let

$$F(X) = \frac{1}{\sigma\sqrt{2\pi}} \exp[-(X - m)^2/(2\sigma^2)] \qquad [12]$$

$$F'(X) = -(X - m)F(X)/\sigma^2. \qquad [13]$$

The mean character value of the selected group is

$$m_S = \frac{1}{\overline{W}} \int_{-\infty}^{\infty} XW(X)F(X)dX. \qquad [14]$$

Using Eqs. **14** and 3, the selection differential, $I$, is

$$I = m_S - m = \frac{1}{\overline{W}} \int_{-\infty}^{\infty} (X - m)W(X)F(X)dX. \qquad [15]$$

From Eqs. **10a**, **13**, and **15** we obtain

$$s_i = \frac{A_i}{\sigma} \cdot \frac{I}{\sigma}, \qquad [16a]$$

or, measuring the excess of the allele $G_i$ in units of the standard deviation of $X$,

$$\frac{s_i}{a_i} = \frac{I}{\sigma} \qquad [16b]$$

in which

$$a_i = \frac{A_i}{\sigma} = \frac{X_i - m}{\sigma}. \qquad [17]$$

Eqs. **16** are well known (ref. 12, p. 228; ref. 13), though customarily derived by assuming truncation selection, and lead directly to the standard formula for change of a metrical character by selection in terms of selection differential and heritability. The requisite assumption is normality on the character scale, which is roughly equivalent to the linearity of regression usually assumed in deriving the heritability formula. Eqs. **16** are equivalent to $s \simeq ig$ of Milkman (1), but our derivation shows that it is not necessary to assume that the fitness function be nondecreasing.

### Optimum phenotype selection

The formulas given above are not restricted to directional selection. As an example of stabilizing selection, assume that the quantitative character has an optimum for fitness at $X = X_{op}$. Assume further that the logarithmic fitness decreases in proportion to the squared deviation from the optimum, so that

$$W(X) = W_{max}e^{-k(X-X_{op})^2}, \qquad [18]$$

in which $k$ is a positive constant (Fig. 1). This is essentially the same model as used by Lande (13), who expressed it as $W(Z) = \exp\{-Z^2/(2w^2)\}$, in which $w$ is the "width" of the fitness function. Assume that $X$ follows the normal distribution with mean $m$ and variance $\sigma^2$. It is convenient to let $x = (X - m)/\sigma$ and $x_{op} = (X_{op} - m)/\sigma$, in which case $dX = \sigma dx$. Making these substitutions in Eqs. **18** and **12** and substituting the results into Eq. 3 gives

$$\overline{W} = \frac{W_{max}}{\sqrt{2\pi}} \int_{-\infty}^{\infty} [\exp\{-k\sigma^2(x - x_{op})^2 - (x^2/2)\}]dx. \quad [19]$$

Integrating this, we obtain

$$\overline{W} = W_{max} \frac{\exp\left\{\dfrac{-k\sigma^2 x_{op}^2}{1 + 2k\sigma^2}\right\}}{\sqrt{1 + 2k\sigma^2}} \qquad [20a]$$

or

$$\overline{W} = W_{max} \frac{\exp\left\{-\dfrac{k(X_{op} - m)^2}{1 + 2k\sigma^2}\right\}}{\sqrt{1 + 2k\sigma^2}} \qquad [20b]$$

If the mean and optimum coincide, $\overline{W} = W_{max}/\sqrt{1 + 2k\sigma^2}$.

The selection differential, $I$, is given by Eq. **15.** Making substitutions from Eqs. **18** and **12** as before and integrating leads to

$$\frac{I}{\sigma} = \frac{2k\sigma^2}{1 + 2k\sigma^2} x_{op} \qquad [21a]$$

or

$$I = \frac{2k\sigma^2}{1 + 2k\sigma^2}(X_{op} - m). \qquad [21b]$$

From Eq. **18** it is apparent that $2k$ is the reciprocal of the variance, $\rho^2$, of the fitness function. Therefore we can write Eq. **21b** as

$$\frac{I}{\sigma} = \frac{R}{1 + R} x_{op} \text{ or } I = \frac{R}{1 + R}(X_{op} - m), \qquad [22]$$

in which $R = \sigma^2/\rho^2$. Then from Eq. **16b**

$$\frac{s_i}{a_i} = \frac{R}{1 + R} x_{op}. \qquad [23]$$

As expected, the ratio of the selective advantage of an allele to its effect on the character increases as the optimum phenotype diverges from the mean. When the optimum and mean coincide, the selection differential, $I$, becomes zero and there is no tendency for the gene $G_i$ either to increase or decrease. Also, from Eq. **23** we see that the selective advantage increases as $R$ increases; the greater the ratio of the variance of the character distribution to that of the fitness distribution, the more effective is selection.

If the mean and optimum nearly coincide, the average fitness of a population is mainly determined by the variance of $X$. The most important selection is not to shift the mean but to remove outliers. The efficiency of selection to reduce the variance with different fitness functions is also of interest and we hope to treat it later.

### Directional selection less extreme than truncation

A natural intermediate between selection in which fitness is simply proportional to the character and the extreme of truncation selection is an S-shaped curve such as $W_2(X)$ in Fig. 1. This can be represented by the integrated normal distribution. Let

$$W(X) = W_{max} \sqrt{\frac{k}{\pi}} \int_{-\infty}^{X} e^{-k(Y-C)^2} dY. \qquad [24]$$

As $X$ increases $W(X)$ approaches $W_{max}$. As $k$ increases this approaches truncation at $X = C$.

$$W'(X) = W_{max} \sqrt{\frac{k}{\pi}} e^{-k(X-C)^2}. \qquad [25]$$

Letting $x = (X - m)/\sigma$, $c = (C - m)/\sigma$, and $dX = \sigma dx$ and substituting into Eq. **10b**, we obtain

$$s_i = A_i \frac{W_{max}}{\overline{W}} \sqrt{\frac{k}{\pi}} \frac{\exp\left\{-\dfrac{k\sigma^2 c^2}{1 + 2k\sigma^2}\right\}}{\sqrt{1 + 2k\sigma^2}} \qquad [26]$$

or

$$\frac{s_i}{a_i} = \frac{W_{max}}{\overline{W}} \cdot \frac{1}{\sqrt{2\pi}} \sqrt{\frac{R}{1 + R}} \exp\left\{-\frac{R}{1 + R}\frac{c^2}{2}\right\}, \qquad [27]$$

in which $R = 2k\sigma^2$ is the ratio of the variance of the frequency function to that of the normal distribution that, when integrated, gives the fitness function.

We can compare the efficiency of truncation selection to less extreme alternatives by noting that with truncation selection $R/(1 + R) = 1$. For example, with truncation selection at $W_{max}/\overline{W} = 2$, which implies that $c = 0$, $s_i/a_i = 0.798$. For $R = 1$, this is 0.564, and for $R = 0.5$ this is 0.461. Thus truncation selection is more efficient, but not greatly so. For truncation selection to be twice as efficient ($s_i/a_i = 0.399$), $R$ must be ⅓; that is, the variance of the integrated normal fitness function must be 3 times that of the frequency function.

### Discussion

Eqs. **10** and the various special cases, Eqs. **11, 16, 23,** and **27,** show how much selection is allocated to each locus when the overall phenotypic selection pattern is given. It is clear that the efficiency of selection in changing gene frequencies depends not only on the magnitude of the phenotypic effect, but on the mean fitness and the shape of the fitness function.

Lande (13) has derived formulas for the rate of change of phenotypic characters with specified intensities of selection and has applied these to analysis of the amount of selection required to account for observed paleontological changes. Our interest is complementary: to consider selection at individual loci.

It is clear that truncation is the most efficient form of directional selection, as measured by $s/a$. It also appears from Eq. **27** that substantial departures from strict truncation selection do not greatly reduce the efficiency. In the following paper (14) we discuss this in more detail and apply the theory to data on viability-reducing mutants in *Drosophila* to see if this type of selection can reduce the mutation load.

Although we considered the average excess of an allele in the above treatment, the present method can be applied to individual genotypes at a locus. Namely, if $s$ is the selection coefficient (not the average excess in fitness) of a genotype $G_iG_j$ relative to another genotype, say $G_oG_o$, then we have

$$s = (G/\sigma) \cdot (I/\sigma). \qquad [28]$$

in which $G = \overline{X}(G_iG_j) - \overline{X}(G_oG_o)$. Furthermore, we can extend this to genotypes determined by two loci if the average excess of the joint effect is small as compared to $\sigma$.

We thank Russell Lande, Joel Cohen, and Sewall Wright for useful discussions. This is contribution no. 1219 from the National Institute of Genetics, Mishima, Japan, and no. 2293 from the Genetics Laboratory, University of Wisconsin-Madison. This work was supported in part by National Institutes of Health Grant GM 22038.

1. Milkman, R. (1978) *Genetics* **88,** 391–403.
2. Wills, C. (1978) *Genetics* **89,** 403–417.
3. King, J. L. (1967) *Genetics* **55,** 483–494.
4. Sved, J. A., Reed, T. E. & Bodmer, W. F. (1967) *Genetics* **55,** 469–481.
5. Milkman, R. (1967) *Genetics* **55,** 493–495.
6. Wallace, B. (1968) *Topics in Population Genetics* (Norton, New York).
7. Wallace, B. (1975) *Evolution* **29,** 465–473.
8. Fisher, R. A. (1930) *The Genetical Theory of Natural Selection* (Clarendon, Oxford); (1958) (Dover, New York), Rev. Ed.
9. Haldane, J. B. S. (1930) *Proc. Cambridge Philos. Soc.* **27,** 131–136.
10. Wright, S. (1969) *Evolution and the Genetics of Populations* (Univ. of Chicago Press, Chicago), Vol. 2.
11. Falconer, D. (1960) *Introduction to Quantitative Genetics* (Oliver and Boyd, Edinburgh, Scotland).
12. Crow, J. F. & Kimura, M. (1970) *An Introduction to Theoretical Population Genetics* (Harper and Row, New York); reprinted by Burgess, Minneapolis, MN.
13. Lande, R. (1976) *Evolution* **30,** 314–334.
14. Crow, J. F. & Kimura, M. (1979) *Proc. Natl. Acad. Sci. USA* **76,** in press.

# Efficiency of truncation selection*

(rank-order selection/fitness potential/mutation load/viability/fitness)

JAMES F. CROW† AND MOTOO KIMURA‡

†Genetics Laboratory, University of Wisconsin, Madison, Wisconsin 53706; and ‡National Institute of Genetics, Mishima, Shizuoka-ken 411, Japan

*Contributed by James F. Crow, September 27, 1978*

**ABSTRACT** Truncation selection is known to be the most efficient form of directional selection. When this is modified so that the fitness increases linearly over a range of one or two standard deviations of the value of the selected character, the efficiency is reduced, but not greatly. When truncation selection is compared to a system in which fitness is strictly proportional to the character value, the relative efficiency of truncation selection is given by $f(c)/\sigma$, in which $f(c)$ is the ordinate of the frequency distribution at the truncation point and $\sigma$ is the standard deviation of the character. It is shown, for mutations affecting viability in *Drosophila*, that truncation selection or reasonable departures therefrom can reduce the mutation load greatly. This may be one way to reconcile the very high mutation rate of such genes with a small mutation load. The truncation model with directional selection is appropriate for this situation because of the approximate additivity of these mutations. On the other hand, it is doubtful that this simple model can be applied to all genes affecting fitness, for which there are intermediate optima and antagonistic selection among components with negative correlations. Whether nature ranks and truncates, or approximates this behavior, is an empirical question, yet to be answered.

In the preceding paper (1) we showed the relation between the average increase in a character caused by a gene substitution and the corresponding increase in fitness, for various fitness functions. It is generally known that truncation selection is the most efficient form of directional selection from the standpoint of maximum change of gene frequency for a given effect of the gene on the character, but we quantify this by comparison with specific alternatives. Then, using *Drosophila* data on mutation rates and selection intensities for viability-reducing mutants, we show that truncation selection can indeed reduce the mutation load to an acceptable value despite a high mutation rate.

## A broken line approximation

We follow the example of Milkman (2) and approximate various alternatives to truncation selection by connected straight lines (Fig. 1). The fitness function increases linearly between $c - d$ and $c + d$, in which $c$ and $d$ are measured in units of $\sigma$. As $d$ becomes 0 we have truncation selection. As $d$ increases so that the slanted line covers the whole range of values of the character, fitness becomes simply proportional to the character value.

As before (1), $X$ is the measured character, $F(X)$ is the frequency function, and $W(X)$ is the fitness function. It is convenient to measure the character as a deviation of $X$ from its mean, $m$, and in units of the standard deviation, $\sigma$; $x = (X - m)/\sigma$. The frequency function on the transformed scale will be designated $f(x)$.

The publication costs of this article were defrayed in part by page charge payment. This article must therefore be hereby marked "*advertisement*" in accordance with 18 U. S. C. §1734 solely to indicate this fact.

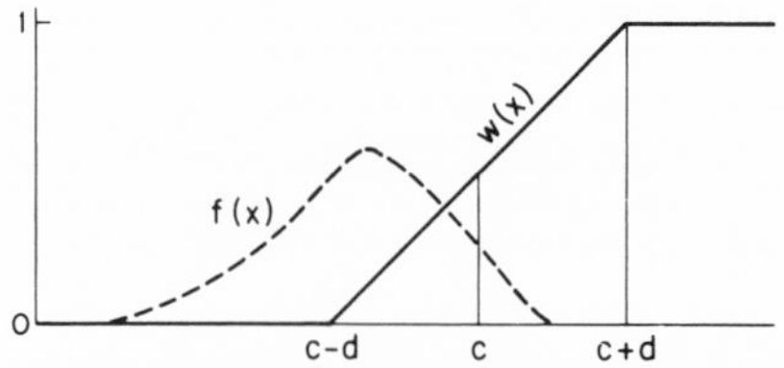


FIG. 1. Broken-line approximation to various fitness functions. As $d$ becomes small the model approaches truncation selection. The $x$ axis is scaled in units of the standard deviation $\sigma$. Fitness $w(x)$ is expressed as a fraction of the maximum fitness, $W_{max}$.

From Fig. 1 and letting $w(x) = W(x)/W_{max}$ we see that

$$w(x) = \begin{cases} 0 & x < c - d \\ (x + d - c)/2d & c - d < x < c + d \\ 1 & x > c + d. \end{cases}$$

From Eqs. **3** and **10a** in the previous paper (1) we have

$$\bar{w} = \int_{-\infty}^{\infty} w(x)f(x)dx \qquad [1]$$

$$\frac{s_i}{a_i} = -\int_{-\infty}^{\infty} w(x)f'(x)dx/\bar{w}. \qquad [2]$$

From Eq. **2** we can compare the efficiency of truncation selection with various broken-line alternatives. Some numerical values are given in Table 1, in which the frequency is normally distributed, $f(x) = (1/\sqrt{2\pi}) \exp(-x^2/2)$. It is clear, as emphasized by Milkman (2), that truncation selection is more efficient than the alternatives, but not by an enormous amount.

The broken line approximation is very close to the more biological integrated normal fitness function discussed in the earlier paper (1). Fig. 2 shows an example for comparison. The variance of the frequency function, assumed to be normally distributed, is taken to be the same as that of the integrated normal fitness function, so $R = 1$. The value of $c$ is assumed to be 0, which implies that $\overline{W}/W_{max} = \bar{w} = 1/2$. From equation **27** in ref. 1 we see that $s_i/a_i = 0.564$. The broken-line model with the same value of $s_i/a_i$ has $d = 1.56$. The two functions are shown in Fig. 2, along with the truncation fitness function, for which $s_i/a_i = 0.798$. Truncation selection is 1.77 times as efficient.

There is no obvious way to decide how large $d$ may become. Is it meaningful for $d$ to be greater than 3–5 standard deviations? In most cases there is not information enough to decide. However, a natural extreme alternative to truncate selection is for fitness to be simply proportional to the character value, $X$. This is reasonable for a trait, such as viability, for which there is a natural maximum and minimum value.

* This is the second of two related papers. Paper no. 1 is ref. 1.

Table 1. Measures of the relative efficiency of truncation selection against a broken-line alternative

| $d$ | $\overline{w}$ = 0.9 | $\overline{w}$ = 0.5 | $\overline{w}$ = 0.1 |
|---|---|---|---|
| 0 | 0.195 (1.00) | 0.798 (1.00) | 1.755 (1.00) |
| 0.5 | 0.187 (0.96) | 0.765 (0.96) | 1.687 (0.96) |
| 1.0 | 0.170 (0.87) | 0.683 (0.86) | 1.535 (0.87) |
| 1.5 | 0.153 (0.79) | 0.578 (0.72) | 1.379 (0.79) |
| 2.0 | 0.139 (0.71) | 0.479 (0.60) | 1.252 (0.71) |
| 3.0 | 0.118 (0.61) | 0.332 (0.42) | 1.063 (0.61) |
| 4.0 | 0.103 (0.53) | 0.250 (0.31) | 0.925 (0.53) |
| 5.0 | 0.091 (0.46) | 0.200 (0.25) | 0.816 (0.46) |

The table gives $s_i/a_i$ for specified values of the proportion selected, $\overline{w}$. The values in parentheses are relative to $d$ = 0, truncation selection.

This is illustrated in Fig. 3. The character value $X$ goes from 0 to 1, as does the fitness value $W(X)/W_{max} = w(X)$. If the fitness is proportional to the character, $w(X) = X$ and

$$s_i = \frac{X_i - m}{\overline{X}} = \frac{A_i}{\overline{w}} \qquad [3]$$

or

$$\frac{s_i}{a_i} = \frac{\sigma}{\overline{w}}. \qquad [4]$$

With truncation selection we use equation **11** from the previous paper (1):

$$s_i = \frac{A_i}{\overline{w}} F(C) = \frac{A_i}{\overline{w}} \cdot \frac{f(c)}{\sigma} \qquad [5]$$

$$\frac{s_i}{a_i} = \frac{f(c)}{\overline{w}}, \qquad [6]$$

in which $c = (C - m)/\sigma$. For a corresponding intensity of selection and phenotypic effect of an allele, the relative efficiency of truncation selection against the alternative of simple proportionality from Eqs. **4** and **6** is

$$RE = \frac{f(c)}{\sigma}. \qquad [7]$$

Some numerical values are given in Table 2. The smaller the standard deviation, $\sigma$, relative to the range of $X$, the greater is the relative efficiency of truncation selection. Although $s_i/a_i$ increases with the intensity of selection (i.e., as $\overline{w}$ decreases), as expected, the relative efficiency of truncation selection is greater as $\overline{w}$ approaches ½. Intermediate models are not shown, but from Table 1 it can be seen that when $D$ is nearly $\sigma$ (or $d$ = 1) the selection is about 85% as efficient as truncation selection. Although it is unreasonable to expect nature to practice perfect truncation selection, a model in which $d$ is 1 or 2 may often be appropriate.

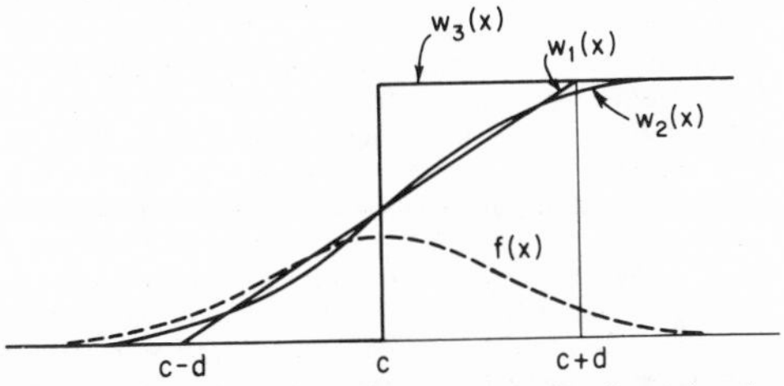


FIG. 2. Close agreement between a broken-line fitness function, $w_1(x)$, and an integrated normal distribution, $w_2(x)$, with the same efficiency of selection. Truncation selection, $w_3(x)$, is 1.77 times as efficient.

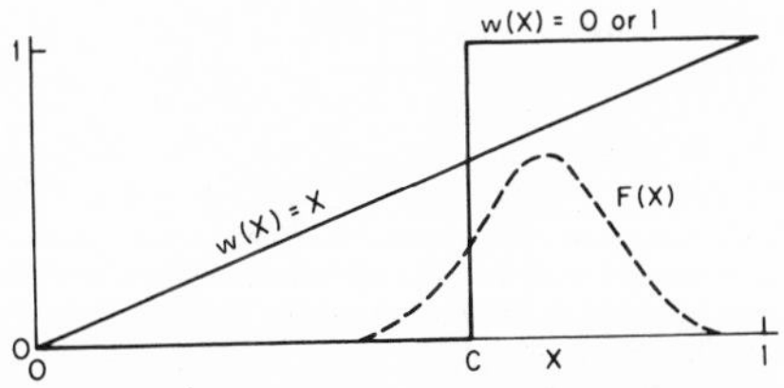


FIG. 3. Truncation selection, $w(X)$ = 0 or 1, compared to fitness proportional to the character value, $w(X) = X$. The character value, $X$, is scaled from 0 to 1.

## Application to viability mutants in *Drosophila*

Mutants causing a small decrease in homozygous viability have been studied extensively by Mukai and others (3–5). The individual effects of the mutants are small, no larger than about 0.03 on a scale for which 1.00 is maximum, and possibly considerably smaller because the procedure provides only a maximum estimate. The mutants are approximately additive over the range of interest here (6). They show considerable dominance, enough that the major impact of the mutants is through heterozyous expression (4, 7, 8). The minimum estimate of the mutation rate is about 0.4 per gamete. If fitness is proportional to viability and the genes act independently, the equilibrium fitness, relative to a maximum of 1, is exp(−0.8), or 0.45; the mutation load, by the Haldane mutation load principle (9, 6), is 0.55. If the mutation rate is twice this high, as is quite reasonable because the estimate is only a minimum, the load is 0.80. Can this large load be reduced by truncation selection?

Viability is restricted to the range 0–1 and, if not normally distributed, is at least unimodal (lethals and mutants causing large effects are not included in the data). Furthermore, the mutants are individually small and approximately additive in their homozygous effects. So this trait fits the requirements of our variable $X$. However, it is clear that natural selection acts not on the homozygous effects of these mutants on viability but on their pleiotropic heterozygous effects on total fitness. That the heterozygous effects of these mutants on "fitness potential" (2) are also approximately additive is suggested by the high correlation between homozygous effect on viability and heterozygous effects on fitness of chromosomes with accumulated mutants (7).

Keeping in mind that the variable $X$ is not viability *per se* but the pleiotropic heterozygous effects of the viability mutants on total fitness, which is assumed to be proportional, we can attempt a rough quantitative interpretation. The value of $s_i$ can be inferred from the inbred viability load in natural populations (10, 11) and the decrement in viability per generation of mutation accumulation (4). The observed ratio of these quantities is 30–40 (7, 8), but there is reason to think that this is a

Table 2. Relative efficiency, $RE = f(c)/\sigma$, of truncation selection against the alternative of fitness being proportional to the measured character

| | Standard deviation, $\sigma$ | | | | |
|---|---|---|---|---|---|
| $\overline{w}$ | 0.01 | 0.02 | 0.05 | 0.10 | 0.167 |
| 0.97, 0.03 | 6.8 | | | | |
| 0.94, 0.06 | 11.9 | 6.0 | | | |
| 0.85, 0.15 | 23.3 | 11.7 | 4.66 | | |
| 0.70, 0.30 | 34.8 | 17.4 | 6.96 | 3.48 | |
| 0.50 | 39.9 | 19.9 | 7.98 | 3.99 | 2.39 |

The values in the upper right corner of the table are omitted to keep the character distribution mainly within the range of 0–1.

slight underestimate, so we use 50.[§] This means that there are about 50 times as many mutants in an equilibrium population as arise in a single generation of mutation, so the mean persistence time for a mutant is 50 generations. The average rate of elimination per generation is roughly the reciprocal of this, or 0.02, so $s_i$ has an average value of about $-0.02$. This value is quite reliable, being based on direct measurements that are reproducible and on simple theory.

Note that if the individual $s_i$ values vary, the value $-0.02$ is the average for mutants in the equilibrium population. It is the harmonic mean of the value for newly arisen mutants, whose arithmetic mean would be somewhat larger. The value that we use, and quantities derived from it, apply to those mutants in an equilibrium population, among which those with longer persistence times are disproportionately represented. However, it is the mutation load of a population at equilibrium in which we are interested, so the $-0.02$ value is appropriate.

If fitness is proportional to viability and $s = -0.02$, then this is the value of $A_i/\overline{w}$. If fitness is subject to truncation selection, then from Eq. 7 this is multiplied by $\sigma/f(c)$.

We can get some idea of the value of $\sigma$ from mutation experiments. The genetic variance of viability from accumulated viability mutations increases at a rate of about 0.0005 per generation per zygote (4). Because of the 50 generations of mean persistence of a mutant in the population, the equilibrium value is 50 times as large, or 0.0250. This is for homozygous effects. Dominant effects have been estimated at about 1/4th of the homozygous effects (4), so the heterozygous variance, $\sigma^2$, is 0.0250/16, or 0.0016. So $\sigma = 0.04$ by this crude procedure.

It is reasonable that the standard deviation be less than 0.1, and probably considerably less, because the whole distribution is restricted to the range 0–1 and there are other causes of viability reduction that we are not including in this mutational analysis (e.g., balancing selection). We shall assume that $\sigma$ lies in the range of 0.02–0.05. For these values, we see from Table 2 that the relative efficiency of truncation selection ranges from 5 to 20. Thus the decrease in fitness per mutant need be only 1/5th to 1/20th of the 0.02 inferred from the mean persistence time of mutants in the population, or 0.001–0.004. In other words, with truncation selection, mutants with an effect on fitness of 0.1–0.4% are eliminated from the population as efficiently as if they had effects of 2.0% but were eliminated by additive selection. Truncation selection eliminates the mutants in groups so that the conventional H. J. Muller principle of "one mutant—one genetic death" is bypassed by picking off several mutants by a single extinction.

Of course nature does not practice perfect truncation selection. But, as shown earlier, there can be a considerable departure therefrom with retention of a large proportion of the benefit. We conclude that truncation selection, or something even very roughly approximating it, can be very effective in reducing the mutation load for mutants having minor deleterious effects on viability.

## Discussion

We have made several assumptions that are uncertain, although not gratuitous. The major assumption is that the effects of these genes on heterozygous fitness, or fitness potential in Milkman's term (2), are additive. This is justified partially because of the high correlation between homozygous and heterozygous viability (4, 7). More to the point is the correlation between homozygous *viability* and heterozygous *fitness* effects, evidenced both by direct experiments (12–14) and by the short persistence time of the viability mutants, which implies a great deal of heterozygous selection on total fitness. Of greater uncertainty are the assumptions about the variance of viability. Although this can be estimated under *laboratory* conditions for *homozygous* chromosomes, there is a large extrapolation to the variance of *heterozygous* effects in a *natural* population. The strength of our conclusions depends on the extent to which these plausible assumptions can be verified experimentally.

We can summarize the principal numerical conclusions as follows:

| | |
|---|---|
| $U = 0.4$ | Mutation rate per gamete for polygenic viability mutants (minimum) |
| $s_i = -0.02$ | Mean selective disadvantage per mutant, the reciprocal of the number of generations that a mutant persists in the population |
| $n_m = 2U/s_i = 40$ | Number of mutants per diploid individual at equilibrium (minimum) |
| $\overline{w} = 0.9$ | Proportion selected (weak truncation selection assumed for illustration) |
| $s_i/a_i = 0.195$ | From Table 1 |
| $a_i = -0.103$ | Mean effect of a mutant on standardized fitness potential (assuming 10% truncation) |
| $\sigma = 0.02$–$0.05$ | Standard deviation of fitness potential $X$ |
| $A_i = a_i\sigma = -0.002$–$-0.005$ | Average effect of a single mutant on fitness potential |
| $1 - n_mA_i = 0.80$–$0.92$ | Mean fitness potential, relative to a maximum of 1, consistent with 10% truncation selection |

The striking point, as already emphasized, is the large difference between $s_i$ and $A_i$. With perfect truncation, a reproductive excess sufficient to permit culling only 10% of the population can sustain the observed rate of mutation and mutant elimination. If fitness effects were strictly additive the mean fitness would be decreased by $-n_ms$, or 80%. Yet with truncation selection the mean fitness potential is reduced by only 10–18%. With more intense selection, say truncation at 20%, the reduction would be less, and a larger mutation rate could easily be balanced.

Experiments of Latter and Robertson (15), Sved and Ayala (16, 17), and Simmons, Mitchell, and others (12–14) have shown that there is much stronger selection against the same chromosome for total fitness than for viability alone. In *Drosophila* the reproductive capacity is so great that there is an enormous opportunity for selection in nature either through viability or fertility differences. Truncation selection, or a close approximation to it, may not be needed, at least for elimination of recurrent mutations, even with the high overall mutation rates implied by Mukai's work (3, 4).

On the other hand, many organisms, such as most mammals, have a limited reproductive rate. There is no reason to think that the mutation rates are less; if any part of the mutation rate is time dependent rather than generation dependent or if the process of evolutionary adjustment of mutation rates to increasing generation length lags far behind such increases, we might expect the mutation rate in long-lived mammals to be higher. Yet, animals with low reproductive capacity are not overtly handicapped by a large mutation load. Furthermore, their rates of morphological and cytological evolution are in many cases higher than those in organisms, such as insects and many invertebrates, with enormously greater reproductive capacity. Clearly, a high reproductive rate is not needed for

[§] The mutation accumulation is measured against a standard chromosome with no accumulated mutations, while the equilibrium population standard is a normal heterozygote, which includes dominant effects. With 25% dominance (4) the ratio should be increased by 33%; hence we take 50 as the value.

elimination of recurrent mutations or for adaptive evolution and maintenance of polymorphisms. Truncation selection provides one mechanism whereby a small selective intensity can be magnified in its effect on gene frequency change without a great requirement for excess fertility. This conclusion is consistent with the pioneering work of King (18), Sved *et al.* (19), and Milkman (20), as well as the more recent studies of Milkman (2) and Wills (21).

Yet, in our view, it is risky to carry this argument much further. The rank-order model assumes that the genes affecting fitness contribute additively, or at least that the rank order does not change with changes of environment or residual genotype. This is reasonable for genes contributing to a single trait such as size and, as we have discussed, is experimentally shown for new mutants reducing fitness. But, for the normal variance in fitness there are often negative correlations among components. Hiraizumi (22) found that, although the fitness components were positively correlated among the less fit genotypes, these correlations were negative for highly fit genotypes. Most traits that are fitness components have an intermediate optimum, thus making the same allele positively selected in some individuals and negatively in others, as emphasized especially by Wright (ref. 23, and many earlier papers referred to in this reference). Lande (24) has discussed the lowered effectiveness of directional selection when two characters with negative correlations are selected simultaneously. With many loci acting on traits that are sometimes negatively correlated, the additivity assumption necessary for a consistent ranking of all genes seems untenable when all components of fitness are considered. Although truncation can increase the efficiency of selection, it cannot solve all the problems.

It is clear that the mutation load can be substantially reduced and the efficiency of selection substantially increased by truncation selection, and almost as much by similar but less extreme selection schemes. In addition to the broken-line model that we have discussed, it is known that, if the deleterious effect of mutant genes is proportional to the square of their number, the mutation load is half as large as with no epistasis (25). The actual amount of epistasis observed for homozygous effects on viability would produce about the same reduction (6). But the actual shape of the fitness curve in nature is not known. The question is: Does nature rank and truncate, or do something approximating this? It will have to be answered empirically.

We thank Roger Milkman for some helpful comments on the manuscript. This work was supported in part by National Institutes of Health Grant GM 22038. This is contribution No. 1220 from the National Institute of Genetics, Mishima, Japan, and No. 2294 from the Genetics Laboratory, University of Wisconsin-Madison.

1. Kimura, M. & Crow, J. F. (1978) *Proc. Natl. Acad. Sci. USA* **75,** 6168–6171.
2. Milkman, R. (1978) *Genetics* **88,** 391–403.
3. Mukai, T. (1964) *Genetics* **50,** 1–19.
4. Mukai, T., Chigusa, S. I., Mettler, L. E. & Crow, J. F. (1972) *Genetics* **72,** 335–355.
5. Ohnishi, O. (1977) *Genetics* **87,** 529–545.
6. Crow, J. F. (1970) in *Mathematical Topics in Population Genetics,* ed. Kojima, K. (Springer, Berlin), Vol. 1, pp. 128–177.
7. Simmons, M. J. & Crow, J. F. (1977) *Annu. Rev. Genet.* **11,** 49–78.
8. Crow, J. F. (1979) *Genetics,* in press.
9. Haldane, J. B. S. (1937) *Am. Nat.* **71,** 337–349.
10. Greenberg, R. & Crow, J. F. (1960) *Genetics* **45,** 1153–1168.
11. Temin, R. G., Meyer, H. U., Dawson, P. S. & Crow, J. F. (1969) *Genetics* **61,** 497–519.
12. Mitchell, J. A. (1977) *Genetics* **87,** 763–774.
13. Mitchell, J. A. & Simmons, M. J. (1977) *Genetics* **87,** 775–783.
14. Simmons, M. J., Sheldon, E. W. & Crow, J. F. (1978) *Genetics* **88,** 575–590.
15. Latter, B. D. H. & Robertson, A. (1962) *Genet. Res.* **3,** 110–138.
16. Sved, J. A. & Ayala, F. J. (1970) *Genetics* **66,** 97–113.
17. Sved, J. A. (1975) *Genet. Res.* **18,** 97–105.
18. King, J. L. (1967) *Genetics* **55,** 483–494.
19. Sved, J. A., Reed, T. E. & Bodmer, W. F. (1967) *Genetics* **55,** 469–481.
20. Milkman, R. (1967) *Genetics* **55,** 493–495.
21. Wills, C. (1978) *Genetics* **89,** 403–417.
22. Hiraizumi, Y. (1960) *Genetics* **46,** 615–624.
23. Wright, S. (1978) *Variability Within and Among Natural Populations,* Evolution and the Genetics of Populations (University of Chicago Press, Chicago), Vol. 4.
24. Lande, R. (1979) *Evolution,* in press.
25. Kimura, M. & Maruyama, T. (1966) *Genetics* **54,** 1337–1351.

# Possibility of extensive neutral evolution under stabilizing selection with special reference to nonrandom usage of synonymous codons

(molecular evolution/quantitative character/natural selection/neutral mutation–random drift hypothesis/mode of evolution)

MOTOO KIMURA

National Institute of Genetics, Mishima, 411 Japan

*Contributed by Motoo Kimura, June 3, 1981*

**ABSTRACT** The rate of evolution in terms of the number of mutant substitutions in a finite population is investigated assuming a quantitative character subject to stabilizing selection, which is known to be the most prevalent type of natural selection. It is shown that, if a large number of segregating loci (or sites) are involved, the average selection coefficient per mutant under stabilizing selection may be exceedingly small. These mutants are very slightly deleterious but nearly neutral, so that mutant substitutions are mainly controlled by random drift, although the rate of evolution may be lower as compared with the situation in which all the mutations are strictly neutral. This is treated quantitatively by using the diffusion equation method in population genetics. A model of random drift under stabilizing selection is then applied to the problem of "nonrandom" or unequal usage of synonymous codons, and it is shown that such nonrandomness can readily be understood within the framework of the neutral mutation–random drift hypothesis (the neutral theory, for short) of molecular evolution.

It is generally accepted that stabilizing selection is the most prevalent type of natural selection at the phenotypic level (1–4). It eliminates phenotypically extreme individuals and preserves those that are near the population mean (5). It is also called centripetal selection (6) or normalizing selection (7), and many examples have been reported. Probably the best example in human populations is the relationship between the birth weights of babies and their neonatal mortality, as studied by Karn and Penrose (8). These authors found that babies whose weight is very near the mean have the lowest mortality. This optimum weight is slightly heavier than the mean, and mortality increases progressively as the birth weight deviates from this optimum (see also ref. 9). Unlike the type of natural selection that Darwin (10) had in mind when he tried to explain evolution, stabilizing selection acts to keep the status quo rather than to cause a directional change. From this, it might appear that stabilizing selection is antithetical to evolutionary change.

In this note, I intend to show that, under stabilizing phenotypic selection, extensive "neutral evolution" can occur. By neutral evolution, I mean accumulation of mutant genes in the species through random genetic drift (due to finite population size) under mutational pressure. Thus, beneath an unchanged morphology, a great deal of cryptic genetic change may be occurring in natural populations of all organisms, transforming even genes of "living fossils" (11). This will substantiate my neutral mutation–random drift hypothesis (the neutral theory, for short) of molecular evolution (ref. 12; for review, see ref. 13). I shall also show that this gives a clue to understanding "nonrandom" or unequal usage of synonymous codons (14–16) based on the neutral theory.

The publication costs of this article were defrayed in part by page charge payment. This article must therefore be hereby marked "*advertisement*" in accordance with 18 U. S. C. §1734 solely to indicate this fact.

**Selection intensity at an individual locus when overall phenotypic selection is given**

Let us consider a quantitative character, such as height, weight, concentration of some substance, or a more abstract quantity that represents Darwinian fitness in an important way. We assume that the character is determined by a large number of loci (or sites), each with a very small effect in addition to being subjected to environmental effects. We also assume that genes are additive with respect to the character. We follow the method used by Bulmer (17, 18) and Kimura and Crow (19). Let $X$ be the measured phenotypic character with the mean $M$ and the variance $\sigma^2$. We denote by $X_{op}$ the optimum phenotypic value and, unless otherwise stated, we shall take this point as the origin. Let $F(X)$ and $W(X)$ be the relative frequency and the fitness of individuals with character value $X$. Two examples of $W(X)$ are shown in Fig. 1.

Consider a particular locus at which a pair of alleles $A_1$ and $A_2$ are segregating with respective frequencies $1 - p$ and $p$. We assume a random-mating diploid population and let $X_{ij}$ be the average phenotypic value of $A_iA_j$ individuals, where $i = 1$ or 2 and $j = 1$ or 2.

It is often convenient to measure various quantities relating to the character value in units of the standard deviation ($\sigma$). For this purpose, lowercase letters will be used such as $x = (X - X_{op})/\sigma$, and the corresponding frequency and fitness functions will be denoted by $f(x)$ and $w(x)$. We also let $m = (M - X_{op})/\sigma$ and $a_{ij} = (X_{ij} - M)/\sigma$. Note that $a_{ij}$ is the deviation of $A_iA_j$ from the population mean in $\sigma$ units.

We assume that the background distribution of the character is the same among different genotypes at this locus and that this is given by $f(x)$ with good approximation, because individual gene effects are assumed to be extremely small. Let $w_{ij}$ be the relative fitness of $A_iA_j$, then

$$w_{ij} = \int_{-\infty}^{\infty} w(x)f(x - a_{ij})dx, \quad [1]$$

as explained in ref. 19. Assuming that $a_{ij}$ is small, we expand $f(x - a_{ij})$ in a Taylor series,

$$f(x - a_{ij}) = f(x) - a_{ij}f'(x) + (a_{ij}^2/2)f''(x) - \dots, \quad [2]$$

as in ref. 19, but here we retain the second-order term, so that we get, from Eq. 1,

$$w_{ij} = b_0 - a_{ij}b_1 + a_{ij}^2 b_2/2, \quad [3]$$

where

$$b_0 = \int_{-\infty}^{\infty} w(x)f(x)dx, \qquad b_1 = \int_{-\infty}^{\infty} w(x)f'(x)dx,$$

and

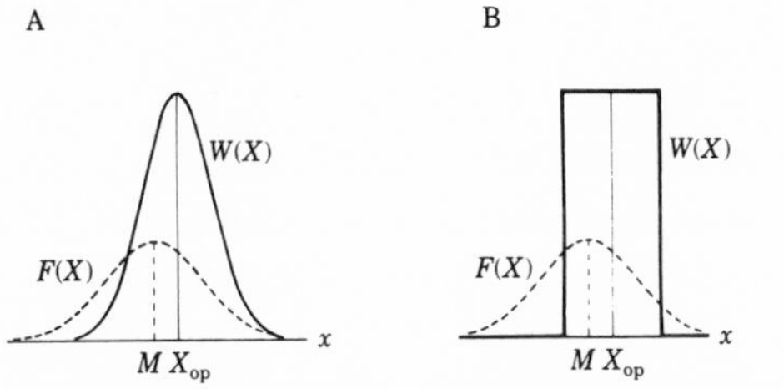


FIG. 1. Examples of the fitness function W(X). (A) Normal distribution. (B) Uniform distribution. ——, fitness function; - - - -, frequency function.

$$b_2 = \int_{-\infty}^{\infty} w(x) f''(x) dx.$$

Here, the prime denotes differentiation.

Let $a$ be the effect of substituting $A_2$ for $A_1$ on the character value $x$. Then, under random mating and assuming an additive gene effect on $x$, we find that $a_{11} = -2ap$, $a_{12} = a(1-2p)$, and $a_{22} = 2a(1-p)$. Then, by using Eq. **3**, we can compute the mean fitness $\bar{w} = w_{11}(1-p)^2 + w_{12}2(1-p)p + w_{22}p^2$ and the average fitness of $A_2$—i.e., $w_2 = w_{21}(1-p) + w_{22}p$—and they turn out to be as follows:

$$\bar{w} = b_0 + a^2 p(1-p) b_2 \quad [4]$$

$$w_2 = b_0 - a(1-p)b_1 + a^2(1-p)b_2/2. \quad [5]$$

The change of the frequency of $A_2$ in one generation is given by $\Delta p = p(w_2 - \bar{w})/\bar{w}$ (see p. 180 of ref. 20). Therefore, by substituting Eqs. **4** and **5** in this expression and neglecting terms involving $a^3$ and higher order terms, we obtain

$$\Delta p = \frac{p(1-p)}{b_0}\left[-ab_1 + a^2 b_2\left(\frac{1}{2} - p\right)\right], \quad [6]$$

in agreement with Bulmer (17). Then, the selection coefficient, $s$, which represents the selective advantage of $A_2$ over $A_1$ is

$$s = -ab_1/b_0 + a^2 b_2(\tfrac{1}{2} - p)/b_0. \quad [7]$$

With this selection coefficient $s$, the change of $p$ by selection per generation is $\Delta p = sp(1-p)$.

In the special, but important, case in which both the frequency and the fitness functions are given by normal distributions,

$$f(x) = \frac{1}{\sqrt{2\pi}} \exp[-(x-m)^2/2], \quad [8]$$

and

$$w(x) = \exp(-kx^2), \quad [9]$$

Eq. **7** reduces to

$$s = -m\lambda a + (\lambda^2 m^2 - \lambda)(\tfrac{1}{2} - p)a^2, \quad [10]$$

where $\lambda = 2k/(1+2k)$. An equivalent result was obtained earlier by Bulmer (18).

Note that, if we use the original scale ($X$) and express the density function of the frequency distribution of the character by $F(X) = (1/\sqrt{2\pi}\sigma)\exp[-(X-M)^2/2\sigma^2]$, the fitness function by $W(X) = \exp(-KX^2)$, and the effect of allele substitution by $A$, then the parameters in Eq. **10** are $m = (M - X_{op})/\sigma$, $\lambda = 2K\sigma^2/(1+2K\sigma^2)$, and $a = A/\sigma$. From Eq. **10**, it may be seen that, if the deviation of the mean from the optimum is much larger than the effect of the allele substitution ($|m| \gg |a|$), we have $s/a \approx -\lambda m$, which agrees with equation **23** of ref. 19. In this case, the situation is similar to truncation selection (see refs. 21 and 22), and natural selection acts very efficiently to change the mean toward the optimum. During this short period of directional selection, extensive shift of gene frequencies is expected to occur at many loci, but this process itself will seldom cause gene substitutions.

If, on the other hand, the mean is at the optimum ($m = 0$), we have $s = \lambda(\tfrac{1}{2} - p)a^2$ from Eq. **10**. In this case, $A_2$ is disadvantageous if $p < \tfrac{1}{2}$ but advantageous if $p > \tfrac{1}{2}$. This selection is frequency dependent, and alleles behave as if negatively overdominant. The change of gene frequency is then given by $\Delta p = \lambda a^2 p(1-p)(p - \tfrac{1}{2})$ in agreement with Robertson (23) and Wright (24). What is pertinent to our evolutionary consideration is that, here, every mutation is deleterious, because $\Delta p < 0$ if $p$ is small. Furthermore, if a large number of loci are segregating, each with a very small effect, $a$ is small so that $a^2$ is extremely small. This applies with particular force if we consider an individual nucleotide site rather than the conventional gene locus, as it has been estimated (25) that the average individual in a large mammalian population is likely to be heterozygous at a million or so nucleotide sites. This substantiates Ohta's claim (26, 27) that the majority of mutants at the molecular level are nearly neutral but very slightly deleterious. As I shall show below, negatively overdominant alleles are far more susceptible to random genetic drift than unconditionally deleterious alleles having the same magnitude of selection coefficient.

As to the intensity of natural selection involved, we can measure it in terms of load ($L$)—i.e., by the fraction of individuals that are eliminated in each generation by natural selection due to deviation of their phenotypic values from the optimum. For the frequency and fitness functions given as Eqs. **8** and **9**, we obtain

$$L = 1 - \sqrt{1-\lambda}\exp(-\lambda m^2/2). \quad [11]$$

For $m = 0$, this reduces to $L = 1 - \sqrt{1-\lambda}$. In the special case in which the fitness function has the same variance as the frequency function—i.e., when $K = 1/(2\sigma^2)$ or $k = \tfrac{1}{2}$, we get $L \approx 0.293$ or $\approx 30\%$ elimination. In general, $L$ is likely to be small for any single character in mammals. For example, according to Haldane (1), the intensity of selection acting on birth weight of babies through their neonatal mortality is $L = 0.027$. If $L$ is small, we have, with good approximation, $L = \lambda/2$.

## Behavior of mutant alleles in a finite population under stabilizing selection

**Probability of Fixation of a Mutant Allele.** We denote by $N_e$ the effective size of the population (for the meaning of $N_e$, see ref. 28). Roughly speaking, $N_e$ is equal to the number of breeding individuals in one generation. This number is likely to be much smaller in most cases than the apparent population size, which we denote by $N$. To simplify expressions, we let $\beta_1 = -\lambda ma$ and $\beta_2 = \lambda(1 - \lambda m^2)a^2/2$, so that Eq. **10** reduces to

$$s = \beta_1 - \beta_2(1-2p), \quad [12]$$

where $p$ is the frequency of $A_2$. Ignoring mutational change for a moment, and denoting the frequency of $A_2$ by $y$, the mean and the variance in the change of $y$ during one generation are

$$M_{\delta y} = [\beta_1 - \beta_2(1-2y)]\, y(1-y) \quad [13]$$

and

$$V_{\delta y} = y(1-y)/2N_e. \quad [14]$$

Let $u(p)$ be the probability that $A_2$ becomes eventually fixed in the population (i.e., reaches 100% in frequency), given that its initial frequency is $p$. Then, $u(p)$ can be expressed in terms of $M_{\delta y}$ and $V_{\delta y}$ by using a general formula for the probability of fixation obtained by Kimura (ref. 29; see also p. 424 of ref. 20, where $x$ is used instead of $y$). We are particularly interested in the probability of fixation of $A_2$ when it is initially singly represented in the population. If we denote this probability by $u$, then this is given by $u(p)$ with $p = 1/(2N)$. We then obtain

$$u = 1\bigg/\left[2N\int_0^1 \exp\{-B_1x + B_2x(1-x)\}dx\right], \qquad [15]$$

where $B_1 = 4N_e\beta_1$ and $B_2 = 4N_e\beta_2$.

In the above treatment, we have assumed that $m$ (the deviation of the mean from the optimum) remains unchanged throughout the process. This assumption appears to be unrealistic because, if $m \neq 0$, one would expect $|m|$ to be reduced with time by the directional component of selection. There is an important possibility, however, that this change is opposed by mutational pressure so that $m$ remains constant under continued stabilizing selection, although $|m|$ at equilibrium is likely to be small. This occurs when the optimum and the mutational equilibrium point do not coincide. We shall elaborate such a case when we discuss the problem of nonrandom synonymous codon usage.

To show that mutants that have negative overdominance (as induced by stabilizing selection) are far more likely to be fixed by random drift than unconditionally deleterious mutants that have comparable selection coefficients, some examples of the probabilities of fixation ($u$) for these two cases are listed in Table 1. In the case of stabilizing selection, we let $m = 0$ and denote $\lambda a^2/2$ by $s_s$ (selection coefficient for stabilizing selection), so that $\Delta p = -s_sp(1-p)(1-2p)$. For the unconditionally deleterious case, we denote the selection coefficient against $A_2$ by $-s'$ ($s' > 0$), so that the probability of fixation is given by

$$u = S'/[2N(e^{S'}-1)], \qquad [16]$$

where $S' = 4N_es'$ (see p. 426 of ref. 20). In both cases, $u$ is tabulated taking the probability of fixation of the completely neutral case as the unit—i.e., it is expressed as a multiple of $u_0 = 1/(2N)$. It is clear from Table 1 that an enormous difference exists between the two cases in fixation probability and that, under stabilizing selection, extensive neutral evolution is possible even when $4N_es_s$ is 8 or more. For $B = 4N_es_s > 8$, it can be shown that $u/u_0 \approx \sqrt{B/\pi}\exp(-B/4)$.

**Gene Frequency Distribution.** We now incorporate mutational pressure and investigate the probability distribution of allelic frequencies at statistical equilibrium attained under stabilizing selection in a finite population. We shall denote by $\phi(p)$ the probability density such that $\phi(p)dp$ represents the probability that the frequency of $A_2$ in the population lies in the range $p \sim p + dp$, where $0 < p < 1$. We assume reversible mutations between the two alleles and let $v_1$ be the mutation rate from $A_1$ to $A_2$ and $v_2$ be the rate in the reverse direction. Then, the mean and the variance in the change of $A_2$ in one generation are, respectively,

$$M_{\delta p} = [\beta_1 - \beta_2(1-2p)]\,p(1-p) - v_2p + v_1(1-p) \qquad [17]$$

and

$$V_{\delta p} = p(1-p)/(2N_e). \qquad [18]$$

By using Wright's (30) formula for the steady-state gene-frequency distribution (see p. 434 of ref. 20), we obtain

$$\phi(p) = Ce^{B_1p - B_2p(1-p)}\,p^{V_1-1}(1-p)^{V_2-1}, \qquad [19]$$

where $B_1 = 4N_e\beta_1$, $B_2 = 4N_e\beta_2$, $V_1 = 4N_ev_1$, $V_2 = 4N_ev_2$, and $C$ is determined so that $\int_0^1 \phi(p)dp = 1$. The probability of $A_2$ being temporarily fixed in the population may be obtained by integrating $\phi(p)$ from $1 - [1/(2N)]$ to 1, and we obtain, with sufficient accuracy,

$$f_2 = C\,e^{B_1}/[V_2(2N)^{V_2}]. \qquad [20]$$

Similarly, the probability of $A_1$ being temporarily fixed in the population (i.e., $A_2$ lost) is

$$f_1 = C/[V_1(2N)^{V_1}]. \qquad [21]$$

Then, the ratio of $f_2$ to $f_1$ is

$$f_2/f_1 = e^{B_1}(V_1/V_2)(2N)^{V_1-V_2}. \qquad [22]$$

In this paper, we shall be mainly concerned with the situation in which both $V_1$ and $V_2$ are much smaller than unity and alleles are fixed most of the time. This situation is particularly pertinent when we consider individual nucleotide sites rather than conventional gene loci, because the mutation rate per site must be of the order of $10^{-8}$ rather than of $10^{-5}$.

In general, for any set of values of $B_1$, $B_2$, $V_1$, and $V_2$, we can compute the mean frequency $\bar{p}$ and the mean heterozygosity $\bar{H}_e$ per locus through numerical integration by using $\bar{p} = E(p) = \int_0^1 p\phi(p)dp$ and $\bar{H}_e = E[2p(1-p)] = \int_0^1 2p(1-p)\phi(p)dp$. If the phenotype is determined by $n$ equivalent loci in addition to environmental effects, we have $M = 2nA\bar{p}$ and $\sigma^2 = nE[2A^2p(1-p)]/\rho^2 = nA^2\bar{H}_e/\rho^2$, where A is the effect of substituting $A_2$ for $A_1$ on the character ($A = a\sigma$) and $\rho^2$ is the fraction of phenotypic variance due to gene segregation—i.e., broad sense heritability. Furthermore, if mutation rates are equal in both directions ($v_1 = v_2 \equiv v$) and the phenotypic mean coincides with the optimum phenotype (i.e., $m = 0$ or $B_1 = 0$, $B_2 = 2N_e\lambda a^2$), the distribution formula is much simplified, and the values of $E[p(1-p)]$ are tabulated by Bulmer (18) for some combinations of values of $4N_ev$ and $v/(ka^2)$. One interesting property of the frequency distribution of alleles under stabilizing selection is that it is more U shaped than the strictly neutral case having the same mutation rate. It may sometimes be convenient to take as the standard the situation in which the minus alleles are fixed at all loci. Then, the range of $X$ lies between 0 and $2nA$ if we assume that $n$ loci are involved and the effect of allele substitution is the same at all the loci. Let

$$X_{op} = 2nAQ_{op}, \qquad [23]$$

where $Q_{op}$ is the position of the optimum when the total range of $X$ is rescaled so that it lies in the interval [0, 1]. If $A_1$ is the minus allele and $A_2$ is the plus allele, so that $A > 0$, then the optimum is less than the mean if $Q_{op} < \bar{p}$ and more than the mean if $Q_{op} > \bar{p}$.

Table 1. Relative probability of fixation ($u/u_0$) of negatively overdominant and unconditionally deleterious mutants

| $\hat{S}$ | Negatively overdominant | Unconditionally deleterious |
|---|---|---|
| 0 | 1.00 | 1.00 |
| 1.0 | 0.84 | 0.58 |
| 8.0 | 0.23 | $2.7 \times 10^{-3}$ |
| 16.0 | 0.042 | $1.8 \times 10^{-6}$ |
| 20.0 | 0.017 | $4.1 \times 10^{-8}$ |
| 30.0 | 0.0017 | $2.8 \times 10^{-12}$ |

$\hat{S}$ stands for $4N_es_s$ for the negatively overdominant case and $4N_es'$ for the unconditionally deleterious case.

**Application to the problem of nonrandom codon usage**

Recently, nonrandom or unequal usage of synonymous codons has been reported in many genes of various organisms (for review, see ref. 14). Indeed, nonrandom codon usage appears to be a rule rather than an exception, and this is often mentioned as evidence against the neutral theory. I shall now show that this can be explained in the framework of the neutral theory. Note that the existence of selective constraint (negative selection) by no means contradicts the neutral theory (see ref. 13).

To simplify the argument, we group nucleotide bases A (adenine) and U (uracil) as $A_1$ and C (cytosine) and G (guanine) as $A_2$. It is known (31) that, at the third position of degenerate codons in mammalian mRNAs, $A_2$ predominate over $A_1$. For globin mRNA, the ratio of $A_2$ to $A_1$ at position 3 is ≈7:3 (32). As shown above, the distribution function $\phi(p)$ (Eq. **19**), when applied to a nucleotide site rather than a gene locus, indicates that either $A_1$ or $A_2$ is fixed most of the time in the course of evolution. This is because the mutation rate per site is exceedingly low, so that the probability of polymorphism per nucleotide site is very small, although this probability may amount to more than 10% when applied to a locus that is comprised of 1000 or so nucleotide sites.

As to the cause of nonrandom codon usage, recent studies of Ikemura (15, 16) are instructive. He found a strong positive correlation between the frequency of synonymous codon usage and abundance of cognate tRNA in *Escherichia coli*. This correlation appears to be related to the translational efficiency (see also ref. 33). If this applies in general to other organisms, the most plausible explanation for preferential codon usage is that it represents the optimum state in which the population of synonymous codons matches that of cognate tRNA available in the cell. This will help to carry out more efficient cell function, leading to higher Darwinian fitness. This appears to be compatible with the genome hypothesis of Grantham *et al.* (14), who claim that a surprising consistency of choices of degenerate bases exists among genes of the same or similar genomes and that "the genome and not the individual gene is the unit of selection."

Let $Q_{op}$ be the optimum proportion of $A_2$ (guanine or cytosine) at position 3 of the codons and assume that mutation rates are equal between $A_1$ and $A_2$—i.e., $V_1 = V_2$—then the mean of $p(\bar{p})$ does not coincide with $Q_{op}$ unless $Q_{op} = 0.5$. So, we assume that stabilizing selection is at work to hold $\bar{p}$ near $Q_{op}$. At individual sites, however, $A_2$ is either fixed or lost most of the time. Let $f_2$ be the probability that $A_2$ is fixed in the population at a given site. Similarly, let $f_1$ be the probability that $A_1$ is fixed ($A_2$ is lost). Then, from Eq. **22**, we have $f_2/f_1 = \exp(B_1)$, where $B_1 = 4N_e\beta_1 = -4N_e\lambda ma$. Thus, we can estimate $B_1$ by the relationship $B_1 = \ln(f_2/f_1)$, and we obtain $B_1 = 0.85$ for $f_2/f_1 = 0.7/0.3$. In most mammalian species, the effective size $N_e$ must be at least $10^4$. Therefore, the intensity of selection that acts at an individual site to produce nonrandom codon usage is an exceedingly weak one, leaving plenty of room for random drift to operate. This is consistent with Latter's (34) claim that mutations responsible for enzyme polymorphisms are very slightly deleterious with "*Ns*" values in the range 1–3.

One important question that remains is the extent to which the rate of evolution in terms of mutant substitution is influenced by such selection. As the relative evolutionary rate (in terms of mutant substitution) under stabilizing selection as compared with the strictly neutral case is given by $u/u_0$ with $u_0 = 1/(2N)$, we have, from Eq. **15**,

$$u/u_0 = 1 \Big/ \int_0^1 \exp[-B_1x + B_2x(1-x)]dx, \qquad [24]$$

where $B_1 = -4N_e\lambda ma$ and $B_2 = 2N_e\lambda(1 - \lambda m^2)a^2$. If we assume that $2N_e\lambda a^2$ is negligibly small, so that $B_2 \approx 0$, then we get

$$u/u_0 \approx 2f_1f_2\ln(f_2/f_1)/(f_2 - f_1). \qquad [25]$$

For $f_2/f_1 = 0.7/0.3$, as we observe at the third position of the codons in globin and other mammalian mRNAs, we get $u/u_0 = 0.89$. In other words, the evolution is retarded by ≈10% from what is expected under complete selective neutrality. Under the more extreme condition $f_2/f_1 = 0.9/0.1$, we get $u/u_0 \approx 0.49$, which means ≈50% retardation.

In actual situations, however, there are four possible "alleles" (bases) per nucleotide site rather than two and, together with other complications due to differences in the speed of translation among different types of genes, etc. (16), we need more careful and detailed analysis to arrive at a more accurate figure for the retardation.

**Discussion**

During its lifetime, an individual is subject to natural selection through a large number of quantitative characters, many of which are mutually correlated. Let us assume, to simplify the treatment, that we can choose a certain number, say $n_c$, of independent characteristics that collectively represent, to a first approximation, the total pattern of selection. Various parameters pertaining to the *i*th character will be expressed by subscript $i$ ($i = 1, 2, \ldots, n_c$).

Because the total selection intensity is limited, the selection intensity, as measured by $L_i$ at each component character is expected to be small if the number of characters involved is large. Let $L_T$ be the total selection intensity, then $(1 - L_T) = \Pi_i(1 - L_i)$, so that $L_T \approx 1 - \exp(-\Sigma_i L_i)$. To simplify the treatment still further, let us suppose that the $L_i$s are all equal among component characters, so that $L_i \approx -(1/n_c)\ln(1 - L_T)$. The selection coefficient per site is then $-\lambda_i(1/2 - p_i)a_i^2$ and, noting that $L_i \approx \lambda_i/2$ and dropping the subscript $i$, we have $s = \ln(1 - L_T)(1 - 2p)a^2/n_c$ (approximately). On the right-hand side of this formula, we note that $a^2/n_c = A^2/(n_c\sigma^2)$, where $n_c\sigma^2$ is the variance of the total phenotype. Let $\bar{h}_e$ be the average heterozygosity per site and, if we denote by $n_{nuc}$ the total number of nucleotide sites concerned, then $n_{nuc}A^2\bar{h}_e = n_c\sigma^2\rho^2$, where $A$ is the effect of substituting one nucleotide on a component phenotype and $\rho^2$ is broad sense heritability. Thus, the coefficient for stabilizing selection $s_s$, as defined by the relationship $s = -s_s(1 - 2p)$, turns out to be $s_s = -[\ln(1 - L_T)]\rho^2/(n_{nuc}\bar{h}_e)$. This represents the selection intensity involved in nucleotide substitution under stabilizing selection (assuming $m = 0$).

Let us assume that the average heterozygosity per enzyme locus with respect to electrophoretically detectable alleles is 0.1 and (rather conservatively) that there is twice as much heterozygosity with respect to silent alleles. Then, if the average number of nucleotide sites that comprises a locus is $10^3$, we get $\bar{h}_e = 3 \times 10^{-4}$. Extrapolating this to the total genome of a mammal that has $3.5 \times 10^9$ nucleotide sites, the average number of heterozygous nucleotide sites per individual is $n_{nuc}\bar{h}_e = 1.05 \times 10^6$. As typical values of genetic load and heritability for a mammal (such as the human species), let us suppose that $L_T = 0.5$ and $\rho^2 = 0.5$, then, we obtain $s_s = 3.3 \times 10^{-7}$. This is a very small selection coefficient for stabilizing selection and shows that the majority of mutations at the molecular level are nearly neutral but very slightly deleterious. This agrees with Ohta's hypothesis of very slightly deleterious mutations (26, 27). However, the fitness of the species does not drift downward in this view as it does in Ohta's hypothesis. Also, in this view, those genes that are substituted by random drift and those that are responsible for phenotypic variability of quantitative traits belong to the same class. It is possible that many, and even most, of the mutants affecting a quantitative trait are regulatory rather than structural. DNA outside the coding region may be more important from this standpoint than translated DNA. The present analysis agrees with Lande (35), who suggests that many polygenic changes can accumulate by random drift because they have little or no net phenotypic effect.

Needless to say, some sites produce much larger phenotypic effects than others and therefore are subject to stronger selection. On the other hand, a certain fraction of sites (presumably a large fraction) produce no phenotypic effects at all and therefore are completely neutral with respect to natural selection.

The picture of evolution that emerges from the present analysis is as follows. From time to time, the position of the optimum shifts due to changes in environment and the species tracks such changes rapidly by altering its mean. But, most of the time, stabilizing selection predominates. Under this selection, neutral evolution (random fixation of alleles by sampling drift) occurs extensively, transforming all genes, including those of living fossils, profoundly at the molecular level.

I thank Drs. Tomoko Ohta and R. Milkman for stimulating discussions in composing the manuscript. Thanks are also due to Drs. J. Maynard Smith, J. F. Crow, and K. Aoki for helpful suggestions and useful comments to improve the presentation. This is contribution no. 1364 from the National Institute of Genetics, Mishima, Shizuoka-ken, 411 Japan.

1. Haldane, J. B. S. (1959) in *Darwin's Biological Work*, ed. Bell, P. R. (Cambridge Univ. Press, Cambridge, England), pp. 101–149.
2. Mather, K. (1973) *Genetical Structure of Populations* (Chapman & Hall, London).
3. Wright, S. (1977) *Evolution and the Genetics of Populations* (Univ. Chicago Press, Chicago), Vol. 3.
4. Parkin, D. T. (1979) *An Introduction to Evolutionary Genetics* (Arnold, London).
5. Mather, K. (1953) *Symp. Soc. Exp. Biol.* **7**, 66–95.
6. Simpson, G. G. (1944) *Tempo and Mode in Evolution* (Columbia Univ. Press, New York).
7. Waddington, C. H. (1957) *The Strategy of the Genes* (Allen & Unwin, London).
8. Karn, M. N. & Penrose, L. S. (1951) *Ann. Eugen.* **16**, 147–164.
9. Cavalli-Sforza, L. L. & Bodmer, W. F. (1971) *The Genetics of Human Populations* (Freeman, San Francisco).
10. Darwin, C. (1859) *The Origin of Species* (Murray, London).
11. Kimura, M. (1969) *Proc. Natl. Acad. Sci. USA* **63**, 1181–1188.
12. Kimura, M. (1968) *Nature (London)* **217**, 624–626.
13. Kimura, M. (1979) *Sci. Am.* **241** (5), 94–104.
14. Grantham, R., Gautier, C., Gouy, M., Mercier, R. & Pavé, A. (1980) *Nucleic Acids Res.* **8**, r49–r62.
15. Ikemura, T. (1980) in *Genetics and Evolution of RNA Polymerase, tRNA and Ribosomes*, ed. Osawa, S., Ozeki, H., Uchida, H. & Yura, T. (Univ. Tokyo Press, Tokyo), pp. 519–523.
16. Ikemura, T. (1981) *J. Mol. Biol.* **146**, 1–21.
17. Bulmer, M. G. (1971) *Heredity* **27**, 157–162.
18. Bulmer, M. G. (1972) *Genet. Res.* **19**, 17–25.
19. Kimura, M. & Crow, J. F. (1978) *Proc. Natl. Acad. Sci. USA* **75**, 6168–6171.
20. Crow, J. F. & Kimura, M. (1970) *An Introduction to Population Genetics Theory* (Harper & Row, New York); reprinted (1977) by Burgess, Minneapolis, MN.
21. Milkman, R. (1978) *Genetics* **88**, 391–403.
22. Crow, J. F. & Kimura, M. (1979) *Proc. Natl. Acad. Sci. USA* **76**, 396–399.
23. Robertson, A. (1956) *J. Genetics* **54**, 236–248.
24. Wright, S. (1935) *J. Genetics* **30**, 243–256.
25. Kimura, M. (1974) *Cold Spring Harbor Symp. Quant. Biol.* **38**, 515–524.
26. Ohta, T. (1973) *Nature (London)* **246**, 96–98.
27. Ohta, T. (1974) *Nature (London)* **252**, 351–354.
28. Kimura, M. & Ohta, T. (1971) *Theoretical Aspects of Population Genetics* (Princeton Univ. Press, Princeton, NJ).
29. Kimura, M. (1962) *Genetics* **47**, 713–719.
30. Wright, S. (1938) *Proc. Natl. Acad. Sci. USA* **24**, 253–259.
31. Jukes, T. H. (1978) *J. Mol. Evol.* **11**, 121–127.
32. Kimura, M. (1981) *Proc. Natl. Acad. Sci. USA* **78**, 454–458.
33. Post, L. E., Strycharz, G. D., Nomura, M., Lewis, H. & Dennis, P. P. (1979) *Proc. Natl. Acad. Sci. USA* **76**, 1697–1701.
34. Latter, B. D. H. (1975) *Nature (London)* **257**, 590–592.
35. Lande, R. (1980) *Am. Nat.* **116**, 463–479.

PART ELEVEN

# Probability and Time of Fixation or Extinction

## Introduction

Fisher, Haldane, Wright, and Robertson all recognized the importance of the process in which a new mutant gene introduced in a natural or artificial population increases its frequency by interaction between selection and random genetic drift and eventually becomes fixed in the population. Malécot (1948, 1952) used a method of moment-generating function and generalized Fisher's (1922) result (see also Haldane 1927) to the case of an arbitrary initial frequency and genic selection. Kimura also had presented a number of significant results on this problem since his early days as a population geneticist, and derived (no. 27) a more general formula for the ultimate fixation probability. The formula allowed him to obtain the probability under various forms of natural selection and to unify all the previous results of Fisher, Haldane, and Wright (Crow 1987). This formula has been crucial to the development of the neutral theory.

Kimura and Ohta (no. 28) were the first to treat separately the mean time of a mutant gene until fixation and that until loss. Although the unconditional mean time, until either of these alternative events, was studied by Watterson (1962) and Ewens (1963), the distinction is important because only mutants that are fixed in a population can play a key role in evolution. However, like alleles at the major histocompatibility complex loci, balanced polymorphic alleles that can persist for a long time without fixation and produce descendants may well be evolutionarily significant (Klein 1986). A neutral gene, initially represented singly in a population of effective size $N_e$, takes about $4N_e$ generations to become fixed. This is much longer than the unconditional mean time because loss of the gene is much more frequent and the mean time for this is very short. This topic of fixation time was later extended to various cases by a number of authors. For instance, Nei and Roychoudhury (1973) and Maruyama (1974, 1977) studied effects of selection in detail, noting that the mean fixation time for a favorable mutant (with no dominance) is the same as that for the corresponding deleterious mutant. This is related to time-reversal properties of stochastic processes, the future being the "mirror-image" of the past (Watterson 1976; Levikson 1977; Sawyer 1977). The standard deviation of the fixation time of a neutral mutant is about $2.15N_e$ (Narain 1970; Kimura and Ohta 1969), and the distribution itself was presented by Kimura (no. 29).

When a locus is duplicated, harmful mutations and drift may lead to loss of gene expression at one of the loci. Under the assumption of free recombination, Kimura and King (no. 30) dealt with the mean time until loss of gene expression and compared the result with experimental data from tetraploid fish. In 1985, Kimura took up the same subject, assuming complete linkage between two loci. Under particular interactions between duplicated genes, recombination affects the time greatly; the problem has some resemblance to the advantage or disadvantage of sexuality (Muller 1932; Crow and Kimura [no. 14]; Maynard Smith 1968; Felsenstein 1974; Haigh 1978). Further studies by computer simulations were made by Bailey, Poulter, and Stockwell (1978), Takahata and Maruyama (1979), and Li (1980). Watterson (1983) developed an approximate formula for the mean time of duplicated gene silencing.

Kimura (1985b; no. 31) treated the average fixation time of selected alleles under continuous mutation pressure. It takes a long time for a deleterious mutation to become fixed. The situation is different in the case of duplicated loci or where mutations at two loci are individually deleterious, but becomes harmless in combination (compensatory neutral). When linkage between two loci is loose, compensatory neutral mutations are unimportant, whereas with the tight linkage they have an ample chance to participate in evolution. Compensatory mutations may be found as double substitutions in relatively conserved molecules such as tRNA. Kimura (no. 31) carried out numerical calculations of the formulated diffusion equations as well as Monte Carlo experiments. At this time, since the institutional computer was not suited for multiple users, he set up a personal 15KB computer (Hewlett Packard 9825A) at his home and first used it to write this paper. The last sentence of Kimura (no. 31) shows his then stance for the neutral theory: "Finally, if the present model of compensatory neutral evolution is realistic, it will not only lend support to Ohta's concept of very slightly deleterious mutations, but also help us to understand some 'nonrandom' evolutionary amino acid substitutions under the framework of the neutral theory." However, as shown by Li and Nei (1977), when $4N_es > 10$ (in which $s$ is the selective disadvantage), the fixation time becomes too long to be of any evolutionary importance.

# ON THE PROBABILITY OF FIXATION OF MUTANT GENES IN A POPULATION[1]

MOTOO KIMURA[2]

*University of Wisconsin, Madison, Wisconsin*

Received January 29, 1962

THE success or failure of a mutant gene in a population is dependent not only on selection but also on chance. This fact was first treated quantitatively by FISHER (1922) who later (1930) worked out the probability of ultimate survival of a mutant gene for the case of genic selection (i.e. no dominance). Equivalent results have been obtained by HALDANE (1927) and WRIGHT (1931). Also the probability was estimated for a recessive mutant gene by HALDANE (1927) and WRIGHT (1942).

The present author (KIMURA 1957) extended these results to include any level of dominance. The probability of eventual fixation, $u(p)$, was expressed in terms of the initial frequency, $p$, the selection coefficients, and the effective population number. This function was used by ROBERTSON (1960) in his theory of selection limits in plant and animal breeding.

The purpose of this note is to present a more general formula for $u(p)$ which encompasses random fluctuations in selection intensity as well as random drift because of small population number. It will also be used to solve a question relating to "quasi-fixation" posed by the author in 1955.

*Derivation of the formula*: Consider a population in which the frequency of the allele $A$ is $p$ $(0 \leqq p \leqq 1)$. We assume that the population is sufficiently large and the change in $p$ per generation sufficiently small that the change in $p$ through time may be satisfactorily approximated by a continuous stochastic process.

Let $u(p,t)$ be the probability that allele $A$ will be fixed (i.e. its frequency becomes 1) during a time interval $t$ (conveniently measured in generations), given that $p$ is the initial frequency of $A$. From the nature of Mendelian mechanism of inheritance, the process of change in $p$ is Markovian, i.e. the frequency distribution of $A$ in the next generation is determined by the frequency of $A$ in the present generation and is not dependent on the way in which the present gene frequency was attained. Thus we have

$$u(p,t+\delta t) = \int f(p,p+\delta p;\delta t)\,u(p+\delta p,t)\,d(\delta p)$$

where $f(p,p+\delta p;\delta t)$ is the probability density of the change from $p$ to $p+\delta p$ during short time interval $\delta t$ and the integration is over all possible values of $\delta p$.

[1] This paper was prepared for publication under Contract No. DA-11-022-ORD-2059, Mathematics Research Center, United States Army, University of Wisconsin. This also constitutes paper number 391 of the National Institute of Genetics, Mishima-shi, Japan.

[2] On leave from the National Institute of Genetics, Mishima-shi, Japan.

Genetics **47**: 713–719 June 1962.

Here we assume that the probability density depends on $p$ and $\delta t$ but not on $t$ (i.e. the process is ~~temporary~~ time homogeneous).

Expanding $u(p+\delta p,t)$ in the right side of the above equation in terms of $\delta p$ and putting

$$\lim_{\delta t\to 0}\frac{1}{\delta t}\int(\delta p)f(p,p+\delta p;\delta t)\mathrm{d}(\delta p)=M$$

$$\lim_{\delta t\to 0}\frac{1}{\delta t}\int(\delta p)^2 f(p,p+\delta p;\delta t)\mathrm{d}(\delta p)=V$$

but neglecting higher order terms involving $(\delta p)^3$, $(\delta p)^4$, etc., we have

$$\frac{\partial u(p,t)}{\partial t}=\frac{V}{2}\frac{\partial^2 u(p,t)}{\partial p^2}+M\frac{\partial u(p,t)}{\partial p} \tag{1}$$

Since time is conveniently measured in generations, we substitute $M_{\delta p}$ and $V_{\delta p}$ for $M$ and $V$, where $M_{\delta p}$ and $V_{\delta p}$ are the mean and the variance of the change of $p$ per generation. The probability $u(p,t)$ may then be obtained by solving the above partial differential equation (known as the Kolmogorov backward equation) with boundary conditions:

$$u(0,t)=0,\ u(1,t)=1.$$

We are especially interested in the present paper in the ultimate probability of fixation defined by

$$u(p)=\lim_{t\to\infty}u(p,t),$$

for which $\partial u/\partial t=0$ and which therefore satisfies the equation

$$\frac{V_{\delta p}}{2}\frac{\mathrm{d}^2u(p)}{\mathrm{d}p^2}+M_{\delta p}\frac{\mathrm{d}u(p)}{\mathrm{d}p}=0 \tag{2}$$

with boundary conditions:

$$u(0)=0,\ u(1)=1.$$

The equation may readily be solved to give

$$u(p)=\frac{\int_0^p G(x)\mathrm{d}x}{\int_0^1 G(x)\mathrm{d}x} \tag{3}$$

where

$$G(x)=e^{-\int\frac{2M_{\delta x}}{V_{\delta x}}\mathrm{d}x} \tag{4}$$

in which $M_{\delta x}$ and $V_{\delta x}$ are the mean and the variance of the change in gene frequency, $x$, per generation. Formula (3) gives the probability of fixation of allele $A$ when its initial frequency is $p$. It has a pleasing simplicity and generality comparable to WRIGHT's (1938, 1949) well-known formula for the frequency distribution of genes at equilibrium. In my notation, this is

$$\varphi(x) = \frac{C}{V_{\delta x}\, G(x)} \tag{5}$$

where $\varphi(x)$ is the probability density of the gene frequency $x$ and $C$ is a constant chosen such that

$$\int_0^1 \varphi(x)\, dx = 1$$

From the derivation it may be seen that formula (3) holds not only for sexually reproducing diploid organisms, but also for haploid or polyploid organisms and asexually reproducing plants.

Finally, the chance of fixation of an individual mutant gene in a randomly mating diploid population is given by

$$u = u\left(\frac{1}{2N}\right), \tag{6}$$

where $N$ is the number of reproducing individuals in the population.

*Some applications:* The simplest case is genic selection in which $A$ has a constant selective advantage $s$ over its alleles in a randomly mating population of size $N$. If the frequency of $A$ is $x$, then the mean and the variance in the rate of change in $x$ per generation are

$$\begin{cases} M_{\delta x} = sx(1-x) \\ V_{\delta x} = x(1-x)/(2N) \end{cases} \tag{7}$$

so that $2M_{\delta x}/V_{\delta x} = 4Ns$, $G(x) = e^{-4Nsx}$ and we obtain from (3)

$$u(p) = \frac{1 - e^{-4Nsp}}{1 - e^{-4Ns}} \tag{8}$$

For $|\,2Ns\,| < \pi$, the right side of the above equation may be expanded in terms of $4Ns$ as follows:

$$\begin{aligned} u(p) &= \sum_{i=1}^{\infty} \frac{\phi_i(p)(-1)^{i-1}}{i!}(4Ns)^{i-1} \\ &= p + 2Nsp(1-p) + \frac{(2Ns)^2}{3} p(p-1)(2p-1) + \ldots \end{aligned} \tag{9}$$

where the $\phi_i(p)$'s are Bernoulli polynomials. Thus for a small value of $2Ns$, $u(p) - p$ is $2N$ times $sp(1-p)$. In other words, the total advance is $2N$ times the change in the first generation as pointed out by ROBERTSON (1960).

The probability of fixation of an individual mutant gene is obtained from (8) by putting $p = 1/(2N)$.

$$u = (1 - e^{-2s})/(1 - e^{-4Ns}) \tag{10}$$

If we assume that $|s|$ is small, we obtain

$$u = (2s)/(1 - e^{-4Ns}) \tag{11}$$

as a good approximation. This agrees with the result obtained by FISHER (1930) and WRIGHT (1931). This formula is good even for negative $s$, though $u$ for such a case is very small unless $|Ns|$ is not large. For a positive $s$ and very large $N$ we obtain the known result that the probability of ultimate survival of an advantageous mutant gene is approximately twice the selection coefficient (HALDANE 1927). On the other hand, if we let $s \to 0$ in (10), we obtain $u = 1/2N$, the result known for a neutral gene.

Next, I shall consider a more general case of zygotic selection under random mating. Let $s$ and $sh$ be, respectively, the selective advantage of mutant homozygote and the heterozygote, then

$$\begin{cases} M_{\delta x} = sx(1-x)[h+(1-2h)x] \\ V_{\delta x} = x(1-x)/(2N) \end{cases} \tag{12}$$

and therefore

$$G(x) = e^{-2cDx(1-x)-2cx}$$

where $c = Ns$ and $D = 2h - 1$. Thus we obtain

$$u(p) = \int_0^p e^{-2cDx(1-x)-2cx}dx \Big/ \int_0^1 e^{-2cDx(1-x)-2cx}dx \tag{13}$$

which is the formula given by the present author in a previous paper (KIMURA 1957).

For a completely recessive gene $h = 0$ or $D = -1$ and we have

$$u(p) = \int_0^p e^{-2cx^2}dx \Big/ \int_0^1 e^{-2cx^2}dx \tag{14}$$

If $s$ is positive and small but $Ns$ is large, the above formula leads approximately to

$$u = \sqrt{\frac{2s}{\pi N}} \approx 1.128\sqrt{\frac{s}{2N}} \tag{15}$$

for $p = \frac{1}{2N}$, giving the probability of fixation of individual mutant genes which is advantageous but completely recessive. It is interesting to note that the more exact value given here in (15) lies between $\sqrt{s/N}$, the value obtained by HALDANE (1927) by treating this as a branching process and $\sqrt{s/(2N)}$, obtained by WRIGHT (1942) with his method of integral equations. WRIGHT's numerical approximation, $1.1\sqrt{s/2N}$, is very close to (15).

I shall now investigate the effect of random fluctuation of selection intensity on the fixation of mutant genes.

The simplest situation is again genic selection in which the mutant gene $A$ has a selective advantage $s$ over its alleles, but $s$ is now a random variable with mean $\bar{s}$ and variance $V_s$. Thus (7) may be replaced by

$$\begin{cases} M_{\delta x} = \bar{s}x(1-x) \\ V_{\delta x} = V_s x^2(1-x)^2 + x(1-x)/(2N) \end{cases} \tag{16}$$

to give

$$G(x) = \left(\frac{\lambda_1 - x}{x - \lambda_2}\right)^{k/\sqrt{1+4r}} \tag{17}$$

where

$$k = 2\bar{s} / V_s$$

and

$$\begin{cases} \lambda_1 = (1 + \sqrt{1+4r})\ /2 \\ \lambda_2 = (1 - \sqrt{1+4r})\ /2 \end{cases}$$

where $r = 1/(2NV_s)$.

There are several interesting special cases. First, if $\bar{s} = 0$, then $G(x) = 1$ and we have

$$u(p) = p$$

a result which might be expected, since the gene is neutral on the average. Secondly, if $\bar{s} \neq 0$, then $k \neq 0$ and I shall consider the following three cases: (1) $r \to \infty$, (2) $r = 2$ and (3) $r \to 0$. In all these cases, we assume that $0 < V_s \ll 1$ and therefore $r \gg 1/(2N)$. We will also assume that $\bar{s}$ is positive but small.

*Case 1.* Consider the situation in which $r$ is very large or equivalently $2NV_s$ is very small. Then, approximately

$$2M_{\delta x} / V_{\delta x} = 4N\bar{s}[1 - 2NV_s x(1-x)].$$

Assuming further that $N\bar{s}$ is large but $8N^2\bar{s}V_s$ is very small, we have

$$u = 2\bar{s} - V_s, \tag{18}$$

namely the probability of fixation of a favorable gene is reduced by $V_s$ due to random fluctuation in $s$.

*Case 2.* Here we consider a special case in which $r = 2$, or equivalently $V_s = 1/(4N)$. For this case

$$G(x) = (2-x)^{k/3}(1+x)^{-k/3}$$

If $k$ is very small, $u$ is approximately $1/(2N)$, while if $k$ is very large it is nearly $2\bar{s}$. For an intermediate value of $k$ such as $k = 3$, we obtain

$$u = [N(3 \log 2 - 1)]^{-1}$$

which is roughly $0.9/N$ or $2.4\bar{s}$.

*Case 3.* We now come to a more interesting case in which $r$ is very small, namely $V_s \gg 1/(2N)$. In this case, as $r$ approaches 0, $\lambda_1$ and $\lambda_2$ approach 1 and 0, respectively. Here the situation is rather delicate and we restrict our consideration to the behavior of $u(p)$ for an intermediate value of $p$. For such a value of $p$ which is neither very close to 0 nor 1, the effect of random sampling of gametes in reproduction may be ignored for a large population ($N \to \infty$), so that we can write

$$G(x) = \left(\frac{1-x}{x}\right)^k$$

keeping in mind that fixation here means strictly "quasifixation" (cf. KIMURA

1954), though in actual cases the prefix "quasi" may be unnecessary, because random sampling of gametes is always at work in the subterminal class.
Then we have

$$u(p) = \int_0^p \left(\frac{1-x}{x}\right)^k dx \bigg/ \int_0^1 \left(\frac{1-x}{x}\right)^k dx \tag{19}$$

It may readily be seen that if $k \geqq 1$, both denominator and numerator of the above formula are divergent, which may be interpreted as showing that $u(p) = 1$ for all $p > 0$. This agrees with the obvious fact that once the advantageous mutant gene has reached the stage where it is represented by sufficiently many individuals, it will almost certainly be led to fixation by natural selection. On the other hand, if $0 \leqq k < 1$, the integral in the denominator converges and we get

$$\int_0^1 \left(\frac{1-x}{x}\right)^k dx = \frac{\pi K}{\sin \pi K}$$

This shows that if $0 \leqq 2\bar{s} < V_s$, then $u(p) < 1$, namely there is a finite chance that gene $A$ will be lost from the population by random fluctuation in selection intensity (plus random sampling of gametes in the neighborhood of $x = 0$) even if the gene is advantageous on the average. This gives an affirmative answer to the question posed by the present author in 1955:

> "When $\bar{s}$ and $V_s$ are of the same order of magnitude, or $\bar{s}$ ($> 0$) is much smaller than $V_s$, is quasi-fixation still possible on the side of the disadvantageous class?"

For a small value of $p$ (still assuming that $N \to \infty$ and $1/N \ll p \ll 1$), we have approximately

$$u(p) = \frac{\sin(\pi k) p^{1-k}}{\pi k (1-k)}$$

As an example, if $\bar{s}$ is 1/10 of $V_s$ and if the initial frequency is one percent, the chance of fixation is about three percent.

Application of a similar method for the case of a completely recessive advantageous gene leads to a rather unexpected result that there is no finite chance of its loss if $p \gg 1/N$, where $N \to \infty$.

## SUMMARY

The probability of ultimate fixation of a gene in a population is treated as a continuous stochastic process and the solution is given as a function of the initial frequency and the mean and variance of the gene frequency change per generation. The formula is given by equation (3).

The formula is shown to include previous results as special cases and is applied to solve problems where there is random fluctuation in selection intensity. It is also used to show some circumstances under which, even in a large population, an advantageous gene may be lost because of fluctuating selective values.

ACKNOWLEDGMENTS

The author would like to express his thanks to DR. J. F. CROW who kindly reviewed the draft and made valuable suggestions for improving the manuscript.

LITERATURE CITED

FISHER, R. A., 1922 On the dominance ratio. Proc. Roy. Soc. Edinburgh. **42:** 321–341.

1930 The distribution of gene ratios for rare mutations. Proc. Roy. Soc. Edinburgh. **50:** 204–219.

HALDANE, J. B. S., 1927 A mathematical theory of natural and artificial selection. Part V: Selection and mutation. Proc. Cambridge Phil. Soc. **23:** 838–844.

KIMURA, M., 1954 Process leading to quasifixation of genes in natural populations due to random fluctuation of selection intensities. Genetics **39:** 280–295.

1955 Stochastic processes and distribution of gene frequencies under natural selection. Cold Spring Harbor Symposia Quant. Biol. **20:** 33–53.

1957 Some problems of stochastic processes in genetics. Ann. Math. Stat. **28:** 882–901.

ROBERTSON, A., 1960 A theory of limits in artificial selection. Proc. Roy. Soc. B. **153:** 234–249.

WRIGHT, S., 1931 Evolution in Mendelian populations. Genetics **16:** 97–159.

1938 The distribution of gene frequencies under irreversible mutation. Proc. Natl. Acad. Sci. U.S. **24:** 253–259.

1942 Statistical genetics and evolution. Bull. Am. Math. Soc. **48:** 223–246.

1949 Adaptation and selection. pp. 365–389. *Genetics, Paleontology, and Evolution.* Edited by G. L. JEPSON, G. G. SIMPSON, and E. MAYR. Princeton Univ. Press, Princeton, N.J.

# THE AVERAGE NUMBER OF GENERATIONS UNTIL FIXATION OF A MUTANT GENE IN A FINITE POPULATION[1]

MOTOO KIMURA AND TOMOKO OHTA[2]

*National Institute of Genetics, Mishima, Japan*

Received July 26, 1968

A mutant gene which appeared in a finite population will eventually either be lost from the population or fixed (established) in it. The mean time until either of these alternative events takes place was studied by WATTERSON (1962) and EWENS (1963). They made use of a method previously announced by DARLING and SIEGERT (1953), and, independently by FELLER (1954). Actually, DARLING and SIEGERT refer to its application to genetics.

From the standpoint of population genetics, however, it is much more desirable to determine separately the mean time until fixation and that until loss. Since the gene substitution in a population plays a key role in the evolution of the species, it may be of particular interest to know the mean time for a rare mutant gene to become fixed in a finite population, excluding the cases in which such a gene is lost from the population.

In the present paper, a solution to this problem will be presented together with Monte Carlo experiments to test some of the theoretical results. Throughout this paper, the senior author (M. K.) is responsible for the mathematical treatments, while the junior author (T. O.) is responsible for the numerical treatments based on computers.

## BASIC THEORY

Let us consider a diploid population consisting of $N$ individuals and having the variance effective number $N_e$, which may be different from the actual number (for the definition of $N_e$, see KIMURA and CROW 1963). Throughout this paper, we will denote by $p$ the frequency of a mutant gene ($A_2$), so that $1-p$ represents the frequency of its allele ($A_1$). Also, we will use the diffusion models (cf. KIMURA 1964) to solve the problem. Let $u(p,t)$ be the probability that the mutant allele $A_2$ becomes fixed (i.e., its frequency becomes unity) by the $t$th generation, given that its frequency is $p$ at the start (i.e., at $t=0$). If we define a quantity $T_1(p)$ by the relation

$$T_1(p) = \int_0^\infty t \frac{\partial u(p,t)}{\partial t} dt, \tag{1}$$

then

$$\bar{t}_1(p) = T_1(p)/u(p) \tag{2}$$

[1] Contribution No. 692 from the National Institute of Genetics, Mishima, Shizuoka-ken, Japan.
[2] Supported by a post-doctoral fellowship from the Japan Soc. for the Promotion of Science.

Genetics **61**: 763–771 March 1969.

represents the average number of generations until the mutant allele with initial frequency $p$ becomes fixed in the population, excluding the cases in which the allele is lost from it. In the above expression, $u(p)$ stands for the probability of ultimate fixation (KIMURA 1957, 1962), such that

$$u(p) = \lim_{t\to\infty} u(p,t). \tag{3}$$

If we denote by $M_{\delta p}$ and $V_{\delta p}$ the mean and the variance of the rate of change in the frequency of $A_2$ per generation, then as shown by KIMURA (1962), $u(p,t)$ satisfies the following partial differential equation

$$\frac{\partial u(p,t)}{\partial t} = \tfrac{1}{2} V_{\delta p} \frac{\partial^2 u(p,t)}{\partial p^2} + M_{\delta p} \frac{\partial u(p,t)}{\partial p} \tag{4}$$

Here we assume that the process of change in gene frequency is time homogeneous, that is, both $M_{\delta p}$ and $V_{\delta p}$ do not depend on time parameter $t$.

Differentiating each term of the above equation (4) with respect to $t$, multiplying each of the resulting terms by $t$, followed by integrating them with respect to $t$ from 0 to $\infty$, we obtain

$$\int_0^\infty t \frac{\partial^2 u(p,t)}{\partial t^2} dt = \tfrac{1}{2} V_{\delta p} \frac{\partial^2}{\partial p^2} T_1(p) + M_{\delta p} \frac{\partial}{\partial p} T_1(p)$$

The left hand side of this equation is reduced to $-u(p)$, since

$$\int_0^\infty t \frac{\partial^2 u(p,t)}{\partial t^2} dt = \left[ t \frac{\partial u(p,t)}{\partial t} \right]_0^\infty - \int_0^\infty \frac{\partial u(p,t)}{\partial t} dt = -u(p,\infty),$$

in which we assume that $t\partial u(p,t)/\partial t$ vanishes at $t = \infty$.

Thus, we have the following ordinary differential equation for $T_1(p)$,

$$T_1''(p) + a(p) T_1'(p) + b(p) = 0, \tag{5}$$

where $a(p) = 2M_{\delta p}/V_{\delta p}$ and $b(p) = 2u(p)/V_{\delta p}$

The above equation can be integrated immediately, and if we determine the two constants involved by the following two boundary conditions,

$$\lim_{p\to 0} \bar{t}_1(p) = \text{finite} \tag{6}$$

and

$$\bar{t}_1(1) = 0, \tag{7}$$

we obtain

$$T_1(p) = u(p) \int_p^1 \psi(\xi) u(\xi) \{1-u(\xi)\} d\xi + \{1-u(p)\} \int_0^p \psi(\xi) u^2(\xi) d\xi. \tag{8}$$

In the above formula,

$$u(p) = \int_0^p G(x)dx \Big/ \int_0^1 G(x)dx, \quad \text{(Kimura 1962)}, \tag{9}$$

is the probability of ultimate fixation, and $\psi(x)$ is given by

$$\psi(x) = 2\int_0^1 G(x)dx \Big/ \{V_{\delta x}\, G(x)\}\ , \tag{10}$$

where

$$G(x) = \exp\left\{ -\int_0^x \frac{2M_{\delta\xi}}{V_{\delta\xi}}\, d\xi \right\}\ , \tag{11}$$

in which exp $\{\cdot\}$ stands for the exponential function.

Of the two boundary conditions, the first, i.e. (6), may need some comments. It reflects the fact that in a finite population a single mutant gene which appeared in the population reaches fixation within a finite time. It is also equivalent to the relation $\lim_{p\to 0} T_1(p) = Kp$, in which $K$ is a constant.

From (8) and (2), the required solution for our problem is

$$\bar{t}_1(p) = \int_p^1 \psi(\xi)u(\xi)\{1-u(\xi)\}d\xi + \frac{1-u(p)}{u(p)} \int_0^p \psi(\xi)u^2(\xi)d\xi\ , \tag{12}$$

where $u$ and $\psi$ are given respectively by (9) and (10). Similarly, we can derive the average number of generations until the mutant gene is lost from the population, excluding the cases in which the mutant gene is ultimately fixed in the population. This is given by

$$\bar{t}_0(p) = \frac{u(p)}{1-u(p)} \int_p^1 \psi(\xi)\{1-u(\xi)\}^2 d\xi + \int_0^p \psi(\xi)\{1-u(\xi)\}u(\xi)d\xi \tag{13}$$

SOME SPECIAL CASES

In this section we assume that the factor causing random fluctuation in gene frequency is the random sampling of gametes alone so that

$$V_{\delta p} = p(1-p)/(2N_e),$$

where $N_e$ is the variance effective number.

In the simplest case of neutral mutations, we have

$$M_{\delta p} = 0.$$

Thus from (12), the average number of generations until fixation (excluding the cases of loss) is

$$\bar{t}_1(p) = -\frac{1}{p}\{4N_e(1-p)\log_e(1-p)\}. \tag{14}$$

At the limit of $p \to 0$, we have

$$\bar{t}_1(0) = 4N_e. \tag{15}$$

This shows that an originally rare mutant gene in a population of effective size $N_e$ takes about $4N_e$ generations until it spreads to the whole population if we disregard the cases in which such a gene is eventually lost from the population. Similarly, from (13), the number of generations until loss (excluding the cases of fixation) for neutral mutations is

$$\bar{t}_0(p) = -4N_e\left(\frac{p}{1-p}\right)\log_e p \tag{16}$$

If the mutant gene $A_2$ has the selective advantage $s/2$ over its allele $A_1$ (case of genic selection) such that

$$M_{\delta p} = \frac{s}{2}p(1-p),$$

then, writing $N_e s = S$, we have, from (12)

$$\bar{t}_1(p) = J_1 + \frac{1-u(p)}{u(p)}J_2 \tag{17}$$

where

$$J_1 = \frac{2}{s(1-e^{-2S})}\int_p^1 \frac{(e^{2S\xi}-1)(e^{-2S\xi}-e^{-2S})}{\xi(1-\xi)}\,d\xi\,,$$

$$u(x) = (1-e^{-2Sx})/(1-e^{-2S})$$

and

$$J_2 = \frac{2}{s(1-e^{-2S})}\int_0^p \frac{(e^{2S\xi}-1)(1-e^{-2S\xi})}{\xi(1-\xi)}\,d\xi\,.$$

A more general case of genotypic selection can also be worked out in a similar way using equation (12).

## NUMERICAL EXAMPLES AND MONTE CARLO EXPERIMENTS

The numerical evaluation of some of the results of the foregoing sections together with Monte Carlo experiments were performed using computers TOSBAC 3400 and IBM 360. The only case which can be evaluated easily from the formula is the one of selectively neutral mutations (formula 14). For other cases, one needs numerical integration. In this section, the numerical examples for the cases of no selection, genic selection (no dominance) and overdominance will be given.

The Monte Carlo experiments were performed for the cases of no selection and genic selection by the following scheme. In each generation, the change of gene frequency by genic selection was carried out deterministically using the formula

$$\Delta p = \frac{s}{2}p(1-p)/(1+sp), \tag{18}$$

where $s$ is the selection coefficient for the mutant homozygote. Sampling of zygotes was performed by generating pseudo-random numbers, $X\,(0<X<1)$, using the subroutines RAND in TOSBAC 3400 and RANDU in IBM 360. Each experiment was continued until fixation or extinction and the number of generations involved was recorded. 400 replicate trials were done for each set of parameters.

Figure 1 shows the comparison of the results of Monte Carlo simulation (dots) and those of analytical solution (curves) for the cases of $N_e=10$ (upper curve) and $N_e=5$ (lower curve). The abscissa represents the initial frequency $p$. As $p$

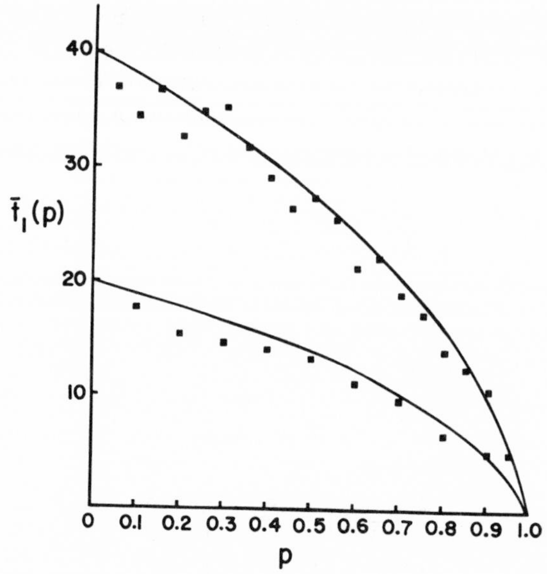


FIGURE 1.—Average number of generations until fixation of a selectively neutral mutant gene as a function of its initial frequency. The theoretical values are represented by curves and those of Monte Carlo simulation by dots. $2N_e=20$ in the upper curve and $2N_e=10$ in the lower one.

changes from 0 to 1, the number of generations until fixation changes from 40 to 0 for the case of $N_e=10$ and from 20 to 0 in the case of $N_e=5$. As it is clear from the figure, the agreement between theMonte Carlo results and the theoretical predictions is satisfactory, although the latter seems to overestimate the true value slightly.

The results for the case of genic selection are given in Figure 2 with a constant initial frequency ($p = 0.1$) and a varying selection intensity. The curve represents the theoretical results by numerical integration, and the dots represent the results of Monte Carlo experiments. The numerical integration was performed by Simpson's rule using the computer IBM 360. Again, the agreement between the two is satisfactory, though the theoretical treatment seems to underestimate the actual value slightly. The slight discrepancy may be due to the existence of the denominator in formula (18) that effectively decreases the value of $s$ in the numerator as compared with the expression $(s/2)p(1-p)$ used in the theoretical treatment. As expected, the selective advantage accelerates the fixation of the advantageous allele. So, it may be of some interest to compare such acceleration with the rate of steady decay (KIMURA 1955). It can be shown that, as $N_es$ increases, the inverse of the rate of steady decay decreases more rapidly as compared with the shortening of the fixation time (with $p = 0.1$). For example, when $N_es = 5$, the time until fixation is about half of that of the neutral case, while the rate of steady decay is about 3.5 as large.

In the case of overdominance between a pair of alleles $A_1$ and $A_2$, the formula

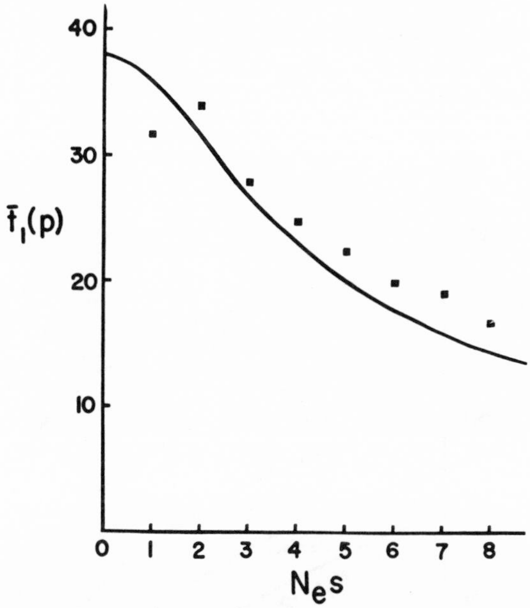


FIGURE 2.—The relationship between time for fixation and selective advantage in the case of genic selection (no dominance). The curve represents the theoretical results by numerical integration and the dots represent the results of Monte Carlo simulation. In this figure, the effective population number $N_e=10$ and the initial frequency $p=0.1$.

(12) contains double integrals. Though no simulation experiments were performed in this case, we studied, using formula (12), the theoretical relationship between the time until fixation and the initial frequency for various values of $N_e s$ assuming that the fitnesses of the three genotypes $A_1A_1$, $A_1A_2$ and $A_2A_2$ are $1-s$, 1 and $1-s$, respectively. In such a case, the equilibrium gene frequency is 0.5 and the overdominance is most effective for retarding gene fixation (ROBERTSON 1962). Now, the time until fixation is the function of $N_e$ and $N_e s$. So, in Figure 3, the results for $N_e s = -2, -1, 1, 2$ and 4 are illustrated in comparison with the selectively neutral case ($N_e s = 0$). From the figure, it may be seen, for example, that the overdominance prolongs the time until fixation almost twice as compared with the selectively neutral case when $N_e s$ is about 2. Let us compare the present results with the rate of steady decay. Again, the increase of fixation time for larger $N_e s$ is slower as compared with the increase of the inverse of the rate of steady decay which is equivalent to the retardation factor of ROBERTSON. For instance, when $N_e s = 2$, the retardation factor is about $1/0.4 = 2.5$ (MILLER 1962). However, the disagreement is not as large as in the case of the genic selection.

DISCUSSION

It now appears (KIMURA 1968) that mutation and random genetic drift play a more important role in determining the genetic structure of Mendelian populations than previously considered, especially when molecular mutations are

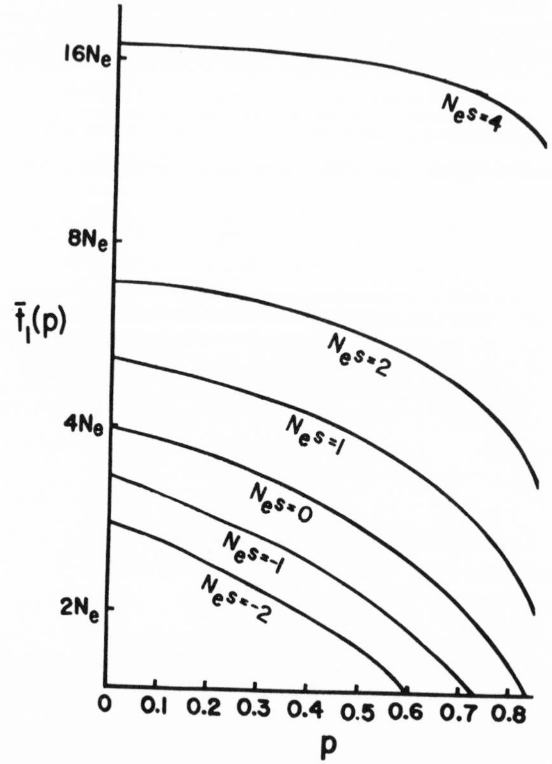


FIGURE 3.—Time for fixation of an overdominant mutant gene as a function of its initial frequency with various intensities of selection. Only the theoretical results obtained by numerical integration are shown. No Monte Carlo experiments were performed. The selectively neutral case is also shown for comparison. The ordinate gives the time for fixation by the logarithmic scale.

taken into account. Furthermore, the majority of such mutations appear to be almost neutral for natural selection. Thus the results obtained in the previous sections assuming $M_{\delta p} = 0$ should be applicable to such mutations.

Now, in a population consisting of $N$ individuals, if we assume that each mutant gene is represented only once at the moment of its occurrence, $p = 1/(2N)$, and from formula (14), the average number of generations until fixation of a neutral mutation becomes

$$\bar{t}_1\left(\frac{1}{2N}\right) = -8NN_e\left(1 - \frac{1}{2N}\right) \log_e\left(1 - \frac{1}{2N}\right) \qquad (19)$$

If $N$ is large, this is very close to $4N_e$, the value given as $\bar{t}_1(0)$ in (15). Since the condition (6), i.e. $\bar{t}_1(0)$ = finite is a crucial one in deriving the formula (8) and therefore formula (12), it may be of some interest to examine the value of $\bar{t}_1(0)$ more in detail. It was shown by KIMURA (1955) that for a neutral mutation

$$u(p,t) = p + \sum_{i=1}^{\infty} (-1)^i \frac{2i+1}{2i(i+1)} (1-r^2) T^{1}_{i-1}(r) e^{-\lambda_i t}, \qquad (20)$$

where $r=1-2p$, $\lambda_i=i(i+1)/(4N_e)$ and $T^{1}_{i-1}(r)$ represents the Gegenbauer polynomial. The above formula enables us to calculate $\bar{t}_1(p)$ directly from (1) and (2), giving

$$\bar{t}_1(p)=4N_e(1+r)\sum_{i=1}^{\infty}(-1)^{i-1}\frac{2i+1}{i^2(i+1)^2}T^{1}_{i-1}(r). \tag{21}$$

At the limit of $p\to 0$ $(r\to 1)$, if we use the relation $T^{1}_{i-1}(1)=i(i+1)/2$, we obtain

$$\bar{t}_1(0)=4N_e\sum_{i=1}^{\infty}(-1)^{i-1}\frac{2i+1}{i(i+1)}=4N_e\,, \tag{22}$$

thus confirming the result given in (15). The above results show that a single mutant gene, if it is neutral, takes about $4N_e$ generations until fixation if we disregard the cases in which it is eventually lost. In this connection, it is interesting to note FISHER's (1930) inference that in the absence of favorable selection the number of individuals having a gene derived from a single mutation cannot greatly exceed the number of generations since its occurrence.

Next, let us consider the number of generations until a neutral mutant gene is lost from the population disregarding the cases in which it is eventually fixed. This is given by formula (16) by putting $p=1/(2N)$. Namely,

$$\bar{t}_0\left(\frac{1}{2N}\right)=\frac{4N_e}{2N-1}\log_e(2N)\approx 2\left(\frac{N_e}{N}\right)\log_e(2N) \tag{23}$$

Since the ratio $N_e/N$ is around 0.8 in man (CROW 1954), a single mutant gene which appeared in a human population will be lost from the population on the average in about $1.6\log_e 2N$ generations. If $N=10^4$, this amounts to about 16 generations. These results show that a great majority (fraction $1-\frac{1}{2N}$) of neutral or nearly neutral mutant genes which appeared in a finite population are lost from the population within a few generations, while the remaining minority (fraction $\frac{1}{2N}$) spread over the entire population (i.e. reach fixation) taking a very large number of generations.

In the present paper, we have studied the average (i.e., the first moment) of the length of time until fixation (and, separately, until loss), but the present method can immediately be adapted to obtain the $n$th moment of the length of time until fixation in terms of $(n-1)$th moment, thus enabling us to obtain the higher moments step by step starting from the first moment.

SUMMARY

In a finite population, a mutant gene is either fixed in the population or lost from it within a finite length of time. A theory was presented which enables us to obtain the average number of generations until fixation, and separately, that until loss, based on the method of diffusion equations. Also, Monte Carlo experiments were performed to test some of the theoretical results.—It was shown that a single mutant gene, if it is selectively neutral, takes about $4N_e$ generations until fixation in a population of effective size $N_e$.

LITERATURE CITED

CROW, J. F., 1954 Breeding structure of populations. II. Effective population number. pp. 543–556. In: *Statistics and Mathematics in Biology*. Iowa State College Press, Ames, Iowa.

DARLING, D. A., and A. J. F. SIEGERT, 1953 The first passage problem for a continuous Markov process. Ann. Math. Statist. **24**: 624–639.

EWENS, W. J., 1963 The mean time for absorption in a process of genetic type. J. Australian Math. Soc. **3**: 375–383.

FELLER, W., 1954 Diffusion processes in one dimension. Trans. Am. Math. Soc. **77**: 1–31.

FISHER, R. A., 1930 *The Genetical Theory of Natural Selection*. Clarendon Press, Oxford.

KIMURA, M., 1955 Solution of a process of random genetic drift with a continuous model. Proc. Natl. Acad. Sci. U.S. **41**: 144–150. —— 1957 Some problems of stochastic processes in genetics. Ann. Math. Statist. **28**: 882–901. —— 1962 On the probability of fixation of mutant genes in a population. Genetics **47**: 713–719. —— 1964 Diffusion models in population genetics. J. Appl. Probability **1**: 177–232. —— 1968 Evolutionary rate at the molecular level. Nature **217**: 624–626.

KIMURA, M., and J. F. CROW, 1963 The measurement of effective population number. Evolution **17**: 279–288.

MILLER, G. G., 1962 The evaluation of eigenvalues of a differential equation arising in a problem in genetics. Proc. Cambridge Phil. Soc. **58**: 588–593.

ROBERTSON, A., 1962 Selection for heterozygotes in small populations. Genetics **47**: 1291–1300.

WATTERSON, G. A., 1962 Some theoretical aspects of diffusion theory in population genetics. Ann. Math. Statist. **33**: 939–957.

# The length of time required for a selectively neutral mutant to reach fixation through random frequency drift in a finite population*

By MOTOO KIMURA

*National Institute of Genetics, Mishima, Japan*

(*Received* 25 *August* 1969)

SUMMARY

Frequency distribution of the length of time until fixation (excluding the cases of eventual loss) of a selectively neutral mutant in a finite population was obtained. With $4N_e$ generations ($N_e$, effective population size) as the unit length of time, the distribution has mean $\mu_1 = 1$, standard deviation $\sigma = 0{\cdot}538$, skewness $\gamma_1 = 1{\cdot}67$ and kurtosis $\gamma_2 = 4{\cdot}51$ (see Fig. 1).

A general theory concerning the average number of generations until a mutant gene becomes fixed (established) in a finite population was developed by Kimura & Ohta (1969*a*) based on the diffusion models. For the special case of selectively neutral mutants, the problem was simplified and it was shown by them that the average length of time until fixation (excluding the cases of eventual loss) is approximately $4N_e$ generations, where $N_e$ is the 'variance' effective number of the population (cf. Kimura & Crow, 1963). In addition it was shown by Narain (1969) and also by Kimura & Ohta (1969*b*) that for this case the standard deviation of the length of time until fixation is about $(2{\cdot}15)N_e$ generations.

The purpose of this note is to show that, for a neutral mutant, the entire probability distribution of the length of time until fixation can readily be derived from my previous results on the process of random genetic drift. Actually, I have shown (Kimura 1955) that if $f(p,1;t)$ is the probability of a selectively neutral allele reaching fixation by the $t$th generation, then

$$f(p,1;t) = p + \sum_{i=1}^{\infty} (2i+1)\, p(1-p)\, (-1)^i \mathrm{F}(1-i, i+2, 2, p)\, e^{-\lambda_i} \quad (t > 0), \tag{1}$$

where $p$ is the initial frequency of the mutant allele and F denotes the hypergeometric function, and $\lambda_i = i(i+1)/4N_e$.

If we restrict our consideration to the cases in which the mutant allele is eventually fixed but disregard those in which it is eventually lost from the population, then the cumulative probability distribution of the length of time until fixation is given by $f(p,1;t)/p$, since the probability of eventual fixation is $f(p, 1; \infty) = p$. The corresponding density function denoted by $y(p, t)$ may then be obtained by differentiating this with respect to $t$, namely

$$y(p,t) = (1-p) \sum_{i=1}^{\infty} (2i+1)\, (-1)^{i+1} \mathrm{F}(1-i, i+2, 2, p)\, \lambda_i e^{-\lambda_i t} \quad (t > 0). \tag{2}$$

* Contribution no. 742 from the National Institute of Genetics, Mishima, Shizuoka-ken, Japan. Aided in part by a Grant-in-Aid from the Ministry of Education, Japan.

We are particularly interested in the case in which the initial frequency of the mutant gene is very low. This corresponds to the situation in which the population is large while the mutant allele is represented only by one or two individuals at the moment of its appearance. The distribution for this case may be approximated by $\lim_{p\to 0} y(p, t)$, which we will denote by $y(t)$. Thus we have

$$y(t) = \sum_{i=1}^{\infty} (2i+1)(-1)^{i+1}\lambda_i e^{-\lambda_i t} \quad (t > 0). \tag{3}$$

Note that F is reduced to unity at $p = 0$. The moments of the length of time until fixation can then be evaluated by using this distribution. Namely, let $\mu'_n$ be the $n$th moment ($n \geqslant 1$) around zero, then

$$\mu'_n = \int_0^\infty t^n y(t)\,dt = (4N_e)^n n! \sum_{i=1}^{\infty} \frac{2i+1}{[i(i+1)]^n}(-1)^{i+1}. \tag{4}$$

In particular, the first four moments turn out to be as follows:

$$\begin{aligned}
\mu'_1 &= 4N_e \\
\mu'_2 &= (4N_e)^2[(\tfrac{1}{3}\pi^2)-2] \approx 1{\cdot}29(4N_e)^2 \\
\mu'_3 &= (4N_e)^3(12-\pi^2) \approx 2{\cdot}13(4N_e)^3 \\
\mu'_4 &= (4N_e)^4[(\tfrac{7}{15}\pi^4)+8\pi^2-120] \approx 4{\cdot}41(4N_e)^4.
\end{aligned}$$

For $n$ larger than 4, we have, approximately,

$$\mu'_n \approx (4N_e)^n n! \frac{3}{2^n},$$

the first term of the series.

Figure 1 illustrates this probability distribution in terms of $T = t/(4N_e)$, namely, taking $4N_e$ generations as the unit length of time. This has mean $\mu_1 = 1$, standard deviation $\sigma = 0{\cdot}538$, skewness $\gamma_1 = \mu_3/\sigma^3 = 1{\cdot}67$ and kurtosis $\gamma_2 = \mu_4/\sigma^4 - 3 = 4{\cdot}51$. It somewhat resembles the gamma distribution.

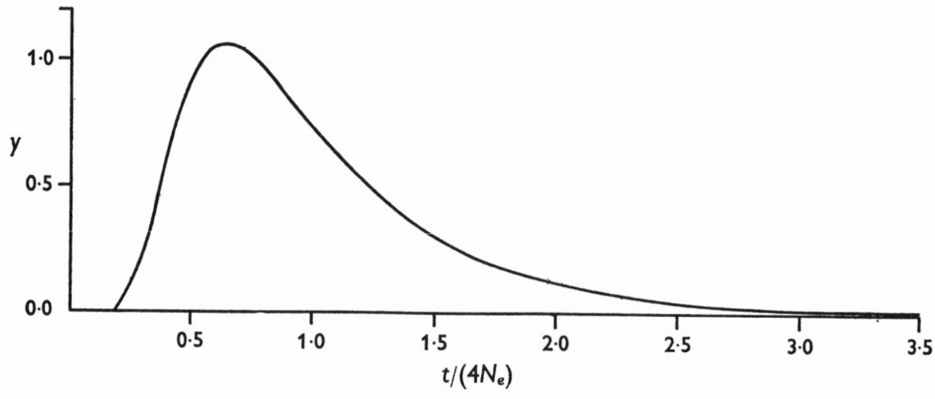


Fig. 1. Probability distribution of the length of time until fixation of a selectively neutral mutant. Abscissa: time measured with $4N_e$ generations as the unit length of time. $N_e$ stands for the effective size of the population. Ordinate: probability density ($y$).

REFERENCES

KIMURA, M. (1955). Solution of a process of random genetic drift with a continuous model. *Proc. Natn. Acad. Sci., U.S.A.* **41**, 144–150.

KIMURA, M. & CROW, J. F. (1963). The measurement of effective population number. *Evolution* **17**, 279–288.

KIMURA, M. & OHTA, T. (1969*a*). The average number of generations until fixation of a mutant gene in a finite population. *Genetics, Princeton* **61**, 763–771.

KIMURA, M. & OHTA, T. (1969*b*). The average number of generations until extinction of an individual mutant gene in a finite population. *Genetics, Princeton* (in the Press).

NARAIN, P. (1969). A note on the diffusion approximation for the variance of the number of generations until fixation of a neutral mutant gene. *Genet. Res., Camb.* (in the Press).

# Fixation of a deleterious allele at one of two "duplicate" loci by mutation pressure and random drift

(population genetics/mutational load/gene duplication/diffusion equation method)

MOTOO KIMURA* AND JACK L. KING†

*National Institute of Genetics, Mishima 411, Japan; and †Department of Biological Science, University of California, Santa Barbara, California 93106

*Contributed by Motoo Kimura, April 4, 1979*

**ABSTRACT** We consider a diploid population and assume two gene loci with two alleles each, *A* and *a* at one locus and *B* and *b* at the second locus. Mutation from wild-type alleles *A* and *B* to deleterious alleles *a* and *b* occurs with mutation rates $v_a$ and $v_b$, respectively. We assume that alleles are completely recessive and that only the double recessive genotype *aabb* shows a deleterious effect with relative fitness $1 - \epsilon$. Then, it can be shown that if $v_a > v_b$ mutant *a* becomes fixed in the population by mutation pressure and a mutation–selection balance is ultimately attained with respect to the *B/b* locus alone. The main aim of this paper is to investigate the situation in which $v_a = v_b$ exactly. In this case a neutral equilibrium is attained and either locus can drift to fixation for the mutant allele. Diffusion models are developed to treat the stochastic process involved whereby the deleterious mutant eventually becomes fixed in one of the two duplicated loci by random sampling drift in finite populations. In particular, the equation for the average time until fixation of mutant *a* or *b* is derived, and this is solved numerically for some combinations of parameters $4N_ev$ and $4N_e\epsilon$, where $v$ is the mutation rate ($v_a = v_b \equiv v$) and $N_e$ is the effective size of the population. Monte Carlo experiments have been performed (using a device termed "pseudo sampling variable") to supplement the numerical analysis.

It is expected that, in natural populations, deleterious alleles are constantly arising by mutations at every gene locus. However, frequencies of such alleles are kept low by natural selection, unless the deleterious effects are extremely small (smaller than the reciprocal of the effective population size).

It is common sense in population genetics that the mutation pressure can not overcome the barrier of negative selection. An interesting exception to this rule occurs when the environment changes in such a way that the wild-type allele at one locus becomes no longer necessary. Then, amorphic mutations at that locus become selectively neutral and mutation pressure, in conjunction with random genetic drift, can lead one of such alleles to eventual fixation in the population. The loss of vitamin C synthesizing ability in several vertebrates can be explained in the light of such considerations (1). Gene duplications also create conditions that enable random drift to operate much more prominently on recessive deleterious mutants than what was possible before duplication (as we shall show later). This allows fixation of mutants that are slightly deleterious for contemporary conditions but which may have other useful effects for adaptation to a new environment. Here again we find that the paradigm of the neutral theory (2) or "non-Darwinian" view (3) gives adequate explanation for phenomena relating to progressive evolution.

The present paper consists of two parts. In the first part, we discuss a situation in which mutation pressure leads to fixation of a deleterious allele. In the second part, which is the main part of this paper, we present a new treatment for the problem of random fixation of a nonfunctional allele at one of two loci after duplication. Since the present paper was submitted, we discovered that the results of the first part had already been obtained by Christiansen and Frydenberg (4), so in this revised version we present our results of this part only in an abbreviated form in order to serve as an introduction to the second part.

The publication costs of this article were defrayed in part by page charge payment. This article must therefore be hereby marked "*advertisement*" in accordance with 18 U. S. C. §1734 solely to indicate this fact.

## FIXATION BY MUTATION PRESSURE

Let us consider a random mating diploid population. We assume that a pair of alleles *A* and *a* are segregating in the first locus and alleles *B* and *b* are segregating in the second locus. We shall refer to *a* and *b* as mutant alleles. To simplify the situation as much as we can, we make additional assumptions. We assume that the population is large enough so that random fluctuation of gene frequencies can be neglected (this assumption will be removed in the next section). We also assume that genes in two loci combine completely at random. Strictly speaking, under epistatic interaction in fitness, nonrandom association or linkage disequilibrium will develop, particularly if linkage is tight. However, if selection coefficients involved are small and if the linkage is loose, quasilinkage equilibrium (5–7) will be realized, and we can neglect the linkage disequilibrium without serious error. Now, we assume that relative fitnesses of various genotypes are as given in Table 1, in which $s$ is the selection coefficient against a single recessive mutation and $\epsilon$ denotes the epistatic effect in fitness ($s \geq 0$, $\epsilon \geq 0$). In other words, $\epsilon$ represents the excess selection against the double homozygote over that expected with multiplicative interaction. Then the differential equations giving the rates of change of mutant allele frequencies are

$$dp/dt = (1-p)\{v_a - p^2[s(1-sq^2) + \epsilon q^2]\} \text{ and}$$
$$dq/dt = (1-q)\{v_b - q^2[s(1-sp^2) + \epsilon p^2]\} \qquad [1]$$

where $t$ denotes time in generations. It is assumed here that the selection is weak and linkage is loose so that assumption of complete random combination of genes is essentially valid. From these equations, we can see that if $s > 0$ the mutation–selection balance will be realized at both loci, unless $s$ is extremely small (at least as small as the mutation rates $v_a$ and $v_b$).

Let us assume, then, that the mutant alleles are completely recessive so that $s = 0$. In this case, mutant alleles are not only completely recessive but are also completely hypostatic to normal alleles at another locus. In other words, only the double recessive *aabb* is deleterious with selection coefficient $\epsilon$. This

Abbreviation: PSV, pseudo sampling variable.

Genetics: Kimura and King *Proc. Natl. Acad. Sci. USA 76 (1979)* 2859

Table 1. Fitness of various genotypes for the diploid model

| | $(1-p)^2$ $AA$ | $2(1-p)p$ $Aa$ | $p^2$ $aa$ | Marginal $w$ |
|---|---|---|---|---|
| $(1-q)^2$ $BB$ | 1 | 1 | $1-s$ | $1-sp^2$ |
| $2(1-q)q$ $Bb$ | 1 | 1 | $1-s$ | $1-sp^2$ |
| $q^2$ $bb$ | $1-s$ | $1-s$ | $(1-s)^2-\epsilon$ | $(1-s)(1-sp^2)-\epsilon p^2$ |
| Marginal $w$ | $1-sq^2$ | $1-sq^2$ | $(1-s)(1-sq^2)$ $-\epsilon q^2$ | $\overline{w}=(1-sp^2)(1-sq^2)-\epsilon p^2q^2$ |

The letter $w$ stands for the selective value.

type of epistasis is known as "duplicate" gene in classical genetics; selfing of the double heterozygote $AaBb$ ($F_1$) leads to 15:1 segregation in $F_2$ (for example, see ref. 8). In this case, Eqs. **1** reduce to

$$dp/dt = (1-p)(v_a - \epsilon p^2q^2) \text{ and}$$

$$dq/dt = (1-q)(v_b - \epsilon p^2q^2). \qquad [2]$$

We can then show that, if $v_a > v_b$, the mutant allele $a$ increases by mutation pressure to reach fixation ($p = 1$); mutation–selection balance will be reached for the allele $b$ ($q = \sqrt{v_b/\epsilon}$). In Fig. 1, courses of the frequencies of mutant alleles in an infinitely large population are illustrated assuming mutation rates $v_a = 2 \times 10^{-5}$, $v_b = 1 \times 10^{-5}$, and selection coefficient $\epsilon = 1 \times 10^{-3}$ and taking $1/\epsilon$ (= 1000 generations) as the unit length of time ($T = t\epsilon$). These are constructed based on the numerical solution of differential Eqs. **2** by the Runge–Kutta method.

In the case of equal mutation rates at both loci ($v_a = v_b \equiv v$), the situation is quite different, and neutral equilibrium will be reached so that $p^2q^2 = v/\epsilon$. Then, mutant frequencies are subject to random drift, and in a finite population the deleterious mutant will eventually become fixed in one of the two loci. We shall investigate the stochastic process involved in the next section.

## FIXATION BY RANDOM DRIFT

Let us consider a diploid population of effective size $N_e$ and use the selection model as shown in Table 1 but with $s = 0$ so that only the double mutant homozygote, *aabb*, is deleterious with

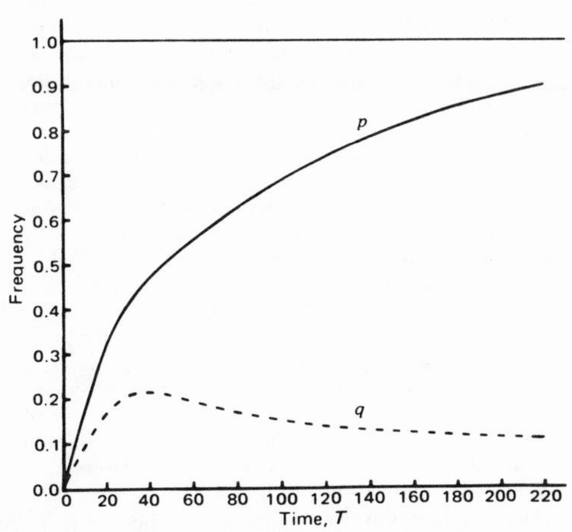


FIG. 1. Courses of change in the frequencies of deleterious alleles by mutation pressure (see text for details). —, For allele $a$; - - -, for allele $b$. Both lines start from $p_0 = q_0 = 0$.

selective disadvantage $\epsilon$. To treat the process of change of mutant frequencies, we make use of the diffusion equation method (9). Because the analytical solution is difficult to obtain for the present case, we resorted to numerical solution of the partial differential equation involved, for which the senior author (M.K.) is responsible. He is also responsible for Monte Carlo experiments performed to supplement the numerical analysis.

Let $u(p, q; t)$ be the probability of fixation of mutant allele $a$ or $b$ by the $t$th generation, given that the initial (i.e., at $t = 0$) frequencies are $p$ and $q$ for $a$ and $b$, respectively. Then, assuming random combination of alleles at both loci, we can show that $u(p, q; t)$ satisfies the following Kolmogorov backward equation:

$$\frac{\partial u}{\partial t} = \tfrac{1}{2} V_{\delta p} \frac{\partial^2 u}{\partial p^2} + \tfrac{1}{2} V_{\delta q} \frac{\partial^2 u}{\partial q^2} + M_{\delta p} \frac{\partial u}{\partial p} + M_{\delta q} \frac{\partial u}{\partial q} \qquad [3]$$

in which

$$V_{\delta p} = p(1-p)/(2N_e),\ V_{\delta q} = q(1-q)/(2N_e),$$

$$M_{\delta p} = (1-p)(v - \epsilon p^2q^2),\ M_{\delta q} = (1-q)(v - \epsilon p^2q^2), \qquad [4]$$

and $u$ stands for $u(p, q; t)$. For the rationale of Eq. **3**, see Crow and Kimura (ref. 10, p. 429). Let $\overline{T}(p, q)$ be the average time until fixation of mutant $a$ or $b$ in the population such that

$$\overline{T}(p, q) = \int_0^\infty t[\partial u(p, q; t)/\partial t]dt. \qquad [5]$$

Then, using Eq. **3**, we can show that $\overline{T}(p, q)$ satisfies the following elliptic equation.

$$\frac{1}{2} V_{\delta p} \frac{\partial^2 \overline{T}}{\partial p^2} + \frac{1}{2} V_{\delta q} \frac{\partial^2 \overline{T}}{\partial q^2} + M_{\delta p} \frac{\partial \overline{T}}{\partial p} + M_{\delta q} \frac{\partial \overline{T}}{\partial q} + 1 = 0. \qquad [6]$$

Let us denote by $y(p, q)$ the average time until fixation measured with $4N_e$ generations as the unit, so that $\overline{T}(p, q) = 4N_e y(p, q)$. Then $y$ satisfies the equation

$$p(1-p)y_{pp} + q(1-q)y_{qq} + (V - Ep^2q^2) \times \{(1-p)y_p + (1-q)y_q\} + 1 = 0 \qquad [7]$$

in which $V = 4N_e v$, $E = 4N_e \epsilon$, and $y_p$, $y_{pp}$, etc., stand for $\partial y/\partial p$, $\partial^2 y/\partial p^2$, etc. Note that $y(p, q)$ depends only on the products $4N_e v$ and $4N_e \epsilon$ but not on $N_e$, $v$, and $\epsilon$ separately. The appropriate boundary conditions are

$$y(1, q) = y(p, 1) = 0 \qquad [8a]$$

and

$$y(0, q) = \text{finite},\ y(p, 0) = \text{finite}. \qquad [8b]$$

To apply the numerical method, we cover the domain ($0 \leq p \leq 1$, $0 \leq q \leq 1$) by $n \times n$ square meshes each with side length $h$ ("mesh size"). Let $p = hi$ and $q = hj$, in which $i$ and $j$ are integers ($i, j = 0, 1, \ldots, n$). Then Eq. **7** may be converted to a set of finite difference equations in which symbol $y_{i,j}$ is used to represent $y(ih, jh)$. Boundary conditions **8** can be incorporated as follows. Condition **8a** is straightforward and we have $y_{n,j} = y_{i,n} = 0$. Condition **8b** is more subtle, but we can replace it by $y_{0,j} = y_{1,j} + (y_{1,j} - y_{2,j})$ and $y_{i,0} = y_{i,1} + (y_{i,1} - y_{i,2})$.

For given values of $V$ and $E$, this set of equations for $y_{i,j}$ can be solved numerically by using a computer (details will be published elsewhere). We chose mesh size $h = 0.1$, and the Gauss–Seidel method was used (for the numerical solution of partial differential equations see, for example, ref. 11, p. 391). The solid curve in Fig. 2 illustrates, for the case $4N_ev = 2$, the average time until fixation of mutant $a$ or $b$, starting from $p = q = 0$, as function of $4N_e\epsilon$ for value of $4N_e\epsilon < 50$. In other words, the ordinate represents $\overline{T}(0, 0)$ or $4N_ey(0, 0)$.

In order to supplement these results, Monte Carlo simulation experiments were performed with assumed various population sizes ($N_e$) ranging from 25 to 2500, $4N_ev = 2$, and $4N_e\epsilon$ ranging from 1 to $10^4$. Note that $\overline{T}/(4N_e)$ depends only on $4N_ev$ and $4N_e\epsilon$. Starting from $p = q = 0$, the average time until fixation of $a$ or $b$ was investigated. (In nature, newly tetraploid populations presumably are the immediate descendants of one or a few individuals and can be assumed to reach the effective population size $N_e$ quickly before accumulating any mutant alleles.) Each solid circle in Fig. 2 represents the average of 100 replicate trials, except for a few cases ($4N_e\epsilon > 1000$) for which each point is the average of 10–50 replicate trials. The broken line at the tail of the solid curve for $4N_ev = 2$ represents values of $\overline{T}(0, 0)$ inferred from these simulation experiments for higher values of $4N_e\epsilon$. Thus, the broken-line curve represents crude approximation values only. Fig. 2 also shows the results of simulation for $4N_ev = 0.4$ (each open circle is the average of 1000 replicate trials). In the simulation experiments, each generation consists of random sampling drift followed by mutation and selection.

To simulate the gene frequency change by random sampling of gametes in one generation, instead of actually sampling gametes $2N_e$ times as is usually done in Monte Carlo experiments in population genetics (for example, see ref. 12), we simply generated a random number (called a "pseudo sampling variable" or PSV) and a realized value of this variable was added to the gene frequency ($p$) to produce the frequency ($p'$)

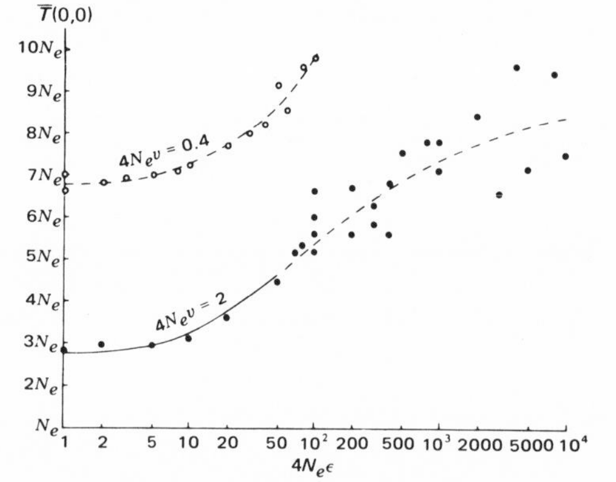


FIG. 2. Average time until fixation of a deleterious mutant at one of two "duplicate loci." Relationship between the average time, $\overline{T}$ (0, 0), and $4N_e\epsilon$ is illustrated for two cases, $4N_ev = 2$ and $4N_ev = 0.4$. $N_e$, effective population size; $\epsilon$, selection coefficient against the double mutant individual; $v$, mutation rate per locus (see text for details).

after sampling drift. The essential point is that it is a uniform random number that has mean = 0 and variance $\sigma^2 = p(1 - p)/(2N_e)$. In other words, if $\xi_{PSV}$ is a PSV, then $\xi_{PSV} = \sqrt{3\sigma^2} \cdot U_1$, in which $U_1$ is a random variable that follows a uniform distribution between $-1$ and $+1$ and is commonly used in Monte Carlo experiments. If $p'$ ($= p + \xi_{PSV}$) becomes negative by chance, which may sometimes happen when $p$ is near zero, then $p'$ is set to zero to continue the experiment. On the other hand, if $p'$ becomes larger than $1 - 1/(2N_e)$, $p'$ is set equal to unity and the run is ended. The reason why PSV can substitute for the actual sampling comes from the nature of the continuous stochastic process—namely, only the mean and the variance (but not the detailed shape of the distribution) of the change per generation determine the process, as long as the higher moments are negligible in magnitude (see ref. 10, p. 374). Note that this scheme of pseudo sampling simulates the diffusion process itself rather than the discrete, binomial sampling process, for which the diffusion model is usually regarded as an approximation.

Finally, we should remark that, because the time until fixation of $a$ or $b$ has a large standard deviation around its mean ($\overline{T}$), this mean time, in very rough sense, may represent the time by which fixation occurs in half of the cases—i.e., $T_{0.5} \approx \overline{T}$. Actually, the distribution of fixation time has a positive skewness of roughly unity, so that $T_{0.5}$ is somewhat smaller than $\overline{T}$.

## DISCUSSION

It is clear that the remarkable phenomenon of a deleterious allele reaching fixation in one of the two loci is possible only when the "duplicate" type epistasis is complete in fitness. In addition, mutation rates ($v_a$ and $v_b$) must be significantly different at the two loci for the mutation pressure to control the process deterministically. Under these conditions, mutant $a$ at the first locus goes to fixation if $v_a > v_b$ and $b$ reaches mutation–selection balance, and vice versa. This confirms the results obtained earlier by Christiansen and Frydenberg (4). The rate of increase of $a$ is roughly proportional to the difference in mutation rates, $v_a - v_b$. In other words, $a$ is pushed by the pressure that comes from the excess mutation rate.

From the standpoint of evolution, the most likely origin of "duplicate" type epistasis is gene duplication, especially as it occurs in the formation of allotetraploids. Considering the prevalence of gene duplication in evolution, it might be expected that "duplicate" genes are commonly found in plants and, to a lesser extent, in animals. Duplicate genes are indeed not uncommon, but loss of duplicate genes and reversion to functional diploidy are certainly common and perhaps more usual (13).

Where duplicate genes persist, it is of course possible that complete hypostasis and recessivity of deleterious mutants may be lacking. Slightly deleterious mutants in *Drosophila* usually show considerable dominance, so that they are mainly selected against in the heterozygous condition (14, 15). For "null" mutants at enzyme loci, on the other hand, it is possible that the heterozygotes with the wild-type (active) allele are so nearly normal that we can regard the mutant alleles, for practical purposes, as being completely recessive. Presumably, such mutants at duplicate enzyme loci might also be completely hypostatic. However, unless we see evidence to the contrary, it seems likely that mutation rates stay the same in duplicated loci, and the effect of unequal mutation rates at duplicate loci may seldom be important in and of itself.

The situation is less clear for unequal epistasis—e.g., where the $B$ allele is completely epistatic but the $A$ allele is not. If multiple alleles are considered, it seems likely that slightly hypomorphic alleles might occur and increase at one or the other locus, through mutation and drift in the absence of strong selection. Once one duplicate locus had such a slightly deleterious allele either fixed or in a significantly high frequency, the symmetry of selection would be ended and additional, more severe, mutant alleles should also increase in frequency at the same locus.

Loss of gene expression in one or the other of duplicated loci has recently been reported in some groups of fish. In this case, fixation of "null" alleles by mutation and drift appears to be the most likely explanation. According to Allendorf (13), species in both salmonid and catostomid fish have lost approximately 50% of the gene duplication produced by tetraploidy. We can show that the time needed for such an evolutionary loss ($T_{0.5}$) depends much on the mutation rate, assuming that the population starts from the state having no null mutants. This assumption may be realistic because the tetraploidization must have started from a single individual or a few related individuals free of null alleles. As shown in Fig. 2, if $2N_e v = 1$ (i.e., one new mutation appears in each generation), it takes roughly $9N_e$ generations if $4N_e \epsilon = 10{,}000$. If the mutation rate for appropriate null alleles is much lower, the time needed for 50% loss must be longer. Our rough estimation based on Monte Carlo experiments suggests that, if $4N_e v = 0.1$ and $4N_e \epsilon = 1000$, the average time ($\overline{T}$) until fixation of a null allele in one of the loci is about $20N_e$ generations which must be somewhat larger than the time needed for 50% loss. These results do not seem to agree with those of Bailey *et al.* (16) who claim that $T_{0.5} \approx 15N_e + v^{-3/4}$. Clearly, $T_{0.5}$ can be much shorter than $15N_e$ if $4N_e v$ is larger than unity. A more detailed study of the problem including the situation in which epistasis is not complete has been done by N. Takahata and T. Maruyama (personal communication), and the results will be published in the near future. In particular, they found that $T_{0.5}$ becomes much larger if null alleles show small deleterious effects in combinations other than the double null homozygote.

We thank Dr. F. W. Allendorf for calling our attention to the paper of Christiansen and Frydenberg (ref. 4). Thanks are also due to Drs. T. Maruyama and N. Takahata for stimulating discussions. This is contribution no. 1247 from the National Institute of Genetics, Mishima, Shizuoka-ken 411, Japan.

1. Jukes, T. H. & King, J. L. (1975) *J. Hum. Evol.* **4**, 85–88.
2. Kimura, M. (1968) *Nature (London)* **217**, 624–626.
3. King, J. L. & Jukes, T. H. (1969) *Science* **164**, 788–798.
4. Christiansen, F. B. & Frydenberg, O. (1977) *Am. J. Hum. Genet.* **29**, 195–207.
5. Kimura, M. (1965) *Genetics* **52**, 875–890.
6. Wright, S. (1967) *Proc. Natl. Acad. Sci. USA* **58**, 165–172.
7. Nagylaki, T. (1974) *Proc. Natl. Acad. Sci. USA* **71**, 526–530.
8. Darlington, C. D. & Mather, K. (1949) *The Elements of Genetics* (Allen & Unwin, London).
9. Kimura, M. (1964) *J. Appl. Probab.* **1**, 177–232.
10. Crow, J. F. & Kimura, M. (1970) *An Introduction to Population Genetics Theory* (Harper & Row, New York).
11. Todd, J., ed. (1962) *Survey of Numerical Analysis* (McGraw-Hill, New York).
12. Kimura, M. & Ohta, T. (1969) *Genetics* **61**, 763–771.
13. Allendorf, F. W. (1978) *Nature (London)* **272**, 76–78.
14. Mukai, T., Chigusa, S. I., Mettler, L. E. & Crow, J. F. (1972) *Genetics* **72**, 335–355.
15. Simmons, M. J. & Crow, J. F. (1977) *Annu. Rev. Genet.* **11**, 49–78.
16. Bailey, G. S., Poulter, R. T. M. & Stockwell, P. A. (1978) *Proc. Natl. Acad. Sci. USA* **75**, 5575–5579.

# Diffusion Models in Population Genetics with Special Reference to Fixation Time of Molecular Mutants under Mutational Pressure

MOTOO KIMURA

*National Institute of Genetics, Mishima 411, Japan*

The method of applying diffusion equations to treat stochastic processes of gene frequencies in Mendelian populations (*i.e.*, reproductive communities) has been quite successful in population genetics. It has played a particularly important role in the development of theoretical population genetics at the molecular level (*1–4*). Although stochastic processes of gene frequencies can be formulated more exactly in terms of finite Markov chains involving multiplication of transition matrices, biologically useful results have seldom emerged from such an approach. On the other hand, the diffusion equation method (or "diffusion models" (*5*) as they are often called) which are continuous approximations of the underlying discrete processes, can deal with many difficult but biologically important problems, often leading to simple and elegant solutions. Furthermore, the approximations are usually quite accurate even for very small population sizes, as can be seen by numerically checking them with more exact discrete treatments and or with extensive Monte Carlo simulation experiments.

Historically, the first use of diffusion models in population genetics appears to go back to Fisher's 1922 paper (*6*). Unfortunately, the result he obtained on the rate of decay of genetic variability in a finite population was incorrect (due to his oversight of a necessary term) but he later corrected this (*7*). Actually the correct result was first obtained by Wright (see ref. *8*). Personally, I was introduced to the world of theoretical population genetics by studying Wright's 1931 paper (*8*) when I was working in Professor Kihara's laboratory in Kyoto a few years after the Second World War. Then in 1949 I moved to Mishima, and by studying Wright's papers (*9*, *10*) I came to know that the partial differential equation called the "Fokker-Planck equation" in physics could be used to describe the processes of gene frequency change in finite populations.

Soon I found, through studying an advanced textbook on probability theory then newly published in Japan (*11*), that two types of partial differential equations, called Kolmogorov forward and the Kolmogorov backward equations, were available to treat continuous stochastic processes. I was quite fascinated by these formulations and tried to apply them to my study of theoretical population genetics. This work was much extended when I went to the United States for study. My 2 years stay at the University of Wisconsin as a graduate student under Dr. James Crow starting from 1954 (now, 30 years ago!) was particularly conducive to the realization of my project of investigating stochastic processes of gene frequency change in terms of diffusion models. At about the same time, Dr. Sewall Wright moved to Madison after his retirement from the University of Chicago to join the Genetics Department of the University of Wisconsin. I was fortunate that I could embark on my life's work in this marvellous environment.

I can perhaps claim that my main life as a scientist has been devoted to the study of diffusion models in population genetics, as influenced by the great work of Sewall Wright. Even my neutral theory of molecular evolution is on this line of development: it is a flower, so to speak, that bloomed on a tree whose main trunk comprises the diffusion models of populations genetics.

In this paper, I would like to exemplify the use of diffusion models by considering the problem: how long does it take for a mutant gene or genes to become fixed in a finite population under continued mutation pressure? I shall treat the single locus case first.

## AVERAGE TIME UNTIL FIXATION UNDER MUTATION PRESSURE AT A SINGLE LOCUS

Let us consider a particular locus, and denote the normal, wild-type allele by $A$. We assume that $A$ mutates irreversibly to its allele $A'$ at the rate $v$ per generation. In reality, the mutant allele $A'$ is not usually a single entity (particularly at the molecular level) but a set of mutant alleles, but we designate them collectively as $A'$. Let $1+s$ and $1+h$ be respectively the relative fitnesses of mutant homozygote ($A'A'$) and heterozygote ($A'A$), taking the fitness of wild-type homozygote ($AA$) as unity (Table I).

We assume a random mating, diploid population of effective size $N_e$; roughly speaking, $N_e$ is equal to the number of breeding individuals in one generation (for more details, see refs. *2* and *12*).

Let $p$ be the frequency of mutant allele $A'$ in the population and denote by $u(p, t)$ the probability that $A'$ becomes fixed in the population by time $t$ (or the $t$-th generation), given that its initial frequency (at time $t=0$) is $p$. Then, $u(p, t)$ satisfies the following Kolmogorov backward equation

$$\frac{\partial u(p,t)}{\partial t} = \frac{1}{2} V_{\delta p} \frac{\partial^2 u(p,t)}{\partial p^2} + M_{\delta p} \frac{\partial u(p,t)}{\partial p}, \tag{1}$$

where

$$M_{\delta p} = p(1-p)[sp + h(1-2p)] + v(1-p) \tag{2}$$

and

$$V_{\delta p} = p(1-p)/(2N_e). \tag{3}$$

In these expressions, $M_{\delta p}$ represents the mean change of $p$ per genera-

TABLE I

Fitnesses and Frequencies of Three Genotypes at a Single Locus

| Genotype | $AA$ | $AA'$ | $A'A'$ |
|---|---|---|---|
| Fitness | 1 | $1+h$ | $1+s$ |
| Frequency | $(1-p)^2$ | $2p(1-p)$ | $p^2$ |

tion due to mutation and natural selection, and we assume that the selection coefficients are small (*i.e.*, $|s| \ll 1$, $|h| \ll 1$).

Let $\overline{T}(p)$ be the average time until fixation of the mutant allele, given that its initial frequency is $p$, so that

$$\overline{T}(p) = \int_0^\infty t \frac{\partial u(p,t)}{\partial t} dt. \tag{4}$$

Then, it can be shown (see ref. *13*) that $\overline{T}(p)$ satisfies the ordinary differential equation

$$\frac{1}{2} V_{\delta p} \frac{d^2\overline{T}(p)}{dp^2} + M_{\delta p} \frac{d\overline{T}(p)}{dp} + 1 = 0, \tag{5}$$

where the appropriate boundary conditions are

$$\overline{T}'(0) = \text{finite} \tag{6a}$$

and

$$\overline{T}(1) = 0. \tag{6b}$$

Note that if the population is finite in size, the mutant allele (even when it is deleterious) becomes fixed in the population under irreversible mutation. In other words,

$$u(p, \infty) = 1,$$

although the time required for such fixation may be very long if the mutant is definitely deleterious.

The solution of Eq. (5) with coefficients given by Eqs. (2) and (3), and with boundary conditions (6a) and (6b) is

$$\overline{T}(p) = 4N_e y(p), \tag{7}$$

where

$$y(p) = \int_p^1 e^{-B(\eta)} \eta^{-V} d\eta \int_0^\eta \frac{e^{B(\xi)} \xi^{V-1}}{1-\xi} d\xi, \tag{8}$$

in which

$$B(\xi) = (S/2)\xi^2 + H\xi(1-\xi), \tag{9}$$

$S=4N_e s$, $H=4N_e h$, and $V=4N_e v$ (*13*).
Note that $y(p)$ depends on the products $N_e s$, $N_e h$, and $N_e v$, but not on $N_e$, $s$, $h$, and $v$ separately.

In what follows we shall be concerned with the fixation time for a slightly deleterious mutant. For the purpose of comparing the single locus case with the two-locus case to be studied in the next section, we particularly consider the semidominant deleterious mutation so that

$$h = -s' \text{ and } s = -2s', \tag{10}$$

where $s'$ ($\geqq 0$) is the selection coefficient against $A'$ (Table II). For this case, $B(\xi)$ of formula (9) reduces to

$$B(\xi) = -S'\xi \tag{11}$$

where $S'=4N_e s'$. An equivalent case was studied by Li and Nei (*14*) in which their $p$ and $s$ correspond to our $1-p$ and $-s'$. We shall also restrict our consideration to the case $p=0$. In other words, we shall

TABLE II
Assignment of Fitnesses to Three Genotypes Involving the "Wild Type" Allele $A$ and the Semidominant Deleterious Allele $A'$

| Genotype | $AA$ | $AA'$ | $A'A'$ |
|---|---|---|---|
| Fitness | 1 | $1-s'$ | $1-2s'$ |
| Frequency | $(1-p)^2$ | $2p(1-p)$ | $p^2$ |

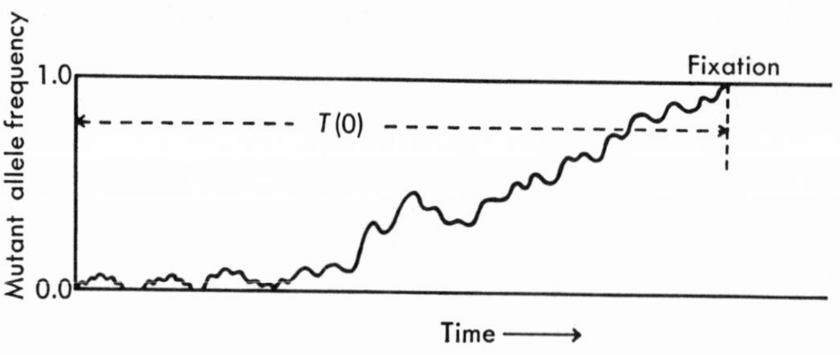


Fig. 1. Diagram illustrating the process of mutant fixation under continued irreversible mutation pressure. $T(0)$ stands for the length of time for the mutant allele $A'$ to become fixed in the population, starting from the state in which the population is free of $A'$. $\bar{T}(0)$ given by Eq. (12) is the average of $T(0)$ when $A'$ is semidominant and deleterious.

investigate the average number of generations until fixation of the mutant allele, starting from a population consisting exclusively of the wild-type allele (see Fig. 1). Under these restrictions, formula (7) reduces to

$$\bar{T}(0) = 4N_e \int_0^1 e^{S'\eta}\eta^{-V} d\eta \int_0^\eta e^{-S'\xi}\xi^{V-1}(1-\xi)^{-1} d\xi, \tag{12}$$

where

$$S' = 4N_e s' \text{ and } V = 4N_e v.$$

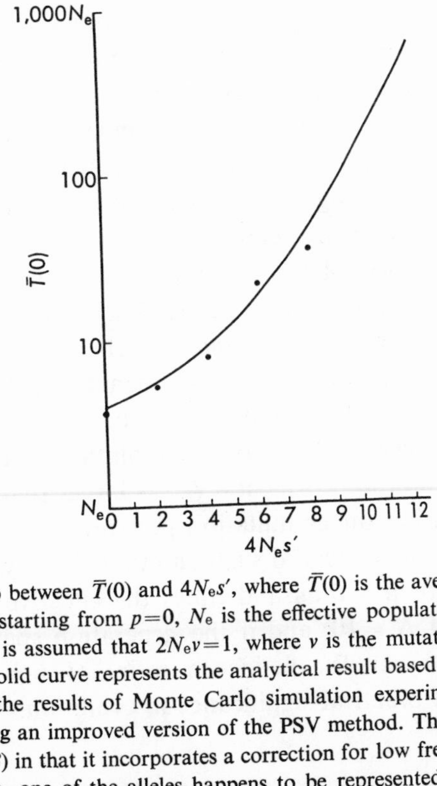


Fig. 2. Relationship between $\bar{T}(0)$ and $4N_es'$, where $\bar{T}(0)$ is the average time until fixation of the mutant allele starting from $p=0$, $N_e$ is the effective population size and $s'$ is selective disadvantage. It is assumed that $2N_ev=1$, where $v$ is the mutation rate for deleterious allele ($A \rightarrow A'$). The solid curve represents the analytical result based on the diffusion model and dots represent the results of Monte Carlo simulation experiments. The experiments were performed using an improved version of the PSV method. This differs from the original PSV method (*13*) in that it incorporates a correction for low frequency classes, namely, if, at any generation, one of the alleles happens to be represented in the population less than five times, a Poisson random variable is used to sample that allele. Each dot represents the average of 50 replicate trials assuming 100 diploid breeding individuals ($N_e=100$) and the mutation rate $v=0.005$ so that one mutation on the average is assumed to occur in each generation ($2N_ev=1$) when the population consists exclusively of the wild type allele.

For a selectively neutral mutant ($S'=0$), the formula is much simplified: if $V=4N_e v \neq 1$, we have

$$\bar{T}(0)=\frac{4N_e}{V-1}\int_0^1 \frac{1-\xi^{V-1}}{1-\xi}\,d\xi=\frac{4N_e}{V-1}\,[\gamma+\psi(V)] \qquad (13)$$

(*13*), where $\psi(\cdot)$ stands for the digamma function and $\gamma=0.577\cdots\cdots$ If, on the other hand, $4N_e v=1$, we have $\bar{T}(0)=4N_e(\pi^2/6)\approx 6.58N_e$.

An interesting special case occurs when $V=2$ or $2N_e v=1$; in this case we have $\bar{T}(0)=4N_e$. In other words, it takes on the average four times the effective population size for a neutral mutant allele to become fixed under continued irreversible mutation pressure. This corresponds to the rate such that one mutant gene is fed into the population in each generation if the population consists exclusively of the wild type allele (assuming that $N_e$ is equal to the actual population size).

In Fig. 2, the average time until fixation $\bar{T}(0)$ is illustrated for various values of $4N_e s'$ ranging from 0 to 12 (abscissa) assuming $2N_e v=1$. The solid curve in this figure represents the theoretical results obtained by numerical integration of Eq. (12) for $V=2$, while dots represent the corresponding results obtained by Monte Carlo simulation experiments (assuming $N_e=100$ and $v=1/200$, and 5 levels of $s'$ values). The agreement between the analytical and experimental results is satisfactory. These results show that deleterious mutations are unlikely to be incorporated into the species in the course of evolution unless $4N_e s'$ is small, say, less than 10. In other words, if $4N_e s'$ is larger than 10, the average time taken for fixation under mutation pressure is so enormously long that we can practically disregard such an event. For example, if $4N_e s'=13$, it takes more than a thousand times the effective population size. This means that, if $N_e=10^5$ and if the generation span is one year, it takes more than $10^8$ years for the incorporation of such a mutant, which is too long to be of practical significance in evolution.

## MUTATION AT TWO LOCI WITH COMPENSATORY INTERACTION IN FITNESS

I shall now consider a more interesting case involving two loci (or

TABLE III
Table of Fitnesses of Four Gene Combinations at Two Loci

| | $A$ | $A'$ |
|---|---|---|
| $B$ | 1 | $1-s'$ |
| $B'$ | $1-s'$ | 1 |

sites) in which mutations are individually deleterious but become harmless (*i.e.*, selectively neutral) in combination. Let $A'$ be the mutant allele which is produced irreversibly at the rate $v$ per generation from its wild type allele $A$ at the first locus (or site). Likewise, let $B'$ be the mutant allele at the second locus (or site) produced at the same rate $v$ from its allele $B$. We denote by $s'$ the selection coefficient against the single mutant ($A'B$ or $AB'$), and assume that the double mutant $A'B'$ has the same fitness as the wild type $AB$ (Table III). To simplify the treatment, we disregard dominance at each locus, so that we adopt the scheme of genic selection or haploid selection model. As to the effective population size, we again denote by $N_e$ the effective size, assuming a random mating, diploid population. Since the selection model is haploid, we can just as well consider a haploid population consisting of $2N_e$ breeding individuals.

As to recombination between the two loci (or sites), I shall consider two extreme situations, namely, 1) the case of free recombination between loci, and 2) that of complete linkage. As we shall see, these two cases give quite different results.

*1. Case of Free Recombination between Loci*

We assume independent assortment of alleles between the two loci. Let $\overline{T}=\overline{T}(p,q)$ be the average time until joint fixation of $A'$ and $B'$ under continued mutation pressure, given that the initial frequencies of $A'$ and $B'$ are $p$ and $q$, respectively. Then, $\overline{T}(p,q)$ satisfies the equation,

$$\frac{1}{2}V_{\delta p}\frac{\partial^2\overline{T}}{\partial p^2}+\frac{1}{2}V_{\delta q}\frac{\partial^2\overline{T}}{\partial q^2}+M_{\delta p}\frac{\partial\overline{T}}{\partial p}+M_{\delta q}\frac{\partial\overline{T}}{\partial q}+1=0, \qquad (14)$$

where

$$V_{\delta p}=p(1-p)/(2N_e),\ V_{\delta q}=q(1-q)/(2N_e),$$

$$M_{\delta p} = -s'p(1-p)(1-2q) + v(1-p)$$

and

$$M_{\delta q} = -s'q(1-q)(1-2p) + v(1-q).$$

The appropriate boundary conditions are

$$\overline{T}(1,q) = \overline{T}_1(q), \quad \overline{T}(p,1) = \overline{T}_1(p),$$

$$\lim_{p\to 0} \frac{\partial}{\partial p} \overline{T}(p,q) = \text{finite, and } \lim_{q\to 0} \frac{\partial}{\partial q} \overline{T}(p,q) = \text{finite},$$

where

$$\overline{T}_1(x) = 4N_e \int_x^1 e^{-S'\eta}\eta^{-V} d\eta \int_0^\eta e^{S'\xi}\xi^{V-1}(1-\xi)^{-1} d\xi. \tag{15}$$

Note that if $A'$ is fixed in the population, mutant $B'$ becomes more advantageous than $B$. Similarly, if $B'$ is fixed in the population, $A'$ becomes more advantageous than $A$. I have not been able to obtain the analytical solution of the partial differential equation (14), but I applied a

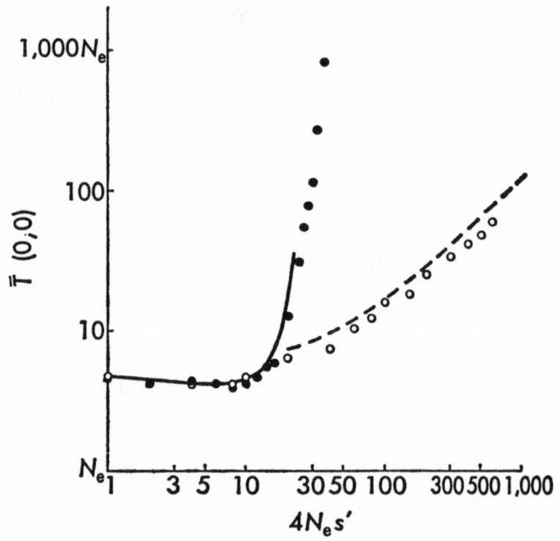


Fig. 3. Relationship between $\overline{T}(0, 0)$, the average time until joint fixation at two loci (with time expressed as multiples of $N_e$), and $4N_e s'$ ($s'$, the selection coefficient against the single mutants), assuming $2N_e v = 1$ ($v$, mutation rate per locus). For details, see text.

numerical method to obtain the values of $\bar{T}(0, 0)$ for some sets of parameter values. More specifically, I used the method of approximating a partial differential equation by a finite difference equation and then applying the Gauss-Seidel iteration procedure (see, *e.g.*, ref. *15*) to obtain the numerical solutions on $n \times n$ square meshes that cover the domain ($0 \leqq p \leqq 1$, $0 \leqq q \leqq 1$), where I usually assumed $n=10$. The solid curve in Fig. 3 illustrates the relationship between $\bar{T}(0, 0)$ and $4N_e s'$ thus obtained assuming $2N_e v=1$. For $4N_e s'=0$ (neutral mutations), it takes about $5.1N_e$ generations until joint fixation of mutants starting from $p=q=0$, *i.e.*, when the initial population consists exclusively of wild type alleles at both loci. (I owe the following analytical demonstration which leads to the same result to Dr. G. Watterson (personal communication): if $2N_e v=1$ (*i.e.*, $\theta=2$ in ref. *16*), the mean time for the $A'$ allele to fix is $E(T_{A'})=4N_e$. Also, according to Table 5 of ref. *16*, $E(T_{A'}|A'$ fixes before $B')=2.9N_e$. But, by symmetry between $A'$ and $B'$, $E(T_{A'})=(1/2)E(T_{A'}|A'$ fixes before $B')+(1/2)E(T_{A'}|A'$ fixes after $B')$. Thus $E(T_{A'}|A'$ fixes after $B')=2E(T_{A'})-E(T_{A'}|A'$ fixes before $B')=2\times 4N_e-2.9N_e=5.1N_e$.) For slightly deleterious mutations in the range $0<4N_e s' \leqq 10$, the joint fixation time is slightly shorter than this. For $4N_e s'=10$, for example, I obtained $\bar{T}(0, 0)\approx 4.5N_e$ by numerical method. For larger values of $4N_e s'$ the fixation time quickly grows to become very large. The result that, for a region of $s'$ with very small values, mutant alleles with more deleterious effect shorten the fixation time was rather unexpected, and an error in numerical solution was suspected. However, the same result was obtained by extensive simulation experiments so this must be valid.

Two types of Monte Carlo experiments were performed, one which made use of the improved pseudo-sampling variable (PSV) method, the other which used the standard method of faithfully sampling $2N_e$ gametes in each generation. In Fig. 3, the results obtained using the standard type simulation method are plotted with solid dots. Each dot represents the average of 50 replicate trials, assuming $N_e=100$ and $v=0.005$ so that $2N_e v=1$. From these results it appears that for compensatory mutations with enough deleterious effect so that $4N_e s'>40$, the average time until joint fixation is so enormously long that they are unlikely to be incorporated in the species in evolution.

TABLE IV
Assignment of Fitness and Frequency Parameters for Four Genotypes under Complete Linkage between the Two Loci

| Genotype | $AB$ | $A'B$ or $AB'$ | $A'B'$ |
|---|---|---|---|
| Fitness | 1 | $1-s'$ | 1 |
| Frequency | $p_0$ | $p_1$ | $p_2$ |

2. *Case of Complete Linkage between Loci*

Since we assume no crossing-over between the two loci (or sites) in this case, we may treat four genotypes as if they were four alleles at a single locus. Furthermore, we may lump $A'B$ and $AB'$ together because of equal fitnesses and denote their frequencies collectively as $p_1$ (see Table IV). Let us denote the frequency of the double mutant $A'B'$ by $p_2$, and that of the wild type by $p_0$ $(=1-p_1-p_2)$. For a full treatment of the process of fixation of $A'B'$, we must consider simultaneously the behavior of the single mutants as well as that of the double mutant. Thus, let $u=u(p_1, p_2; t)$ be the probability that the double mutant $A'B'$ becomes fixed in the population by the $t$-th generation, given that the initial frequencies of the single and the double mutants are $p_1$ and $p_2$, respectively. Then $u$ satisfies the equation

$$\frac{\partial u}{\partial t}=\frac{p_2(1-p_2)}{4N_e}\frac{\partial^2 u}{\partial p_2^2}-\frac{p_1p_2}{2N_e}\frac{\partial^2 u}{\partial p_2 \partial p_1}+\frac{p_1(1-p_1)}{4N_e}\frac{\partial^2 u}{\partial p_1^2}$$
$$+M_{\delta p_2}\frac{\partial u}{\partial p_2}+M_{\delta p_1}\frac{\partial u}{\partial p_1}, \tag{16}$$

where

$$M_{\delta p_2}=vp_1+s'p_1p_2 \tag{17a}$$

and

$$M_{\delta p_1}=2vp_0-s'p_1(1-p_1)-vp_1. \tag{17b}$$

The analytical solution of Eq. (16) appears to be very difficult to obtain, but for the biologically interesting case in which

$$s' \gg v>0, \tag{18}$$

we may apply the following shortcut approximation. It is reasonable to assume that in this case single mutants, because of their selective disadvantage, remain in low frequencies throughout the process and that the quasi-equilibrium

$$M_{\delta p_1}=0 \tag{19}$$

holds approximately. Then, if we disregard the second order term in $p_1^2$, we have

$$2v(1-p_1-p_2)-s'p_1-vp_1=0. \tag{20}$$

This gives

$$p_1=\frac{2v}{s'+3v}(1-p_2), \tag{21}$$

and therefore we have

$$M_{\delta p_2}=(v+s'p_2)\frac{2v}{s'+3v}(1-p_2) \tag{22}$$

from Eq. (17a). Under the assumption of quasi-equilibrium of $p_1$, it is only necessary to consider the process of change of $p_2$. Writing $p$ for $p_2$, let us denote by $u(p;t)$ the probability that $A'B'$ becomes fixed in the population by the $t$-th generation given that its frequency is $p$ at time 0. Then, we can apply Eq. (1) with the mean and the variance of change per generation given by

$$M_{\delta p}=\frac{2s'v}{s'+3v}p(1-p)+\frac{2v^2}{s'+3v}(1-p) \tag{23}$$

and

$$V_{\delta p}=p(1-p)/(2N_e). \tag{24}$$

Formally, this corresponds to the case of an advantageous mutant at a single locus with selection coefficient

$$s_1=\frac{2s'v}{s'+3v} \tag{25a}$$

and the mutation rate

$$v_1 = \frac{2v^2}{s' + 3v} \,. \tag{25b}$$

Note that these are very small quantities, particularly when $s'$ is very much larger than $v$. Then, the average time until fixation of the double mutant $A'B'$ starting from the population consisting exclusively of the wild type $(AB)$ is given by Eq. (7) by putting $p=0$ and assuming $S=8N_e s_1=4AB/(B+3A)$, $H=S/2$, and $V=4N_e v_1=2A^2/(B+3A)$, in which $A=4N_e v$ and $B=4N_e s'$.

In Fig. 3, the average fixation time computed by this approximation method is plotted as a function of $4N_e s'$ ($\geqq 20$) with a broken line. In order to check the validity of this approximation, Monte Carlo simulation experiments were performed, and the results are plotted by open circles. Each circle is the average of 50 replicate trials assuming 250 breeding individuals ($N_e=250$) with mutation rate $v=0.002$ so that

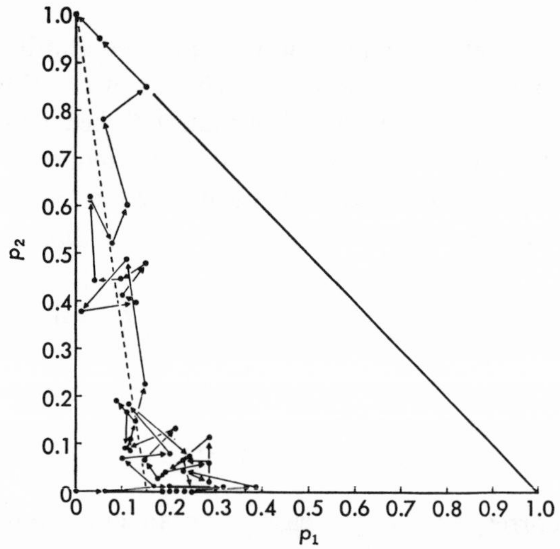


Fig. 4. An example of a sample path obtained from the Monte Carlo simulation experiments as illustrated with intervals of 5 generations. Parameter values are $N_e=50$, $s'=0.1$, and $v=0.01$ so that $2N_e v=1$ and $4N_e s'=20$. The broken line represents the quasi-equilibrium or "moving equilibrium" computed using Eq. (21).

$2N_e v=1$. The agreement between the approximate treatment and the results of simulation experiments is satisfactory, although there is a possibility that the approximation method slightly overestimates the true value. Figure 4 illustrates an example of sample paths drawn from the simulation experiments in order to explain the meaning of the quasi-equilibrium assumed.

These results, when compared with those obtained in the previous subsection, bring to light a remarkable point that the average fixation time is very much shorter under complete linkage than with free recombination when the deleterious effect of single mutants is large such that $4N_e s'>20$. Indeed, the difference between these two cases grows very rapidly as $4N_e s'$ becomes larger. For example, when $4N_e s'=30$, the mean fixation time is about $8N_e$ generations under complete linkage, whereas it is over $100N_e$ generations under free recombination (still assuming the mutation rate so that $2N_e v=1$).

The remarkable effect of very tight linkage in reducing the fixation time of the compensatory deleterious mutants is evident from the fact that, even for $4N_e s'=400$, the average fixation time under complete linkage is about $50N_e$, which is only 10 times as long as the selectively neutral case. This means that if this model of compensatory effect in fitness is realistic, deleterious mutants such that $4N_e s'=400$ still have an ample chance to participate in evolution by random drift under continued mutation pressure.

I shall now discuss the implication of such a finding in the context of considering the mechanism of molecular evolution.

## DISCUSSION

The problem of whether evolution can proceed through an intermediate deleterious state has often been debated in the literature of evolution. Usually the issue is settled by claiming that the presumed deleterious state is misconceived and that it is, in fact, advantageous or at least neutral.

The present model of "compensatory neutral evolution" is rather unusual in that a marked deleterious intermediate state can easily be overcome by mutational pressure and random drift under complete linkage. Probably the most appropriate circumstance to which this model applies is the coupled substitutions of amino acids within molecules, which Ohta (*17*, *18*) referred to in relation to her hypothesis that very slightly deleterious as well as neutral mutations play an important role in molecular evolution. Wyckoff (*19*), in his comparison of rat and bovine pancreatic ribonucleases, noted that "a number of changes are paired." For example, in bovine RNase, amino acid positions 57 and 79 are occupied respectively by valine and methionine, while in rat RNase, these positions are occupied by isoleucine and leucine. What is important is that these two amino acid sites are close to each other in the three-dimensionally folded structure, although they are relatively far apart in the linear sequence.

A similar example was found by Tsukihara *et al.* (*20*) in their study of [2Fe-2S] ferredoxins isolated from various plants and algae. According to them, in *Equisetum* (horsetail) species, when two duplicated genes (I and II) of this protein are compared in terms of amino acid sequence, a change from threonine to arginine at position 25 correlates with a change from arginine to glutamine at position 42 in the molecule. Again, these two amino acid positions are close to each other in the three-dimensionally folded structure.

That such physical proximity of sites within a folded protein is the basis of fitness interaction is strongly suggested by the mutation studies of Yanofsky *et al.* (*21*) on tryptophan synthetase A protein. For example, position 210 in the wild type protein is occupied by glycine but if this is replaced by glutamic acid by mutation, the enzyme becomes nonfunctional. However, if a further change occurs at position 174, changing tyrosine of the wild type to cysteine, the activity of the enzyme is recovered, although a mutation (Tyr→Cys) at the second position alone causes loss of function. By extensive reversion studies of this sort, these authors came to the conclusion that the interacting amino acid sites (as in positions 210 and 174) are close to each other in the folded protein.

Now, we can envisage two sites, A and B, assumed in our model in the previous section (Tables III and IV) as representing two amino acid sites within a protein (Fig. 5). These two codon positions, being in the same cistron, must be very tightly linked, approximating the complete linkage assumed in our model of Table IV.

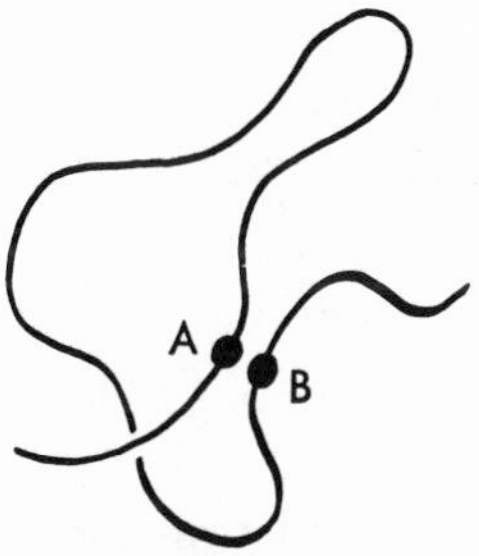


Fig. 5. A diagram illustrating two interacting amino acid sites (A and B) within a folded protein whose evolutionary pattern may conform to the model of "compensatory neutral evolution" investigated in the previous subsection.

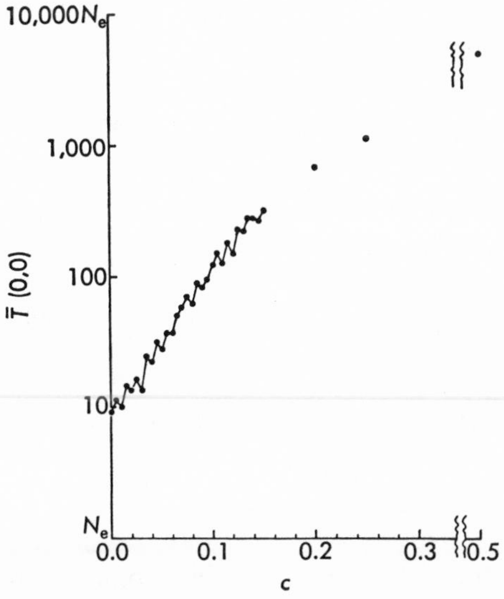


Fig. 6. The effect of crossing-over on retarding the average fixation time of individually deleterious but compensatory neutral mutants, as studied by Monte Carlo simulation experiments. Each dot represents the average of 50 replicate trials assuming a diploid population consisting of 100 breeding individuals ($N_e=100$) with mutation and selection parameters $v=0.005$, $s'=0.1$. The abscissa ($c$) stands for the recombination fraction between the two loci and the ordinate, $\bar{T}(0, 0)$, represents the average time until fixation of the double mutants $A'B'$ starting from the population consisting exclusively of wild type alleles at both loci ($AB$).

A problem which immediately arises is how tight the linkage must be for the model to be valid. To investigate this, a series of Monte Carlo simulation experiments were performed. The results are illustrated in Fig. 6, where the average fixation time $\bar{T}(0, 0)$ is plotted by dots for various values of recombination fraction $c$, assuming $N_e=100$, $s'=0.1$, $v=0.005$ so that $2N_ev=1$ and $4N_es'=40$. Each dot represents the average of 50 replicate trials.

In addition, an approximation method has been developed which can predict quite well the simulation results for a low crossing-over value ($c$). The approximation method is based on the premise that for a small $c$ such that

$$v \leqq c \ll s', \tag{26}$$

the effect of crossing-over is mainly to decrease the frequency of double mutant $A'B'$ through its crossing-over with the wild type $AB$. Thus we have, as a good approximation,

$$M_{\delta p_2} = vp_1 + s'p_1p_2 - p_0p_2c. \tag{27a}$$

This corresponds to Eq. (17a) in the case of no crossing-over. For the change of frequency of single mutants, the crossing-over contributes to increasing it by the amount $2cp_0p_2$, so that we have

$$M_{\delta p_1} = 2vp_0 - s'p_1(1-p_1) - vp_1 + 2cp_0p_2, \tag{27b}$$

which reduces to Eq. (17b) in the case of $c=0$. Then, the assumption of quasi-equilibrium, *i.e.*, $M_{\delta p_1}=0$, leads to the following expression

$$p_1 = \frac{(2v+2cp_2)(1-p_2)}{s'+3v}, \tag{28}$$

where a small term involving $cp_2$ has been omitted from the numerator. Substituting this in Eq. (27a) and writing $p$ for $p_2$, we get

$$M_{\delta p} = p(1-p)[s_2p + h_1(1-2p)] + v_1(1-p), \tag{29}$$

where

$$v_1 = \frac{2v^2}{s'+3v}, \tag{29a}$$

$$h_1 = \frac{cv + 2s'v - cs'}{s' + 3v} \tag{29b}$$

and

$$s_2 = \frac{2c^2 + 2cv + 4s'v}{s' + 3v}\,. \tag{29c}$$

Note that Eq. (29) is equivalent to Eq. (2). Then, Eqs. (7)–(9) are available to obtain the average time until joint fixation of $A'$ and $B'$, namely, we can obtain from these formulae the approximate value for $\bar{T}(0, 0)$ by putting $p=0$ and assuming $S=4N_e s_2=(2C^2+2CA+4BA)/(B+3A)$, $H=4N_e h_1=(CV+2BA-CB)/(B+3A)$ and $V=4N_e v_1=2A^2/(B+3A)$, in which $A=4N_e v$, $B=4N_e s'$, and $C=4N_e c$. In Fig. 7, the analytical results computed by using this approximation method are illustrated by a

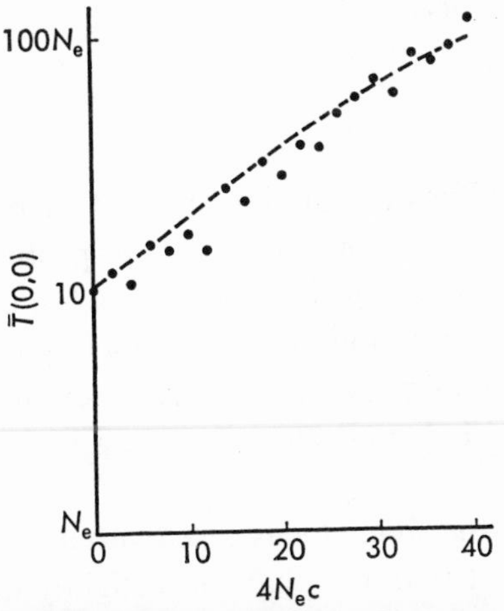


Fig. 7. An experimental check on the validity of the approximation theory that incorporates the effect of low rate crossing-over on retardation of the average fixation time of individually deleterious but compensatory neutral mutants. The broken curve represents the theoretical results giving the relationship between the average fixation time, $\bar{T}(0, 0)$, and $4N_e c$, where $N_e$ is the effective population size and $c$ is the recombination fraction between the two loci. Other parameters assumed are $2N_e v=1$ and $4N_e s'=40$, where $v$ is the mutation rate for the deleterious allele per locus per generation and $s'$ is the selection coefficient against the individual mutant. The dots represent the corresponding results taken from the series of Monte Carlo simulation experiments which are shown more extensively in the previous figure (Fig. 6).

broken curve for $A=4N_ev=2$, $B=4N_es'=40$ and assuming values $C=4N_ec=0$–$40$. In the same figure, the corresponding results from the Monte Carlo simulation experiments are plotted by dots. The agreement between the analytical and experimental results is excellent, suggesting the validity of the approximation theory employed.

Using this theory, I have computed the value of $4N_ec$ which retards the average fixation time by a factor of two for a few values of $4N_es'$ (still assuming $2N_ev=1$). It turned out this is $4N_ec\approx 5$ for $4N_es'=400$ and $4N_ec\approx 6.3$ for $4N_es'=100$. Although more study is needed, it looks as if the intra-cistronic crossing-over of $10^{-4}$ or less per generation does not spoil the effect of the present model.

Finally, if the present model of compensatory neutral evolution is realistic, it will not only lend support to Ohta's concept (*17*, *18*) of very slightly deleterious mutations, but will also help us to understand some "non-random" evolutionary amino acid substitutions under the framework of the neutral theory.

## SUMMARY

The use of the diffusion equation method in population genetics is demonstrated through its application to the problem of the average time until fixation of a mutant gene or genes in a finite population under continued (irreversible) mutation pressure. I mainly investigate the situation in which the initial population consists exclusively of the wild type allele (or alleles). I denote by $v$ the mutation rate per locus per generation.

The treatment for the single locus case (with a pair of alleles $A$ and $A'$) is straightforward. It is shown that for a slightly deleterious mutant the average time taken for fixation is too long to be of practical significance in evolution unless $4N_es'<10$, where $N_e$ is the effective population size and $s'$ is the selection coefficient against the mutant allele ($A'$).

For the two locus case (with a pair of alleles $A$ and $A'$ at the first locus, and $B$ and $B'$ at the second), I investigate the situation in which mutations are individually deleterious but become harmless (*i.e.*, selectively neutral) in combination: I assign relative fitnesses 1, $1-s'$, $1-s'$

and 1 respectively to $AB$, $A'B$, $AB'$, and $A'B'$. Two extreme cases, i) free recombination and ii) complete linkage between the loci are considered. Assuming $2N_e v=1$, where $v$ is the mutation rate per locus, the following results are obtained.

i) In the case of free recombination, the average time until fixation ($\bar{T}$) is about $5N_e$ generations for neutral mutations ($s'=0$). For slightly deleterious mutations in the range $0<4N_e s'\leqq 10$, $\bar{T}$ is slightly shorter but not very much (*e.g.*, $\bar{T}=4.5N_e$ for $4N_e s'=10$). If $4N_e s'$ is much larger, the fixation time quickly becomes very large, and for mutations with $4N_e s'>40$, $\bar{T}$ becomes so enormous that such mutations are unlikely to play a part in evolution.

ii) In the case of complete linkage, single mutations with much larger deleterious effects are allowed to participate in "compensatory neutral evolution," namely, joint fixation of the selectively neutral double mutant ($A'B'$) occurs without having to wait an unrealistically long time. In fact, even for $4N_e s'=400$, the average fixation time is only 10 times as long as the completely neutral case. The bearing of this finding on molecular evolution is discussed with special reference to coupled substitutions at interacting amino acid (or nucleotide) sites within a folded protein (or RNA) molecule.

*Acknowledgments*

I thank Dr. Tomoko Ohta for enlightening discussions in the course of this work and much help in carrying out simulation experiments. Thanks are also due to Dr. Tomitake Tsukihara for stimulating discussions, and to Dr. Kenichi Aoki for having gone over the manuscript to suggest improved presentation.

REFERENCES

1 Kimura, M. (1971) *Theor. Pop. Biol.* **2**, 174–208.
2 Kimura, M. and Ohta, T. (1971) *Theoretical Aspects of Population Genetics*, Princeton Univ. Press, Princeton.
3 Nei, M. (1975) *Molecular Population Genetics and Evolution*, North-Holland, Amsterdam.
4 Kimura, M. (1983) *The Neutral Theory of Molecular Evolution*, Cambridge Univ. Press, Cambridge.

5 Kimura, M. (1964) *J. Appl. Probabl.* **1**, 177–232.
6 Fisher, R.A. (1922) *Proc. Roy. Soc. Edinburgh* **42**, 321–341.
7 Fisher, R.A. (1930) *Proc. Roy. Soc. Edinburgh* **50**, 205–220.
8 Wright, S. (1931) *Genetics* **16**, 97–159.
9 Wright, S. (1945) *Proc. Natl. Acad. Sci. U.S.* **31**, 382–389.
10 Wright, S. (1949) In *Genetics, Paleontology, and Evolution* (Jepsen, G.L. *et al.*, eds.), pp. 365–389, Princeton Univ. Press, Princeton.
11 Kunisawa, K. (1951) *Modern Probability Theory*, Iwanami Shoten, Tokyo (in Japanese).
12 Crow, J.F. and Kimura, M. (1970) *An Introduction to Population Genetics Theory*, Harper & Row, New York.
13 Kimura, M. (1980) *Proc. Natl. Acad. Sci. U.S.* **77**, 522–526.
14 Li, W.-H. and Nei, M. (1977) *Genetics* **86**, 901–914.
15 Ortega, J.M. and Poole, W.G., Jr. (1981) *An Introduction to Numerical Methods for Differential Equations*, Pitman Pub., Marshfield, Mass.
16 Watterson, G.A. (1983) *Genetics* **105**, 745–766.
17 Ohta, T. (1973) *Nature* **246**, 96–98.
18 Ohta, T. (1974) *Nature* **252**, 351–354.
19 Wyckoff, H.W. (1968) *Brookhaven Symp. Biol.* **21**, 252–257.
20 Tsukihara, T., Kobayashi, M., Nakamura, M., Katsube, Y., Fukuyama, K., Hase, T., Wada, K., and Matsubara, H. (1982) *Biosystems* **15**, 243–257.
21 Yanofsky, C., Horn, V., and Thorpe, D. (1964) *Science* **146**, 1593–1594.

PART TWELVE

# Age of Alleles and Reversibility

## Introduction

Maruyama and Kimura (no. 32) discovered a general procedure for finding sojourn times and such quantities as the total number of heterozygotes during the time the allele is moving toward fixation. This is a technical paper showing how to obtain an arbitrary moment of the sojourn time and related quantities in a one-dimensional diffusion equation with exit boundaries. Two events of gene frequency change, fixation and extinction, are distinguished.

Fisher (1930) argued that the number of individuals having a gene derived from a single neutral mutation cannot greatly exceed the number of generations since its occurrence. Such a retrospective view is important and useful in understanding the nature of extant genetic variations. Kimura and Ohta (no. 33) took it up for the first time and opened up the whole topic of retrospective properties of stochastic processes arising in population genetics (Ewens 1979, p. 112). With a wealth of DNA sequence data that have been accumulating since 1980, much of the theory has been concerned with the ancestral history of genes in a phylogeny. One such theory, the coalescent process of Kingman (1982a,b), has become widely used to understand DNA variations and test the neutral theory (Hudson 1983; Hudson, Kreitman, and Aguadé 1987; Tajima 1989b).

Maruyama and Kimura (no. 34) pointed out an interesting property related to reversibility (identity of prospective and retrospective behaviors) of stochastic process arising in population genetics. Although the property was recognized in mathematics (Dynkin 1965), the application to genetics was new. It is counterintuitive, but true, that the average time required for an advantageous mutant allele to increase its frequency on the way to fixation is equal to that for the same allele to decrease its frequency on the way to extinction against selection. The reversibility argument was extended by Sawyer (1977) to the case where there is no stationary distribution.

An extension of an earlier work on the age of a mutant allele was made so as to compute any moment of the sum of an arbitrary function of the gene frequency along sample paths between two specified frequencies (Maruyama and Kimura, no. 35). An application of Nagylaki's formula (1974b) to the moments led them to a set of second order differential and difference equations (see also Maruyama 1977).

# SOME METHODS FOR TREATING CONTINUOUS STOCHASTIC PROCESSES IN POPULATION GENETICS[1]

TAKEO MARUYAMA AND MOTOO KIMURA

National Institute of Genetics, Misima 411

*Received October 21, 1971*

The theory of stochastic processes of the change of gene frequencies in finite populations is occupying an increasingly important place in the mathematical theory of population genetics. It is particularly important for our consideration of the mechanism of evolution at the molecular level (Kimura 1971). The most powerful method so far developed for treating such processes is the method of diffusion equations or "diffusion models" (Kimura 1964; see also Wright 1969, chapters 13 and 14; Crow and Kimura 1970, chapters 8 and 9).

In the present paper, we consider a single locus with a pair of alleles $A$ and $A'$ segregating in a finite Mendelian population. We shall denote the frequency of $A$ by $x$ and that of its allele ($A'$) by $1-x$. We assume no mutations so that $x=0$ and $x=1$ act as "absorbing barriers". In what follows, we regard the stochastic process in the change of gene frequency as a collection of sample paths $\{\omega\}$ and denote by $x(\omega, t)$ the position of a particular path $\omega$ at time $t$. Let $x(\omega, 0) \equiv p$ be the position where the path starts at time 0. We denote by $\tau(\omega)$ the time when the path $\omega$ exits from the interval $(0, 1)$. Note that the exit time $\tau(\omega)$ depends on the path $\omega$. Since we assume no mutations, $\tau(\omega)$ is always finite, and $x(\omega, \tau(\omega))=0$ or 1, depending on the exit.

Now, consider an arbitrary function $f(x)$, and let

$$F^{(n)}(p) \equiv E\left\{\left[\int_0^{\tau(\omega)} f(x(\omega, t))\, dt\right]^n\right\}, \tag{1}$$

where $E\{\ \}$ indicates the expectation with respect to the paths. Then $F^{(n)}(p)$ represents the expectation of the $n$-th moment of the sum of the quantity given by $f(x)$ from $t=0$ until the exit time, given that the initial gene frequency is $p$. For example, if we want to measure the total heterozygosity, we may put $f(x)=2x(1-x)$. This problem was investigated by Kimura (1969) for $n=1$. In the expression for $F^{(n)}(p)$ given by (1), the sum contains two kinds of paths, one going to $x=0$ and the other to $x=1$. We may however want to distinguish them and calculate such quantity only for those paths going to one destination, say, $x=1$. A typical example is the time until fixation as first

1) Contribution No. 855 from the National Institute of Genetics, Misima, Shizuoka-ken 411, Japan. Aided in part by a Grant-in-Aid from the Ministry of Education, Japan.

worked out by Kimura and Ohta (1969). Now let

$$F_1^{(n)}(p) \equiv u(p)E\left\{\left[\int_0^{\tau(\omega)} f(x(\omega, t))dt\right]^n \middle| x(\omega, \tau(\omega))=1\right\}, \tag{2}$$

where $u(p)$ is the probability of ultimate fixation, that is, the probability of the path reaching $x=1$. The quantity of our particular interest is $F_1^{(n)}(p)/u(p)$ which represents the $n$-th moment of the sum of $f(x)$ under the condition that the path's ultimate destination is the fixation. The crucial point in the present treatment is that $F^{(n)}(p)$ and $F_1^{(n)}(p)$ satisfy appropriate differential equations to be discussed below. First, consider $F^{(1)}(p)$. It can be shown that it satisfies

$$LF^{(1)}(p)+f(p)=0, \tag{3}$$

where

$$L=\frac{V_{\delta p}}{2}\frac{d^2}{dp^2}+M_{\delta p}\frac{d}{dp} \tag{4}$$

is a second order differential operator, and $M_{\delta p}$ and $V_{\delta p}$ are respectively the mean and variance of the change of gene frequency $p$ per unit time (generation). In a typical situation, if $A$ has selective advantage $s$ in homozygotes and $sh$ in heterozygotes over its allele $A'$, $M_{\delta p}=sp(1-p)\{h+(1-2h)p\}$, and if the "variance" effective size of the population is $N_e$, $V_{\delta p}=p(1-p)/(2N_e)$. The appropriate boundary conditions for equation (3) are $F^{(1)}(0)=F^{(1)}(1)=0$.

Similarly, it can be shown that $F^{(n)}(p)$ for $n\geqq 2$ satisfies the differential equation

$$LF^{(n)}(p)+nf(p)F^{(n-1)}(p)=0 \tag{5}$$

with boundary conditions $F^{(n)}(0)=F^{(n)}(1)=0$. A rigorous proof of these theorems can be found in Dynkin (1965, chapter 13). Second, consider $F_1^{(1)}(p)$ of (2). A slight modification of Dynkin's method proves that $F_1^{(1)}(p)$ satisfies

$$LF_1^{(1)}(p)+u(p)f(p)=0 \tag{6}$$

with $F_1^{(1)}(0)=F_1^{(1)}(1)=0$, and more generally $F_1^{(n)}(p)$, $n\geqq 2$, satisfies

$$LF_1^{(n)}(p)+nf(p)F_1^{(n-1)}(p)=0, \tag{7}$$

with boundary conditions $F_1^{(n)}(0)=F_1^{(n)}(1)=0$. Finally, the fixation probability $u(p)$ satisfies

$$Lu(p)=0 \tag{8}$$

with boundary conditions $u(0)=0$, $u(1)=1$ (cf. Kimura 1962).

In general, we are concerned with an ordinary second order differential equation of the form

$$LY(p)+K(p)=0, \tag{9}$$

where $K(p)$ is a given function of $p$ and $Y(p)$ satisfies the boundary conditions

$$Y(0)=Y(1)=0. \tag{10}$$

Equation (9) can readily be integrated and we have

$$Y(p)=\{1-u(p)\}\int_0^p \psi_K(\xi)\,u(\xi)\,d\xi+u(p)\int_p^1 \psi_K(\xi)\{1-u(\xi)\}d\xi\,, \tag{11}$$

where

$$\psi_K(\xi)=2K(\xi)\int_0^1 G(\lambda)\,d\lambda\Big/\{V_{\delta\xi}G(\xi)\}\,,$$

$$u(p)=\int_0^p G(\xi)\,d\xi\Big/\int_0^1 G(\xi)\,d\xi$$

and

$$G(\xi)=\exp\left\{-2\int_0^\xi (M_{\delta x}/V_{\delta x})\,dx\right\}.$$

As an example, let us calculate the sum of heterozygosity for the paths leading to fixation (but excluding the paths leading to loss), which we denote by $H_1^{(1)}(p)$. This corresponds to $K(p)=u(p)f(p)=u(p)2p(1-p)$. To simplify calculations, we shall assume that $A$ is selectively neutral ($s=0$). Then, $M_{\delta\xi}=0$ and $V_{\delta\xi}=\xi(1-\xi)/(2N_e)$ so that $G(\xi)=1$, $u(p)=p$, and $\psi_K(\xi)=8N_e\xi$. Thus, from (11), we obtain $H_1^{(1)}(p)=F_1^{(1)}(p)/p=(4N_e/3)(1-p^2)$. Similarly, the expectation of the second moment of the sum of heterozygosity from $x=p$ until $x=1$ (denoted by $H_1^{(2)}(p)$) can be calculated. To check the validity of the formulae for $H_1^{(1)}(p)$ and $H_1^{(2)}(p)$, we performed Monte Carlo experiments using the computer TOSBAC 3400, and the agreement between theoretical predictions and the experimental results was satisfactory.

The above analyses can be generalized as follows. By choosing two arbitrary values $x_0<x_1$ in (0, 1) we ask what will be the total sum of a certain quantity before a path reaches either $x_0$ or $x_1$ for the first time. The answer to this problem is given by the solution of equation (3) and (5), with boundary conditions $F(x_0)=F(x_1)=0$. Similarly, if we want to measure the quantity for those paths which reach $x_1$ before $x_0$ is reached, then the solution is given by equations (6) and (7), but with boundary conditions $F(x_0)=F(x_1)=0$. In addition the *first exit* probability through $x_1$ is given by the solution of (8) with conditions $u(x_0)=0$, $u(x_1)=1$.

Finally, if we put $f(x)=\delta(x-y)$, where $\delta(\cdot)$ is the Dirac delta function, then, the unique solution, denoted by $\Phi(p, y)$ of equation (3) with boundary conditions $\Phi(0, y)=\Phi(1, y)=0$ gives the total sojourn time at $y$ for the paths starting from $p$. This is equivalent to "the transient function" $t(x)$ of Ewens (1969), and also to the steady flux distribution with $\nu_m=1$ given by Kimura (1969). Similarly, let $\Phi_1(p, y)$ be the sojourn time at $y$ of the paths starting from $p$ and going to fixation (excluding the paths leading to extinction). Then, this is given by $F_1^{(1)}(p)/u(p)$ where $F_1^{(1)}(p)$ is the solution of (6) with $f(p)=\delta(p-y)$. Thus, using (11), we obtain

$$\left.\begin{aligned}\Phi_1(p, y)&=2u(y)\{1-u(y)\}/\{V_{\delta y}u'(y)\}\,, &&(y\geq p)\,,\\ \Phi_1(p, y)&=2\{1-u(p)\}u^2(y)/\{u(p)V_{\delta y}u'(y)\}\,, &&(y<p)\,,\end{aligned}\right\} \tag{12}$$

where $u'(y)=du(y)/dy$. In a special but important case in which a single mutant gene is destined to eventual fixation, the average number of generations which the mutant allele spends in the interval between $x$ and $x+dx$ before fixation is given by $\Phi_1(x)dx$, where

$$\Phi_1(x)=2u(x)\{1-u(x)\}/\{V_{\delta x}u'(x)\}, \qquad (x\geqq 1/2N). \tag{13}$$

For a neutral mutant, this reduces to $\Phi_1(x)=4N_e$, showing that the mutant spends on the average $2N_e/N$ generations at each frequency class until fixation in the population, where $N$ is the actual size of the population.

The present method can also be extended to obtain the moment generating functions. For example, let

$$\phi(\lambda, p)\equiv\sum_{n=0}^{\infty}(\lambda^n/n!)F_1^{(n)}(p), \text{ then } \phi \text{ satisfies}$$

$$L\phi(\lambda, p)+\lambda\phi(\lambda, p)f(p)=0$$

with condition $F_1^{(0)}(p)=u(p)$. In the special case of a selectively neutral mutant, we can obtain the moment generating function of the probability distribution of the time until fixation by putting $L=\{p(1-p)/4N_e\}d^2/dp^2$, $f(p)=1$ and $u(p)=p$ in these equations.

## LITERATURE CITED

Crow, J. F., and M. Kimura, 1970 An Introduction to Population Genetics Theory. Harper and Row, New York.

Dynkin, E. B., 1965 Markov Processes. Springer-Verlag, Berlin.

Ewens, W. J., 1969 Population Genetics. Methuen, London.

Kimura, M., 1962 On the probabillty of fixation of mutant genes in a population. Genetics **47**: 713–719.

Kimura, M., 1964 Diffusion models in population genetics. J. Appl. Probability **1**: 177–232.

Kimura, M., 1969 The number of heterozygous nucleotide sites maintained in a finite population due to steady flux of mutations. Genetics **61**: 893–903.

Kimura, M., 1971 Theoretical foundation of population genetics at the molecular level. Theoretical Population Biology **2**: 174–208.

Kimura, M., and T. Ohta, 1969 The average number of generations until fixation of a mutant gene in a finite population. Genetics **61**: 763–771.

Wright, S., 1969 Evolution and the Genetics of Populations. Vol. 2. The Theory of Gene Frequencies. University of Chicago Press, Chicago.

# THE AGE OF A NEUTRAL MUTANT PERSISTING IN A FINITE POPULATION*

MOTOO KIMURA AND TOMOKO OHTA

*National Institute of Genetics, Mishima, Japan*

Manuscript received May 11, 1972
Revised copy received February 16, 1973

ABSTRACT

Formulae for the mean and the mean square age of a neutral allele which is segregating with frequency $x$ in a population of effective size $N_e$ have been obtained using the diffusion equation method, for the case of $4N_e v < 1$ where $v$ is the mutation rate. It has been shown that the average ages of neutral alleles, even if their frequencies are relatively low, are quite old. For example, a neutral mutant whose current frequency is 10% has the expected age roughly equal to the effective population size $N_e$ and the standard deviation $1.4N_e$ (in generations), assuming that this mutant has increased by random drift from a very low frequency. Also, formulae for the mean "first arrival time" of a neutral mutant to a certain frequency $x$ have been presented. In addition, a new, approximate method has been developed which enables us to obtain the condition under which frequencies of "rare" polymorphic alleles among local populations are expected to be uniform if the alleles are selectively neutral. ——It was concluded that exchange of only a few individuals on the average between adjacent colonies per generation is enough to bring about such a uniformity of frequencies.

IN one of our previous papers (KIMURA and OHTA 1969a), we presented a theory on the average number of generations until a mutant gene becomes fixed in a finite population (excluding the cases of loss). The theory can be extended, as outlined in MARUYAMA and KIMURA (1971), to obtain the average number of generations until a mutant gene reaches a certain frequency for the first time starting from a lower frequency (i.e., the mean first arrival time). We need such a theory when we try to understand the evolutionary process consisting of a sequence of mutant substitutions in each of which an originally rare mutant increases its frequency and finally reaches fixation in the population.

On the other hand, in order to understand the nature of extant variations, we need to know the ages of mutant alleles within a population. In other words, we have to consider the problem of how many generations a mutant allele has persisted in the population since it appeared by mutation.

In the present paper, we present a solution to this problem for the case of selectively neutral alleles using the method of diffusion equations. It will be shown that the expected age of such a mutant is quite old; it is much older than one might suppose on the common-sense ground. We shall also discuss the bearing

* Contribution No. 887 from the National Institute of Genetics, Mishima, Shizuoka-ken, 411 Japan.

Genetics **75:** 199–212 September, 1973.

of the present finding on our neutral mutation-random drift hypothesis of molecular polymorphisms (KIMURA and OHTA 1971).

## BASIC THEORY

We use the diffusion model (KIMURA 1964; see also CROW and KIMURA 1970, p. 371), and denote by $\phi(p, x; t)$ the probability density that the frequency of the mutant allele becomes $x$ at time $t$ ($t$-th generation) given that it is $p$ at the start ($t = 0$). We first consider the case in which mutations are so rare that further mutations can be neglected. Such a treatment should be realistic if we consider mutants at the molecular level, that is, at each nucleotide site. If the mutant is selectively neutral and if the "variance effective size" of the population is $N_e$, then the transition probability density satisfies the partial differential equation

$$\frac{\partial\phi(p,x;t)}{\partial t} = \frac{1}{4N_e}\frac{\partial^2}{\partial x^2}\{x(1-x)\phi(p,x;t)\} \tag{1}$$

with the initial condition $\phi(p,x;0) = \delta(x-p)$, where $\delta(\cdot)$ stands for Dirac's delta function.

Our main aim is to evaluate the mean and the variance of the time interval in generations since an allele which now has intermediate frequency $x$ had a lower frequency $p$.

Let

$$T_i = \int_0^\infty t^i\phi(p,x;t)\,dt \tag{2}$$

be the $i$-th moment ($i = 0, 1, 2, \ldots$) of $t$, then the mean time interval is given by

$$\bar{t}(p, x) = T_1/T_0, \tag{3}$$

while the mean square time is given by

$$\overline{t^2(p, x)} = T_2/T_0, \tag{4}$$

from which the variance can readily be obtained. We first derive an equation for $T_0$ as follows: Integrating both sides of equation (1) with respect to $t$ from $t = 0$ to $t = \infty$, we obtain

$$\int_0^\infty \frac{\partial\phi}{\partial t}dt = \frac{1}{4N_e}\frac{\partial^2}{\partial x^2}\{x(1-x)\int_0^\infty \phi dt\}, \tag{5}$$

where $\phi$ stands for $\phi(p,x;t)$. This yields

$$\phi(p,x;\infty) - \phi(p,x;0) = \frac{1}{4N_e}\frac{d^2}{dx^2}\{x(1-x)T_0\},$$

but the two terms on the left-hand side vanish if we assume $1 > x > p \geqq 0$, because $\phi$ is asymptotically proportional to $\exp\{-t/(2N_e)\}$ for a large $t$ (cf.

KIMURA 1955) while it is equal to $\delta(x-p)$ for $t=0$. Then, by integrating twice the resulting equation, i. e.,

$$\frac{1}{4N_e}\frac{d^2}{dx^2}\{x(1-x)T_0\}=0,$$

with respect to $x$, we obtain

$$T_0=\frac{4N_e}{x(1-x)}\{(C_1-1)x+C_2\}, \tag{6}$$

where $C_1$ and $C_2$ are constants. These constants can be determined from the consideration that as $x$ approaches unity, $T_0$, as defined by (2), must approach $4N_ep$, because $\phi(p,1;t)/(4N_e)$ represents the amount of fixation during the $t$-th generation, and the sum of this quantity over all generations must be equal to $p$, the probability of ultimate fixation. This leads to $C_1-1=-C_2$ and $C_2=p$.

Thus we obtain

$$T_0=4N_ep/x. \tag{7}$$

The equations for $T_1$ and $T_2$ can be obtained in a similar way, so we shall derive a general equation for $T_i$ $(i=1,2,\ldots)$.

Multiplying $t^i$ to both sides of equation (1) and then integrating them with respect to $t$ from $t=0$ to $\infty$, we obtain

$$\int_0^\infty t^i\frac{\partial\phi}{\partial t}dt=\frac{1}{4N_e}\frac{\partial^2}{\partial x^2}\{x(1-x)\int_0^\infty t^i\phi dt\}. \tag{8}$$

The left-hand side of this equation yields

$$\Big[\,t^i\phi\,\Big]_{t=0}^{t=\infty}-i\int_0^\infty t^{i-1}\phi dt,$$

the first term of which vanishes because $t^i\phi$ vanishes both at $t=0$ and $t=\infty$. Thus we obtain the ordinary differential equation for $T_i$ as follows:

$$\frac{1}{4N_e}\frac{d^2}{dx^2}\{x(1-x)T_i\}+iT_{i-1}=0, \tag{9}$$

where $i\geqq 1$.

In the special case of $i=1$, by putting $T_0=4N_ep/x$, equation (9) reduces to

$$\frac{1}{4N_e}\frac{d^2}{dx^2}\{x(1-x)T_1\}+\frac{4N_ep}{x}=0. \tag{10}$$

Then, integrating this equation twice with respect to $x$, we obtain

$$T_1=\frac{4N_e}{x(1-x)}\{(C_1+4N_ep)x-4N_epx\log_ex+C_2\},$$

where $C_1$ and $C_2$ are constants. In determining these constants, we note that as $x$ approaches unity, $T_1/T_0$ should approach

$$\bar{t}_1(p)=-\frac{1}{p}\{4N_e(1-p)\log_e(1-p)\},$$

the average number of generations until fixation (KIMURA and OHTA 1969a). This leads to

$$C_1 + 4N_e p = -C_2 = 4N_e\{(1-p)\log_e(1-p) + p\}.$$

Thus we obtain the formula for the mean time interval (in generations)

$$\bar{t}(p,x) \equiv \frac{T_1}{T_0} = 4N_e\{-\frac{1-p}{p}\log_e(1-p) - \frac{x}{1-x}\log_e x - 1\}. \quad (11)$$

Note that this is different from the mean first arrival time which we denote by $\bar{t}_x(p)$ and on which we later present a formula in the discussion. Whereas $\bar{t}_x(p)$ represents the average number of generations until a mutant allele happens to reach a certain frequency $x$ for the first time starting from a lower frequency $p$, $\bar{t}(p,x)$ represents the average number of generations which an allele having frequency $x$ at present has persisted in the population since it had a lower frequency $p$ in the past.

Similarly, we can obtain the following formula for the mean square age by solving equation (9) for the case of $i = 2$ under the condition that as $x$ approaches unity $T_2/T_0$ should approach the mean square time until fixation as given by KIMURA and OHTA (1969b).

$$\begin{aligned}\overline{t^2(p,x)} = \frac{32N_e^2}{p}\{[(1-p)\log_e(1-p) + 2p]\frac{x\log_e x}{1-x} \\ - p\int_0^x \frac{\log_e z}{1-z}dz + 2(1-p)\log_e(1-p) \\ + 2p + p\int_{1-p}^1 \frac{\log_e z}{1-z}dz\}, \quad (x > p). \quad (12)\end{aligned}$$

The variance of the age is then given by

$$\sigma_t^2(p,x) = \overline{t^2(p,x)} - \{\bar{t}(p,x)\}^2.$$

For a mutant allele which is represented only once at the moment of appearance in a population consisting of $N$ individuals, we may put $p = 1/(2N)$ in the above formulae. As $N$ gets large, $p$ approaches zero, and at the limit we have

$$\bar{t}(0,x) = 4N_e\left(-\frac{x}{1-x}\log_e x\right) \quad (13)$$

and

$$\overline{t^2(0,x)} = 32N_e^2\left(\frac{x}{1-x}\log_e x - \int_0^x \frac{\log_e z}{1-z}dz\right) \quad (14)$$

respectively for the mean and the mean square ages. These formulae should be valid for molecular mutants which are selectively neutral and which are subject to random frequency drift in a large population.

However, when we consider each gene locus (cistron) as our basic unit, rather than each nucleotide site, we must take into account the possibility that further mutation occurs before the mutant allele reaches a high frequency. In addition there is the possibility that a mutant allele that has once become fixed in the population eventually has its frequency decreased by further mutation in conjunction with random drift. Also it is possible if the mutation rate is sufficiently high that an allele can never reach complete fixation.

Although the complete treatment taking all these possibilities into account is difficult, we have worked out the average age of the mutant for the case $4N_ev < 1$, where $v$ is the mutation rate per locus per generation. Following KIMURA and CROW (1964), we assume that the number of possible allelic states per locus is so large that whenever mutation occurs it leads to a new (not a pre-existing) allele. The treatment using this model is more complicated than that without mutation, so that we shall only summarize the results.

If $4N_ev < 1$, it can be shown that the probability of a mutant allele's reaching fixation $(x = 1)$ is

$$u(p) = 1 - (1-p)^{1-4N_ev},$$

disregarding the possibility that its frequency later decreases by mutation and random drift. It can also be shown that for $4N_ev < 1$, the average age of a mutant having current frequency $x$ is

$$\bar{t}_{\uparrow}(0, x) = \frac{4N_e}{1-4N_ev}\left\{ \log_e x + \int_0^1 \frac{1-(1-z)^{1-4N_ev}}{z}\,dz \right.$$
$$\left. + \frac{1}{(1-x)^{1-4N_ev}} \int_x^1 \frac{(1-z)^{1-4N_ev}}{z}\,dz \right\}. \tag{13a}$$

It is assumed that the mutant allele increased from a very low frequency sometime in the past rather than decreased from the fixed state. As $4N_ev$ approaches 0, equation (13a) reduces to (13), but in general, numerical integration will be required to compute $\bar{t}_{\uparrow}(0, x)$ from this equation.

On the other hand, if the allele in question decreased from the fixed state—that is, from a frequency of 100%—by mutation and random drift, rather than directly increasing from a low frequency, then the corresponding formula becomes

$$\bar{t}_{\downarrow}(1, x) = \frac{4N_e}{1-4N_ev}\left\{ \frac{(1-x)^{1-4N_ev}}{1-(1-x)^{1-4N_ev}} \int_0^x \frac{[1-(1-z)^{1-4N_ev}]^2}{(1-z)^{1-4N_ev}z}\,dz \right.$$
$$\left. + \int_x^1 \frac{1-(1-z)^{1-4N_ev}}{z}\,dz \right\}. \tag{13b}$$

This gives the expected "age" of the allele counted from $p = 1 - 1/(2N)$ to the present frequency $x$ assuming that $N$ is large. When $4N_ev = 0$, this agrees with $\bar{t}_{\uparrow}(0, 1-x)$, as it should. It does not include the length of time during which the allele remained fixed in the population before the frequency $1 - 1/(2N)$ was reached. In order to estimate this length of time (the waiting time), let $u_x(1 - 1/2N)$ be the probability that the allele frequency goes down to $x$ from $1 - 1/(2N)$ without previously going back to unity. Then, it can be shown that

$u_x(1 - 1/2N) = \{2N(1-x)\}^{4N_ev-1}$. Also, the probability that one mutant is produced during a short time interval of length $\Delta t$ in the population of fixed state is $2Nv\Delta t$. Combining these two probabilities, we find that for the allele whose current frequency is $x$ and which had once been fixed in the population, the average length of time during which it was fixed is

$$\bar{t}_{fix} = \frac{1}{2Nvu_x(1-1/2N)} = \frac{\{2N(1-x)\}^{1-4N_ev}}{2Nv}.$$

Note that this is the waiting time until a successful mutation first occurs in the population leading to the downward journey reaching $x$. When $4N_ev = 0$ and $x = 0$, this reduces to $\bar{t}_{fix} = 1/v$.

Finally, we can show that if $4N_ev \geqq 1$, the probability is zero that a new mutant reaches complete fixation (assuming that $N_e$ is large). In other words, complete fixation is prevented by the opposing mutation pressure.

Figure 1 illustrates for several values of $4N_ev$ the relationship between $x$ and $\bar{t}_{\uparrow}(0, x)$ with solid lines, and that between $x$ and $\bar{t}_{\downarrow}(1, x)$ with dotted lines. They were obtained by numerical integration of formulae (13a) and (13b) using a computer. Note that $4N_ev = 0.2$ corresponds to a heterozygosity of about 16%, the value observed in man and Drosophila. Note also that with this level of $4N_ev$, the expected age of the mutant is not much influenced by mutation.

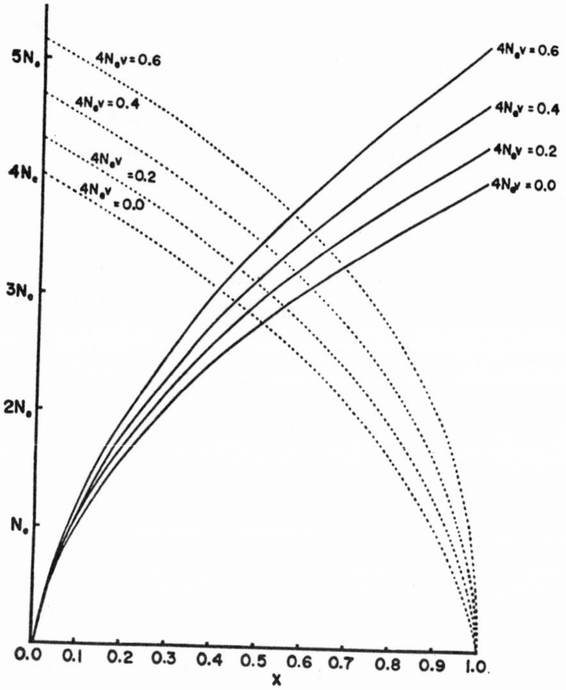


FIGURE 1.—Relationship between $x$ and $\bar{t}\uparrow(0, x)$ as shown by solid lines and that between $x$ and $\bar{t}_\downarrow(1, x)$ as shown by dotted lines for some values of $4N_ev$. The abscissae represent the gene frequencies and the ordinates represent the corresponding ages in generations.

## MONTE CARLO EXPERIMENTS

In order to check the validity of the above treatment (especially, formulae 13, 13a and 13b), we performed Monte Carlo experiments using TOSBAC 3400 in our institute. The procedure of the experiments follows the one used by HILL and ROBERTSON (1966). Assuming no selection, sampling of gametes is carried out by generating pseudo-random numbers that follow the uniform distribution (RAND 20 in TOSBAC 3400). Namely, if $x$ is the frequency of the mutant allele and if a random number is less than $x$, one mutant gamete is sampled, while if it is larger than $x$, one gamete with normal allele is sampled. Sampling is repeated $2N_e$ times to obtain the total of gametes to form the next generation. Each experiment starts by having a mutant allele represented only once in the population and whenever loss or fixation of the mutant occurs, a new mutant is again supplied to continue the experiment. In each generation, the age and the frequency of the mutant allele are recorded.

Figure 2 illustrates the results of the experiments to check formula 13 for the frequency classes up to 0.1. The abscissa represents frequencies of mutant alleles and the ordinate, the corresponding ages. The curve in the figure shows the theoretical prediction for the mean age (assuming no mutation), while the two types of dots represent observed values; the square dots are for the case of $N_e = 100$ and the circular dots for $N_e = 200$. The observed values are the outcome of $10^4$ gen-

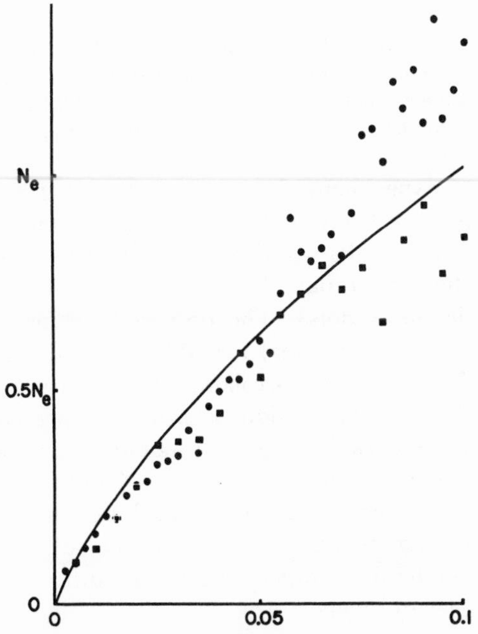


FIGURE 2.—Results of Monte Carlo experiments on the average age of new, neutral mutants. The abscissae represent frequencies of mutant alleles, while the ordinates represent the corresponding ages. In the figure, the curve gives the theoretical values, the square dots give the results for $N_e = 100$ and the circular dots the results for $N_e = 200$.

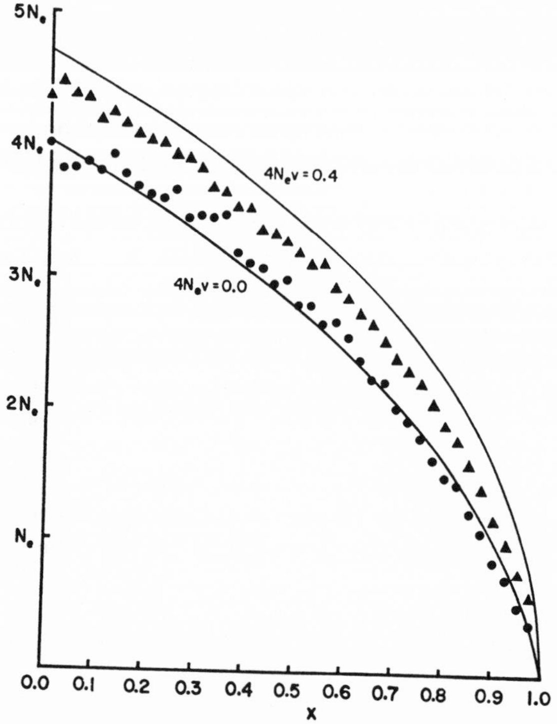


FIGURE 3.—Results of Monte Carlo experiments to check formula 13b for $4N_ev = 0$ and $4N_ev = 0.4$. Circular dots represent the outcome of 10,000 runs assuming $4N_ev = 0$, while triangular dots represent the outcome of 50,000 runs assuming $4N_ev = 0.4$. The abscissae represent the frequencies and the ordinates represent the corresponding ages. $N_e = 20$ is assumed.

erations of simulation experiments for the case of $N_e = 100$, but $3 \times 10^4$ generations for $N_e = 200$. As seen from the figure, the agreement between the theoretical and the experimental results is satisfactory. Figure 3 illustrates the results of the experiments to check formula 13b for the cases of $4N_ev = 0$ (circular dots) and $4N_ev = 0.4$ (triangular dots). The observed values are the outcomes of 10,000 runs (rather than generations) for $4N_ev = 0$ and 50,000 runs for $4N_ev = 0.4$, assuming $N_e = N = 20$. They corroborate the theoretical prediction that the expected age gets older under mutation pressure irrespective of the direction of mutation. We should remark here that systematic deviation of experimental results from theoretical prediction for $4N_ev = 0.4$ must be due to a very small population number, i.e., 20, employed in the simulation experiments; the pressure of mutation in preventing fixation does not become sufficiently effective unless the population number becomes much larger. The diffusion treatment is really adequate for such larger numbers.

## DISCUSSION

In the present paper we have derived, under the assumption of no further mutation, formula (11) which gives the average number of generations which a neutral allele segregating in a population with frequency $x$ has persisted in the population since it had the initial frequency $p$ ($< x$) in the past. In this general form, it also gives the average number of generations which a neutral allele takes when its frequency increases from $p$ to $x$ through random genetic drift ($p < x$). Therefore, $\bar{t}(0, x)$ given by formula 13 is appropriate to express the expected age of a neutral allele with frequency $x$, if the allele has increased its frequency in the population since it first appeared by mutation. On the other hand, if the allele in question has decreased from the previously fixed state, $\bar{t}_{\downarrow}(1, x)$ is appropriate to express its age, where $\bar{t}_{\downarrow}(1, x) = \bar{t}_{\uparrow}(0, 1-x)$ for $4N_e v = 0$. When we try to apply these formulae to actual situations, one difficulty that we encounter is that we cannot know which of these two alternative events has actually occurred. However, we can attach a probability statement to them (as pointed out to us by the referee). Namely, the probabilities of these two alternative events (assuming no further mutation) are $1 - x$ and $x$, respectively. This follows from the consideration that probability is $p/x$ that a mutant allele with initial frequency $p$ subsequently reaches a higher frequency $x$ before it is either lost from the population or fixed in it. Similarly the probability is $p/(1-x)$ that the frequency of the allele decreases to $x$.

We have also studied the effect of further mutation on the age of neutral alleles and have found that the effect is relatively minor if $4N_e v$ is small (see Figure 1).

These results should be compared with the mean first arrival time, that is, the average number of generations until a neutral allele reaches frequency $x$ for the first time starting from a lower frequency $p$. This is given by

$$\bar{t}_x(p) = 4N_e \left\{ \frac{1-x}{x} \log_e(1-x) - \frac{1-p}{p} \log_e(1-p) \right\}. \tag{15}$$

At the limit $p \to 0$, this reduces to

$$\bar{t}_x(0) = 4N_e \left\{ \frac{1-x}{x} \log_e(1-x) + 1 \right\}. \tag{16}$$

When $x$ is much smaller than unity, we have

$$\bar{t}_x(0) \approx 4N_e x. \tag{17}$$

Equation (15) is a special case of a more general equation (A1) which can hold when the mutant is selected as well as when it is neutral (see APPENDIX).

In Figure 4, the mean age $\bar{t}(0, x)$ and the mean first arrival time $\bar{t}_x(0)$ are plotted for frequencies up to 0.1. From the figure it may be seen that expected age of a neutral mutant whose current frequency is 10% is roughly equal to the effective population size, $N_e$, and it is about five times the corresponding first arrival time. The standard deviation of the age is roughly $1.4N_e$ which is slightly larger than the mean. Since the expected age of a neutral mutant whose frequency is 50% is about $2.8N_e$ generations (see Figure 1), this example suggests that even "rare" polymorphic alleles whose current frequencies are a few percent

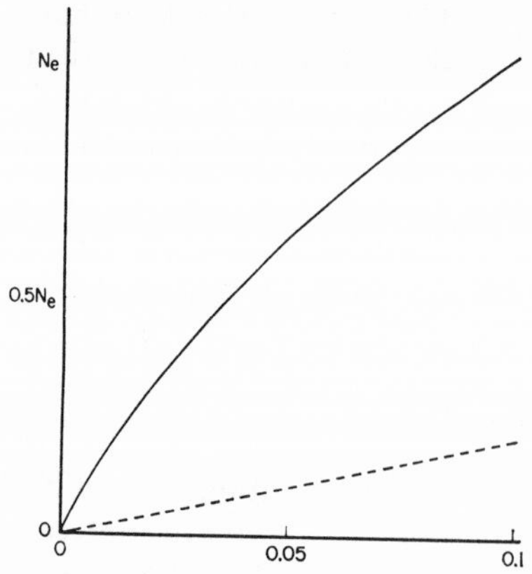


FIGURE 4.—Comparison between the average age (solid line) and the average first arrival time (broken line). The abscissae represent frequencies and the ordinates the ages or times in generations.

have quite old ages if they are neutral and if the population is very large. In fact, a neutral allele whose frequency is only 3.5% has the expected age of about half the population size, i.e. $0.5N_e$ if it has increased from a very low frequency. Furthermore, if this allele happens to be the remnant of a previously fixed allele, expected age becomes still larger. In general, we can make the following probability statement regarding these alternative events (as suggested to us by DR. ALAN ROBERTSON): If $4N_ev$ is small so that there are never more than two alleles segregating simultaneously, the probability that the allele with frequency $x$ has increased from a very small frequency is $1-x$, while that it has decreased from a previously fixed state is $x$. This means that if we combine these two alternative possibilities, the average age of the polymorphism with two alleles segregating with respective frequencies of $x$ and $1-x$ is

$$-4N_e\{x\log_e x + (1-x)\log_e(1-x)\}.$$

One additional property of neutral alleles which may be of interest from a mathematical standpoint is that the average age of a mutant allele having current frequency $x$ (assuming $p = 0$) is equal to the average time until extinction of the same allele (excluding the cases of its eventual fixation). This may be evident by comparing equation (13) of this paper with equation (16) of KIMURA and OHTA (1969a). The same applies to the mean square age, as may be seen by comparing equation (14) of this paper with equation (A7) of KIMURA and OHTA (1969b).

Let us now consider the bearing of the present findings on the spatial pattern of genetic variation. Here we are particularly concerned with the question: how much migration is required to keep the frequencies of a "rare" polymorphic allele essentially uniform among localities when the allele is selectively neutral? First, consider a one-dimensional habitat forming a circle of radius $r$. Let $N_T$ be the total number of breeding individuals in one generation and assume that they are distributed uniformly with density $\delta$ so that $N_T = 2\pi r\delta$. If we denote by $\sigma^2$ the mean square distance of individual migration (assumed to be isotropic) in one generation, then the distance of migration during $t$ generations should follow the normal distribution with mean zero and variance $\sigma^2{}_t = t\sigma^2$ when $t$ is large. If the abscissa of this distribution is wrapped around the circle of radius $2\pi r$, and if the resulting (superimposed) probability distribution on the circle is approximately uniform, then the frequencies among localities of a mutant allele having age $t$ will become essentially uniform. On the other hand, if the superimposed probability distribution on the circle markedly deviates from the uniform distribution, clear local differentiation of allelic frequencies should result. Figure 5 illustrates two such contrasting cases ($\pi r = \sigma_t$ and $\pi r = 3\sigma_t$) together with an intermediate case ($\pi r = 2\sigma_t$).

If we substitute $\bar{t}(0, x)$ for $t$ in the above reasoning and if we note that the superimposed distribution is essentially uniform when

$$\pi r \leqq \sigma_t\,, \tag{18}$$

then we obtain

$$N_T \leqq 2\delta\sigma\sqrt{\bar{t}(0, x)} \tag{19}$$

as a condition for the uniform distribution of allelic frequencies among localities.

Next, we consider a two-dimensional habitat extending over a sphere of radius $r$. Let us assume that the individuals are distributed uniformly with density $\delta$. Let $\sigma^2$ be the mean square distance of individual migration in one generation, and

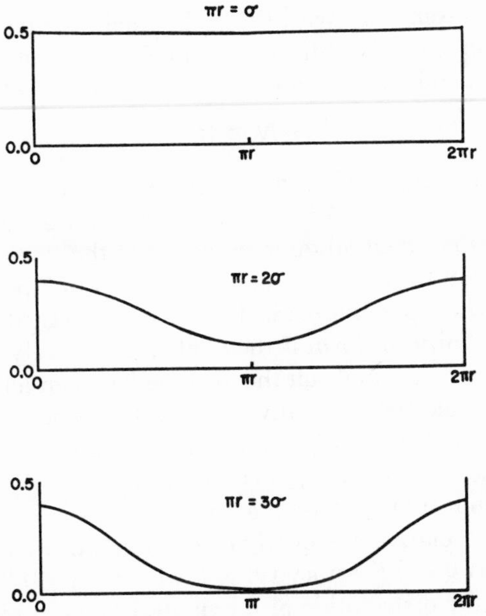


FIGURE 5.—Diagrams illustrating the frequency distribution produced by wrapping one-dimensional normal distribution around a circle of radius $r$ for three cases $\pi r = \sigma$, $2\sigma$ and $3\sigma$.

assume that the migration is isotropic. Wrapping the sphere by the two-dimensional normal distribution for the distance of migration in $t$ generations and considering the resulting probability distribution on the sphere similar to the one considered above, the condition (18) leads to

$$\pi N_T \leqq 4\delta\sigma^2 t \tag{20}$$

for the two-dimensional case, since $N_T = 4\pi r^2\delta$ in this case.

For a mutant allele whose average frequency in the whole population happens to be 0.1, the condition for uniform distribution reduces roughly to

$$\sigma^2\delta \geqq 1 \tag{21}$$

if we put $t = \bar{t}(0, 0.1) \approx N_e$ in (20).

If the total population is subdivided into colonies (subpopulations) in each of which the mating is at random, and migration in one generation is restricted to neighboring colonies (two-dimensional stepping stone model, cf. KIMURA and WEISS 1964), condition (21) becomes

$$mN \geqq 1, \tag{22}$$

where $m$ is the rate at which each colony exchanges individuals with four surrounding colonies each generation and $N$ is the effective size of each colony. However, since the age $t(0, x)$ at $x = 0.1$ has a relatively large standard deviation, it may be safer to use the first arrival time $\bar{t}_x\ (0)$ at $x = 0.1$ for $t$ in (20) to derive the required condition. This leads approximately to

$$mN \geqq 5, \tag{23}$$

These results agree essentially with those obtained by KIMURA and MARUYAMA (1971) based on an entirely different method. For a neutral mutant having $x = 0.05$, the corresponding formula becomes approximately

$$mN \geqq 10. \tag{24}$$

This means that uniform distribution of frequencies among colonies can be attained if exchange of at least 2.5 individuals occurs on the average between two neighboring colonies per generation irrespective of the size of each colony. Thus we conclude that when the average frequency of a "rare" polymorphic allele is a few percent in the whole population, its frequencies among different localities are expected to be essentially uniform if the allele is selectively neutral and if there is migration of a few individuals on the average between adjacent colonies each generation. It is possible that in many Drosophila species, with their enormous population size and with their high individual mobility in addition to the possibility of long range migration by airborne individuals, virtual panmixia are usually attained even if subpopulations are very far apart.

In addition, the associative overdominance at a neutral locus caused by linked selected loci (cf. OHTA and KIMURA 1970, 1971; OHTA 1971) will contribute at least partly to keep the distribution of the neutral alleles uniform among localities. The reason for this is that the associative overdominance creates a sort of inertia so that whenever a local frequency is temporarily disturbed it tends to go back to the original frequency, although there are no definite equilibrium frequencies for the neutral allele to settle on the long term basis (OHTA 1973).

PRAKASH, LEWONTIN and HUBBY (1969), in their studies on the pattern of genetic variation among subpopulations of *Drosophila pseudoobscura*, rejected the model of neutral isoalleles on the ground that frequencies of rarer alleles at several loci (such as malic dehydrogenase locus) are essentially identical among widely separated subpopulations, and that the isoallelic hypothesis cannot explain such identical allelic configurations. We hope that the above treatment has made it clear that their observations are not incompatible with our neutral mutation-random drift hypothesis of molecular polymorphisms.

Also we would like to point out that if the observed uniformity of the frequencies of rare alleles among localities is due to weak "balancing selection" rather than migration, the effective size of the local population (not the whole species) has to be probably at least the order of a million, not mentioning the fact that the selection coefficients have to be equal among localities. This is because, as first discovered by ROBERTSON (1962) for overdominant alleles, the balancing selection actually accelerates fixation rather than retards it if the equilibrium frequencies lie outside the range 0.2–0.8 unless $N_e\ (s_1 + s_2)$ is very large, where $s_1$ and $s_2$ are the selection coefficients against the two homozygotes. In fact, if the equilibrium frequency is 5%, $N_e\ (s_1 + s_2)$ has to be about 2100 in order to retard fixation by a factor of 100 as compared with the completely neutral case (see also CROW and KIMURA 1970, p. 414).

We would like to thank DR. ALAN ROBERTSON for constructive criticism.

## APPENDIX

### *A general formula for the average number of generations until a mutant allele first reaches a frequency x starting from a lower frequency p (the mean "first arrival time")*

Let $M_{\delta x}$ and $V_{\delta x}$ be, respectively, the mean and the variance of the change in one generation of the frequency of a mutant allele having frequency $x$. We assume that the stochastic process of change in gene frequency is time homogenous—in other words, the selection coefficient of the mutant remains constant with time even if it may be frequency-dependent.

Then it can be shown using the diffusion equation method that the average number of generations until the mutant reaches frequency $x$ for the first time starting a lower frequency $p$ is given by

$$\bar{t}_x(p) = \int_p^x \psi_x(\xi) u_x(\xi)\{1 - u_x(\xi)\} d\xi + \frac{1-u_x(p)}{u_x(p)} \int_0^p \psi_x(\xi) u_x^2(\xi) d\xi, \quad \text{(A1)}$$

where

$$\psi_x(\xi) = \frac{2\int_0^x G(\lambda)d\lambda}{V_{\delta\xi} G(\xi)}, \quad u_x(p) = \frac{\int_0^p G(\lambda)d\lambda}{\int_0^x G(\lambda)d\lambda}$$

and

$$G(\xi) = \exp\left\{ -\int_0^\xi \frac{2M_{\delta x}}{V_{\delta x}} dx \right\} .$$

When $x = 1$, $\bar{t}_x(p)$ reduces to $\bar{t}_1(p)$ (the average time until fixation) given by KIMURA and OHTA (1969a). Also, in the special case of no selection (neutral allele), $M_{\delta x} = 0$ and $V_{\delta x} = x(1-x)/(2N_e)$, so that (A1) reduces to

$$\bar{t}_x(p) = 4N_e\left\{ \frac{1-x}{x} \log_e(1-x) - \frac{1-p}{p} \log_e(1-p) \right\} , \qquad \text{(A2)}$$

where $N_e$ is the "variance" effective size of the population.

## LITERATURE CITED

CROW, J. F. and M. KIMURA, 1970 *An Introduction to Population Genetics Theory*. Harper and Row, New York.

HILL, W. G. and ALAN ROBERTSON, 1966 The effect of linkage on limits to artificial selection. Genet. Res. **8:** 269–294.

KIMURA, M., 1955 Solution of a process of random genetic drift with a continuous model. Proc. Nat. Acad. Sci. U.S. **41:** 144–150. ——, 1964 Diffusion models in population genetics. Jour. Applied Probability **1:** 177–232.

KIMURA, M. and J. F. CROW, 1964 The number of alleles that can be maintained in a finite population. Genetics **49:** 725–738.

KIMURA, M. and T. MARUYAMA, 1971 Pattern of neutral polymorphism in a geographically structured population. Genet. Res. **18:** 125–131.

KIMURA, M. and T. OHTA, 1969a The average number of generations until fixation of a mutant gene in a finite population. Genetics **61:** 763–771. ——, 1969b The average number of generations until extinction of an individual mutant gene in a finite population. Genetics **63:** 701–709. ——, 1971 Protein polymorphism as a phase of molecular evolution. Nature **229:** 467–469.

KIMURA, M. and G. H. WEISS, 1964 The stepping stone model of population structure and the decrease of genetic correlation with distance. Genetics **49:** 561–576.

MARUYAMA, T. and M. KIMURA, 1971 Some methods for treating continuous stochastic processes in population genetics. Japanese Jour. Genetics **46:** 407–410.

OHTA, T., 1971 Associative overdominance caused by linked detrimental mutations. Genet. Res. **18:** 277–286. ——, 1973 Effect of linkage on behavior of mutant genes in a finite population. Theoret. Pop. Biol. **4:** 145–162.

OHTA, T. and M. KIMURA, 1970 Development of associative over-dominance through linkage disequilibrium in finite populations. Genet. Res. **16:** 165–177. ——, 1971 Behavior of neutral mutants influenced by associated overdominant loci in finite populations. Genetics **69:** 247–260.

PRAKASH, S., R. C. LEWONTIN and J. L. HUBBY, 1969 A molecular approach to the study of genic heterozygosity in natural populations. IV. Patterns of genic variation in central, marginal and isolated populations of *Drosophila pseudoobscura*. Genetics **61:** 841–858.

ROBERTSON, A., 1962 Selection for heterozygotes in small populations. Genetics **47:** 1291–1300.

Corresponding Editor: R. ALLARD

# A NOTE ON THE SPEED OF GENE FREQUENCY CHANGES IN REVERSE DIRECTIONS IN A FINITE POPULATION[1]

TAKEO MARUYAMA AND MOTOO KIMURA
*National Institute of Genetics, Mishima, Japan*

Received March 6, 1973

When we consider evolutionary processes from the standpoint of population genetics, the probability and the time involved for a mutant gene to become fixed (i.e., to spread to the whole population) play important roles (cf., Kimura and Ohta, 1971). Following the earlier works of Haldane (1927), Fisher (1930) and Wright (1931), a general theory of the probability of gene fixation in finite populations was worked out by Kimura (1957, 1962), while the mean time required for fixation was first obtained by Kimura and Ohta (1969*a*) (see also Kimura and Ohta, 1969*b*; and Narain, 1970). For selectively neutral mutants the distribution of the fixation time has also been obtained (Kimura 1970). Furthermore, the "sojourn time" at each gene frequency class in the course of fixation was found by Maruyama and Kimura (1971).

What is crucial in these studies, is that we must distinguish two types of paths in the change of gene frequency, one leading to fixation and the other leading to extinction. The theory developed by us (Maruyama and Kimura, 1971) is general in that it enables us to calculate moments of arbitrary order of a quantity associated with a restricted class of paths, from the start of the process to the "first exit time". The purpose of the present note is to show, using this theory, that the sojourn (as well as the mean) times are the same for the two processes, one going from a low initial frequency to fixation and the other going from very high initial frequency to extinction

[1] Contribution No. 926 from the National Institute of Genetics, Mishima, Shizuoka-ken, 411 Japan.

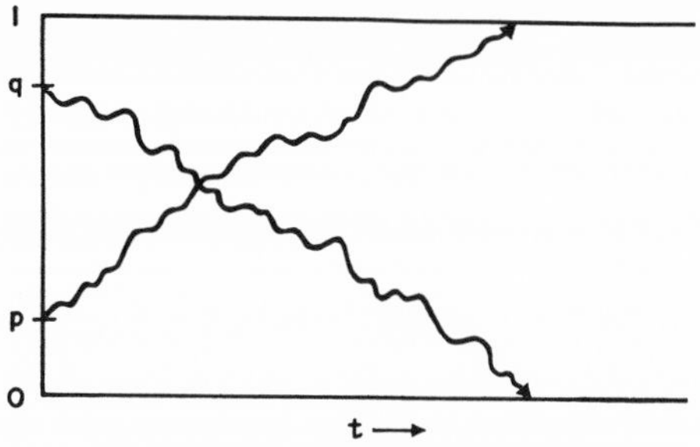


FIG. 1. Diagram illustrating two sample paths, one leading to fixation starting from initial frequency $p$ and the other leading to extinction starting from initial frequency $q$ $(q > p)$. The abscissa is time; the ordinate is gene frequency.

under otherwise the same conditions. In other words, these two reverse processes are reciprocal to each other.

Throughout this paper, we consider one locus at which a pair of alleles, $A$ and $A'$, are segregating. We shall denote the frequency of allele $A$ by such symbols as $p$, $q$, $x$ and $y$. Let $M_{\delta p}$ and $V_{\delta p}$ be respectively the mean and variance of the change of gene frequency p per generation. In a typical situation, if $A$ has selective advantage s in homozygotes and sh in heterozygotes over its allele $A'$, and if the mutation rate from $A$ to $A'$ is $v$ and the rate in reverse direction is $v'$, we have $M_{\delta p} = sp(1-p)\{h + (1-2h)p\} + v' - (v + v')p$, and $V_{\delta p} = p(1-p)/(2N_e)$, where $N_e$ is the "variance" effective size of the population. We assume that both end frequencies of the process can be reached within a finite length of time.

Let $\Phi_1(p,y)$ be the density of average sojourn time of allele $A$ at frequency $y$ in the path starting from $p$ and going to fixation, and let $\Phi_0(p,y)$ be the density of average sojourn time at $y$ of the path starting from $p$ and going to extinction of allele $A$. Let $u_1(p)$ and $u_0(p) = 1 - u_1(p)$ be respectively the probabilities of ultimate fixation and ultimate extinction of allele $A$, given that its initial frequency is $p$:

$$u_1(p) = \int_0^p G(x)\,dx \Big/ \int_0^1 G(x)\,dx, \qquad \text{(Kimura, 1962)}$$

where

$$G(x) = \exp\{-2 \int (M_{\delta x}/V_{\delta x})\,dx\}.$$

Then, from equation (12) of Maruyama and Kimura (1971), we have, for the process leading to fixation,

$$\Phi_1(p, y) = 2u_1(y)u_0(y)/\{V_{\delta y}u_1'(y)\} \quad \text{for } y \geqq p, \qquad (3)$$

and

$$\Phi_1(p, y) = \frac{2u_0(p)[u_1(y)]^2}{u_1(p)V_{\delta y}u_1'(y)} \quad \text{for } y < p, \qquad (4)$$

where $u_1'(y) = du_1(y)/dy$. Essentially the same set of equations as (3) and (4) were later obtained by Ewens (1973) using a different method.

Similarly, for the process leading to extinction

$$\Phi_0(q, y) = \frac{2u_1(q)[u_0(y)]^2}{V_{\delta y}u_1'(y)u_0(q)} \quad \text{for } y \geqq q, \qquad (5)$$

and

$$\Phi_0(q, y) = 2u_1(y)u_0(y)/\{V_{\delta y}u_1'(y)\} \quad \text{for } y < q. \qquad (6)$$

Let us suppose that $p < q$ and consider the behavior of the allele in the interval $(p,q)$. Since $\Phi_1(p,y)$ of (3) and $\Phi_0(q,y)$ of (6) have the same expression and they are independent of $p$ and $q$, we note that the path starting from $p$ and going to fixation, and the path starting from $q$ and going to extinction have the identical mean sojourn time in the interval $(p,q)$. In these two formulae, (3) and (6), if we let $p$ go to 0 and $q$ go to 1, we have the situations where a single mutant gene spreads into the entire population and where the allele frequency decreases to zero from near unity. Quite remarkably the features of the both processes are the same, despite the possible great difference in the probabilities of occurrence of these two events.

The average total length of time required for a single mutant gene to reach fixation is given by

$$\lim_{p \to 0} \bar{t}_1(p) = \int_0^1 \Phi_1(p, y)\,dy$$

and this is equal to

$$\lim_{q \to 1} \bar{t}_0(q) = \int_0^1 \Phi_0(q, y)\,dy$$

which is the average time for the reverse of the above process. This result can also be derived from the equations of Kimura and Ohta (1969*a*), (see also Crow and Kimura, 1970, p. 430, formulae 8.9.1 and 8.9.2).

We can similarly obtain the higher moments of these quantities, using the method given in Maruyama and Kimura (1971). For example, the second moments denoted by $\Phi_1^{(2)}(p,y)$ and $\Phi_0^{(2)}(q,y)$, of the sojourn times are

$$\Phi_1^{(2)}(p, y) = \Phi_0^{(2)}(q, y) = \frac{8[u_1(y)u_0(y)]^2}{[V_{\delta y}u_1'(y)]^2} \quad \text{for } p < y < q.$$

It might be rather difficult to perceive, from common sense, that the average time which a selectively advantageous mutant allele (say $A$) takes when it increases its frequency from $p$ to $q$, on the way to fixation under random drift in conjunction with favorable selection, is equal to the average time which the same allele takes when its frequency decreases from $q$ to $p$ against selection toward eventual extinction. In both cases, the average time is shortened by selection. This property was noticed by Maruyama (1972) for the special case of genic selection ($h = 1/2$). Needless to say, the probabilities of occurrence of these two events can be very different especially when selection coefficient is large. On the other hand, assuming no selection, it can be shown that under irreversible mutation from $A$ to $A'$, the average time until fixation of $A$ is prolonged as compared with the case of no mutation. This might be easily understood by considering that the frequency of $A$ is decreased at the rate $v$ per generation thereby prolonging the time for its frequency to reach unity. However, it might be rather difficult to accept (using only common sense) that the reverse process of $A$ decreasing its frequency from near unity to zero is equally prolonged under the same mutation pressure. Kimura and Ohta (1973) confirmed this fact both analytically (by the diffusion model) and experimentally (by Monte Carlo method).

## Summary

It is shown that in finite populations the average length of time which it takes for an allele to increase its frequency from $p$ to $q$ ($> p$) on the way to fixation is equal to the average length of time which the same allele takes when its frequency decreases from $q$ to $p$ on the way to extinction, although the probabilities of occurrence of these two events can be very different.

## Literature Cited

Crow, J. F., and M. Kimura. 1970. An introduction to population genetics theory. Harper and Row, New York.

Ewens, W. J. 1973. Conditional diffusion processes in population genetics. Theoret. Pop. Biol. 4:21–30.

Fisher, R. A. 1930. The distribution of gene ratios for rare mutations. Proc. Roy. Soc. Edin. 50:205–220.

Haldane, J. B. S. 1927. A mathematical theory of natural and artificial selection. Part V: Selection and mutation. Proc. Camb. Phil. Soc., 23:833–844.

Kimura, M. 1957. Some problems of stochastic processes in genetics. Ann. Math. Stat. 28: 882–901.

——. 1962. On the probability of fixation of mutant genes in a population. Genetics 47: 713–719.

——. 1970. The length of time required for a selectively neutral mutant to reach fixation through random frequency drift in a finite population. Genet. Res. 15:131–133.

Kimura, M., and T. Ohta. 1969*a*. The average number of generations until fixation of a mutant gene in a finite population. Genetics 61:763–771.

—— and ——. 1969*b*. The average number of generations extinction of an individual mutant gene in a finite population. Genetics 63:701–709.

—— and ——. 1971. Theoretical aspects of population genetics. Princeton University Press, Princeton.

—— and ——. 1973. The age of a neutral mutant persisting in a finite population. Genetics (*in press*).

Maruyama, T. 1972. The average number and the variance of generation at particular gene frequency in the course of fixation of a mutant gene in a finite population. Genet. Res. 19:109–114.

Maruyama, T., and M. Kimura. 1971. Some methods for treating continuous stochastic processes in population genetics. Jap. J. Genetics 46:407–410.

Narain, P. 1970. A note on the diffusion approximation for the variance of the number of generations until fixation of a neutral mutant gene. Genet. Res. 15:251–255.

Wright, S. 1931. Evolution in Mendelian populations. Genetics 16:97–159.

# Moments for Sum of an Arbitrary Function of Gene Frequency along a Stochastic Path of Gene Frequency Change

(population genetics/behavior of mutant/stochastic process/diffusion models)

TAKEO MARUYAMA AND MOTOO KIMURA

National Institute of Genetics, Mishima, Japan

*Contributed by Motoo Kimura, January 27, 1975*

**ABSTRACT** A diffusion model is developed to compute any moment of the sum of an arbitrary function of the gene frequency along sample paths between any two specified frequencies. This is used to calculate the mean age of a mutant of frequency $x$, including or excluding the possibility of its having been at value $x = 1$ during the process, and the total frequency of heterozygotes involving an allele since its origin.

Recent molecular discoveries have increased the likelihood that mutations responsible for amino-acid substitutions in evolution and for protein polymorphisms are selectively neutral, or so nearly neutral that chance plays an important role in determining their frequencies (1–8). Although stochastic processes in gene frequency change in finite populations can often be formulated by use of matrix algebra (in terms of Markov chains), answers in actually usable form seldom come from such a formulation. The most powerful method so far has been to use diffusion equations (9). Using this method, it has been possible to obtain such quantities as the probability of fixation (10, 11), sojourn times (12–14), average time to fixation or loss (15), conditional time to fixation or loss (16, 17), total number of individuals affected by a mutant gene (18, 19), total number of heterozygotes for a mutant allele (20), and number of heterozygous nucleotide sites under a steady flux of mutations in a finite population (21). More generally, using a theorem of Dynkin (22), we can compute the sum of an arbitrary function of gene frequency, $f(x)$, such as the fraction of heterozygotes, along a sample path that is destined to fixation starting from an initial frequency $p$ (12, 23).

In this paper we consider a related, but somewhat different question: what is the expected value of the sum of a function of gene frequency along a sample path in the past, given the present frequency. It is an extension of earlier work (24, 25) on the age of a mutant allele persisting in a finite population.

We consider a stochastic process of change in gene frequency which we assume to be time-homogeneous, and let $\phi(p,x;t)$ be the transition probability density that the frequency of a mutant allele is $p$ at time 0 and becomes $x$ at time $t$ $(0 < p < 1,\ 0 < x < 1)$. Then, $\phi(p,x;t)$ satisfies the Kolmogorov backward equation

$$\frac{\partial \phi}{\partial t} = L(\phi), \qquad [1]$$

where

$$L = \frac{V_{\delta p}}{2}\frac{\partial^2}{\partial p^2} + M_{\delta p}\frac{\partial}{\partial p} \qquad [2]$$

is the differential operator, in which $M_{\delta p}$ and $V_{\delta p}$ are the mean and variance of the change of gene frequency $p$ per unit time (generation).

Let $\xi_{\omega,\tau}$ be the frequency of the mutant allele on a given sample path $\omega$ at time $\tau$ $(\leqq t)$. We are interested in the expected value, and more generally, the moments of the integral of $f(\xi)$ taken along the sample path that starts from $p$ at time 0 and reaches $x$ sometime later.

Let us start by considering the conditional first moment,

$$E\left\{\int_0^t f(\xi_{\omega,\tau})d\tau \,|\, \xi_{\omega,t} = x\right\},$$

where $E$ stands for the expectation with respect to sample paths. In terms of the transition probability, this can be expressed as

$$F_x^{(1)}(p)/\Phi(p,x),$$

where

$$\Phi(p,x) = \int_0^\infty \phi(p,x;t)dt \qquad [3]$$

and

$$F_x^{(1)}(p) = \int_0^1 \Phi(p,\xi)f(\xi)\Phi(\xi,x)d\xi. \qquad [4]$$

In other words, $F_x^{(1)}(p)/\Phi(p,x)$ represents the mean value of the sum of a function $f(\xi)$ of allelic frequency $\xi$ in the past, given that the initial frequency was $p$ and that the present frequency is $x$.

Throughout this paper, we assume that $x = 0$ or $x = 1$ or both are absorbing barriers, so that the mutant allele which appeared in a population becomes either fixed $(x = 1)$ or lost $(x = 0)$ from the population within a finite time. Then using [1], [2], and [3] and noting $\phi(p,x;0) = \delta(x - p)$, it can readily be shown that $\Phi(p,x)$ satisfies the differential equation

$$L\{\Phi(p,x)\} = -\delta(x - p), \qquad [5]$$

where $\delta(\cdot)$ is the Dirac delta function. Note that $\Phi(p,x)$ is the density of mean sojourn time at $x$ starting from

$p$ (see ref. 12). The equation satisfying the first moment $F_x^{(1)}(p)$ can be obtained by applying the operator $L$ to both sides of [4] and noting [5]:

$$L\{F_x^{(1)}(p)\} + f(p)\Phi(p,x) = 0. \quad [6]$$

More generally, if we denote the conditional $n$th moment $(n \geqq 1)$ by $F_x^{(n)}(p)/\Phi(p,x)$ so that

$$F_x^{(n)}(p) = \Phi(p,x)\cdot E\left\{\left(\int_0^t f(\xi_{\omega,\tau})d\tau\right)^n \Big| \xi_{\omega,t} = x\right\} \quad [7]$$

then, we can show, following the method given by Nagylaki (23), that

$$F_x^{(n)}(p) = n!\left(\prod_{i=1}^{n}\int_0^1 d\xi_i\right) \times \left(\prod_{i=1}^{n}\Phi(\xi_{i-1}, \xi_i)f(\xi_i)\right)\Phi(\xi_n, x), \quad [8]$$

where $\xi_0 = p$. Applying the operator $L$ to both sides of this equation and noting [5], we obtain

$$L\{F_x^{(n)}(p)\} + nf(p)F_x^{(n-1)}(p) = 0. \quad [9]$$

Since we are, in general, concerned with an ordinary second order differential equation of the form

$$L\{Y(p)\} + K(p) = 0, \quad [10a]$$

or

$${}^1/_2V_{\delta p}Y''(p) + M_{\delta p}Y'(p) + K(p) = 0, \quad [10b]$$

it may be convenient here to present its formal solutions under two genetically appropriate boundary conditions: For the boundary conditions,

$$Y(0) = Y(1) = 0 \quad [11]$$

the solution is given by

$$Y(p) = \{1 - u(p)\}\int_0^p \psi_K(\xi)u(\xi)d\xi + u(p)\int_p^1 \psi_K(\xi)\{1 - u(\xi)\}d\xi \quad [12]$$

where

$$\psi_K(\xi) = 2K(\xi)\int_0^1 G(\lambda)d\lambda \Big/ \{V_{\delta\xi}G(\xi)\},$$

$$u(p) = \int_0^p G(\xi)d\xi \Big/ \int_0^1 G(\xi)d\xi$$

and

$$G(\xi) = \exp\left\{-2\int_0^\xi (M_{\delta\eta}/V_{\delta\eta})d\eta\right\}.$$

On the other hand, for the boundary conditions

$$Y(0) = 0,\ Y'(1) = \text{finite}, \quad [13]$$

the appropriate solution is

$$Y(p) = \int_0^p G(\xi)d\xi\int_\xi^1 \frac{2K(\eta)}{V_{\delta\eta}G(\eta)}d\eta. \quad [14]$$

(See ref. 23 for [13]).

The problem of evaluating the mean age of an allele segregating within a finite population is the special case where $f(x) = 1$. If we use a multiple allelic mutation model (26) assuming that every mutation at a locus leads to a new, not preexisting allele, then boundary condition [11] gives the age of a mutant of frequency $x$ that has never reached frequency 0 or 1 (age before fixation), while boundary condition [13] includes the possibility that the mutant may have been at frequency 1 in the past (age including fixation).

In particular, if the mutant alleles are selectively equivalent (neutral), and if the mutation rate is $v$ per generation, then, for a population with the effective size $N_e$, we have

$$M_{\delta p} = -vp,\ V_{\delta p} = \frac{p(1-p)}{2N_e}, \quad [15]$$

and therefore $G(\xi) = (1-\xi)^{-4N_ev}$ and $u(p) = 1 - (1-p)^{1-4N_ev}$, where we assume $4N_ev \leqq 1$. Using Eq. [12], the mean age before fixation at the limit $p \to 0$ turns out to be as follows:

$$\lim_{p\to 0}\frac{F_x^{(1)}(p)}{\Phi(p,x)} = \frac{4N_e}{1-4N_ev}\left\{\int_0^x \frac{1-(1-\xi)^{1-4N_ev}}{\xi}d\xi + \frac{1-(1-x)^{1-4N_ev}}{(1-x)^{1-4N_ev}}\int_x^1 \frac{(1-\xi)^{1-4N_ev}}{\xi}d\xi\right\}. \quad [16]$$

On the other hand, the mean age, including fixation, at the limit $p \to 0$ is

$$\lim_{p\to 0}\frac{F_x^{(1)}(p)}{\Phi(p,x)} = \frac{4N_e}{1-4N_ev}\left\{\int_0^x \frac{1-(1-\xi)^{1-4N_ev}}{\xi(1-\xi)^{1-4N_ev}}d\xi + [1-(1-x)^{1-4N_ev}]\int_x^1 \frac{d\xi}{\xi(1-\xi)^{1-4N_ev}}\right\}. \quad [17]$$

These formulae should give good approximations to the ages (before fixation and including fixation) of a neutral mutant allele that appeared in very low frequency in the population sometime in the past and whose present frequency is $x$. In fact, Eqs. [16] and [17] agree with Eqs. [10] and [12] of Maruyama (25), who denoted these two types of ages by $A_1\left(\frac{1}{2N}, y\right)$ and $A_2\left(\frac{1}{2N}, y\right)$ where $y$ is the present frequency of the allele in question, and $1/(2N)$ is the initial frequency of the allele. Eq. [16] also agrees with Eq. [13a] of Kimura and Ohta (24), who designated the age as $\bar{t}_{\uparrow}(0, x)$.

Let us now consider the mean sojourn time of the process, assuming that both the starting frequency $p$ and the present frequency $x$ are given. If we denote the mean density, at frequency $z$, of such a sojourn time by $\Phi_x(p, z)$, then we obtain

$$\Phi_x(p, z) = \frac{\Phi(p, z)\Phi(z, x)}{\Phi(p, x)}, \quad [18]$$

where $\Phi(p, z)$ is the unconditional mean sojourn time that satisfies Eq. [**5**]. This formula follows immediately by noting that $\Phi_x(p, z)$ is a special case of $F_x^{(1)}(p)/\Phi(p, x)$ with $f(\xi) = \delta(\xi - z)$ in Eq. [**4**]. We note here that the distribution of the conditional sojourn time in this paper may be obtained by combining Nagylaki's Eq. **37** (ref. 23) with the mean conditional sojourn time given above, which, following his notation, we may denote as $\bar{\tau}_x(p, z)$.

As an application of formula [**18**], consider the process of pure random drift in a finite population, and let

$$M_{\delta p} = 0 \text{ and } V_{\delta p} = \frac{p(1-p)}{2N_e}. \qquad [19]$$

We assume here that the mutant allele is selectively neutral and no further mutation takes place during the process. Then, we obtain

$$\left.\begin{aligned} &\Phi_x(0, z) = 4N_e && \text{for } z \leqq x \\ \text{and} \\ &\Phi_x(0, z) = 4N_e \frac{x(1-z)}{(1-x)z} && \text{for } z > x. \end{aligned}\right\} \qquad [20]$$

The theory developed in this paper allows us to compute such a quantity as the total heterozygosity, or the sum of the fraction of heterozygotes over all generations in the past, due to a mutant allele whose present frequency is known. For a randomly mating population, this may be obtained by putting $f(x) = 2x(1 - x)$. Assuming selective neutrality and no mutation in the process, the total heterozygosity in the past, which we denote by $H_x(p)$, is given by $F_x^{(1)}(p)/\Phi(p,x)$ with $M_{\delta p} = 0$, $V_{\delta p} = p(1 - p)/(2N_e)$, and $f(p) = 2p(1 - p)$. For $p < x$, the expected total heterozygosity in the past for a mutant allele whose initial frequency is $p$ and whose present frequency is $x$ is

$$H_x(p) = \frac{4N_e}{3}(2x - x^2 - p^2), \qquad (p < x). \qquad [21]$$

In order to check this formula, we have performed Monte Carlo experiments (using computer TOSBAC 3400) simulating a random mating population consisting of 10 breeding individuals ($N_e = 10$). In each experiment, we started with a population in which the mutant allele is represented only once ($p = 1/20$). The experiments were repeated 100,000 times (giving $10^5$ sample paths). The agreement between the theoretical predictions of [**21**] and the experimental outcomes is satisfactory; the average difference between the two for $x \geqq 0.1$ was about 1.3%.

We thank Dr. J. F. Crow and Dr. T. Nagylaki for reading the manuscript and offering suggestions for improving the presentation. This research was supported in part by U.S. Public Health Service Research Grants GM19513 and GM20293. This is Contribution no. 1040 from the National Institute of Genetics, Mishima, Shizuoka-ken, 411, Japan.

1. Kimura, M. (1968) *Nature* **217,** 624–626.
2. King, J. L. & Jukes, T. H. (1969) *Science* **164,** 788–798.
3. Crow, J. F. (1969) *Proc. Int. Congr. Genet. XII* (Tokyo) **3,** 105–113.
4. Crow, J. F. (1972) *J. Hered.* **63,** 306–316.
5. Kimura, M. & Ohta, T. (1971) *Nature* **229,** 467–469.
6. Kimura, M. & Ohta, T. (1971) *Theoretical Aspects of Population Genetics* (Princeton University Press, Princeton, N.J.).
7. Kimura, M. & Ohta, T. (1974) *Proc. Nat. Acad. Sci. USA* **71,** 2848–2852.
8. Ohta, T. (1974) *Nature* **252,** 351–354.
9. Kimura, M. (1964) *J. Appl. Probabil.* **1,** 177–232.
10. Kimura, M. (1957) *Ann. Math. Stat.* **28,** 882–901.
11. Kimura, M. (1962) *Genetics* **47,** 713–719.
12. Maruyama, T. & Kimura, M. (1971) *Jap. J. Genet.* **46,** 407–410.
13. Maruyama, T. (1972) *Genet. Res.* **19,** 109–113.
14. Ewens, W. J. (1973) *Theor. Pop. Biol.* **4,** 21–30.
15. Ewens, W. J. (1969) *Population Genetics* (Methuen, London).
16. Kimura, M. & Ohta, T. (1969) *Genetics* **61,** 763–771.
17. Kimura, M. & Ohta, T. (1969) *Genetics* **63,** 701–709.
18. Nei, M. (1971) *Theor. Pop. Biol.* **2,** 426–430.
19. Li, W. & Nei, M. (1972) *Amer. J. Human Genet.* **24,** 667–679.
20. Maruyama, T. (1972) *Genet. Res.* **20,** 141–149.
21. Kimura, M. (1969) *Genetics* **61,** 893–903.
22. Dynkin, E. B. (1965) in *Markov Processes* (Springer-Verlag, Berlin), Vol. II, pp. 46–53.
23. Nagylaki, T. (1974) *Proc. Nat. Acad. Sci. USA* **71,** 746–749.
24. Kimura, M. & Ohta, T. (1973) *Genetics* **75,** 199–212.
25. Maruyama, T. (1974) *Genet. Res.* **23,** 137–143.
26. Kimura, M. & Crow, J. F. (1964) *Genetics* **49,** 725–738.

P A R T   T H I R T E E N

# Intergroup Selection

## Introduction

Kimura's preference for diffusion models is typified in his 1983 paper (no. 36). He incorporated intergroup selection into a diffusion model, in addition to conventional individual selection, mutation, migration, and random drift, and obtained the condition for group selection to prevail over counteracting individual selection (Wilson 1975; Uyenoyama 1979; Aoki 1982). Unlike ordinary diffusion equations, the equation has killing and creating terms of a sample path. Those killing and creating terms are respectively proportional to the average frequency of a mutant in the whole population and its frequency in a particular deme. The problem is also related to the third phase of Wright's (1970, 1977) shifting balance theory (Crow, Engels, and Denniston 1990; Crow 1990b for review; Kimura 1983a for criticism) in which intergroup selection by differential migration facilitates the establishment of favorable gene combinations in other demes. Shimakura (1985) proved the existence and uniqueness of solutions in Kimura's diffusion.

# Diffusion model of intergroup selection, with special reference to evolution of an altruistic character

(population genetics/sociobiology/stochastic process)

MOTOO KIMURA

National Institute of Genetics, Mishima, 411 Japan

*Contributed by Motoo Kimura, July 19, 1983*

**ABSTRACT** Assume a diploid species consisting of an infinite number of competing demes, each having $N_e$ reproducing members and in which mating is at random. Then consider a locus at which a pair of alleles $A$ and $A'$ are segregating, where $A'$ is the "altruistic allele," which has selective disadvantage $s'$ relative to $A$ with respect to individual selection, but which is beneficial for a deme in competition with other demes; namely, a deme having $A'$ with frequency $x$ has the advantage $c(x - \bar{x})$ relative to the average deme, where $c$ is a positive constant and $\bar{x}$ is the average of $x$ over the species. Let $\phi = \phi(x;t)$ be the distribution function of $x$ among demes in the species at time $t$. Then, we have $\partial\phi/\partial t = L(\phi) + c(x - \bar{x})\phi$, where $L$ is the Kolmogorov forward differential operator commonly used in population genetics [i.e., $L = (1/2)(\partial^2/\partial x^2)V_{\delta x} - (\partial/\partial x)M_{\delta x}$], and $M_{\delta x}$ and $V_{\delta x}$ stand for the mean and variance of the change in $x$ per generation within demes. As to migration, assume Wright's island model and denote by $m$ the migration rate per deme per generation. By investigating the steady state, in which mutation, migration, random drift, and intra- and interdeme selection balance each other, it is shown that the index $D = c/m - 4N_e s'$ serves as a good indicator for predicting which of the two forces (i.e., group selection or individual selection) prevails; if $D > 0$, the altruistic allele predominates, but if $D < 0$, it becomes rare and cannot be established in the species.

In population genetics, diffusion models have been used quite successfully in treating the change of gene frequencies in finite populations (1–3). Particularly, diffusion models have proved themselves to be invaluable in developing theoretical population genetics at the molecular level (refs. 4 and 5; for review, see ref. 6). In all these treatments, it is customary to assume that natural selection acts through survival and reproduction of individuals rather than through competition of subpopulations.

Although the significance of group selection in evolution is controversial, it is an essential ingredient in the now-famous shifting-balance theory of evolution by Wright (see refs. 7 and 8 for review). Previously, Wright (9) had pointed out that it is difficult to see how socially advantageous but individually disadvantageous mutations can be fixed without some form of intergroup selection. Despite his emphasis on interdeme selection, however, Wright has not produced any quantitative theory treating the process of such selection.

The incentive of this paper derives from the recent work of Aoki (10), who investigated in quantitative terms the conditions for group selection to prevail over counteracting individual selection. In what follows, I shall develop a diffusion model for treating intergroup selection in addition to conventional individual selection, mutation, migration, and random drift. I shall then propose an index (to be denoted by $D$) which serves as a good indicator for predicting which of the two—that is, group selection or individual selection—prevails when they counteract each other.

The publication costs of this article were defrayed in part by page charge payment. This article must therefore be hereby marked "*advertisement*" in accordance with 18 U.S.C. §1734 solely to indicate this fact.

**Diffusion model incorporating group selection**

Let us assume a hypothetical population (species) consisting of an infinite number of competing subgroups (demes), each with the effective size $N_e$. Consider a particular gene locus and assume a pair of alleles $A$ and $A'$. I shall refer to $A'$ as the "altruistic allele." This is an allele that lowers the Darwinian fitness of individuals containing it but is helpful for demes to win in competition and proliferate.

We denote the frequency (proportion) of the altruistic allele $A'$ within a deme by $x$ and consider the frequency distribution of $x$ $(0 \leqq x \leqq 1)$ among the entire collection of demes making up the species. Let $\phi(x;t)$ be the distribution function of $x$ at time $t$ such that $\phi(x;t)\Delta x$ represents the fraction of demes whose frequency of $A'$ lies in the range $(x, x + \Delta x)$.

We assume that, within each deme, mating takes place at random among diploid members, and mutation and individual selection occur as treated in standard population genetic theory (2): in each generation, mutation occurs from $A$ to $A'$ at the rate $v'$, and, in the reverse direction, at the rate $v$. In other words, $v'$ is the rate at which the altruistic allele $A'$ is produced, and $v$ is the back mutation rate. Therefore, the rate of change in $x$ by mutation is $v'(1 - x) - vx$. $A'$ is assumed to have selective disadvantage $s'$ $(>0)$ relative to $A$, and selection is "genic" (i.e., "no dominance"), so that the rate of change in $x$ by individual selection is $-s'x(1 - x)$.

Migration is assumed to occur following Wright's island model (11)—namely, each deme contributes emigrants to the entire gene pool of the species at the rate $m$ and receives immigrants from that pool at the same rate. Thus, if $\bar{x}$ is the average frequency of $A'$ in the entire species, the rate of change in $x$ in a given deme by migration is $m(\bar{x} - x)$.

Let us now consider the effect of interdeme selection. We shall denote by $c(x)$ the coefficient of interdeme selection. This represents the rate at which the number of demes belonging to the gene frequency class $x$ change through interdeme competition. This means that during a short time interval of length $\Delta t$, the change of $\phi(x;t)$ is

$$\Delta\phi = \frac{1 + c(x)\Delta t}{1 + \overline{c(x)}\Delta t}\,\phi - \phi,$$

where $\phi$ stands for $\phi(x;t)$ and $\overline{c(x)}$ is the average of $c(x)$ over the entire array of demes in the species. At the limit $\Delta t \to 0$, we have

$$\Delta\phi = [c(x) - \overline{c(x)}]\,\phi\Delta t. \qquad [1]$$

In this paper, we shall treat the case in which $c(x)$ is linear so that Eq. 1 becomes

$$\Delta\phi = c(x - \bar{x})\phi\Delta t, \qquad [1a]$$

where $c$ is a positive constant.

By combining all the above evolutionary factors, the equation for $\phi(x;t)$ becomes

$$\frac{\partial\phi}{\partial t} = \frac{1}{2}\frac{\partial^2}{\partial x^2}\{V_{\delta x}\phi\} - \frac{\partial}{\partial x}\{M_{\delta x}\phi\} + c(x - \bar{x})\phi, \qquad [2]$$

where

$$V_{\delta x} = x(1 - x)/(2N_e) \qquad [2a]$$

is the variance of change in $x$ per generation due to random sampling of gametes;

$$M_{\delta x} = v'(1 - x) - vx + m(\bar{x} - x) - s'x(1 - x) \qquad [2b]$$

is the mean change per generation of $x$ within demes due to the joint effect of mutation, migration, and individual selection; $c$ is the coefficient of interdeme competition; and $\bar{x}$ is the mean of $x$ over the species—i.e.,

$$\bar{x} = \int_0^1 x\phi dx, \qquad [2c]$$

in which $\phi = \phi(x;t)$.

Note that Eq. 2, except for the last term $c(x - \bar{x})\phi$, is the standard form of the Kolmogorov forward equation commonly used in population genetics (cf. page 372 of ref. 2).

For the purpose of deriving expressions for the moments of $x$, the following equation is useful. Let $f(x)$ be a polynomial of $x$ or a suitable continuous function of $x$. Then, using the same method as used by Ohta and Kimura (12) to derive their equation **A3**, we obtain

$$\frac{\partial E\{f(x)\}}{\partial t} = E\left\{\frac{V_{\delta x}}{2}\frac{\partial^2 f(x)}{\partial x^2} + M_{\delta x}\frac{\partial f(x)}{\partial x} + c(x - \bar{x})f(x)\right\}, \qquad [3]$$

where $E$ stands for the operator of taking expectation with respect to the distribution $\phi$.

For example, if we put $f(x) = x$, we have $f'(x) = 1$ and $f''(x) = 0$, and we obtain the following equation that gives the rate of change of the mean.

$$\frac{d\bar{x}}{dt} = v' - (v' + v)\bar{x} + (s' + c)\sigma_x^2 - s'\bar{x}(1 - \bar{x}), \qquad [3a]$$

where $\bar{x} = E(x)$ is the mean and $\sigma_x^2 = E\{(x - \bar{x})^2\}$ is the variance of the distribution. From this we have, as the condition for $\bar{x}$ to increase when mutations are neglected, $d\bar{x}/dt = (s' + c)\sigma_x^2 - s'\bar{x}(1 - \bar{x}) > 0$ in agreement with Aoki (10).

**Distribution at steady state**

We shall now investigate the steady-state distribution, to be denoted by $\phi(x)$, which will be realized when a statistical equilibrium is reached under the joint effects of recurrent mutation, migration, intra- and interdeme selection, and random genetic drift. At this state, $\partial\phi/\partial t = 0$ and Eq. 2 reduces to

$$L\{\phi(x)\} + c(x - \bar{x})\phi(x) = 0, \qquad [4]$$

where $L$ is the Kolmogorov forward differential operator,

$$L = \frac{1}{2}\frac{\partial^2}{\partial x^2}V_{\delta x} - \frac{\partial}{\partial x}M_{\delta x}. \qquad [4a]$$

Let $\phi(x) = \phi_0(x)\psi(x)$, where we choose $\phi_0(x)$ so that $L\{\phi_0(x)\} = 0$. More precisely, we let $\phi_0(x)$ represent the steady-state distribution attained if the group selection were absent, i.e., if $c = 0$. This satisfies the "flux zero" condition (see page 170 of ref. 3), that is,

$$-\frac{1}{2}\frac{\partial}{\partial x}\{V_{\delta x}\phi_0(x)\} + M_{\delta x}\phi_0(x) = 0, \qquad [5]$$

and it leads to the well-known distribution due to Wright (11):

$$\phi_0(x) = \text{Const. } e^{-S'x}x^{V'+M\bar{x}-1}(1 - x)^{V+M(1-\bar{x})-1}, \qquad [6]$$

where $S' = 4N_e s'$, $V' = 4N_e v'$, $V = 4N_e v$, and $M = 4N_e m$, $(0 < x < 1)$. Because the constant term (Const.) is simply a normalizing factor so that all of the frequency classes add up to unity, I shall neglect this in the following treatment.

Under condition 5, Eq. 4 yields

$$(V_{\delta x}/2)\psi''(x) + M_{\delta x}\psi'(x) + c(x - \bar{x})\psi(x) = 0, \qquad [7]$$

where primes denote differentiation and

$$\bar{x} = \frac{\int_0^1 x\phi_0(x)\psi(x)dx}{\int_0^1 \phi_0(x)\psi(x)dx}. \qquad [7a]$$

Because we must have $\psi(x) =$ constant for $c = 0$, we attempt to solve Eq. 7 under the condition that $\psi(x)$ remains finite and continuous throughout the closed interval $[0,1]$, including the two singular points $x = 0$ and 1. In particular we let $\psi(0) = 1$.

Substituting Eqs. **2a** and **2b** in Eq. **7** but neglecting mutation terms $v'$ and $v$, we have

$$x(1 - x)\frac{d^2\psi}{dx^2} + [M(\bar{x} - x) - S'x(1 - x)]\frac{d\psi}{dx} + C(x - \bar{x})\psi = 0, \qquad [8]$$

where $M = 4N_e m$, $S' = 4N_e s'$, $C = 4N_e c$, $0 \leqq x \leqq 1$, and $\psi$ stands for $\psi(x)$.

Let $\psi(x) = e^{y(x)}$, then Eq. 8 becomes

$$x(1 - x)\left[\frac{d^2y}{dx^2} + \left(\frac{dy}{dx}\right)^2 - S'\frac{dy}{dx}\right] + \left(M\frac{dy}{dx} - C\right)(\bar{x} - x) = 0. \qquad [9]$$

Letting $x \to 0$ and noting that $d^ny/dx^n$ remains finite at $x = 0$ $(n = 1, 2, 3, \ldots)$ we obtain

$$y'(0) = R,\ y''(0) = \frac{R(S' - R)}{1 + M\bar{x}}, \ldots, \qquad [10]$$

where $R = C/M = c/m$. This means that in the neighborhood of $x = 0$, we have

$$y(x) = Rx + \frac{1}{2}\frac{R(S' - R)}{1 + M\bar{x}}x^2 + \cdots. \qquad [11]$$

Similarly, letting $x \to 1$, we obtain

$$y'(1) = R,\ y''(1) = \frac{R(S' - R)}{1 + M(1 - \bar{x})}, \ldots, \qquad [12]$$

so that, at the neighborhood of $x = 1$, we have

$$y(x) = y(1) - R(1 - x) + \frac{1}{2}\frac{R(S' - R)}{1 + M(1 - \bar{x})}(1 - x)^2 + \cdots. \qquad [13]$$

Of these two Taylor series, the former, that is, Eq. **11** is particularly useful in deriving an approximate distribution function and also an approximate value for $\bar{x}$, as I shall explain below.

Eq. **11** (together with higher order terms in the series) suggests that $y(x) = Rx$ is a good approximation for the solution of Eq. **9** when $|S' - R|$ is small. In fact, it can be shown that $y(x) = Rx$ is the exact solution of Eq. **9** when $S' - R = 0$. In the following treatment, I shall denote $R - S'$ by $D$, so that

$$D = R - S' = c/m - 4N_e s'. \qquad [14]$$

Because the required solution of the original Eq. **4** is

$$\phi(x) = \text{Const. } e^{y(x) - S'x}\{x^{\alpha-1}(1-x)^{\beta-1}\}, \qquad [15]$$

where $\alpha = V' + M\bar{x}$ and $\beta = V + M(1 - \bar{x})$, we may use (neglecting the constant term)

$$\phi(x) = e^{Dx}x^{\alpha-1}(1-x)^{\beta-1} \qquad [15a]$$

as a good approximation to the distribution when $|D|$ is small. Substituting this for $\phi_0(x)\psi(x)$ in Eq. **7a** and incorporating further approximations, such as $e^{Dx} = 1 + Dx$, we obtain

$$\bar{x} = \frac{\alpha}{\alpha + \beta}\left[1 + D\frac{\beta}{(\alpha+\beta)(1+\alpha+\beta)}\right].$$

If we further assume that the mutation rates are much lower than the migration rate, i.e., $v' + v \ll m$, we obtain

$$\bar{x} = \frac{1}{2BD}\left[BD - 1 + \sqrt{(1 - BD)^2 + 4\hat{x}_0 BD}\right], \qquad [16]$$

where $B = M/[(M + 1)(V + V')]$ and $\hat{x}_0 = V'/(V + V')$. Note that $\hat{x}_0$ is the equilibrium frequency of $A'$ attained if the alleles were selectively neutral.

In order to obtain more exact values for $\bar{x}$, we must resort to the numerical solution of the differential equation (Eq. **9**). This was done as follows. Note that $x = 0$ and $x = 1$ are singular points and that we must be careful in the neighborhood of these two points. For a small value of $x$, say $x < 0.05$, Eq. **11** was used to obtain values of $y(x)$. Beyond such a small value of $x$, Lunge–Kutta methods were used to integrate Eq. **9** step by step up to a certain point near 1, say $x = 0.9$, beyond which Eq. **13** was incorporated to make sure of the process of numerical integration.

In terms of this function $y(x)$, the required solution of the original Eq. **4** is given by Eq. **15**. Note that in this equation, the terms inside the braces represents the gene frequency distribution when natural selection is absent (i.e., alleles are neutral). It is evident, therefore, that the exponent $y(x) - S'x$ represents the joint effect of inter- and intrademe selection. Note also that, in the neighborhood of $x = 0$, we have

$$y(x) - S'x = Dx\left\{1 - \frac{1}{2}\cdot\frac{R}{1 + M\bar{x}}x + \cdots\right\}. \qquad [17]$$

As will be demonstrated in the next section, $D$ decides which of the two types of selection predominates: if $D > 0$, interdeme competition wins over counteracting individual selection; if $D < 0$, individual selection predominates. Transition between these two states is rather rapid and, if $D = 0$, the alleles behave as if completely neutral with respect to natural selection.

## Results of numerical studies

Using the above theory, we can compute the frequency distribution $\phi(x)$ in the following way. For a given set of parameters, $v$, $v'$, $m$, $s'$, $c$, and $N_e$, we first give a trial value for $\bar{x}$ (the mean of the frequency distribution), and compute $y(x)$ through numerically solving differential Eq. **9**. Then, combining $\psi(x) = e^{y(x)}$ and $\phi_0(x)$ of Eq. **6**, we compute a new value of $\bar{x}$ by using formula **7a**. This process is repeated until the trial value of $\bar{x}$ and the computed value of $\bar{x}$ become sufficiently close. The final trial value $\bar{x}$ thus obtained serves to derive the steady-state distribution $\phi(x)$. The variance and the higher moments can be obtained from this distribution. They also can be obtained from equations derived from Eq. **3** by setting $f(x) = x^n$ ($n = 1, 2, 3, \ldots$) and assuming $\partial E\{f(x)\}/\partial t = 0$ if the correct value of $\bar{x}$ is known. Note that in diffusion models, parameters giving mutation, migration, and selection all enter as products with $N_e$, namely, as $4N_e v'$, $4N_e v$, $4N_e m$, $4N_e s'$, and $4N_e c$.

Some of the results obtained by the numerical integration are illustrated by solid curves in Fig. 1 *A* and *B*, where the abscissae represent the index $D$ ($D = R - S'$) and the ordinates represent the average frequency ($\bar{x}$) of the altruistic allele.

In Fig. 1*A*, the solid curve represents the case in which the intensity of individual selection is varied ($4N_e s'$ is changed from 1.0 to 0.04) while holding other parameters constant—i.e., $4N_e v = 4N_e v' = 0.004$, $4N_e m = 1$, and $4N_e c = 0.4$. This means that $D$ ranges from $-0.6$ to 0.36. Note that this corresponds to mutation rates $v' = v = 10^{-4}$, selection coefficients $s' = 0.025 \sim 0.001$, $c = 0.01$, and migration rate $m = 0.025$ if the effective population size ($N_e$) is 10, but $v' = v = 10^{-5}$, $s' = 0.0025 \sim 0.0001$, $c = 0.001$, and $m = 0.0025$ if $N_e = 100$. In the same figure (Fig. 1*A*), the dotted line represents the approximate relationship between $D$ and $\bar{x}$ as computed by using Eq. **16**. It is remarkable that the agreement between the approximate values (dotted curve) and the more exact values (solid curve) are so close to each other as to be indistinguishable over most of the range studied. Note that in this case $D = 0$ corresponds to $\bar{x} = 0.5$ because the mutation rates are equal in both directions.

Fig. 1*B* illustrates the case in which the mutation rate at which the altruistic allele $A'$ is produced (i.e., $v'$) is only 1/10 as high as the back mutation rate ($v$). More precisely, it is assumed that $4N_e v' = 0.0004$ and $4N_e v = 0.004$. To construct the solid curve, we assumed $4N_e c = 4N_e s' = 0.4$, and $4N_e m$ is varied from 2.0 to 0.4 ($D$ ranges from $-0.2$ to 0.6). In this case, the mean frequency corresponding to $D = 0$ is $\bar{x} = v'/(v' + v) = 1/11 \approx 0.091$ because of the lower mutation rate to $A'$. In the same figure (Fig. 1*B*), the dotted curve is obtained by using the approximation equation (Eq. **16**), assuming $V' = 0.0004$, $V = 0.004$, and $M = 1$, and changing $D$ from $-0.2$ to 0.6. In this case, too, the approximate values are very close to the exact values over most of the range.

These results also demonstrate remarkably rapid transition from $\bar{x} \approx 0$ to $\bar{x} \approx 1$ as $D$ changes from a negative to a positive value. This must be valid as long as the deme size ($N_e$) is small, so that both $4N_e v$ and $4N_e v'$ are much smaller than unity.

## Discussion

The diffusion model used in this paper is an extension and elaboration of a model proposed in my earlier paper (13) in reference to Wright's theory of evolution. In that paper, I argued as follows (see page 47 of ref. 13). Suppose "the total population is subdivided into small and mutually competing subgroups, each of which has finite probability of splitting and multiplying into two or more groups as well as of being exterminated." Then, "$\psi(\mathbf{x},t)$, the quantity proportional to the probability of the group whose gene frequency is $\mathbf{x}$ at time $t$, $\cdots$, will satisfy the equation

$$\partial\psi/\partial t = L(\psi) + C\psi,$$

where $L$ is the differential operator" of the type given in Eq.

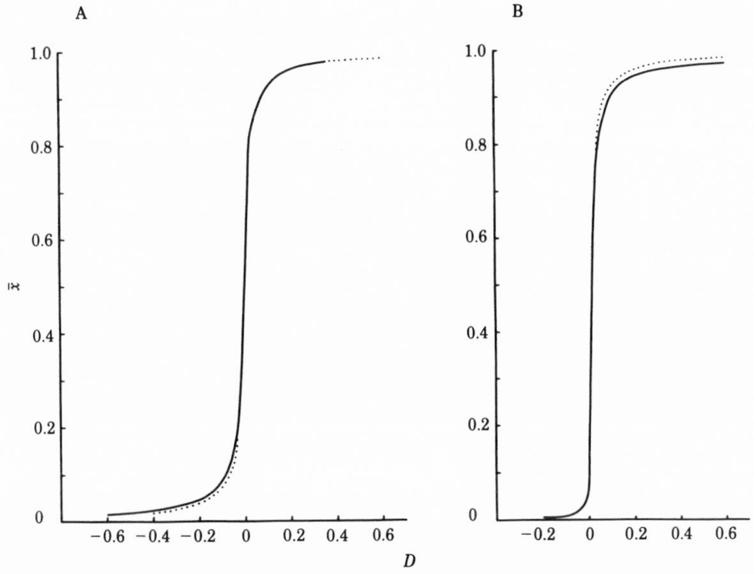


FIG. 1. Relationship between the average frequency of the altruistic allele ($\bar{x}$) and the index $D(D = c/m - 4N_e s')$. The solid curves were obtained by numerical integration of relevant equations, whereas the dotted curves were derived by using the approximation equation (Eq. **16**). In each case, the agreement between the two curves is satisfactory. Note an extremely rapid transition from $\bar{x} \approx 0$ to $\bar{x} \approx 1$ as $D$ changes from a negative value to a positive value.

**4a** of this paper (actually, its multidimensional form) and $C$ is "the rate of increase of the number of subgroups with given genic constitution due to intergroup competition."

More recently, a similar approach was used by Boorman and Levitt (14) to investigate group selection. They assign extinction function $-E(x)$ to $C$, so that, at the limit $t = \infty$, all the frequency classes vanish—i.e., $\psi(x,\infty) = 0$ for all $x$. In other words, the cohort eventually becomes extinct.

On the other hand, in the present model, in each generation a certain fraction of demes become extinct while some of the other demes multiply, so that the total number of demes remains constant. This is clearly seen by setting $f(x) = 1$ in Eq. **3**: the right hand side of this equation vanishes, and we have

$$\frac{\partial}{\partial t} E\{1\} = \frac{\partial}{\partial t}\int_0^1 \phi(x;t)dx = 0.$$

Because evolution is a consequence of survival rather than extinction, our model is probably more suitable than that of Boorman and Levitt to treat evolution of altruistic traits through group selection.

As shown in the preceding sections, the index $D$ serves as a good indicator for predicting whether group or individual selection prevails over the other when the two counteract. This means that the group selection prevails if

$$D = \frac{c}{m} - 4N_e s' > 0. \qquad [\mathbf{18}]$$

This also may be expressed as

$$4N_e m < c/s'. \qquad [\mathbf{18a}]$$

It turns out that this is essentially equivalent to Aoki's (10) condition, $nm \lesssim k/(2s)$, where $n$ is the size of each deme, $k$ is a measure of the intensity of group selection, $m$ is the migration rate, and $s$ is the selection coefficient against the altruistic allele. By taking into account the fact that individuals are haploid in Aoki's model and therefore his $n$ corresponds to $2N_e$, his condition may be expressed in our terminology as $2N_e m \lesssim c/(2s')$, which agrees essentially with our condition **18a**. In addition to confirming Aoki's condition, the present formulation (Eq. **18**) has the merit of showing in concrete terms how isolation (smaller $m$) enhances the effect of intergroup selection and how smaller deme size ($N_e$) diminishes the effect of individual selection.

The main purpose of this paper is to propose a mathematical model of group selection and to show its utility in treating evolution of altruistic characters. For biological aspects of the problem involved, readers are invited to consult chapter 5 ("Group selection and altruism") of Wilson's celebrated book (15) on sociobiology and also more recent writings such as those by Maynard Smith (16), Uyenoyama (17), Wade (18), Crow and Aoki (19), and Leigh (20).

I thank Dr. K. Aoki for stimulating discussions and helpful comments in composing this manuscript. This is contribution 1,518 from the National Institute of Genetics, Mishima, Shizuoka-ken, 411 Japan. This work is supported in part by a Grant-in-Aid from the Ministry of Education, Science and Culture of Japan.

1. Kimura, M. (1964) *J. Appl. Probab.* **1**, 177–232.
2. Crow, J. F. & Kimura, M. (1970) *An Introduction to Population Genetics Theory* (Harper & Row, New York); reprinted (1977) by Burgess, Minneapolis, MN.
3. Kimura, M. & Ohta, T. (1971) *Theoretical Aspects of Population Genetics* (Princeton Univ. Press, Princeton, NJ).
4. Kimura, M. (1971) *Theor. Pop. Biol.* **2**, 174–208.
5. Nei, M. (1975) *Molecular Population Genetics and Evolution* (North-Holland, Amsterdam).
6. Kimura, M. (1983) *The Neutral Theory of Molecular Evolution* (Cambridge Univ. Press, Cambridge, England), in press.
7. Wright, S. (1970) in *Mathematical Topics in Population Genetics*, ed. Kojima, K. (Springer, Berlin), pp. 1–31.

8. Wright, S. (1977) *Evolution and the Genetics of Populations* (Univ. of Chicago Press, Chicago), Vol. 3, pp. 443–473.
9. Wright, S. (1945) *Ecology* **26,** 415–419.
10. Aoki, K. (1982) *Evolution* **36,** 832–842.
11. Wright, S. (1931) *Genetics* **16,** 97–159.
12. Ohta, T. & Kimura, M. (1971) *Genetics* **68,** 571–580.
13. Kimura, M. (1955) *Cold Spring Harbor Symp. Quant. Biol.* **20,** 33–53.
14. Boorman, S. A. & Levitt, P. R. (1973) *Theor. Pop. Biol.* **4,** 85–128.
15. Wilson, E. O. (1975) *Sociobiology* (Belknap, Cambridge).
16. Maynard Smith, J. (1976) *Q. Rev. Biol.* **51,** 277–283.
17. Uyenoyama, M. K. (1979) *Theor. Pop. Biol.* **15,** 58–85.
18. Wade, M. J. (1982) *Evolution* **36,** 949–961.
19. Crow, J. F. & Aoki, K. (1982) *Proc. Natl. Acad. Sci. USA* **79,** 2628–2631.
20. Leigh, E. G., Jr. (1983) *Proc. Natl. Acad. Sci. USA* **80,** 2985–2989.

PART FOURTEEN

# Infinite Allele, Infinite Site, and Ladder Models

## Introduction

One may imagine that Kimura would not have proposed the neutral theory without writing the celebrated paper on the infinite allele model—more grammatically, the model of infinitely many alleles or the infinite-number-of-alleles model (Kimura and Crow, no. 37). Such a guess seems only partly correct, for it is clearly Wright (1931), Haldane (1957), and Muller (1958) that were essential in proposing the neutral theory. Kimura and Crow stated: "It is not the purpose of this article to discuss the plausibility of many very nearly neutral, highly mutable multiple isoalleles, or the evidence for and against. Instead, we propose to examine some of the population consequences of such a system if it does exist. The possibility seems great enough to warrant such an inquiry." Crow (1989b) recalled how they came to the consideration of such a system of isoalleles. There is no indication of the neutral theory in this paper. In fact, the infinite allele model was applied not only to neutral but also to overdominant alleles, and the time was not yet ripe for the neutral theory. Nevertheless, one can still perceive some link in scientific thought.

The infinite allele model has been extensively used in bringing refinements of the neutral theory. The mathematical refinements culminated in the sampling theory (Ewens 1972), the line-of-descent process (Griffiths 1980; Watterson 1984; Tavaré 1984), and measure-valued diffusions (Ethier and Griffiths 1984), all of which assume the infinite allele model. It also served to develop useful statistical methods for testing the neutral theory. In particular, the Houston group headed by Nei (1987; Kimura 1983a for review) has made significant contributions and supported the neutral theory.

Using a Kolmogorov backward equation, Kimura (no. 38) obtained the expected number of heterozygous nucleotide sites in a particular range of allele frequencies. This number, under a steady flux of mutations, is mathematically equivalent to the so-called Green function, from which other quantities can be derived. The main interest was the average number of heterozygous sites per individual, which under random mating is the same as the number of segregating sites between two haploid genomes that are randomly sampled from a population (Watterson 1975). Kimura assumed no linkage among those heterozygous sites (cf. Ohta and Kimura, no. 13), while Watterson assumed complete linkage. This mutation model, proposed in 1969, was called the infinite site model (Kimura 1971). Under neutrality, Kimura estimated the average number of heterozygous sites per individual to be 80,000 in human populations.

The theoretical results of the original neutral theory rely much on the infinite allele model (Kimura and Crow, no. 37) and the infinite site model (no. 38; Kimura 1971). One argument against the neutral theory was that the extent of electrophoretically detectable polymorphisms is rather low. Since the electrophoretic method does not meet the requirement in the infinite allele model, Ohta and Kimura (no. 39) considered the charge state, stepwise mutation, or ladder model of electrophoretic mutations. The model received much attention from the mathematical point of view (see Moran 1975, 1976 for the distribution of allele frequencies; Kingman 1976, 1977), but the biological consequence does not differ much from that obtained from the infinite allele model (Li 1976; Takahata 1980; chapter 8 in Kimura 1983a). Also, the appropriateness of the infinite allele model was supported experimentally by Fuerst and Ferrell (1980), who used hemoglobin variants (Brown, Marshall, and Weir 1981 for criticism).

# THE NUMBER OF ALLELES THAT CAN BE MAINTAINED IN A FINITE POPULATION[1]

MOTOO KIMURA[2] AND JAMES F. CROW

*University of Wisconsin, Madison*

Received January 3, 1964

IT has sometimes been suggested that the wild-type allele is not a single entity, but rather a population of different isoalleles that are indistinguishable by any ordinary procedure. With hundreds of nucleotides, each presumably capable of base substitutions and with additional permutations possible through sequence rearrangements, gains, and losses, the number of possible gene states becomes astronomical. It is known that a single nucleotide substitution can have the most drastic consequences, but there are also mutations with very minute effects and there is the possibility that many are so small as to be undetectable.

It is not the purpose of this article to discuss the plausibility of such a system of isoalleles, or the evidence for and against. Instead, we propose to examine some of the population consequences of such a system if it does exist. The probability seems great enough to warrant such an inquiry.

If a large number of different states can arise by mutation, this doesn't necessarily mean that a large fraction of these would coexist in a single population. Some will be lost by random drift and others may be selectively disadvantageous. On the other hand, some may persist by being beneficial in heterozygous combinations.

We shall consider three possibilities: (1) A system of selectively neutral isoalleles whose frequency in the population is determined by the mutation rate and by random drift. (2) A system of mutually heterotic alleles. (3) A mixture of heterotic and harmful mutants.

## 1. *Selectively Neutral Isoalleles*

To isolate the essential problem, we consider an extreme situation in which the number of possible isoallelic states at a locus is so large that each new mutant is a state not preexisting in the population. This provides an estimate of the upper limit for the number of different alleles maintained in the population.

The distribution in successive generations of the descendants of an individual mutant gene was solved by FISHER (1930) and under less restricted conditions,

[1] Paper No. 922 from the Division of Genetics; also Contribution No. 487 from the National Institute of Genetics, Mishima-shi, Japan. Sponsored by the Mathematics Research Center, U.S. Army, University of Wisconsin (Contract No. DA-11-022-ORD-2059) and supported in part by the National Institutes of Health (RG-7666).

[2] On leave from the National Institute of Genetics, Mishima-shi, Japan.

Genetics **49**: 725–738 April 1964.

though less exactly, by HALDANE (1939). An approximate solution to our problem was in fact given by HALDANE, but we present the following more elementary and more exact procedure:

Let $u$ be the average rate of mutation of the alleles existing in a diploid population, so that in a population of size $N$ ($2N$ genes) there will be $2Nu$ new mutants introduced per generation, each new mutant being regarded as different from any allele preexisting in the population.

In a randomly mating population of effective size $N_e$, the probability of two uniting gametes carrying alleles that are identical in the sense of being descended from the same allele in some common ancestor is

$$F_t = \frac{1}{2N_e} + \left(\frac{2N_e - 1}{2N_e}\right)F_{t-1} \tag{1}$$

where $F_t$ is the inbreeding coefficient in generation $t$ (WRIGHT 1931; MALÉCOT 1948).

The two alleles will be in identical states only if neither of them has mutated since the previous generation. The probability that neither has mutated is $(1-u)^2$. Thus we can generalize the formula (as MALÉCOT did) to include mutation by writing

$$F_t = \left[\frac{1}{2N_e} + \left(\frac{2N_e - 1}{2N_e}\right)F_{t-1}\right](1-u)^2 \tag{2}$$

To specify the equilibrium condition when the loss of alleles by random drift exactly balances the gain of new alleles by mutation, let $F_t = F_{t-1} = F$. The solution, ignoring terms containing $u^2$, is

$$F = \frac{1-2u}{4N_e u - 2u + 1} \approx \frac{1}{4N_e u + 1} \tag{3}$$

In this context, $F$ is the probability that an individual will be homozygous. If all the alleles were equally frequent, the proportion of homozygotes would be the reciprocal of the number of alleles at this locus maintained in the population. If there are variations in allele frequencies, the proportion of homozygotes will be greater than this. Therefore, $n = 1/F$ may be used as a measure of the *effective number* of alleles maintained in the population, which in general will be less than the actual number.

Some numerical values of $F$ and $n$ are given in Table 1 and the relations are shown graphically in Figure 1. If $4N_e$ is much less than the reciprocal of the mutation rate, $F$ approaches 1 and all the genes in the population will usually be the descendants of a single mutant. If $4N_e$ is larger than $1/u$, more than one allele will usually be maintained and as $N_e$ gets larger more individuals will be heterozygous.

The effective number, $N_e$, is usually smaller than the actual number. It is of course much nearer the number of sexually mature individuals than the number counted at immature stages, particularly if there is heavy pre-adult mortality

TABLE 1

*The average proportion of homozygosity, F, (upper figure) and the effective number of alleles per locus, n, (lower figure) in a randomly mating population of effective size $N_e$. The alleles are selectively neutral and the mutation rate of any allele is u. The number of possible mutant states is assumed to be large enough so that each new mutant is different from the others in the population.*

| | Effective population number, $N_e$ | | | | | |
|---|---|---|---|---|---|---|
| Mutation rate, $u$ | $10^2$ | $10^3$ | $10^4$ | $10^5$ | $10^6$ | $10^7$ |
| $10^{-4}$ | .96 | .71 | .20 | .024 | .0025 | .00025 |
| | 1.04 | 1.4 | 5.0 | 41 | 401 | 4001 |
| $10^{-5}$ | .996 | .96 | .71 | .20 | .024 | .0025 |
| | 1.004 | 1.04 | 1.4 | 5.0 | 41 | 401 |
| $10^{-6}$ | .9996 | .996 | .96 | .71 | .20 | .024 |
| | 1.0004 | 1.004 | 1.04 | 1.4 | 5.0 | 41 |
| $10^{-7}$ | .99996 | .9996 | .996 | .96 | .71 | .20 |
| | 1.00004 | 1.0004 | 1.004 | 1.04 | 1.4 | 5.0 |

(WRIGHT 1931; FISHER 1939; CROW and MORTON 1955). If the expectation of progeny is not the same for all individuals in the population the effective number for monoecious diploids is given by

$$N_e = \frac{N\bar{k} - 1}{\bar{k} - 1 + V/\bar{k}} \qquad \text{(KIMURA and CROW 1963)} \qquad (4)$$

where $\bar{k}$ = mean number of progeny per parent, $V$ = variance in number of progeny per parent, and $N$ = population number in the parent generation. There is a slight modification for a bisexual population (see KIMURA and CROW 1963). The special case of a population of stable size, $\bar{k} = 2$, was first given by WRIGHT (1938a). In this case (4) becomes

$$N_e = \frac{4N - 2}{2 + V} \qquad (5)$$

HALDANE's (1939) approximate solution for the minimum number of genes expected in a stable sized population of $N$ individuals is (in our terminology) $16Nu/(V + 2)$, in rough agreement with (3) and (5) when $N$ is large compared with $u^{-1}$.

The general conclusion of this section is that, for selectively neutral alleles, if the effective population number is much less than the reciprocal of the mutation rate almost all the genes in the population at a given locus will be descended from a single mutant.

## 2. *Mutually Heterotic Alleles*

It has been known since the early work of FISHER (1922) that, in an infinite population, heterozygote superiority in fitness for a pair of alleles leads to a stable polymorphism. With more than two alleles the necessary and sufficient conditions for maintaining a stable equilibrium are more delicate. The conditions were given by KIMURA (1956) and confirmed for a discontinuous model by MANDEL (1959). The complexity of the conditions, however, does not change the general conclusion that overdominance is a potent factor for maintaining a polymorphism in a large population.

Recently the behavior of overdominant genes in a finite population has been investigated by ROBERTSON (1962) utilizing some mathematical results of MILLER (1962). ROBERTSON showed that when the equilibrium allele frequency is outside the range 0.2 to 0.8 there are some circumstances where heterozygote advantage actually accelerates the rate of fixation and loss of alleles by random drift, rather than retarding it as might have been expected. This suggests that if there are a large number of mutually heterotic alleles, they may under some circumstances be lost by random drift more rapidly than if they were neutral.

In a system of mutually heterotic alleles, the population fitness will be greatest when the number of heterozygotes is maximized. In general, the larger the number of alleles the greater the proportion of heterozygotes. Hence, if the requisite mutations occur the population can reduce the segregation load (CROW 1958) by increasing the number of alleles that are maintained. On the other hand, the effect of random drift in reducing the number of alleles increases greatly with increase in the number of alleles in the population, being roughly proportional to square of the number of alleles (KIMURA 1955). A larger number can be maintained if the homozygotes are more disadvantageous, but this increases the segregation load.

Therefore, with a population of a certain size and mutation rate there must be, for a given pattern of homozygote disadvantage, a maximum number of alleles that can be maintained. This will correspond to the minimum segregation load.

We are interested in considering such an extreme situation where the segregation load is minimum. To make the mathematics more manageable, we assume that each homozygote has the same disadvantage, $s$, with respect to the heterozygotes, all of which are assumed to have the same fitness. In an infinite population each allele would be of equal frequency at equilibrium; in a finite population there will be departures because of random drift. We need to obtain the distribution of allele frequencies at equilibrium under the joint influence of mutation, selection, and random drift.

As in Section 1, we assume that the number of possible mutant alleles is so large that no mutation is repeated in a finite population. Using WRIGHT's (1937) general distribution formula and incorporating some of FISHER's (1958) inventive methods the average homozygosity and the effective number of alleles can be expressed in terms of $s$, $u$, and $N_e$.

*Mathematical methods:* In a randomly mating population of effective size $N_e$, let $\Phi(x)dx$ be the expected number of alleles whose frequency is in the range $x$ to $x + dx$. The value of $x$ may change from generation to generation by mutation, selection, and random drift, but at equilibrium a stable distribution will be reached, the formula for which can be obtained from an equation given by WRIGHT (1938b):

$$\Phi(x) = \frac{C}{V_{\delta x}} e^{2\int \frac{M_{\delta x}}{V_{\delta x}} dx} \tag{6}$$

where $C$ is a constant, $M_{\delta x}$ and $V_{\delta x}$ are respectively the mean and variance of the rate of change of $x$ per generation.

We let $u$ be the rate of mutation from the allele under consideration to all other allelic states. As stated before, we assume that each new mutation is unique. For simplicity, we assume that $u$ is the same for all alleles. We designate by $F$ the sum of the squares of the allele frequencies; i.e.

$$F = \sum_i x_i^2$$

where $x_i$ is the frequency of the $i$th allele, $A_i$, in the population.

Since the rate of change in the frequency of a particular allele with frequency $x$ by mutation is $-ux$ and by selection is

$$-sx(x - F), \tag{7}$$

we have

$$M_{\delta x} = -ux - sx(x - F). \tag{8}$$

As stated before, $s$ is the selective advantage of a heterozygote over a homozygote. This is most conveniently measured in MALTHUSIAN parameters (FISHER 1930, 1958). With discrete, nonoverlapping generations the change in $x$ caused by selection is

$$-\frac{sx(x - F)}{1 - sF} \tag{9}$$

Since we are considering circumstances where $sF$ is very small, this is not appreciably different from (7). The variance of the rate of change in $x$ is given by

$$V_{\delta x} = \frac{x(1 - x)}{2N_e} \tag{10}$$

A great mathematical simplification is possible if we replace this by

$$V_{\delta x} = \frac{x}{2N_e} \tag{11}$$

which introduces no significant error, since we are mainly concerned with large numbers of alleles, which have individually low frequencies. Substituting (8) and (11) into the distribution formula (6) leads to

$$\Phi(x) = C\, e^{-2S(x-F)^2-4Mx} x^{-1} \tag{12}$$

where

$$S = N_e s \text{ and } M = N_e u \tag{13}$$

In deriving equation (12), $F$ was assumed to be a constant, and is interpreted as the expected value of the sum of squares of the allele frequencies, or more simply as the reciprocal of the effective number of alleles maintained in the population. The treatment of $F$ as a constant will be shown later to be satisfactory as an approximation.

The constant $C$ is determined by the condition that the allele frequencies add up to unity, $\Sigma x_i = 1$, or

$$\int_0^1 x\Phi(x)dx = 1. \tag{14}$$

Note that this is different from the usual way by which $C$ in WRIGHT's formula is evaluated. The reason is that in the present instance $\Phi(x)$ is related to the number of different genes in a population rather than the probability of a certain gene frequency in a population.

From (14) we obtain

$$C^{-1} = \int_0^1 e^{-2S(x-F)^2-4Mx}dx. \tag{15}$$

Putting $y = x - F + M/S$, we get

$$C^{-1} = e^{-4MF + 2M^2/S}\int_{-F+M/S}^{1-F+M/S} e^{-2Sy^2}dy \tag{16}$$

At equilibrium, when the random extinction of alleles is exactly balanced by new mutations, we have the following condition at the subterminal class (cf. WRIGHT 1931; FISHER 1958):

$$2Nu = \frac{1}{2}\,\Phi\left(\frac{1}{2N}\right)\frac{1}{2N} \tag{17}$$

or

$$4M = C\, e^{-2S\left(\frac{1}{2N}-F\right)^2-2u} \tag{18}$$

In any population, the expected number of alleles maintained is much smaller than the total number of individuals; thus $1/2N$ is very small compared to $F$ and, since $u$ is very small, (18) is simplified to

$$4M = C\, e^{-2SF^2} \tag{19}$$

Thus, from the two relations (16) and (19), $F$ may be determined as a function of $M$ and $S$.

In equation (12), $F$ was used as the expected value of the sum of squares of the allele frequencies. This can be demonstrated by evaluating

$$\int_0^1 x^2\Phi(x)dx \tag{20}$$

which turns out to be very nearly $F$.

It is only necessary that

$$u << sFe^{2S(1-2F)} + 4M$$

because

$$\int_0^1 x^2\,\Phi\,(x)dx = F - \frac{u}{s}e^{-2S(1-2F)-4M}$$

*The effective number of alleles maintained:* If all alleles were of equal frequency, the number of alleles, $n$, would be given by

$$n = 1/F. \tag{21}$$

Therefore, we define $n$ as the effective number of alleles. The segregation load will be given by

$$L_s = sF = s/n \tag{22}$$

In order to get a solution for $F$, we first eliminate $C$ from (16) and (19). This leads to

$$e^{-Z^2/2} = r\int_{-Z}^{2\sqrt{S}-Z} e^{-y^2/2}dy \tag{23}$$

where

$$Z = 2\sqrt{S}(F - \frac{M}{S}) \text{ and } r = \frac{2M}{\sqrt{S}} \tag{24}$$

For any given value of $M$ and $S$, the corresponding value of $Z$ may be obtained from (23) and then $F$ is calculated from

$$F\sqrt{S} = \frac{r+Z}{2}. \tag{25}$$

The relation between $2M/\sqrt{S}$ and $F\sqrt{S}$ is given in Table 2 for various equally spaced values of $Z$ between $-3$ and $+3$. Numerical calculation is facilitated by the fact that, as seen from (23), $r$ is the ratio of the ordinate of the normal curve with zero mean and unit variance at $Z$ to the area under the curve from $-Z$ to $2\sqrt{S} - Z$. Since $\sqrt{S}$ is 10 or more in most cases of interest, the area is practically equivalent to integration from $-Z$ to $+\infty$.

For example, with $N_e = 10^5$, $s = 10^{-3}$, and $u = 10^{-5}$, we have $S = 100$ and $M = 1$, so that $r = 2M/\sqrt{S} = 0.2$. Table 2 gives $n/\sqrt{S} = 1.35$ or $n = 13.5$.

For values of $r$ outside the range tabulated, the following approximations are satisfactory:

1. *Small values of* r. For this, use

$$F\sqrt{S} = \frac{1}{2}\sqrt{4.6\log_{10}(0.4/r)} \tag{26}$$

For example, with $N_e = 10^5$, $s = 0.1$, and $u = 0.5 \times 10^{-5}$, we have $\sqrt{S} = 100$, $2M = 1$, so that $r = 0.01$. From (26), $F = 0.0136$ and $n = 73.6$, as compared with 73.8 from Table 2.

2. *Large values of* r. For this, use

$$n = 4M(1 + \frac{1}{r^2}). \tag{27}$$

When $s = 0$, $r = \infty$, leading to $n = 4M = 4N_e u$. This is in approximate agreement with the more exact value derived in Section 1, $n = 4N_e u + 1$. This can also be verified directly by integrating (20) for the case

$$\Phi(x) = 4M(1-x)^{4M-1}\,x^{-1}$$

TABLE 2

*Factors for computing the effective number of alleles* (n), *the proportion of homozygous loci* (F), *and the segregation load* (sF) *in a population of effective size* $N_e$, *mutation rate* u, *and selective disadvantage of homozygotes* s. $M = N_e u$ *and* $S = N_e s$. *The table is accurate when* $\sqrt{S} > 4$

| $r = \frac{2M}{\sqrt{S}}$ | $\frac{n}{\sqrt{S}}$ | $b = \sqrt{S}F$ |
|---|---|---|
| 0.0044 | 0.666 | 1.502 |
| 0.0105 | 0.738 | 1.355 |
| 0.0176 | 0.794 | 1.259 |
| 0.0360 | 0.895 | 1.120 |
| 0.0553 | 0.973 | 1.025 |
| 0.0984 | 1.112 | 0.899 |
| 0.139 | 1.220 | 0.819 |
| 0.204 | 1.375 | 0.727 |
| 0.288 | 1.553 | 0.644 |
| 0.389 | 1.755 | 0.570 |
| 0.509 | 1.982 | 0.505 |
| 0.646 | 2.233 | 0.441 |
| 0.798 | 2.507 | 0.399 |
| 0.964 | 2.803 | 0.357 |
| 1.141 | 3.120 | 0.321 |
| 1.329 | 3.456 | 0.289 |
| 1.525 | 3.809 | 0.263 |
| 1.729 | 4.177 | 0.239 |
| 1.939 | 4.560 | 0.219 |
| 2.110 | 4.876 | 0.205 |
| 2.373 | 5.359 | 0.187 |
| 2.552 | 5.685 | 0.176 |
| 2.823 | 6.194 | 0.161 |
| 3.006 | 6.603 | 0.153 |
| 3.283 | 7.105 | 0.141 |

Equation (27) shows that when $r$ is large, the number of alleles is determined almost entirely by effective population size and mutation rate, since overdominance increases the number of alleles only by the fraction $1/r^2$.

*Results of the calculations:* Figures 1 to 5 show the values of $F$ (the proportion of homozygous loci), $n$ (the effective number of alleles maintained), and $L_s$ (the segregation load) for a number of values of effective population number, mutation rate, and selective disadvantage of homozygotes. Corresponding to each selection coefficient, population size, and mutation rate there is a certain average homozygosity and a corresponding segregation load.

For example, with $u = 10^{-5}$ and $s = .001$, a population of effective number 10,000 has an effective number of alleles of less than five and a segregation load somewhat larger than .0002 per locus (Figure 2). If $s$ is increased the number of alleles maintained is increased, but so is the load. If $s = .01$, $n = 8$, and $L_s = .0012$ (Figure 3); if $s = 0.1$, $n = 22$, and $L_s = .0045$ (Figure 4); if $s = 1$, a balanced lethal condition, the number of alleles is almost 60 but the load has increased to .016 per locus (Figure 5).

With lethal homozygotes the situation is almost the same as with self-sterility alleles, a situation thoroughly investigated by WRIGHT (1939, 1960) and FISHER (1958).

WRIGHT's (1939) graph shows some 80 to 90 self-sterility alleles maintained by a mutation rate of $10^{-5}$ in a population of $10^4$ compared with our effective number of about 60 for the same situation. This is as expected: because the alleles will drift away from equal frequencies, the effective number of alleles is smaller than the actual number, the former being $1/\Sigma x^2$ and the latter being $1/\bar{x}$, where $\bar{x}$ is the mean frequency of an allele. For example, with three alleles with frequencies 2/3, 1/6, and 1/6, $\bar{x} = 1/3$ and $\Sigma x^2 = 1/2$. Thus the number of alleles is three, but the effective number is two; i.e. two alleles of equal frequency would produce the same proportion of heterozygotes.

For estimating the actual number of different alleles in the population, the average number as used by WRIGHT is appropriate. For assessing such things as the fraction of incompatible pollinations, the effective number is the quantity needed. This is the quantity that is estimated by the ordinary procedure of allelism tests.

*Mixed Heterotic and Harmful Mutants*

The model that we have discussed is artificial in assuming only overdominant mutants with equal homozygote fitness. Under this system, it would be advantageous for the mutation rate to be high, for this would lower the segregation load. On the other hand, if there are both overdominant mutants and deleterious mutants the situation would be different.

Consider first a situation where some loci produce only over dominant mutants of the type we have discussed and the remainder of the loci produce mutants that are deleterious in both homozygous and heterozygous state. If we let the proportion of heterotic loci be $P$ and the proportion of loci producing deleterious mutants be $Q$, the average total load per locus will be

$$\bar{L}_T = \sqrt{\frac{s}{N_e}}\,(Pr + Qb) \tag{28}$$

where $r = 2M/\sqrt{S}$ and $b = \sqrt{S}\,F$. The values of $r$ and $b$ are given by the first and third columns of Table 2.

Given $P$ and $Q$, values of $r$ and $b$ can be determined to minimize the total load. For example, if $P = Q = ½$, inspection of Table 2 shows that the average of columns 1 and 3 is minimum when $r$ is approximately 0.2. The average is .47 and therefore the load per locus is $.47\sqrt{s/N_e}$. The segregation load is about 7/2 the mutation load. For $N_e = 10^5$ and $s = 10^{-3}$, the mutation rate that minimizes $\bar{L}_T$ is $10^{-5}$.

It is probable that any locus that produces heterotic mutants also gives rise to deleterious mutants as well. If a fraction $p$ of the mutants are deleterious and a

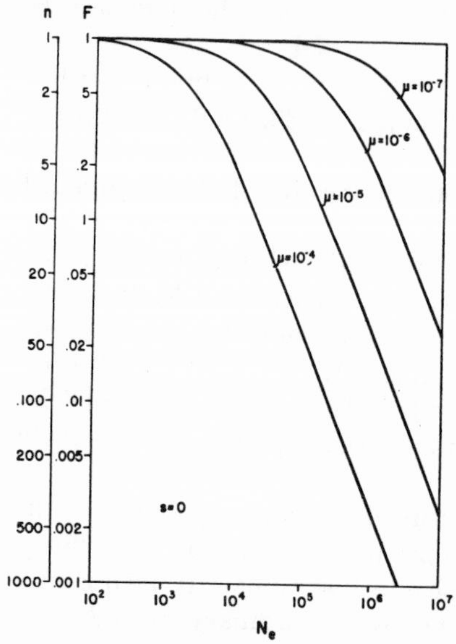


FIGURE 1.—The probability of homozygosity ($F$) and the effective number of alleles ($n$) maintained by a mutation rate ($\mu$) in a population of effective number $N_e$. The mutants are assumed to be selectively neutral and each mutant allele is of a type not already existing in the population.

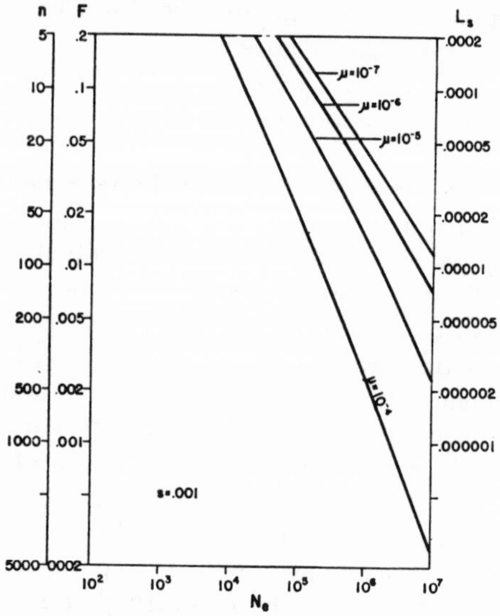


FIGURE 2.—The probability of homozygosity ($F$), the effective number of alleles maintained ($n$), and the segregation load ($L_s$) in a population of effective number $N_e$ and mutation rate $\mu$. The selective disadvantage of homozygotes ($s$) is .001. Because of the approximations used, the values near the top of the graph may be inaccurate.

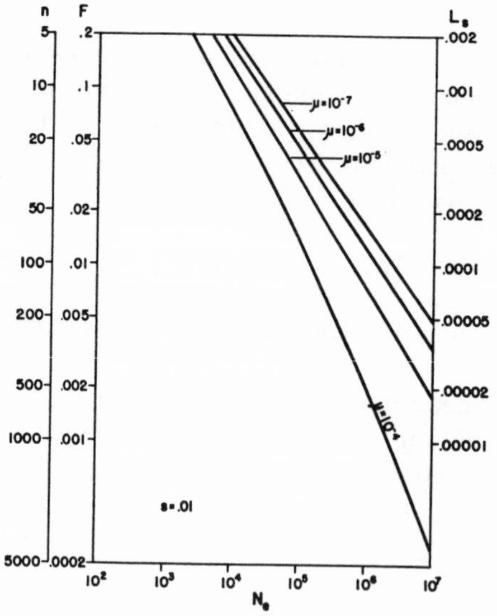


FIGURE 3.—Same as Figure 2, but with $s = .01$.

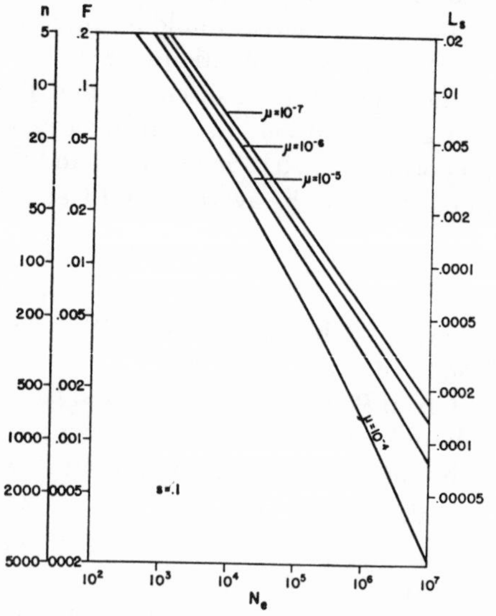


FIGURE 4.—Same as Figure 2, but with $s = .1$.

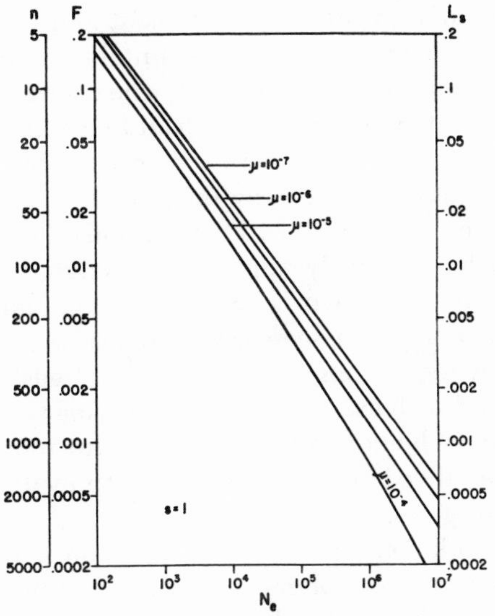


FIGURE 5.—Same as Figure 2, but with $s = 1$.

fraction $q = 1 - p$ are heterotic, then the total load per locus is

$$L_T = 2\,pu + F's \tag{29}$$

where $F'$ is the same function of $qu$ as $F$ is of $u$. Equations (28) and (29) can of course be combined, if the total load is to be determined for a number of loci, some which are giving rise only to deleterious mutants and others are mixed.

## DISCUSSION

The model chosen for discussion is unrealistic, except for very special cases. Yet it can help to provide some insight as to what situations are possible or likely in a natural population. The first case discussed, $s = 0$, shows the maximum heterozygosity per locus that can be maintained in a population by mutation alone, in the absence of any selective advantage of heterozygotes or other selective mechanism that maintains intermediate allele frequencies. The critical quantity is $4N_e u$. If this quantity is larger than one, less than half the individuals in the population will be homozygous for this locus; if less than one, more than half will be homozygous. Of course, if some of the mutants are selectively disadvantageous, if the mutation rates to different alleles are different, or if some mutants are duplicates of preexisting alleles, the proportion of homozygosity will be higher; hence these calculations represent an upper limit for heterozygosity in a population of given effective size with no selection favoring heterozygotes.

The second model discussed, the rather artificial one where each mutant is equally deleterious when homozygous and with all heterozygotes equal in fitness, provides some insight into the minimum genetic load required to maintain such a polymorphism. For example, when $s = .01$ and $u = 10^{-5}$, a population of effective size $10^4$ will have a segregation load of about .0012 (Figure 3). Under this circumstance the effective number of alleles maintained is about eight. If the selection intensity were increased to .1 (Figure 4) the number of alleles is raised to about 22, but the segregation load is .0045, about four times as large. Corresponding to a given value of $s$, $N_e$ and $u$ there is a certain load required to maintain the alleles in the population, as given by the graphs.

It has frequently been pointed out by WRIGHT and others that the total amount of selection that can be effectively applied to a population is limited. The fact that a certain amount of selection is required to maintain a polymorphism is shown by the calculation of these segregation loads. A large population can maintain a great many segregating loci, perhaps hundreds or thousands, provided these are of the type (if such exist) where there are many possible mutants, each slightly deleterious as a homozygote, but which are mutually heterotic in all combinations. On the other hand, any departure from these conditions reduces the number of heterozygotes.

Although these calculations, based as they are on a rather artificial model that favors the development of polymorphisms, do not place very severe limitations on the number of segregating loci they do cast doubt on some suggested models of population structure. One of the most extreme possibilities is that suggested by WALLACE (1958) who tentatively concluded that "on the average an individual member of the Drosophila population studied is heterozygous for genes at 50 percent or more of all loci". We suspect that the effective population number in Drosophila may well be $10^4$ or less. A mutation in order to be detected in WALLACE's experiment would have to have had a substantial viability effect. If $s$ is as small as .01, $L_s$ on our model would be $1.2 \times 10^{-3}$. If there are 10,000 loci, and half are segregating, the load would be 5000 $L_s$ or 6, and with independently acting loci the average fitness of the population would be only $e^{-6}$ or .002, compared with a hypothetical Drosophila heterozygous at all loci.

These calculations make the unlikely assumption that the requisite number of heterotic mutants for minimum load exist at all relevant loci. If the assumption is not true, the necessary reduction in fitness would be greater. For these reasons we think it is more likely that the typical Drosophila is homozygous for the majority of its genes, though the segregating minority may still be hundreds of loci. Furthermore, the segregation load although it probably depends on a minority of loci, may still exceed the mutation load as has been repeatedly suggested (e.g. CROW 1952). That the absolute number of polymorphisms may be large is indicated by the many new ones that are being discovered in man as new techniques are introduced. In very large populations, the possibility of many very nearly neutral, highly mutable multiple isoalleles cannot be ruled out, although there is no experimental evidence for the existence of such systems.

The present analysis is obviously unsatisfactory because of the various approximations and the restrictive nature of the assumptions. We have not been able to handle mathematically the situation when $s$ is different for different alleles. In an infinite population it is sufficient to replace $s$ by the harmonic mean of the $s$'s in determining the segregation load, but the situation in finite populations is not clear, nor is the effect of different fitnesses of different heterozygotes. We hope that a more general and accurate treatment will be possible.

We should like to thank ETAN MARKOWITZ and JOSEPH FELSENSTEIN for aid in computer programming and calculations.

SUMMARY

For a locus where two or more alleles are maintained by selective superiority of the heterozygotes the average fitness of the population is increased with a larger number of alleles. On the other hand, the effect of random drift in reducing the number of alleles increases greatly as the number of alleles increases, being roughly proportional to the square of the allele number. Therefore, with a population of a certain effective number and mutation rate there must be, for a given level of heterozygote advantage, a maximum number of alleles maintained. This will correspond to the minimum segregation load.

The effective number of alleles maintained in the population ($n$), the probability that a randomly chosen individual will be homozygous for this locus ($F$), and the segregation load ($L$) are given graphically for various population sizes and selection coefficients. It is assumed that all homozygotes are equally deleterious, and that each new mutant is an allele that does not already exist in the population.

When there is no selection at all, the number of isoalleles maintained in the population is approximately $4N_eu + 1$, where $N_e$ is the effective population number and $u$ is the mutation rate. Thus, if $4N_e$ is much less than the reciprocal of the mutation rate, most individuals in the population will be homozygous for this locus.

LITERATURE CITED

CROW, J. F., 1952 Dominance and overdominance. pp. 282–297. *Heterosis*. Edited by J. W. GOWEN. Iowa State College Press, Ames, Iowa. —— 1958 Some possibilities for measuring selection intensities in man. Human Biol. **30**: 1–13.

CROW, J. F., and N. E. MORTON, 1955 Measurement of gene frequency drift in small populations. Evolution **9**: 202–214.

FISHER, R. A., 1922 On the dominance ratio. Proc. Roy. Soc. Edinburgh **42**: 321–341. —— 1930, 1958 *The Genetical Theory of Natural Selection*. The Clarendon Press, Oxford. Second Revised Edition, Dover Press, New York. —— 1939 Stage of enumeration as a factor influencing the variance in the number of progeny, frequency of mutants and related quantities. Ann. Eugenics **9**: 406–408.

HALDANE, J. B. S., 1939 The equilibrium between mutation and random extinction. Ann. Eugenics **9**: 400–405.

KIMURA, M., 1955 Random drift in a multi-allelic locus. Evolution **9**: 419–435. —— 1956 Rules for testing the stability of a selective polymorphism. Proc. Natl. Acad. Sci. U.S. **42**: 336–340. —— 1961 Some calculations on the mutational load. Japan. J. Genet. **36** (suppl.): 179–190.

KIMURA, M., and J. F. CROW, 1963 The measurement of effective population number. Evolution **17**: 279–288.

MALÉCOT, G., 1948 *Les mathématiques de l'hérédité*. Masson, Paris.

MANDEL, S. P. H., 1959 The stability of multiple allelic system. Heredity **13:** 289–302.

MILLER, G. F., 1962 The evaluation of eigenvalues of a differential equation arising in a problem in genetics. Proc. Cambridge Phil. Soc. **58:** 588–593.

ROBERTSON, A., 1962 Selection for heterozygotes in small populations. Genetics **47:** 1291–1300.

WALLACE, B., 1958 The average effect of radiation-induced mutations on viability in *Drosophila melanogaster*. Evolution **12:** 532–552.

WRIGHT, S., 1931 Evolution in Mendelian populations. Genetics **16:** 97–159. —— 1937 The distribution of gene frequencies in populations. Proc. Natl. Acad. Sci. U.S. **23:** 307–320. —— 1938a. Size of population and breeding structure in relation to evolution. Science **87:** 430–431. —— 1938b The distribution of gene frequencies under irreversible mutation. Proc. Natl. Acad. Sci. U.S. **24:** 253–259. —— 1939 The distribution of self-sterility alleles in populations. Genetics **24:** 538–552. —— 1960 On the number of self-incompatible alleles maintained in equilibrium by a given mutation rate in a population of given size: a re-examination. Biometrics **16:** 61–85.

# THE NUMBER OF HETEROZYGOUS NUCLEOTIDE SITES MAINTAINED IN A FINITE POPULATION DUE TO STEADY FLUX OF MUTATIONS[1]

MOTOO KIMURA

*National Institute of Genetics, Mishima, Japan*

Received September 10, 1968

IN natural populations, it is expected that there is a constant supply of mutations in each generation. These mutations may have different persistence depending on their fitnesses, but collectively, they constitute the ultimate source of genetic variability in the populations.

Since the maintenance of genetic variability is an important subject of study in population genetics, it may be worthwhile to investigate, using various models, the effect of mutation on the genetic variability. For example, KIMURA and CROW (1964) studied the number of alleles maintained in a finite population, assuming that each mutant is an allele not preexisting in the population.

In the present paper I will use a different model and will investigate the number of heterozygous sites per individual and some related quantities that represent the statistical properties of the mutant frequency distribution, assuming that a very large number of independent sites are available for mutation. In this paper, "site" refers to a single nucleotide pair, although the theory is still appropriate to a small group of nucleotides, such as a codon.

## THE NUMBER OF HETEROZYGOUS SITES

Throughout this paper, I will consider a Mendelian population consisting of $N$ diploid individuals, each of which has a chromosome set comprising a very large number of sites. Since the effective number of the population may be different from the actual number $N$, the letter $N_e$ will be used to represent the "variance" effective number (cf. KIMURA and CROW 1963).

Let us assume that in the entire population in each generation mutants appear on the average in $v_m$ sites. We will also assume that the total number of sites per individual is so large and the mutation rate per site is so low that whenever a mutant appears, it represents a mutation at a previously homoallelic site.

Now, consider a particular site in which a mutant has appeared. We will denote by $p$ the frequency of the mutant form. Let $\phi(p,x;t)$ be the probability density that the frequency of the mutant form in the population becomes $x$ after

[1] Contribution No. 694 from the National Institute of Genetics, Mishima, Shizuoka-ken 411 Japan. Aided in part by a Grant-in-Aid from the Ministry of Education, Japan.

Genetics .61: 893–903 April 1969.

$t$ generations, given that it is $p$ at $t=0$. Then, it can be shown (KIMURA 1964) that $\phi$ satisfies the following partial differential equation

$$\frac{\partial\phi(p,x;t)}{\partial t}=\frac{1}{2}V_{\delta p}\frac{\partial^2\phi(p,x;t)}{\partial p^2}+M_{\delta p}\frac{\partial\phi(p,x;t)}{\partial p} \tag{1}$$

where $M_{\delta p}$ and $V_{\delta p}$ stand for the mean and the variance of the change of mutant frequency $p$ per generation. More precisely, the mean and the variance of the amount of change in mutant frequency $p$ during a short time interval from $t$ to $t+\delta t$ are $M_{\delta p}\,\delta t$ and $V_{\delta p}\,\delta t$ respectively. The above equation is a time homogeneous form of the Kolmogorov backward equation and is valid only when both $M_{\delta p}$ and $V_{\delta p}$ are independent of the time parameter $t$. Except for such a restriction, the equation is quite general. In a typical situation which we will investigate more in detail later, we will assume that in each site the mutant has a selective advantage $s$ in homozygotes and $sh$ in heterozygotes over the preexisting form so that

$$M_{\delta p}=sp(1-p)\{h+(1-2h)p\}, \tag{2}$$

and that the sole factor causing random fluctuation in the mutant frequency is random sampling of gametes so that

$$V_{\delta p}=p(1-p)/(2N_e). \tag{3}$$

We will assume that the parameters $p$, $s$ and $sh$ are the same for mutations at different sites. However, if both $s$ and $sh$ vary from site to site, we may use their means $\bar{s}$ and $\overline{sh}$ in the following treatments.

Since any mutant that appears in a finite population is either lost from the population or fixed in it within a finite length of time (cf. KIMURA and OHTA 1969), under continued production of new mutations over many generations, a balance will be reached between production of new mutants and their random extinction or fixation. In such a state of statistical equilibrium there is a stable frequency distribution among mutant forms at different sites, if we consider only the sites in which the mutants are neither fixed nor lost. The main aim of the present section is to obtain the average number of heterozygous sites per individual in such an equilibrium population.

Let us consider the function $\phi(p,x;t)$ in equation (1). Since $v_m$ is the number of sites in which new mutations appear in the population in each generation, $v_m\phi(p,x;t)dx$ represents the contribution made by mutants which appeared $t$ generations earlier with initial frequency $p$ to the present frequency class in which the mutant frequencies are in the range $x\sim x+dx$ (i.e. from $x$ to $x+dx$). Thus, considering all the contributions made by mutations in the past, the expected number of sites in which the mutants are in the frequency range $x\sim x+dx$ in the present generation is

$$[v_m\int_0^\infty\phi(p,x;t)dt]dx \tag{4}$$

which we will denote by $\Phi(p,x)dx$, where $0<x<1$.

Now, under random mating, the frequency of the heterozygote is $2x(1-x)$

for a site having the mutant with frequency $x$. So, assuming random mating, the total number of heterozygous sites per individual is

$$H(p) = \int_0^1 2x(1-x)\Phi(p,x)dx$$
$$= v_m \int_0^1 2x(1-x)dx \int_0^\infty \phi(p,x;t)dt \qquad (5)$$

We note that the integral with respect to $x$ is strictly over the open interval $(0<x<1)$ since we consider only sites in which the mutant frequency is neither 0 nor 1. Actually it would be more appropriate to write the limit of integration as $1/(2N)$ and $1-1/(2N)$ rather than 0 and 1, but for the sake of simplicity I will use the latter limits unless this causes the integral to diverge. In order to obtain an equation for $H(p)$, we multiply each term of equation (1) by $v_m 2x(1-x)$, and then integrate each of the resulting terms first with respect to $x$ over the interval (0, 1) and then with respect to $t$ over $(0,\infty)$. This yields

$$\int_0^\infty \frac{\partial}{\partial t}\{v_m \int_0^1 2x(1-x)\phi(p,x;t)dx\}dt$$
$$= \frac{1}{2} V_{\delta p} \frac{\partial^2}{\partial p^2} H(p) + M_{\delta p} \frac{\partial}{\partial p} H(p). \qquad (6)$$

The left hand side of this equation becomes

$$v_m \int_0^1 2x(1-x)\phi(p,x;\infty)dx$$
$$- v_m \int_0^1 2x(1-x)\phi(p,x;0)dx\,,$$

which is further reduced to $-2p(1-p)v_m$ by applying the conditions

$$\phi(p,x;\infty) = 0 \quad (0<x<1) \qquad (7)$$

and

$$\phi(p,x;0) = \delta(x-p), \qquad (8)$$

where $\delta(\cdot)$ is Dirac delta function. The first condition (7) follows from the fact that the mutant form either becomes fixed or lost within a finite length of time. The second condition (8) is simply an expression of the fact that the initial frequency of the mutant is $p$.

Thus, we obtain the ordinary differential equation for $H(p)$,

$$\frac{1}{2} V_{\delta p} H''(p) + M_{\delta p} H'(p) + 2v_m p(1-p) = 0. \qquad (9)$$

The solution which satisfies the boundary conditions

$$H(0) = H(1) = 0 \qquad (10)$$

is

$$H(p) = \{1-u(p)\} \int_0^p \psi_H(\xi)u(\xi)d\xi$$
$$+ u(p) \int_p^1 \psi_H(\xi)\{1-u(\xi)\}d\xi. \qquad (11)$$

In the above formula,

$$\psi_H(\xi) = 4v_m\xi(1-\xi) \int_0^1 G(x)dx \Big/ \{V_{\delta\xi}G(\xi)\}, \qquad (12)$$

and

$$u(p) = \int_0^p G(x)dx \Big/ \int_0^1 G(x)dx \quad \text{(Kimura 1962)} \tag{13}$$

is the probability of ultimate fixation, in which

$$G(x) = \exp\left\{-2\int_0^x \frac{M_{\delta\xi}}{V_{\delta\xi}}\, d\xi\right\}, \tag{14}$$

where exp $\{\cdot\}$ denotes the exponential function.

In the special case of no dominance for which $h=\frac{1}{2}$, formula (2) gives $M_{\delta p} = \frac{1}{2}sp(1-p)$. Combining this with the formula for $V_{\delta p}$ given in (3), we have $G(x)=e^{-2N_e sx}$ or $G(x)=e^{-2Sx}$ if we put $S=N_e s$. Then, $\psi_H(\xi)=4N_e v_m e^{2S\xi}(1-e^{-2S})/S$, and $u(p)=(1-e^{-2Sp})/(1-e^{-2S})$. Thus formula (11) yields

$$H(p) = \frac{4N_e v_m}{S}\left(\frac{1-e^{-2Sp}}{1-e^{-2S}} - p\right), \tag{15}$$

where $S=N_e s$. This may also be expressed as

$$H(p) = \frac{4v_m}{s}\,[u(p)-p]. \tag{15$'$}$$

At the limit of $s \to 0$, we have

$$H(p) = 4N_e v_m p(1-p). \tag{16}$$

In a population consisting of $N$ individuals, if the mutant form in each site is represented only once at the moment of its occurrence, $p=1/(2N)$ and the number of heterozygous sites per individual is given by $H(1/2N)$ in the above formulas. Thus, for the case of no dominance, we have approximately

$$H(1/2N) \approx 4v_m(N_e/N) \tag{17a}$$

if the mutant is advantageous such that $2N_e s >> 1$,

$$H(1/2N) \approx 2v_m/(Ns') \tag{17b}$$

if it is deleterious such that $2N_e s' >> 1$ in which $s'=-s$, and,

$$H(1/2N) \approx 2v_m(N_e/N) \tag{17c}$$

if it is almost neutral such that $|2N_e s| << 1$.

These results suggest that mutations having a definite advantage or disadvantage can not contribute greatly to the heterozygosity of an individual because of the rare occurrence of advantageous mutations and rapid elimination of deleterious ones. They also show that in a finite population the total number of heterozygous sites per individual is determined by the number of mutations per gamete and the population numbers, and not by the total number of sites.

## STATISTICAL PROPERTIES OF THE EQUILIBRIUM DISTRIBUTION UNDER STEADY FLUX OF MUTATIONS

The number of heterozygous sites $H(p)$ studied in the previous section is but one property of the equilibrium distribution $\Phi(p,x)$. Namely, it is the expectation of $2x(1-x)$ with respect to this distribution. Here $\Phi(p,x)$ represents the stable frequency distribution of mutant forms among segregating sites $(0<x<1)$ such that $\Phi(p,x)dx$ gives the expected number of sites having mutants in the frequency range $x \sim x+dx$.

Now, let us study more generally the expectation of an arbitrary function $f(x)$ with respect to this distribution. We will denote such an expectation (functional) by $I_f(p)$, that is,

$$I_f(p) \equiv \int_0^1 f(x)\Phi(p,x)dx$$
$$= v_m \int_0^\infty [\int_0^1 f(x)\phi(p,x;t)dx]dt \tag{18}$$

Again, the integral with respect to $x$ is over the open interval $(0,1)$ and actually it is more appropriate if we use $1/(2N)$ and $1-1/(2N)$ as the limit of the integration, especially when the value of the integral changes significantly by including $x=0$ and 1. Using the same procedure that was used to derive (9) from (1) except that in this case each term of (1) is multiplied by $v_m f(x)$ rather than by $v_m 2x(1-x)$, we obtain the following ordinary differential equation for $I_f(p)$:

$$\tfrac{1}{2} V_{\delta p} I_f''(p) + M_{\delta p} I_f'(p) + v_m f(p) = 0\,. \tag{19}$$

This corresponds to (9) which is a special case of $f(p)=2p(1-p)$. Furthermore, since the "mutations" at $p=0$ and $p=1$ do not contribute to the segregating sites, $\phi(0,x;t)=\phi(1,x;t)=0$ for $0<x<1$. Therefore we have the boundary conditions

$$I_f(0) = I_f(1) = 0. \tag{20}$$

The solution of equation (19) which satisfies the boundary conditions (20) is

$$I_f(p) = \{1-u(p)\}\int_0^p \psi_f(\xi)u(\xi)d(\xi) + u(p)\int_p^1 \psi_f(\xi)\{1-u(\xi)\}d\xi\,, \tag{21}$$

where $u(p)$ is the probability of fixation given by (13) and

$$\psi_f(\xi) = 2v_m f(\xi)\int_0^1 G(x)dx \Big/ \{V_{\delta\xi} G(\xi)\}$$
$$= 2v_m f(\xi) \,/\, \{V_{\delta\xi} u'(\xi)\}\,, \tag{22}$$

in which $u'(\xi)=du(\xi)/d\xi$. We note here that $H(p)$ in the previous section is a special case of $I_f(p)$ in which $f(x)=2x(1-x)$, as comparison of (21) with (11) clearly shows. Furthermore, there are several other quantities of genetic interest that may be derived by assigning various functions of $x$ to $f$ in the above formula (21).

The total number of segregating sites in the population at any given moment may be obtained by taking $f(x)=1$ in (21). If there is no dominance and the random change in mutant frequency is due to random sampling of gametes, that is, if

$$M_{\delta p} = \frac{S}{2} p(1-p) \text{ and } V_{\delta p} = p(1-p)/(2N_e)\,, \tag{23}$$

we have

$$I_1(p) = \frac{2N_e v_m}{S}\left\{\frac{1-e^{-2Sp}}{1-e^{-2S}}\int_p^1 \frac{1-e^{-2S(1-\xi)}}{\xi(1-\xi)}d\xi \right.$$
$$\left. + \frac{e^{-2Sp}-e^{-2S}}{1-e^{-2S}}\int_0^p \frac{e^{2S\xi}-1}{\xi(1-\xi)}d\xi\right\}, \tag{24}$$

where $S = N_e s$. If the mutant is represented only once at the moment of its occurrence, $p = 1/(2N)$, and the above formula reduces approximately to

$$I_1\left(\frac{1}{2N}\right) = \frac{2v_m}{1-e^{-2S}}\left(\frac{N_e}{N}\right)\left\{\log_e(2N) - e^{-2S}\int_{S/N}^{2S}\frac{e^{\lambda}}{\lambda}d\lambda + \int_0^{2S-S/N}\frac{1-e^{-\lambda}}{\lambda}d\lambda + \left(1-\frac{S}{N}-e^{-2S}\right)\right\}, \tag{25}$$

assuming that $|s|$ is small and $N$ is large. The integrals in the right hand side of the above formula may be evaluated by using the exponential integrals

$$E_1(x) = \int_x^{\infty}\frac{e^{-\lambda}}{\lambda}d\lambda \text{ and } E_i(x) = \int_{-\infty}^{x}\frac{e^{\lambda}}{\lambda}d\lambda, \quad (x>0)$$

for which fairly extensive tabulations are available (cf. ABRAMOWITZ and STEGUN 1964). Thus, if the mutant is advantageous such that $S = N_e s >> 1$, we obtain

$$I_1(1/2N) \approx 2v_m(N_e/N)\{\log_e(4NN_e s) + \gamma + 1\}, \tag{26}$$

where $\gamma$ is EULER's constant $0.5772\cdots$. On the other hand, if the mutant is deleterious $(s<0)$, writing $-s = s'$ and assuming $N_e s' >> 1$, we obtain

$$I_1(1/2N) \approx 2v_m(N_e/N)\{-\log_e(N_e s'/N) - \gamma + 1\}. \tag{27}$$

If the mutation is neutral $(s=0)$, formula (24) reduces to

$$I_1(p) = -4N_e v_m\{p\log_e p + (1-p)\log_e(1-p)\}, \tag{28}$$

from which we obtain

$$I_1(1/2N) \approx 2v_m(N_e/N)\{\log_e(2N) + 1)\} \tag{29}$$

Going back to the general formula (21), the mean and the variance of the number of mutants per individual is given by $I_f(p)$ with $f=2x$ and $2x(1-x)$ respectively. The variance of the number of heterozygous sites per individual may be obtained from $H(p)-K(p)$, where $K(p)=I_f(p)$ with $f=\{2x(1-x)\}^2$. For the case of no dominance corresponding to (23), we have, assuming $s\neq 0$,

$$K(p) = \frac{8N_e v_m}{S}\left\{u(p)\left(\frac{1}{6}-\frac{1}{2S^2}\right) - \left(\frac{p^2}{2}-\frac{p^3}{3}\right) + \frac{p(1-p)}{2S} + \frac{2p}{(2S)^2}\right\}, \tag{30}$$

where $S=N_e s$ and $u(p)=(1-e^{-2Sp})/(1-e^{-2S})$. On the other hand, if $s=0$, we have

$$K(p) = \frac{4}{3}N_e v_m p(1-p)(1+p-p^2). \tag{31}$$

Thus, for neutral mutations, the variance in the number of heterozygous sites per individual is

$$\sigma_H^2(p) = \frac{4}{3}N_e v_m p(1-p)(2-p+p^2). \tag{32}$$

If $p=1/(2N)$, this gives

$$\sigma_H(1/2N) = \left\{\frac{4}{3}\left(\frac{N_e}{N}\right)v_m\right\}^{1/2} \tag{33}$$

approximately. The substitutional load in a finite population studied by KIMURA and MARUYAMA (1969) is given by $I_f(p)$ with $f = s - \{sp^2 + sh2p(1-p)\}$ assuming that $s \geqq sh > 0$ in (2).

Finally, as seen from the definition (18), the distribution function $\Phi$ itself may be obtained from $I_f(p)$ of (21) by putting $f(x) = \delta(x-y)$, where $\delta(\cdot)$ is the Dirac delta function. In this case

$$\psi_f(\xi) = 2v_m\delta(\xi - y)/\{V_{\delta\xi}u'(\xi)\},$$

and the first integral in the right hand side of (21) vanishes if $y > p$ because in the integral $\xi \leqq p$ and therefore $\delta(\xi - y) = 0$. On the other hand, the second integral vanishes if $y < p$ because in that integral $\xi \geqq p$ and therefore $\delta(\xi - y) = 0$. Thus, we obtain

$$\Phi(p,y) = 2v_m u(p)\{1-u(y)\}/\{V_{\delta y}u'(y)\} \tag{34}$$

for $p \leqq y < 1$, and

$$\Phi(p,y) = 2v_m\{1-u(p)\}u(y)/\{V_{\delta y}u'(y)\} \tag{35}$$

for $0 < y \leqq p$. The case which may be of the most genetic significance is the one in which each mutant is represented only once at the moment of its occurrence so that $p = 1/(2N)$. In this case, only (34) is needed to express the mutant frequency distribution among segregating sites. Thus writing $\Phi(y)$ for $\Phi(1/2N, y)$ and using the letter $x$ rather than $y$ to represent the mutant frequency, we obtain the distribution

$$\Phi(x) = 2v_m u\left(\frac{1}{2N}\right)\{1-u(x)\}/\{V_{\delta x}u'(x)\}, \tag{36}$$

in which $1/(2N) \leqq x \leqq 1 - 1/(2N)$. Since from (13), we have approximately

$$u\left(\frac{1}{2N}\right) = \left(\frac{1}{2N}\right)\Big/\int_0^1 G(x)dx,$$

the above distribution (36) may also be expressed as

$$\Phi(x) = \frac{2v}{V_{\delta x}G(x)}\,\frac{\int_x^1 G(x)dx}{\int_0^1 G(x)dx}, \tag{37}$$

where $v = v_m/(2N)$ is the mutation rate per gamete per generation, still assuming that whenever a mutation occurs it represents a new mutation at a different site and that each mutant is represented only once at the moment of its occurrence. This agrees with the formula obtained by KIMURA (1964) as an extension of WRIGHT's distribution for irreversible mutation, except that $v$ there stands for the mutation rate per locus.

## DISCUSSION

We should start our discussion by examining the adequacy of the present model. The basic assumptions of the model are that: (i) a very large (practically infinite) number of sites are available for mutation and (ii) whenever a mutant appears, it represents a mutation at a new (different) site. However, since we are only considering segregating sites, the second assumption may be weakened and replaced by the assumption that whenever a mutation occurs it takes place at a site in which a previous mutation is not still segregating. This means that the present model is adequate to represent reality if the total number of sites available for mutation is very much larger than $I_1(p)$ the number of temporarily segregating sites.

There are never more than four "alleles" corresponding to four kinds of nucleotides. However only very rarely will more than two types be present simultaneously so two-allele theory is adequate.

In mammals, the number of nucleotide pairs making up the haploid chromosome set is estimated to be $3\sim4\times10^9$ and this is sufficient to code for $2\times10^6$ polypeptides each consisting of 500 amino acids. In other words, the total number of cistrons may be as large as two million. On the other hand, the effective number of population ($N_e$) is probably tens of thousands or less in most cases.

If in each generation, one advantageous mutant gene appears within the population ($v_m=1$) consisting of $N=10^4$ individuals and having an effective number half as large ($N_e/N=0.5$), then, assuming $s=0.01$, we have, from (26), $I_1(1/2N) \approx 16.1$. This is very much smaller than two million and the model is clearly adequate to treat such a situation. Mutant genes with definitely deleterious effects such as $s<-0.1$ must be much more common. So, if we take $-s=s'=0.1$ and $v=v_m/(2N)=0.1$, that is, 10% selective disadvantage and the mutation rate per gamete of 0.1, we have, from (27), $I_1(1/2N)\approx6.8\times10^3$, which is still much smaller than two million. There is some possibility that neutral or nearly neutral mutations occur at a considerably higher rate of roughly 2 per gamete per generation (KIMURA 1968a). If we take $v=v_m/(2N)=2$, we obtain, from (29), $I_1(1/2N) \approx4.4\times10^5$. This is a large number amounting to about 22% of the total number cistronic loci, a fraction too large to be neglected. However, the model is appropriate if we consider the total number of nucleotide sites ($4\times10^9$) rather than the cistrons. Actually, the present model is most pertinent if we take the individual nucleotide site as the unit of mutation. Then $I_1(1/2N)$ represents the number of nucleotide sites in which mutant forms are segregating in the population.

Similarly, $H(1/2N)$ represent the number of heterozygous nucleotide sites per individual. Assuming that the majority of molecular mutations due to base substitution is almost neutral for natural selection and that they occur at the rate of 2 per gamete per generation ($v=v_m/2N=2$), we have, from (17c),

$$H(1/2N)\approx8N_e.$$

Thus, in a population of effective size 10,000, the average number of heterozygous nucleotide sites per individual is about $8\times10^4$. Furthermore, from (33), the standard deviation of this number is about 230.

The probability of a particular site being heterozygous for a selectively neutral mutant is $4N_eu$, where $u$ (i.e. $u=v$/total number of sites) is the mutation rate per site. More accurately, if mutation rates are equal in all directions, the proportion of heterozygous sites is $4N_eu/(1+16N_eu/3)$ (cf. KIMURA 1968b). However, $u$ is of the order $10^{-9}$ whereas $N_e$ is probably less than $10^5$, so $4N_eu$ is completely adequate.

A cistron of 1000 sites will be heterozygous at one or more sites with probability $1-(1-4N_eu)^{1000}\approx1-e^{-4000N_eu}$. For example, if $u=10^{-9}$ and $N_e=10^5$, the heterozygosity per nucleotide is $4\times10^{-4}$ and the proportion of heterozygous cistrons is $1-e^{-0.4}=0.33$.

On the other hand, KIMURA and CROW (1964) and KIMURA (1968b) showed that for a model in which each new mutant per cistron is not previously represented in the population—a model that should be almost equivalent to the present model—the heterozygosity is $4N_eU/(1+4N_eU)$, where $U$ is the mutation rate per cistron. For a cistron of 1000 nucleotides, then, $U=10^{-6}$ and $4N_eU/(1+4N_eU)$ is $0.4/(1+0.4)=0.29$.

The lack of correspondence between $1-e^{-4N_eU}$ and $4N_eU/(1+4N_eU)$ is because the first permits each site to come to equilibrium independently whereas the second regards all sites as completely linked. The truth must usually be somewhere in between the two models. If intra-cistronic recombination is frequent the present model is more appropriate; probably this is so low that the second formula is more correct. However, if the number of heterozygous sites per individual is of interest, the formula of the present paper is appropriate.

There is another interpretation of $H(1/2N)$ that is of particular use in assessing the substitutional load based on competition. Since $I_f(p)$ with $f=2x(1-x)$ gives the variance of the number of mutants per individual, $\sigma_m=\sqrt{H(1/2N)}$ is equal to the standard deviation, if each mutant is represented only once at the moment of its occurrence. Now, let us assume that in each generation definitely advantageous mutations with respect to competitive ability occur at $v_m$ of the sites. Then, at statistical equilibrium in which the gene substitution is proceeding at a constant rate, the difference in the number of mutant sites between an average individual and the one having the most probable largest number of mutants within the population is

$$\tilde{\bar{x}}_{N,1}\approx\sqrt{2\log_e(0.4N)} \tag{38}$$

times of $\sigma_m$. The above asymptotic formula (38) is FRANK's formula giving "the most probable largest normal value" (cf. GUMBEL 1958).

Let $K$ be the average number of gene substitutions in the population per generation so that $K=v_mu(1/2N)$ (cf. KIMURA and MARUYAMA 1968). Then the substitutional load measured in Malthusian parameters with respect to competitive ability may be given by

$$\tilde{L}_e=\frac{s}{2}\,\tilde{x}_{N,1}\,\sigma_m. \tag{39}$$

Assuming no dominance and enough selective advantage ($N_es>>1$), we have approximately $u(1/2N)=s(N_e/N)$ and $\sigma_m=\sqrt{4v_m(N_e/N)}$, and therefore,

$$\tilde{L}_e=\sqrt{2Ks\log_e(0.4N)} \tag{40}$$

where

$$K=v_ms(N_e/N) \tag{41}$$

For example, let us consider a population of $N=25{,}000$ in which gene substitution is being carried out at the rate of 2 per generation ($K=2$). If the selection coefficient of the advantageous mutant gene is $s/2=0.1$, we have $\tilde{L}_e\approx2.7$ from (40), namely, disregarding environmental effects an individual carrying the largest number of advantageous mutant genes in the population must have about

$e^{2.7}$ or 14.9 times as many offspring as the average individual. In this case, the actual number $\nu_m$ of advantageous mutations appearing in each generation is 20 from (41) assuming that $N_e/N=0.5$. On the other hand, if the selection coefficient is one hundredth as large ($s/2=0.001$), the load becomes 1/10 as large ($\tilde{L}_e=0.27$) but the number of advantageous mutations must be 100 times more frequent ($\nu_m=2{,}000$), such that one out of every 25 gametes carries a new advantageous mutation in each generation. This is a very high rate of production of advantageous mutations comparable to that of recessive lethal genes.

The mathematical treatment in the present paper enables us to obtain not only the average number of heterozygous nucleotide sites but also various statistical properties of the mutant frequency distribution attained under a steady flux of mutations including the frequency distribution itself. The gene frequency distribution obtained by FISHER (1930) assuming a supply of one mutation in each generation and also the distribution obtained by WRIGHT (1945) assuming irreversible mutations were both the solutions of the appropriate forward equations under the condition of constant probability flux. The present treatment shows that they are special cases of equation (37) derived by assigning a special function to $f(x)$ in (21).

I believe that the present treatment has brought some refinement and extension to the great work of WRIGHT (1938, 1942 and 1945) on the distribution of gene frequencies under irreversible mutation. By so doing I hope to penetrate into the domain of population genetics at the molecular level.

I would like to express my thanks to DR. J. F. CROW for reading the manuscript and making valuable suggestions.

## SUMMARY

A theoretical treatment was presented which enables us to obtain the average number of heterozygous nucleotide sites per individual and related quantities that describe the statistical property of the mutant frequency distribution attained under steady flux of mutations in a finite population.—The main assumptions of the model are that (i) a very large (practically infinite) number of sites are available for mutation and (ii) whenever a mutant appears, it represents a mutation at a new (different) site. Such a model may be particularly realistic if we consider the individual nucleotide site rather than the conventional genetic locus as a unit of mutation.—In a population consisting of $N$ individuals and having the variance effective number $N_e$, the average number of heterozygous sites per individual due to neutral or nearly neutral mutations is $2\nu_m N_e/N$, where $\nu_m$ is the number of sites in which new mutations appear in the population in each generation.—In a mammalian species having the variance effective number of 10,000, if the majority of molecular mutations due to base substitutions is almost neutral for natural selection and if they occur at the rate of 2 per gamete per generation ($\nu_m/2N=2$), then the average number of heterozygous nucleotide sites per individual becomes about $8\times10^4$ with the standard deviation of about 230.

LITERATURE CITED

ABRAMOWITZ, M., and I. A. STEGUN, (ed.) 1964 *Handbook of Mathematical Functions with Formulas, Graphs, and Mathematical Tables.* U.S. Department of Commerce, Washington, D.C.

FISHER, R. A., 1930 *The Genetical Theory of Natural Selection.* Oxford, Clarendon Press.

GUMBEL, E. J., 1958 *Statistics of Extremes.* New York, Columbia University Press.

KIMURA, M., 1962 On the probability of fixation of mutant genes in a population. Genetics **47:** 713–719. —— 1964 Diffusion models in population genetics. J. Appl. Probab. **1:** 177–232. —— 1968a Evolutionary rate at the molecular level. Nature **217:** 624–626. —— 1968b Genetic variability maintained in a finite population due to mutational production of neutral and nearly neutral isoalleles. Genet. Res. **11:** 247–269.

KIMURA, M., and J. F. CROW, 1963 The measurement of effective population number. Evolution **17:** 279–288. —— 1964 The number of alleles that can be maintained in a finite population. Genetics **49:** 725–738.

KIMURA, M., and T. MARUYAMA, 1969 The substitutional load in a finite population. Heredity **24:** 101–114.

KIMURA, M., and T. OHTA, 1969 The average number of generations until fixation of a mutant gene in a finite population. Genetics **61:** 763–771.

WRIGHT, S., 1938 The distribution of gene frequencies under irreversible mutation. Proc. Natl. Acad. Sci. U.S. **24:** 253–259. —— 1942 Statistical genetics and evolution. Bull. Amer. Math. Soc. **48:** 223–246. —— 1945 The differential equation of the distribution of gene frequencies. Proc. Natl. Acad. Sci. U.S. **31:** 382–389.

# A model of mutation appropriate to estimate the number of electrophoretically detectable alleles in a finite population*

By TOMOKO OHTA and MOTOO KIMURA
*National Institute of Genetics, Mishima, Japan*

(*Received* 20 *February* 1973)

SUMMARY

A new model of mutational production of alleles was proposed which may be appropriate to estimate the number of electrophoretically detectable alleles maintained in a finite population. The model assumes that the entire allelic states are expressed by integers $(\ldots, A_{-1}, A_0, A_1, \ldots)$ and that if an allele changes state by mutation the change occurs in such a way that it moves either one step in the positive direction or one step in the negative direction (see also Fig. 1). It was shown that for this model the 'effective' number of selectively neutral alleles maintained in a population of the effective size $N_e$ under mutation rate $v$ per generation is given by

$$n_e = \sqrt{(1+8N_e v)}.$$

When $4N_e v$ is small, this differs little from the conventional formula by Kimura & Crow, i.e. $n_e = 1+4N_e v$, but it gives a much smaller estimate than this when $4N_e v$ is large.

Since a model of isoalleles with infinite states was proposed by Kimura & Crow (1964) it has been used extensively to estimate the number of selectively neutral isoalleles that can be maintained in a finite population under a given mutation rate. In this model it is assumed that the number of possible allelic states at a locus is so large that whenever mutation occurs it represents a new, not pre-existing allele. This model could be applied directly to actual situations if individual variants were identified at the level of nucleotide or amino acid sites. At present, however, our experimental analyses of the genetic variability of natural populations are at much cruder level of identifying electrophoretically detectable variants. In other words, a gene mutation can be detected only when it leads to a replacement of amino acid which causes change in electric charge of the molecule. Not only such variants occupy a relatively small fraction of the entire variants at the molecular level, but also they are identified only as a discrete spectrum of broad bands on the electrophoresis gels. This means that the electrophoretic method does not have the resolving power which the model of Kimura & Crow presupposes.

The purpose of the present note is to propose a model which may be more appropriate to estimate the number of electrophoretically detectable alleles, allowing us to compare theoretical predictions with actual observations. Let us assume that the entire sequence of allelic states are expressed by integers as shown in Fig. 1, and that if an allele changes its state by a single step mutation, the change occurs in such a way that it moves either one step in the positive direction or one step in the negative direction. In other words, it can

* Contribution no. 922 from the National Institute of Genetics, Mishima, Shizuoka-ken, 411, Japan.

mutate only to one of the two adjacent states. In this model, one positive and one negative changes (in charge) cancel each other, leading the allele back to the original state. An actual example of this type of change is afforded by mutant proteins A 11 and A 46 of tryptophan synthetase of *E. coli*. According to Henning & Yanofsky (1963), A 11 moves toward the negative direction after electrophoresis on cellulose acetate while A 46 moves toward the positive direction. The mobility of the double mutant A 11–46 protein was found to be identical with that of the wild-type A protein.

Consider a diploid population with the effective size $N_e$, and let $v$ be the mutation rate per locus per generation. To simplify the treatment, we shall assume that under mutation, changes toward the positive and the negative directions occur with equal frequencies (i.e. each with $\frac{1}{2}v$ as shown in Fig. 1). Let $x_i$ be the frequency of the $i$th allele $A_i$ ($i$ = integer), and also let

$$C_k = E\{\sum_i x_i x_{i+k}\} \quad (k = 0, 1, \ldots), \tag{1}$$

where $E$ stands for the operator for taking expectation. The summation is over all relevant alleles in the population. Note that $C_0$ is the expected value of the sum of squares of allelic frequencies, so that it gives the average homozygosity under random mating. The correlation between frequencies of alleles that are $k$ steps apart may be given by $C_k/C_0$.

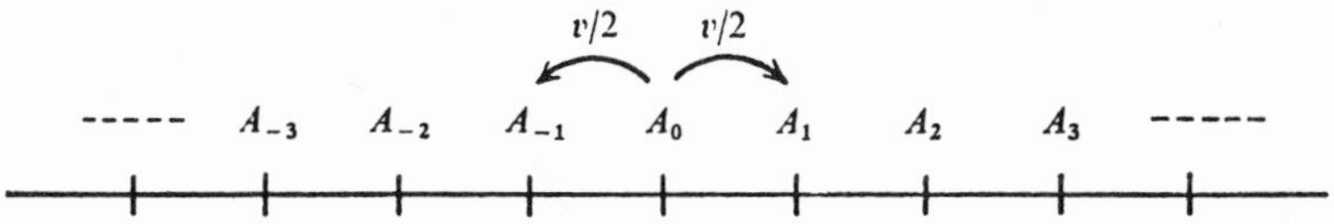


Fig. 1. Diagram illustrating the model of production of electrophoretically detectable alleles.

In order to obtain a set of equations giving the rate of change in $C_k$, we use the basic equation for generating the moments (Ohta & Kimura, 1971). This equation takes the following form:

$$(d/dt)\,E(f) = E\{L(f)\}, \tag{2}$$

where $L$ is the differential operator of the Kolmogorov backward equation (see equation A 5 of Ohta & Kimura, 1971) and $f$ is an arbitrary continuous function of $x_i$'s. Assuming that the alleles are selectively neutral, the mean ($M$), the variance ($V$) and the covariance ($W$) of gene frequency changes per generation are

$$M_{\delta x_i} = \tfrac{1}{2}v(x_{i-1}+x_{i+1}) - vx_i,$$

$$V_{\delta x_i} = \frac{x_i(1-x_i)}{2N_e} \tag{3}$$

and

$$W_{\delta x_i \delta x_j} = -\frac{x_i x_j}{2N_e} \quad (i \neq j).$$

Let $f = x_i^2$ in formula (2), then

$$\frac{d}{dt}E(x_i^2) = E\left\{v(x_{i-1}+x_{i+1})\,x_i - 2vx_i^2 + \frac{x_i(1-x_i)}{2N_e}\right\}$$

or

$$\frac{d}{dT}E(x_i^2) = E\{2N_e v(x_{i-1}x_i + x_i x_{i+1}) - (4N_e v + 1)\,x_i^2 + x_i\},$$

where $T\,(= t/(2N_e))$ is time measured in the unit of $2N_e$ generations. By summing up the above equation for all $i$, and noting (1), we obtain

$$dC_0/dT = -(1+4N_e v)\,C_0 + 4N_e vC_1 + 1, \tag{4}$$

since $\sum_i x_i = 1$. Similarly, by letting $f = x_i x_{i+k}$, we obtain the following equation for $C_k$:

$$\frac{dC_k}{dT} = 2N_e vC_{k-1} - (1+4N_e v)\, C_k + 2N_e vC_{k+1} \quad (k \geqslant 1). \tag{5}$$

At equilibrium, we have $dC_k/dt = 0$ for all $k$, and the appropriate equilibrium solution for the set of equations (4) and (5) which vanishes at $k = \infty$ is given by

$$C_k = H_0 \lambda_1^k, \tag{6}$$

where

$$H_0 = 1/\surd(1+8N_e v) \tag{7}$$

and

$$\lambda_1 = \frac{1+4N_e v - \surd(1+8N_e v)}{4N_e v}. \tag{8}$$

To check this solution, we considered a finite ($n$) set of allelic states arranged on a circle, and derived a set of equations transforming $C_k$'s ($k = 0, 1, \ldots, n$) from one generation to the

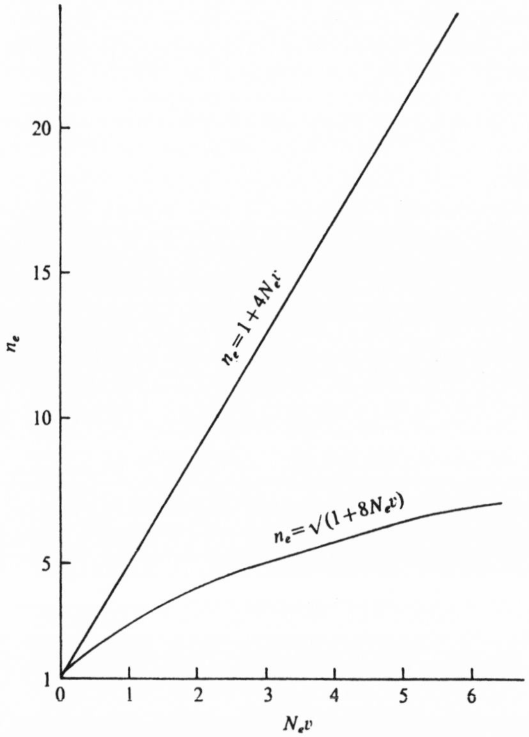


Fig. 2. Relationship between the effective number of alleles ($n_e$) and $N_e v$ under the two models (the model of Kimura & Crow and the present model).

next. Then we multiplied the matrix corresponding to the finite set of transformations a large number of times by a computer assuming $n = 10, 20$ and $40$ allelic states and $N_e v = 0{\cdot}5$ and $2{\cdot}0$. It was found that formula (6) gives good approximation to the equilibrium values of $C_0$, $C_1$, etc., obtained by the matrix multiplication.

The 'effective' number of alleles is given by the reciprocal of the average homozygosity $H_0$. Thus, we obtain

$$n_e = \surd(1+8N_e v). \tag{9}$$

Monte Carlo experiments were performed to check this formula for various combinations of values of $N_e$ and $v$ and the results were satisfactory. The above formula for $n_e$ should be compared with the corresponding formula,

$$n_e = 1 + 4N_e v, \tag{10}$$

obtained by Kimura & Crow (1964). Fig. 2 illustrates the relationship between $N_e v$ and $n_e$ for these two models.

It may be seen from the figure that for a small value of $N_e v$ these two formulae differ rather little. For example, if $4N_e v = 0{\cdot}2$, we have $n_e = 1{\cdot}18$ from (9) but $n_e = 1{\cdot}20$ from (10). The former gives the average heterozygosity of 15·5 % while the latter gives 16·7 %. However, these two formulae give very different estimates for $n_e$ when $N_e v$ is large. For example, if $N_e v = 100$, the present model (9) gives $n_e = 28{\cdot}3$, while the conventional model (10) gives $n_e = 401$. Recently, Ayala *et al.* (1972) in criticizing our neutral theory of protein polymorphism (Kimura & Ohta, 1971), point out that in *D. willistoni*, which is estimated to have $10^9$ breeding flies per generation, if we take a lower estimate of $10^{-7}$ for mutation rate, we should have $4N_e v = 400$, while the observed average heterozygosity per locus is about 18 %. In other words, the theoretical value of $n_e$ obtained from Kimura & Crow's formula overestimates the true value by the factor of some 400. Such a marked discrepancy is reduced, however, if we use the present formula, although the theoretical value is still some 20 times greater. In addition, there are two possibilities which have to be taken into account and both of which reduce our theoretical estimate based on the neutral theory. First, as pointed out by Kimura & Ohta, the rate $v = 10^{-7}$ may be appropriate for *the neutral mutation rate per year*, but not per generation. This means we should take much smaller value for the neutral mutation rate per generation for fruit flies that breed all the year round in tropical forests. Secondly, there is possibility that $N_e = 10^9$ is an overestimate, not as the number of breeding flies per generation, but as the number applicable to the formula of effective allele number. Not only this number is controlled by the minimum population size when population number fluctuates from generation to generation, but also it takes the length of time in the order of the population size for the equilibrium state in the distribution of allelic frequencies to be established. (Treatment of this subject will be published elsewhere.) It is quite likely that the effect of small population number during the last glaciation still remains in the genetic composition of tropical fruit flies. In this connexion we would like to point out that neutral alleles behave quite differently from lethal genes (having short life-span) and inversion polymorphisms (many of which may be subject to 'balancing selection').

## REFERENCES


AYALA, F. J., POWELL, J. R., TRACEY, M. L., MOURÃO, C. A. & PÉREZ-SALAS, S. (1972). Enzyme variability in the *Drosophila willistoni* group. IV. Genetic variation in natural populations of *Drosophila willistoni*. *Genetics* **70**, 113–139.

HENNING, U. & YANOFSKY, C. (1963). An electrophoretic study of mutationally altered A proteins of the tryptophan synthetase of *Escherichia coli*. *J. Molecular Biology* **6**, 16–21.

KIMURA, M. & CROW, J. F. (1964). The number of alleles that can be maintained in a finite population. *Genetics* **49**, 725–738.

KIMURA, M. & OHTA, T. (1971). Protein polymorphism as a phase of molecular evolution. *Nature* **229**, 467–469.

OHTA, T. & KIMURA, M. (1971). Linkage disequilibrium between two segregating nucleotide sites under the steady flux of mutations in a finite population. *Genetics* **68**, 571–580.

PART FIFTEEN

# Molecular Evolution

## Introduction

Those who currently study DNA sequences for the study of systematics may be surprised to find how little information was available until quite recently. Kimura and Ohta (no. 40) compared four DNA sequences of 5S ribosomal RNA genes from humans, yeast, and two bacteria, and dated the eukaryote-prokaryote divergence as about 2 billion years ago. Now the number of 5S rRNA sequences alone is more than 620 (Hori and Osawa 1987), and many other genes have become available (see Woese 1991 for a review of 16S rRNA). The LASL-GDB (version 78.0, 1993) released from GenBank/EMBL consists of 133,970 entries, the total number of nucleotides stored being 153,217,959. This is one of a very few papers in which Kimura did not discuss the mechanism of molecular evolution in any detail and simply applied the molecular clock to the phylogenetic analysis. Ironically, the paper may mark the beginning of the present bifurcation between molecular evolution and population genetics (long-term vs. short-term evolution). This may not have been Kimura's intention, but at the deepest level the neutral theory indeed demanded a bifurcation of molecular evolution from phenotypic evolution (Provine 1990).

As soon as amino acid sequence data were generated around 1960, molecular evolution became popular. With the introduction of DNA technology into evolutionary biology about a decade ago, it soon became clear that the genome has been evolving in a complex manner due to its own mechanisms, such as unequal crossing over, gene conversion, duplicative transposition, and so forth. Stimulated by these molecular evolutionary mechanisms, many population geneticists have gotten involved in quantifying them. Ohta (1976) was one of the first to quantify unequal crossing over on the basis of population genetics considerations. Kimura and Ohta (no. 41) analyzed the correlation between two members of a multigene family subjected to intrachromosomal unequal crossing over and ordinary interchromosomal crossing over. The spread of a new mutant in the family is often called concerted evolution, but the extent of homogenization may depend on the distance between members. This homogenization process is stochastic, and under some simplified assumptions it resembles the spread of a mutant gene in a randomly mating population (e.g., Ohta 1980).

# Eukaryotes-Prokaryotes Divergence estimated by 5S Ribosomal RNA Sequences

DATING the principal events in the history of life on the Earth is an interesting subject in evolutionary studies. Here we estimate the time of divergence of the eukaryotes and the prokaryotes through comparative studies of 5S ribosomal RNA sequences, coupled with those of cytochrome *c*. By prokaryotes we mean primitive forms having no true nucleus (bacteria and blue-green algae), while by eukaryotes we mean higher nucleated organisms such as plants and animals including yeasts and fungi. The principle we use in our estimation is that the rate of nucleotide substitutions in the course of evolution is constant per year per site for each informational macromolecule as long as the structure and function of the molecule remain unaltered.

To estimate the evolutionary distances (number of mutant substitutions) among 5S rRNA sequences, we made the alignment shown in Fig. 1, using published data[1–3] on human, yeast and bacterial (*Escherichia coli* and *Pseudomonas fluorescens*) sequences. To arrive at this alignment, previous attempts[1,3] involving two or three sequences were helpful. The alignment is made in such a way that the number of matches between sequences is maximized while keeping the gaps inserted as few as possible. It involves a trial and error process, shifting various regions, and counting the number of nucleotides by which the two sequences agree with each other, followed by calculation of probability that this or better agreement occurs by chance. Figure 1 shows clearly (as was noted already by others) the marked conservative nature of this molecule as shown by the fact that only a small number of gaps need be inserted to obtain homology. The observed differences between sequences in terms of the fraction of different sites are given in Table 1. The mutational distance, that is, the average number of nucleotide substitutions per site, was estimated using the formula

$$K = -3/4 \ln(1-(4/3)\lambda)$$

where $\lambda$ is the fraction of sites by which two homologous

Reprinted by permission from *Nature*, vol. 243, pp. 199–200. 

**Table 1** Fraction of Different Sites between 5S rRNA Sequences

| | Yeast | *E. coli* | *P. fluorescens* |
|---|---|---|---|
| Human | 0.395 | 0.457 | 0.478 |
| Yeast | | 0.474 | 0.565 |
| *E. coli* | | | 0.319 |

sequences differ. The formula was derived under the assumptions that in the course of evolution nucleotide substitutions occur spatially at random and with uniform probabilities and that each of the four bases (A, C, G, U) mutates to any of the remaining three with equal probability. (For details see ref. 4.) The equivalent formula has been previously derived by Jukes and Cantor[5]. We should also note that Dayhoff's[3] empirical relationship tabulated in her Table 11–3 is practically equivalent to this formula especially for the purpose of comparing different $K$ values.

The average mutational distance between the eukaryotes (man, yeast) and the prokaryotes (*E. coli*, *P. fluorescens*) turned out to be

$$K_{eu-pro} = 0.817 \pm 0.158$$

where the standard error was obtained from four observations (comparisons). On the other hand, the corresponding quantity between the human and yeast was

$$K_{h-y} = 0.561 \pm 0.095$$

where the error is a theoretical one computed by equation (5) in ref. 4. It may be interesting to note here that yeast is more closely related to man than to the bacteria, supporting the thesis (compare ref. 6) that the division between the eukaryotes and prokaryotes is more basic than divisions within eukaryotes. The remoteness of the eukaryotes-prokaryotes divergence relative to the human-yeast divergence can be estimated by the ratio $K_{eu-pro}/K_{h-y}$ which is approximately 1.5. This is much lower than the corresponding estimate of McLaughlin and Dayhoff[7] who obtained the ratio 2.6 using data on cytochrome $c$, $c_2$ and tRNA sequences. It is also lower than the corresponding estimate, ~2, obtained by Jukes (1969) (quoted in ref. 7). Hoping to resolve this discrepancy we calculated the evolutionary distances using the same data[3] on tRNA as McLaughlin and Dayhoff[7] but restricting our treatment only to paired regions. The reason for doing this is that it seems as if there is no excess of highly conserved regions (as inferred from our statistical analysis of the frequency distribution of the number of evolutionary changes per site in the alignment of

| | |
|---|---|
| Human | – G– UCUACGGCC– AUACCACCCUGAACGCGCCCGAUCUCGUCUGAU– CUCGGAAGCUAAGCAG |
| Yeast | – G– GUUGCGGCC– AUACCAUCUAGAAAGCACCGUUCUCCGUCCGAUAACCUGUAGUUAAGCUG |
| *E. coli* | UGCCUGGCGGCC– GUAGCGCGGUGGUCCCACCUGACCCCAUGCCGAACUCAGAAGUGAAACGC |
| *P. fluorescens* | UGUUCUUUGACGAGUAGUGGCAUUGGAACACCUGAUCCCAUCCCGAACUCAGAGGUGAAACGA |
| | |
| Human | GGUCGGGCCUG– GUUAGUACUUGGAUGGGAGACCGCCUGGGAAUACCGGGUGCUGUAG– GCUU |
| Yeast | GUAAGAGCCUGACCGAGUAGUGUAGUGGGUGACCAUACGCGAAACCUAGGUGCUGCA– – AUCU |
| *E. coli* | CGUAGCGCC– – – GAUGGUAGUGUG– – GGGUCUCCCCAUGCGAGAGUAGGGAACUGCCAGGCAU |
| *P. fluorescens* | UGCAUCGCC– – – GAUGGUAGUGUG– – GGGUUUCCCCAUGUCAAGAUCUCG– ACCAUAGAGCAU |

**Fig. 1** Alignment of 5S rRNA sequences.

3

Fig. 1), so that evolutionary change appears to be uniform over the entire sequence and in this respect 5S rRNA might be more similar to the paired than unpaired regions of tRNA. The evolutionary distances turned out to be $K_1=0.836\pm0.136$ for the eukaryotes-prokaryotes divergence but $K_2=0.420\pm0.029$ for the average of rat-yeast ($tRNA^{Ser}$) and wheat-yeast ($tRNA^{Phe}$) divergences. It may be seen that although $K_1$ is comparable to $K_{eu-pro}$, $K_2$ is clearly lower than $K_{h-y}$, so that the ratio $K_1/K_2=1.99$ is still considerably higher than $K_{eu-pro}/K_{h-y}=1.46$. It is possible that the difference is due to sampling error, with the true value lying somewhere between these two. At any rate we should take these estimates as tentative (including the problem that might arise because of the multiplicity of ribosomal genes). There is some reason to believe, however, that our estimate of 1.5 is consistent with the fossil records as explained below.

From comparative studies of cytochrome *c* sequences among eukaryotes, we can estimate the remoteness of the human-yeast divergence relative to the mammal-fish divergence. This allows us to estimate the absolute time of the human-yeast divergence as it is known from classical palaeontological studies that the common ancestor of the fish and the mammals goes back to some 400 m.y. (compare refs 8 and 9). From a number of comparisons involving various species of fish and mammals (data taken from ref. 3), we obtained the results that the mutational distance (in terms of the number of amino acid substitutions) between mammals and yeast is about three times that between mammals and the fish. This puts the time of the human-yeast divergence back to about $1.2\times10^9$ yr. This agrees with Dickerson[10] who obtained $1{,}200\pm75$ m.y. as the estimated date of the branch point for animals/plants/protists. We should note here that we avoided using the cytochrome $c_2$ sequence of the bacterium *Rhodospirillum rubrum* to estimate the prokaryotes-eukaryotes divergence, because there seems to be some difference in function between cytochromes $c$ and $c_2$ (compare ref. 7).

Multiplying $1.2\times10^9$ yr by the ratio $K_{eu-pro}/K_{y-h}\approx1.5$, we arrive at the result that the divergence between the eukaryotes and prokaryotes goes back to some $1.8\times10^9$ yr. Recent studies on Precambrian fossils (compare ref. 11) suggest that the eukaryotes evolved from prokaryotes at some point between the Bitter Springs formation ($10^9$ yr old) and the Gunflint formation ($2\times10^9$ yr old). With additional relevant data forthcoming (for example, ref. 12) we hope that the studies of molecular evolution will soon supply a more accurate date.

We tentatively conclude that the eukaryotes diverged from prokaryotes nearly $2\times10^9$ yr ago, thus opening up the way toward "higher organisms".

We thank Dr K. Miura and Mr H. Komiya for calling our attention to the relevant literature on 5S rRNA sequences and Drs T. Maruyama, S. Takemura and S. Kondo for helpful discussions.

MOTOO KIMURA
TOMOKO OHTA

*National Institute of Genetics,*
*Mishima, Shizuoka-ken* 411,
*Japan*

Received December 29, 1972; revised March 5, 1973.

[1] DuBuy, B., and Weissman, S. M., *J. Biol. Chem.*, **246**, 747 (1971).
[2] Hindley, J., and Page, S. M., *FEBS Lett.*, **26**, 157 (1972).
[3] Dayhoff, M. O., *Atlas of Protein Sequence and Structure 1972* (National Biomedical Research Foundation, Washington, DC, 1972).
[4] Kimura, M., and Ohta, T., *J. Mol. Evol.*, **2**, 87 (1972).
[5] Jukes, T. H., and Cantor, C. R., in *Mammalian Protein Metabolism* (edit. by Munro, H. N.), 21 (Academic Press, New York, 1969).
[6] Margulis, L., *Origin of Eukaryotic Cells* (Yale Univ. Press, New Haven and London, 1970).
[7] McLaughlin, P. J., and Dayhoff, M. O., *Science*, **168**, 1469 (1970).
[8] Romer, A. S., *The Procession of Life* (Weidenfeld and Nicolson, London, 1968).
[9] McAlester, A. L., *The History of Life* (Prentice-Hall, Englewood Cliffs, 1968).
[10] Dickerson, R. E., *J. Mol. Evol.*, **1**, 26 (1971).
[11] Barghoorn, E. S., *Sci. Amer.*, **224**, 30 (1971).
[12] Brownlee, G. G., Cartwright, E., McShane, T., and Williamson, R., *FEBS Lett.*, **25**, 8 (1972).

# Population genetics of multigene family with special reference to decrease of genetic correlation with distance between gene members on a chromosome

(coincidental evolution/unequal crossing-over/diffusion model/repeated genes)

MOTOO KIMURA AND TOMOKO OHTA

National Institute of Genetics, Mishima, Shizuoka-ken 411, Japan

*Contributed by Motoo Kimura, May 15, 1979*

**ABSTRACT** A mathematical method is developed which enables us to treat exactly the process of coincidental evolution under mutation, unequal intrachromosomal crossing-over as well as ordinary crossing-over between homologous chromosomes in a finite population of the effective size $N$. It makes use of finite difference equations involving two quantities denoted by $f_i$ and $\phi_i$, in which $f_i$ is the identity coefficient of two gene members that are $i$ steps apart on the same chromosome and $\phi_i$ is that of two members $i$ steps apart on two homologous chromosomes. When the number of genes ($n$) per family is large, the finite difference equations can be approximated by ordinary second-order differential equations which can then be solved analytically. Results obtained by the present method are compared with the corresponding results previously obtained by one of us (T.O.) using conventional diffusion models of gene frequency changes in population genetics. It is shown that the previous results obtained by T.O. regarding second-order statistics are essentially valid, and they give good approximations particularly when $N\beta$ is small, where $\beta$ is the rate of ordinary interchromosomal crossing-over within the multigene family.

Tandemly repeated genes on a chromosome forming a multigene family, such as ribosomal RNA and immunoglobulin genes, provide an interesting new subject of study in evolutionary genetics. Its main features are contraction and expansion of the gene number and coincidental (or horizontal) evolution among members (1, 2). To account for these phenomena, Smith (3) proposed a model that assumes continuous process of unequal crossing-over in the germ line and showed that mutant genes may spread or become extinct through unequal crossing-over in a family on one chromosome. Recently, one of us (T.O.) (4–6) developed a population genetics theory for treating evolution of multigene families by taking into account unequal crossing-over. In her treatment, it is assumed that a multigene family is evolving under mutation, random drift, sister chromatid unequal crossing-over, and ordinary interchromosomal crossing-over. The last is assumed to occur at meiosis and may include unequal crossing-over in addition to regular equal crossing-over. Also, the stochastic process of spreading and extinction of mutant genes in a family was approximated by conventional diffusion models that are used to treat gene frequency changes in a finite population.

In the present paper we shall develop a more exact treatment of the problem with special reference to the relationship between genetic correlation among gene members of a family and their distance on the chromosome. The results will then be compared with those previously obtained by T.O.

## Basic theory

We consider a multigene family evolving under mutation, random drift, and equal and unequal crossing-over in a random mating diploid population with effective size $N$. Let us assume that the family consists of $n$ genes, and that intrachromosomal (that is, between-sister-chromatid) unequal crossing-over occurs with shift by *one* gene unit (Fig. 1). Thus, we assume that one gene is either duplicated or deleted each with probability 1/2 when one unequal crossing-over occurs. We denote by $\gamma$ the rate at which such unequal crossing-over occurs per gene family per generation. We also consider interchromosomal crossing-over, but we assume that this always involves equal crossing-over; no shift occurs at chromosome pairing. We denote by $\beta$ the rate at which such equal, interchromosomal crossing-over occurs per family per generation. Let us further assume that mutation occurs at the rate $v$ per gene per generation following the infinite allele model of Kimura and Crow (7)—namely, every mutation leads to a new, not preexisting allele.

In the following, we attempt to derive the genetic correlation (in terms of identity coefficient) between members of a gene family as a function of distance on the chromosome. We first consider a simple situation in which no interchromosomal crossing-over occurs ($\beta = 0$). Let us denote by $f_i$ the coefficient of identity between two genes that are $i$ steps apart on the chromosome (see Fig. 1). This is the probability that two randomly chosen genes that are $i$ steps apart are identical by descent and, therefore, identical in allelic state. Under this definition, we have $f_0 = 1$. After one generation of unequal crossing-over and mutation, identity coefficient $f_i$ changes to $f_i'$ according to the relation

$$f_i' = (1 - v)^2 \left[\left(1 - \frac{\gamma i}{n}\right) f_i + \frac{\gamma i}{n} \frac{f_{i-1} + f_{i+1}}{2}\right], \qquad [1]$$

in which $i \geqq 1$. In this formula, $\gamma i/n$ is the probability that unequal crossing-over occurs between two genes that are $i$ steps apart. When this occurs, the distance of these two genes either increases or decreases by one gene unit each with probability 1/2. In addition, the term $(1 - v)^2$ is added from the consideration that two identical genes remain identical only when neither mutates in one generation. Note that when we assume $\beta = 0$, $f_i$ is not influenced by random sampling of gametes. At statistical equilibrium in which $f_i' = f_i$, we obtain

The publication costs of this article were defrayed in part by page charge payment. This article must therefore be hereby marked "*advertisement*" in accordance with 18 U. S. C. §1734 solely to indicate this fact.

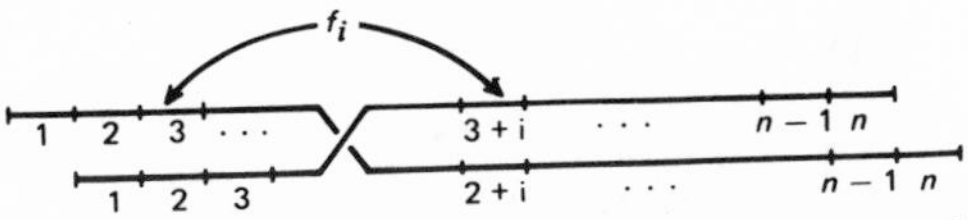


FIG. 1. Diagram illustrating an intrachromosomal unequal crossing-over and the meaning of identity coefficient $f_i$.

$$2vf_i = \frac{\gamma i}{n}\,\frac{f_{i+1} - 2f_i + f_{i-1}}{2}, \qquad [2]$$

where we neglect terms involving $v^2$ in Eq. 1. Letting $x = i/n$, writing $f(x)$ for $f_i$, and noting $f(x \pm h) = f(x) \pm hf'(x) + h^2f''(x)/2 + \ldots$, in which $h = 1/n$, we obtain

$$2vf(x) = \gamma x\,\frac{f\left(x + \frac{1}{n}\right) - 2f(x) + f\left(x - \frac{1}{n}\right)}{2} \approx \frac{\gamma x}{2n^2}\,\frac{d^2f(x)}{dx^2},$$

in which we assume $1/n << 1$. This leads to the following ordinary differential equation:

$$x\,\frac{d^2}{dx^2}f(x) - \frac{4n^2v}{\gamma}f(x) = 0, \ (0 \leq x \leq 1). \qquad [3]$$

The solution of this equation which satisfies the condition $f(0) = 1$ and which is bounded is

$$f(x) = 2\sqrt{ax}\,K_1\,(2\sqrt{ax}), \qquad [4]$$

in which $a = 4n^2v/\gamma$ and $K_1(\cdot)$ denotes a modified Bessel function. In Fig. 2, this solution is illustrated for various values of $a$ as a function of $x$ or $i/n$.

Next, we consider a more general situation in which interchromosomal crossing-over also takes place ($\beta > 0$). In this case we need another identity coefficient denoted by $\phi_i$. This is the identity probability between two randomly chosen genes on homologous chromosomes that are $i$ steps apart from each other on the chromosome (Fig. 3). Then the transition of $f_i$ and $\phi_i$ from one generation to the next can be computed through several steps as follows:

(*i*) By intrachromosomal unequal crossing-over, which occurs at the rate $\gamma$ per gene family, $f_i$ and $\phi_i$ change, respectively, to

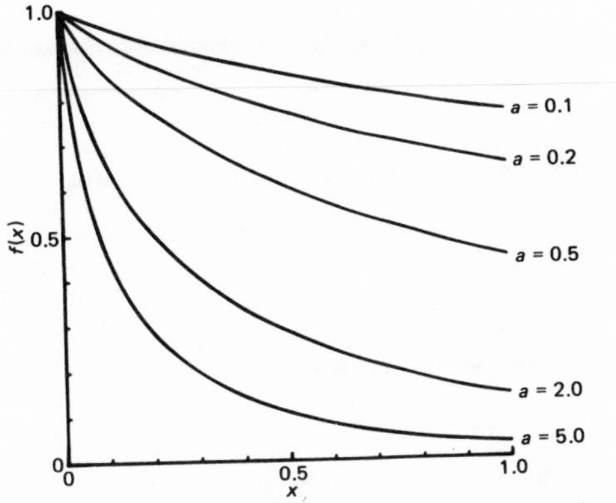


FIG. 2. Relationship between identity coefficient, $f(x)$, and the distance, $x$, on the chromosome between two genes for various values of $a$.

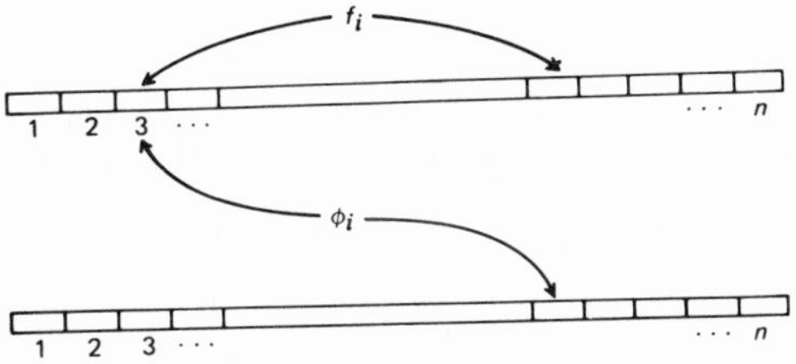


FIG. 3. Diagram illustrating the meaning of two identity coefficients $f_i$ and $\phi_i$.

$$f_i' = \left(1 - \frac{i\gamma}{n}\right)f_i + \frac{i\gamma}{n}\,\frac{f_{i-1} + f_{i+1}}{2} \qquad [5]$$

and

$$\phi_i' = \phi_i,$$

in which $i \geqq 1$. Note that the expected value of $\phi_i$ for a given $i$ does not change by intrachromosomal unequal crossing-over.

(*ii*) After an ordinary interchromosomal crossing-over, which occurs at the rate $\beta$ per gene family, $f_i'$ and $\phi_i'$ changes to $f_i''$ and $\phi_i''$;

$$f_i'' = \left(1 - \frac{i\beta}{n}\right)f_i' + \frac{i\beta}{n}\,\phi_i'$$

$$\phi_i'' = \left(1 - \frac{1}{2N-1}\cdot\frac{i\beta}{n}\right)\phi_i' + \frac{1}{2N-1}\cdot\frac{i\beta}{n}\,f_i'. \qquad [6]$$

The second equation is based on the following consideration. After an ordinary interchromosomal crossing-over, two genes (which we call the first and the second genes) that are $i$ steps apart and that are located on one particular chromosome become separated into two different chromosomes with probability $i\beta/n$. There are $2N - 1$ ways of matching the first gene on this particular chromosome with the second gene on any other chromosome, so that $f_i'$ contributes to $\phi_i''$ for the fraction $(i\beta/n)/(2N - 1)$. On the other hand, the first and the second genes each on two different chromosomes come to the same chromosome after an ordinary crossing-over with probability $(i\beta/n)/(2N - 1)$, so that $\phi_i'$ is decreased by this amount. Essentially the same reasoning was used by Ohta (5, 6).

(*iii*) Next, by random sampling of gametes in reproduction, $f_i''$ and $\phi_i''$ change to $f_i'''$ and $\phi_i'''$ according to the relation

$$f_i''' = f_i''$$

$$\phi_i''' = \left(1 - \frac{1}{2N}\right)\phi_i'' + \frac{1}{2N}f_i''. \qquad [7]$$

Note that the probability of two randomly sampled chromosomes being descended from a single chromosome is $1/(2N)$.

(*iv*) Finally, we consider the effect of mutation on reducing the identity coefficients. Let $v$ be the mutation rate per gene per generation, then

$$f_i^{(\mathrm{IV})} = (1 - v)^2 f_i'''$$

$$\phi_i^{(\mathrm{IV})} = (1 - v)^2 \phi_i'''. \qquad [8]$$

Combining these four steps, (*i*)–(*iv*), we get

$$f_i^{(\mathrm{IV})} = (1 - v)^2\left\{\left(1 - \frac{i\beta}{n}\right) \times \left[f_i + \frac{i\gamma}{2n}(f_{i-1} - 2f_i + f_{i+1})\right] + \frac{i\beta}{n}\,\phi_i\right\} \qquad [9]$$

Genetics: Kimura and Ohta Proc. Natl. Acad. Sci. USA 76 (1979) 4003

and

$$\phi_i^{(IV)} = (1-v)^2 \left\{ \left[\left(1-\frac{1}{2N}\right)\frac{1}{2N-1}\cdot\frac{i\beta}{n} + \frac{1}{2N}\left(1-\frac{i\beta}{n}\right)\right] \times \left[f_i + \frac{i\gamma}{2n}(f_{i-1}-2f_i+f_{i+1})\right] + \left[\left(1-\frac{1}{2N}\right)\left(1-\frac{1}{2N-1}\frac{i\beta}{n}\right)+\frac{1}{2N}\frac{i\beta}{n}\right]\phi_i\right\}. \quad [\mathbf{10}]$$

At equilibrium, we have $f_i^{(IV)} = f_i$ and $\phi_i^{(IV)} = \phi_i$. Writing $f(x)$ for $f_i$ and $\phi(x)$ for $\phi_i$, where $x = i/n$, and after neglecting small terms such as $\beta x/(2N)^2$, $v^2$, and $v/N$, we obtain the following set of equations from [**9**] and [**10**]:

$$\frac{\gamma}{2n^2}(1-\beta x)x\frac{d^2}{dx^2}f(x) - (2v+\beta x)f(x) + \beta x\phi(x) = 0 \quad [\mathbf{11}]$$

$$\frac{\gamma}{2n^2}x\frac{d^2}{dx^2}f(x) + f(x) - (1+4Nv)\phi(x) = 0. \quad [\mathbf{12}]$$

From these two equations, we immediately obtain

$$\phi(x) = \frac{(1+2v)f(x)}{1+(1-\beta x)4Nv}. \quad [\mathbf{13}]$$

If we assume $\beta \ll 1$, which is a realistic assumption for $\beta$, Eq. **13** reduces to

$$\phi(x) = f(x)/(1+4Nv) \quad [\mathbf{14}]$$

with good approximation. If we take averages each for $\phi(x)$ and $f(x)$ over the whole region ($0 \leqq x \leqq 1$), then the resulting equation agrees with the second formula in equations 13 of Ohta (5), who derived it by an entirely different method. From Eqs. **11** and **12**, we can easily derive the equation for $f(x)$. If we assume $\beta \ll 1$, this becomes as follows

$$x\frac{d^2}{dx^2}f(x) - a(1+bx)f(x) = 0, \ (0 \leqq x \leqq 1), \quad [\mathbf{15}]$$

in which $a = 4n^2v/\gamma$ and $b = 2N\beta/(1+4Nv)$.

Let $x = z/(2\sqrt{ab})$ and $f(x) = W(z)$, then Eq. **15** reduces to Whittaker's differential equation (see ref. 8, p. 337)

$$\frac{d^2W}{dz^2} + \left\{-\frac{1}{4} + \frac{k}{z} + \frac{(1/4)-m^2}{z^2}\right\}W = 0, \quad [\mathbf{16}]$$

in which $k = -(1/2)\sqrt{a/b}$ and $m = \pm\,½$. A pertinent solution of this equation, in terms of Whittaker's function, is given by $W_{k,1/2}(z)$. Thus, we obtain

$$f(x) = \Gamma(1-k)W_{k,1/2}(2\sqrt{ab}\,x). \quad [\mathbf{17}]$$

It can then be shown that, at the limit $\beta \to 0$ or $b \to 0$, the right-hand side of Eq. **17** reduces to $2\sqrt{ax}\,K_1\,(2\sqrt{ax})$ so that [**17**] reduces to [**4**], as it should (for mathematical properties of Whittaker's function, see ref. 9, p. 505). We can also show that

$$f(x) = e^{-\sqrt{abx}}(2\sqrt{ab}\,x)\int_0^{\infty} e^{-2\sqrt{ab}\,xt}\left(\frac{t}{1+t}\right)^{(1/2)\sqrt{a/b}}dt, \quad [\mathbf{18}]$$

in which $a = 4n^2v/\gamma$ and $b = 2N\beta/(1+4Nv)$. This integral representation is quite useful for numerical computations of $f(x)$.

Fig. 4 illustrates $f(x)$ as a function of $x$ for several combinations of values of $a$ and $b$. Note that $f(x)$ drops sharply if $b$ is large.

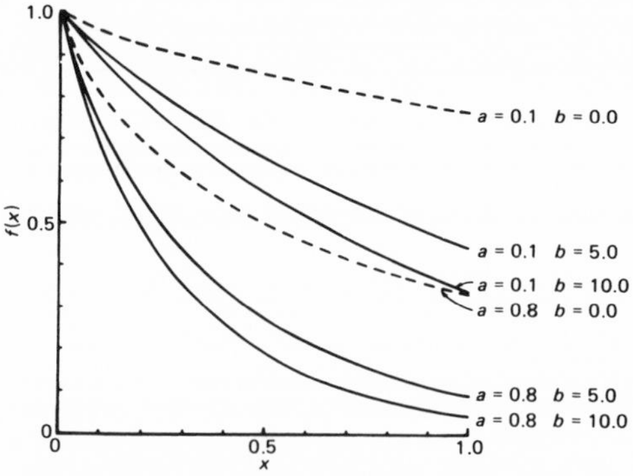


FIG. 4. Relationship between identity coefficient, $f(x)$, and the distance, $x$, on the chromosome when interchromosomal crossing-over is involved. Cases with $b = 0.0$ (no interchromosomal crossing-over) are plotted by broken lines.

## Average gene identity

We can now compute the average identity coefficient, which is the probability of two randomly chosen genes from the multigene family being identical. This is given by

$$\bar{f} = \frac{2}{n(n+1)}\{nf_0 + (n-1)f_1 + \cdots + f_{n-1}\}. \quad [\mathbf{19}]$$

When $n$ is large, this may be approximated by

$$\bar{f} = 2\int_0^1 (1-x)f(x)dx. \quad [\mathbf{20}]$$

In the case of no interchromosomal crossing-over ($\beta = 0$), if we apply Eq. **4**, we get

$$\bar{f} = \frac{1}{a}\{2 - L^2K_2(L)\} - \frac{1}{4a^2}\{16 - 2L^3K_3(L) - L^4K_2(L)\}, \quad [\mathbf{21}]$$

in which $K_2(\cdot)$ and $K_3(\cdot)$ stand for modified Bessel functions and $L = 2\sqrt{a}$ and $a = 4n^2v/\gamma$.

In her previous studies, Ohta (5, 6) assumed that the gene lineages are arranged randomly on the chromosome so that $f_i$ is constant. This assumption was needed to treat the effect of interchromosomal crossing-over. Thus, her result is likely to be biased when $\beta$ is large. However, when $\beta = 0$, it should give an approximately correct value, as shown by Perelson and Bell (10), who applied the birth–death process to the problem. Table 1 gives a comparison between the result obtained by the present method (Eqs. **4** and **20**) and that by the first formula in equations 13 of ref. 5, assuming $\beta = 0$. The table shows that the assumption of constant $f_i$ gives a slight overestimation. Note that the parameter $\kappa$ (effective number of cycles of unequal crossing-over) in ref. 5 corresponds to $\gamma/2$, and thus $\alpha$ in equations 13 of ref. 5 is equal to $\gamma/n^2$ in our notation.

Exact values of $f_i$ and $\phi_i$ can be obtained numerically by repeatedly using Eq. **1** and Eqs. **9** and **10**. When $n$ is small, this procedure gives the most exact results. Values of $f_i$ and $\phi_i$ at equilibrium can be obtained by repeating calculations over $2/v$ generations starting from a homogeneous gene family (i.e., $f_i = \phi_i = 1$ at the first generation) for a given set of parameters $a$ and $b$. Then the average identity coefficients can be computed by using Eq. **19**. Table 2 gives values of $\bar{f}$ thus obtained

Table 1. Average identity coefficient for various values of $a$, assuming $b = 0$

| $a$ | Eqs. 4 and **20** | First formula of eqs. 13 of ref. 5 |
|---|---|---|
| 0.05 | 0.9397 | 0.9756 |
| 0.1 | 0.8993 | 0.9524 |
| 0.2 | 0.8378 | 0.9091 |
| 0.3 | 0.7898 | 0.8696 |
| 0.4 | 0.7498 | 0.8333 |
| 0.5 | 0.7155 | 0.8000 |
| 0.8 | 0.6340 | 0.7143 |
| 1.0 | 0.5916 | 0.6667 |
| 2.0 | 0.4505 | 0.5000 |
| 5.0 | 0.2698 | 0.2857 |
| 8.0 | 0.1933 | 0.2000 |

for various $n$ as well as those calculated from Eqs. **4, 18,** and **20.** Approximate results of Ohta (using first the formula of equations 13 of ref. 5) are also given in the last column. They slightly overestimate the true value when $b$ is small, but underestimate it when $b$ is large. This is reasonable because the effect of interchromosomal crossing-over is exaggerated under the assumption of constant $f_i$. In addition, it may be seen from Table 2 that when $n$ is small, exact $\bar{f}$ is slightly larger than the corresponding values obtained by Eqs. **4, 18,** and **20.**

## Discussion

The method developed in the present paper provides a theoretical basis for studying genetic correlation between gene members of a multigene family. Although analytical solutions are obtained only for the case in which unequal crossing-over involves a shift by one gene unit, we can extend the method for numerically generating the exact genetic correlation for more general cases. The treatments of such cases will be reported in our next paper.

A simple way of grasping the process of unequal crossing-over leading to homogeneity of gene members is to regard it as "contraction-expansion process" of gene lineages in a family (1). Fig. 5 illustrates a situation in which the gene lineage 4 expands by chance through repeated unequal crossing-over. The process may be approximated by treating it as diffusion of gene lineages in the family (4, 5). Our present formulation also leads to a different type of diffusion equation and provides a basis for assessing the adequacy and limitation of the previous approximate treatments (5) in which evolution of a multigene family is treated as a double diffusion process: diffusion of a gene lineage in a family and that of a gene family type in a population. The interchromosomal crossing-over further complicates the situation through exchange of gene members between homologous families on different chromosomes. This contributes to increasing genetic variation, and it may be responsible for gene diversity of the immunoglobulin family. The amino acid diversity of reported sequences of variable regions of immunoglobulins fits the prediction of the simple model of

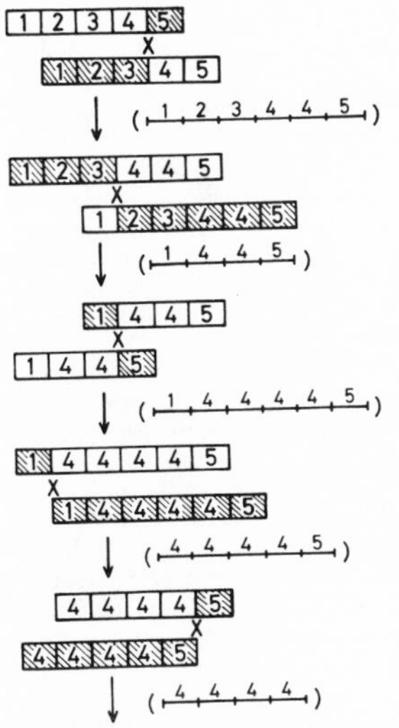


FIG. 5. Diagram illustrating the process of expansion by chance of a gene lineage (in this case, gene lineage 4) on the chromosome through continued unequal intrachromosomal crossing-over. Sequences inside parentheses represent one of two products of unequal crossing-over that happen to survive for the next division.

unequal crossing-over and that interchromosomal crossing-over contributes much to increasing amino acid diversity (11). On the other hand, there is little interchromosomal exchange in *Drosophila* satellite DNA (12), and this may be one cause of high homogeneity of repeated members of this gene family.

Negative selection against mutants that disturb the basic functions may also be responsible for the maintenance of homogeneity of gene members in the family. This type of natural selection will reduce "effectively neutral mutation rate" (13), thus making $v$ in our model smaller, with the result that homogeneity increases. In the multiplicational multigene families such as ribosomal RNA and transfer RNA genes (1), the spacer region is often more variable than the transcribed region (14), and this observation fits this prediction.

As to observations on the decrease of genetic correlation with chromosomal distance, only a few data are available. Wellauer *et al.* (15) suggested that the distribution of length heterogeneity of spacer regions of ribosomal DNA is random on the chromosome. On the other hand, Buongiorno-Nardelli (16) found that the neighboring genes are more similar in length than random members. Length heterogeneity may be attributed to unequal crossing-over between internal repeating sequences inside the spacer of this gene family. Then unequal crossing-over plays a role similar to mutation in generating new variations; thus the rate of production of variants must be fairly high. Our theory may be used to estimate the mean number of gene

Table 2. Average coefficient of gene identity ($a = 0.8$)

| $b$ | Numerical result | | | Eqs. 4, 18, and **20** | First formula of eqs. 13 of ref. 5 |
|---|---|---|---|---|---|
| | $n = 10$ | $n = 20$ | $n = 50$ | | |
| 0.0 | 0.7088 | 0.6919 | 0.6841 | 0.6340 | 0.7143 |
| 0.8333 | 0.6564 | 0.6380 | 0.6305 | 0.5884 | 0.6618 |
| 4.1667 | 0.5569 | 0.5290 | 0.5144 | 0.4912 | 0.5114 |
| 8.3333 | 0.4973 | 0.4517 | 0.4476 | 0.4278 | 0.3982 |

shift at unequal crossing-over when more detailed information on correlation of length heterogeneity with distance becomes available.

Another interesting observation concerns the presence of pseudogene in the 5S RNA gene family (17). Each repeating unit contains one gene and one pseudogene. Somehow, in an original unit that spread in the family, the gene appears to have duplicated to give the pseudogene by unequal crossing-over. After the duplication, the gene and the pseudogene diverged from each other, while the homogeneity among the units has been retained through a continuous process of unequal crossing-over.

Ohta (11) applied her simple model of double diffusion process (unequal crossing-over and random genetic drift) to analyze sequence variability of the variable region of immunoglobulins. Her results show that the sequence variability, including the hypervariability at the antigen binding sites, is in accord with the prediction of the simple model and thus the amino acid diversity is considered to be the result of evolutionary accumulation of mutant genes in this gene family. The effectively neutral mutation rate and, consequently, the rate of mutant substitution is roughly 3 times higher in the hypervariable region than in the framework (nonhypervariable) region. Therefore the amino acid diversity is much more pronounced within the species at the hypervariable region than at the framework region. Information on the arrangement of the variable region genes on the chromosome is not available and, therefore, we do not know the relationship between genetic similarity and the distance on the chromosome. It would be very interesting to apply our theory when the data on this relationship are obtained.

We thank Drs. T. Maruyama and N. Takahata for useful suggestions regarding solutions of differential equations in the present paper. This is contribution no. 1262 from the National Institute of Genetics (Mishima, Japan).

1. Hood, L., Campbell, J. H. & Elgin, S. C. R. (1975) *Annu. Rev. Genetics* **9**, 305–353.
2. Tartof, K. D. (1975) *Annu. Rev. Genetics* **9**, 355–385.
3. Smith, G. P. (1974) *Cold Spring Harbor Symp. Quant. Biol.* **38**, 507–513.
4. Ohta, T. (1976) *Nature (London)* **263**, 74–76.
5. Ohta, T. (1978) *Genet. Res.* **31**, 13–28.
6. Ohta, T. (1979) *Genetics* **91**, 591–607.
7. Kimura, M. & Crow, J. F. (1964) *Genetics* **49**, 725–738.
8. Whittaker, E. T. & Watson, G. N. (1935) *A Course of Modern Analysis* (Cambridge Univ. Press, Cambridge, England), 4th Ed.
9. Abramowitz, M. & Stegun, I. A., eds. (1964) *Handbook of Mathematical Functions with Formulas, Graphs, and Mathematical Tables.* (U.S. Dept. Commerce, Washington, DC).
10. Perelson, A. S. & Bell, G. I. (1977) *Nature (London)* **265**, 304–310.
11. Ohta, T. (1978) *Proc. Natl. Acad. Sci. USA* **75**, 5108–5112.
12. Yamamoto, M. & Miklos, G. L. G. (1978) *Chromosoma* **66**, 71–98.
13. Kimura, M. (1979) *Proc. Natl. Acad. Sci. USA* **76**, 3440–3444.
14. Brown, D. D. & Sugimoto, K. (1974) *Cold Spring Harbor Symp. Quant. Biol.* **38**, 501–505.
15. Wellauer, P. K., Reeder, R. H., Dawid, I. B. & Brown, D. D. (1976) *J. Mol. Biol.* **105**, 487–505.
16. Buongiorno-Nardelli, M. (1977) *J. Mol. Biol.* **110**, 105–117.
17. Fedoroff, N. V. & Brown, D. D. (1978) *Cold Spring Harbor Symp. Quant. Biol.* **42**, 1195–1200.

# PART SIXTEEN

# Nucleotide Substitutions

## Introduction

When data on nucleotide sequences began to accumulate at an accelerated pace and began to be used for comparative studies of molecular evolution, it became clear that there are substitutional biases among four nucleotides. At present, there are a number of statistical models (usually based upon stationary Markov processes) which take such biases into account. This line of statistical study was initiated by Kimura (no. 42), who distinguished transitional and transversional (type I and II) differences between two homologous sequences. The proposed model was applied to pseudogenes and to each codon position in relation to the neutral theory.

However, the model cannot account for unequal composition of nucleotides in DNA sequences. To explain this, different substitution rates among four nucleotides other than transitions and transversions are required, and hence Takahata and Kimura (no. 44) made one such extension. Kimura (no. 43) proposed two other substitution models; the three-substitution-type (3ST) and the two-frequency-class (see Lanave et al. 1984 for a more general model and chapter 5 in Nei 1987 for a review). The problem in using these elaborated models is that we must determine various unknown parameters, and some of the formulas are often inapplicable because of sampling errors. More seriously, the actual substitution process is sometimes much more complicated than what is ordinarily assumed by time-homogeneous Markov processes (Uzzell and Corbin 1971; Holmquist et al. 1983; Takahata 1991b). For instance, substitutions occur in such a way that different lineages evolved toward different base compositions (Li, Luo, and Wu 1985; Li and Tanimura 1987; Easteal 1991). In addition, it is very difficult to make an accurate correction for extensive multiple-hit substitutions. If such Markov models need to be reexamined, so does the maximum likelihood method of constructing phylogenetic trees (Felsenstein 1981; Hasegawa and Kishino 1989).

# A Simple Method for Estimating Evolutionary Rates of Base Substitutions Through Comparative Studies of Nucleotide Sequences

Motoo Kimura

National Institute of Genetics, Mishima 411, Japan

**Summary.** Some simple formulae were obtained which enable us to estimate evolutionary distances in terms of the number of nucleotide substitutions (and, also, the evolutionary rates when the divergence times are known). In comparing a pair of nucleotide sequences, we distinguish two types of differences; if homologous sites are occupied by different nucleotide bases but both are purines or both pyrimidines, the difference is called type I (or "transition" type), while, if one of the two is a purine and the other is a pyrimidine, the difference is called type II (or "transversion" type). Letting P and Q be respectively the fractions of nucleotide sites showing type I and type II differences between two sequences compared, then the evolutionary distance per site is $K = -(1/2) \ln \{(1 - 2P - Q) \sqrt{1 - 2Q}\}$. The evolutionary rate per year is then given by $k = K/(2T)$, where T is the time since the divergence of the two sequences. If only the third codon positions are compared, the synonymous component of the evolutionary base substitutions per site is estimated by $K'_S = -(1/2) \ln (1 - 2P - Q)$. Also, formulae for standard errors were obtained. Some examples were worked out using reported globin sequences to show that synonymous substitutions occur at much higher rates than amino acid-altering substitutions in evolution.

**Key words**: Molecular evolution – Evolutionary distance estimation – Synonymous substitution rate

During the last few years, rapid sequencing of DNA has become feasible, and data on nucleotide sequences of various parts of the genome in diverse organisms have started to accumulate at an accelerated pace. Each new report on such sequences

Contribution No. 1330 from the National Institute of Genetics, Mishima, 411 Japan invites evolutionary considerations through comparative studies. Therefore, it is desirable if good statistical methods are established for estimating evolutionary distances between homologous sequences in terms of the number of nucleotide base substitutions.

Recently, Miyata and Yasunaga (1980) proposed a new method for this purpose. Their method consists in tracing, through successive one step changes, all the possible paths (restricted by the assumption of "the minimum substitution number") for each pair of codons compared, giving each step its due weight based on relative acceptance rates of amino acid substitutions. Using this method they succeeded in revealing some interesting properties of synonymous base substitutions. One drawback of their method, however, is that it is too tedious.

The purpose of this note is to derive some simple formulae which enable us to obtain reliable estimates on the evolutionary distances between two nucleotide sequences compared. The present method of estimation is not only handy but also has the merit of incorporating the possibility that sometimes "transition" type substitutions may occur more frequently than "transversion" type substitutions.

Let us compare two homologous sequences and suppose that n nucleotide sites are involved. In what follows, we express sequences in terms of RNA codes, so that the four bases are designated by letters U, C, A and G.

Now, we fix our attention on one of the n sites, and investigate how the homologous sites in two species differentiate from each other in the course of evolution, starting from a common ancestor T years back. Since there are 4 possibilities with respect to a base occupied at each site, there are 16 combinations of base pairs when the homologous sites in two species are compared. These are listed in Table 1. For example, UU in the first line represents the case in which homologous sites in the first and second species are both occupied by base U. Similarly, UC in the second line represents the case in which homologous sites in the first and second species are occupied by U and C.

**Table 1.** Types of nucleotide base pairs occupied at homologous sites in two species. Type I difference includes four cases in which both are purines or both are pyrimidines (line 2). Type II difference consists of eight cases in which one of the bases is a purine and the other is a pyrimidine (lines 3 and 4).

| | | | | | |
|---|---|---|---|---|---|
| Same (Frequency) | UU ($R_1$) | CC ($R_2$) | AA ($R_3$) | GG ($R_4$) | Total (R) |
| Different, Type I (Frequency) | UC ($P_1$) | CU ($P_1$) | AG ($P_2$) | GA ($P_2$) | Total (P) |
| Different, Type II | UA ($Q_1$) | AU ($Q_1$) | UG ($Q_2$) | GU ($Q_2$) | Total |
| (Frequency) | CA ($Q_3$) | AC ($Q_3$) | CG ($Q_4$) | GC ($Q_4$) | (Q) |

When the homologous sites are occupied by different bases, we shall distinguish two types of differences, namely, type I and type II differences. As shown in Table 1, the type I difference includes 4 cases in which both are either purines or pyrimidines, and the type II difference includes 8 cases in which one is a purine and the other is a pyrimidine.

Let $\alpha$ and $\beta$ be the rates of base substitutions as shown in Fig. 1. In other words, $\alpha$ is the rate of transition type substitutions, and $2\beta$ is that of transversion type substitutions, so that the total rate of substitutions per site per unit time (year) is $k = \alpha + 2\beta$. Note that $\alpha$ and $\beta$ refer to evolutionary rates of mutant substitutions in the species rather than the ordinary mutation rates at the level of an individual. To simplify the following treatments, we assume that these rates are equal for all bases as shown in Fig. 1.

We now define two probabilities denoted by P and Q, where P is the probability of homologous sites showing a type I difference, while Q is that of these sites showing a type II difference. More detailed specifications of probabilities of individual combinations of bases are given in Table 1. For example, the probability of UC (and also CU) is denoted by $P_1$. Let T be the time since divergence of the two species (measured in years), and denote by P(T) and Q(T) the probabilities of type I and type II differences at time T. Note that the probability of identity at homologous sites which we denote by R(T) is equal to $1 - P(T) - Q(T)$.

Then, we can derive the equations for P and Q at time $T + \Delta T$ in terms of P, Q, and R at time T as follows, where $\Delta T$ stands for the length of a short time interval. Consider a particular base pair of type I, say UC, which occurs with probability $P_1(T + \Delta T)$ (see second line in Table 1). We can distinguish three ways by which UC at time $T + \Delta T$ is derived from various base pairs at time T.

(i) UC is derived from UC (which occurs with probability $P_1(T)$) when both U and C remain unchanged. Since the probability of occurrence of substitution per site during

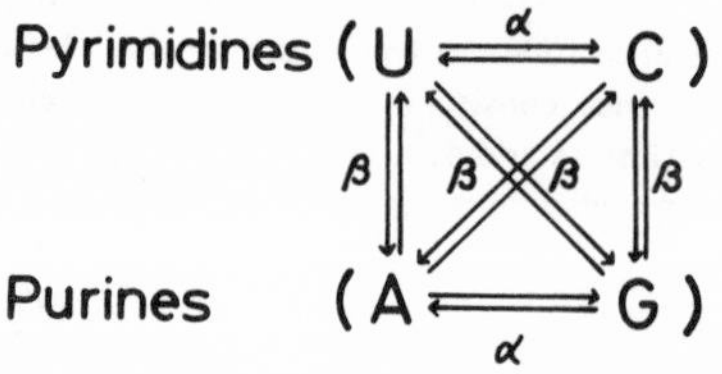


**Fig. 1.** Scheme of evolutionary base substitutions and their rates per unit time

a short time interval of length $\Delta T$ is $(\alpha + 2\beta)\Delta T$, the probability of no change occurring at both homologous sites is $[1 - (\alpha + 2\beta)\Delta T]^2$ so that the contribution coming from this class is $[1 - (\alpha + 2\beta)\Delta T]^2 P_1(T)$, which reduces to $[1 - 2(\alpha + 2\beta)\Delta T]P_1(T)$ if we neglect small terms of the order $(\Delta T)^2$. We also ignore the rare possibility that CU changes to UC by a double change.

(ii) UC is derived from UU when U in the second species is replaced by C, and from CC when C in the first species is replaced by U (see the first line in Table 1). Since UU and CC occur with respective frequencies $R_1(T)$ and $R_2(T)$, and since each changes to UC at the rate $\alpha$, the total contribution coming from this class is $\alpha\Delta T\,[R_1(T) + R_2(T)]$. We neglect cases in which substitutions occur simultaneously at both sites because such probabilities are of the order of $(\Delta T)^2$.

(iii) UC can also be derived from UA, UG, AC and GC (see the third line in Table 1) which occur with respective frequencies $Q_1(T)$, $Q_2(T)$, $Q_3(T)$ and $Q_4(T)$, and each of which changes to UC at the rate $\beta$. The contribution of this class to UC at $T + \Delta T$ is $\beta\Delta T\,[Q_1(T) + Q_2(T) + Q_3(T) + Q_4(T)]$ or $\beta\Delta T \cdot Q(T)/2$.

Combining all these contributions coming from various base pair classes at time T, we get

$$P_1(T + \Delta T) = [1 - (2\alpha + 4\beta)\,\Delta T]\,P_1(T) + \alpha\Delta T[R_1(T) + R_2(T)] + \beta\Delta T \cdot Q(T)/2 \;.$$

Similarly, for the base pair AG,

$$P_2(T + \Delta T) = [1 - (2\alpha + 4\beta)\,\Delta T]\,P_2(T) + \alpha\Delta T[R_3(T) + R_4(T)] + \beta\Delta T \cdot Q(T)/2$$

Summing these two equations, and noting $P(T) = 2P_1(T) + 2P_2(T)$ and $R(T) = R_1(T) + R_2(T) + R_3(T) + R_4(T) = 1 - P(T) - Q(T)$, and writing $\Delta P(T) = P(T + \Delta T) - P(T)$ we get

$$\Delta P(T)/\Delta T = 2\alpha - 4(\alpha + \beta)\,P(T) - 2(\alpha - \beta)\,Q(T) \tag{1}$$

Carrying out a similar series of calculations for base pairs of type II, we obtain

$$\Delta Q(T)/\Delta T = 4\beta - 8\beta Q(T) \;. \tag{2}$$

From these two finite difference equations (Eqs. 1 and 2), we obtain the following set of differential equations

$$\frac{dP(T)}{dT} = 2\alpha - 4(\alpha + \beta)\,P(T) - 2(\alpha - \beta)\,Q(T)$$

$$\frac{dQ(T)}{dT} = 4\beta - 8\beta Q\,(T) \tag{3}$$

The solution of this set of equations which satisfies the condition

$$P(0) = Q(0) = 0 \;, \tag{4}$$

i.e., no base differences exist at T = 0, is as follows.

$$P(T) = \frac{1}{4} - \frac{1}{2}e^{-4(\alpha+\beta)T} + \frac{1}{4}e^{-8\beta T} \tag{5}$$

$$Q(T) = \frac{1}{2} - \frac{1}{2}e^{-8\beta T} \;. \tag{6}$$

Writing $P_T$ and $Q_T$ for P(T) and Q(T), we get, from these two equations,

$$4(\alpha + \beta)T = -\log_e(1 - 2P_T - Q_T) \tag{7}$$

and

$$8\beta T = -\log_e(1 - 2Q_T) \;, \tag{8}$$

so that

$$4\alpha T = -\log_e(1 - 2P_T - Q_T) + (1/2)\log_e(1 - 2Q_T) \;. \tag{9}$$

Since the rate of evolutionary base substitutions per unit time is

$$k = \alpha + 2\beta \;,$$

the total number of substitutions (including revertant and superimposed changes) per site which separate the two species (and therefore involve two branches each with length T) is

$$K = 2Tk = 2\alpha T + 4\beta T ,$$

where $\alpha T$ and $\beta T$ are given by Eqs. (8) and (9). Then, omitting the subscript T from $P_T$ and $Q_T$, we obtain

$$K = -\frac{1}{2} \log_e \{(1 - 2P - Q)\sqrt{1 - 2Q}\} . \quad (10)$$

It is remarkable that, as can be seen from Eqs. (3), at equilibrium $2P = Q = 1/2$, even when $\alpha \neq \beta$.

This equation may be used to estimate the evolutionary distance between two sequences in terms of the number of base substitutions per site that have occurred in the course of evolution extending over T years. In this equation, $P = n_1/n$ and $Q = n_2/n$, where $n_1$ and $n_2$ are respectively the numbers of sites for which two sequences differ from each other with respect to type I ("transition" type) and type II ("transversion" type) substitutions and n is the total number of sites compared.

In the special case of $\alpha = \beta$, Eqs. (5) and (6) reduce to

$$P_T = Q_T/2 = (1 - e^{-8\alpha T})/4 . \quad (11)$$

Then, substituting $P = Q/2$ in (10), we get

$$K = -\frac{3}{4} \log_e (1 - \frac{4}{3}\lambda) , \quad (12)$$

where $\lambda = P + Q = 3Q/2$ is the fraction of sites for which two sequences differ from each other. This formula is well-known (see Kimura and Ohta 1972), and it was first obtained by Jukes and Cantor (1969). However, in actual situations, particularly when the third positions of codons are compared, P is often larger than Q, and therefore the assumption of $\alpha = \beta$ or $P = Q/2$ is not always realistic. This is one reason why the new formula (10) is better than (12). Eq. (10) also has a desirable property in that as P and Q get small, it converges to $K = P + Q$ independent of $\alpha$ and $\beta$.

Since a large fraction of substitutions at the third positions are synonymous, that is, they do not cause amino acid changes, it would be interesting to estimate the synonymous component of the substitution rate at this position.

As is evident from the standard RNA code table, for a given pair of bases in the first and the second codon positions, roughly speaking there are two situations; either the third position is completely synonymous (four-fold degeneracy) or synonymy is restricted within purines or pyrimidines (two-fold degeneracy). These two situations occur roughly in equal numbers. Thus, the synonymous component of substitutions at the third position which we denote by $k_S'$ may be estimated by

$$k_S' = \frac{1}{2}(\alpha + 2\beta) + \frac{1}{2}\alpha = \alpha + \beta . \quad (13)$$

Let $K_S' = 2Tk_S' = 2(\alpha + \beta)T$ be the synonymous component of the distance, then, we get

$$K_S' = -\frac{1}{2} \log_e (1 - 2P - Q) , \quad (14)$$

if we apply Eq. (7), so that the synonymous component of substitution rate per unit time may be estimated from $k_S' = K_S'/(2T)$. Writing $k'_{nuc(S)}$ rather than $k_S'$ to emphasize that this refers to the rate per nucleotide site, we get, using Eq. 14, the following formula.

$$k'_{nuc(S)} = -\frac{1}{4T}\log_e(1-2P-Q) \quad . \tag{15}$$

The corresponding formula for amino acid-altering substitutions may be obtained by

$$k_{nuc(A)} = K''/(2T) \quad , \tag{16}$$

where K'' is the evolutionary distance computed by applying Eq. (10) to a set of the first and second positions of codons (i.e., by excluding the third positions in the sequence comparison).

We can also derive formulae for the error variances of the estimates K and $K_S'$. Let $\delta K$, $\delta P$ and $\delta Q$ be respectively small changes in K, P and Q. Then

$$\delta K = a\delta P + b\delta Q$$

where

$$a = \frac{1}{1-2p-Q} \tag{17}$$

and

$$b = \frac{1}{2}\left(\frac{1}{1-2P-Q} + \frac{1}{1-2Q}\right) \quad , \tag{18}$$

so that

$$\sigma_K^2 = E\,\{(\delta K)^2\} = a^2E\,\{(\delta P)^2\} + 2abE\,\{\delta P\delta Q\} + b^2E\,\{(\delta Q)^2\} \quad ,$$

where E stands for the expectation operator. Then noting that the sampling variances and the covariance are $E\,\{(\delta P)^2\} = P(1-P)/n$, $E\,\{(\delta Q)^2\} = Q(1-Q)/n$ and $E\,\{\delta P\delta Q\} = -PQ/n$. The standard error of K is then

$$\sigma_K = \frac{1}{\sqrt{n}}\left|\sqrt{(a^2P + b^2Q) - (aP + bQ)^2}\right| \quad . \tag{19}$$

In a similar manner, we can derive the standard error of $K_S'$ which turns out to be as follows.

$$\sigma_{K'S} = \frac{\sqrt{4P + Q - (2P + Q)^2}}{2(1-2P-Q)\sqrt{n}} \quad . \tag{20}$$

As an example, let us compare the nucleotide sequence of the rabbit $\beta$ globin (Efstratiadis and Kafatos 1977) with that of chicken $\beta$ globin (Richards et al. 1979). There are 438 nucleotide sites that can be compared, corresponding to 146 amino acid sites (codons). Among these sites, we find that there are 58 sites for which these two sequences have type I differences, and 63 sites with type II differences. Thus, P = 0.132, Q = 0.144 and we obtain K = 0.348. Mammals and birds probably diverged during the Carboniferous period (see Romer 1968), so we tentatively take $T = 300 \times 10^6$ years. The evolutionary rate per site is then $k_{nuc} = K/(2T) = 0.58 \times 10^9$ per year. This is the overall rate per site, but it is much more interesting to estimate separately the evolutionary rates for the three codon positions. For the first position, there are 146 nucleotide sites compared, and we find P = 15/146 and Q = 21/146, giving $K_1 = 0.300$, where subscript 1 denotes that it refers to the first codon position. Similarly, for the second position, we find P = 7/146 and Q = 18/146 so that $K_2 = 0.195$. Finally, for the third position, P = 36/146 and Q = 24/146, and we get $K_3 = 0.635$, which is much higher than the corresponding estimates for the first and second positions. We can also estimate the synonymous component of the evolutionary distance per third codon position. From Eq. (14), this turns out to be $K_S' = 0.535$.

In Table 2, results of similar calculations are listed for various comparisons involving the human β (Marotta et al. 1977) and the mouse β globin sequences (Konkel et al. 1978), in addition to the chicken and rabbit β-globin sequences. Except for the last two comparisons involving the abnormal, globin like α-3 gene (Nishioka et al. 1980) it is clear that the relationship $K_2 < K_1 < K_3$ holds generally; the evolutionary mutant substitutions are most rapid at the third position, and this is followed by the first position and then, at the second position the substitutions are the slowest.

This can be readily interpreted by the neutral theory of molecular evolution (Kimura 1968; King and Jukes 1969; see also Kimura 1979) as follows. Among the three codon positions, base substitutions at the second positions tend to produce more drastic changes in the physico-chemical properties of amino acids than those at the first positions. Take for example a codon for Pro (CCN). Substitutions for base C at the first position of U, A, and G, lead respectively to Ser, Thr and Ala. In terms of Miyata's distance (based on polarity and volume differences between an amino acid pair; see Miyata et al. 1979), they are 0.56, 0.87 and 0.06 units apart from Pro, with the average distance of about 0.5. On the other hand, the corresponding average distance resulting from substitutions for C at the second position of the codon turns out to be about 2.5.

This means that mutational changes at the first position have a higher chance of not being harmful (i.e., selectively neutral or equivalent) than those at the second position, and therefore, have a higher chance of being fixed in the species by random drift (Kimura and Ohta 1974). This type of reasoning applies more forcibly to the third position (as compared with the first and the second positions), since a majority of mutational changes at this position do not cause amino acid changes.

The ratio per site of synonymous to amino acid-altering substitutions as measured by $2K_S'/(K_1 + K_2)$, is 4.17 for the human β-rabbit β comparison ($T = 8 \times 10^7$ years), 2.16 for the chicken β-rabbit β comparison ($T = 3 \times 10^8$ years) but only 1.41 for the rabbit α-rabbit β comparison ($T = 5 \times 10^8$ years). It looks as if synonymous substitutions occur more frequently during the later (i.e. more recent) stages of globin evolution than its early stages, but such a tendency is probably more apparent than real. In my opinion, this is due to lower detectability of synonymous substitutions for more remote comparisons; as more and more synonymous substitutions accumulate at the third positions of codons, it becomes progressively difficult to detect all of them. In fact, as compared with the first and the second positions, the third position shows marked deviation of the base composition from equality (e.g., G 36%, C 30%, U 27%, A 7% in human β) so that the present method of estimation may become imprecise for a very large value of T.

**Table 2.** Evolutionary distances in terms of the number of base substitutions estimated for several comparisons of globin sequences. $K_1$, $K_2$, and $K_3$ respectively denote the number of base substitutions at the first, second, and third positions of codons, while $K_S'$ stands for the estimated number of substitutions due to synonymous changes in the third position. Estimated values of these parameters together with their standard errors are listed. The primary sequence of rabbit α-globin is taken from Heindell et al. (1978)

| Comparison | Evolutionary distances per nucleotide site | | | |
|---|---|---|---|---|
| | $K_1$ | $K_2$ | $K_3$ | $K_S'$ |
| Human β vs. Mouse β | 0.17 ± 0.04 | 0.13 ± 0.03 | 0.34 ± 0.06 | 0.28 ± 0.05 |
| Rabbit β vs. Mouse β | 0.16 ± 0.04 | 0.13 ± 0.03 | 0.43 ± 0.07 | 0.36 ± 0.07 |
| Human β vs. Rabbit β | 0.06 ± 0.02 | 0.06 ± 0.02 | 0.28 ± 0.06 | 0.25 ± 0.05 |
| Chicken β vs. Rabbit β | 0.30 ± 0.05 | 0.19 ± 0.04 | 0.64 ± 0.11 | 0.53 ± 0.10 |
| Rabbit α vs. Rabbit β | 0.54 ± 0.09 | 0.44 ± 0.07 | 0.90 ± 0.15 | 0.69 ± 0.13 |
| Rabbit α vs. Mouse α-1 | 0.12 ± 0.03 | 0.11 ± 0.03 | 0.54 ± 0.09 | 0.47 ± 0.09 |
| Rabbit α vs. Mouse α-3 | 0.27 ± 0.06 | 0.28 ± 0.06 | 0.69 ± 0.13 | 0.56 ± 0.12 |
| Mouse α-1 vs. Mouse α-3 | 0.16 ± 0.04 | 0.20 ± 0.05 | 0.30 ± 0.06 | 0.22 ± 0.05 |

The last two comparisons in Table 2 involve the globin-like α-3 gene recently sequenced in the mouse (Nishioka et al. 1980). This gene completely lacks two intervening sequences normally present in all the α and β globin genes, and it does not encode globin. In other words, it is inactive in the production of stable mRNA. However, from a comparison of this sequence with the normal, adult mouse α globin (α-1) and the rabbit α globin nucleotide sequences, it is evident that this gene evolved from a normal ancestral α globin gene through duplication and subsequent loss of its intervening sequences. This must have occurred after the mouse and the rabbit diverged from their common ancestor some 80 million years ago. This α-globin-like gene acquired in its coding region a number of insertions and deletions of nucleotides. In making sequence comparisons, therefore, I chose only those codons of the α-3 gene which do not contain such changes and which are either identical with or differ only through base substitutions from the corresponding (homologous) codons of the mouse α-1 and the rabbit α genes.

As seen from the last three lines of Table 2, this unexpressed α-3 gene evolved at a much faster rate than its normal counterpart (mouse α-1 gene), particularly with respect to the first and the second codon positions. This is easy to understand from the neutral theory. Under a normal situation, each gene is subject to a selective constraint coming from the requirement that the protein which it produces must function normally. Evolutionary changes are restricted within such a set of base substitutions. However, once a gene is freed from this constraint, as is the case for this globin-like α-3 gene, practically all the base substitutions in it become indifferent to Darwinian fitness, and the rate of base substitutions should approach the upper limit set by the mutation rate (This holds only if the neutral theory is valid, but not if the majority of base substitutions are driven by positive selection; see Kimura 1977). If the rates of synonymous substitutions are not very far from this limit (Kimura 1977) we may expect that the rates of evolution of a "dead gene" are roughly equal to those of synonymous substitutions. Recently Miyata (personal communication) computed the evolutionary rate of a mouse pseudo alpha globin gene ($\psi\alpha$ 30.5) which was sequenced by Vanin et al. (1980) and which appears to be essentially equivalent to the α-3 gene. He also obtained a result supporting this prediction.

Finally, it would be interesting to estimate base substitution rates in non-coding regions such as introns. I use data presented by van Ooyen et al. (1979) who investigated similarity between the nucleotide sequences of rabbit and mouse β-globin genes. They list (see their Table 2) separately the numbers of "transition" and "transversion" type differences between the homologous parts of these sequences. For the small introns excluding 5 gaps that amount to 6 nucleotides, $P = 27/113$, $Q = 18/113$ and $n = 113$. Using Eqs. 10 and 19, we get $K = 0.60 \pm 0.12$. This is not significantly different from the substitution rate at the third position $K_3 = 0.43 \pm 0.07$. The large introns of rabbit and mouse β globin genes differ considerably in length, being separated from each other by 14 gaps (determined by optimization of alignment of the sequences) which amount to 109 nucleotides. Excluding these parts, $P = 113/557$ and $Q = 179/557$, from which we get $K = 0.90 \pm 0.07$. This value is significantly larger than $K_3$. It is likely, as pointed out by van Ooyen et al. (1979) that insertions and deletions occur rather frequently in this part in addition to point mutations, and that they inflate the estimated value of the "nucleotide substitution rate," since a majority of these changes may also be selectively neutral and subject to random fixation by genetic drift.

*Aknowledgments.* I thank Drs. Y. Tateno, N. Takahata, and J.F. Crow for reading carefully the first draft of this paper and for giving many useful suggestions for improving its presentation.

**References**

Efstratiadis A, Kafatos FC (1977) Cell 10:571–585
Heindell HC, Liu A, Paddock GV, Studnicka GM, Salser WA (1978) Cell 15:43–54
Jukes TH, Cantor CR (1969) Evolution of protein molecules. In: Munro HN (ed), Mammalian protein metabolism, II. New York, Academic Press, p 21
Kimura M (1968) Nature 217:624–626
Kimura M (1977) Nature 267:275–276
Kimura M (1979) Sci Am 241 (No.5, Nov.):94–104
Kimura M, Ohta T (1972) J Mol Evol 2:87–90
Kimura M, Ohta T (1974) Proc Natl Acad Sci USA 71:2848–2852
King JL, Jukes TH (1969) Science 164:788–798
Konkel DA, Tilghman SM, Leder P (1978) Cell 15:1125–1132
Marotta CA, Wilson JT, Forget BG, Weissman SM (1977) J Biol Chem 252:5040–5053
Miyata T, Miyazawa S, Yasunaga T (1979) J Mol Evol 12:219–236
Miyata T, Yasunaga T (1980) J Mol Evol 16:23–36
Nishioka Y, Leder A, Leder P (1980) Proc Natl Acad Sci USA 77:2806–2809
Richards RI, Shine J, Ullrich A, Wells JRE, Goodman HM (1979) Nucleic Acids Res 7: 1137–1146
Romer AS (1968) The procession of life. Weidenfeld and Nicolson, London.
Vanin EF, Goldberg GI, Tucker PW, Smithies O (1980). Nature 286:222–226
van Ooyen A, van den Berg J, Mantel N, Weissmann C (1979) Science 206:337–344

Received July 24, 1980

# Estimation of evolutionary distances between homologous nucleotide sequences

(molecular evolution/comparison of base sequences/base substitution rate/neutral mutation-random drift hypothesis)

MOTOO KIMURA

National Institute of Genetics, Mishima, 411 Japan

*Contributed by Motoo Kimura, October 3, 1980*

**ABSTRACT** By using two models of evolutionary base substitutions—"three-substitution-type" and "two-frequency-class" models—some formulae are derived which permit a simple estimation of the evolutionary distances (and also the evolutionary rates when the divergence times are known) through comparative studies of DNA (and RNA) sequences. These formulae are applied to estimate the base substitution rates at the first, second, and third positions of codons in genes for presomatotropins, preproinsulins, and $\alpha$- and $\beta$-globins (using comparisons involving mammals). Also, formulae for estimating the synonymous component (at the third codon position) and the standard errors are obtained. It is pointed out that the rates of synonymous base substitutions not only are very high but also are roughly equal to each other between genes even when amino acid-altering substitution rates are quite different and that this is consistent with the neutral mutation-random drift hypothesis of molecular evolution.

Data on nucleotide sequences of various parts of the genome in diverse organisms are appearing at an accelerated, almost explosive, rate. Many of these sequences are of interest for studies of molecular evolution. Before long, comparative studies of amino acid sequences, which have played a major role during the last 15 years or so, will be superseded by studies of nucleotide sequences. Already it has become increasingly evident that a preponderance of synonymous and other silent base substitutions is a general but remarkable feature of molecular evolution and that this is consistent with the neutral theory of molecular evolution (1–8).

In estimating the evolutionary distances between homologous sequences in terms of the number of base substitutions, corrections for multiple and revertant changes at homologous sites are essential. This is because only four kinds of bases exist in nucleotide sequences and even two random sequences show a 25% average match at individual sites. In this paper, I derive some formulae which are useful for estimating evolutionary distances between nucleotide sequences by using two models of evolutionary base substitutions.

## THREE-SUBSTITUTION-TYPE (3ST) MODEL

Consider a pair of homologous sites in two sequences being compared. We investigate how these sites have diverged from each other during their descent from a common ancestor $T$ years back. At each individual site, bases are successively substituted one after another in the course of time. To formulate this, we assume a model of evolutionary base substitutions as shown in Fig. 1*a*. Throughout this paper we use RNA codes so that the four bases are expressed by letters U, C, A, and G. Let $\alpha$, $\beta$, and

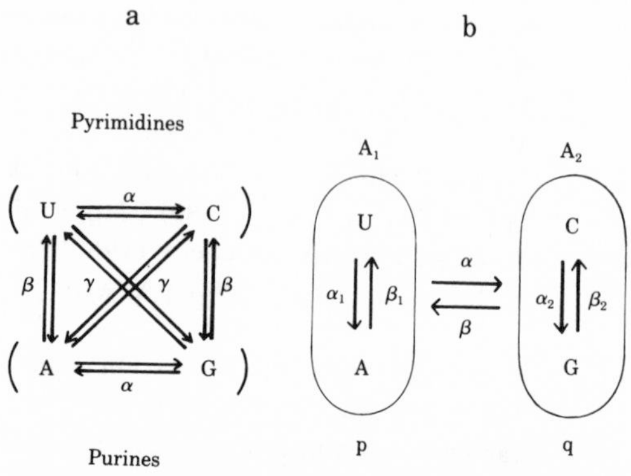


FIG. 1. Two models of evolutionary base substitutions: (*a*) "three-substitution-type" (3ST) model; (*b*) "two-frequency-class" (2FC) model.

$\gamma$ be the rates of base substitutions as indicated in Fig. 1*a*, $\alpha$ being the rate of "transition" type substitutions and $\beta$ and $\gamma$ being rates of "transversion" type substitutions. The total rate of base substitutions per unit time (year) is $k = \alpha + \beta + \gamma$.

It is important to note here that $\alpha$, $\beta$, and $\gamma$ refer to evolutionary rates by which bases are substituted in the species rather than ordinary mutation rates at the level of individuals. The total number of base substitutions per site which separate the two sequences and therefore involve two branches each with length $T$ is given by $2Tk$ which we denote by $K$.

$$K = 2Tk = 2(\alpha + \beta + \gamma)T. \quad [1]$$

When we compare homologous sites of sequences 1 and 2, we note that there are 12 combinations of different bases (Table 1). Let $P$ be the probability (relative frequency) that, at time $T$, homologous sites are occupied by base pair UC, CU, AG, or GA. In other words, $P$ is the probability of homologous sites showing the transition type base differences. Similarly, let $Q$ be the probability of homologous sites being occupied by pair UA, AU, CG, or GC, and $R$ be the probability of UG, GU, CA, or AC. Thus, $Q + R$ represents the probability of homologous sites showing transversion type differences.

Then it can be shown that $P$, $Q$, and $R$ satisfy the following set of differential equations (details of the derivation will be published elsewhere).

The publication costs of this article were defrayed in part by page charge payment. This article must therefore be hereby marked "*advertisement*" in accordance with 18 U. S. C. §1734 solely to indicate this fact.

Abbreviations: 3ST model, three-substitution-type model; 2FC model, two-frequency-class model.

*Proc. Natl. Acad. Sci. USA 78 (1981)*    455

$$dP/dT = 2\alpha - 2(2\alpha + \beta + \gamma)P$$
$$- 2(\alpha - \gamma)Q - 2(\alpha - \beta)R$$
$$dQ/dT = 2\beta - 2(\beta - \gamma)P \qquad [2]$$
$$- 2(\alpha + 2\beta + \gamma)Q - 2(\beta - \alpha)R$$
$$dR/dT = 2\gamma - 2(\gamma - \beta)P$$
$$- 2(\gamma - \alpha)Q - 2(\alpha + \beta + 2\gamma)R.$$

The solution of this set of equations that satisfies the condition

$$P = Q = R = 0 \text{ at } T = 0, \qquad [3]$$

i.e., the two sequences are identical at the start, is

$$\left.\begin{array}{l} P = (1 - e^{\lambda_1 T} - e^{\lambda_2 T} + e^{\lambda_3 T})/4 \\ Q = (1 - e^{\lambda_1 T} + e^{\lambda_2 T} - e^{\lambda_3 T})/4 \\ R = (1 + e^{\lambda_1 T} - e^{\lambda_2 T} - e^{\lambda_3 T})/4 \end{array}\right\} \qquad [4]$$

in which $\lambda_1 = -4(\alpha + \beta)$, $\lambda_2 = -4(\alpha + \gamma)$, and $\lambda_3 = -4(\beta + \gamma)$. From these equations, we get

$$4(\alpha + \beta)T = -\ln[1 - 2(P + Q)]$$
$$4(\alpha + \gamma)T = -\ln[1 - 2(P + R)] \qquad [5]$$
$$4(\beta + \gamma)T = -\ln[1 - 2(Q + R)].$$

Because the evolutionary distance in terms of the number of base substitutions between the two sequences is given by Eq. 1, we obtain

$$K = -(1/4)\ln[(1 - 2P - 2Q)(1 - 2P - 2R)(1 - 2Q - 2R)]. \qquad [6]$$

This formula has the desirable property that, as $P$, $Q$, and $R$ approach zero, it converges to $K = P + Q + R$.

If the divergence time $T$ is known, the base substitution rate per year is then given by

$$k_{nuc} = K/(2T), \qquad [7]$$

in which the subscript nuc means that the estimate refers to the rate per nucleotide site.

In the special case in which two types of transversion substitutions occur equally frequently so that $\gamma = \beta$, it can be shown that Eq. 6 reduces to

$$K = -(1/2)\ln[(1 - 2P - Q')\sqrt{1 - 2Q'}], \qquad [8]$$

in which $Q' = Q + R$ is the total proportion of transversion differences (9). This formula is useful when only two types of differences (i.e., transition and transversion) are distinguished in comparative studies of sequences (as in ref. 10). In a still simpler situation in which $\alpha = \beta = \gamma$, Eq. 6 reduces to

Table 1.    Combinations of bases

| Type of difference | Transition type | Transversion type | |
|---|---|---|---|
| Sequence 1 | U C A G | U A C G | U G C A |
| Sequence 2 | C U G A | A U G C | G U A C |
| Frequency | $P$ | $Q$ | $R$ |

Various types of different base pairs at homologous nucleotide sites in two sequences compared.

$$K = -(3/4)\ln[1 - (4/3)\lambda], \qquad [9]$$

in which $\lambda = P + Q + R$ is the proportion of sites that differ in the two sequences. This formula was obtained by Jukes and Cantor (11), and a formula for the large sample standard error of this estimator was given by Kimura and Ohta (12).

It is known that a large fraction of base substitutions at the third position in the codons are synonymous (i.e., do not lead to amino acid changes). So, it may be of interest to derive a formula for estimating the synonymous component of the number of base substitutions at position 3. From the code table we note that, roughly speaking, for a given combination of bases in the first and second codon positions, base substitutions at the third position are either completely synonymous or synonymous within purines or pyrimidines. Since these two situations occur approximately in equal frequencies, we can estimate the synonymous component of the number of substitutions per year at the third position of codons by

$$k_S' = \frac{1}{2}(\alpha + \beta + \gamma) + \frac{1}{2}\alpha = \frac{1}{2}(\alpha + \beta) + \frac{1}{2}(\alpha + \gamma). \qquad [10]$$

The corresponding distance is then given by $K_S' = 2Tk_S' = (1/4)[4(\alpha + \beta)T + 4(\alpha + \gamma)T]$, and noting Eqs. 5, we obtain

$$K_S' = -(1/4)\ln[(1 - 2P - 2Q)(1 - 2P - 2R)]. \qquad [11]$$

In the special case $\gamma = \beta$, this reduces to $K_S' = -(1/2)\ln(1 - 2P - Q')$, where $Q' = Q + R$ (ref. 9).

It is desirable to have a formula for the error variance (due to sampling) of the estimated value of $K$. If $n$ is the number of nucleotide sites for which the two sequences are compared, then it can be shown that the large sample variance of $K$ is

$$\sigma_K^2 = (1/n)[a^2P + b^2Q + c^2R - (aP + bQ + cR)^2], \qquad [12]$$

in which $a = (C_{12} + C_{13})/2$, $b = (C_{12} + C_{23})/2$, and $c = (C_{13} + C_{23})/2$ in which $C_{12} = 1/(1 - 2P - 2Q)$, $C_{13} = 1/(1 - 2P - 2R)$ and $C_{23} = 1/(1 - 2Q - 2R)$.

Similarly, for the estimate of the synonymous component $K_S'$, the error variance is

$$\sigma_{K_S'}^2 = (1/n)[a_S^2P + b_S^2Q + c_S^2R - (a_SP + b_SQ + c_SR)^2], \qquad [13]$$

in which $a_S = (C_{12} + C_{13})/2$, $b_S = C_{12}/2$, and $c_S = C_{13}/2$.

As an example, let us compare the nucleotide sequence of human presomatotropin (13) with that of rat presomatotropin (14). Excluding insertions or deletions ("gaps") that amount to three codons, there are 214 homologous codon positions that can be compared. For the first codon positions, we find $P = 28/214$, $Q = 9/214$, and $R = 10/214$, and, from Eqs. 6, we obtain $K = 0.264$. It is likely that the human and the rat diverged from each other late in the Mesozoic, some 80 million years ago, so we may take $T = 8 \times 10^7$. The evolutionary rate per site at codon position 1 for presomatotropin is then $k_{nuc} = K/(2T) = 1.65 \times 10^{-9}$ per year. From Eq. 12, the error variance of $K$ becomes (taking $n = 214$) $\sigma_K^2 = 1.34 \times 10^{-3}$ so that the standard error is $\sigma_K = 3.66 \times 10^{-2}$. We can calculate the corresponding estimates of $K$ for positions 2 and 3, and also for the synonymous component, as shown in the first line in Table 2. The table also lists (in the lines marked 3ST) estimates of evolutionary distances similarly computed by using data on the human preproinsulin gene (15, 16), rat preproinsulin gene I (17, 18), rabbit $\beta$-globin (19), mouse $\beta$-globin (20), rabbit $\alpha$-globin (21), and mouse $\alpha$-1-globin genes (22). Note that in the first four comparisons the diver-

Table 2. Estimates of $K$

| Comparison | Model | Evolutionary distance per nucleotide site $K_1$ | $K_2$ | $K_3$ | $K'_S$ |
|---|---|---|---|---|---|
| Human vs. rat presomatotropins | 3ST | 0.26 ± 0.04 | 0.18 ± 0.03 | 0.53 ± 0.07 | 0.44 ± 0.07 |
| | 2FC | 0.28 ± 0.05 | 0.18 ± 0.04 | 0.75 ± 0.20 | — |
| Human vs. rat I preproinsulins: A + B chains | 3ST | 0.04 ± 0.03 | 0.00* | 0.46 ± 0.12 | 0.38 ± 0.12 |
| | 2FC | 0.04 ± 0.04 | 0.00 | 0.60 ± 0.39 | — |
| C peptide | 3ST | 0.18 ± 0.06 | 0.27 ± 0.10 | 0.95 ± 0.46 | 0.77 ± 0.51 |
| | 2FC | 0.15 ± 0.08 | 0.30 ± 0.14 | —† | — |
| Rabbit vs. mouse β-globins | 3ST | 0.16 ± 0.03 | 0.13 ± 0.03 | 0.43 ± 0.07 | 0.36 ± 0.07 |
| | 2FC | 0.17 ± 0.05 | 0.14 ± 0.04 | 0.49 ± 0.11 | — |
| Rabbit vs. mouse‡ α-globins | 3ST | 0.12 ± 0.03 | 0.12 ± 0.03 | 0.54 ± 0.09 | 0.47 ± 0.09 |
| | 2FC | 0.13 ± 0.04 | 0.13 ± 0.05 | 0.65 ± 0.17 | — |
| Rabbit α- vs. rabbit β-globins | 3ST | 0.60 ± 0.08 | 0.44 ± 0.04 | 0.90 ± 0.14 | 0.68 ± 0.13 |
| | 2FC | 0.64 ± 0.12 | 0.53 ± 0.14 | 1.19 ± 0.38 | — |

Evolutionary distances per site (together with standard errors) as estimated by using two models (3ST and 2FC). $K_i$, ($i$ = 1, 2, 3), denotes the number of base substitutions at codon position $i$ that separates the two sequences compared, and $K'_S$ is the synonymous component at position 3.

* No observed changes among 51 codons.

† Inapplicable case.

‡ Mouse α-1-globin gene of ref. 22

gence time may be taken as $T = 8 \times 10^7$ years and that the evolutionary rates per year can be obtained by dividing these values by $2T = 1.6 \times 10^8$.

## TWO-FREQUENCY-CLASS (2FC) MODEL

This model is motivated by the observation that, in mammalian mRNAs, bases C and G are much higher in frequency than U and A at the third codon positions. For example, the average base composition at the third positions in several mammalian globin sequences (computed from data in table 2 of ref. 5) are: C, 40%; G, 32%; A, 6%; and U, 22%.

Let us group the four bases into two classes, U + A in one class (called $A_1$) and C + G in the other (called $A_2$). Let $\alpha$ and $\beta$ be, respectively, the substitution rate of $A_2$ for $A_1$ and vice versa as shown in Fig. 1*b*. We denote by $X$ and $Y$ the respective frequencies of $A_1A_1$ and $A_2A_2$ pairs, and by $Z$ the frequencies of the sum of $A_1A_2$ and $A_2A_1$ pairs when two homologous sequences are compared ($X + Y + Z = 1$). Then it can be shown that $X$, $Y$, and $Z$ satisfy the differential equations

$$dX/dT = -2\alpha X + \beta Z$$
$$dY/dT = -2\beta Y + \alpha Z \qquad [14]$$
$$dZ/dT = 2\alpha X + 2\beta Y - (\alpha + \beta)Z.$$

To simplify the analysis, we assume that frequencies of $A_1$ and $A_2$ are in equilibrium so that they do not change with time. This means that the frequencies of $A_1$ and $A_2$ are given by $p$ and $q = 1 - p$,

$$p = \beta/(\alpha + \beta). \qquad [15]$$

Under this assumption, the evolutionary distance between two sequences with respect to substitutions between $A_1$ and $A_2$ is $K = 2T(p\alpha + q\beta) = 4pq(\alpha + \beta)T$, and, incorporating the relevant solution of Eqs. **14**, this leads to

$$K = -\theta \ln(1 - Z/\theta), \qquad [16]$$

in which $\theta = 2pq$, $p$ is the frequency of base group $A_1$, and $q = 1 - p$ is that of $A_2$. Also, $Z$ is the fraction of sites by which the two sequences differ from each other (i.e., $A_1A_2$ and $A_2A_1$). Note that this formula has the desirable property of converging to $K = Z$ as $Z$ approaches zero, irrespective of the value of $\theta$. If $\theta$ is in the range 0.4–0.5, then $K$ does not depend much on $\theta$ if $Z$ is less than 0.2. Note also that Eq. **9** is equivalent to this formula when $\theta = 3/4$. In applying this formula it may be desirable to estimate $p$ not simply from the two sequences being compared but from a number of related sequences (if they are available). For example, for the comparison of globin sequences, we take $p = 0.28$ which is the average frequency of U + A at the third codon positions for six globins (rabbit α-, mouse α-, human β-, rabbit β-, mouse β-, and chicken β-). Let us suppose then that, in general, $p$ is estimated by a sample of size $N$, and $Z$ is estimated from a sample of size $n$. Then it can be shown that the standard error of $K$ is given by

$$\sigma_K = \sqrt{a^2\sigma_\theta^2 + b^2\sigma_Z^2}, \qquad [17]$$

in which $\sigma_\theta^2 = 4(1 - 2p)^2p(1 - p)/N$, $\sigma_Z^2 = Z(1 - Z)/n$, $a = K/\theta - Z/(\theta - Z)$, and $b = \theta/(\theta - Z)$. Note that, for $p = 0.28$, which we assume for codon position 3 of globins, we have $\theta = 0.4$. If we take conservatively $N = 500$ (because, the six globins used to estimate $p$ are not wholly independent), then $\sigma_\theta^2 = 3.12 \times 10^{-4}$. Because $\theta$ is not very sensitive to the change of $p$ at the neighborhood of 0.5, we may take $\theta = 0.5$ unless $p$ and $q$ differ greatly from each other. Note that, for $\theta = 0.5$, we have $\sigma_\theta^2 = 0$ so that $N$ is irrelevant for computing $\sigma_K^2$.

In order to estimate the total distance $K$ by using this model, we first estimate the component $K_b$ ("between-class component") by applying Eq. **16**, classifying the four bases into two groups $A_1$ and $A_2$. Next, we apply Eq. **16** to the first class $A_1$, proceeding as if the two bases U and A make up 100%. This yields an estimate for the component $K_{w1}$ ["within-class $A_1$ component," corresponding to $2T(\alpha_1 + \beta_1)$ of Fig. 1*b*]. Similarly, we obtain $K_{w2}$ ("within-class $A_2$ component"). Then the total distance is obtained by

$$K = K_b + pK_{w1} + qK_{w2}. \qquad [18]$$

For codon position 3 of globins, we take $\theta = 0.4$, but for codon positions 2 and 3, we take $\theta = 0.5$. The difference between the estimate obtained by using the 2FC model and that obtained by the 3ST model becomes significant only when the base composition deviates greatly from equality and at the same time the evolutionary distance involved is large. This is evident when we compare values estimated by these two methods for $K_1$ and $K_2$ as listed in Table 2.

At the third codon positions, and particularly when the distance is large, however, the difference may become large. As an example, let us compare the $\alpha$- and $\beta$-globins of the rabbit. Excluding insertions and deletions (gaps) that amount to 9 codons, there are 139 codons that can be compared ($n = 139$). For the third positions of these codons, we find $nX = 8$ (UU 4, AA 2, UA 1, AU 1), $nY = 82$ (CC 29, GG 32, CG 16, GC 5), and $nZ = 49$. Applying Eqs. **16** and **17** with $\theta = 0.4$ ($p = 0.28$), $Z = 49/139$, $n = 139$, and $N = 500$, we obtain $K_b = 0.836 \pm 0.334$. Also, applying these equations within classes $A_1$ (consisting of U and A) and $A_2$ (C + G), we get $K_{w1} = 0.346 \pm 0.306$ and $K_{w2} = 0.359 \pm 0.099$. Altogether we get $K_3 = 1.19 \pm 0.38$. On the other hand, the corresponding estimate obtained by using the 3ST model turns out to be $K_3 = 0.90 \pm 0.14$, which is likely to be an underestimate (see bottom line in Table 2).

## CORRECTION FOR EXCLUDING INAPPLICABLE CASES

Equations for $K$, such as Eqs. **6**, **11**, and **16**, are derived by deterministic methods which are based on the assumption that the lengths of sequences involved are infinite. In other words, the sampling effect due to finite number of codons is disregarded. On the other hand, the actual sequences are all finite in length, and the observed numbers of differences are subject to statistical fluctuation. The most serious consequence of such fluctuation is that cases arise, particularly when the true value of $K$ is large and $n$ is small, for which the equations cannot be applied. I shall explain this using Eq. **16.** Let $n$ be the total number of homologous sites and let $j$ be the observed number of sites for which the two sequences differ from each other—that is, the number of $A_1A_2$ plus $A_2A_1$ pairs ($j = 0, 1, \ldots, n$). Then, $j$ follows the binomial distribution

$$f(j) = \binom{n}{j} Z^j(1 - Z)^{n-j}, \qquad [19]$$

in which $Z = \theta[1 - \exp(-K/\theta)]$. If $j$ happens to become equal to or larger than $n\theta$, then Eq. **16** cannot be used to estimate $K$ by letting $Z = j/n$ in this equation, because $(1 - Z/\theta)$ becomes negative. If we exclude such "inapplicable cases," the estimate of $K$ becomes biased and a correction will be required. Let $\tilde{K}$ be the average value of $K$ obtained under the condition that inapplicable cases are excluded—i.e., $\tilde{K} = E\{K|j < n\theta\}$.

$$\tilde{K} = -\sum_{j=0}^{L} \theta \ln\{1 - j/(n\theta)\}f(j) \Big/ \sum_{j=0}^{L} f(j), \qquad [20]$$

in which $L$ is the maximum integer such that $L < n\theta$. Fig. 2 depicts the relationship between the true distance $K$ and the conditional distance $\tilde{K}$, assuming $\theta = 0.4$ and $n = 140$. The graph suggests the possibility of a serious underestimate for $K$ when its estimated value (applying Eq. **16** to position 3 of globins) turns out to be larger than about 1.0.

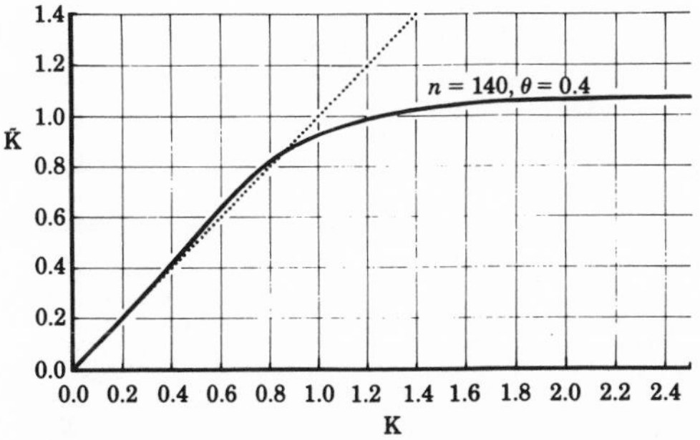


**FIG. 2. Relationship between the true distance $K$ and the conditional distance $\tilde{K}$. For details, see text.**

## DISCUSSION

Table 2 shows that the evolutionary rates of synonymous base substitutions at the third positions of codons are not only high but also are roughly equal (the two seemingly higher values are for C peptide, which has a large standard error, and the bottom comparison which involves a much longer time period, probably $T = 5 \times 10^8$). This is particularly evident if we contrast the evolutionary distances in presomatotropin with those of insulin (preproinsulin A + B chains), both involving human vs. rat comparisons. In presomatotropin hormone, the distance due to amino acid-altering substitutions per site, as estimated by $(K_1 + K_2)/2$, is 0.22, but the corresponding distance is only 0.02 in insulin. This means that amino acid-altering substitutions proceed some 10 times faster in presomatotropin than in insulin. On the other hand, the synonymous component at position 3, as estimated by $K_S'$, is roughly equal in these two proteins. Furthermore, in $\alpha$- and $\beta$-globins, the rates of synonymous substitutions are about equal to those of presomatotropin and insulin, although their amino acid-altering substitutions are intermediate between those of somatotropin and insulin. Note that the divergence time of rabbit and mouse is approximately the same as that of man and rat. Such uniform rate of synonymous substitutions has also been brought out by Miyata *et al.* (7).

These observations can be explained readily by the neutral mutation-random drift hypothesis of molecular evolution (the neutral theory, in short; see ref. 2). Unlike the Darwinian paradigm, this theory states that the majority of evolutionary mutant substitutions in the species are caused by random fixation of selectively neutral (i.e., selectively equivalent, but not necessarily functionally equivalent) mutants rather than by positive Darwinian selection. Although favorable mutations no doubt occur, the theory assumes that they are so rare as to be neglected in calculating rates of molecular evolution. The neutral theory predicts that the probability of a mutation being selectively neutral (that is, not harmful) is larger the less the mutation disrupts the existing structure and function of the molecule. At the limit in which all the mutations are selectively neutral, the rate of evolution per site ($k$) becomes equal to the total mutation rate ($v$) per site. In my opinion (see ref. 1), synonymous mutations are not very far from this limit and therefore the evolutionary rates of synonymous substitutions per site are nearly equal between different molecules.

Recently, an opposing view was proposed by Perier *et al.* (23). They claimed that the driving force for fixation is positive natural selection operating on some fraction of amino acid-altering ("replacement") changes and, that such a selected fixation carries along with it neutral alterations (including changes at silent sites) that have accumulated in that region of the DNA. In other words, they invoke the "hitchhiking" effect to explain fixation of synonymous changes.

I would like to point out that, unless we ignore the principles of population genetics, such an explanation cannot account for actual observations. In fact, such hitchhiking cannot bring about substitutions of neutral mutants at a very high rate when the selected changes occur at a very low rate. For example, take the histone H4 gene. The rate of replacement changes is almost zero, yet synonymous base substitutions occur at a rate comparable to that of replacement changes in fibrinopeptides, one of the most rapidly evolving molecules (1).

We can treat the problem in more detail. Because the hitchhiking effect extends only over short distances around a selectively driven gene, particularly in bringing associated mutations to fixation in the population, we consider a small segment of DNA, such as a gene locus, within which the incidence of crossing over is so low as to be neglected. Let us suppose that a new, advantageous, mutant allele at this gene locus appeared, first singly represented, in the population. In order that this selected mutant can bring other unselected (neutral) mutants to fixation by hitchhiking, the gene copy in which this advantageous mutant appeared must also contain at the same time a number of neutral mutants. Furthermore, in order to make the rate of substitution of neutral mutants per site $m$ times higher than that of selectively driven mutations (in this case, amino acid-altering changes), each gene copy in the population must contain on the average $m$ neutral mutants, irrespective of whether an advantageous mutation happened to occur in it or not. This factor $m$ must be very large, probably 1000 or more in histone H4. On the other hand, if each gene copy contains a large number of neutral mutants, the corresponding (homologous) genes in different individuals differ from each other in so many bases that there is no such thing as a species-specific nucleotide sequence of a particular gene, say histones, hemoglobins, etc. In other words, every individual in the species would have quite different homologous sequences. This is contrary to observations.

Furthermore, the hitchhiking theory cannot explain the observation that, when genes of different proteins are studied, the evolutionary rates of synonymous substitutions are not only high, but also they are roughly equal to each other, even when their amino acid-altering substitution rates differ greatly.

I thank Drs. T. H. Jukes and K. M. Weiss for reading the manuscript and correcting the English. Thanks are also due to Dr. J. F. Crow who read the first draft carefully and made many useful suggestions to improve the presentation and to Dr. N. Takahata for stimulating discussions. Contribution no. 1339 from the National Institute of Genetics, Mishima, Shizuoka-ken, 411 Japan.

1. Kimura, M. (1977) *Nature (London)* **267**, 275–276.
2. Kimura, M. (1979) *Sci. Am.* **241 (5)**, 94–104.
3. Jukes, T. H. (1978) *J. Mol. Evol.* **11**, 267–269.
4. Jukes, T. H. & King, J. L. (1979) *Nature (London)* **281**, 605–606.
5. Jukes, T. H. (1980) *Naturwissenschaften*, **67**, 534–539.
6. Jukes, T. H. (1980) *Science*, **210**, 973–978.
7. Miyata, T., Yasunaga, T. & Nishida, T. (1980) *Proc. Natl. Acad. Sci. USA* **77**, 7328–7332.
8. Nichols, B. P. & Yanofsky, C. (1979) *Proc. Natl. Acad. Sci. USA* **76**, 5244–5248.
9. Kimura, M. (1980) *J. Mol. Evol.*, in press.
10. van Ooyen, A., van den Berg, J., Mantel, N. & Weissmann, C. (1979) *Science* **206**, 337–344.
11. Jukes, T. H. & Cantor, C. R. (1969) in *Mammalian Protein Metabolism, III*, ed. Munro, H. N. (Academic, New York), pp. 21–132.
12. Kimura, M. & Ohta, T. (1972) *J. Mol. Evol.* **2**, 87–90.
13. Martial, J. A., Hallewell, R. A., Baxter, J. D. & Goodman, H. M. (1979) *Science* **205**, 602–607.
14. Seeburg, P. H., Shine, J., Martial, J. A., Baxter, J. D. & Goodman, H. M. (1977) *Nature (London)* **270**, 486–494.
15. Bell, G. I., Pictet, R. L., Rutter, W. J., Cordell, B., Tischer, E. & Goodman, H. M. (1980) *Nature (London)* **284**, 26–32.
16. Sures, I., Goeddel, D. V., Gray, A. & Ullrich, A. (1980) *Science* **208**, 57–59.
17. Cordell, B., Bell, G., Tischer, E., DeNoto, F. M., Ullrich, A., Pictet, R., Rutter, W. J. & Goodman, H. M. (1979) *Cell* **18**, 533–543.
18. Lomedico, P., Rosenthal, N., Efstratiadis, A., Gilbert, W., Kolodner, R. & Tizard, R. (1979) *Cell* **18**, 545–558.
19. Efstratiadis, A., Kafatos, F. C. & Maniatis, T. (1977) *Cell* **10**, 571–585.
20. Konkel, D. A., Tilghman, S. M. & Leder, P. (1978) *Cell* **15**, 1125–1132.
21. Heindell, H. C., Liu, A., Paddock, G. V., Studnicka, G. M. & Salser, W. A. (1978) *Cell* **15**, 43–54.
22. Nishioka, Y., Leder, A. & Leder, P. (1980) *Proc. Natl. Acad. Sci. USA* **77**, 2806–2809.
23. Perier, F., Efstratiadis, A., Lomedico, P., Gilbert, W., Kolodner, R. & Dodgson, J. (1980) *Cell* **20**, 555–566.

# A MODEL OF EVOLUTIONARY BASE SUBSTITUTIONS AND ITS APPLICATION WITH SPECIAL REFERENCE TO RAPID CHANGE OF PSEUDOGENES*

NAOYUKI TAKAHATA AND MOTOO KIMURA

*National Institute of Genetics, Mishima, 411 Japan*

Manuscript received February 2, 1981
Revised copy received May 18, 1981

ABSTRACT

A model of evolutionary base substitutions that can incorporate different substitutional rates between the four bases and that takes into account unequal composition of bases in DNA sequences is proposed. Using this model, we derived formulae that enable us to estimate the evolutionary distances in terms of the number of nucleotide substitutions through comparative studies of nucleotide sequences. In order to check the validity of various formulae, Monte Carlo experiments were performed. These formulae were applied to analyze data on DNA sequences from diverse organisms. Particular attention was paid to problems concerning a globin pseudogene in the mouse and the time of its origin through duplication. We obtained a result suggesting that the evolutionary rates of substitution in the first and second codon positions of the pseudogene were roughly 10 times faster than those in the normal globin genes; whereas, the rate in the third position remained almost unchanged. Application of our formulae to histone genes H2B and H3 of the sea urchin showed that, in each of these genes, the rate in the third codon position is tremendously higher than that in the second position. All of these observations can easily and consistently be interpreted by the neutral theory of molecular evolution.

RECENT developments in DNA-sequencing techniques (MAXAM and GILBERT 1977; SANGER, NICKLEN and COULSON 1977), together with methods for amplifying gene copies in a bacterial plasmid, have made possible rapid determinations of DNA sequences of genes. Because data on DNA sequences are obtainable only from living organisms, it is necessary to develop mathematical models to estimate the number of evolutionary nucleotide substitutions through comparison of DNA sequences of homologous genes in related species.

Prior to a recent flood of data on nucleotide sequences, there already existed a large body of data on amino acid sequences in diverse organisms, and many mathematical models have been proposed to treat protein evolution in terms of amino acid substitutions, (see, for example, ZUCKERKANDL and PAULING 1965; FITCH and MARGOLIASH 1967; JUKES and CANTOR 1969; OHTA and KIMURA 1971; NEI 1975 for review). However, for comparative studies of nucleotide sequences, different mathematical models have to be employed. Sev-

* Contribution No. 1354 from The National Institute of Genetics, Mishima, Shizuoka-ken, 411 Japan.

Genetics 98: 641–657 July, 1981

eral authors have considered the problem of estimating the evolutionary distance of the homologous part of the genome between related species (JUKES and CANTOR 1969; KIMURA and OHTA 1972; MIYATA and YASUNAGA 1980; HOLMQUIST 1980; HOLMQUIST and PEARL 1980).

Recently, KIMURA (1980; 1981) developed three different models that can partially incorporate different substitutional rates between four bases and applied them to analyze data on various nucleotide sequences from several organisms. He showed that a preponderance of synonymous and other silent nucleotide substitutions is a general feature of molecular evolution and that this is consistent with the neutral theory (KIMURA 1968; see KIMURA 1979 for review).

In this paper, we extend the mathematical models of KIMURA (1980, 1981), and we derive appropriate formulae for estimating evolutionary distances in terms of the number of nucleotide substitutions per site. In addition to an analytical treatment, we used simulation methods to check the validity of the formulae and determine their range of applicability. This is necessary because formulae for estimating evolutionary distances generally do not have high resolving power when they are applied to evolutionarily distant organisms; they are accompanied by large error variances. Therefore, we conducted extensive Monte Carlo experiments in which the values of the parameters involved were greatly altered.

We have applied our formulae to analyze actual data on nucleotide sequences. Particular attention was paid to the evolution of the globin pseudogene in the mouse (NISHIOKA, LEDER and LEDER 1980; VANIN *et al.* 1980). Although similar analyses have recently been carried out by KIMURA (1980), PROUDFOOT and MANIATIS (1980) and, in more detail, by MIYATA and YASUNAGA (1981), we re-examined the problem to estimate the time of its origin and to discuss the evolutionary implications.

## MODEL AND ANALYSIS

Let us consider a model of base substitutions, as shown in Figure 1, in which the four bases are represented by the letters U, A, C and G in terms of mRNA codes. The rates of base substitutions per unit time (say, year) between the four bases are designated by $\alpha$, $\beta$, $\gamma$, $\delta$ and $\varepsilon$. For instance, $\alpha$ is the rate of transition-type substitutions from U to C or A to G, while $\beta$ is the rate for the reverse directions. The rates of transversion-type substitutions are denoted by $\gamma$, $\delta$ and $\varepsilon$. In comparing two homologous sequences, there are 16 combinations of base pairs at each site. The possible combinations and their relative frequencies (probabilities) are listed in Table 1. They represent the expected relative frequencies of their occurrence in two homologous sequences. For example, $S$ (the sum of $S_i$'s) stands for the probability that the bases at a homologous site are identical and $P$ $(= 2P_1 + 2P_2)$ the probability of their showing transition-type differences, while $Q$ $(= 2Q_1 + 2Q_2)$ and $R$ $(= 2R_1 + 2R_2)$ are the probabilities of transversion-type differences.

We denote the probabilities at time $T$ of the four bases by $U(T)$, $A(T)$, $C(T)$ and $G(T)$. Starting from a common ancestor $(T = 0)$, we can express these

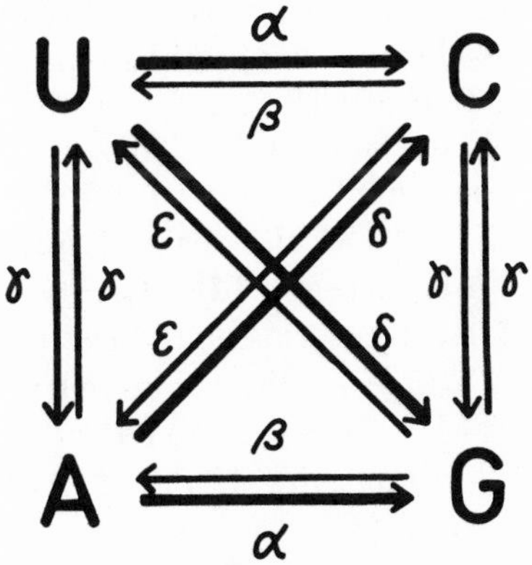


FIGURE 1.—Scheme of base substitutions and their rates per unit time.

probabilities at time $T + \Delta T$, using their probabilities at time $T$, and the rates of base substitutions, where $\Delta T$ stands for a short time interval. Neglecting small quantities involving $(\Delta T)^2$ and higher order terms, we have, for example,

$$\mathrm{U}(T+\Delta T) = \{1-(\alpha+\delta+\gamma)\Delta T\}\mathrm{U}(T) + \beta\Delta T\mathrm{C}(T) + \varepsilon\Delta T\mathrm{C}(T) + \gamma\Delta T\mathrm{A}(T).$$

As seen from Figure 1, the first term in the right-hand side of this equation corresponds to the probability of no change, and the last three terms are the probabilities that U came from the remaining three bases during the short time interval $\Delta T$. The corresponding probabilities for other bases can be obtained in a similar manner, and we get the following set of ordinary differential equations by letting $\Delta T \to 0$;

$$\frac{d\mathrm{U}(T)}{dT} = -(\alpha+\delta+\gamma)\mathrm{U}(T) + \beta\mathrm{C}(T) + \varepsilon\mathrm{G}(T) + \gamma\mathrm{A}(T) \,,$$

$$\frac{d\mathrm{A}(T)}{dT} = -(\alpha+\delta+\gamma)\mathrm{A}(T) + \beta\mathrm{G}(T) + \varepsilon\mathrm{C}(T) + \gamma\mathrm{U}(T) \,,$$

$$\frac{d\mathrm{C}(T)}{dT} = -(\beta+\varepsilon+\gamma)\mathrm{C}(T) + \alpha\mathrm{U}(T) + \delta\mathrm{A}(T) + \gamma\mathrm{G}(T) \,,$$

$$\frac{d\mathrm{G}(T)}{dT} = -(\beta+\varepsilon+\gamma)\mathrm{G}(T) + \alpha\mathrm{A}(T) + \delta\mathrm{U}(T) + \gamma\mathrm{C}(T) \,.$$

Likewise, we can derive the equations for changes of probabilities of the base

TABLE 1

*Types of nucleotide base pairs occurring at homologous sites in two sequences and their probabilities (relative frequencies)*

| Types | U C A G<br>U C A G | U C A G<br>C U G A | U A C G<br>A U G C | U G A C<br>G U C A |
|---|---|---|---|---|
| Probabilities | $\underbrace{S_1\ S_2\ S_3\ S_4}_{S}$ | $\underbrace{2P_1\ 2P_2}_{P}$ | $\underbrace{2Q_1\ 2Q_2}_{Q}$ | $\underbrace{2R_1\ 2R_2}_{R}$ |

pairs listed in Table 1, although the procedures involved are more complicated. As an example, let us consider the change of base pair UC. Noting that the probability of UC is equal to that of CU, *i.e.*, $P_1$, we first get the probability of no change occurring in either nucleotide. This is $[1-(\alpha+\gamma+\varepsilon)\Delta T]\,[1-(\beta+\gamma+\delta)\Delta T]$, so that the contribution from this class is $[1-(\alpha+\gamma+\varepsilon)\Delta T]\times[1-(\beta+\gamma+\delta)\Delta T]P_1(T)$. UC is also derived from UU and CC with probabilities $[1-(\alpha+\gamma+\varepsilon)\Delta T]\alpha\Delta T$ and $[1-(\beta+\gamma+\delta)\Delta T]\beta\Delta T$, respectively. Then, the contribution from these classes is $[1-(\alpha+\gamma+\varepsilon)\Delta T]\alpha\Delta TS_1(T)+[1-(\beta+\gamma+\delta)\Delta T]\beta\Delta TS_2(T)$. Additional contributions come from pairs UA and GC with probabilities $[1-(\alpha+\gamma+\varepsilon)\Delta T]\delta\Delta TQ_1(T)$ and $[1-(\beta+\gamma+\delta)\Delta T]\varepsilon\Delta TQ_2(T)$, and also from pairs UG and AC with probabilities $[1-(\alpha+\gamma+\varepsilon)\Delta T]\gamma\Delta TR_1(T)$ and $[1-(\beta+\gamma+\delta)\Delta T]\gamma\Delta TR_2(T)$. Combining all these contributions, and neglecting $(\Delta T)^2$ and higher order terms, as before, we have

$$\begin{aligned}P_1(T+\Delta T)=&[1-(\alpha+\beta+2\gamma+\delta+\varepsilon)\Delta T]P_1(T)+\alpha\Delta TS_1(T)\\&+\beta\Delta TS_2(T)+\delta\Delta TQ_1(T)+\varepsilon\Delta TQ_2(T)+\gamma\Delta T\{R_1(T)+R_2(T)\}.\end{aligned}$$

Continuing these calculations for other base pairs and taking the limit $\Delta T\to 0$, we get a complete set of differential equations (equations 1). It is interesting to note that a more convenient derivation of equations (1) is possible if we use the differential equations for $\mathrm{U}(T)$, $\mathrm{A}(T)$, $\mathrm{C}(T)$ and $\mathrm{G}(T)$ and combine them through relationships such as $\dfrac{dP_1(T)}{dT}=\mathrm{U}(T)\dfrac{d\mathrm{C}(T)}{dT}+\mathrm{C}(T)\dfrac{d\mathrm{U}(T)}{dT}$ and $P_1(T)=\mathrm{U}(T)\mathrm{C}(T)$ for the case of the UC pair. This is true because bases in one species change independently from those in other species. We can verify by direct calculation that both derivations give the same set of equations. Thus, we obtain a complete set of differential equations as follows:

$$\left.\begin{aligned}
\frac{dS_1(T)}{dT}&=-2(\alpha+\gamma+\delta)S_1(T)+2\beta P_1(T)+2\gamma Q_1(T)+2\varepsilon R_1(T)\\
\frac{dS_2(T)}{dT}&=-2(\beta+\gamma+\varepsilon)S_2(T)+2\alpha P_1(T)+2\gamma Q_2(T)+2\delta R_2(T)\\
\frac{dS_3(T)}{dT}&=-2(\alpha+\gamma+\delta)S_3(T)+2\beta P_2(T)+2\gamma Q_1(T)+2\varepsilon R_2(T)\\
\frac{dS_4(T)}{dT}&=-2(\beta+\gamma+\varepsilon)S_4(T)+2\alpha P_2(T)+2\gamma Q_2(T)+2\delta R_1(T)\\
\frac{dP_1(T)}{dT}&=-(\alpha+\beta+2\gamma+\delta+\varepsilon)P_1(T)+\alpha S_1(T)+\beta S_2(T)\\
&\qquad+\delta Q_1(T)+\varepsilon Q_2(T)+\gamma[R_1(T)+R_2(T)]\\
\frac{dP_2(T)}{dT}&=-(\alpha+\beta+2\gamma+\delta+\varepsilon)P_2(T)+\alpha S_3(T)+\beta S_4(T)\\
&\qquad+\delta Q_1(T)+\varepsilon Q_2(T)+\gamma[R_1(T)+R_2(T)]\\
\frac{dQ_1(T)}{dT}&=-2(\alpha+\gamma+\delta)Q_1(T)+\gamma[S_1(T)+S_3(T)]\\
&\qquad+\varepsilon[P_1(T)+P_2(T)]+\beta[R_1(T)+R_2(T)]
\end{aligned}\right\}\quad(1)$$

$$\left.\begin{aligned}\frac{dQ_2(T)}{dT} &= -2(\beta+\gamma+\delta)Q_2(T)+\gamma[S_2(T)+S_4(T)]\\ &\quad +\delta[P_1(T)+P_2(T)]+\alpha[R_1(T)+R_2(T)]\\ \frac{dR_1(T)}{dT} &= -(\alpha+\beta+2\gamma+\delta+\varepsilon)R_1(T)+\delta S_1(T)+\varepsilon S_4(T)\\ &\quad +\gamma[P_1(T)+P_2(T)]+\alpha Q_1(T)+\beta Q_2(T)\\ \frac{dR_2(T)}{dT} &= -(\alpha+\beta+2\gamma+\delta+\varepsilon)R_2(T)+\delta S_3(T)+\varepsilon S_2(T)\\ &\quad +\gamma[P_1(T)+P_2(T)]+\alpha Q_1(T)+\beta Q_2(T).\end{aligned}\right)$$

To solve equations (1), we define six variables

$$\left.\begin{aligned}X_\pm(T) &= S_1(T)+S_3(T)\pm 2Q_1(T)\\ Y_\pm(T) &= S_2(T)+S_4(T)\pm 2Q_2(T)\\ Z_\pm(T) &= P(T)\pm R(T)\ ,\end{aligned}\right\} \tag{2}$$

where we take the same sign for the subscripts of $X$, $Y$ and $Z$ as that in the right-hand side. Then, from (1), we can derive two sets of equations,

$$\frac{d}{dT}\begin{pmatrix}X_+(T)\\ Y_+(T)\\ Z_+(T)\end{pmatrix}=\begin{pmatrix}-2(\alpha+\delta) & 0 & \beta+\varepsilon\\ 0 & -2(\beta+\varepsilon) & \alpha+\delta\\ 2(\alpha+\delta) & 2(\beta+\varepsilon) & -(\alpha+\beta+\delta+\varepsilon)\end{pmatrix}\begin{pmatrix}X_+(T)\\ Y_+(T)\\ Z_+(T)\end{pmatrix} \tag{3}$$

and

$$\frac{d}{dT}\begin{pmatrix}X_-(T)\\ Y_-(T)\\ Z_-(T)\end{pmatrix}=\begin{pmatrix}-2(\alpha+2\gamma+\delta) & 0 & \beta-\varepsilon\\ 0 & -2(\beta+2\gamma+\varepsilon) & \alpha-\delta\\ 2(\alpha-\delta) & 2(\beta-\varepsilon) & -(\alpha+\beta+4\gamma+\delta+\varepsilon)\end{pmatrix}\times\begin{pmatrix}X_-(T)\\ Y_-(T)\\ Z_-(T)\end{pmatrix}. \tag{4}$$

In these equations, the transformation matrices have a common form

$$M=\begin{pmatrix}-2c & 0 & b\\ 0 & -2d & a\\ 2a & 2b & -(c+d)\end{pmatrix}. \tag{5}$$

Note that $a=c$ and $b=d$ hold in (3). As we can easily calculate the eigenvalues and projection operators for the matrix of (5), we can solve the initial value problems of equations (3) and (4). Let $\lambda_i$'s ( $(i=1, 2$ and $3)$ be the eigenvalues and $p_i$'s be the corresponding projection operators. Then, we have

$$\lambda_1=-(c+d),\ \lambda_2=-(c+d-g)\text{ and }\lambda_3=-(c+d+g), \tag{6}$$

and

$$p_1=\frac{1}{g^2}\begin{pmatrix}2ab & -2b^2 & -b(d-c)\\ -2a^2 & 2ab & a(d-c)\\ -2a(d-c) & 2b(d-c) & (d-c)^2\end{pmatrix}$$

$$\left.\begin{aligned} p_2 &= \frac{1}{2g^2}\begin{pmatrix} (d-c)(d-c+g)+2ab & 2b^2 & b(d-c+g) \\ 2a^2 & (d-c)(d-c-g)+2ab & -a(d-c-g) \\ 2a(d-c+g) & 2b(-d+c+g) & 4ab \end{pmatrix} \\ p_3 &= \frac{1}{2g^2}\begin{pmatrix} (d-c)(d-c-g)+2ab & 2b^2 & b(d-c-g) \\ 2a^2 & (d-c)(d-c+g)+2ab & -a(d-c+g) \\ 2a(d-c+g) & 2b(d-c-g) & 4ab \end{pmatrix} \end{aligned}\right\} \quad (7)$$

where $g = \sqrt{(d-c)^2+4ab}$.

By using these formulae, we obtain the solutions $X(T)$ at time $T$ under an arbitrary initial condition of $\mathbf{X}(0)$. Thus,

$$\mathbf{X}(T) = \{e^{\lambda_1 T}p_1 + e^{\lambda_2 T}p_2 + e^{\lambda_3 T}p_3\}\mathbf{X}(0) \quad (8)$$

where $\mathbf{X}(\cdot)$ is a column vector that can be either $(X_+, Y_+, Z_+)^t$ or $(X_-, Y_-, Z_-)^t$, in which the superscript $t$ denotes the transpose.

We assume that the frequency ($\omega$) of U + A does not change with time, and also that $\mathrm{U}(T) = \mathrm{A}(T)$ and $\mathrm{C}(T) = \mathrm{G}(T)$ for all $T$. Then, $S_i(T)$, $P_i(T)$, $Q_i(T)$ and $R_i(T)$ can all be expressed in terms of $X_\pm(T)$, $Y_\pm(T)$ and $Z_\pm(T)$. The evolutionary rate of base substitutions per unit time is

$$k = (\alpha + \gamma + \delta)\omega + (\beta + \gamma + \varepsilon)(1-\omega)\ , \quad (9)$$

or

$$k = \gamma + 2\omega(1-\omega)(\alpha + \beta + \delta + \varepsilon)\ . \quad (9a)$$

These equations are derived from the consideration that U or A each with the frequency $\omega/2$ changes to the other bases at the rate of $\alpha + \gamma + \delta$, and C or G each with $(1-\omega)/2$ changes at the rate of $\beta + \gamma + \varepsilon$. Note that we have $\omega = (\beta + \varepsilon)/(\alpha + \beta + \delta + \varepsilon)$. Therefore, the expected number of substitutions per site between two species with divergence time $T$ is given by

$$K = 2Tk\ .$$

If we use formula (9a) for $k$, then

$$K = 2\gamma T + 4\omega(1-\omega)(\alpha + \beta + \delta + \varepsilon)T\ . \quad (10)$$

Before deriving an expression for $K$ in terms of $X_\pm$, $Y_\pm$ and $Z_\pm$, we shall obtain the explicit expression for the eigenvalues and the functional forms of those quantities under the assumption of the steady state of U + A content. As the initial conditions are now

$$\mathbf{X}_\pm(0) = (\omega, 1-\omega, 0)^t \quad (11)$$

for both cases, the solutions for $\mathbf{X}_+(T)$ are expressed in a simple form,

$$\left.\begin{aligned} X_+(T) &= \omega\{\omega + (1-\omega)e^{\lambda_0 T}\} \\ Y_+(T) &= (1-\omega)(1-\omega+\omega e^{\lambda_0 T}) \\ Z_+(T) &= 2\omega(1-\omega)(1-e^{\lambda_0 T}) \end{aligned}\right\} \quad (12)$$

where

$$\lambda_0 = -2(\alpha + \beta + \delta + \varepsilon)\ . \tag{13}$$

On the other hand, the eigenvalues of equation (4) are

$$\begin{aligned} \lambda_1 &= -(\alpha + \beta + \delta + \varepsilon + 4\gamma) \\ \lambda_2 &= \lambda_1 + g \\ \lambda_3 &= \lambda_1 - g \end{aligned} \tag{14}$$

where $g^2 = \{\alpha + \delta - (\beta + \varepsilon)\}^2 + 4(\alpha - \delta)(\beta - \varepsilon)$. Using (7) and (14), the solutions for $\mathbf{X}_-(T)$ are, for $g \neq 0$,

$$\left.\begin{aligned} X_-(T) &= \frac{1}{g^2}\,[2b\{a\omega - b(1-\omega)\}e^{\lambda_1 T} + \{\xi\omega + b^2(1-\omega)\}e^{\lambda_2 T} \\ &\qquad + \{\eta\omega + b^2(1-\omega)\}e^{\lambda_3 T}] \\ Y_-(T) &= \frac{1}{g^2}\,[-2a\{a\omega - b(1-\omega)\}e^{\lambda_1 T} + \{a^2\omega + \eta(1-\omega)\}e^{\lambda_2 T} \\ &\qquad + \{a^2\omega + \xi(1-\omega)\}e^{\lambda_3 T}] \\ Z_-(T) &= \frac{1}{g^2}\,[-2(d-c)\{a\omega - b(1-\omega)\}e^{\lambda_1 T} + \{a(d-c+g)\omega \\ &\qquad - b(d-c-g)(1-\omega)\}e^{\lambda_2 T} + \{a(d-c-g)\omega \\ &\qquad - b(d-c+g)(1-\omega)\}e^{\lambda_3 T}] \end{aligned}\right\} \tag{15}$$

in which $a = \alpha - \delta$, $b = \beta - \varepsilon$, $c = \alpha + \delta + 2\gamma$, $d = \beta + \varepsilon + 2\gamma$, $\xi = \frac{1}{2}(d-c)(d-c+g) + ab$ and $\eta = \frac{1}{2}(d-c)\ (d-c-g) + ab$. In this case, however, it does not seem feasible to derive a simple formula for $2\lambda_1 T = (\lambda_2 + \lambda_3)T = -2(\alpha + \beta + \delta + \varepsilon + 4\gamma)$ as a function of $\mathbf{X}_-$(T) and $\omega$.

A great simplification is possible if we assume that $\delta = \theta\alpha$ and $\varepsilon = \theta\beta$, where $\theta$ is a constant. Then, equations (15) are much simplified (see 15a below), although equations (12) remain the same. Furthermore, $2\gamma T$ in (10) can be expressed in terms of $X_-(T)$, $Y_-(T)$, $Z_-(T)$ and $\omega$, which can be estimated from observations. The solutions for $\mathbf{X}_-(T)$ become, for $\alpha \neq \beta$ and $\theta \neq 1$,

$$\begin{aligned} X_-(T) &= \frac{\omega}{2g}\,\{(d-c+g)e^{\lambda_2 T} - (d-c-g)e^{\lambda_3 T}\} \\ Y_-(T) &= \frac{1-\omega}{2g}\,\{-(d-c-g)e^{\lambda_2 T} + (d-c+g)e^{\lambda_3 T}\} \\ Z_-(T) &= \frac{a\omega + b(1-\omega)}{g}\,(e^{\lambda_2 T} - e^{\lambda_3 T})\ , \end{aligned} \tag{15a}$$

where $\omega = \dfrac{\beta}{\alpha + \beta}$. Note that, under the above assumption, if $\alpha = \beta$, the model reduces to the "three-substitution-type" (3ST) model of KIMURA (1981). Using equations (15a), we get

$$X_-(T)Y_-(T) - \left(\frac{Z_-(T)}{2}\right)^2 = \omega(1-\omega)e^{(\lambda_2+\lambda_3)T}\ . \tag{16}$$

Combining this with equations (12) and (13), we get

$$\gamma T = -\frac{1}{8}\ln\left\{\frac{X_{-}(T)Y_{-}(T) - \left(\frac{Z_{-}(T)}{2}\right)^2}{\omega(1-\omega) - \left(\frac{Z_{+}(T)}{2}\right)}\right\}, \tag{17}$$

and therefore we obtain an appropriate equation for $K$ as follows.

$$K = -\frac{1}{4}\ln\left[\left\{\frac{X_{-}(T)Y_{-}(T) - \left(\frac{Z_{-}(T)}{2}\right)^2}{\omega(1-\omega)}\right\}\left\{1 - \frac{Z_{+}(T)}{2\omega(1-\omega)}\right\}^{8\omega(1-\omega)-1}\right], \tag{18}$$

or more explicitly

$$K = -\frac{1}{4}\ln\left[\left\{\frac{(S_1+S_3+2Q_1)(S_2+S_4-2Q_2) - \left(\frac{P-R}{2}\right)^2}{\omega(1-\omega)}\right\} \times \left\{1 - \frac{P+R}{2\omega(1-\omega)}\right\}^{8\omega(1-\omega)-1}\right], \tag{18a}$$

where $\omega$ is the fraction of the sum of two bases U and A, and $S_1$, etc., are as defined in Table 1. In addition, we can estimate the unknown parameter $\theta$ by using the relationship

$$\frac{1-\theta}{1+\theta} = \frac{\left(\omega - \frac{1}{2}\right)(P-R)}{(1-\omega)(S_1+S_3-2Q_1) - \omega(S_2+S_4-2Q_2)}, \tag{19}$$

while the ratio of $\beta$ to $\alpha$ can be determined from the assumption that $\omega$ does not change with time and that

$$\omega = \frac{\beta}{\alpha+\beta} = S_1 + S_3 + 2Q_1 + \frac{1}{2}(P+R). \tag{20}$$

Formula (19) may be verified by substituting equations (15a) for the right-hand side of equation (19), noting at the same time equations (2). Estimated evolutionary distances (denoted by $\tilde{K}$) for several comparisons are shown in Table 2 together with values of $\theta$ and $\omega$. The table also contains the estimates of the standard error of $\tilde{K}$ that were obtained using a procedure similar to the one used by KIMURA (1980, 1981) but assuming that the estimation of $\omega$ is not accompanied by sampling error. It is also based on the assumption of a multinomial distribution of the variables in the right-hand side of (18). This seems to give a good estimate in the light of the results of Monte Carlo experiments.

## MONTE CARLO EXPERIMENTS

In order to check the validity and the range of applicability of our formula (18), we performed Monte Carlo experiments. The procedures were as follows.

TABLE 2

*Evolutionary distances per nucleotide site at the first, second and third codon position estimated by using the 3ST and the present model* (TK)

| Comparison | | Evolutionary distances per nucleotide site $\tilde{K}_1$ | $\tilde{K}_2$ | $\tilde{K}_3$ |
|---|---|---|---|---|
| Chicken β *vs.* Rabbit β | (3ST) | 0.300 | 0.195 | 0.636 |
| | (TK) | 0.299 ± 0.044 | 0.237 ± 0.037 | 0.691 ± 0.160 |
| | ω | 0.384 | 0.637 | 0.267 |
| | θ | 0.489 | 1.048 | — 0.079 |
| Human *vs.* Rat pregrowth hormone | (3ST) | 0.265 | 0.177 | 0.531 |
| | (TK) | 0.269 ± 0.038 | 0.178 ± 0.032 | 0.692 ± 0.121 |
| | ω | 0.430 | 0.609 | 0.258 |
| | θ | 0.592 | 0.135 | 0.468 |
| Human *vs.* Rat insulin C peptide | (3ST) | 0.182 | 0.274 | 0.947 |
| | (TK) | 0.178 ± 0.273 | 0.289 ± 0.104 | ∞ |
| | ω | 0.081 | 0.581 | 0.226 |
| | θ | 2.63 | 0.821 | 1.00 |
| Human *vs.* Rat insulin A + B peptide | (3ST) | 0.040 | 0.000 | 0.461 |
| | (TK) | 0.042 ± 0.026 | 0.000 ± 0.019 | 0.786 ± 0.300 |
| | ω | 0.480 | 0.628 | 0.275 |
| | θ | 0.964 | 1.00 | 0.535 |
| Rabbit β *vs.* Mouse β | (3ST) | 0.160 | 0.127 | 0.427 |
| | (TK) | 0.171 ± 0.031 | 0.135 ± 0.030 | 0.463 ± 0.080 |
| | ω | 0.373 | 0.644 | 0.339 |
| | θ | 0.954 | 0.757 | 0.398 |
| Rabbit α *vs.* Rabbit β | (3ST) | 0.600 | 0.437 | 0.903 |
| | (TK) | 0.600 ± 0.087 | 0.517 ± 0.062 | 1.29 ± 1.05 |
| | ω | 0.389 | 0.633 | 0.234 |
| | θ | 0.071 | 1.09 | 0.290 |
| | (3ST) | 0.124 | 0.115 | 0.544 |
| | (TK) | 0.126 ± 0.029 | 0.131 ± 0.027 | 0.774 ± 0.205 |

TABLE 2—Continued

| Comparison | | Evolutionary distances per nucleotide site $\tilde{K}_1$ | $\tilde{K}_2$ | $\tilde{K}_3$ |
|---|---|---|---|---|
| Mouse α-1 *vs.* Rabbit α (Exons 1, 2 and 3) | | | | |
| | $\omega$ | 0.390 | 0.575 | 0.231 |
| | $\theta$ | 0.539 | 0.966 | 0.182 |
| | (3ST) | 0.008 | 0.008 | 0.470 |
| | (TK) | 0.008 | 0.008 | 0.470 |
| | | ± 0.013 | ± 0.008 | ± 0.088 |
| *S. purpuratus vs. P. miliaris* H3 | | | | |
| | $\omega$ | 0.380 | 0.493 | 0.530 |
| | $\theta$ | 0.000 | 0.000 | − 0.160 |
| | (3ST) | 0.086 | 0.020 | 0.479 |
| | (TK) | 0.087 | 0.020 | 0.484 |
| | | ± 0.032 | ± 0.016 | ± 0.104 |
| *S. purpuratus vs. P. miliaris* H2B | | | | |
| | $\omega$ | 0.303 | 0.436 | 0.535 |
| | $\theta$ | 1.00 | 0.000 | 0.536 |
| | (3ST) | 0.307 | 0.371 | 0.768 |
| | (TK) | 0.303 | 0.374 | 0.862 |
| | | ± 0.062 | ± 0.060 | ± 0.259 |
| Mouse ψα *vs.* Rabbit α | | | | |
| | $\omega$ | 0.381 | 0.587 | 0.258 |
| | $\theta$ | − 0.458 | 0.867 | 0.115 |
| | (3ST) | 0.133 | 0.121 | 0.658 |
| | (TK) | 0.129 | 0.138 | 0.883 |
| Mouse α-1 *vs.* Rabbit α | | | | |
| | $\omega$ | 0.383 | 0.576 | 0.232 |
| | $\theta$ | − 9.22 | 0.952 | 0.009 |
| | (3ST) | 0.209 | 0.300 | 0.337 |
| | (TK) | 0.207 | 0.294 | 0.358 |
| | | ± 0.045 | ± 0.114 | ± 0.074 |
| Mouse α-1 *vs.* Mouse ψα | | | | |
| | $\omega$ | 0.352 | 0.564 | 0.376 |
| | $\theta$ | 0.527 | 2.57 | 1.18 |

First, we prepared a random nucleotide sequence to be used as a common ancestor. It consisted of $n$ sites, with frequencies of U, A, C and G being given by $\omega/2$, $\omega/2$, $(1-\omega)/2$ and $(1-\omega)/2$. From this sequence, two descendent sequences were derived by independent nucleotide substitutions according to the scheme shown in the previous section. In simulation experiments, we assigned values of substitutional rates so that $k$ in equation (9) is of the order of $10^{-3}$. The experiments were continued until one substitution per site had occurred on the average in each lineage. We compared the two sequences through time and counted the actual number of nucleotide substitutions involved in two lineages. The total number of nucleotide subsütitutions, $K$, was monitored by summing the actual numbers of substitutions observed until a given time $T$. On the other hand, the expected number of nucleotide substitutions over $T$, as denoted by $K_E$, was calculated by $2kT$ for a given value of $k$. Note that, strictly speaking, $K_E$ is different from $K$, although no significant differences between these two were observed in the simulations experiments. We also observed the relative frequencies of the various classes in Table 1 at specified times and calculated the esüimtaed evolutionary distances using equation (18) or (18a). In a similar way, we obtained the estimate $\tilde{K}_{JC} = -\frac{3}{4}\ln(1-\frac{4}{3}\lambda)$ as a reference point; where $\lambda$ is the fraction of different sites between two nucleotide sequences (see JUKES and CANTOR 1969; KIMURA and OHTA 1972). These processes were repeated 100 times, assuming $n = 100$. Each quantity of concern was obtained by taking the average.

The results are illustrated through Figures 2 to 4. As typical situations, we assigned the parameter values similar to the ones derived from comparisons of DNA sequences of exons 1 and 2 in the mouse and the rabbit $\alpha$-globin genes, together with a mouse pseudogene. The parameters are determined separately at different codon positions. Figures 2, 3 and 4, respectively, represent situations at

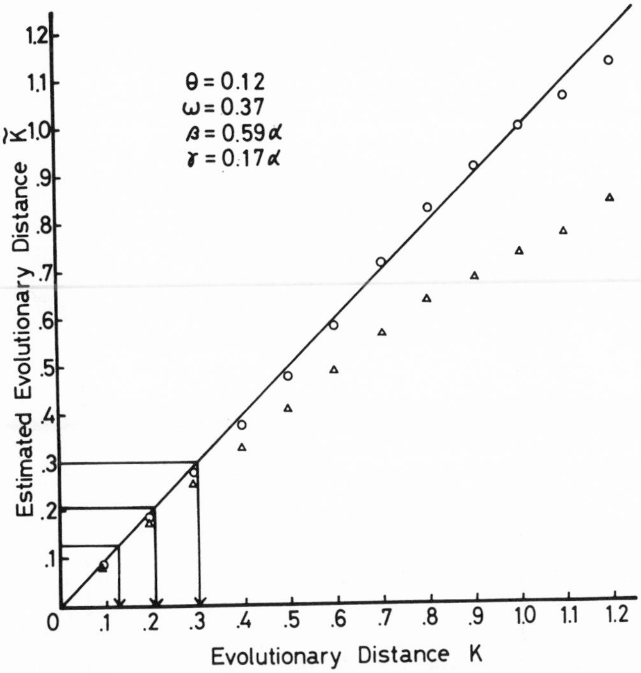


FIGURE 2 to 4.—Relationship between the actual evolutionary distance $K$ and the estimated evolutionary distance $\tilde{K}$ based on the formula (18) (the broken lines). The solid lines represent $K = \tilde{K}$. The parameters used in these figures are determined by taking the average for the data on nucleotide sequences of the $\alpha$ globin gene and the $\alpha$ pseudogene ($\alpha 3$) of the mouse, and the $\alpha$-globin gene of the rabbit. FIGURE 2: First codon position.

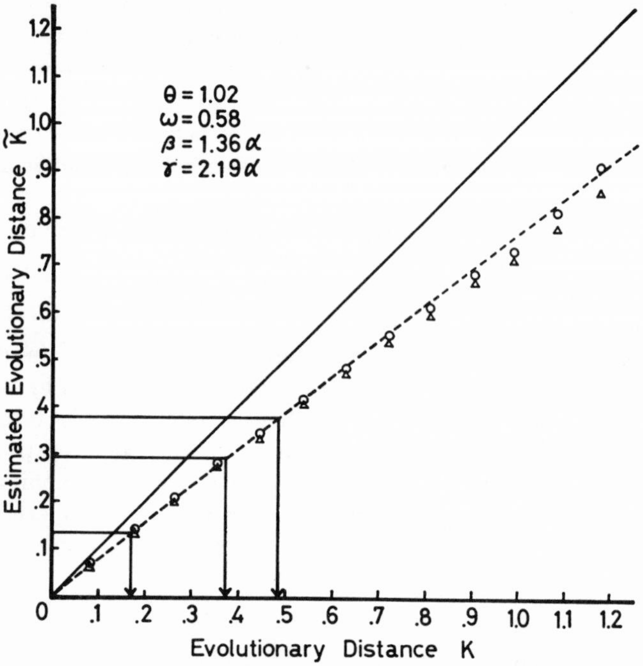


FIGURE 3.—Second codon position.

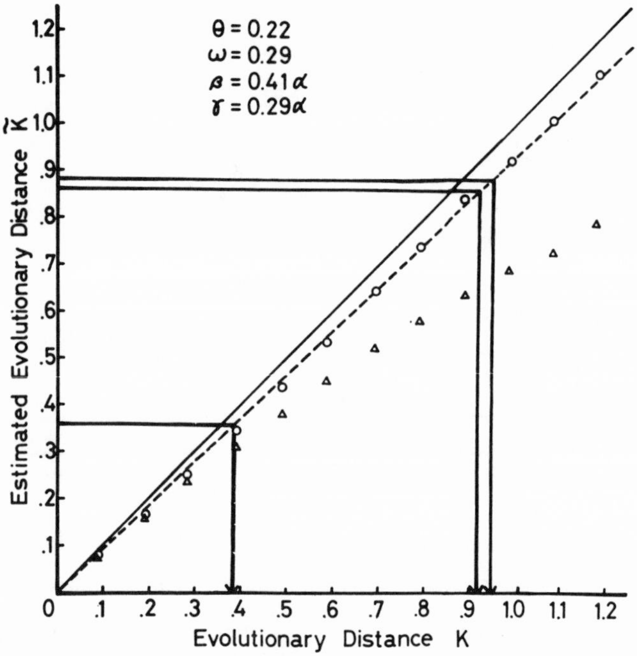


FIGURE 4.—Third codon position.

the first, second and third positions of codons within the above two exons. The abscissa stands for the actual evolutionary distance $K$ (in terms of the number of base substitutions), while the ordinate represents the estimated evolutionary distance $\tilde{K}$. Because the difference between $K$ and $K_E$ turned out to be quite small in all experiments, we will not discriminate between them. Estimates obtained by using equation (18) are marked by open circles, while those estimated by using the fomula of JUKES and CANTOR (1969) are indicated open triangles. It is evident from Figure 3 that both sets of estimates are close to each other when $\theta$ is near 1 and $\omega$ is about 0.5 and that they need appropriate corrections when $K$ is large. Such a discrepancy for the second codon position seems to have been caused by the relatively high transversion type substitution rate $\gamma$ assumed. On the other hand, when $\theta$ and $\omega$ are, respectively, less than 1 and 0.5, marked difference between the two sets of estimators occur (see Figures 2 and 4). Fortunately, equation (18) provides good estimates of the actual evolutionary distance if $K$ does not exceed unity, and the linearity is almost completely preserved in this range of $K$ values. The simulation experiments show that the present formula (18) is useful, especially when the rates of transversion-type substitutions are low and the frequencies of two classes of bases differ greatly from each other. Although not perfectly proven, if the real pattern of base substitutions is something like this, the present method is the most accurate, followed by the 3ST (KIMURA 1981) and JUKES and CANTOR methods. On the other hand, if substitution rates are equal in all directions, all three methods of these give almost the same result.

Note that cases arise in which we cannot estimate the evolutionary distances from these formulae. This occurs when arguments of logarithms become zero or negative. Such situations should occur frequently when a great many substitutions are involved. Because we excluded such inapplicable cases from our calculations, the estimated evolutionary distance $\tilde{K}$ gives an underestimate for $K$. In some cases, the fraction of such inapplicable cases for formula (18) became more than 50% if $K \geqq 1$. When we analyze actual data, this difficulty often arises if we compare the sequences in which many substitutions have occurred. One example is afforded by the third position of codons in the C peptide of insulin when comparison is made between human and rat (see Table 2). Such a problem arises because the true nature of substitutional processes is stochastic; whereas, our present treatment is deterministic. As the number of nucleotide sites actually compared is finite, sometimes less than 100, the small sample size creates a large sampling error. Particularly, when $K > 2$, no estimator seems to provide good information on the actual evolutionary distance, as simulation experiments show. We also note that in such cases $\tilde{K}$ has a very large error variance.

## RESULTS AND DISCUSSION

Using equation (18), we calculated the evolutionary distances between various DNA sequences, as shown in Table 2. This table also contains evolutionary distances estimated using the "three-substitution-type" (3ST) model of KIMURA (1981) for the purpose of comparison (this corresponds to the case where $\alpha = \beta$

and $\varepsilon = \delta$ in Figure 1). It can be seen from this table, that equation (18) often provides larger estimates than the corresponding estimates obtained from the 3ST model. Particularly, when the estimated value of $\theta$ from equation (19) is small and that of $\omega$ from (20) differs noticeably from 0.5, the discrepancy between the two sets of estimates becomes large. Such a dependence of equation (18) on $\omega$ seems to represent a favorable property as an estimator of evolutionary dis-ance because, as pointed out by Kimura (1981), the U and A content, particularly at the third position of codons, is often much less than 0.5. For instance, the value of $\omega$ is about 0.23 at the third codon positions in rabbit $\alpha$ and $\beta$ globins. This bias often results in unreliable estimates of the evolutionary distance, as we mentioned before.

On the other hand, the value of $\theta$ is quite sensitive to changes of observed values of various classes involved in equation (19). In most cases, $\theta$ is less than unity, but, in some cases, it happens to be negative or exceed unity. The estimated value of $\theta$ for each case may not be reliable, but the results suggest that transition-type substitutions can occur more frequently than those of transversion types (*i.e.*, from U to G and A to C, or *vice versa*). It is fortunate that equation (18) does not contain $\theta$.

It may be clear from Table 2 that evolutionary base substitution is faster (roughly 2.5 ~ 5.9 times) at the third position of codons than at the second positions in the functional globin genes. This characteristic is particularly conspicuous in histone genes. The ratio per site of the rates of third to the second positions is about 59 for the comparison of *S. purpuratus* and *P. miliaris* H3 sequences, while it is about 24 for the H2B sequences in the same comparison (for data, see SURES, LOWRY and KEDES 1978; SCHAFFNER *et al.* 1978). These species probably diverged between $6 \times 10^7$ and $16 \times 10^7$ years ago (DURHAM 1966; KEDES 1979); therefore, the rate $k_3$ per site per year is $(1.5 \sim 3.9) \times 10^{-9}$ for the H3 sequences and $(1.5 \sim 4.0) \times 10^{-9}$ for the H2B sequences. These values are very similar; moreover, the rates $k_3$ estimated for other genes using several other comparisons show roughly the same values. For example, we have $4.3 \times 10^{-9}$ for the human and rat pregrowth hormone comparison (with the divergence time $T = 8 \times 10^7$ years) and $1.2 \times 10^{-9}$ for the chicken and rabbit $\beta$ globin comparison ($T = 3 \times 10^8$ years). In some cases, formula (18) gives 1.4 to 1.7 times higher estimates for $k_3$ than does the 3ST model, but whether or not the model can decrease the estimated variance of $k_3$ is still uncertain until more data are available. The rough equality of the evolutionary rates at the third position of codons among genes is in sharp contrast to wide differences of the rate at the second position, where most substitutions alter amino acids. At any rate, we can confirm the conclusion of KIMURA (1980, 1981) and MIYATA, YASUNAGA and NISHIDA (1980) that the rates of nucleotide substitutions at the third position of codons are not only very high but also roughly equal to each other between genes even when amino acid altering substitutional rates are quite different.

Now, let us examine the history of the $\alpha$ globin pseudogene in the mouse, as studied by VANIN *et al.* (1980) and NISHIOKA, LEDER and LEDER (1980). We apply equation (18) to estimate evolutionary distances. The sequences of the normal $\alpha$-globin genes, including the noncoding regions, have also been determined in the mouse and the rabbit (see KONKEL, MAIZEL and LEDER 1979; HARDISON *et al.* 1979, and references therein). Thus, we can compare DNA sequences of these three genes. Our aim is to estimate the time of occurrence of duplication leading to the $\alpha$ pseudogene in the mouse line and the relative evolutionary rates at each codon position in the pseudogene relative to those in the normal $\alpha$-globin genes. Although several models concerning the appearance of pseudogenes are conceivable (see for example PROUDFOOT and MANIATIS 1980; MIYATA and YASUNAGA 1981; LI, personal communication), we assume here a simple one.

Let us assume that the duplication occurred $T_d$ years ago, and thereafter a duplicated gene became "dead" and started to evolve at the rate $k'_i$ instead of $k_i$, where $i$ (= 1, 2 or 3) denotes the codon position. At the incipient stage, the mouse population must have been polymorphic with respect to the number of $\alpha$-globin genes per individual. However, it is likely that the duplicate gene could accumulate mutations at a higher rate than the normal gene, due to its multiplicity. Let $T_0$ be the divergence time of the mouse and the rabbit, and let $\tilde{K}_i(X-Y)$ be the evolutionary distance in the $i$th codon position of homologous genes between species $X$ and $Y$. For example, $\tilde{K}_i(M\psi\alpha\text{—}R\alpha)$ denotes the evolutionary distance in the $i$th position between the mouse $\alpha$ pseudogene and the rabbit $\alpha$ gene. In the following study, we make no correction for $\tilde{K}$ and compare only the part including exon 1 and 2, excluding the exon 3 region from the calculation because of an unusual characteristic of this region, as pointed out by MIYATA and YASUNAGA (1981). Then, $T_d$ and $k_i'$, relative to $T_0$ and $k_i$, can be calculated for each codon position by

$$\frac{T_d}{T_0} = \frac{\tilde{K}_i(M\alpha\text{—}M\psi\alpha) - \tilde{K}_i(M\psi\alpha\text{—}R\alpha) + \tilde{K}_i(M\alpha\text{—}R\alpha)}{\tilde{K}_i(M\alpha\text{—}R\alpha)}$$

and

$$\frac{k_i'}{k_i} = \frac{\tilde{K}_i(M\alpha\text{—}M\psi\alpha) + \tilde{K}_i(M\psi\alpha\text{—}R\alpha) - \tilde{K}_i(M\alpha\text{—}R\alpha)}{\tilde{K}_i(M\alpha\text{—}M\psi\alpha) - \tilde{K}_i(M\psi\alpha\text{—}R\alpha) + \tilde{K}_i(M\alpha\text{—}R\alpha)}$$

Substituting the values of Table 2 in the above equations, the ratios of $T_d/T_0$ are respectively, 0.26, 0.42 and 0.43 for the first, second and third codon positions, while $k_1'/k_1 = 11.5$, $k_2'/k_2 = 13.9$ and $k_3'/k_3 = 0.9$. Roughly speaking, this means that the duplication responsible for the mouse $\alpha$ pseudogene occurred about $(0.3 \sim 0.4)T_0$ years ago. If we take $8.0 \times 10^7$ as $T_0$, $T_d$ becomes about 20 ~ 30 million years. On the other hand, the rates in the first and second positions in the pseudogene turn out to be roughly 10 times faster than those of normal genes; whereas, the rate in the third positions remain unaltered. The estimated values of $2k_1'T_0$ (= 1.48) and $2k_2'T_0$ (= 1.92) are both about 2 times greater than the estimated value of $2k_3T_0$ (= 0.883). This might indicate that there are some selective constraints even against the changes in the third positions in the normal gene. Another possibility is that equation (18) still gives an underestimate for the evolutionary distance. Considering the fact that estimated values of $k_3$ and $k_3'$ are similar, the latter might be more probable in the light of the results of Monte Carlo experiments. We could make some correction for the values of $\tilde{K}$ based on Monte Carlo experiments, but we did not take such an approach here because it seemed unlikely that we can get more precise estimates of these values because of the inevitable large sampling errors.

In the above analysis, we have tacitly assumed that the $\alpha$ pseudogene is fixed in the mouse population. In fact, it is likely that several million years are sufficient for such a nonfunctional pseudogene to become fixed in a population (see MARUYAMA and TAKAHATA 1981; TAKAHATA 1981). Therefore, it is highly probable that the $\alpha$ pseudogene is fixed in the mouse population. This conclusion, however, is tentative in the sense that we ignored the effect of recurrent unequal crossing over. As pointed out by OHTA (1981), it is possible that unequal crossing over plays a prominent role in the evolution of duplicate genes, even when a small number of them are tightly linked (*i.e.*, multigene family of small size). If unequal crossing over occurs frequently in the course of evolution, the fixation of a pseudogene at a specific locus may be considered transient. However, a preliminary study incorporating such a mechanism still supports the view that all individuals carry a pseudogene in their genome for several million years, although its location on a chromosome may vary from individual to individual or in time, (the details will be published elsewhere).

We conclude that a duplicate gene leading to the $\alpha$ pseudogene in the mouse line was introduced $20 \sim 30$ million years ago by unequal crossing over and became fixed in the population several million years after the duplication occurred, and that many nucleotide substitutions have accumulated at a high rate, irrespective of codon position, due to the loss of selective constraints.

We thank K. AOKI for his helpful comments in composing the manuscript and T. OHTA for stimulating discussion.

## LITERATURE CITED

DURHAM, J. W., 1966 *Echinoides*. pp. 270–295. In: *Treatise on Invetebrate Paleontology, Part U, Echinodermata 3*. Edited by R. C. MOORE. Univ. Kansas Press, Lawrence, Kansas.

FITCH, W. M. and E. MARGOLIASH, 1967 Construction of phylogenetic trees. Science **155:** 279–284.

HARDISON, R. C., E. T. BUTLER III, E. LACY, T. MANIATIS, N. ROSENTHAL and A. EFSTRATIDIS, 1979 The structure and transcription of four linked rabbit $\beta$-like globin genes. Cell **18:** 1285–1297.

HOLMQUIST, R., 1980 Evolutionary analysis of $\alpha$ and $\beta$ hemoglobin genes by REH theory under the assumption of equiprobability of genetic events. J. Mol. Evol. **15:** 149–159.

HOLMQUIST, R. and D. PEARL, 1980 Theoretical foundations for quantitative paleogenetics. III. The molecular divergence of nucleic acids and proteins for the case of genetic events of unequal probability. J. Mol. Evol. **16:** 211–267.

JUKES, T. H. and C. H. CANTOR, 1969 Evolution of protein molecules. pp. 21–123. *Mammalian Protein Metabolism*. Edited by H. N. MUNRO. Academic Press, New York.

KEDES, L. H., 1979 Histone genes and histone messengers. Ann. Rev. Biochem. **48:** 837–870.

KIMURA, M., 1968 Evolutionary rate at the molecular level. Nature **217**: 624–626. ——, 1979 The neutral theory of molecular evolution. Scientific American **241**(5): 98–126. ——, 1980 A simple method for estimating evolutionary rates of base substitutions through comparative studies of nucleotide sequences. J. Mol. Evol. **16**: 111–120. ——, 1981 On estimation of evolutionary distances between homologous nucleotide sequences. Proc. Natl. Acad. Sci. U.S. **78**: 454–458.

KIMURA, M. and T. OHTA, 1972 On the stochastic model for estimation of mutational distance between homologous proteins. J. Mol. Evol. **2**: 87–90.

KONKEL, D. A., J. V. MAIZEL, JR. and P. LEDER, 1979 The evolution and sequence comparison of two mouse chromosomal β-globin genes. Cell **18**: 865–873.

MARUYAMA, T. and N. TAKAHATA, 1981 Numerical studies of the frequency trajectories in the process of fixation of null genes at duplicated loci. Heredity **46**: 49–57.

MAXAM, A. and W. GILBERT, 1977 A new method for sequencing DNA. Proc. Natl. Acad. Sci. U.S. **74**: 560–564.

MIYATA, T. and T. YASUNAGA, 1980 Molecular evolution of mRNA: A method for estimating evolutionary rates of synonymous amino acid substitutions from homologous nucleotide sequences and its application. J. Mol. Evol. **16**: 23–36. ——, 1981 Rapid evolving mouse alpha globin-related pseudogene and its evolutionary history. Proc. Natl. Acad. Sci. U.S. **78**: 450–453.

MIYATA, T., T. YASUNAGA and T. NISHIDA, 1980 Nucleotide sequence divergence and functional constraint in mRNA evolution. Proc. Natl. Acad. Sci. U.S. **77**: 7328–7332.

NEI, M., 1975 *Molecular Population Genetics and Evolution.* North-Holland Publishing Company: Amsterdam, Oxford.

NISHIOKA, Y., A. LEDER and P. LEDER, 1980 Unusual α-globin-like gene that has clearly lost both globin intervening sequences. Proc. Natl. Acad. Sci. U.S. **77**: 2806–2809.

OHTA, T., 1981 Genetic variation in small multigene families. Genet. Research **37**: 133–149.

OHTA, T. and M. KIMURA, 1971 On the constancy of the evolutionary rate of cistrons. J. Mol. Evol. **1**: 18–25.

PROUDFOOT, N. J. and T. MANIATIS, 1980 The structure of a human α-globin pseudogene and its relationship to α-globin gene duplication. Cell **21**: 537–544.

SANGER, F., S. NICKLEN and A. R. COULSON, 1977 DNA sequencing with chain-terminating inhibitors. Proc. Natl. Acad. Sci. U.S. **74**: 4563–4567.

SCHAFFNER, W., G. KUNZ, H. DAETWYLER, J. TELFORD, H. O. SMITH and M. L. BIRNSTIEL, 1978 Genes and spacers of cloned sea urchin histone DNA analyzed by sequencing. Cell **14**: 655–671.

SURES, I., J. LOWRY and L. H. KEDES, 1978 The DNA sequence of sea urchin (S. *pupuratus*) H2A, H2B and H3 histone coding and spacer regions. Cell **15**: 1033–1044.

TAKAHATA, N., 1981 On the disappearance of duplicate gene expression. In: *Molecular Evolution, Protein Polymorphism and The Neutral Theory.* Edited by M. KIMURA. Japan Scientific Societies Press, Tokyo. (In press).

VANIN, E. F., G. I. GOLDBERG, P. W. TUCKER and O. SMITHIES, 1980 A mouse α-globin-related pseudogene lacking intervening sequences. Nature **286**: 222–226.

ZUCKERKANDL, E. and L. PAULING, 1965 Evolutionary divergence and convergence in proteins. pp. 97–166. In: *Evolving Genes and Proteins.* Edited by V. BRYSON and H. J. VOGEL. Academic Press: New York and London.

Corrsponding editor: M. NEI

## PART SEVENTEEN

# Molecular Clock

## Introduction

Kimura (no. 45) regarded the molecular clock (rough constancy of molecular evolutionary rate) of Zuckerkandl and Pauling (1965) as evidence for the neutral theory (Kimura and Ohta, no. 46), and this raised a heated debate on the concept and its accuracy. It is wrong, however, to say that the molecular clock is expected only under neutrality or that the neutral theory always leads to a molecular clock. The existence of the clock depends strictly on how new mutations, selected or unselected, arise. The usual subsidiary assumption in the neutral theory is that new mutations occur by a Poisson process so that the variance to mean ratio of the number of substitutions must be one (Ohta and Kimura 1971; Kimura 1983a; no. 47). If advantageous mutations occur under the same assumption, the Poisson clock also exists.

Kimura (no. 45) could not finish this paper without predicting that living fossils would have accumulated amino acid changes at the same rate as rapidly evolving species. Living fossils are morphologically in *status quo* for a long time so that they provide an excellent example of the bifurcation between molecular and phenotypic evolution.

To Kimura, the constancy of the molecular evolutionary rate always concerns uniformity over diverse lineages. He originally thought that the neutral theory can account for it only by assuming that mutation rates per year, rather than per generation, are uniform among lineages. However, he was not certain, and later, instead of asking why this is so (Wilson, Carlson, and White 1977; Wu and Li 1985; Britten 1986 for the generation time effect), Kimura and Ohta argued (no. 46) that "the species with short generation time tends to have small body size and attain a large population number, while the species which takes many years for one generation tends to have a small population number." This was the first inkling of the negative correlation between generation time and effective population size ($N_e$).

Once a slightly deleterious mutation arises in a population, the chance for it to contribute to molecular evolution is larger in a smaller population. The supposition on the relationship between $N_e$ and generation time, on the other hand, implies a high per-year rate of slightly deleterious mutations in a large population, provided that the mutation rate is nearly constant *per generation.* It follows that the substitution rate of slightly deleterious mutations may become more or less constant *per year.* This is the argument in Ohta and Kimura (1971), and subsequent refinements of the neutral random drift theory along this line were made most extensively by Ohta (1973, 1976, 1977). One may note, however, that there is little quantitative evidence for the assumption of the constant mutation rate per generation. Recently, Kimura has returned to his original explanation that the mutation rate must be constant per year if the rate of molecular evolution is uniform among diverse lineages (Kimura 1991a,b). He reasoned this on the basis that the per-generation constancy is based on traditional mutation studies on visible characters and viability polygenes, and that these may not be applied to neutral or near-neutral mutations.

However, it is still not clear why neutral mutations are constant per year rather than per generation, the problem of the so-called generation time effect (Wilson, Carlson, and White 1977; Gillespie 1987). In a special issue of the *Journal of Molecular Evolution,* "Molecular Evolutionary Clock," Kimura (no. 47) called for some experiments to settle this issue. According to the neutral theory, there are two sources responsible for irregular rates of molecular clock: the change in the mutation rate per year and changes in selective constraint (see Takahata 1987 for a quantitative discussion of these possibilities, but see Gillespie 1984 for an alternative explanation). The largest deviation from the molecular clock was found in the hystricognath rodent insulin gene and was attributed to loss of selective constraint. Graur, Hide, and Li (1991) suggested, however, that the guinea pig diverged before the separation of the primates and the artiodactyls from the myomorph rodents (rats and mice), and this revised time scale supported the molecular clock.

# THE RATE OF MOLECULAR EVOLUTION CONSIDERED FROM THE STANDPOINT OF POPULATION GENETICS

By Motoo Kimura*

DEPARTMENT OF BIOLOGY, PRINCETON UNIVERSITY

*Communicated by James F. Crow, May 26, 1969*

*Abstract.*—The rate of amino acid substitutions in the evolution of homologous proteins is remarkably constant. Furthermore, estimated rates of amino acid substitutions based on comparisons of the alpha hemoglobin chains of various mammals with that of the carp are about the same as those based on comparisons of the carp alpha and mammalian beta or the alpha and beta chains in mammals. These uniformities are regarded as evidence for the hypothesis that a majority of amino acid substitutions that occurred in these proteins are the result of random fixation of selectively neutral or nearly neutral mutations.

Two implications of this possibility are discussed: (*a*) Random gene frequency drift is playing an important role in determining the genetic structure of biological populations and (*b*) genes in "living fossils" may be expected to have undergone as many DNA base (and therefore amino acid) substitutions as corresponding genes (proteins) in more rapidly evolving species.

---

Although since Darwin a great deal of knowledge has accumulated concerning evolution at the phenotypic level, the molecular changes are only beginning to be understood. Recent developments in the comparative study of protein sequence, however, have provided a powerful tool by which evolution at the molecular level may be investigated.

A few years ago, Zuckerkandl and Pauling[1] claimed that the evolutionary rate is approximately constant in most polypeptide chains. In hemoglobins, according to their estimate, about one substitution occurs in every 800 million years per amino acid site. Although the constancy of the molecular evolutionary rate has been accepted in many writings other than Zuckerkandl and Pauling, no systematic check on this thesis has been made taking due account of the statistical variations involved. For many biologists who are accustomed to think of evolution in terms of natural selection, this thesis may not readily be accepted.

Since extensive information is now available concerning amino acid arrangements in various proteins, as admirably compiled by Dayhoff and Eck,[2] I shall first examine in some detail the rate of amino acid change in hemoglobin $\alpha$ and $\beta$ chains. The method of estimation is as follows: Let $n_{aa}$ be the total number of amino acid sites in two polypeptide chains compared with each other (preferably excluding deletions and insertions), and let $d_{aa}$ be the number of sites in which they are different. If we denote by $K_{aa}$ the mean number of substitutions per amino acid site over the whole evolutionary period that separated these two polypeptides, then assuming independence of substitution, we have

$$d_{aa} = n_{aa}\,(1 - e^{-K_{aa}}),$$

so that $K_{aa}$ may be estimated from

$$K_{aa} = -2.30 \log_{10} (1 - p_d), \qquad (1)$$

where $p_d = d_{aa}/n_{aa}$ is the fraction of different sites. An equivalent formula has also been used by Zuckerkandl and Pauling.[1] The standard error of this estimate may be obtained by

$$\sigma_K = \sqrt{\frac{p_d}{(1 - p_d)n_{aa}}}. \quad (2)$$

The rate of substitution per amino acid site *per year* may then be obtained from

$$k_{aa} = K_{aa}/(2T), \quad (3)$$

where $T$ is the number of years that have elapsed since divergence from a common ancestor.

For example, if we compare the hemoglobin $\alpha$ chain of the carp with that of man, excluding insertions or deletions amounting to 3 amino acids, we have $n_{aa} = 140$ and $d_{aa} = 68$. Thus, from formulas (1) and (2), we obtain $K_{aa} = 0.665 \pm 0.082$. From paleontological evidence, we are reasonably sure that the common ancestor of the carp and man lived in the Devonian period, the age of fishes (about 350–400 million years ago). During this period, most of the basic differentiation of fishes occurred, and in its later part crossopterygians gave rise to amphibians (see, for example, refs. 3 and 4). So, we may assume that divergence of lines leading to the carp and man occurred around the middle of Devonian (Fig. 1), and take $2T = 750 \times 10^6$. This gives the rate of substitution $k_{aa} = 8.9 \times 10^{-10}$ per amino acid site per year. Table 1 lists the result of similar calculations in which the $\alpha$ chain of the carp is compared with the human, mouse, rabbit, horse, and bovine $\alpha$ chains. The mammals probably diverged from their common ancestor some 80 million years ago (see Fig. 1).

The table reveals a remarkable fact that these chains have differentiated in relation to carp's $\alpha$ chain roughly to the same extent. Actually, most of the $K_{aa}$ values agree with each other within the limit of statistical error. The standard errors attached to the values of $K_{aa}$ are valid for comparisons of independent estimates, but, when two different $K_{aa}$ values in the table are compared with each other, we must note that they are correlated in the sense that they share a common ancestor about 800 million years ago. Thus, the standard error appropriate to such comparisons is

$$\sigma_K^* = \sigma_K\sqrt{1 - r}, \quad (4)$$

where $r = (2T - T_o)/(2T)$ in which $T_o$ is the number of years after divergence. Then the standard error of the difference is $\sqrt{2}\sigma_K^*$, which is approximately 0.05. The agreement of the $K_{aa}$ values for different mammals as compared with the carp appears to be even more remarkable if we consider the fact that these $\alpha$ chains of mammals also differ from each other roughly in 20 amino acid sites on the average. The results given in Table 1 are combined and summarized in the first row of Table 2 as comparison 1. Assuming $2T = 7.5 \times 10^8$ as before, and, disregarding the error involved in the estimation of the time parameter, we obtain the rate of substitution; $k_{aa} = (8.9 \pm 0.5) \times 10^{-10}$ per amino acid site per year.

Comparisons 2 and 3 in the same table summarize the results of similar analysis with respect to $\alpha$ chains among mammals, taking the human $\alpha$ chain as the reference point in the former and that of the mouse in the latter. They give $k_{aa} = (8.8 \pm 0.9) \times 10^{-10}$ and $k_{aa} = (10.9 \pm 0.9) \times 10^{-10}$. These values are not only alike but they also agree well with the corresponding value derived above in comparison 1 in which the carp $\alpha$ is compared with various mammalian $\alpha$'s.

Similar calculations for $\beta$ chains are summarized as comparisons 4 and 5 in Table 2.

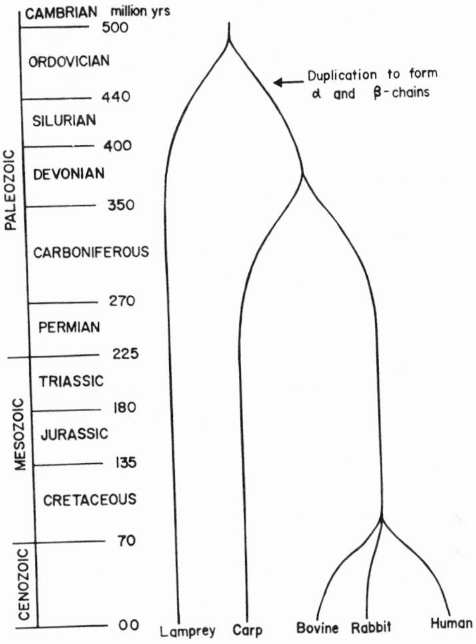


FIG. 1.—A phylogenetic tree of the vertebrate evolution.

Much more interesting and significant results are obtained by comparing the $\alpha$ and $\beta$ chains. It is generally accepted that they have originated by gene duplication in the remote past. In Table 3, the $\beta$ chain of man is compared with human, mouse, rabbit, horse, bovine, and carp $\alpha$ chains. In these comparisons, nonmatching parts are excluded between the $\alpha$ and $\beta$ chains due to deletion or insertion that amount to 9 or 10 amino acids. The table shows that each of these $\alpha$ chains had diverged from the human $\beta$ chain almost to the same extent. Actually, their $K_{aa}$ values (expected numbers of substitutions per amino acid site) agree with each other within the limit of statistical error. In Table 2, they are combined to give $K_{aa} = 0.799 \pm 0.038$ as comparison 6. They indicate that the two structural genes corresponding to the $\alpha$ and $\beta$ chains, after their origination by duplication, have diverged from each other independently and to the same extent on whatever evolutionary line they are placed so that the amount of divergence is the same, irrespective of whether the compared $\alpha$ and $\beta$ chains are taken from the same organism (man) or from two different organisms, such as man and carp, which have evolved independently over 350 million years. To my knowledge, the significance of this remarkable fact in relation to the

TABLE 1. *Comparison of the hemoglobin $\alpha$ chain of carp with those of various mammalian species, showing the number of different amino acid sites.*

| Comparison | $d_{aa}$ | $n_{aa}$ | $K_{aa}$ |
|---|---|---|---|
| Carp $\alpha$—Human $\alpha$ | 68 | 140 | $0.665 \pm 0.082$ |
| " —Mouse $\alpha$ | 68 | 140 | $0.665 \pm 0.082$ |
| " —Rabbit $\alpha$ | 72 | 140 | $0.722 \pm 0.087$ |
| " —Horse $\alpha$ | 67 | 140 | $0.651 \pm 0.081$ |
| " —Bovine $\alpha$ | 65 | 140 | $0.624 \pm 0.079$ |

$K_{aa}$ stands for the mean number of substitutions per amino acid site that separate two chains compared. Note smaller errors appropriate for comparisons between different mammals as described in the text.

TABLE 2. *Summary of the results of comparisons involving hemoglobin $\alpha$, $\beta$, and globin chains in various vertebrate species regarding amino acid substitution in evolution.*

| No. | | Comparison | $K_{aa}$ | $2T \times 10^{-8}$ | $k_{aa} \times 10^{10}$ |
|---|---|---|---|---|---|
| 1 | Carp $\alpha$ vs. | Human, mouse, rabbit, horse and bovine $\alpha$'s | $0.665 \pm 0.037$ | 7.5 | $8.9 \pm 0.5$* |
| 2 | Human $\alpha$ vs. | Horse, bovine, pig, and sheep $\alpha$'s | $0.141 \pm 0.014$ | 1.6 | $8.8 \pm 0.9$ |
| 3 | Mouse $\alpha$ vs. | Human, horse, bovine, pig, rabbit and sheep $\alpha$'s | $0.175 \pm 0.015$ | 1.6 | $10.9 \pm 0.9$ |
| 4 | Human $\beta$ vs. | Horse, pig, sheep and bovine $\beta$'s | $0.190 \pm 0.016$ | 1.6 | $11.9 \pm 1.0$ |
| 5 | Mouse $\beta$ vs. | Human, rabbit, horse, pig and bovine $\beta$'s | $0.225 \pm 0.019$ | 1.6 | $14.0 \pm 1.2$ |
| 6 | Human $\beta$ vs. | Human, mouse, rabbit, horse, bovine and carp $\alpha$'s | $0.799 \pm 0.038$ | 9.0 | $8.9 \pm 0.4$ |
| 7 | Rabbit $\beta$ vs. | Human, mouse, rabbit, horse, bovine and carp $\alpha$'s | $0.829 \pm 0.039$ | 9.0 | $9.2 \pm 0.4$ |
| 8 | Human $\beta$ vs. | Lamprey globin | $1.281 \pm 0.135$ | 10.0 | $12.8 \pm 1.4$ |

$T$ stands for the length of time elapsed since the evolutionary divergence of the two chains compared. For details, see text.

* Error due to inaccuracy of time parameter is not included.

mechanism of evolution has never been stressed in print before. The amino acid substitutions have proceeded at the same rate throughout the diverse lines of vertebrate evolution. This supports the hypothesis that the changes are largely fortuitous.

As to the time of gene duplication that formed the $\alpha$ and $\beta$ chains, it certainly must have occurred before the evolutionary divergence of the human and carp lines, but after the divergence of the human and lamprey lines, since the lamprey globin found in the blood is a monomer. Recently, Ohno *et al.*[5] suggested, based on cytological evidence, that gene duplication started to occur widely at the jawless stage of vertebrate evolution. So, we may assume that the formation of $\alpha$ and $\beta$ chains occurred about 450 million years ago toward the end of Ordovician period (Fig. 1). Assuming this and taking $2T = 9 \times 10^8$ (years), we obtain $k_{aa} = (8.9 \pm 0.4) \times 10^{-10}$ per amino acid site per year, in good agreement with the previous values.

The same clear indication of constancy in the rate of amino acid substitution in evolution is obtained if we compare the rabbit $\beta$ chain with human, mouse, rabbit, horse, bovine, and carp's $\alpha$ chains. It gives $k_{aa} = (9.2 \pm 0.4) \times 10^{-10}$ per amino acid site per year (Table 2, comparison 7).

I shall conclude this type of analysis by comparing human $\beta$ hemoglobin with lamprey globin. Excluding insertions or deletions amounting to 15 amino acids, $n_{aa} = 144$ and $d_{aa} = 104$. This gives $K_{aa} = 1.281 \pm 0.135$ as listed in the last row of Table 2. If we assume that the divergence of lines leading to lamprey

TABLE 3. *Comparison of the $\beta$ chain of man with the $\alpha$ chains of various vertebrate species.*

| Comparison | $d_{aa}$ | $n_{aa}$ | $K_{aa}$ |
|---|---|---|---|
| Human $\beta$—Human $\alpha$ | 75 | 139 | $0.776 \pm 0.092$ |
| " —Mouse $\alpha$ | 75 | 139 | $0.776 \pm 0.092$ |
| " —Rabbit $\alpha$ | 79 | 139 | $0.840 \pm 0.098$ |
| " —Horse $\alpha$ | 77 | 139 | $0.807 \pm 0.094$ |
| " —Bovine $\alpha$ | 76 | 139 | $0.791 \pm 0.093$ |
| " —Carp $\alpha$ | 77 | 139 | $0.807 \pm 0.094$ |

(a jawless fish of today) and man occurred in the earlier part of Ordovician period some 500 million years ago, $2T = 10^9$ (years), we obtain $k_{aa} = (12.8 \pm 1.4) \times 10^{-10}$ per amino acid site per year, giving once again a similar value for the rate of substitution. From the uniformity in the rates of substitution as shown in this table we may take $k_{aa} = 10^{-9}$ (per amino acid site per year) as a representative figure for hemoglobins.

I believe that the present analysis supports the hypothesis that at least in the hemoglobin $\alpha$ and $\beta$ chains, amino acid substitutions and the underlying nucleotide substitutions have proceeded at a constant rate and in a fortuitous manner throughout the diverse lines of vertebrate evolution during the past 500 million years. It is extraordinary that they mainly depend on *time measured in years* but are almost independent of generation time, living conditions, or even the genetic background.

In my previous paper[6] on the rate of molecular evolution, in order to estimate the rate of nucleotide substitution per genome per generation in mammalian species, I computed the average rate of amino acid substitution, using three proteins: hemoglobins, cytochrome *c*, and triosephosphate dehydrogenase. This gave the rate of about one substitution in $28 \times 10^6$ years for a polypeptide chain consisting of 100 amino acids, that is, roughly $\bar{k}_{aa} = 0.4 \times 10^{-9}$ per amino acid site per year. Recently, King and Jukes[7] obtained a more reliable figure, $\bar{k}_{aa} = 1.6 \times 10^{-9}$, by averaging seven proteins. According to them, the rates of substitution differ among various proteins, ranging from $0.33 \times 10^{-9}$ (insulins) to $4.29 \times 10^{-9}$ (fibrinopeptide A).

In this context, it might be convenient to coin a new word, for example, *pauling*, for a unit of evolutionary rate at the molecular level, defined as the rate of substitution of $10^{-9}$ per amino acid site per year. Accordingly, the hemoglobin rate is nearly one pauling, while the rates for the seven proteins range from 33 centipaulings (insulins) to 4.3 paulings (fibrinopeptide A). This unit is analogous to the *darwin*, the unit of evolutionary rate at the phenotypic level, representing an increase or decrease of quantitative measurement, such as the length of a tooth, at the rate of 1/1000 per 1000 years.[8]

Because of the estimated high rate of nucleotide substitution per genome per generation and from the consideration of the accompanying substitutional load, I concluded[6] that the majority of molecular mutations due to base substitution must be neutral or almost neutral for natural selection. The present analysis on the rate of amino acid substitution in hemoglobins seems to offer additional support for this conclusion. The remarkable constancy per year is most easily understood by assuming that in diverse vertebrate lines the rate of production of neutral mutations per individual per year is constant.

The above inference is based on a simple principle[6] that for neutral mutations the rate of gene substitution in a population is equal to the rate of production of new mutations per gamete, because for such a mutation, the probability of gene fixation is equal to the initial frequency. Also, it was shown by Kimura and Ohta[9] that it takes about $4N_e$ generations for a neutral mutant to reach fixation through random drift in a population of effective size $N_e$, if we exclude the cases in which it is lost from the population by chance.

The above arguments bring us to inquire about the mutation rate of the hemoglobin genes at the molecular level. In other words, what is the mutation rate per codon per year in hemoglobin $\alpha$ and $\beta$ cistrons? The assumption of neutral substitutions requires that it is at least about $10^{-9}$ per year. From the observed frequencies of rare hemoglobin variants and by statistical treatment based on the theory of the number of generations until extinction of mutant genes,[10] the estimated mutation rate per amino acid site per generation in man turns out to be $u_{aa} = 4.4 \times 10^{-8}$. (Details will be published elsewhere.[11])

Within the past 15 generations or so, average generation time must have been roughly 20 years for man, so that the mutation rate per year per amino acid site amounts approximately to $U_{aa} = 2.2 \times 10^{-9}$. Considering many uncertainties involved in the process of estimation both of the rate of amino acid substitution in evolution and the rate of mutation per amino acid site per year, we may consider that the estimated mutation rate $U_{aa} = 2.2 \times 10^{-9}$ is at least as high as the rate of substitution $k_{aa} = 10^{-9}$. This indicates that the observed mutation rate is sufficiently high to accommodate the assumed neutral mutations.

Other evidence for a considerable degree of randomness in DNA base-pair substitution during evolution is the observation that the amino acid composition of proteins can be predicted from a knowledge of the code and DNA base ratios.[12] This observation also suggests that "genic" and "nongenic" DNA do not differ in composition.

As to the differences in rates of evolution among various proteins such as insulin and fibrinopeptide A, it may be that different fractions of the amino acid substitutions are neutral, as pointed out by King and Jukes.[7] One must also admit the possibility of intrinsic differences in mutation rates. Alpha and beta hemoglobins should be similar in both respects, hence the similar rate of evolution.

In man, the total number of nucleotide pairs making up the haploid chromosome set is estimated to be about $3 \sim 4 \times 10^9$ (cf. refs. 13 and 14). This number is roughly the same among different species of mammals. If we assume that the average rate of amino acid substitution is $\bar{k}_{aa} = 1.6 \times 10^{-9}$ (per amino acid site per year), as estimated by King and Jukes[7] for seven proteins, and note that about 20 per cent of base substitutions are synonymous,[12] that is, code for the same amino acid, then we obtain the figure that in the evolutionary history of mammals nucleotide substitution has proceeded at the rate of about 2 or 2.5 per year. This is some four or five times higher than the corresponding figure obtained in the previous report.[6] In the line leading to man, average generation time was probably longer than 10 years. This gives at least the rate of nucleotide substitution per generation of about 20, making the contrast still larger with Haldane's estimate[15] of 1/300 per generation as the standard rate of gene substitution in evolution. Considering the amount of selective elimination that accompanies the process of gene substitution (i.e., the cost of natural selection or the substitutional load[6]), the most natural interpretation is, I believe, that a majority of molecular mutations due to DNA base substitution are almost neutral in natural selection.

In this connection it is appropriate to mention the remarkable experiment conducted by Cox and Yanofsky[16] using Treffers' mutator gene in *E. coli*. This gene is known to cause preferentially the transversion from an AT pair to a CG pair in the genome of *E. coli*. According to them, the estimated rate of mutation is $3.5 \times 10^{-6}$ per AT pair per generation. In a strain containing this gene, they observed an 0.2–0.5 per cent increase in the GC content of the DNA after 80 subcultures, which corresponds to 1200–1600 cell generations. On the other hand, the expected increase in GC content based on the above mutation rate and assuming some 50 per cent GC in the genome of *E. coli* turns out to be about 0.21–0.28 per cent. Thus, the agreement between the observed and the expected amount of increase in GC content is satisfactory. It shows that the rate of mutation per genome is approximately equal to the rate of mutant (nucleotide pair) substitution in the population and suggests that a majority of the base substitutions are selectively neutral, though in this case most of the mutants that are destined to reach fixation may have not yet been fixed. This agrees also with their observation that the strain is fully viable after accumulation of more than 7000 base substitutions in the course of the experiment.

Under such a circumstance, Sueoka's[17] "effective base conversion rates" are indeed equal to mutation rates.

It should be noted, however, that it is at present unknown what fraction of the total DNA in the genome of a higher organism is used for protein synthesis. If some (conceivably a large proportion) of the base pairs do not so act, substitutions among these might have even a larger probability of being neutral, and the number of changes would be at least as great.

These observations suggest that, particularly if all the DNA is taken into account, random drift is a much more important factor in evolution than has commonly been believed.

If amino acid changes are often due to chance, then these should be established as frequently in evolutionary conservative species as in those that undergo rapid changes in morphology. While we recognize that a constant morphology does not necessarily reflect a constant internal physiology, it is nonetheless likely that living fossils such as coelacanths, horseshoe crabs, and *Lingula* probably have fewer changes in internal function than more rapidly evolving animals. It would support the hypothesis of this paper if hemoglobins and other proteins show the same rate of amino acid substitution in such living fossils as in rapidly evolving species.

None of the above arguments are intended to imply that natural selection is not important in evolution. What we have postulated is that, surrounding the adaptive changes that occur by selection, there is a great deal of random noise from near-neutral random changes.

I would like to express my thanks to Dr. J. F. Crow for valuable suggestions and help in composing the manuscript. Thanks are also due Drs. E. Dempster and J. L. King for many helpful and stimulating discussions.

* On leave from the National Institute of Genetics, Mishima, Japan. This paper constitutes Contribution no. 720 from the National Institute of Genetics, Mishima, Shizuoka-ken, Japan. Aided in part by a grant-in-aid from the Ministry of Education, Japan. This work was also supported in part by the National Institutes of Health (GM 15422).

[1] Zuckerkandl, E., and L. Pauling, in *Evolving Genes and Proteins*, ed. V. Bryson and H. J. Vogel (New York: Academic Press, 1965), pp. 97–166.

[2] Dayhoff, M. O., and R. V. Eck, *Atlas of Protein Sequence and Structure 1967–68* (Silver Spring: National Biomedical Research Foundation, 1968).
[3] Simpson, G. G., C. S. Pittendrigh, and L. H. Tiffany, *Life: An Introduction to Biology* (London: Routledge and Kegan Paul, 1958).
[4] de Beer, G., *Atlas of Evolution* (London: Thomas Nelson and Sons, 1964).
[5] Ohno, S., U. Wolf, and N. B. Atkin, *Hereditas*, **59**, 169–187 (1968).
[6] Kimura, M., *Nature*, **217**, 624–626 (1968).
[7] King, J. L., and T. H. Jukes, *Science*, **164**, 788–798 (1969).
[8] Haldane, J. B. S., *Evolution*, **3**, 51–56 (1949).
[9] Kimura, M., and T. Ohta, *Genetics*, **61**, 763–771 (1969).
[10] Kimura, M., and T. Ohta, *Genetics*, in press.
[11] Kimura, M., in preparation.
[12] Kimura, M., *Genet. Res.*, **11**, 247–269 (1968).
[13] Muller, H. J., *Bull. Am. Math. Soc.*, **64**, 137–160 (1958).
[14] Vogel, F., *Nature*, **201**, 847 (1964).
[15] Haldane, J. B. S., *J. Genet.*, **55**, 511–524 (1957).
[16] Cox, E. C., and C. Yanofsky, these PROCEEDINGS, **58**, 1895–1902 (1967).
[17] Sueoka, N., these PROCEEDINGS, **48**, 582–592 (1962).

# Protein Polymorphism as a Phase of Molecular Evolution

MOTOO KIMURA & TOMOKO OHTA

National Institute of Genetics, Mishima, Shizuoka-ken, Japan

**It is proposed that random genetic drift of neutral mutations in finite populations can account for observed protein polymorphisms.**

SINCE one of us[1] put forward the theory that the chief cause of molecular evolution is the random fixation of selectively neutral mutants, some have supported the theory[2-5] and others have criticized it[6-8].

## Rate of Evolution

Probably the strongest evidence for the theory is the remarkable uniformity for each protein molecule in the rate of mutant substitutions in the course of evolution. This is particularly evident in the evolutionary changes of haemoglobins[5], where, for example, the number of amino-acid substitutions is about the same in the line leading to man as in that leading to the carp from their common ancestor. Similar constancy is found on the whole for cytochrome c, although the rate is different from that of the haemoglobins. The observed rate of amino-acid substitution for the haemoglobins is very near to one pauling ($10^{-9}$/amino-acid site/yr) over all vertebrate lines[5]. The rate for cytochrome c is roughly 0.3, while the average rate for several proteins is about 1.6 times this figure[2].

If we define the rate, $k$, of mutant substitution in evolution as the long term average of the number of mutants that are substituted in the population at a cistron per unit time (year, generation and so on), then under the neutral mutation–random drift theory, we have a simple formula

$$k = u \tag{1}$$

where $u$ is the mutation rate per gamete for neutral mutants per unit time at this locus. Note that this rate $k$ is different from the rate at which an individual mutant increases its frequency within a population. The latter depends on effective population size.

The uniformity of the rate of mutant substitution per year for a given protein may be explained by assuming constancy of neutral mutation rate per year over diverse lines. Moreover, the difference of the evolutionary rates among different molecules can be explained by assuming that the different fraction of mutants is neutral depending on the functional requirement of the molecules.

On the other hand, it can be shown that if the mutant substitution is carried out principally by natural selection

$$k = 4N_e s_1 u \tag{2}$$

where $N_e$ is the effective population number of the species, $s_1$ is the selective advantage of the mutant and $u$ is the rate at which the advantageous mutants are produced per gamete per unit time[9]. In this case we must assume that in the course of evolution three parameters $N_e$, $s_1$ and $u$ are adjusted in such a way that their product remains constant per year over diverse lines. The mere assumption of constancy in the "internal environment" is, however, far from being satisfactory to explain such uniformity of evolutionary rate. In our example of carp–human divergence, we must assume that $N_e s_1 u$ is kept constant in two lines which have been separate for some 400 million years in spite of the fact that the evolutionary rates at the phenotypic level (likely to be governed by natural selection) are so different.

## Polymorphism in Sub-populations

Kimura[1] also suggested that the widespread enzyme polymorphisms in *Drosophila*[10] and man[11] as detected by electrophoresis are selectively neutral and that the high level of heterozygosity at such loci can be explained by assuming that most mutations at these cistrons are neutral. This suggestion, however, has been much criticized[12-14]. One of the chief objections is that the same alleles are found in similar frequencies among different sub-populations of a species and that some kind of balancing selection must therefore be involved.

Robertson[15] suggested that if a large fraction of mutations at a locus is selectively neutral, we find either very many alleles segregating in large populations, or a small number of different set of alleles in different isolated small populations. He considered that because neither of these alternatives is found, most polymorphisms have at some time been actively maintained by selection.

Actually, both the situations suggested by Robertson are typical of the heterochromatic pattern of chromosomes in wild populations of the perennial plant *Trillium kamtschaticum*. Extensive cytological studies of this plant by Haga and his associates[16,17] have shown that several chromosome types are segregating within a large population, while different types are fixed in small isolated populations. Indeed, Robertson's suggestion is pertinent if isolation between sub-populations is nearly complete and if the mutation rate for neutral isoallelic variations is sufficiently high that more than one new mutant appears within a large population each generation. The chromosome polymorphism in *Trillium* can be explained by assuming a relatively high mutation rate per chromosome and very low migration rate per generation for this plant.

## Mutation and Mobility

On the other hand, it is possible that in animal species such as *Drosophila*, mouse and man well able to migrate, no local population is sufficiently isolated to prevent the entire species or subspecies from forming effectively one panmictic population. In his study on "isolation by distance", Wright[18], using his model of continuum over an area, has concluded that the total species differ little from a single panmictic population if the size of the "neighbourhood" from which parents come is more than 200.

Recently, Maruyama[19,20] made an extensive mathematical analysis of the stepping stone model of finite size. He worked out the exact relationship between local differentiation of gene frequencies and the amount of migration. His results show that in the two dimensional stepping stone model, if $N$ is the effective size of each colony and $m$ is the rate at which each colony exchanges individuals with four surrounding colonies per generation, then marked local differentiation is possible only when $Nm$ is smaller than unity (assuming a large number of colonies arranged on a torus). This is a very severe restriction for migration between colonies because the number of individuals which each colony exchanges with surrounding colonies must be less than an average of one per generation, irrespective of the size of each colony. For the model of continuous distribution of individuals over an area, this condition is equivalent to $N_\sigma < \pi$, where $N_\sigma$ is the average number of individuals within a circle of radius $\sigma$, the standard deviation of the distance of individual migration in one direction per generation. If, on the other hand, there is more migration, the whole population tends to become effectively panmictic. The transition from marked local differentiation to practical panmixis is very rapid for the distribution over an area, and it can be shown that if $N_\sigma > 12$, the whole populations behave as if it were a single panmictic population.

This means that when two or more alleles happen to be segregating within a species, their frequencies among different localities far apart from each other are nearly the same. For animals with separate sexes, it is expected that at least several individual males and females usually exist within a circle of radius $\sigma$ and the condition $N_\sigma > 12$ is therefore almost always met by widely distributed and actively moving animals. Maruyama also showed that when isolation is more complete and different alleles tend to fix in different local populations, they are connected by zones of intermediate frequencies. The overall pattern then mimics a gene frequency cline resulting from selection, even if alleles are in fact neutral.

## Heterozygosity and Probability of Polymorphism

Let us now consider the number of neutral isoalleles maintained in a finite population. Kimura and Crow[21] have shown that if $u$ is the mutation rate per locus (cistron) for neutral mutants, the effective number of alleles maintained in a population of effective size $N_e$ at equilibrium is

$$n_e = 4N_eu + 1 \quad (3)$$

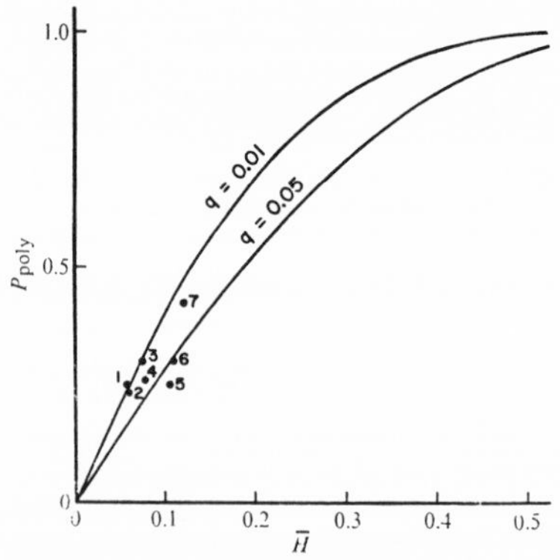


**Fig. 1** Relationship between the probability of polymorphism ($P_{poly}$) and the average heterozygosity $H$. The two curves represent the theoretical relationship based on the neutral polymorphism theory; the dots represent observed values. 1, *Limulus polyphemus*; 2, *Peromyscus polionotus*; 3, *Homo sapiens*; 4, *Mus musculus* (Denmark); 5, *Drosophila persimitis*; 6, *Mus musculus* (California); 7, *Drosophila pseudoobscura*.

In deriving this formula, it was assumed that the possible number of allelic states per locus is so large that whenever a mutant appears it represents a new, not pre-existing allele. The effective number of alleles given by formula (3) is equal to the reciprocal of the average homozygosity and is the number estimated by the ordinary procedure of allelism test. Then the average heterozygosity is

$$\bar{H} = 1 - 1/n_e = 4N_eu/(4N_eu + 1) \quad (4)$$

This is the mean frequency of the heterozygotes averaged over all cases including monomorphic and polymorphic cases.

We shall call a population "monomorphic" if the sum of the frequencies of "variant" alleles is $q$ or less. Then it can be shown[22] that the probability of a population being monomorphic is

$$P_{mono} = q^{n_e - 1} \quad (5)$$

when $n_e$ is the effective number of alleles. This formula has been derived under the same condition as formulae (3) and (4). Note that the value of $q$ is arbitrary but a reasonable value is 0.01. If, however, a sample from each locality consists of only a dozen or so individuals, $q = 0.05$ may be more appropriate. The probability that a population is polymorphic $(1 - P_{mono})$ is then

$$P_{poly} = 1 - q^{\bar{H}/(1-\bar{H})} \quad (6)$$

Fig. 1 illustrates the relationship between the probability of polymorphism and the average heterozygosity at two levels of $q$ (0.01, 0.05) together with some observed values compiled by Selander *et al.*[23] in their Table 3. The agreement between theoretical and observed values is satisfactory.

From this discussion it can be seen that for actively moving animals such as *Drosophila*, mouse and man, the frequency and the observed pattern of polymorphism can be explained by assuming effective migration such that

$$Nm > 4 \quad (7)$$

between adjacent local populations and also assuming a low mutation rate per cistron for neutral isoalleles such that

$$4N_eu \approx 0.1 \quad (8)$$

In formulae (7) and (8), $N$ refers to the effective size of the local population (colony) while $N_e$ refers to the effective size of the total Mendelian population such as species or subspecies.

When these two conditions are met, the effective number of alleles is on the average 1.1 and at each of the polymorphic loci (constituting roughly 0.3 of all loci) a particular allele takes a high frequency such as 0.8 while the remaining alleles exist in frequencies less than 0.2 in all. Furthermore, frequencies of those alleles among different local populations are about the same throughout the species.

## Relative Neutral Mutation Rate

As mentioned already, we know from studies of molecular evolution the average rate of amino-acid substitution. Then, using equation (1), we can infer that the mutation rate for neutral isoalleles is $u_{aa}=1.6\times10^{-9}$ per amino-acid site per year. If the average cistron responsible for isozyme polymorphisms consists of 300 amino-acids and if 0.3 of amino-acid changes lead to change of the electric charge

$$u=1.6\times10^{-9}\times300\times(0.3)=1.6\times10^{-7}$$

per year. If the fraction of neutral mutants is less among mutants that can be detected by electrophoresis than among those that cannot be so detected, this figure is an overestimate. The same applies if the changes that can be detected by electrophoresis are restricted to amino-acids that are exposed to the surface of the protein molecule. It is therefore possible that the true mutation rate is much lower than this and in the following treatment we take $u=10^{-7}$.

For species such as the mouse, with possibly two generations per year, the mutation rate per generation for neutral isoalleles detectable by electrophoresis is half as large, while for man it should be some twenty times as large. Then the effective population number that satisfies formula (8) is $N_e\approx0.5\times10^6$ for the mouse and $N_e\approx1.3\times10^4$ for man. The effective number here refers to the species or subspecies in the course of evolution.

We note that if the mutation rate $u$ is constant per year, then the product $N_eu$ should be less variable among different organisms than its components $N_e$ and $u$, because the species with short generation time tends to have small body size and attain a large population number, while the species which takes many years for one generation tends to have a small population number. At any rate, $u=10^{-7}$ per year is much lower than the standard figure of $10^{-5}$ per generation even for man and this suggests that, in general, neutral mutants constitute a small fraction of all the mutants at a cistron. Thus, we consider this as one important revision to earlier work[1] in which it was assumed that the neutral mutation rate per locus was high.

We must emphasize, however, that most mutants that spread into the species are neutral, even if the neutral mutants constitute a small fraction of the total mutants at the time of occurrence. Those mutants that are destined to spread to the species take a long time until fixation and on their way take the form of "protein polymorphism".

If most protein polymorphisms constitute a phase of molecular evolution, then the behaviour of molecular mutants in a population is crucial for an understanding of the polymorphism. It was shown by Kimura and Ohta[24] that for a selectively neutral mutant, it takes about $4N_e$ generations to reach fixation in the population (excluding the cases of eventual loss). We can also compute, using the solution of Kimura[25], the average number of generations that have elapsed since the appearance of a neutral allele which happens to have reached frequency 0.5. It turns out that this is about $(1.25)N_e$ generations.

One further factor we must consider is "associative overdominance". When truly overdominant loci are distributed over the genome, they will cause neutral loci to behave as if they were overdominant[26]. As we have shown (unpublished), this will somewhat prolong the time spent by a neutral mutant at intermediate frequencies, but this has no effect on the rate of mutant substitution in evolution. The associative overdominance, however, will play an important role when a small number of chromosomes are extracted from natural populations and rapidly multiplied for an experiment. In this case, spurious "balancing selection" will be observed.

## Polymorphism in Living Fossils

Returning to the problems of evolutionary time, we note that the average number of generations between two consecutive fixations of mutants at a given locus (cistron) in the course of evolution is $1/u$. This is roughly ten times as long as the time taken for an individual mutant to reach fixation if $4N_eu=0.1$. It may be interesting to ask, then, how long it takes until new mutants accumulate in the species causing detectable change in the fraction $P_d$ of proteins. If we denote by $T_d$ the average length of time for such change, then

$$T_d=-(1/u)\log_e(1-P_d) \quad (9)$$

According to Selander *et al.*[27], two Danish subspecies of the house mouse differ at 32% of their loci. Putting $P_d=0.32$ and assuming $u=10^{-7}$ per year, we obtain $T_d\approx3.9\times10^6$ yr. The time since divergence of these two subspecies from their common ancestor is given by $T_d/2$ or roughly 2 million years.

In our view, protein polymorphism and molecular evolution are not two separate phenomena, but merely two aspects of a single phenomenon caused by random frequency drift of neutral mutants in finite populations. If this view is correct, we should expect that not only genes in "living fossils" have undergone as many DNA base (and therefore amino-acid) substitutions as corresponding genes in more rapidly evolving species as predicted by Kimura[5], but also they are equally polymorphic and heterozygous at the protein level. A study by Selander *et al.*[23] on the variation of the horseshoe crab at the protein level seems to support this view.

At the moment, our observations are limited to a few organisms, and we do not know how typical their heterozygosities are. It is possible that for organisms with short generation time and small effective population number, the level of heterozygosity is much lower (because of a very small $N_eu$).

The neutral mutation-random drift theory allows us to make a number of definite quantitative as well as qualitative predictions by which the theory can be tested. We hope that through this process we will be able to gain deeper understanding of the mechanism of evolution at the molecular level and will be emancipated from a naive pan-selectionism.

Received October 29, 1970.


1 Kimura, M., *Nature*, **217**, 624 (1968).
2 King, J. L., and Jukes, T. H., *Science*, **164**, 788 (1969).
3 Crow, J. F., *Proc. Twelfth Intern. Cong. Genet.*, **3**, 105 (1969).
4 Arnheim, N., and Taylor, C. E., *Nature*, **223**, 900 (1969).
5 Kimura, M., *Proc. US Nat. Acad. Sci.*, **63**, 1181 (1969).
6 Maynard Smith, J., *Nature*, **219**, 1114 (1968).
7 Richmond, R. C., *Nature*, **225**, 1025 (1970).
8 Clarke, B., *Science*, **168**, 1009 (1970).
9 Kimura, M., and Ohta, T., *J. Mol. Evol.* (in the press).
10 Lewontin, R. C., and Hubby, J. L., *Genetics*, **54**, 595 (1966).
11 Harris, H., *Proc. Roy. Soc.*, B, **164**, 298 (1966).
12 Prakash, S., Lewontin, R. C., and Hubby, J. L., *Genetics*, **61**, 841 (1969).
13 Petras, M. L., Reimer, J. D., Biddle, F. G., Martin, J. E., and Linton, R. S., *Canad. J. Genet. Cytol.*, **11**, 497 (1969).
14 Maynard Smith, J., *Amer. Nat.*, **104**, 231 (1970).
15 Robertson, S., in *Population Biology and Evolution* (edit. by Lewontin, R.), 5 (Syracuse University Press, New York, 1968).
16 Haga, T., and Kurabayashi, M., *Cytologia*, **18**, 13 (1953).
17 Haga, T., in *Chromosomes Today*, **2** (edit. by Darlington, C. D., and Lewis, K. R.), 207 (Oliver and Boyd, Ltd, Edinburgh, 1969).
18 Wright, S., *Annals of Eugenics*, **15**, 323 (1951).
19 Maruyama, T., *Theoretical Population Biology*, **1**, 101 (1970).
20 Maruyama, T., *Japan. J. Genet.*, **45**, 481 (1970).
21 Kimura, M., and Crow, J. F., *Genetics*, **49**, 725 (1964).
22 Kimura, M., *Theoretical Population Biology* (in the press).
23 Selander, R. K., Yang, S. Y., Lewontin, R. C., and Johnson, W. E., *Evolution*, **24**, 402 (1970).
24 Kimura, M., and Ohta, T., *Genetics*, **61**, 763 (1969).
25 Kimura, M., *Proc. US Nat. Acad. Sci.*, **41**, 144 (1955).
26 Ohta, T., and Kimura, M., *Genet. Res.*, **16**, 165 (1970).
27 Selander, R. K., Hunt, W. G., and Yang, S. Y., *Evolution*, **23**, 379 (1969).

# Molecular Evolutionary Clock and the Neutral Theory

Motoo Kimura

National Institute of Genetics, Mishima, 411 Japan

**Summary.** From the standpoint of the neutral theory of molecular evolution, it is expected that a universally valid and exact molecular evolutionary clock would exist if, for a given molecule, the mutation rate for neutral alleles *per year* were exactly equal among all organisms at all times. Any deviation from the equality of neutral mutation rate per year makes the molecular clock less exact. Such deviation may be due to two causes: one is the change of the mutation rate per year (such as due to change of generation span), and the other is the alteration of the selective constraint of each molecule (due to change of internal molecular environment). A statistical method was developed to investigate the equality of evolutionary rates among lineages. This was used to analyze protein data to demonstrate that these two causes are actually at work in molecular evolution. It was emphasized that departures from exact clockwise progression of molecular evolution by no means invalidates the neutral theory. It was pointed out that experimental studies should be done to settle the issue of whether the mutation rate for nucleotide change is more constant per year or per generation among organisms whose generation spans are very different.

**Key words:** Protein evolution — Population genetics — Mutation rate

## Introduction

One of the most remarkable features of molecular evolution is its clocklike progression, namely, for a given protein or sequence of DNA, the rate of amino acid or nucleotide substitution is roughly constant among diverse lineages as well as within lineages over time. This is known as the "molecular evolutionary clock," and its existence was first suggested by Zuckerkandl and Pauling (1965). Since then, this property has been used extensively to reconstruct phylogenies, particularly when fossil records are missing or incomplete. Recent studies of Osawa's group on the evolution of a wide range of organisms using 5S rRNA sequences (see Hori et al. 1985), and those of Sibley's group on bird phylogenies using DNA–DNA hybridization (Sibley and Ahlquist 1986) are particularly successful examples.

At the same time, there has been much dispute regarding the accuracy and the underlying mechanism of the "clock"; some authors have even doubted the validity of the rate constancy hypothesis. In retrospect, it is interesting to note that Zuckerkandl (1965) made a suggestion contradicting the rate constancy hypothesis, that a great many polypeptide molecules of "living fossils" might possibly differ very little from those of their ancestors of millions of years ago. If we accept the viewpoint that evolution at the molecular level is caused by accumulation of slightly advantageous mutations as suggested by Zuckerkandl and Pauling (1965), such a prediction might very well be valid.

On the other hand, from the standpoint of the neutral theory, I predicted (Kimura 1969) that "genes in 'living fossils' may be expected to have undergone as many base (and therefore amino acid) substitutions as corresponding genes (proteins) in more rapidly evolving species." Since then, ample evidence has accumulated supporting my prediction based on the neutral theory (see Kimura 1983).

The rate-constancy hypothesis or the concept of the molecular evolutionary clock has been particularly controversial when used to estimate the date

Reprinted with permission from *Journal of Molecular Evolution,* vol. 26. © by Springer-Verlag, 1987.

of the human–ape divergence. Wilson and Sarich (1969) made a bold suggestion, based on molecular data and assuming clocklike regularity of molecular evolution, that the date of human–ape divergence is only 4 or 5 million years (Myr) ago. Since this differed so much from the then accepted date of at least 20 or 30 Myr (see for example, de Beer 1964, who placed the date about 30 Myr), it caused a great deal of controversy. It now appears that the molecular approach gives much more reliable information on this problem than the paleontological one (Pilbeam 1984), particularly if a "generation time effect" is taken into account.

Some authors (e.g., Wilson et al. 1977), while strongly supporting the validity of the molecular clock concept, do not consider the underlying mechanism to be important. I have a different view, and in this paper, I intend to discuss the accuracy of the molecular evolutionary clock from the standpoint of the neutral theory.

**Population Genetics of Mutant Substitutions**

For our discussion of the mechanism underlying the molecular evolutionary clock, we must make a clear distinction between mutation at the individual level and substitution of molecular mutants (such as different amino acids) at the population level.

The process of molecular evolution consists of a sequence of events in which rare mutants increase their frequencies and spread through the species, finally reaching the state of fixation (100% in frequency). A great majority of such mutants, including those having a small selective advantage, are lost from the population by chance within a small number of generations. Only a tiny fraction can spread through the whole population to reach fixation, for which a very long time is required.

In the case of selectively neutral mutants (i.e., selectively equivalent alleles), it takes on the average $4N_e$ generations until fixation (Kimura and Ohta 1969), where $N_e$ is the effective size of the population; $N_e$ is roughly equal to the number of breeding individuals in one generation.

Let us now consider the sequence of events by which mutant genes are substituted one after another in the population (species) in the course of time. We denote by $k_g$ the rate of mutant substitution per generation. Since each substitution may take a very long time, the rate should be measured as a long term average. The rate of substitution thus defined is independent of how quickly individual mutants spread into the population. What matters is the average interval between consecutive fixations.

Consider a gene locus (or a corresponding protein), and let v be the mutation rate per generation for a particular class of mutants (such as selectively neutral amino acid changes). Further, let u be the probability of ultimate fixation of an individual mutant. This is the probability that an individual mutant which appeared in the population eventually spreads in the population, reaching 100% in frequency.

In a population consisting of N diploid individuals, if we assume that each mutant is represented only once at the moment of appearance, the rate of evolution per generation is

$$k_g = 2Nvu. \qquad (1)$$

This formula is based on the consideration that in a population of N diploid individuals, 2Nv new mutants appear per generation, of which the fraction u eventually spread through the whole population. (For a haploid population, 2N in the above formula should be replaced by N.) As mentioned already, this rate is in the sense of a long-term average; to measure the evolutionary rate we must take a much longer time than required for individual mutant substitution.

If the mutant is selectively neutral, the probability of ultimate fixation is equal to its initial frequency, that is, $u = 1/(2N)$ in the diploid population, and therefore, from Eq. (1), we have

$$k_g = v. \qquad (2)$$

In other words, for neutral alleles, the rate of evolution is equal to the mutation rate.

The above formulation of the evolutionary rate at the molecular level is implicit in my first paper on the neutral theory (Kimura 1968), but it was presented more explicitly in later writings (see Crow and Kimura 1970; Kimura and Ohta 1971). Note that for a haploid population, there are N genes so that we have $k_g = Nvu$ and $u = 1/N$. Therefore, we again obtain $k_g = v$, the same result as the diploid case. Note also that Eq. (2) is independent of the population size, and it holds even when the population size fluctuates from time to time. Furthermore, it is independent of the mode of reproduction; it is equally applicable to sexual and asexual organisms (Kimura and Ohta 1971). The formula also implies that the rate of evolution should be independent of the environment where organisms are placed, as long as the mutation rate remains the same.

Let g be the generation span measured in years, and let $k_1$ be the evolutionary rate per year so that $k_1 = k_g/g$. Then, from Eq. (2), we obtain, for the selectively neutral case, the following formula for the evolutionary rate per year:

$$k_1 = v_0/g. \qquad (3)$$

In this formula, $v_0$ is the mutation rate per generation for selectively neutral alleles. (Subscript 0 in $v_0$ denotes that it refers to neutral alleles.)

Since the neutral theory assumes that a certain fraction, say $f_0$, of the mutations are selectively neutral, while the rest (i.e., $1 - f_0$) are sufficiently deleterious to be eliminated from the population, the above formula (3) can also be expressed as follows:

$$k_1 = f_0(v_T/g), \quad (4)$$

where $v_T$ is the total mutation rate per generation so that $v_0 = f_0 v_T$.

Then, the existence of the molecular evolutionary clock means, if the neutral theory is valid, that $v_T/g$ remains the same (i.e., constant) among diverse lineages and over time for a given gene (or a protein). Here, we assume that $f_0$ represents the level of selective constraint, and that $f_0$ is constant for a given molecule or a part of one molecule: generally speaking, $f_0$ is smaller for functionally more important molecules which are subject to stronger selective constraint.

The situation is much more complicated if mutant substitutions are caused exclusively by positive Darwinian selection. Let $v_A$ be the mutation rate per generation for advantageous mutants, and let s be the selective advantage of an individual mutant over the preexisting allele. If we assume that s is small but $4N_e s$ is large so that $0 < s \ll 1$ and $\exp(-4N_e s) \ll 1$ for any of the advantageous mutants, then we have approximately $u = 2s(N_e/N)$ (Kimura 1964). Substituting this in Eq. (1), we obtain

$$k_g = 4N_e s v_A. \quad (5)$$

If s varies from mutation to mutation, then the mean selective advantage $\bar{s}$ ($>0$) may be substituted for s in the above equation. Furthermore, if we denote by $f_A$ the fraction of advantageous mutants, so that $v_A = f_A v_T$, the equation which corresponds to (4) becomes as follows:

$$k_1 = 4N_e \bar{s} f_A v_T/g. \quad (6)$$

This means that the rate of evolution under natural selection depends on various factors, namely, the effective population size ($N_e$), the fraction of advantageous mutants ($f_A$) and their average selective advantage ($\bar{s}$), as well as the total mutation rate per year ($v_T/g$). Then, in order to explain the constancy of the evolutionary rate among diverse lineages, we must assume that the product of all these parameters remains constant among lineages. However, it is likely that these parameters tend to differ from species to species. Particularly, $f_A$ should depend strongly on the environment where the species is placed, being high for a species offered a new ecologic opportunity but low for those kept long in a stable environment. It seems to me that a highly complicated and arbitrary set of assumptions must be invoked in order to explain the clocklike behavior of molecular evolution from the standpoint of the selectionists. Merely assuming that the amount of selection is equal among lines by no means leads to the constancy (equality) of the evolutionary rate, contrary to the statement of some authors. In fact, adaptive evolutionary change (due to mutant substitutions for which $4N_e s \gg 1$) should be characterized by marked differences in its rate among lineages over time, as evidenced by evolution at the phenotypic level revealed by paleontological studies (see, for example, Simpson 1944).

On the other hand, if evolutionary change is largely controlled by mutation pressure and random drift, as claimed by the neutralists, the constancy of evolutionary rate can be explained by assuming constancy of mutation rate per amino acid (or nucleotide) site among diverse lineages. This assumption has to be checked by future experiments.

**Some Statistical Analyses of Protein Evolution**

The evolutionary rate per year (i.e., $k_g/g$) is sometimes found to be inconstant, undermining the clocklike progression of molecular evolution. From the standpoint of the neutral theory, this can be explained as being due to two causes.

One is the change in the mutation rate per year (i.e., $v_T/g$), and the other is the alteration of the selective constraint ($f_0$) of each molecule. The former (change of $v_T/g$) is likely to be brought about by the change of the generation span (g), while the latter (change of $f_0$) is caused by the alteration of internal molecular environment such as gene duplication, or through change of interacting molecules, details of which are mostly unknown at present.

In this section, I shall present a few examples suggesting that these two causes are really at work in mammalian evolution. Before I go into data analysis, I shall develop a simple statistical method which will be useful for our analysis.

Let us assume a phylogenetic tree in which several branches diverge from a common ancestor A, as shown in Fig. 1. We denote by L the number of branches or lines involved ($L \geqq 3$). We consider a particular molecule, such as the $\alpha$ chain of hemoglobin, and let $n_{aa}$ be the total number of amino acid sites compared between two lines. For simplicity, we assume that $n_{aa}$ is constant for all pairs of comparisons. We denote by $D_{ij}$ the observed number of amino acid differences between the i-th and j-th lines ($i = 1, 2, \ldots, L$; $j = 1, 2, \ldots, L$). Let $X_i$ be the number of amino acid substitutions which have accumulated in the i-th line since divergence from the common ancestor A. These L parameters ($X_1, X_2, \ldots, X_L$) may be estimated from $L(L-1)/2$ observed numbers of amino acid differences by using the least square method as follows.

Let

$$S = \sum_{i=1}^{L} \sum_{\substack{j=1 \\ (j \neq i)}}^{L} (\tilde{D}_{ij} - X_i - X_j)^2 \qquad (7)$$

be the sum of squares of errors in fitting the observed numbers ($\tilde{D}_{ij}$) by the corresponding linear estimates ($X_i + X_j$). Note that the observed numbers here refer not to the raw data ($D_{ij}$) but rather to the corrected values ($\tilde{D}_{ij}$) in which multiple substitutions and back mutational changes are taken into account. In this paper, I use the well-known formula:

$$\tilde{D}_{ij} = -n_{aa} \log_e(1 - D_{ij}/n_{aa}). \qquad (7a)$$

[Note: for DNA sequence data, a different correction should be used (see Kimura 1980, 1981; see also Kimura 1983).]

Then, we determine $X_i$'s so that S is minimized. Differentiating Eq. (7) with respect to $X_i$'s and setting each of the resulting equations equal to zero, i.e., $\partial S/\partial X_i = 0$ ($i = 1, 2, \ldots, L$), we obtain

$$\hat{X}_i = \frac{1}{L-2}\left[\tilde{D}_{i\cdot} - \frac{\tilde{D}_{\cdot\cdot}}{2(L-1)}\right], \qquad (8)$$

where

$$\tilde{D}_{i\cdot} = \sum_j{}' \tilde{D}_{ij} \qquad (8a)$$

and

$$\tilde{D}_{\cdot\cdot} = \sum_{i=1}^{L} \tilde{D}_{i\cdot} \qquad (8b)$$

In these expressions, the hat (^) on $X_i$ means that this is the least square estimate and the primed summation in Eq. (8a) indicates that the sum is over all j values excluding $j = i$. Furthermore, we assume that $\tilde{D}_{ij} = \tilde{D}_{ji}$. The large-sample standard error $\sigma_i$ for $\hat{X}_i$ is given by the formula,

$$\sigma_i = \frac{1}{L-2} \sqrt{\left[\left(\frac{L-2}{L-1}\right) V(D_i) + \frac{1}{4(L-1)^2} \sum_{i=1}^{L} V(D_i)\right]}, \qquad (9)$$

where

$$V(D_i) = n_{aa} \sum_j{}' \left(\frac{d_{ij}}{1 - d_{ij}}\right), \qquad (9a)$$

and

$$d_{ij} = D_{ij}/n_{aa}. \qquad (9b)$$

**Table 1.** The observed number of amino acid differences between hemoglobin α-chains of six mammals: $n_{aa} = 141$, $L = 6$ (data from Dayhoff 1978)

| | Mouse | Rabbit | Dog | Horse | Bovine |
|---|---|---|---|---|---|
| Human | 18 | 25 | 23 | 18 | 17 |
| Mouse | | 27 | 25 | 24 | 19 |
| Rabbit | | | 28 | 25 | 25 |
| Dog | | | | 27 | 28 |
| Horse | | | | | 18 |

**Table 2.** The number of amino acid differences between hemoglobin β-chains of six mammals: $n_{aa} = 146$, $L = 6$ (data from Dayhoff 1978)

| | Mouse | Rabbit | Dog | Horse | Bovine |
|---|---|---|---|---|---|
| Human | 27 | 14 | 15 | 25 | 25 |
| Mouse | | 28 | 30 | 36 | 39 |
| Rabbit | | | 21 | 25 | 30 |
| Dog | | | | 30 | 28 |
| Horse | | | | | 30 |

The standard error here is the probable error caused by sampling a restricted number, namely $n_{aa}$, of amino acid sites.

An additional quantity which will be useful for investigating if the rates of evolution among lines are uniform is the statistic R. This is the ratio of the observed variance of $X_i$'s to the theoretically expected variance under the assumption of uniformity, and if $X_i$'s follow Poisson distribution, it can be shown that $(L-1)R$ follows the $\chi^2$ (chi-square) distribution with $L-1$ degrees of freedom. In terms of observed differences,

$$R = \frac{(L+1)V_D}{(L-1)\bar{D}}, \qquad (10)$$

where $\bar{D}$ and $V_D$ are the mean and variance of $\tilde{D}_{ij}$ values (see p. 77 of Kimura 1983). Note that "the observed variance" here is not exactly identical to the variance of $\hat{X}_i$'s, since the latter does not contain the residual error variance (usually small) of least square estimation.

Let us now apply these methods to the analysis of actual data. Table 1 lists observed number of amino acid differences ($D_{ij}$'s) between six mammals (human, mouse, rabbit, dog, horse, and bovine) when their hemoglobin α-chains are compared. It is likely that these mammals diverged from each other late in the Mesozoic about 80 million years ago.

Noting that $n_{aa} = 141$ and $L = 6$, and applying Eq. (10) we obtain $R = 1.26$. With this R value, $\chi^2 = (L-1)R = 6.3$, and from a table of the $\chi^2$ distribution with five degrees of freedom, we find

28

that deviation of R from unity is not statistically significant. So we cannot reject the hypothesis that the evolutionary rates among six mammals are the same with respect to $\alpha$-hemoglobin. The estimated numbers of substitutions for these six lines are as follows:

Human: $\hat{X}_1 = 8.3 \pm 2.7$
Mouse: $\hat{X}_2 = 11.9 \pm 2.9$
Rabbit: $\hat{X}_3 = 16.9 \pm 3.1$
Dog: $\hat{X}_4 = 17.2 \pm 3.1$
Horse: $\hat{X}_5 = 11.6 \pm 2.8$
Bovine: $\hat{X}_6 = 10.1 \pm 2.8$

A similar analysis can be made for hemoglobin $\beta$ with respect to the same set of animals, using data as listed in Table 2. In this case, noting $n_{aa} = 146$ and $L = 6$, we obtain $R = 3.1$. This gives $\chi^2 = 15.5$, and with five degrees of freedom, we find from a table of $\chi^2$ that the deviation of R from unity is significant at about the 1% level. The number of amino acid substitutions for the six lines as estimated by using Eq. (8) turns out to be as follows:

Human: $\hat{X}_1 = 6.4 \pm 2.8$
Mouse: $\hat{X}_2 = 22.9 \pm 3.6$
Rabbit: $\hat{X}_3 = 9.9 \pm 3.0$
Dog: $\hat{X}_4 = 11.8 \pm 3.0$
Horse: $\hat{X}_5 = 18.4 \pm 3.7$
Bovine: $\hat{X}_6 = 20.3 \pm 3.5$

By inspection of these values, we immediately note that the human line accumulated much fewer amino acid substitutions, and the mouse line many more substitutions, as compared with the average mammals (i.e., about 15.0). If we remove these two extreme lines, and test the uniformity of the number of substitutions among the remaining four lines ($L = 4$), we obtain $R = 1.1$, a value expected under the assumption of uniformity. On the other hand, if we test the uniformity for a group consisting of mouse, human, and dog ($L = 3$), we obtain $R = 6.6$. These results are consistent with recent reports showing that the evolutionary rate in rodents is higher than the average mammal, while the rate of the human line is lower (Kikuno et al. 1985; Wu and Li 1985).

The difference in the results with respect to human and mouse lines for hemoglobulin $\alpha$ and $\beta$ chains is likely to be due to statistical fluctuations, since the mutation rates and the levels of selective constraint appear to be the same for the $\alpha$ and $\beta$ chains of these mammals. Therefore, it is desirable to repeat the statistical analysis after combining the data on the $\alpha$ and $\beta$ chains, i.e., for $n_{aa} = 141 + 146 = 287$ and $L = 6$. This gives the following results:

Human: $\hat{X}_1 = 14.6 \pm 3.8$
Mouse: $\hat{X}_2 = 34.5 \pm 4.5$
Rabbit: $\hat{X}_3 = 26.9 \pm 4.3$
Dog: $\hat{X}_4 = 29.1 \pm 4.3$
Horse: $\hat{X}_5 = 30.0 \pm 4.4$
Bovine: $\hat{X}_6 = 30.2 \pm 4.4$

The R value turns out to be 1.9, giving $\chi^2 = 9.7$. With five degrees of freedom, the deviation of R from unity is significant at about the 5% level. In particular, it is clear that the human line definitely accumulated many fewer mutations than the mouse line.

The most likely cause of such differences in the evolutionary rate is the "generation time effect": If the mutation rates among organisms having widely different generation spans are roughly equal to each other when measured taking one generation as the unit, the mutation rate should be much higher, when measured per year, for an organism having a shorter generation span (and vice versa). Based on the neutralist principle that the rate of evolution is equal to the mutation rate for neutral alleles, the evolutionary rate of amino acid (or nucleotide) substitutions per year must be much higher in rodent than in man.

The intrinsic mutation rate for neutral alleles may also change in some lines due to change (usually loss) of selective constraint. One of the best examples suggesting this is the exceptionally rapid evolutionary change of insulin observed in the guinea pig (*Cavia*) and also in its relative the coypu or nutria (*Myocastor*). (These animals belong to the group of hystricognath rodents.)

Table 3 lists the number of amino acid substitutions among seven mammals including guinea pig and coypu. First, if we exclude the coypu, which is more closely related to the guinea pig than to the other five mammals (human, elephant, sheep, sperm whale, and rabbit), and apply the above statistical method, that is, Eqs. 8, 9, and 10 (noting $n_{aa} = 51$ and $L = 6$), we obtain

Human: $\hat{X}_1 = 1.2 \pm 1.7$
Elephant: $\hat{X}_2 = 1.1 \pm 1.6$
Sheep: $\hat{X}_3 = 1.7 \pm 1.7$
Sperm whale: $\hat{X}_4 = 1.7 \pm 1.7$
Rabbit: $\hat{X}_5 = 1.5 \pm 1.7$
Guinea pig: $\hat{X}_6 = 20.5 \pm 3.1$

and

$$R = 13.2.$$

This very high value of R clearly comes from the fact that the guinea pig line accumulated more than 10 times as many amino acid substitutions as the rest of the lines since divergence from the common ancestor. The R value becomes even higher if we

**Table 3.** The number of amino acid differences between insulin (A and B peptides) of seven mammals (data from Dayhoff 1978)

| | Elephant | Sheep | Sperm whale | Rabbit | Guinea pig | Coypu |
|---|---|---|---|---|---|---|
| Human | 2 | 4 | 3 | 1 | 18 | 18 |
| Elephant | | 2 | 4 | 3 | 17 | 19 |
| Sheep | | | 2 | 4 | 18 | 20 |
| Sperm whale | | | | 3 | 18 | 20 |
| Rabbit | | | | | 18 | 18 |
| Guinea pig | | | | | | 15 |

Excluding insertions and deletions, 51 amino acid sites are compared ($n_{aa}$ = 51)

select three lines, namely, human, elephant, and coypu lines. For this set of comparisons (L = 3), we obtain R = 18.4.

On the other hand, if we exclude from Table 3, guinea pig and coypu, and test the uniformity of substitution rates among the remaining five lines (human, elephant, sheep, sperm whale, and rabbit), we get R = 0.62, indicating strong uniformity. The average number of substitutions for these five lines turns out to be $\bar{X}$ = 1.44.

From the data given in Table 3, we can also estimate the numbers of amino acid substitutions in various branches including guinea pig and coypu as illustrated in Fig. 2. It is interesting to note that the rates of amino acid substitutions in the two lines, one leading to guinea pig and the other leading to coypu from their nearest common ancestor B, are roughly equal to each other even if both have very high rates. Assuming that the ancestor A goes back to 80 million ($8 \times 10^7$) years ago, the average rate of amino acid substitutions in the mammalian line, excluding guinea pig and coypu, turns out to be 1.44/($51 \times 8 \times 10^7$) or $0.35 \times 10^{-9}$/aa/year (per amino acid site per year). On the other hand, the average rate for these hystricognath rodents since their descent from the ancestor A is $5.2 \times 10^{-9}$/aa/year. From the standpoint of the neutral theory, a possible explanation, first proposed by Kimura and Ohta (1974), is that the insulin of these hystricognath rodents lost its original selective constraint at an early stage of evolution so that most amino acid changes in that molecule became neutral, i.e., nonharmful (see also Jukes 1979). This is consistent with the observation that in the guinea pig, zinc is absent from the insulin-producing cells, coinciding with the loss of the usually invariant histidine B10 (for more detailed discussion and relevant references, see p. 114 of Kimura 1983).

It is concluded that due to loss of selective constraint, as evidenced by a drastic change in the molecular environment, mutations started to accumulate in guinea pig and coypu insulins at a very high rate. In this connection, investigation of the

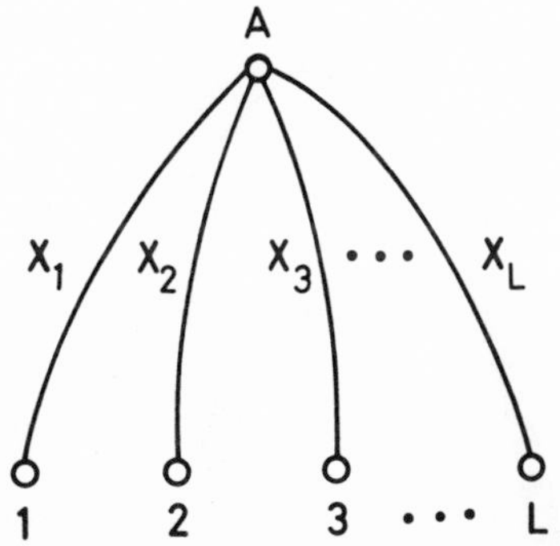


**Fig. 1.** A phylogenetic tree in which L lineages (L ≧ 3) diverge from their common ancestor A. Letters $X_1$, $X_2$, etc. denote the number of amino acid substitutions accumulated in the individual lineages

evolutionary rates of the middle segment (C peptide) of proinsulin in mammals, including guinea pig, is of interest. The C peptide is the part which is removed when the active insulin (A and B peptides) is formed from its precursor, proinsulin. It is known that the C peptide generally has a much higher rate of amino acid substitution in evolution than insulin.

In Table 4, the fraction of amino acid differences, i.e., $D_{ij}/n_{aa}$ ($D_{ij}$ = the observed number of differences; $n_{aa}$ = the number of amino acid sites compared), is listed for 10 comparisons involving five mammals (guinea pig, rat, human, horse, and bovine). The C peptide consists of about 30 amino acid sites, but insertions or deletions are found in several comparisons, so I excluded these changes and counted only the number of amino acid substitutions. This means that $n_{aa}$ may differ from comparison to comparison.

From these data, the number of amino acid substitutions *per sites,* namely $\hat{X}_i/n_{aa}$, can be estimated, and the following estimates have been obtained:

Guinea pig: 0.26 ± 0.11
Rat: 0.25 ± 0.10
Human: 0.13 ± 0.09
Horse: 0.16 ± 0.09
Bovine: 0.24 ± 0.11

These values show that the evolutionary rate of the guinea pig is not significantly higher than other mammals with respect to the C peptide of proinsulin. In fact, R = 0.66, obtained from this set of comparisons, strongly suggests that the rates of evolution of the C peptide are uniform among these five animals. The evolutionary rate per amino acid site per year turns out to be $k_{aa} = 3.3 \times 10^{-9}$ for guinea pig, and the average rate for these five animals is $\bar{k}_{aa} = 2.6 \times 10^{-9}$ (assuming 80 million years of evolution since the common ancestor).

30

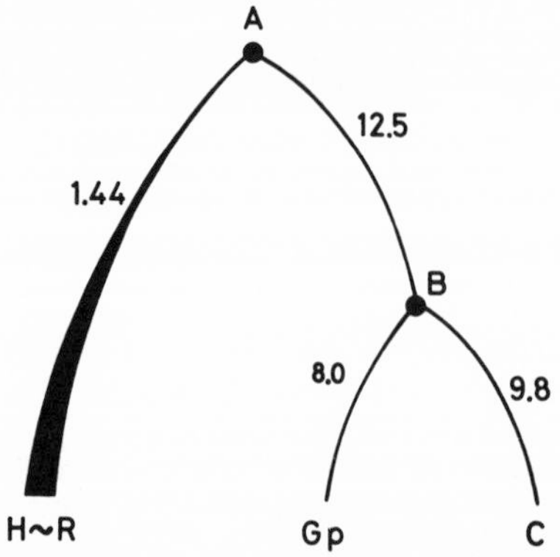


**Fig. 2.** A phylogenetic tree and the numbers of amino acid substitutions in various branches in the evolution of insulin of mammals including guinea pig (Gp) and coypu (C). In this figure, H ~ R denotes a group of five mammals consisting of human, elephant, sheep, sperm whale, and rabbit. The number of amino acid sites compared: $n_{aa} = 51$.

It is interesting to note that the remarkably high evolutionary rate, $k_{aa} = 5.2 \times 10^{-9}$, as found for insulin (A and B peptides) of guinea pig is rather near to $k_{aa} = 3.3 \times 10^{-9}$ obtained for the proinsulin C peptide of this animal. Note also that $5.2 \times 10^{-9}$ is considerably lower than the corresponding amino acid substitution rate for pseudoglobin genes which is roughly, $10^{-8}$ or slightly higher (Li et al. 1981; Miyata and Yasunaga 1981). This means that insulin of guinea pig (and coypu) has not been totally liberated from selective constraint.

Using the numbers of amino acid substitutions given in Fig. 2, and assuming the uniformly high evolutionary rate of insulin in the hystricognath rodents since their descent from the common ancestor A (which we assume to be 80 million years ago), we can estimate the divergence point B for the guinea pig and the coypu. This turns out to be 33.3 million years ago. According to Romer (1968), the guinea pig group suddenly appeared in South America in the Oligocene (25–40 million years ago), so the above estimate is consistent with paleontological observations.

**Table 4.** The fraction ($D_{ij}/n_{aa}$) of amino acid differences between proinsulin C peptide of five mammals including guinea pig

| | Rat | Human | Horse | Bovine |
|---|---|---|---|---|
| Guinea pig | 11/29 | 9/29 | 10/29 | 11/26 |
| Rat | | 10/31 | 12/31 | 9/26 |
| Human | | | 7/31 | 9/26 |
| Horse | | | | 8/26 |

This table was constructed using data compiled in Dayhoff (1978)

## Discussion

As we have seen in the previous section, the generation time effect is one of the possible and (probably important) causes which make the molecular evolutionary clock less exact, reducing its foolproof reliability. However, the magnitude of this effect does not appear to be very large even among organisms whose generation spans are very different. In fact, in their attempt to test whether the generation span has any effect on the molecular evolutionary rate per year, Wilson et al. (1977) could not detect any such effect.

However, recent works by Wu and Li (1985) and Kikuno et al. (1985) show that rodents (having a short generation span) evolve faster than human beings (which have a long generation span), particularly with respect to silent sites, suggesting a generation time effect. Also, Britten (1986) reports, based mainly on DNA–DNA hybridization data, that the evolutionary rate of DNA divergence of rodents is some five times higher than that of higher primates. He suggests that changes in the DNA repair mechanism are the primary cause of such difference. However, the possibility cannot be excluded, I think, that this represents the generation time effect.

Moreover, Sibley and his associates (Catzeflis et al. 1987) carried out DNA–DNA hybridization studies of some muroid rodents, and they came to the conclusion that the average evolutionary rate in these rodents is roughly 10 times as high as that of the hominoid primates. Note that even this much difference is not as large as the actual difference of the generation spans of these two animal groups; they must differ by a factor of at least 40 in the actual generation length.

Sibley and his associates have also discovered the generation time effect in birds (Sibley, personal communication). They found that birds that have longer generation times, i.e., whose first breeding age is higher (such as albatrosses that take 6–9 years until sexual maturity), tend to show less DNA divergence than those that mature more rapidly.

If the generation time effect is really important, particularly when organisms having widely different generation spans are compared, this means that, from the standpoint of the neutral theory, the nucleotide mutation rate is not quite constant per year but is higher for organisms having a shorter generation span (and vice versa). In other words, the number of mutational changes of nucleotides is dependent (albeit mildly) on the number of generations. Experimental studies bearing on this problem are much needed.

The second agent that interferes with the constancy of the evolutionary rate is an alteration of functional constraint of the molecule [leading to change in $f_0$ in Eq. (4)]. This is exemplified by the guinea pig and coypu insulins (see previous section). This is probably the most dramatic example known on this topic. Another possible example is the rapid evolution of the opossum hemoglobin $\alpha$ studied by Stenzel (1974). In this case, the estimated rate of amino acid substitution is $1.7 \times 10^{-9}$ per year, definitely higher than the standard value ($1.2 \times 10^{-9}$) of this molecule in mammals. A change of selective constraint is suggested by the fact that in the opossum hemoglobin $\alpha$, *Gln* occurs at position 58. This is the position which is almost invariably occupied by *His* (bound to heme) in the $\alpha$ hemoglobin of vertebrates. Note that the myoglobin sequence of the opossum is also known, but for this molecule, there is nothing unusual in its sequence (see p. 82 of Kimura 1983). Although we do not know at present what agents initiated such changes (mostly reduction) of functional constraint, it is possible that gene duplication is often responsible.

There may be other factors which also interfere with the accuracy of the molecular evolutionary clock, such as: (i) fixation of very slightly deleterious mutants during population bottlenecks (Ohta 1973, 1974, 1976), (ii) participation of compensatory neutral mutations in molecular evolution (Kimura 1985a,b), and (iii) fluctuation of the neutral space in the course of mutant substitutions (Kimura and Takahata, manuscript in preparation). Whether these factors are responsible for making the molecular clock less accurate to a significant extent remains to be seen.

Ohta and Kimura (1971) were probably the first to investigate statistically the variation of evolutionary rates among lines for a few molecules. They used the *F*-test to see if the observed variance is significantly higher than the variance expected by pure chance, and found that the *F* value is highly significant in $\beta$-type hemoglobins and cytochrome c, but not in hemoglobin $\alpha$. For example, in cytochrome c, the ratio of the observed to expected variances turned out to be about 3.3.

This is consistent with the report of Jukes and Holmquist (1972) that rattlesnake cytochrome c has evolved three or four times as rapidly as turtle cytochrome c. In this case, however, the factor or factors that are responsible for such differences in evolutionary rates among lines is not clear.

It is often stated that, when the variance of evolutionary rates among lineages for a given protein (or a DNA region) is found to be significantly higher than the variance expected by pure chance, or when the estimated rates in different parts of a phylogenetic tree differ significantly, such an observation disproves the neutral theory.

Ayala (1986) analyzed the evolutionary rates of copper–zinc superoxide dismutase (SOD) among eight organisms, and he discovered that the estimated rates become progressively lower as more and more remote comparisons are made. This molecule (SOD) consists of about 150 amino acids, of which 55 are common to all sequences. Let us suppose that these 55 amino acids represent the "invariant sites," and fix our attention on the variable sites consisting of about 100 amino acids. If we use Ayala's (1986) Table IV, which lists the number of amino acid differences, and apply the method of the previous section to estimate the number of amino acid replacements accumulated in four mammalian lines (human, rat, horse, and cow) in the course of evolution from the common ancestor, we obtain the following estimates:

| | |
|---|---|
| Human: | $16.4 \pm 5.2$ |
| Rat: | $13.0 \pm 5.0$ |
| Horse: | $19.2 \pm 5.4$ |
| Cow: | $13.0 \pm 4.9$ |

We also obtian $R = 0.59$, suggesting a strong uniformity of the evolution of SOD among mammalian lineages. The average rate of evolution is $k_{aa} = 1.9 \times 10^{-9}$/aa/year, which is more than six times as high as the corresponding rate of cytochrome c (which Ayala mentions as a protein that exhibits more regular behavior over the whole geological span considered). Assuming this high rate, the number of amino acid substitutions that separate mammals and *Drosophila* should be about 2.3 per variable site. On the other hand, the corresponding observed number is about 0.6, and, even after corrections are made for hidden substitutions using Eq. (7a), the resulting figure is 0.92, which is only 40% of the value expected under the assumption of rate constancy. The discrepancy between the expected and observed values becomes much higher when we make a more remote comparison, such as human and yeast: the expected number of amino acid substitutions per site is 4.6, while the observed number (with a Poisson correction) is 1.2, which is only 26% of the expected value assuming rate constancy.

There are two explanations for such discrepancies. The first is that the evolutionary rate has progressively increased along various lineages since their descent from the common ancestor that lived 1.2 billion years ago. The second explanation, which I think more plausible, is that the observed diminution of the rate in the backward direction is simply an artifact caused by the fact that it becomes progressively difficult to recover hidden substitutions by ordinary statistical means as more and more amino acid substitutions accumulate between the two sequences compared. This may be called the "apparent saturation effect." A similar case was reported by Perler et al. (1980) with respect to nucleotide changes in silent sites in the evolution of preproinsulin gene.

Based on his analysis of SOD data, Ayala (1986) claims that the observed variance in evolutionary rate is much too large and thus inconsistent with the neutral theory. I do not think, however, that his claim is warranted unless the neutral mutation rate per year is shown to be equal among lineages over time and at the same time the molecular evolution of SOD has really accelerated.

Similarly, Gillespie (1984, 1986) criticizes the neutral theory on the ground that the observed value of R [such as computed by Eq. (10)] often turns out to be significantly higher than unity. For example, Ohta and Kimura (1971) analyzed three sets of data involving the $\alpha$- and $\beta$-type hemoglobins and cytochrome c, and found that the observed variance is about three times higher on the average than expected by pure chance. Also, Langley and Fitch (1974) made a similar but more extensive analysis, and obtained results suggesting that the observed variance among lines is about 2.5 times as large as that expected from chance fluctuations.

Gillespie claims that such observations are incompatible with the neutral theory and that a model of evolution by natural selection, which he calls "episodic model," can fit the data better. His model is based on the idea that molecular evolution is episodic, with short bursts of rapid substitutions being separated by long periods of no substitutions. According to him, each environmental change presents a challenge to the species that may be met by amino acid substitutions caused by natural selection.

More specifically, he considers a phylogeny as shown in Fig. 1 (which he calls a "star phylogeny"), and assumes that the number of episodes per lineage follows a Poisson distribution, and that the number of substitutions per episode also follows a certain probability distribution which is the same for all episodes in all lineages. He then states that this model has the remarkable property of predicting that the values of R are in a very restricted range, say from 1.0 to ~3.5, just as observed in the data.

I think it highly unrealistic to assume that the numbers of episodes in different lineages (which must experience different environments) follow the same probability distribution. Also, it is a moot point why natural selection acts in such a way that the number of substitutions per episode follows the same probability distribution for all episodes in all lineages. Here, natural selection is invoked arbitrarily to fit the results, while neglecting all the effects of the mutation rate, population size, selection coefficients, etc., as shown by Eq. (6). If, in reality, the relatively large variation (i.e., $R > 1$) in the number of amino acid substitutions among lineages is caused by some of the lineages having different intrinsic evolutionary rates, as discussed in the previous section, Gillespie's model breaks down completely. (For $R < 1$, his theory does not work anyway.)

From the standpoint of the neutral theory, the clocklike progression of molecular evolution can be explained by assuming that the rate of production of neutral mutations per year is nearly constant among related organisms for a given molecule over time. This means that, if the neutral theory is valid, the rate of evolution at the molecular level is directly proportional to the mutation rate. A dramatic example showing this is a very rapid evolutionary change observed in RNA viruses, which are known to have very high mutation rates: RNA viruses show evolutionary rates roughly a million times as high per year as that of DNA organisms.

Saitou and Nei (1986), who investigated the evolution and polymorphism of influenza A virus genes, found that the rate of nucleotide substitution is of the order of $10^{-3}$/site/year for most genes studied. Since the mutation rate is estimated to be 0.01/site/year, and since this is much higher than the average substitution rate, the authors conclude that most influenza genes are subject to negative selection (i.e., $f_0 \ll 1$). An extremely high substitution rate and the clocklike progression of substitutions in this virus were also reported by Hayashida et al. (1985).

Similarly, Gojobori and Yokoyama (1985) estimated the evolutionary rates of both a retroviral oncogene (v-*mos* gene) and its cellular homolog and found that the former evolves nearly 0.8 million times faster than the latter. They pointed out that the rapid evolution of the former is caused by the correspondingly high mutation rate which is due to a lack of the proofreading enzymes that ensure accurate replication, coupled with the very high replication rates.

What is really remarkable in the evolution of these two types of RNA viruses is that not only do the nucleotide substitutions occur at extraordinarily high rates in clocklike fashion, but also synonymous substitutions (that do not cause amino acid changes) predominate over amino acid altering substitutions, indicating the typical pattern of neutral evolution.

Finally, for those readers who want to know more about the neutral theory of molecular evolution, I would recommend recent papers by Kimura (1986) and Jukes and Kimura (1984), and also, my book (Kimura 1983) on the subject, and the symposium volume by Ohta and Aoki (1985), which contains a number of relevant articles.

*Acknowledgments.* I thank Drs. K. Aoki, C. Sibley, and J.F. Crow for having gone over the manuscript to suggest improved presentation. Thanks are also due to Drs. T. Ohta and N. Takahata for stimulating discussions. This is contribution No. 1716 from the National Institute of Genetics, Mishima, Shizuoka-ken,

411 Japan. This work was supported in part by a Grant-in-Aid from the Ministry of Education, Science and Culture of Japan.

## References


Ayala FJ (1986) On the virtues and pitfalls of the molecular evolutionary clock. J Hered 77:226–235

Britten RJ (1986) Rates of DNA sequence evolution differ between taxonomic groups. Science 231:1393–1398

Catzeflis FM, Sheldon FH, Ahlquist JE, Sibley CG (1987) DNA–DNA hybridization evidence of the rapid rate of muroid rodent DNA evolution. Mol Biol Evol (in press)

Crow JF, Kimura M (1970) An introduction to population genetics theory. Harper & Row, New York

Dayhoff MO (ed) (1978) Atlas of protein sequence and structure, vol 5, suppl 3. National Biomedical Research Foundation, Washington DC

De Beer G (1964) Atlas of evolution. Thomas Nelson & Sons Ltd, London

Gillespie JH (1984) The molecular clock may be an episodic clock. Proc Natl Acad Sci USA 81:8009–8013

Gillespie JH (1986) Natural selection and the molecular clock. Mol Biol Evol 3:138–155

Gojobori T, Yokoyama S (1985) Rates of evolution of the retroviral oncogene of Moloney murine sarcoma virus and of its cellular homologues. Proc Natl Acad Sci USA 82:4198–4201

Hayashida H, Toh H, Kikuno R, Miyata T (1985) Evolution of influenza virus genes. Mol Biol Evol 2:289–303

Hori H, Lim B-L, Ohama T, Kumazaki T, Osawa S (1985) Evolution of organisms deduced from 5S rRNA sequences. In: Ohta T, Aoki K (eds) Population genetics and molecular evolution. Japan Scientific Societies Press, Tokyo/Springer, Berlin, pp 369–384

Jukes TH (1979) Dr. Best, insulin, and molecular evolution. Can J Biochem 57:455–458

Jukes TH, Holmquist R (1972) Evolutionary clock: nonconstancy of rate in different species. Science 177:530–532

Jukes TH, Kimura M (1984) Evolutionary constraints and the neutral theory. J Mol Evol 21:90–92

Kikuno R, Hayashida H, Miyata T (1985) Rapid rate of rodent evolution. Proc Japan Acad 61(B):153–156

Kimura M (1964) Diffusion models in population genetics. J Appl Probab 1:177–232

Kimura M (1968) Evolutionary rate at the molecular level. Nature 217:624–626

Kimura M (1969) The rate of molecular evolution considered from the standpoint of population genetics. Proc Natl Acad Sci USA 63:1181–1188

Kimura M (1980) A simple method for estimating evolutionary rates of base substitutions through comparative studies of nucleotide sequences. J Mol Evol 16:111–120

Kimura M (1981) Estimation of evolutionary distances between homologous nucleotide sequences. Proc Natl Acad Sci USA 78:454–458

Kimura M (1983) The neutral theory of molecular evolution. Cambridge University Press, Cambridge, England

Kimura M (1985a) Diffusion models in population genetics with special reference to fixation time of molecular mutants under mutational pressure. In: Ohta T, Aoki K (eds) Population genetics and molecular evolution. Japan Scientific Societies Press, Tokyo/Springer, Berlin, pp 19–39

Kimura M (1985b) The role of compensatory neutral mutations in molecular evolution. J Genet 64:7–19

Kimura M (1986) DNA and the neutral theory. Philos Trans R Soc Lond [Biol] 312:343–354

Kimura M, Ohta T (1969) The average number of generations until fixation of a mutant gene in a finite population. Genetics 61:763–771

Kimura M, Ohta T (1971) On the rate of molecular evolution. J Mol Evol 1:1–17

Kimura M, Ohta T (1974) On some principles governing molecular evolution. Proc Natl Acad Sci USA 71:2848–2852

Langley CH, Fitch WM (1974) An examination of the constancy of the rate of molecular evolution. J Mol Evol 3:161–177

Li W-H, Gojobori T, Nei M (1981) Pseudogenes as a paradigm of neutral evolution. Nature 292:237–239

Miyata T, Yasunaga T (1981) Rapidly evolving mouse $\alpha$-globin-related pseudogene and its evolutionary history. Proc Natl Acad Sci USA 78:450–453

Ohta T (1973) Slightly deleterious mutant substitutions in evolution. Nature 246:96–98

Ohta T (1974) Mutational pressure as the main cause of molecular evolution and polymorphism. Nature 252(29):351–354

Ohta T (1976) Role of very slightly deleterious mutations in molecular evolution and polymorphism. Theor Popul Biol 10(3):254–275

Ohta T, Aoki K (eds) (1985) Population genetics and molecular evolution. Japan Scientific Societies Press, Tokyo/Springer, Berlin

Ohta T, Kimura M (1971) On the constancy of the evolutionary rate of cistrons. J Mol Evol 1:18–25

Perler F, Efstratiadis A, Lomedico P, Gilbert W, Kolodner R, Dodgson J (1980) The evolution of genes: the chicken preproinsulin gene. Cell 20:555–566

Pilbeam D (1984) The descent of hominoids and hominids. Sci Am 250(3):60–69

Romer AS (1968) The procession of life. Weidenfeld & Nicholson, London

Saitou N, Nei M (1986) Polymorphism and evolution of influenza A virus genes. Mol Biol Evol 3:57–74

Sibley CG, Ahlquist JE (1986) Reconstructing bird phylogeny by comparing DNA's. Sci Am 254(2):68–78

Simpson GG (1944) Tempo and mode in evolution. Columbia University Press, New York

Stenzel P (1974) Opossum Hb chain sequence and neutral mutation theory. Nature 252:62–63

Wilson AC, Carlson SS, White TJ (1977) Biochemical evolution. Annu Rev Biochem 46:573–639

Wilson AC, Sarich VM (1969) A molecular time scale for human evolution. Proc Natl Acad Sci USA 63:1088–1093

Wu C-I, Li W-H (1985) Evidence for higher rates of nucleotide substitution in rodents than in man. Proc Natl Acad Sci USA 82:1741–1745

Zuckerkandl E (1965) The evolution of hemoglobin. Sci Am 212(5):110–118

Zuckerkandl E, Pauling L (1965) Evolutionary divergence and convergence in proteins. In: Bryson V, Vogel HJ (eds) Evolving genes and proteins. Academic Press, New York, pp 97–166


Received November 13, 1986/Revised February 27, 1987

PART EIGHTEEN

# Neutral Theory

## Introduction

Kimura's best-known contribution is the neutral mutation–random drift theory of molecular evolution, for brevity called the neutral theory. It was first presented in 1968 (no. 48). The theory was rejected uncritically by many evolutionists and accepted, equally uncritically, by a number of molecular biologists. Now, although not without controversy, it is a part of the standard body of information on molecular evolution.

The theory was independently proposed by King and Jukes (1969) under the provocative title "Non-Darwinian evolution." Written in early 1968, King and Jukes's paper was at first rejected by *Science,* but later accepted after additional reviews; however, it was not published until 1969. Although the formal publication priority belongs to Kimura, this is clearly an example of an idea being independently thought of by more than one person. Although King and Jukes share with Kimura the credit for proposing the basic idea, further developments have been almost entirely due to Kimura. The early history of the theory has been reviewed by Jukes (1991).

Although Kimura's theory was presented *de novo,* there were nevertheless earlier arguments for neutrality. Darwin himself suggested that some traits might be neutral. Biochemists noted that proteins often function equally well if some amino acids are changed; but one must realize that a change of a magnitude far too small to detect by physiological or chemical methods could still be selected over long time periods. Several people, following the discovery of considerable isozyme polymorphism by Lewontin and Hubby (1966), mentioned the possibility of neutral polymorphisms, but without much conviction and without developing the idea.

The closest approaches to the neutral theory in the pre-Kimura days were those of Sueoka (1962) and Freese (1962). Both were concerned that DNA differences among species of bacteria varied greatly, whereas the amino acid differences were much less. Both suggested that mutation pressure was responsible for the different base ratios, and that many amino acids were interchangeable. Remarkably, these papers were written before the degeneracy of the code was established. These authors clearly had the germ of the neutral idea, but neither paper had much impact on the biological community. Probably this was because most evolutionists paid little attention to bacteria, and neither author developed the idea further.

It was not until Kimura's forceful advocacy and his presentation of one argument after another in successive papers that the theory made its major impact on evolutionary thinking.

Stimulated by Wright's (1966) work and the finding of degeneracy in genetic codes, Kimura (no. 49) presented the full treatment of the so-called $K$ allele model for neutral mutations, the infinite allele model being the limit of $K = \infty$. From the complete solution of the transient distribution of gene frequency expressed by a series of hypergeometric functions, he deduced the stationary distribution, the probability of temporal loss and fixation, and the effective and actual number of alleles. For $K = \infty$, the distribution is called the expected number of alleles in a given gene frequency class or the frequency spectrum (Ewens 1979). The discussion and appendix 1 of Kimura's paper argued that neutral or nearly neutral mutations at the DNA level must be more common than previously thought. As a whole, this paper contains a more thorough argument for the neutral theory than his 1968 *Nature* paper.

In those days, Kimura used to check the validity of theoretical formulas by computer simulation. The detailed description of simulation methods in this paper (no. 49) may still be very instructive to students. Nowadays there are those who rely on computers exclusively. We have to be careful about such results, because of many unexpected errors in programming, too few runs, etc. Even conventional numerical methods for solving a continuum model can create spurious solutions and may lead to nonsense on unconventional problems (Stewart 1992). The cross-checking of simulation with theoretical results, as Kimura did, is always preferable.

If the amino acid substitution rate estimated from protein sequences is extrapolated to the rate in the whole genome, the rate becomes unacceptably high, which is in sharp contrast to the horotelic evolution of Haldane (1957). Muller's (1958) estimate ($4 \times 10^9$) of the total number of nucleotides per genome was particularly important. He measured the DNA content of human sperm and divided it by the average weight of a nucleotide. Kimura (no. 48) came to the conclusion that most mutations produced by nucleotide replacements are almost neutral in natural selection (see Crow 1972) and discussed the extent of polymorphism then newly revealed by electrophoresis (Harris 1966; Hubby and Lewontin 1966; Lewontin 1991). The conclusion of this landmark paper is epitomized by Kimura's remark: "To emphasize the founder principle but deny the importance of random genetic drift . . . is . . . rather similar to assuming a great flood to explain the formation of deep valleys but rejecting a gradual but long lasting process of erosion by water as insufficient to produce such a result."

Two figures in Kimura and Ohta (no. 50) illustrated some important consequences of the neutral theory. Particularly interesting is the qualitative description about the ancestral relationships of alleles and the number of mutations by which they differ from each other. When Wright read this manuscript, he pointed out that the contrast between the neutral and positive Darwinian theories becomes much sharper in interpreting rapidly evolving molecules: under neutrality, the upper limit of evolutionary rate is set by mutation rate, while under positive Darwinian selection there is no such limit. The amino acid sequence analysis of α and β hemoglobins makes clear the points of the paper.

The major features of molecular evolution were enumerated as five principles in Kimura and Ohta (no. 51). Three of them are: (1) the constant rate and its usefulness in reconstructing phylogenetic trees, (2) the important role of negative selection or functional constraints (without which the neutral theory cannot account for different evolutionary rates of different moleculates or different parts of a molecule), and therefore (3) the conservative nature, in contrast to opportunistic ways, of phenotypic or adaptive evolution, and faster rates for substitutions that are less disruptive to the established structures and functions of molecules (see Ohno [1970] and chapter 5 in Kimura [1983a] for further discussion). One example of these principles was soon demonstrated by Kimura (no. 53), when mRNA sequence data began to accumulate. It was immediately clear that most changes occur at the third codon positions so that they are largely synonymous. Although such sequence data for evolutionary studies were very limited at that time, he quickly noticed the high synonymous rate even in the most conserved histone gene (Grunstein, Schedl, and Kedes 1976), which argued for the neutral theory. Kimura (no. 52) eloquently reviewed the status of the neutral theory. It was the dawn of a flood of DNA sequence information. A decade later, he presented to the Royal Society a review paper entitled "DNA and the Neutral Theory" (no. 57).

Some arguments often raised against the original neutral theory are the constancy of molecular evolutionary rate per year, rather low extents of heterozygosity and of its variance, and an excess of rare variants (Nei 1975). To overcome these problems, Kimura (no. 54) proposed a model of effectively neutral mutations. This model is a modification of Ohta's (1973, 1976) slightly deleterious model, but allows greater possibility of effectively neutral mutations even in a large population. Both models predict lower levels of heterozygosity and specific relationships between substitution rates and population sizes. Kimura and Takahata (no. 55) showed by a new simulation method that under the effectively neutral model the variance of heterozygosity can be smaller than that expected from the original neutral theory (Nei, Fuerst, and Chakraborty 1976; Fuerst, Chakraborty, and Nei 1977).

A difficult in the effectively neutral model as well as in the slightly deleterious model is, however, that lowering heterozygosity tends to slow down or even stop the substitution process (Nei 1975). To account for the constant rate per year, it is also necessary to assume specific relationships between generation time and population size, which are, however, necessarily indirect and crude. Kimura reviewed the model in *Scientific American* (Kimura 1979), *Molecular Evolution, Protein Polymorphism and the Neutral Theory,* which he edited (Kimura 1982), and his book *The Neutral Theory of Molecular Evolution* (1983a). He also briefly mentioned it in chapter 11 in *Evolution of Genes and Proteins* (Kimura 1983b, in Nei and Koehn 1983). But he has not mentioned the effectively neutral model in any detail since 1986. He has returned to the original neutral theory. It seems that his change was based on DNA sequence data which could be better interpreted by two distinct classes of mutations: completely neutral or definitely deleterious. We may also imagine that Kimura recognizes the role of a scientific model in the way that Russell (*Sceptical Essays;* see pp. 64–65 in Medawar 1979), Popper (1975), Wimsatt (1987), and others do (no. 48; Kimura 1991a,b). As pointed out by Crow (1985), the (original) neutral theory has, in fact, several attributes of a good scientific theory.

If protein polymorphism is maintained largely by neutral mutation and random extinction of existing alleles, one may wish to know the proportion of neutral mutation rate to the total. Using rare variant as well as polymorphic alleles simultaneously, Kimura (no. 56) estimated the average proportion to be 14% (cf. Nei 1977). The proportion varies from species to species (humans, fruit flies, Japanese monkeys, plaice) or between substrate-specific and nonspecific enzymes, but the 14% neutral mutation is in good agreement with an estimate from globin DNA sequence data. Kimura's estimator was shown to be biased and affected much by deleterious alleles and changes in population size (Watterson 1987; Ewens and Li 1980). Nevertheless, it was a pivotal question to ask what fraction of the total mutation is neutral.

To answer such a question, DNA sequence data rather than allele frequency data provide a more direct means. Kimura (no. 57) reviewed several findings at the DNA level in relation to the neutral theory, and stressed the theory's value as a scientific hypothesis. The neutral theory makes various testable predictions and therefore enables us to examine observations in quantitative terms. These include (1) high substitution rates at the synonymous sites in the protein coding regions and at all sites in introns as well as pseudogenes, (2) rough constancy of synonymous rates in various genes, (3) a preponderance of synonymous substitutions over amino acid replacements even in rapidly evolving RNA viruses (e.g., Gojobori, Moriyama, and Kimura 1990), and (4) unequal usage of synonymous codons (e.g., Ikemura 1985). Kimura (no. 57) has become more and more confident of the neutral theory in its simplest form: "A certain fraction (presumably a large fraction) of nucleotide sites produce no phenotypic effects at all, and therefore are completely neutral while a certain fraction (presumably a very small fraction) of nucleotide or amino acid substitutions are definitely advantageous for the species in adapting to new environments, and therefore they are subjected to straightforward positive natural selection." In this view, there is little or no room for slightly deleterious mutations as being important in molecular evolution.

# Evolutionary Rate at the Molecular Level

## Calculating the rate of evolution in terms of nucleotide substitutions seems to give a value so high that many of the mutations involved must be almost neutral ones

by

MOTOO KIMURA

National Institute of Genetics, Mishima, Japan

COMPARATIVE studies of haemoglobin molecules among different groups of animals suggest that, during the evolutionary history of mammals, amino-acid substitution has taken place roughly at the rate of one amino-acid change in $10^7$ yr for a chain consisting of some 140 amino-acids. For example, by comparing the α and β chains of man with those of horse, pig, cattle and rabbit, the figure of one amino-acid change in $7 \times 10^6$ yr was obtained[1]. This is roughly equivalent to the rate of one amino-acid substitution in $10^7$ yr for a chain consisting of 100 amino-acids.

A comparable value has been derived from the study of the haemoglobin of primates[2]. The rate of amino-acid substitution calculated by comparing mammalian and avian cytochrome *c* (consisting of about 100 amino-acids) turned out to be one replacement in $45 \times 10^6$ yr (ref. 3). Also by comparing the amino-acid composition of human triosephosphate dehydrogenase with that of rabbit and cattle[4], a figure of at least one amino-acid substitution for every $2{\cdot}7 \times 10^6$ yr can be obtained for the chain consisting of about 1,110 amino-acids. This figure is roughly equivalent to the rate of one amino-acid substitution in $30 \times 10^6$ yr for a chain consisting of 100 amino-acids. Averaging those figures for haemoglobin, cytochrome *c* and triosephosphate dehydrogenase gives an evolutionary rate of approximately one substitution in $28 \times 10^6$ yr for a polypeptide chain consisting of 100 amino-acids.

I intend to show that this evolutionary rate, although appearing to be very low for each polypeptide chain of a size of cytochrome *c*, actually amounts to a very high rate for the entire genome.

First, the DNA content in each nucleus is roughly the same among different species of mammals such as man, cattle and rat (see, for example, ref. 5). Furthermore, we note that the G–C content of DNA is fairly uniform among mammals, lying roughly within the range of 40–44 per cent[6]. These two facts suggest that nucleotide substitution played a principal part in mammalian evolution.

In the following calculation, I shall assume that the haploid chromosome complement comprises about $4 \times 10^9$ nucleotide pairs, which is the number estimated by Muller[7] from the DNA content of human sperm. Each amino-acid is coded by a nucleotide triplet (codon), and so a polypeptide chain of 100 amino-acids corresponds to 300 nucleotide pairs in a genome. Also, amino-acid replacement is the result of nucleotide replacement within a codon. Because roughly 20 per cent of nucleotide replacement caused by mutation is estimated to be synonymous[8], that is, it codes for the same amino-acid, one amino-acid replacement may correspond to about 1·2 base pair replacements in the genome. The average time taken for one base pair replacement within a genome is therefore

$$28 \times 10^6 \text{ yr} \div \left(\frac{4 \times 10^9}{300}\right) \div 1{\cdot}2 \doteqdot 1{\cdot}8 \text{ yr}$$

This means that in the evolutionary history of mammals, nucleotide substitution has been so fast that, on average, one nucleotide pair has been substituted in the population roughly every 2 yr.

This figure is in sharp contrast to Haldane's well known estimate[9] that, in horotelic evolution (standard rate evolution), a new allele may be substituted in a population roughly every 300 generations. He arrived at this figure by assuming that the cost of natural selection per generation (the substitutional load in my terminology[10]) is roughly 0·1, while the total cost for one allelic substitution is about 30. Actually, the calculation of the cost based on Haldane's formula shows that if new alleles produced by nucleotide replacement are substituted in a population at the rate of one substitution every 2 yr, then the substitutional load becomes so large that no mammalian species could tolerate it.

Thus the very high rate of nucleotide substitution which I have calculated can only be reconciled with the limit set by the substitutional load by assuming that most mutations produced by nucleotide replacement are almost neutral in natural selection. It can be shown that in a population of effective size $N_e$, if the selective advantage of the new allele over the pre-existing alleles is $s$, then. assuming no dominance, the total load for one gene substitution is

$$L(p) = 2\left\{\frac{1}{u(p)} - 1\right\}$$

$$\int_0^{4Sp} \frac{e^y - 1}{y}\,dy - 2e^{-4S}\int_{4Sp}^{4S} \frac{e^y}{y}\,dy + 2\log_e\left(\frac{1}{p}\right) \qquad (1)$$

where $S = N_e s$ and $p$ is the frequency of the new allele at the start. The derivation of the foregoing formula will be published elsewhere. In the expression given here $u(p)$ is the probability of fixation given by[11]

$$u(p) = (1 - e^{-4Sp})/(1 - e^{-4S}) \qquad (2)$$

Now, in the special case of $|2N_e s| \ll 1$, formulae (1) and (2) reduce to

$$L(p) = 8N_e s \log_e(1/p) \qquad (1')$$

$$u(p) = p + 2N_e s p(1-p) \qquad (2')$$

Formula (1′) shows that for a nearly neutral mutation the substitutional load can be very low and there will be no limit to the rate of gene substitution in evolution. Furthermore, for such a mutant gene, the probability of fixation (that is, the probability by which it will be established in the population) is roughly equal to its initial frequency as shown by equation (2′). This means that new alleles may be produced at the same rate per individual as they are substituted in the population in evolution.

This brings the rather surprising conclusion that in mammals neutral (or nearly neutral) mutations are occurring at the rate of roughly 0·5 per yr per gamete. Thus, if we take the average length of one generation in the history of mammalian evolution as 4 yr, the mutation rate per generation for neutral mutations amounts to roughly two per gamete and four per zygote ($5 \times 10^{-10}$ per nucleotide site per generation).

Such a high rate of neutral mutations is perhaps not surprising, for Mukai[12] has demonstrated that in *Drosophila* the total mutation rate for "viability polygenes" which on the average depress the fitness by about 2 per cent reaches at least some 35 per cent per gamete. This is a much higher rate than previously considered. The fact that neutral or nearly neutral mutations are occurring at a rather high rate is compatible with the high frequency of heterozygous loci that has been observed recently by studying protein polymorphism in human and *Drosophila* populations[13–15].

Lewontin and Hubby[15] estimated that in natural populations of *Drosophila pseudoobscura* an average of about 12 per cent of loci in each individual is heterozygous. The corresponding heterozygosity with respect to nucleotide sequence should be much higher. The chemical structure of enzymes used in this study does not seem to be known at present, but in the typical case of esterase-5 the molecular weight was estimated to be about $10^5$ by Narise and Hubby[16]. In higher organisms, enzymes with molecular weight of this magnitude seem to be common and usually they are "multimers"[17]. So, if we assume that each of those enzymes comprises on the average some 1,000 amino-acids (corresponding to molecular weight of some 120,000), the mutation rate for the corresponding genetic site (consisting of about 3,000 nucleotide pairs) is

$$u = 3 \times 10^3 \times 5 \times 10^{-10} = 1{\cdot}5 \times 10^{-6}$$

per generation. The entire genome could produce more than a million of such enzymes.

In applying this value of $u$ to *Drosophila* it must be noted that the mutation rate per nucleotide pair per generation can differ in man and *Drosophila*. There is some evidence that with respect to the definitely deleterious effects of gene mutation, the rate of mutation per nucleotide pair per generation is roughly ten times as high in *Drosophila* as in man[18,19]. This means that the corresponding mutation rate for *Drosophila* should be $u = 1{\cdot}5 \times 10^{-5}$ rather than $u = 1{\cdot}5 \times 10^{-6}$. Another consideration allows us to suppose that $u = 1{\cdot}5 \times 10^{-5}$ is probably appropriate for the neutral mutation rate of a cistron in *Drosophila*. If we assume that the frequency of occurrence of neutral mutations is about one per genome per generation (that is, they are roughly two to three times more frequent than the mutation of the viability polygenes), the mutation rate per nucleotide pair per generation is $1/(2 \times 10^8)$, because the DNA content per genome in *Drosophila* is about one-twentieth of that of man[20]. For a cistron consisting of 3,000 nucleotide pairs, this amounts to $u = 1{\cdot}5 \times 10^{-5}$.

Kimura and Crow[21] have shown that for neutral mutations the probability that an individual is homozygous is $1/(4N_e u + 1)$, where $N_e$ is the effective population number, so that the probability that an individual is heterozygous is $H_e = 4N_e u/(4N_e u + 1)$. In order to attain at least $H_e = 0{\cdot}12$, it is necessary that at least $N_e = 2{,}300$. For a higher heterozygosity such as $H_e = 0{\cdot}35$, $N_e$ has to be about 9,000. This might be a little too large for the effective number in *Drosophila*, but with migration between subgroups, heterozygosity of 35 per cent may be attained even if $N_e$ is much smaller for each subgroup.

We return to the problem of total mutation rate. From a consideration of the average energy of hydrogen bonds and also from the information on mutation of *rIIA* gene in phage $T_4$, Watson[22] obtained $10^{-8} \sim 10^{-9}$ as the average probability of error in the insertion of a new nucleotide during DNA replication. Because in man the number of cell divisions along the germ line from the fertilized egg to a gamete is roughly 50, the rate of mutation resulting from base replacement according to these figures may be $50 \times 10^{-8} \sim 50 \times 10^{-9}$ per nucleotide pair per generation. Thus, with $4 \times 10^9$ nucleotide pairs, the total number of mutations resulting from base replacement may amount to $200 \sim 2{,}000$. This is 100–1,000 times larger than the estimate of 2 per generation and suggests that the mutation rate per nucleotide pair is reduced during evolution by natural selection[18,19].

Finally, if my chief conclusion is correct, and if the neutral or nearly neutral mutation is being produced in each generation at a much higher rate than has been considered before, then we must recognize the great importance of random genetic drift due to finite population number[23] in forming the genetic structure of biological populations. The significance of random genetic drift has been deprecated during the past decade. This attitude has been influenced by the opinion that almost no mutations are neutral, and also that the number of individuals forming a species is usually so large that random sampling of gametes should be negligible in determining the course of evolution, except possibly through the "founder principle"[24]. To emphasize the founder principle but deny the importance of random genetic drift due to finite population number is, in my opinion, rather similar to assuming a great flood to explain the formation of deep valleys but rejecting a gradual but long lasting process of erosion by water as insufficient to produce such a result.

Received December 18, 1967.

1 Zuckerkandl, E., and Pauling, L., in *Evolving Genes and Proteins* (edit. by Bryson, V., and Vogel, H. J.), 97 (Academic Press, New York, 1965).
2 Buettner-Janusch, J., and Hill, R. L., in *Evolving Genes and Proteins* (edit. by Bryson, V., and Vogel, H. J.), 167 (Academic Press, New York, 1965).
3 Margoliash, E., and Smith, E. L., in *Evolving Genes and Proteins* (edit. by Bryson, V., and Vogel, H. J.), 221 (Academic Press, New York, 1965).
4 Kaplan, N. O., in *Evolving Genes and Proteins* (edit. by Bryson, V., and Vogel, H. J.), 243 (Academic Press, New York, 1965).
5 Sager, R., and Ryan, F. J., *Cell Heredity* (John Wiley and Sons, New York, 1961).
6 Sueoka, N., *J. Mol. Biol.*, **3**, 31 (1961).
7 Muller, H. J., *Bull. Amer. Math. Soc.*, **64**, 137 (1958).
8 Kimura, M., *Genet. Res.* (in the press).
9 Haldane, J. B. S., *J. Genet.*, **55**, 511 (1957).
10 Kimura, M., *J. Genet.*, **57**, 21 (1960).
11 Kimura, M., *Ann. Math. Stat.*. **28**, 882 (1957).
12 Mukai, T., *Genetics*, **50**, 1 (1964).

[13] Harris, H., *Proc. Roy. Soc.*, B, **164**, 298 (1966).
[14] Hubby, J. L., and Lewontin, R. C., *Genetics*, **54**, 577 (1966).
[15] Lewontin, R. C., and Hubby, J. L., *Genetics*, **54**, 595 (1966).
[16] Narise, S., and Hubby, J. L., *Biochim. Biophys. Acta*, **122**, 281 (1966).
[17] Fincham, J. R. S., *Genetic Complementation* (Benjamin, New York, 1966).
[18] Muller, H. J., in *Heritage from Mendel* (edit. by Brink, R. A.), 419 (University of Wisconsin Press, Madison, 1967).
[19] Kimura, M., *Genet. Res.*, **9**, 23 (1967).
[20] *Report of the United Nations Scientific Committee on the Effects of Atomic Radiation* (New York, 1958).
[21] Kimura, M., and Crow, J. F., *Genetics*, **49**, 725 (1964).
[22] Watson, J. D., *Molecular Biology of the Gene* (Benjamin, New York, 1965).
[23] Wright, S., *Genetics*, **16**, 97 (1931).
[24] Mayr, E., *Animal Species and Evolution* (Harvard University Press, Cambridge, 1965).

# Genetic variability maintained in a finite population due to mutational production of neutral and nearly neutral isoalleles*

By MOTOO KIMURA

*National Institute of Genetics, Mishima, Japan*

(*Received* 28 *July* 1967)

## 1. INTRODUCTION

It is well known that populations of sexually reproducing organisms such as man and *Drosophila* contain a large amount of genetic variability. Ubiquity of lethal and detrimental genes has been demonstrated in various species of *Drosophila*. Inbreeding studies suggest that the same situation is met with also in man and other organisms. The existence of genetic variability in quantitative characters has been amply demonstrated by selection experiments with diverse plants and animals. Moreover, recent studies on enzyme polymorphism in man and *Drosophila* (Harris, 1966; Lewontin & Hubby, 1966) strongly suggest that genetic variability is quite pronounced at the protein level. It is probable that at the level of genetic material, or in terms of nucleotide sequence, variability within a population is still greater.

Since each gene is made up of a sequence of at least hundreds or thousands of nucleotide pairs and since some base substitutions may have very little effect, it is possible, as reasoned by Kimura & Crow (1964), that the wild-type gene is not a single entity, but a set of different isoalleles that are indistinguishable by any ordinary procedure. They investigated the population consequences of such a system, assuming neutral and overdominant mutations. More recently, Wright (1966) discussed the evolutionary implications of such a system under the term 'polyallelic random drift'.

The purpose of the present paper is to present a fuller treatment of this system for neutral mutations. Also, some discussion on the nearly neutral mutations will be presented. The recent findings of 'degeneracy' of DNA code, that is, existence of two or more base triplets coding for the same amino acid, seem to suggest that neutral mutations may not be as rare as previously considered. Furthermore, some amino acid substitution in a polypeptide chain may have very little effect on the biological activity of the protein, still adding to the possibility of neutral or nearly neutral mutations.

* Contribution No. 648 from the National Institute of Genetics, Mishima, Shizuoka-ken, Japan. Aided in part by a Grant-in-Aid from the Ministry of Education, Japan, and also by a Grant from Toyo Rayon Foundation.

## 2. AVERAGE HOMOZYGOSITY AND THE EFFECTIVE NUMBER OF ALLELES IN A POPULATION

Throughout this paper, I will consider a population of $N$ diploid individuals and designate by $N_e$ the effective population number (cf. Kimura & Crow, 1963), which may be different from the actual number $N$.

Let us consider a particular locus and assume that there are $K$ possible allelic states $A_1, A_2, \ldots, A_K$ and that each allele mutates with rate $u/(K-1)$ to one of the remaining $(K-1)$ alleles, so that $u$ is the mutation rate per gene per generation and this is equal to all the alleles. Then, if $x_i$ is the frequency of $A_i$ in a population, the amount of change in one generation of $x_i$ denoted by $\delta x_i$ has mean and variance

$$M(\delta x_i) = \frac{u}{K-1}(1-Kx_i), \tag{2.1}$$

$$V(\delta x_i) = x_i(1-x_i)/(2N_e). \tag{2.2}$$

Thus, it can be shown with rather elementary but exact calculation that, at equilibrium in which the mutation and random sampling of gametes balance each other, the frequency distribution of $x_i$ has the first and the second moments about zero as follows:

$$\mu_1' = 1/K, \tag{2.3}$$

$$\mu_2' = \left\{\frac{4N_e u}{K(K-1)} - \frac{2N_e u^2}{(K-1)^2} + \frac{1}{K}\right\} \Big/ \left\{\frac{4N_e uK}{K-1} - \frac{2N_e u^2 K^2}{(K-1)^2} + 1\right\}. \tag{2·4}$$

Therefore the average homozygosity, or the expectation of the sum of squares of allelic frequencies is

$$\bar{H}_0 = E\left(\sum_{i=1}^{K} x_i^2\right) = K\mu_2', \tag{2.5}$$

with $\mu_2'$ given by (2.4).

The effective number of alleles ($n_e$) as defined by Kimura & Crow (1964) is the reciprocal of $\bar{H}_0$, so that

$$n_e \equiv 1/\bar{H}_0 = \left\{4N_e u\left(\frac{K}{K-1}\right) - 2N_e u^2\left(\frac{K}{K-1}\right)^2 + 1\right\} \Big/ \left\{\frac{4N_e u}{K-1} - \frac{2N_e u^2 K}{(K-1)^2} + 1\right\}. \tag{2.6}$$

If $2N_e u^2$ is much smaller than unity, we have, with good approximation,

$$n_e = \left\{4N_e u\left(\frac{K}{K-1}\right) + 1\right\} \Big/ \left\{4N_e u\left(\frac{1}{K-1}\right) + 1\right\}. \tag{2.7}$$

If, in addition, the number of allelic states is indefinitely large ($K = \infty$), the above reduces to

$$n_e = 1/\bar{H}_0 = 4N_e u + 1, \tag{2.8}$$

a result derived by Kimura & Crow (1964) using a different method. Actually, the formula is valid as long as the number of allelic states $K$ is much larger than $4N_e u + 1$.

On the other hand, if $2N_e u^2$ is not necessarily very small but $K = \infty$, (2.6) reduces to

$$n_e = 4N_e u + 1 - 2N_e u^2. \tag{2.9}$$

Since formula (2.6) is exact and no restrictions are placed on mutation rate $u$, effective population number $N_e$ and the number of possible allelic states $K$, it is reassuring to find here that formula (2.8), i.e. $n_e = 4N_e u + 1$ is valid under rather mild restrictions

$$2N_e u^2 \ll 4N_e u + 1 \ll K. \tag{2.10}$$

It is sometimes remarked that a formula like (2.8) is valid only for $u$ up to $1/N_e$, but no such restriction is needed.

### 3. PROBABILITY DISTRIBUTION OF ALLELIC FREQUENCIES AND THE AVERAGE NUMBER OF ALLELES IN A POPULATION

In this section, we will investigate the distribution of allelic frequencies using the method of diffusing approximation (cf. Kimura, 1964). Let $\phi(p, x; t)$ be the probability density that the frequency of $A_i$ becomes $x$ at the $t$th generation given that it is $p$ at the zero (initial) generation. In the following, in order to simplify expressions, letter $x$ rather than $x_i$ will be used to represent the frequency of $A_i$, still assuming that there are $K$ possible allelic states. Since, from (2.1) and (2.2), the mean and the variance of $\delta x$ per generation are respectively

$$M_{\delta x} = \frac{Ku}{K-1}\left(\frac{1}{K} - x\right) \tag{3.1}$$

and

$$V_{\delta x} = \frac{x(1-x)}{2N_e}, \tag{3.2}$$

$\phi(p, x; t)$ satisfies the following Kolmogorov forward equation

$$\frac{\partial \phi}{\partial t} = \frac{1}{4N_e}\frac{\partial^2}{\partial x^2}\left\{x(1-x)\phi\right\} - \bar{m}\frac{\partial}{\partial x}\left\{(\bar{x} - x)\phi\right\}, \tag{3.3}$$

where $\bar{m} = Ku/(K-1)$ and $\bar{x} = 1/K$.

The above equation represents a continuous stochastic process in the change of gene frequency $x$ due to linear evolutionary pressures (mutation, migration) and random sampling of gametes. The solution of this process for arbitrary values of $p$, $\bar{m}$, $\bar{x}$ and $N_e$ was obtained by the present author through the study of the moments of the distribution (cf. Crow & Kimura, 1956). It is given by

$$\phi(p, x; t) = \sum_{1=0}^{\infty} X_i(x) \exp\left\{-i\left(\bar{m} + \frac{i-1}{4N_e}\right)t\right\}, \tag{3.4}$$

where $X_i(x) = x^{B-1}(1-x)^{(A-B)-1}F(A+i-1, -i, A-B, 1-x)$

$$\times F(A+i-1, -i, A-B, 1-p)\frac{\Gamma(A-B+i)\Gamma(A+2i)\Gamma(A+i-1)}{i!\Gamma^2(A-B)\Gamma(B+i)\Gamma(A+2i-1)},$$

in which $A = 4N_e\bar{m}$, $B = 4N_e\bar{m}\bar{x}$ and $F(.,.,.,.)$ represents the hypergeometric function. For the present case $A = 4N_e uK/(K-1)$ and $B = 4N_e u/(K-1)$.

At the limit $t \to \infty$, the above distribution converges to

$$\phi(p, x; \infty) = X_0(x),$$

which is independent of the initial frequency $p$. We will denote this distribution by $\phi(x)$. Thus we obtain

$$\phi(x) = \frac{\Gamma(\alpha+\beta)}{\Gamma(\alpha)\Gamma(\beta)}(1-x)^{\alpha-1}x^{\beta-1}, \qquad (3{\cdot}5)$$

where $\alpha = A-B = 4N_e u$ and $\beta = B = 4N_e u/(K-1)$.

The first and the second moments about zero of this distribution are

$$\mu_1' = 1/K \qquad (3.6)$$

and

$$\mu_2' = \frac{1}{K}\left\{4N_e u\left(\frac{1}{K-1}\right)+1\right\}\Big/\left\{4N_e u\left(\frac{K}{K-1}\right)+1\right\} \qquad (3.7)$$

respectively.

The distribution given by (3.5) is the steady-state distribution which is realized when the effects of mutation (or migration) and random sampling of gametes balance each other. It can also be derived by using Wright's formula for the gene frequency distribution at steady state, namely,

$$\phi(x) = \frac{C}{V_{\delta x}}\exp\left(2\int\frac{M_{\delta x}}{V_{\delta x}}dx\right) \qquad \text{(Wright, 1938}a\text{)}, \quad (3.8)$$

in which constant $C$ is determined such that

$$\int_0^1 \phi(x)dx = 1. \qquad (3.9)$$

Going back to the general solution (3.4), we note, as pointed out earlier (Crow & Kimura, 1956), that

$$\int_0^1 \phi(p, x; t)dx = 1. \qquad (3.10)$$

This means that, for the present case, the procedure (3.9) of determining $C$ is not an arbitrary statistical procedure, but is the one intrinsically determined by the process. Also, it means that with the present formulation no probability mass exists at any time strictly at the boundaries, i.e. at both $x = 0$ and $x = 1$. On the other hand, in an actual population, especially when it is small, we should expect a considerable possibility of an allele being temporarily lost or fixed in the population.

It looks, then, as if the above approach based on the diffusion approximation is inadequate to obtain the probability of temporary loss or fixation. Fortunately, however, it turns out that the required probability mass lies in the intervals

(0, $1/2N$) and ($1-1/2N$, 1). Thus, the probability that allele $A_i$ is temporarily lost from the population may be obtained from

$$f(0) = \int_0^{1/(2N)} \phi(x)dx \tag{3.11}$$

and the probability that it is temporarily fixed in the population from

$$f(1) = \int_{1-1/(2N)}^{1} \phi(x)dx. \tag{3.12}$$

The probability that both $A_i$ and its alleles (collectively denoted by $A_i'$) co-exist in the population is

$$\Omega = \int_{1/(2N)}^{1-1/(2N)} \phi(x)dx. \tag{3.13}$$

If we substitute the distribution formula (3.5) into (3.11), we obtain

$$f(0) = \frac{\Gamma(\alpha+\beta)}{\Gamma(\alpha)\Gamma(\beta)}\int_0^{1/(2N)} (1-x)^{\alpha-1}x^{\beta-1}\,dx \tag{3.14}$$

as the probability that $A_i$ is temporarily lost from the population. Since $u$ and $1/(2N)$ are generally very small, the above reduces, with good approximation, to

$$f(0) = \frac{\Gamma(\alpha+\beta)}{\Gamma(\alpha)\Gamma(\beta+1)}\left(\frac{1}{2N}\right)^{\beta}, \tag{3.15}$$

where $\alpha = 4N_e u$ and $\beta = 4N_e u/(K-1)$. Similarly, the probability that $A_i$ is temporarily fixed is

$$f(1) = \frac{\Gamma(\alpha+\beta)}{\Gamma(\alpha+1)\Gamma(\beta)}\left(\frac{1}{2N}\right)^{\alpha}. \tag{3.16}$$

Now, formula (3.15) may be expressed in the form

$$\tfrac{1}{2}\beta\left(\frac{N}{N_e}\right)f(0) = \frac{1}{2}\left(\frac{N}{N_e}\right)\frac{\Gamma(\alpha+\beta)}{\Gamma(\alpha)\Gamma(\beta)}\left(\frac{1}{2N}\right)^{\beta-1}\frac{1}{2N},$$

but, since

$$\phi\left(\frac{1}{2N}\right) = \frac{\Gamma(\alpha+\beta)}{\Gamma(\alpha)\Gamma(\beta)}\left(\frac{1}{2N}\right)^{\beta-1}$$

approximately, it may also be expressed as

$$2N\left(\frac{u}{K-1}\right)f(0) = \frac{1}{2}\left\{\phi\left(\frac{1}{2N}\right)\frac{1}{2N}\right\}\left(\frac{N}{N_e}\right). \tag{3.17}$$

The left-hand side of the above equation represents the number of populations which have no $A_i$ genes which move to the class having one or more of them, since $2Nu/(K-1)$ is the expected number of $A_i$ genes produced per generation in a population where $A_i$ is absent. On the other hand, the right-hand side of the equation is half the frequency of the subterminal class ($x = 1/2N$) multiplied by the factor $N/N_e$ and it represents the number of populations containing one or more $A_i$

genes which move to the class where $A_i$ is absent (cf. Kimura, 1964, p. 12). At statistical equilibrium in which mutational production of an allele is balanced by random extinction of that allele, the above two numbers should be equal and this justifies equation (3.17) and therefore (3.15). A similar argument applies to (3.16).

I will now proceed to derive the effective and the average number of alleles maintained in a population, using the above approach.

The effective number of alleles, as defined in the previous selection, is the reciprocal of the sum of squares of the allelic frequencies. The latter is

$$\bar{H}_0 = E\left(\sum_{i=1}^{K} x_i^2\right) = K\mu_2' = K\int_0^1 x^2\phi(x)dx = \left\{4N_e u\left(\frac{1}{K-1}\right)+1\right\}\Big/\left\{4N_e u\left(\frac{K}{K-1}\right)+1\right\} \tag{3.18}$$

and this gives the effective allele number

$$n_e = 1/\bar{H}_0 = \left\{4N_e u\left(\frac{K}{K-1}\right)+1\right\}\Big/\left\{4N_e u\left(\frac{1}{K-1}\right)+1\right\}, \tag{3.19}$$

which agrees with (2.7) of the previous section. At the limit $K \to \infty$, where the number of possible allelic states is infinite, this reduces to $n_e = 4N_e u + 1$.

The average number of alleles denoted by $n_a$ is equal to the reciprocal of the mean frequency of alleles existing within a population. The mean here is different from the unconditional mean ($\mu_1'$) in that temporarily lost alleles are not taken into account. The frequency of a particular allele, say $A_i$, averaged over all cases in which it is represented at least once in a population is

$$\bar{x}[x \neq 0] = \mu_1'/\{1-f(0)\}. \tag{3.20}$$

In the present model of assuming equal mutation rates, this value is the same for all the alleles and therefore the average number of alleles turns out to be as follows:

$$n_a = K\{1-f(0)\} = \frac{K\Gamma(\alpha+\beta)}{\Gamma(\alpha)\Gamma(\beta)}\int_{1/(2N)}^{1}(1-x)^{\alpha-1}x^{\beta-1}\,dx, \tag{3.21}$$

where $\alpha = 4N_e u$ and $\beta = 4N_e u/(K-1)$.

Figure 1 illustrates the relation between $n_a$ and $K$ assuming $4N_e u = 1$ and $N = 10^4$, together with the relation between $n_e$ and $K$. At the limit $K \to \infty$, the above formula for $n_a$ converges to

$$n_a = 4N_e u\int_{1/(2N)}^{1}(1-x)^{4N_e u-1}x^{-1}\,dx. \tag{3.22}$$

This can also be derived immediately by the frequency distribution given by Kimura & Crow (1964),

$$\Phi(x) = 4M(1-x)^{4M-1}x^{-1}, \tag{3.23}$$

where $M = N_e u$, in which $u$ is the mutation rate to new (not pre-existing) alleles. This distribution has a different meaning from the one so far considered, such as (3.5), in that $\Phi(x)dx$ represents the *expected number of alleles* whose frequency is in the range $x$ to $x+dx$ within the population, rather than representing the probability that *a particular allele* lies in the frequency range $x$ to $x+dx$. Thus, integrating (3.23) from $x = 1/(2N)$ to $x = 1$, we obtain

$$n_a = 4M \int_{1/(2N)}^{1} (1-x)^{4M-1} x^{-1}\, dx, \tag{3.24}$$

which agrees with (3.22). In the special case of $N = N_e$, this reduces a formula given by Ewens (1964, 1966), who derived it by considering the fate of a new allele. Essentially the same formula as that of Ewens was obtained earlier by Wright

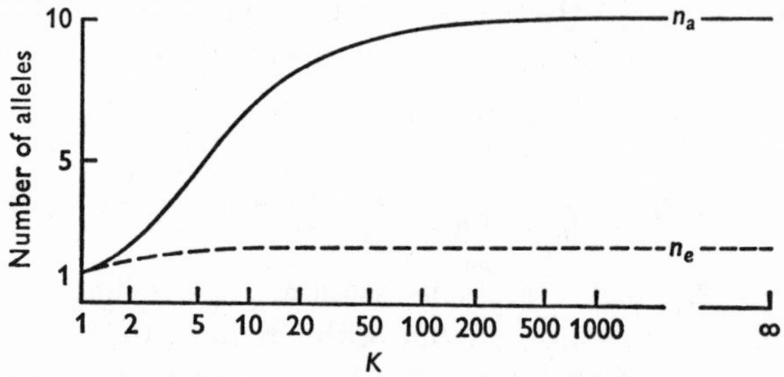


Fig. 1. Relationship between the number of alleles ($n_a$, $n_e$) and the number of possible allelic states ($K$) in a population of $N = 10\,000$, assuming $N_e u = 0{\cdot}25$ ($N_e$; effective population number, $u$; mutation rate). The solid line represents the relationship between the actual number of alleles ($n_a$) and $K$, while the broken line represents the relationship between the effective number of alleles ($n_e$) and $K$.

(1949). It is important to note, however, that in natural as well as controlled populations, the actual population number ($N$) may be considerably different from the effective population number ($N_e$).

The effective number of alleles is obtained from the above distribution (3.23) by evaluating the reciprocal of

$$\int_0^1 x^2 \Phi(x)\, dx,$$

giving

$$n_e = 4M+1 = 4N_e u+1 \tag{3.25}$$

(Kimura & Crow, 1964). This agrees with formula (2.8) in the previous section. Table 1 lists the average number of alleles obtained by numerical integration of formula (3.24), together with the effective number of alleles derived from (3.25). Also, the relation between the number of alleles and $N_e u$ is illustrated in Fig. 2.

## 4. SOME SIMULATION STUDIES

In order to check the validity of the foregoing treatments, simulation studies were carried out by using computer IBM 7090. Two programmes (both in Fortran II) were written that differ essentially in the mode of production of mutant genes. In the first program, a pre-determined number of new mutant alleles are introduced into the population in each generation (deterministic mutation). Also, it is so written that all members of the population contribute equally to the gene pool, from which $2N$ gametes are randomly sampled to form the next generation. Thus, the program simulates a monoecious population whose effective number is equal to the actual number, i.e. $N = N_e$. In the second program, mutation to a new allele is induced with a given probability ($u$) at each step of gamete sampling (random mutation). The program is so written that the population consists of an

Table 1. *The average number* ($n_a$) *and the effective number* ($n_e$) *of alleles in a population of actual size N and effective size* $N_e$

| $N_e u$ \ $N$ | $5\times10^2$ | $10^3$ | $5\times10^3$ | $10^4$ | $5\times10^4$ | $10^5$ | $5\times10^5$ | $10^6$ | Effective number of alleles $n_e$ |
|---|---|---|---|---|---|---|---|---|---|
| 0·001 | 1·028 | 1·030 | 1·037 | 1·040 | 1·046 | 1·049 | 1·055 | 1·058 | 1·004 |
| 0·010 | 1·274 | 1·301 | 1·366 | 1·394 | 1·458 | 1·486 | 1·550 | 1·578 | 1·040 |
| 0·025 | 1·675 | 1·745 | 1·906 | 1·975 | 2·136 | 2·205 | 2·367 | 2·436 | 1·100 |
| 0·050 | 2·324 | 2·462 | 2·784 | 2·923 | 3·245 | 3·384 | 3·705 | 3·844 | 1·200 |
| 0·100 | 3·557 | 3·834 | 4·478 | 4·755 | 5·399 | 5·676 | 6·320 | 6·597 | 1·400 |
| 0·250 | 6·908 | 7·601 | 9·210 | 9·903 | 11·51 | 12·21 | 13·82 | 14·51 | 2·000 |
| 0·500 | 11·82 | 13·20 | 16·42 | 17·81 | 21·03 | 22·41 | 25·63 | 27·02 | 3·000 |
| 1·000 | 20·31 | 23·08 | 29·51 | 32·28 | 38·72 | 41·49 | 47·93 | 50·70 | 5·000 |
| 2·000 | 34·58 | 40·09 | 52·95 | 58·50 | 71·36 | 76·91 | 89·78 | 95·33 | 9·000 |
| 4·000 | 57·67 | 68·64 | 94·30 | 105·4 | 131·1 | 142·2 | 168·0 | 179·0 | 17·000 |
| 6·000 | 76·72 | 93·08 | 131·5 | 148·1 | 186·7 | 203·3 | 242·0 | 258·6 | 25·00 |
| 8·000 | 93·18 | 114·9 | 166·0 | 188·1 | 239·6 | 261·7 | 313·2 | 335·4 | 33·00 |
| 10·000 | 107·7 | 134·7 | 198·5 | 226·1 | 290·4 | 318·1 | 382·5 | 410·2 | 41·00 |

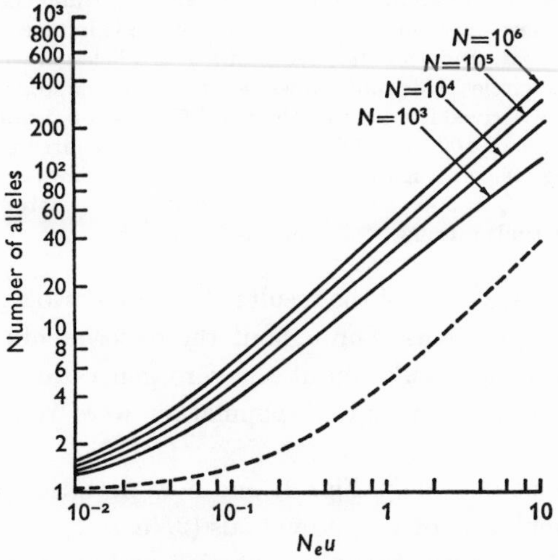


Fig. 2. Graphs showing the relationship between the number of alleles and $N_e u$, where $N_e$ is the effective population number and $u$ is the mutation rate. The solid lines give the average number of alleles ($n_a$) corresponding to four levels of population number, $N = 10^3$, $10^4$, $10^5$ and $10^6$, while the broken line gives the effective number of alleles ($n_e$), which is independent of $N$.

equal number of males and females and that the numbers of breeding males and females may be made smaller than the actual numbers of males and females. Thus it simulates a dioecious population whose effective number may be smaller than the actual number. The effective population number here is given by Wright's formula (Wright, 1938*b*),

$$N_e = 4N_m N_f/(N_m + N_f),$$

where $N_m$ and $N_f$ are the numbers of breeding males and females.

In both programs, sampling of gametes and occurrence of mutation are simulated by generating pseudo-random numbers (using subroutine RAND1). Also, each mutation is treated as a state not pre-existing in the population, so that three formulas, (3.23), (3.24) and (3.25), are relevant in comparing theoretical expectations with computer results. Outputs of both the actual and effective allele numbers were given at pre-assigned intervals. Also, frequency distribution of various alleles within a population was printed out.

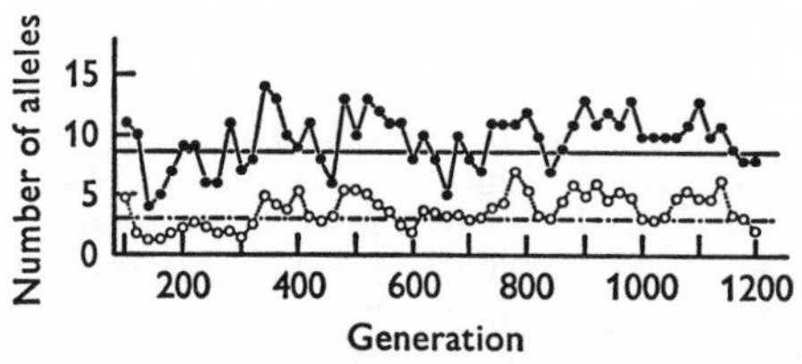


Fig. 3

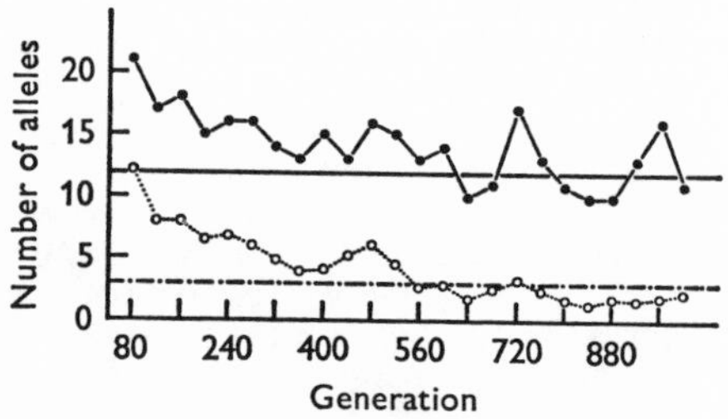


Fig. 4

Figs. 3 and 4. Results of Monte Carlo experiments regarding the number of neutral alleles. In both these experiments, one new mutation is induced in each generation. The average (actual) and the effective number of alleles are plotted respectively by solid and open circles. Horizontal lines, solid and broken, represent corresponding theoretical values derived from the method of diffusion approximation. In Fig. 3, the population consists of 100 monoecious individuals who contribute with equal probability to leaving offspring, namely, $N = N_e = 100$. The mutation rate ($u$) is 0·005. In Fig. 4, $N = N_e = 500$, $u = 0{\cdot}001$. Neutral alleles deterministic mutation: ●—, average (actual) number, ○ - - -, effective number.

Figures 3–6 illustrate some of the results of Monte Carlo experiments performed by using these two programs. Throughout the experiments, the initial condition was set up such that a population at the zero generation contained $2N$ alleles, that is, all the genes in the initial population were represented by different alleles.

In the experiment shown in Fig. 3, one new mutation was induced in each generation in a population of 100 individuals ($2Nu = 1$, $N = N_e = 100$). Starting with 200 alleles, the balance between mutation and random extinction of alleles has been reached well before generation 100. Actually, a few trials indicated that the majority of the initial 200 alleles are lost within the first 20 generations. Both the average (actual) and the effective numbers of alleles are plotted in the figure at intervals of 20 generations from generation 100 through generation 1200

(56 outputs for each of these two allele numbers). Averaged over these 56 outputs, the average and the effective numbers of alleles turned out to be as follows:

$$n_a = 9{\cdot}68, \quad n_e = 3{\cdot}13 \quad \text{(observed)}.$$

The former was obtained by taking the arithmetic mean of the 56 observed values for the actual allele number, while the latter was obtained by taking the harmonic mean of the 56 observed values for the effective allele number. The corresponding values for $n_a$ and $n_e$, derived from equations (3.24) and (3.25) by putting $N = 100$ and $N_e u = 0{\cdot}5$, are as follows:

$$n_a = 8{\cdot}61, \quad n_e = 3{\cdot}00 \quad \text{(theoretical)}.$$

These are shown by the horizontal lines in the figure.

Figure 4 illustrates a result of a similar experiment assuming a population of 500 individuals in which one new mutant allele is introduced in each generation. Starting with 1000 different alleles in the zero generation, the balance between mutation and random extinction of alleles is reached well before generation 200. Actually, a majority of the initial 1000 alleles are lost by generation 50. Note that in a very large population the chance of survival of a single neutral gene for $t$ generations is approximately $2/t$ when $t$ is large (Fisher, 1930). Averaged over 21 outputs (from generation 200 through generation 1000 at intervals of 40 generations), the average and the effective numbers of alleles were as follows:

$$n_a = 13{\cdot}43, \quad n_e = 2{\cdot}79 \quad \text{(observed)}.$$

The corresponding values derived from diffusion approximations are

$$n_a = 11{\cdot}82, \quad n_e = 3{\cdot}00 \quad \text{(theoretical)}.$$

Thus, the two experiments assuming deterministic mutation have given results that agree fairly well with theoretical predictions. The diffusion approximation, however, tends to underestimate $n_a$ slightly. The remaining two experiments which are illustrated in Figs. 5 and 6 were carried out by using the second program (random mutation).

In the experiment shown in Fig. 5, the population consists of 50 males and 50 females ($N = 100$), of which only 25 males and 25 females actually participate in breeding ($N_e = 50$). In each generation, 100 male and 100 female gametes are randomly chosen from these 25 breeding males and 25 breeding females to form the next generation. Mutation to a new, not pre-existing, allele is induced in each gamete with probability 0·005 prior to the formation of zygotes ($u = 0{\cdot}005$). The initial population was set up such that it contained 200 different alleles. The balance between mutation and random extinction of alleles was reached well before generation 100. Actually, the majority of the original 200 alleles were lost by generation 20. The figure depicts the course of fluctuation of the average and the effective numbers of alleles in the population at intervals of 40 generations, from generation 120 through generation 2080. The actual computer outputs were at intervals of 20 generations and gave 100 observed pairs over generations 120–2100, from which the average and the effective numbers of alleles came out as follows:

$$n_a = 6{\cdot}05, \quad n_e = 2{\cdot}07 \quad \text{(observed)}.$$

On the other hand, from equations (3.24) and (3.25), the corresponding values are:

$$n_a = 5{\cdot}30, \quad n_e = 2{\cdot}00 \quad \text{(theoretical)}.$$

In the experiment illustrated in Fig. 6, the actual population number is 100, while the effective population number is 18 (5 breeding males and 45 breeding females). The mutation rate is the same as before. The simulation was carried out until generation 1300 and outputs of both the average and the effective numbers of

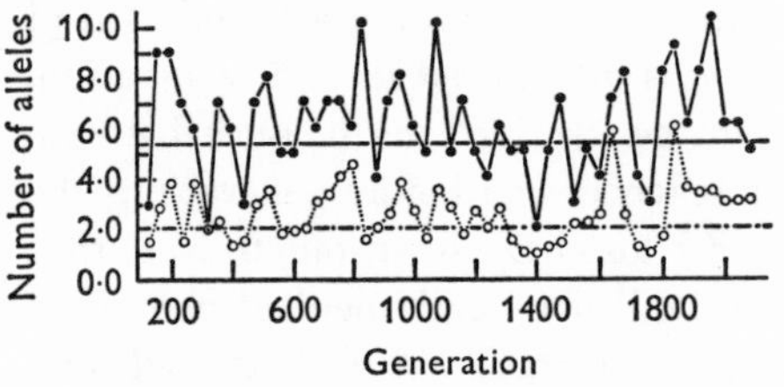


Fig. 5

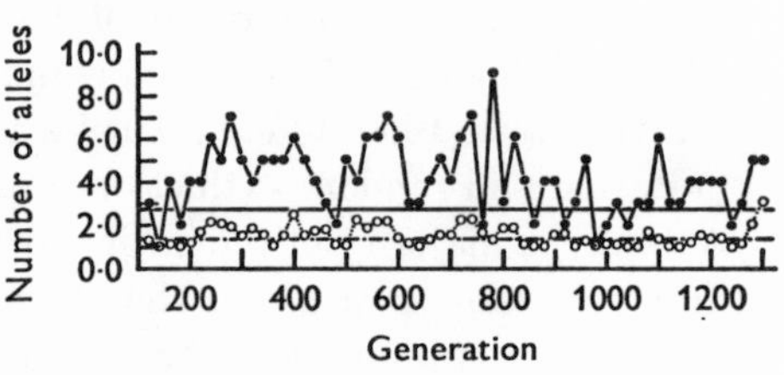


Fig. 6

Figs. 5–6. Results of Monte Carlo experiments as regards the number of neutral alleles. In these 2 experiments, mutation to a new, not pre-existing, allele is induced at each gamete sampling with probability $u = 0{\cdot}005$. The average and the effective numbers of alleles are plotted respectively by solid and open circles. Horizontal lines, solid and broken, represent corresponding theoretical values derived from the diffusion approximations. In Fig. 5, the population consists of 50 males and 50 females, 25 of each sex participating in breeding, so that $N = 100$, $N_e = 50$. In Fig. 6, $N = 100$ but $N_e = 18$. For details, see text. Neutral alleles random mutation: ●—, average (actual) number, ○ - - -, effective number.

alleles are given at intervals of 20 generations starting from generation 120. This yielded 60 pairs of outputs, from which the following values were obtained:

$$n_a = 4.12, \quad n_e = 1{\cdot}38 \quad \text{(observed)}.$$

The corresponding values derived from equations (3.24) and (3.25) by putting $N = 100$, $N_e = 18$, $u = 0{\cdot}005$ are:

$$n_a = 2{\cdot}74, \quad n_e = 1{\cdot}36 \quad \text{(theoretical)}.$$

Additional results of Monte Carlo experiments together with those already mentioned are summarized in Table 2. Despite the smallness of the population number assumed in these experiments, agreement between observed and expected values is fairly good, except that the diffusion approximation tends to underestimate $n_a$.

These Monte Carlo experiments also gave the frequency distribution of various alleles within a population at equilibrium. An example is given in Fig. 7 in which observed values are plotted with the squared dots. They were derived from the same experiment from which Fig. 5 was constructed and they were the averages of 100 actual distributions observed from generation 120 through generation

Table 2. *Summary of the results of Monte Carlo experiments regarding the number of neutral alleles in a population*

In the experiments No. 1 and No. 2, mutation is deterministic, but in the remaining experiments, mutation is stochastic. The numbers inside parentheses indicate the numbers of outputs from which $n_a$ and $n_e$ were computed.

| Expt. no. | Population size | | | | Mutation rate | Output | Observed means | | Diffusion approximations | |
|---|---|---|---|---|---|---|---|---|---|---|
| | $N$ | $Ne$ | ♂ | ♀ | | | $n_a$ | $n_e$ | $n_a$ | $n_e$ |
| 1 (Fig. 3) | 100 | 100 | / | / | 0·005 | 100–1200 (56) | 9·68 | 3·13 | 8·61 | 3·00 |
| 2 (Fig. 4) | 500 | 500 | / | / | 0·001 | 200–1000 (21) | 13·43 | 2·79 | 11·82 | 3·00 |
| 3 (Fig. 5) | 100 | 50 | 25 | 25 | 0·005 | 120–2100 (100) | 6·05 | 2·07 | 5·30 | 2·00 |
| 4 | 100 | 100 | 50 | 50 | 0·005 | 120–2100 (100) | 9·34 | 2·26 | 8·61 | 3·00 |
| 5 (Fig. 6) | 100 | 18 | 5 | 45 | 0·005 | 120–1300 (60) | 4·12 | 1·38 | 2·74 | 1·36 |
| 6 | 100 | 100 | 50 | 50 | 0·005 | 100–1200 (23) | 10·91 | 3·22 | 8·61 | 3·00 |
| 7 | 100 | 50 | 25 | 25 | 0·005 | 100–1200 (23) | 5·52 | 1·93 | 5·30 | 2·00 |
| 8 | 50 | 50 | 25 | 25 | 0·01 | 40–400 (19) | 9·32 | 3·67 | 7·23 | 3·00 |
| 9 | 100 | 50 | 25 | 25 | 0·01 | 50–500 (19) | 10·42 | 3·13 | 8·61 | 3·00 |
| 10 | 200 | 200 | 100 | 100 | 0·01 | 140–1120 (50) | 34·74 | 10·66 | 27·30 | 9·00 |
| 11 | 500 | 167 | 50 | 250 | 0·001 | 220–900 (18) | 6·78 | 1·99 | 5·07 | 1·67 |

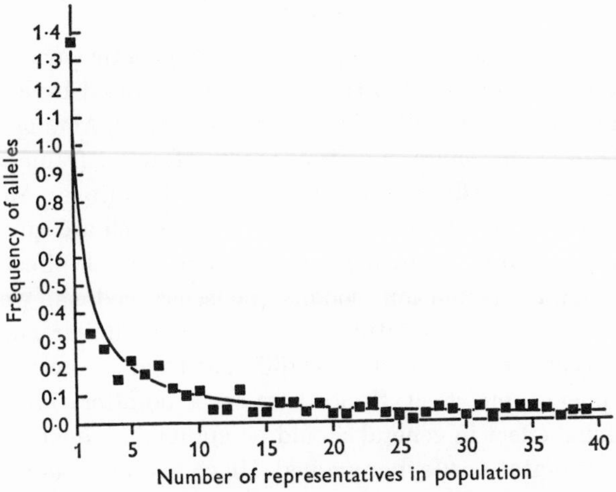


Fig. 7. Frequency distribution of alleles at equilibrium in a population consisting of 50 males and 50 females, of which only 25 males and 25 females actually participate in breeding. The mutation to a new, not pre-existing, allele is induced at each gamete sampling with probability 0·005. Squared dots represent the observed values from a Monte Carlo experiment and the solid curve represents the theoretical distribution obtained from the diffusion approximation. The ordinate stands for the frequency (in absolute number) of alleles having 1, 2, 3, etc., representatives in the population. Frequency distribution of neutral alleles in a population: $N = 100$, $N_e = 50$, $u = 0{\cdot}005$; —, theoretical distributions, ■, observed frequency.

2100 at intervals of 20 generations. The ordinate in the figure stands for the frequency (in absolute numbers) of alleles that have 1, 2, 3, etc., representatives in the population. To make the meaning of the above distribution clearer, let us suppose that a small population consists of 5 individuals, $A_1A_1$, $A_1A_2$, $A_2A_3$, $A_3A_4$, $A_5A_6$. In this case, allele $A_1$ has 3 representatives, alleles $A_2$ and $A_3$ have 2 representatives each, and alleles $A_4$, $A_5$ and $A_6$ have a single representative each. Thus the frequencies (in absolute number) of alleles that have 1, 2 and 3 representatives are 3, 2, and 1 respectively. This population contains 6 different alleles but the sum of squares of allelic frequencies (in proportion) is 0·2 and therefore $n_a = 6$, $n_e = 5$.

Going back to the distribution in Fig. 7, the solid curve represents the theoretical distribution derived from $\Phi(x)dx$ by replacing $dx$ by $1/(2N)$. Since, for the present experiment $M = N_e u = 50 \times \cdot 005 = 0{\cdot}25$, formula (3.23) gives

$$\Phi(x)dx = 1/(2Nx),$$

where $2Nx$ is the number of representatives in the population.

The agreement between the observed and the theoretical distributions is fairly good except that the diffusion approximation tends to underestimate the frequency of alleles represented only once in the population. The same tendency was observed also in the experiment performed with $N = 100$, $N_e = 100$, $u = 0{\cdot}005$ (random mutation).

## 5. DISCUSSION

(1) *Nature of mutant alleles.* The ultimate source of genetic variability in a population is mutation. It is now known that mutation is caused by changes in DNA base arrangements, namely, substitutions, gains and losses. Among them, addition or loss of a single base pair causes a shift of reading frame ('frame shift') and will produce far more drastic effects than single-base alterations. Among the base substitutions, some lead to alteration of amino acids which are quite dissimilar in chemical properties, thus producing marked mutational effects. Especially, changes to the chain termination codons (nonsense codons) would be most damaging. Those leading to substitution of chemically similar amino acids at a position of the polypeptide chain which is different from the active site may produce very little phenotypic effect. Still others cause no alteration of amino acids and their mutational effect in general should be minimal. Sonneborn (1965) called the last category of mutations 'synonymous'. He conjectured that it would not be surprising if 20% or more of all single-base mutations were synonymous. Since a more complete dictionary of the genetic code is now available, an attempt was made to obtain the probability that a mutation is synonymous, giving due weight to the frequencies of various codons in a haploid chromosome set. The method of calculation is given in detail in Appendix I. The results support Sonneborn's conjecture. That is, the probability is about 0·34 if the base pair replacement is exclusively of transitional type, but is roughly 0·23 if all types of single-base substitution occur with equal frequency.

Sonneborn also suggested a possibility of 'recombinational pseudomutation', that is, production of a codon for a different amino acid by recombination of two synonymous codons. For example, UGC (Cys) would be produced from UCU (Ser) and AGC (Ser) by recombination. This event could take place only in a population which is polymorphic with respect to synonymous codons at a given site on a chromosome. Since the mutation rate per nucleotide pair per replication is estimated to be about $10^{-11}$ in *Drosophila* and $10^{-12}$ in man (Kimura, 1967), the mutation rate ($u_c$) per codon per generation is probably of the order of $10^{-8}$ in *Drosophila* and $10^{-9}$ in man even if nearly neutral mutations are included. Thus even for a population of $N_e = 10^4$, $N_e u_c$ is at most of the order of $10^{-4}$. This means that at each site of DNA triplet, synonymous polymorphism must be extremely rare and, accordingly, recombinational pseudomutation is probably a very rare phenomenon in nature.

It is important to note here that probably not all synonymous mutations are neutral, even if most of them are nearly so.

Mutations which lead to substitution of somewhat similar amino acids also produce little or no change in biological activity, depending on their position in the polypeptide chain, and such mutations might be called *imperfectly synonymous*. For example, substitution of serine for glycine, in position 47 of the *E. coli* tryptophan synthetase A, leaves the enzymic activity intact (cf. Watson, 1965). Some of the imperfectly synonymous mutations may produce enzymes that have different electrophoretic property, yet differ little in biological activity. Some of the isozyme polymorphisms must be caused by such mutations.

Thus we have a wide mutation spectrum with respect to fitness: the (recessive) lethal mutations damage the developmental processes so drastically that individuals carrying them in homozygous condition cannot survive to maturity. In the second chromosome of *Drosophila melanogaster*, some 500 loci (possibly about one-eighth of the total) are capable of producing lethal mutations and the total rate amounts to about 0·5 %. Mutations causing less deleterious effect, on what Mukai (1964) called viability polygenes, appear to be much more numerous. The mutation rate for such genes is estimated to be about 14 % per second chromosome. This means that the total mutation rate per individual may reach at least 70 %. Probably, the mutational load due to such viability polygenes is reduced by 'reinforcing type' epistasis (Kimura & Maruyama, 1966). Mutations causing still less deleterious effect are difficult to detect, except possibly those found to be isozyme mutations. In addition, a recent analysis of the genetic variation concerning the number of sternopleural bristles in *Drosophila* suggests that the genes responsible for the character are nearly neutral (Robertson, 1967). It is probable that the same situation will be met with when many other quantitative characters are concerned, and, as suggested earlier by Clayton & Robertson (1955), their existing variation could well be maintained by the equilibrium between inbreeding and mutation.

An important problem confronting us now is what the rate is of occurrence of neutral and nearly neutral mutations. According to Robertson (1967), 'apparently fewer than 30% of the single amino acid substitutions compatible with the genetic code would cause a change in the electrical charge of the protein molecule' (see also Shaw, 1965). This means that mutations that cannot be detected by electrophoresis occur twice as frequently as those that can be so detected. Substitution of similar amino acids such as leucine with isoleucine will not cause a change in the electrical charge.

In discussing neutral or nearly neutral mutations, the fact that substitution of amino acids in many parts of a polypeptide chain often causes no change of its catalytic activity (cf. Watson, 1965) is probably significant.

If the effects of amino acid substitution on the activity of a polypeptide chain were thoroughly known, it would be possible to assess the frequency of neutral or nearly neutral mutations by estimating the frequencies of synonymous or imperfectly synonymous mutations and the average size of the active site or sites in a chain.

The following is a preliminary (and admittedly crude) attempt along this line. According to Goldberg & Wittes (1966), 20 amino acids may be divided into 8 groups of similar amino acids, namely: {Pro}, {Try, Tyr, Phe, Ileu, Leu, Met, Val}, {Cys}, {Thr, Ser}, {Gly, Ala}, {GluN, AspN}, {Glu, Asp} and {His, Arg, Lys}. For each group, they calculated the probability that a single-base substitution would not lead to a change of group, assuming that all the codons have the same frequency. Using these probabilities but giving due weight to each group of amino acids and their expected frequencies (cf. Table A1, column 3), the average probability ($p_s$) was calculated. It turned out that $p_s$ is about 0·43, if all types of a single-base substitution occur with equal frequency. If the base substitution is exclusively of transitional type, the corresponding probability is about 0·46. Let $a$ be the fraction of length which the active site or sites occupy on a polypeptide chain. Since the value of $a$ must vary from molecule to molecule, we will take its average value for $a$. Then, among all mutations due to a single-base substitution with respect to a polypeptide chain, the fraction of mutations that are synonymous or imperfectly synonymous on the non-active site is $(1-a)p_s$. The actual value of $a$ is not known but it is probable that $a$ is at the most 10%. So, the above fraction should be roughly equal to $p_s$. Since there are always some (more drastic but less frequent) mutations due to DNA base additions or losses, $p_s$ will impart an upper limit to the fraction of neutral or nearly neutral mutations among all mutations. In conclusion, it seems probable that neutral or nearly neutral mutations might reach some 40% of all mutations.

If this conclusion turns out to be correct, and, if the remaining 60% of all mutations are detected as viability polygenes, lethals and semi-lethals, the total mutation rate per gamete in *Drosophila* may reach some 60% per generation.

(2) *A definition of neutrality.* For any species, there is an upper limit to the total genetic load, or the amount of selective elimination due to genotypic differences. This is because the reproductive capacity of each species is limited. Furthermore, there is always death or sterility due to environmental causes. Thus, as pointed out earlier by Wright (1931) and others, selection intensity per locus depends on

the total number of segregating loci in a population. The former must decrease as the latter increases.

In the last decade, much emphasis has been laid on the possibility of non-existence of neutral mutations. It may be true that the 'visible' mutations, in the sense that they are discernible by the human eye, do have almost always some selective difference. However, the number of polymorphisms due to such mutations must be small as compared with the total number of loci. On the other hand, recent studies of enzyme polymorphisms suggest that a large number of loci are segregating in a population. In Harris's (1966) study, among the 10 arbitrarily chosen enzymes of man, 3 were found to be polymorphic. In the case of *Drosophila pseudoobscura*, Lewontin & Hubby (1966) studied 18 loci responsible for enzymes and other proteins; they found that the average population is polymorphic for about 30 % of the loci. They estimated that each individual is heterozygous on the average for 12 % of all loci. Since the total gene number in *Drosophila* is estimated to be about 10000 (cf. Muller, 1967), the above findings mean that in this organism each individual is heterozygous for over 1000 loci on the average. The same situation may be met with in man.

This brings us to the problem of natural selection toward holding these polymorphisms. A consideration of the genetic load leads us to conclude that the natural selection acting on the majority of loci at any one time must be small. Kimura & Crow (1964) have shown that if polymorphisms are maintained in thousands of loci by overdominance, each with appreciable selection coefficients, the total load becomes intolerably large.

In considering the effect of selection on each locus, an important quantity is $N_e s$, namely, the product of selection coefficient ($s$) and effective population number ($N_e$). A mutant gene may be called *almost neutral* if $|2N_e s|$ is much smaller than unity. Under this definition, neutrality depends not only on $s$ but also on $N_e$. Thus a gene is almost neutral in a small population but not so in a large one. In this connexion it should be noted that mild overdominance which is efficient enough to maintain a polymorphism in a large population has very little effect in maintaining the polymorphism in a small population.

To see this point more clearly, let us consider a pair of overdominant alleles $A_1$ and $A_2$ and assume that the relative fitnesses of the three genotypes $A_1A_1$, $A_1A_2$ and $A_2A_2$ are $1-s_1$, 1 and $1-s_2$ respectively. Accordingly, in a population of effective size $N_e$, assuming the most favourable condition $s_1 = s_2 \equiv s$ for maintaining polymorphism, it has been shown by the present author (cf. Robertson, 1962) that the probability of co-existence of both alleles decreases at the rate

$$\lambda_0 = \frac{1}{2N_e}\left\{1 - \frac{2N_e s}{5} + \frac{2^4 . 3}{5^4 . 7}(N_e s)^2 - \frac{2^5}{5^5 . 7}(N_e s)^3 - \quad \ldots\right\} \tag{5.1}$$

per generation. The above power series is valid for $N_e s$ up to about 4.

Since the corresponding rate for a strictly neutral pair of alleles is

$$\lambda_0 = 1/(2N_e),$$

the above formula (5.1) shows that for a small value of $N_e s$, the rate of decay of variance is reduced only by the fraction

$$\tfrac{2}{5}N_e s.$$

Thus for a pair of alleles with 1% overdominance ($s = 0{\cdot}01$), if an experimental population is kept in a culture bottle with 50 parents ($N_e = 50$) in each generation, the above fraction becomes 0·2. This value will become much less if $s_1 \neq s_2$ and also if $N_e$ is sometimes reduced in the course of breeding. Thus, no appreciable effects of overdominance should be observed under such conditions. On the other hand, in a population of $N_e = 10^4$, the overdominant alleles with $s = 0{\cdot}01$ will be kept in the population almost indefinitely.

(3) *Population structure and migration.* Usually, a species which occupies a wide territory and consists of a large number of individuals does not form a single panmictic unit, but comprises a number of subgroups or 'demes', mating taking place within each deme nearly at random. However, there is always some migration between the subgroups, so that as a whole the species forms a single reproductive community. It has been advocated by Wright (cf. Wright, 1951) that such a subdivided structure is conducive to the maintenance of genetic variability and therefore is favourable to rapid evolutionary progress. He studied the local differentiation of gene frequencies by assuming a continuous population structure.

The problem of local differentiation in gene frequencies was also studied by Kimura & Weiss (1964), who used 'the stepping stone model' of population structure, in which the entire population is subdivided into colonies and migration is restricted to nearby colonies. In this model, if $N_e$ is the effective number of each colony and $m_1$ is the rate of migration between adjacent colonies, then, assuming mutation as in §2 ($K$ possible allelic states and mutation rates equal in all directions), the gene frequency distribution corresponding to formula (3.5) is approximately

$$\phi(x) = \frac{\Gamma(\bar{A})}{\Gamma(\bar{A}-\bar{B})\,\Gamma(\bar{B})}x^{\bar{B}-1}(1-x)^{(\bar{A}-\bar{B})-1}, \qquad (5{\cdot}2)$$

where

$$\bar{A} = 4N_e m', \quad \bar{B} = 4N_e m'\bar{x},$$

in which

$$m' = \frac{Ku}{K-1}+m_1(1-r_1), \quad \bar{x} = 1/K.$$

In the above expressions, $r_1$ is the correlation coefficient of gene frequencies between two adjacent colonies (i.e. colonies 'one step' apart) and its actual value depends on the number of dimensions as well as the rates of migration and mutation (Kimura & Weiss, 1964; Weiss & Kimura, 1965).

From the above distribution (5.2),

$$\bar{H}_0 = K\int_0^1 x^2\phi(x)dx = \left(\frac{4N_e m'}{K}+1\right)\Big/(4N_e m'+1). \qquad (5.3)$$

For the one-dimensional stepping-stone model, if $u \ll m_1 \ll 1$, it can be shown that

$$1-r_1 = \sqrt{[\{2Ku/(K-1)\,m_1\}]}.$$

Thus, if $N_e = 10^4$, $u = 10^{-5}$, $K = 10$ and $m_1 = 10^{-1}$, we have $r_1 \approx 0{\cdot}985$, and the average homozygosity is approximately

$$\bar{H}_0 = 0{\cdot}11 \quad \text{(heterozygosity of about 89\%)}.$$

On the other hand if $m_1 = 0$ but under otherwise the same condition

$$\bar{H}_0 = 0{\cdot}71 \quad \text{(heterozygosity of about 29\%)}.$$

For the two-dimensional stepping-stone model, if $u \ll m_1 \ll 1$, it can also be shown that

$$1-r_1 = \frac{\pi}{2}\left(\log_e \frac{4}{\sqrt{[2Ku/\{(K-1)\,m_1\}]}}\right)^{-1}.$$

Thus, if $N_e = 10^4$, $u = 10^{-5}$, $K = 10$ and $m_1 = 10^{-1}$, we have $r_1 \approx 0{\cdot}72$ and the average homozygosity is approximately

$$\bar{H}_0 = 0{\cdot}10.$$

On the other hand, if $m_1 = 0$, $\bar{H}_0 = 0{\cdot}71$, as before. These examples show that under subdivided structure and migration, much higher heterozygosity is expected.

When the number of allelic states is infinite, $r_k$ is proportional to the probability that two homologous genes taken one from each of the two colonies $k$ steps apart have the same allelic state. Thus $r_k$ gives the fraction of alleles (in terms of the effective allele number) that are shared by these two colonies.

The advantage of population subdivision in keeping a large number of alleles may best be seen by comparing the average number of alleles maintained in a species under panmixis and under subdivision. For example, in a species of $N = N_e = 10^6$, if $u = 10^{-5}$ and if every mutation is to a new, not pre-existing, allele, then $n_a = 410{\cdot}2$ when the species forms a single random mating unit (see Table 1). On the other hand, $n_a = 1{\cdot}301 \times 1000 \approx 1300$ when the species is subdivided into 1000 completely isolated colonies of size $N = N_e = 10^3$. With a small amount of migration, this number will be reduced but may still be large as compared with the panmictic population.

## SUMMARY

1. The average and the effective numbers of alleles maintained in a finite population due to mutational production of neutral isoalleles were studied by mathematical analysis and computer simulation.

2. The exact formula was derived for the effective number ($n_e$) of alleles maintained in a population of effective size $N_e$, assuming that there are $K$ possible allelic states and mutation occurs with equal frequency in all directions. If the number of allelic states is so large that every mutation is to a new, not pre-existing,

allele, we have $n_e = 4N_e u + 1 - 2N_e u^2$, where $u$ is the mutation rate. Thus, the approximation formula, $n_e = 4N_e u + 1$, given by Kimura & Crow (1964) is valid as long as $2N_e u^2 \ll 1$.

3. The formula for the average number of alleles ($n_a$) maintained in a population of actual size $N$ and effective size $N_e$ was derived by using the method of diffusion approximation. If every mutation is to a new, not pre-existing, allele, we obtain

$$n_a = 4M \int_{1/(2N)}^{1} (1-x)^{4M-1} x^{-1}\, dx,$$

where $M = 4N_e u$. The average number of alleles as a function of $M$ and $N$ is listed in Table 1.

4. In order to check the validity of the diffusion approximations, Monte Carlo experiments were carried out using the computer IBM 7090. The experiments showed that the approximations are satisfactory for practical purposes.

5. It is estimated that among the mutations produced by DNA base substitutions, synonymous mutations, that is, those which cause no alterations of amino acids, amount roughly to 0·2–0·3 in vertebrates. *Incompletely synonymous mutations*, that is, those which lead to substitution of chemically similar amino acids at a different position of the polypeptide chain from the active site and therefore produce almost no phenotypic effects, must be very common. Together with synonymous mutations, they might constitute at least some 40 % of all mutations. These considerations suggest that neutral and nearly neutral mutations must be more common than previously considered.

I would like to express my thanks to Dr Takeo Maruyama for stimulating discussions in the course of the present work. Thanks are also due to Dr Alan Robertson for reading the manuscript and making valuable suggestions.

## APPENDIX I

### *Probability that a mutation is synonymous*

Two or more codons are said to be synonyms (Muller, 1963, cited in Muller, 1967) if they code for the same amino acid. In order to calculate the relative frequency of synonymous mutations among all the mutations produced in an individual by DNA base substitutions, we must know (1) the relative frequencies of various codons in a haploid chromosome set and (2) the frequency of synonymous mutations for each codon. Since the relative frequencies of various codons in the haploid set are not known, we must estimate them either from the frequencies of the four DNA bases, adenine (A), thymine (T), guanine (G) and cytosine (C), or from the frequencies of 16 dinucleotides obtained by the nearest-neighbour analysis (Josse, Kaiser & Kornberg, 1961). Throughout the present calculation, the RNA code dictionary used was the one given by Crick (1966), with UGA included as a nonsense triplet, following Brenner, Barnett, Katz & Crick (1967). We note here that uracil (U) in RNA code corresponds to thymine (T) in DNA code. Note also that the messenger RNA is complementary to one of the strands of DNA from which the former is 'transcribed'. According to Sueoka (cf. 1965), the G–C content of DNA obtained from various vertebrate species lies within the range 40–44 %. So, in the first calculation, the relative frequencies of A, T, G and C are assumed to be 0·285, 0·285, 0·215 and 0·215, respectively. Actually, the transcription is made from only one of the two strands of DNA and the frequencies of A and G may not necessarily be equal respectively to T and C in this strand. However, we will assume that in

Table A1. *Observed and expected frequencies of various amino acids in the proteins of vertebrates (for details, see text)*

| | | Relative frequency (%) | |
|---|---|---|---|
| Amino acid | Observed | Expected (1) | Expected (2) |
| Glu, GluN | 10·13 | 6·51 | 7·89 |
| Asp, AspN | 9·13 | 7·57 | 8·28 |
| Gly | 7·96 | 4·91 | 4·88 |
| Leu | 7·92 | 10·82 | 12·26 |
| Ser | 7·67 | 9·76 | 9·10 |
| Ala | 7·53 | 4·91 | 1·43 |
| Lys | 6·70 | 4·31 | 5·58 |
| Val | 6·68 | 6·50 | 9·53 |
| Arg | 6·28 | 8·16 | 7·34 |
| Thr | 5·87 | 6·50 | 6·78 |
| Pro | 5·21 | 4·91 | 4·46 |
| Ileu | 4·05 | 6·77 | 5·17 |
| Phe | 3·81 | 4·31 | 5·97 |
| Tyr | 3·55 | 4·31 | 3·92 |
| Cys | 2·77 | 3·25 | 2·08 |
| His | 2·37 | 3·25 | 3·01 |
| Met | 1·52 | 1·85 | 1·15 |
| Try | 0·85 | 1·40 | 1·17 |
| Total | 100·00 | 100·00 | 100·00 |

higher organisms, A = T, and G = C hold approximately in each strand of DNA. Using the above frequencies of A, T, G, C and assuming independence of base arrangements, relative frequencies of 64 codons may be obtained. For example, the frequency of AAA is $(0{\cdot}285)^3$ or about 0·02315, from which RNA codon UUU is derived. However, three codons, UAA (Ochre), UAG (Amber) and UGA lead to chain termination in polypeptide synthesis. So the relative frequencies of the remaining 61 codons are recalculated after removing these 3 nonsense codons. In order to test the validity of this approach, the relative frequencies of occurrence of various amino acids in proteins were predicted, using the frequencies of those codons. For example, lysine is coded by AAA and AAG. Thus the estimated frequency of this amino acid is the sum of the frequencies of the two codons, which turns out to be 4·312 %. The third column in Table A1 lists the frequencies of amino acids computed in this way. They should be compared with the corresponding values (second column) actually observed in proteins from vertebrates. The latter values are averages from 61 proteins of vertebrate origin listed in Smith's paper (1966), who compiled the amino acid composition of 80 proteins including those of non-vertebrate origin. Agreement between the observed and the expected frequencies is only fair, but it does indicate that the method is sound as a first approximation in predicting frequencies of various codons. (Prediction is somewhat poor for Ala and Gly but this may be due to some unknown functions of these amino acids.) A similar calculation was carried out using dinucleotide frequencies obtained by Josse *et al.* (1961) for calf thymus DNA (they are as follows: AA 0·089, TT 0·087, CA 0·080, TG 0·076, GA 0·064, TC 0·067, CT 0·067, AG 0·072, GT 0·056, AC 0·052, GG 0·050, CC 0·054, TA 0·053, AT 0·073, CG 0·016, GC 0·044). This second method of calculation assumes that the frequency of codon TTC, for example, is proportional to the product of the frequencies of TT and TC. The last column of Table A1 lists the relative frequencies of amino acids predicted by using frequencies of 61 sense codons thus calculated. The agreement between the

Table A2. *Probability that a mutation is synonymous*

| Type of base substitution | Predicted by mononucleotide frequencies | Predicted by dinucleotide frequencies |
|---|---|---|
| Transition only, equal in both directions | 0·341 | 0·349 |
| All single-base substitutions with equal frequency | 0·233 | 0·231 |

observed and the expected frequencies is less satisfactory than in the previous case.

For each codon, the probability that it still codes for the same amino acid after single-base substitution depends on the type and frequency of DNA base replacement. So, two cases were studied. In the first case, only transition ($\frac{G}{C} \rightleftharpoons \frac{A}{T}$) was considered and this was assumed to occur with equal frequency in both directions. Thus, for example, in terms of RNA codon, GCA (Ala) changes to ACA (Thr), GUA (Val) and GCG (Ala) with equal frequency, and therefore the probability is $\frac{1}{3}$ that GCA still codes for the same amino acid after single-base substitution. This probability was calculated for each of the 61 sense codons, and the resulting probabilities were averaged by giving weight to the frequencies of codons predicted by the two different methods mentioned above. The final results are listed in Table A2. From the figures in the upper row of the table, it will be seen that the probability that a mutation is synonymous is about 0·34.

In the second case studied, every one of the four nucleotides is assumed to change to one of the remaining three nucleotides with equal probability. Thus, for example, UUA (Leu) changes to UUU (Phe), UUC (Phe), UUG (Leu), UCA (Ser), UAA (Ochre), UGA (nonsense), CUA (Leu), AUA (Ileu) and GUA (Val) with equal frequency, and therefore the probability is $\frac{2}{9}$ that UUA still codes for the same amino acid after single-base substitution. In this way, the probability was calculated for each of the 61 sense codons and weighted averages were calculated as in the previous case. The results are given in the bottom row of Table A2. They show that with this type of base substitutions, the probability that a mutation is synonymous is about 0·23.

Summing up, it is estimated that in vertebrate species, the probability that a mutation is synonymous is about 0·34 if the base substitution is exclusively of transition type, but is roughly 0·23 if all types of single-base substitution occur with equal frequency.

## REFERENCES


BRENNER, S., BARNETT, L., KATZ, E. R. & CRICK, F. H. C. (1967). UGA: A third nonsense triplet in the genetic code. *Nature, Lond.* **213** (5075), 449–450.

CLAYTON, G. & ROBERTSON, A. (1955). Mutation and quantitative variation. *Am. Nat.* 89, 151–158.

CRICK, F. H. C. (1966). The genetic code: III *Scient. Am.* **215** (4), 55–62.

CROW, J. F. & KIMURA, M. (1956). Some genetic problems in natural populations. *Proc. Third Berkeley Symp. on Math. Stat. and Prob.* **4**, 1–22.

EWENS, W. J. (1964). The maintenance of alleles by mutation. *Genetics* **50**, 891–898.

EWENS, W. J. & EWENS, P. M. (1966). The maintenance of alleles by mutation—Monte Carlo results for normal and self-sterility populations. *Heredity* **21**, 371–378.

FISHER, R. A. (1930). *The Genetical Theory of Natural Selection.* Oxford: Clarendon Press.

GOLDBERG, A. L. & WITTES, R. E. (1966). Genetic code: Aspects of organization. *Science, N.Y.* **153**, 420–424.

HARRIS, H. (1966). Enzyme polymorphism in man. *Proc. Roy. Soc.* B **164**, 298–310.

JOSSE, J., KAISER, A. D. & KORNBERG, A. (1961). Enzymatic synthesis of deoxyribonucleic acid VIII. Frequencies of nearest neighbour base sequences in deoxyribonucleic acid. *J. Biol. Chem.* **236**, 864–875.

KIMURA, M. (1964). Diffusion models in population genetics. *J. appl. Probability* **1**, 177–232.

KIMURA, M. (1967). On the evolutionary adjustment of spontaneous mutation rates. *Genet. Res.* **9**, 23–34.

KIMURA, M. & CROW, J. F. (1963). The measurement of effective population number. *Evolution* **17**, 279–288.

KIMURA, M. & CROW, J. F. (1964). The number of alleles that can be maintained in a finite population. *Genetics* **49**, 725–738.

KIMURA, M. & MARUYAMA, T. (1966). The mutational load with epistatic gene interactions in fitness. *Genetics* **54**, 1337–1351.

KIMURA, M. & WEISS, G. H. (1964). The stepping stone model of population structure and the decrease of genetic correlation with distance. *Genetics* **49**, 561–576.

LEWONTIN, R. C. & HUBBY, J. L. (1966). A molecular approach to the study of genic heterozygosity in natural populations. II. Amount of variation and degree of heterozygosity in natural populations of *Drosophila pseudoobscura. Genetics* **54**, 595–609.

MUKAI, T. (1964). The genetic structure of natural populations of *Drosophila melanogaster*. I. Spontaneous mutation rate of polygenes controlling viability. *Genetics* **50**, 1–19.

MULLER, H. J. (1967). The gene material as the initiator and the organizing basis of life. In *Heritage from Mendel* (ed. R. A. Brink), pp. 419–447. Madison: Univ. of Wisconsin Press.

ROBERTSON, A. (1962). Selection for heterozygotes in small populations. *Genetics* **47**, 1291–1300.

ROBERTSON, A. (1967). The nature of quantitative genetic variation. In *Heritage from Mendel*, pp. 265–280. (ed. R. A. Brink). Madison: Univ. of Wisconsin Press.

SHAW, C. R. (1965). Electrophoretic variation in enzymes. *Science, N.Y.* **149**, 936–943.

SMITH, M. H. (1966). The amino acid composition of proteins. *J. Theoret. Biol.* **13**, 261–282.

SONNEBORN, T. M. (1965). Degeneracy of the genetic code: Extent, nature, and genetic implications. In *Evolving Genes and Proteins*, pp. 377–397 (ed. Bryson, V. and Vogel, H. J.). New York: Academic Press.

SUEOKA, N. (1965). On the evolution of informational macromolecules. In *Evolving Genes and Proteins* (ed. Bryson, V. and Vogel, H. J.), pp. 479–496. New York: Academic Press.

WATSON, J. D. (1965). *Molecular Biology of the Gene.* New York: Benjamin.

WEISS, G. H. & KIMURA, M. (1965). A mathematical analysis of the stepping stone model of genetic correlation. *J. appl. Probability* **2**, 129–149.
WRIGHT, S. (1931). Evolution in Mendelian populations. *Genetics* **16**, 97–159.
WRIGHT, S. (1938*a*). The distribution of gene frequencies under irreversible mutation *Proc. Natn. Acad. Sci. U.S.A.* **24**, 253–259.
WRIGHT, S. (1938*b*). Size of population and breeding structure in relation to evolution. *Science, N.Y.* **87**, 430–431.
WRIGHT, S. (1949). Genetics of populations. *Encyclopaedia Britannica* **10**, 111–112.
WRIGHT, S. (1951). The genetical structure of populations. *Ann. Eugen.* **15**, 323–354.
WRIGHT, S. (1966). Polyallelic random drift in relation to evolution. *Proc. Natn. Acad. Sci. U.S.A.* **55**, 1074–1081.

# MUTATION AND EVOLUTION AT THE MOLECULAR LEVEL[1]

MOTOO KIMURA AND TOMOKO OHTA

*National Institute of Genetics, Mishima, Japan*

ABSTRACT

Some consequences of the neutral mutation-random drift hypothesis of molecular evolution and polymorphism were worked out with special reference to the relationships among evolutionary rates, mutation rates, protein electrophoretic mobilities, functional constraints and molecular structure. A graphical method was devised to show the pattern of extinction, multiplication and mutational steps involved for selectively neutral mutants in a finite population in the course of evolution. Using the observed mammalian evolutionary rate of protein-encoding cistrons and average heterozygosity for human isozyme polymorphisms, and assuming that the latter are transient states of neutral evolution, it is inferred that, very roughly, a new allele is substituted in each cistron every 10 million years, each individual substitution taking an average of 2 million years. The upper limit of the evolutionary rate is the mutation rate. Using data on the incidence of rare hemoglobin variants in Japanese populations, it was estimated that the mutation rate per amino acid site is about $9 \times 10^{-8}$ per generation. It was shown that in hemoglobins, the rate of amino acid substitutions at the surface parts of the molecules is about 10 times that in the heme pocket in the course of mammalian evolution, although there is no indication that mutation rates differ at different parts of the molecule.

IN this new era of molecular biology, it should not be surprising if evolutionary theory is subject to some revision. On the orthodox Neo-Darwinian view, the speed and direction of evolution are determined almost exclusively by selection and ecological opportunity and the role of mutation as a rate-determining process is essentially nil, provided that the rate of mutation is sufficient to provide the requisite genetic variability. There is no reason to question this view in so far as morphological and physiological traits, or other measures that are related to survival and fertility, are concerned. But there is increasing evidence that it is inconsistent with observations at the molecular level. It now appears likely that mutation and random frequency drift rather than positive Darwinian selection are the main agents controlling the process of nucleotide (and amino acid) substitution in evolution.

As we have pointed out (KIMURA and OHTA 1971b, 1972), the observed uniformity of the rate of amino acid substitutions for each molecule, a very high overall rate when extrapolated to the whole DNA content, and apparent randomness in the pattern of substitutions can all be understood most readily by the neutral mutation-random drift hypothesis put forward by one of us (KIMURA 1968). Essentially the same hypothesis was proposed independently by KING and JUKES (1969) under the more provocative title "non-Darwinian evolution."

[1] Contribution No. 902 from the National Institute of Genetics, Mishima, Shizuoka-ken, 411 Japan.

Genetics Supplement **73**: 19–35 April, 1973.

Although the hypothesis has found some acceptance among molecular biologists, it has encountered a great deal of resistance from traditional evolutionists, who have found it incredible that selectively neutral mutants are common and that natural selection, which has proved to be so potent in producing adaptive change at the morphological and physiological level, can not be the overwhelming factor in producing change at the molecular level.

A corollary of this theory is that at least some protein polymorphisms represent the transient phase of such neutral molecular changes, and KIMURA and OHTA (1971a) have proposed that the great majority of protein polymorphisms are of this type. This hypothesis, too, has been criticized by many population geneticists, but we believe that the various observations on protein polymorphism can be explained satisfactorily by taking account of migration (cf. KIMURA and MARUYAMA 1971) and associative overdominance due to linked selected loci (OHTA and KIMURA 1970, 1971c; OHTA 1971a,b).

As a scientific hypothesis, the neutral theory has the desirable feature that its consequences can be worked out relatively easily by the mathematical theory of population genetics, allowing us to make a number of predictions by which the theory can be tested. In the present paper, we shall make a further effort to relate the theoretical predictions with observations at the molecular level.

## BEHAVIOR OF MOLECULAR MUTANTS IN A POPULATION

Throughout this paper we shall consider mainly mutants produced by DNA base substitutions, the most common type of mutation at the molecular level. In treating the behavior of these mutants within a population, one very important point that we have to keep in mind is that the traditional model of assuming two allelic states (say $A$ and $a$) with reversible mutations at comparable rates is inadequate. This is because each gene (cistron) usually consists of hundreds of nucleotide sites, and therefore the possible number of allelic states is astronomical. In addition, for each allele, the probability of reverting to its previous allelic state by mutation is very much smaller than the probability of mutating further to a new allelic state. As an example, consider the cistron coding for the human hemoglobin $\alpha$ chain with 141 amino acids. This cistron has $3 \times 141$ or 423 nucleotide sites, and each site may be occupied by any one of four bases (A, T, G, C), so that the possible number of allelic states is $4^{423}$ or roughly $10^{254}$. In addition, each allele can mutate to produce $3 \times 423$ or 1269 alleles through single base substitutions alone. If a further base substitution occurs, the probability of the allele's reverting back to the original state is only one in 1269, assuming that mutation is equal in all directions.

Therefore, at the cistron level, a model of irreversible mutation with an effectively infinite array of possible alleles (KIMURA and CROW 1964; KIMURA 1971) is much more realistic at the molecular level than the classical model of reversible mutations between two allelic states. It is also likely that each mutant type is represented only once at the moment of occurrence, since the mutation rate per site is very low, say $10^{-8}$ per generation rather the traditional figure of $10^{-5}$ per gene locus. In a very large population the same mutation may occur several times

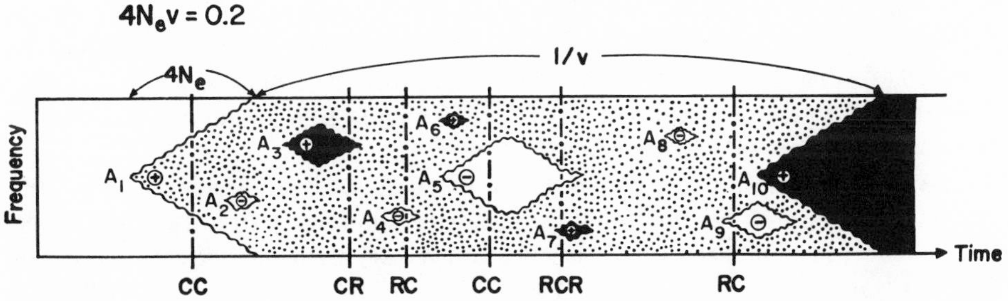


FIGURE 1.—A typical pattern of extinction and multiplication of selectively neutral mutants in a finite population when they occur at the rate of one mutation every ten generations ($4N_ev = 0.2$). See text for details.

within a few generations, but it is very unlikely that these interact in any way since each is ordinarily lost within a few generations.

Although a large number of mutants arise in each generation in any natural population of appreciable size, they are lost from the population within a few generations through random sampling of gametes in reproduction (FISHER 1930). This applies not only to deleterious and selectively neutral mutants, but also to advantageous mutants unless the advantage is very large (cf. KIMURA and OHTA 1971). Only a very small fraction of them can reach an appreciable frequency such as 10%.

Figure 1 represents a typical pattern of extinction and multiplication of selectively neutral mutants. The figure is constructed by taking into account the following basic properties of the population dynamics of neutral mutants:

1) If $p$ is the initial frequency (proportion) of an allele in the population, then the probability of the allele's eventually spreading to the entire population (i.e., the fixation probability) is $p$, while the probability of eventual loss is $1 - p$. More generally, if $p$ is the present frequency of an allele in the population, then the probability is $p/x$ that it subsequently reaches a higher frequency $x$ before it becomes either lost from the population or fixed in it ($0 < p \leqq x \leqq 1$).

2) If a new allele is produced at the rate of $v$ per cistron per gamete in each generation, then the average length of time between consecutive substitutions of alleles in the population is $1/v$ generations (CROW 1969; CROW and KIMURA 1970, p. 369). Thus $v$ is the rate of mutant substitutions in the course of evolution. Note that this rate is independent of the population size. Assuming that each allele is represented only once at the moment of appearance, $2Nv$ new alleles will appear on the average each generation within a population consisting of $N$ diploid individuals, but each has a probability of only $1/(2N)$ of eventually reaching fixation.

3) For each mutant allele destined to reach fixation, it takes on the average $4N_e$ generations from its first appearance until fixation, where $N_e$ is the effective number of the population (KIMURA and OHTA 1969a). More generally, it spends on the average $2N_e/N$ generations at each frequency class in the course of fixation (MARUYAMA and KIMURA 1971).

4) If $p$ is the present frequency of an allele in the population, the average number of generations until extinction (excluding the cases of fixation) is $-4N_e[p/(1-p)]\log_e p$ generations (KIMURA and OHTA 1969a,b).

In addition to these dynamic properties, we would like to add one basic property from the statics of neutral mutants:

5) At equilibrium, when the production of new alleles is balanced by the random extinction of existing alleles, the average homozygosity of an individual is $1/(4N_ev+1)$. The reciprocal of this quantity is called the "effective number of alleles" by KIMURA and CROW (1964). This is the number estimated by the ordinary allelism tests and is usually considerably smaller than the number of alleles actually contained in a population. The average proportion of heterozygotes with respect to this locus is $\bar{H} = 4N_ev/(4N_ev+1)$.

In Figure 1, only mutants that have attained frequencies higher than about 10% are depicted, leaving out an enormous number of mutant alleles that become lost before they reach this frequency. In this example $N$ and $N_e$ are assumed to be equal and $4N_ev$ is taken to be 0.2. Therefore, the effective number of alleles is 1.2 and the average interval between two consecutive fixations of mutants is 5 times as long as the time which each successful mutant takes from its appearance until its fixation in the population. In such a population, the average heterozygosity of an individual is roughly 16%. This is approximately equal to the actual heterozygosity observed in man (HARRIS 1966; see also HARRIS 1971) and Drosophila (LEWONTIN and HUBBY 1966) using electrophoretic techniques.

Continuing the example, if the average neutral mutation rate per cistron (including those mutants that can not be detected by electrophoresis) is roughly $10^{-7}$ per year in mammals, a new allele is substituted at each locus (cistron) every ten million years, each individual substitution taking on the average two million years.

## RARE AND COMMON ELECTROPHORETIC ALLELES

Figure 1 also enables us to make a prediction with respect to the arrangement of "rare" ($R$) and "common" ($C$) alleles detected among proteins characterized on electrophoresis gels.

Let us assume that when a single step mutation (base replacement) occurs at a cistron leading to a change in the electric charge of the protein, then the change occurs in such a way that the band moves either one step in the positive direction or one step in the negative direction on the gel. In Figure 1, such changes are marked by plus and minus signs. In this model, we assume that one positive and one negative charge cancel each other. At various times (abscissae), the sequence on the electrophoresis gels of common (C) and rare (R) alleles are expressed by such representation as CC, CR, RCR, etc. It will be seen that although RCR may occur, CRR, RRC and CRC are unlikely to occur. In particular, CRC presupposes two common alleles separated by at least two mutational steps with a rare intermediate allele between them, and it is clear that such an event is extremely unlikely when the effective number of alleles is 1.2. Recently, BULMER (1971), who analysed the data of PRAKASH *et al.* (1969), found an interesting property of the electrophoretic band pattern. For example, when there are two rare alleles and one common allele, the gel sequence is always RCR and never CRR or RRC. More generally, he found that there is a marked tendency for the rare alleles to occur at the beginning or the end of a gel sequence and for the common alleles to occur in the middle of a sequence. From such observations he claims that the alleles involved are not selectively neutral because, if they were, one would expect all possible orders of sequences to be equally likely. The above prediction using Figure 1 shows clearly that Bulmer's claim regarding the equal chance for all sequences is unfounded, and his observation is indeed compatible with the neutral mutation-random drift hypothesis, as has also been pointed out by MAYNARD-SMITH (1972). In our model, we have assumed that one negative and one positive charge cancel each other. Clearly, this is an oversimplification, and it is possible that under more sensitive tests, the difference between two variants having charges of the same sign can still be discriminated. Under such a circumstance, we should expect that CRR may be found to occur, giving an exception to BULMER's rule.

Figure 2 illustrates a typical pattern of extinction and fixation of alleles when mutations occur more frequently, that is, when $4N_ev = 2$. In this case, the effective number of alleles is 3, comparable to the esterase-5 locus (cf. YAMAZAKI 1971).

In such a population, RCCRR may occur even if we assume that one positive and one negative charge cancel each other. We can also see from the figure that coexisting alleles are often separated from each other by more than two mutational steps (nucleotide replacements).

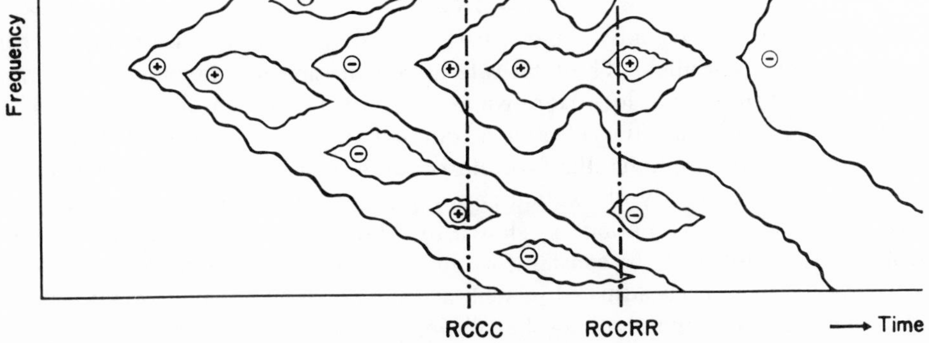

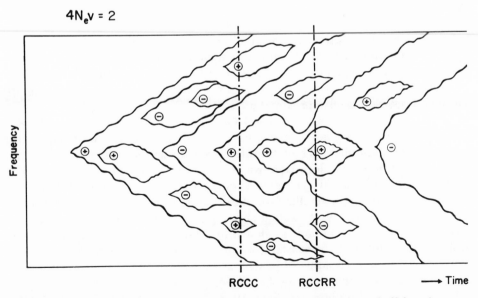

FIGURE 2.—A typical pattern of extinction and multiplication of neutral alleles when new mutations occur ten times as frequently as in Figure 1, that is, when $4N_ev = 2$.

We would like to mention the possibility that some of the rare alleles (frequencies less than 10%) represent nearly neutral but slightly deleterious mutants that could increase their frequencies by random drift (especially when protected by heterozygosity) up to a few percent. This interpretation is still in accord with neutral theory in its extended form (OHTA 1972).

### NEARLY NEUTRAL ALLELES

So far we have restricted our consideration mainly to selectively neutral mutants, but it is likely that in reality there is a continuous transition between definitely advantageous and definitely deleterious mutations. Thus, if we denote by $v(s)$ and $u(s)$ the mutation rate and the fixation probability of mutants having selection coefficients $s$, then the rate of mutant substitution in the course of evolution is given by

$$k = 2N \int v(s)u(s)ds,$$

where the same unit of time is used to measure $v(s)$ and $k$. Of particular importance from the standpoint of molecular evolution is the class of mutants whose selection coefficients are so small that their fate is largely controlled by random drift. When the values of $s$ are restricted to the neighborhood of zero such that $-2 \leqq N_e s \leqq 2$, the mutations are nearly neutral. Therefore, for such mutants, OHTA and KIMURA (1971) called the value of $k$ in the above expression "the effective neutral mutation rate". We should note also that if the selection coefficient $s$ of a mutant fluctuates randomly from generation to generation with mean $\bar{s}$ and the variance $V_s$, and if $|\bar{s}|/V_s$ is not large, say less than ½, the mutant behaves as if neutral even if $|N_e\bar{s}|$ is much larger than unity (OHTA 1972).

In a special but important case in which the mutants are selectively neutral, the above integral reduces to

$$k = v.$$

In other words, for neutral mutants, the rate of gene substitution in evolution is equal to the mutation rate per gamete, as mentioned before. Constancy of $k$ per year, then, may be explained by assuming constancy of $v$ per year.

On the other hand, if the mutants have a definite selective advantage, the integral reduces, with good approximation, to

$$k = 4N_e v s_1,$$

where $s_1$ is the selective advantage of each mutant (assuming semi-dominance) over its preexisting allele. The total rate of substitution per genome is the sum of the rates at individual loci. Here we assume that mutants at different loci behave independently and the selective advantage of each mutant is constant.

Under these assumptions, the formula is valid even for a subdivided population since the probability of fixation is $2s$ regardless of population structure (at least to a good approximation) as shown by MARUYAMA (1970). Thus, for the evolutionary rate ($k$) to be constant, we must assume that the three parameters $N_e$, $v$, and $s_1$ have been adjusted in such a way that their product remains constant in diverse lines in the course of evolution.

Recently, several authors (cf. SVED 1968, MAYNARD SMITH 1968) advocated a truncation selection model to avoid the heavy load that is entailed when the mutant substitutions are carried out by the positive Darwinian selection that acts independently at each locus (KIMURA 1968). With the truncation model, the selection coefficients at each locus are not independent but are determined by the fraction of individuals that survive and contribute to the next generation ("the proportion saved"). Let us suppose that each mutant adds a certain amount $y$ to the underlying fitness score on which the truncation selection is based. Then, if $\sigma^2_T$ is the total variance of this fitness score between individuals, we have

$$s_1 = \frac{y}{\sigma_T} \cdot \frac{z}{S} ,$$

where $S$ is the proportion saved and $z$ is the ordinate of the normal distribution $N(0, 1)$ at the truncation point. The rate of substitution at each locus is then

$$k = 4N_e v \frac{y}{\sigma_T} \cdot \frac{z}{S} .$$

Since $S$, the proportion saved, applies to all the loci concerned, we must assume that in order for the evolutionary rate to be constant per year, $S$ is adjusted such that $N_e vz/S$ is kept constant over all the lineages in the course of evolution. Thus, even if this model avoids the difficulty of the heavy substitutional load, it has no special advantage in explaining the observed constancy of the evolutionary rate per year.

## MOLECULAR STRUCTURE AND EVOLUTIONARY RATES

As pointed out by KING and JUKES (1969), KIMURA (1969), KIMURA and OHTA (1971) and others, the approximately constant evolutionary rate for each molecule constitutes very strong evidence for the neutral mutation-random drift hypothesis. It is remarkable to note that in hemoglobins, for example, the rates of amino acid substitution are approximately equal between the two lines, one leading to the present day carp and the other leading to man from their common ancestor back some 400 million years, while the rates of evolution at the phenotypic level are utterly different between these two lines. However, we must admit that we are rather ignorant of the reason why the rate of mutation in terms of nucleotide replacements seems to be more nearly constant with respect to absolute time than with respect to the number of generations.

Furthermore, according to this hypothesis, the different evolutionary rates among different classes of molecules is mainly due to the difference in the fraction of neutral mutants (depending on the functional requirement of the molecule) rather than to the difference in the intrinsic mutation rates. Thus, the majority of mutations are neutral in a rapidly evolving molecule such as fibrinopeptide A or B, while a large fraction of mutations are deleterious and rejected from the population in a slowly evolving molecule such as cytochrome c or histone IV. In other words, the more stringent the functional requirement of the molecule, the slower is its evolutionary change.

The data of DICKERSON (1971) may be interpreted as supporting this view. His detailed study on the three-dimensional structure of cytochrome c enables us to see the relation between evolutionary conservatism and function in more con-

TABLE 1

*Evolutionary rates of amino acid substitutions at various parts of the hemoglobin molecule*

| Class | Comparison Lamprey—Human $\alpha$ | Lamprey—Human $\beta$ |
|---|---|---|
| Surface | 1.37 | 2.21 |
| Exposed | 1.00 | 1.00 |
| Buried | 0.69 | 0.99 |
| Haem pocket | 0.49 | 0.49 |
| Total | 1.02 | 1.33 |

Rates are expressed per $10^9$ amino acid sites per year, i.e. in paulings, based upon the data of HENDRICKSON and LOVE (1971).

crete physical terms. In this section, we examine the evolutionary rates among different parts of the hemoglobin molecule with special reference to molecular structure and pathology.

Hemoglobin is one of the proteins whose structure is best known. There are several helical and non-helical regions. As with other globular proteins, the inside of the molecule is filled with hydrophobic amino acids and is relatively conservative in evolution. Particularly the haem pocket, which surrounds the haem, appears to be functionally the most rigid. HENDRICKSON and LOVE (1971) analyzed in detail the structure of the hemoglobin of the sea lamprey, *Petromyzon marinus*, and classified its amino acid sites into four groups: surface, exposed, buried and haem pocket. They examined the sequence homologies between the lamprey hemoglobin and some other globins. Using their Table 1, we have calculated the evolutionary rates of amino acid substitutions for the above four classes, assuming that the divergence of lamprey and human occurred 500 million years ago. Table 1 shows the substitution rates in paulings, i. e., per $10^9$ amino acid sites per year. From the table we can see that the evolutionary rate in the haem pocket is only $\frac{1}{3}$ of that at the surface.

In order to examine this relationship in more detail, we analyzed the evolutionary rates of these two parts in the hemoglobin $\alpha$ and $\beta$ chains, mainly among mammals. In the $\alpha$ chain, there are nineteen amino acid sites that are known to form the haem pocket (residues 32, 39, 42, 43, 45, 46, 58, 62, 83, 86, 87, 91, 93, 97, 98, 101, 129, 132, 136). This part plays an important role in the normal functioning of the hemoglobin molecule, and replacements of amino acids in this part tend to cause abnormalities such as unstable hemoglobin (PERUTZ and LEHMAN 1968). From the alignment of hemoglobin $\alpha$ sequences compiled by DAYHOFF (1969, table D-218), we find four amino acid substitutions at these sites. By considering the evolutionary period of divergence among all the organisms compared (human, gorilla, rhesus monkey, mouse C578L, mouse NB, rabbit, horse, bovine, sheep, goat A, goat B, llama, pig, carp), we get 0.165 paulings for the haem pocket in the $\alpha$ chain. Similarly, 21 amino acid sites are known to be situated in the heme pocket (residues 31, 38, 41, 42, 44, 45, 63, 66, 67, 70, 71, 88, 91, 92, 96, 98, 102, 103, 106, 137, 141) of the $\beta$ chain. We find three amino acid substitutions in these sites in our comparison of known hemoglobin $\beta$ sequences (human,

TABLE 2

*Comparison of evolutionary rates between the surface and the haem-pocket in hemoglobin chains*

| Region | Hemoglobin $\alpha$ | Hemoglobin $\beta$ |
|---|---|---|
| Surface | 1.35 | 2.73 |
| Haem pocket | 0.165 | 0.236 |

The listed values represent the rates of amino acid substitution in paulings ($10^{-9}$/amino acid site/year).

gorilla, spider monkey, rhesus monkey, rabbit, mouse SEC, mouse AKR, horse, llama, pig, fetal bovine, sheep A, sheep B, sheep C, goat A, Barbary sheep C, goat C), excluding $\delta$, $\gamma$ and incomplete sequences (DAYHOFF 1969, table D-219). Considering the divergence periods of the organisms involved, we get 0.236 paulings in the heme pocket of the $\beta$ chain. These estimates are considerably smaller than the corresponding values obtained above from the comparison between lamprey and human hemoglobins.

We now examine the surface of the molecule, which is functionally least rigid. It is known that, in contrast to amino acids in the haem pocket, those at the surface usually play no special role in the function. Since there is not always a clear-cut distinction between surface and non-surface sites, we picked out those amino acid sites which are known to be external and to show no clinical symptoms in mutant heterozygotes found in human populations. PERUTZ and LEHMAN (1968) list 18 such sites in hemoglobin $\alpha$ and 23 sites in hemoglobin $\beta$ (see also YANASE and YAMAOKA 1972). By using the same method of calculation as before, the evolutionary rates of these sites turned out to be 1.35 paulings in the $\alpha$ chain and 2.73 paulings in the $\beta$ chain. Thus the evolutionary rates at the surface are almost ten times those of the heme pocket (Table 2). This ratio is greater than those obtained from the comparison between lamprey and human hemoglobins (Table 1), in which the evolutionary rate at the surface is three times that in the haem pocket. This perhaps reflects the fact that the lamprey hemoglobin molecule is a monomer or a dimer whereas the mammalian hemoglobin is a tetramer consisting of two $\alpha$ and two $\beta$ chains. The differentiation has occurred in the functionally important haem pocket more often between lamprey and human hemoglobins than among $\alpha$ or $\beta$ chains. As to the reason why the evolution is faster for the $\beta$ chain than for the $\alpha$, one could perhaps argue that since $\beta$ is produced mainly post natally it has less rigid functional requirements.

## MUTATION RATES AND EVOLUTIONARY RATES

It is desirable to estimate and compare the mutation rate per amino acid site independently of the evolutionary rate for the same molecule. In this section, we intend to show that such a comparison is possible for hemoglobins.

In human populations a number of molecular variants have been detected in hemoglobins. In particular, some 120 variants are known that are due to single amino acid substitutions in the $\alpha$ and $\beta$ chains (see the table compiled by YANASE and YAMAOKA 1972). Also, through a large-scale survey of the incidence of rare hemoglobin variants in Japan, we have fairly good information on their frequencies.

According to YANASE *et al.* (1968), nine different hemoglobin variants due to single amino acid substitution (five variants for $\alpha$ and four for $\beta$) represented by sixteen individuals were found through electrophoretic analysis of approximately 50,000 blood samples. In another survey of hemoglobin variants in Shikoku and Chugoku districts in Japan (personal communication, DR. M. HANADA), ten different variants represented by fifteen individuals due to amino acid substitution in the $\alpha$ or $\beta$ chains were found in a sample of 101,022.

Altogether, the incidence of rare hemoglobin variants due to amino acid substitution in the $\alpha$ or $\beta$ chain is approximately 1/5,000 or, in terms of gene frequency, $10^{-4}$. Similarly, according to IUCHI's (1968) report, 35 variants have been detected from 44 families in the survey covering 279,600 individuals in middle and western parts of Japan during the period 1957–1967. This gives an incidence of about 1/6,400, comparable to the above figure.

Since only about ⅓ of the single amino acid changes can be detected by electrophoresis, our estimate, $10^{-4}$, of gene frequency should be multiplied by 3 to give $3 \times 10^{-4}$. In order to obtain the mutation rate, we must further multiply this value by the fraction which new mutants in each generation occupy among the mutants currently in the population. According to DR. HANADA (personal communication), out of 31 different hemoglobin variants studied (such as Hbs Hikari, Hiroshima, Tagawa I, etc.), the hemoglobins of both parents were successfully determined in 18 cases, while in the remaining 13 cases determination was impossible because of various causes such as death of the parents. Of the former 18 cases, two were found to be due to new mutations, while in 16 cases variants were found to be transmitted from one of the parents. For the two cases of new mutation, one (Hb M Kurume) is certainly due to single amino acid substitution, but for the other (Hb Ube 1) there is some doubt. Thus the fraction of new mutations (in terms of number of different alleles) is either 1/18 or 2/18; so we tentatively take 1.5/18 or 1/12 as a rough estimate. This leads to $2.5 \times 10^{-5}$ as an estimate for the mutation rate for $\alpha$ and $\beta$ chains per generation. Since these chains contain 287 amino acids (141 for the $\alpha$ and 146 for the $\beta$ chain), we obtain the approximate mutation rate $v_{aa} = 9 \times 10^{-8}$ per amino acid site per generation. This is several times higher than the estimate of MOTULSKY (1969), who concluded that the nucleotide mutation rate probably lies somewhere between $10^{-8}$ and $10^{-9}$ per generation.

With $v_{aa} = 9 \times 10^{-8}$/generation and assuming 20 years as the average generation time in the past 15 generations or so in man, the mutation rate per year per amino acid site turns out to be approximately $4.5 \times 10^{-9}$, or $v_{aa} = 4.5 \times 10^{-9}$/year. This is about four times the average evolutionary rate of hemoglobins ($10^{-9}$ per amino acid site per year).

Our tentative figure of 1/12 for the fraction of new variants that appear by mutation in each generation is based on an inevitably small number of observations, but there is a theoretical ground to justify such a figure. The reciprocal of this figure, that is, 12, should represent the average number of generations until extinction ($\bar{t}_0$) of a mutant (assuming that each mutation leads to a new allele) in a finite population. It can be shown (cf. KIMURA and OHTA 1971c, p. 15, table 1.1.) that in a population of $2N = 10^6$ and $N_e = 0.5N$, if the mutant is neutral, $\bar{t}_0 = 13.8$, while if it has 1% selective disadvantage, $\bar{t}_0 = 5.7$. If the mutant is completely recessive for fitness with 1% selective disadvantage in homozygotes, $\bar{t}_0 = 10.3$. So, our tentative figure based on actual observations is consistent with these theoretical values, although in view of the uncertainty of the data, other reasonable values giving less consistent results could be chosen.

Our estimate of the mutation rate per amino acid site, $v_{aa} = 4.5 \times 10^{-9}$/year, is also consistent with FITCH's (1972) finding that in hemoglobins the fraction of "covarions" (concomitantly variable codons in the course of evolution) is about ⅓. Under the neutral mutation-random drift theory, this corresponds to the fraction of neutral mutants, so that about $4.5 \times 10^{-9} \times 1/3$ or $1.5 \times 10^{-9}$ should be equal to the rate of amino acid substitution in evolution. The observed rate of about $10^{-9}$ per amino acid site per year is in good agreement with this prediction.

In the above derivation of mutation rate from the observed frequency of rare variants, we used the multiplying factor 1/12 which is the fraction of new mutants among all the existing mutant alleles. However, it may be more appropriate to use a multiplying factor giving the fraction of new independent mutations among individuals possessing rare variants. Although each allele is usually represented by a single individual in the sample, in some cases, one allele is represented by more than one individual. For example, in the survey by YANASE *et al.* (1968), four hemoglobin variants (Hb Kokura, Hb Umi, Hb Yukuhashi II, Hb Tagawa II) turned out to be the change Asp → Gly at position 47 in the $\alpha$ chain, so that they could actually represent a single mutational event. If we take such a fact into account, the multiplying factor should be reduced by ⅓. With this correction, we have $v_{aa} = 3 \times 10^{-9}$/year, for the mutation rate estimate per amino acid, and $10^{-9}$ for the predicted mutation rate for neutral alleles.

## DISCUSSION

We have reviewed data showing that the rate of evolution is different among different parts of the hemoglobin molecule. In particular, the rate of amino acid substitutions at the surface part of the molecule is about ten times the rate in the haem pocket. We have also estimated the mutation rate per amino acid site, which turned out to be about three times the average evolutionary rate per amino acid site in hemoglobins. With the admittedly insufficient data we have relied on, the present estimate should be considered tentative.

Under the neutral mutation-random drift hypothesis (the neutral theory, for short), the different evolutionary rates among different parts of a molecule can be explained by assuming that the fractions of neutral mutants are different among different parts (depending on the functional requirement of the molecule), rather than representing differences in the intrinsic mutation rates. In particular, the stronger is the functional constraint, the larger is the chance that mutations are deleterious and therefore the smaller is the fraction of selectively neutral mutants.

VOGEL (1969) analyzed the distribution of amino acid replacements in the $\alpha$ and $\beta$ chains using reported hemoglobin variants in human populations. He obtained a result suggesting that there are no codons within the $\alpha$ and $\beta$ cistrons with an increased tendency to mutate. It appears as if mutations occur in the functionally rigid haem pocket just as frequently as in the functionally less rigid surface parts.

Particularly noteworthy is the contrast between the evolutionary conservatism and spontaneous mutability of the two histidines ("distal" and "proximal" histidyl residues) that are linked to the haem. They occupy positions 58 and 87 in the $\alpha$ chain, and positions 63 and 92 in the $\beta$ chain. It is known that these two histidines in the globin molecule are completely invariant (cf. JUKES 1971). In other words, no mutant substitutions have ever occurred in the entire history of vertebrate evolution extending some 500 million years and including a great many lines. On the other hand, in man, it is well known that substitution of tyrosine for one of these histidine residues causes hemoglobin M disease. As shown in Table 3, hemoglobin M variants are reported from various parts of the

TABLE 3

*Hemoglobin M variants involving changes at the distal and proximal histidyl residues*

| Name | Position and change | Symptom |
|---|---|---|
| M Boston<br>M Osaka<br>M Gothenberg<br>M Leipzig-II<br>M Köln<br>M Morin | $\alpha$58, His → Tyr | Cyanosis |
| M Iwate<br>M Reserve<br>M Kankakee II<br>M Oldenberg | $\alpha$87, His → Tyr | Cyanosis |
| M Saskatoon<br>M Elberfeld<br>M Arhus<br>M Emory<br>M Kurume<br>M Yonago<br>M Radom<br>M Chicago<br>M Leipzig<br>M Horlein-Weber<br>M Hamburg | $\beta$63, His → Tyr | Cyanosis (unstable hemoglobin) |
| M Hyde Park<br>M Akita | $\beta$92, His → Tyr | Cyanosis (unstable hemoglobin) |

world under various names (for details, see YANASE and YAMAOKA 1972). In addition, Hb Zürich ($\beta$63, His → Arg) is known. Even within Japan, all the four types of hemoglobin M variants have been found (i. e., M Osaka, M Iwate, M Kurume, M Akita). It appears, then, that although rare, they are by no means excessively rare as mutations compared with other hemoglobin variants that are caused by amino acid replacements in other functionally less critical sites.

The reason that these two histidine sites are invariant in evolution is that mutations at these two sites are highly deleterious or practically lethal in homozygous condition, even if they are viable in heterozygous condition. Note that only about 30% of the hemoglobin molecules of the blood of M Iwate and other Hb M heterozygotes are abnormal (met Hb); the rest are normal (SHIBATA *et al.* 1966). Thus, those mutants have never been able to spread to any species in the course of vertebrate evolution.

Similarly, according to HAYASHI and STAMATOYANNOPOULOS (1972), the penultimate tyrosine at position HC2 in vertebrate hemoglobins is evolutionarily invariant. They studied chemical properties of the variant Hb Bethesda and Hb Rainier, in which this tyrosine is replaced respectively by histidine and by cysteine in the $\beta$ chain. They found that, as expected from the PERUTZ model, these substitutions are associated with extreme disorganization of the normal hemoglobin oxygenation function. Thus, mutations at these amino acid sites, despite repeated occurrence in the course of vertebrate evolution, have never been able to spread to the whole population in any species.

The fact that the rate of evolution is very low at functionally critical sites may also be explained readily by the traditional theory based on positive Darwinian selection. According to this theory the majority of amino acid substitutions in evolution are the result of natural selection acting on definitely advantageous mutations. Under this theory, the observed low evolutionary rate at functionally critical sites can be explained by saying that the fraction of definitely advantageous mutants is very small in such critical sites. In other words, the more stringent the functional requirements, the lower the chance that mutations turn out to be advantageous.

The contrast between the neutral and the positive Darwinian theories, however, becomes much sharper, as pointed out by DR. SEWALL WRIGHT (personal communication 1971), when we try to interpret the rapidly evolving molecules based on these two theories.

Under the neutral theory, the less stringent the functional constraint, the larger the fraction of mutations that are selectively neutral (not harmful) and therefore the higher the evolutionary rate. Thus the upper limit of the evolutionary rate is set by the mutation rate. Furthermore, we should expect that in such molecules as fibrinopeptides A and B, which have very high evolutionary rates the functional constraint is so slight that virtually all mutations are neutral. This allows us to estimate the intrinsic mutation rate per nucleotide site by utilizing the observed evolutionary rates of such molecules (OHTA and KIMURA 1971a). On the other hand, such an interpretation is meaningless under the positive Darwinian theory. Rather, we should expect that rapidly evolving molecules have important functional significance and are undergoing very rapid adaptive improvements by accumulating many advantageous mutations. The rates of evolution in such molecules should have no clear-cut relationship with the intrinsic mutation rate.

We believe that evidence is accumulating that the neutral theory gives much more consistent interpretation and prediction than the positive Darwinian theory as far as observations at the molecular level are concerned. On either a Darwinian or neutral theory those parts of the molecule that are more permissive as to amino acid replacements are expected to evolve faster, but the neutral theory also predicts that the upper limit of the evolutionary rate is the mutation rate. The rate of evolution at such sites is independent of the rate of change of form and function.

The most rapidly evolving molecules known so far are the fibrinopeptides, with the estimated rate $4 \sim 9 \times 10^{-9}$/amino acid site/year. As pointed out by DICKERSON (1971), as well as by KING and JUKES (1969), the fibrinopeptides have little known function after they are removed from fibrinogen when the latter is converted to fibrin in a blood clot. It is worthwhile to note (OHTA and KIMURA 1971a; KING 1972) that the rate of nucleotide substitution estimated from the observed rate of amino acid substitution in fibrinopeptides is roughly equal to the corresponding rate estimated directly from DNA divergence among mammals based on DNA hybridization techniques (cf. KOHNE 1970). Probably the most instructive example is the middle segment (C) of the proinsulin molecule. This middle segment (⅓ of the total length) is removed when the active insulin is formed. It is now clear that this middle segment evolves some ten times as fast as insulin itself in terms of amino acid substitution, and that the rate of evolution at this part is not very different from those of fibrinopeptides (DICKERSON 1971; KIMURA and OHTA 1972; KING 1972). The similarity of evolutionary rates between the middle segment C of proinsulin and the fibrinopeptides is obvious from the neutral theory. In addition, a recent study by FITCH (1971) shows that a rapidly evolving molecule such as fibrinopeptide A has a larger fraction of "covarions" than a slowly evolving molecule such as cytochrome c, and yet, the rates of evolution are approximately the same among fibrinopeptide A, hemoglobin $\alpha$, and cytochrome c when the rate is calculated per covarion.

The following two examples also accord with the principle that functionally less important molecules evolve faster than more important ones. According to *Nature New Biology* (*Correspondent* 1971), it is now well established that although cistrons coding for ribosomal RNA are strongly conserved in evolution, spacer DNA is not. The recent report by BOYER *et al.* (1971) on primate hemoglobins shows that among primates the rate of amino acid substitutions and the incidence of polymorphic variations are higher in the $\delta$ chain forming the minor component $A_2$ (i. e., $\alpha_2\delta_2$) than in the $\beta$ chain forming the major component $A$ (i. e., $\alpha_2\beta_2$) of adult hemoglobins.

The neutral mutation-random drift hypothesis of molecular evolution allows us to make a number of quantitative predictions and interpretations much more straightforwardly than does the Darwinian hypothesis. Therefore, as more and more data on the structures, functions, evolutionary rates and patterns of informational molecules accumulate, we should be able to test its validity unambigously.

Meanwhile, we are confident that mutation is playing a much more direct and important role in controlling the rate and direction of evolution at the molecular level than we have been accustomed to think through decades of evolutionary studies at the phenotypic level, particularly under the flag of neo-Darwinism.

We would like to thank Dr. M. Hanada for supplying valuable data on hemoglobin variants in Japan, and Drs. T. Yanase and S. Shibata for helpful discussions on molecular pathology of abnormal hemoglobins. Thanks are also due to Dr. J. F. Crow who kindly went over the manuscript and offered many useful suggestions to improve the presentation.

## LITERATURE CITED

Boyer, S. H., E. F. Crosby, A. N. Noyes, G. F. Fuller, S. E. Leslie, L. J. Donaldson, G. R. Vrablik, E. W. Schaefer, Jr. and T. F. Thurmon, 1971 Primate hemoglobins: Some sequences and some proposals concerning the character of evolution and mutation. Biochemical Genetics **5:** 405–448.

Bulmer, M. G., 1971 Protein polymorphism. Nature **234:** 410–411.

*Correspondent* 1971 Circles, spacers and satellites on the Riviera. Nature New Biology **231:** 68.

Crow, J. F., 1969 Molecular genetics and population genetics. Proc. XII Intern. Congr. Genetics Vol. **3:** 105–113.

Crow, J. F. and M. Kimura, 1970 *An Introduction to Population Genetics Theory*. Harper and Row, New York.

Dayhoff, M. O. (Ed.), 1969 *Atlas of Protein Sequence and Structure 1969*. National Biomedical Research Foundation, Silver Spring.

Dickerson, R. E., 1971 The structure of cytochrome c and the rates of molecular evolution. J. Molec. Evolution **1:** 26–45.

Fisher, R. A., 1930 *The Genetical Theory of Natural Selection*. Clarendon Press, Oxford.

Fitch, W. F., 1972 Evolutionary variability in hemoglobins. In: *Haematologie und Bluttransfusion* Edited by H. Martin. J. F. Lehmanns Verlag, Munich.

Harris, H., 1966 Enzyme polymorphisms in man. Proc. Roy. Soc. **164B:** 298–310. ——, 1971 Annotation: Polymorphism and protein evolution. The neutral mutation-random drift hypothesis. Jour. Medical Genetics **8:** 444–452.

Hayashi, A. and G. Stamatoyannopoulos, 1972 Role of penultimate tyrosine in haemoglobin $\beta$ subunit. Nature New Biology **235:** 70–72.

Hendrickson, W. A. and W. E. Love, 1971 Structure of lamprey haemoglobin. Nature New Biology **232:** 197–203.

Iuchi, I., 1968 Abnormal hemoglobin in Japan: Biochemical and epidemiologic characters of abnormal hemoglobin in Japan. Acta Haematologica Japonica **31:** 842–851.

Jukes, T. H., 1971 Comparisons of the polypeptide chains of globins. Jour. Molec. Evolution **1:** 46–62.

Kimura, M., 1968 Evolutionary rate at the molecular level. Nature **217:** 624–626. ——, 1969 The rate of molecular evolution considered from the standpoint of population genetics. Proc. Nat. Acad. Sci. **63:** 1181–1188. ——, 1971 Theoretical foundation of population genetics at the molecular level. Theoretical Population Biology **2:** 174–208.

KIMURA, M. and J. F. CROW, 1964 The number of alleles that can be maintained in a finite population. Genetics **49**: 725–738.

KIMURA, M. and T. MARUYAMA, 1971 Pattern of neutral polymorphism in a geographically structured population. Genet. Res. Camb. **18**: 125–131.

KIMURA, M. and T. OHTA, 1969a The average number of generations until fixation of a mutant gene in a finite population. Genetics **61**: 763–771. ——, 1969b The average number of generations until extinction of an individual mutant gene in a finite population. Genetics **63**: 701–709. ——, 1971a Protein polymorphism as a phase of molecular evolution. Nature **229**: 467–469. ——, 1971b On the rate of molecular evolution. Jour. Molecular Evolution **1**: 1–17. ——, 1971c *Theoretical Aspects of Population Genetics*. Princeton University Press, Princeton. ——, 1972 Population genetics, molecular biometry, and evolution. Proc. 6th Berkeley Symposium on Math. Stat. and Probability. **5**: 43–68.

KING, J. L., 1972 The role of mutation in evolution. Proc. 6th Berkeley Symposium on Math. Stat. and Probability. **5**: 69–100.

KING, J. L. and T. H. JUKES, 1969 Non-Darwinian evolution: Random fixation of selectively neutral mutations. Science **164**: 788–798.

KOHNE, D., 1970 Evolution of higher-organism DNA. Quarterly Reviews of Biophysics **33**: 327–375.

LEWONTIN, R. C. and J. L. HUBBY, 1966 A molecular approach to the study of genic heterozygosity in natural populations. II. Amount of variation and degree of heterozygosity in natural populations of *Drosophila pseudoobscura*. Genetics **54**: 595–609.

MARUYAMA, T., 1970 On the fixation probability of mutant genes in a subdivided population. Genet. Res. **15**: 221–225.

MARUYAMA, T. and M. KIMURA, 1971 Some methods for treating continuous stochastic processes in population genetics. Japanese Jour. Genetics **46**: 407–410.

MAYNARD SMITH, J., 1968 "Haldane's dilemma" and the rate of evolution. Nature **219**: 1114–1116. ——, 1972 Protein polymorphism. Nature, New Biology **237**: 31.

MOTULSKY, A. G.. 1969 Some evolutionary implications of biochemical variants in man. Proc. 8th International Congress Anthropological and Ethnological Sci., Tokyo. pp. 364–365.

OHTA, T., 1971a Linkage disequilibrium and associative overdominance due to random genetic drift. Japanese Jour. Genet. **46**: 195–206. ——, 1971b Associative overdominance caused by linked detrimental mutations. Genet. Res. Camb. **18**: 277–286. ——, 1972 Evolutionary rate of cistrons and DNA divergence. Jour. Molecular Evolution **1**: 150–157. ——, 1972 Fixation probability of a mutant influenced by random fluctuation of selection intensity. Genet. Res. Camb. **19**: 33–38.

OHTA, T. and M. KIMURA, 1970 Development of associative overdominance through linkage disequilibrium in finite populations. Genet. Res. Camb. **16**: 165–177. ——, 1971a Functional organization of genetic material as a product of molecular evolution. Nature **233**: 118–119. ——, 1971b On the constancy of the evolutionary rate of cistrons. Jour. Molecular Evolution **1**: 18–25. ——, 1971c Behavior of neutral mutants influenced by associated overdominant loci in finite populations. Genetics **69**: 247–260.

PERUTZ, M. F. and H. LEHMAN, 1968 Molecular pathology of human haemoglobin. Nature **219**: 902–909.

PRAKASH, S., R. C. LEWONTIN and J. L. HUBBY, 1969 A molecular approach to the study of genic heterozygosity in natural populations IV. Patterns of genic variation in central, marginal and isolated populations of *Drosophila pseudoobscura*. Genetics **61**: 841–858.

SHIBATA, S., I. IUCHI and T. MIYAJI, 1966 Abnormal hemoglobins discovered in Japan. Acta Haematologica Japonica **29**: 115–127.

SVED, J. A., 1968 Possible rates of gene substitution in evolution. Amer. Natur. **102**: 283–292.

VOGEL, F., 1969 Point mutations and human hemoglobin variants. Humangenetik **8**: 1–26.

YAMAZAKI, T., 1971 Measurement of fitness at the esterase-5 locus in *Drosophila pseudoobscura*. Genetics **67**: 579–603.

YANASE, T., M. HANADA, M. SEITA, T. OHYA, Y. OHTA, T. IMAMURA, T. FUJIMURA, K. KAWASAKI and K. YAMAOKA, 1968 Molecular basis of morbidity—From a series of studies of hemoglobinopathies in Western Japan. Jap. Jour. Human Genet. **13**: 40–53.

YANASE, T. and K. YAMAOKA, 1972 Amino acid substitutions at the variant and invariant sites in human hemoglobin, and functional integrity. Jap. Jour. Human Genet. **17** (in press)

# On Some Principles Governing Molecular Evolution*

**(population genetics/mutational pressure/negative selection/random drift)**

MOTOO KIMURA† AND TOMOKO OHTA†

National Institute of Genetics, Mishima, Japan

*Contributed by Motoo Kimura, May 1, 1974*

**ABSTRACT** **The following five principles were deduced from the accumulated evidence on molecular evolution and theoretical considerations of the population dynamics of mutant substitutions: (*i*) for each protein, the rate of evolution in terms of amino acid substitutions is approximately constant/site per year for various lines, as long as the function and tertiary structure of the molecule remain essentially unaltered. (*ii*) Functionally less important molecules or parts of a molecule evolve (in terms of mutant substitutions) faster than more important ones. (*iii*) Those mutant substitutions that disrupt less the existing structure and function of a molecule (conservative substitutions) occur more frequently in evolution than more disruptive ones. (*iv*) Gene duplication must always precede the emergence of a gene having a new function. (*v*) Selective elimination of definitely deleterious mutants and random fixation of selectively neutral or very slightly deleterious mutants occur far more frequently in evolution than positive Darwinian selection of definitely advantageous mutants.**

Recent development of molecular genetics has added a new dimension to the studies of evolution. Its impact is comparable to that of Mendelism and cytogenetics in the past. Accumulated evidence suggests (1–8) that, as causes of evolutionary changes at the molecular (genic) level, mutational pressure and random gene frequency drift in Mendelian populations play a much more important role than the orthodox view of neo-Darwinism could lead us to believe.

In the present paper, we intend to enumerate some basic principles that have emerged from recent evolutionary studies of informational macromolecules. Of these, the first four are empirical, while the last one, which is theoretical, enables us to interpret the four empirical principles in a unified way.

(i) *For each protein, the rate of evolution in terms of amino acid substitutions is approximately constant per year per site for various lines, as long as the function and tertiary structure of the molecule remain essentially unaltered.* In their influential paper on the evolution of "informational macromolecules," Zuckerkandl and Pauling (9), noting that the mean evolutionary rates of globins are approximately equal per year among different lineages, suggested the existence of a molecular evolutionary clock. Actually, the idea of such a clock was implicit in the earlier writings of Ingram (10) and Jukes (11). The approximate constancy of the evolutionary rate in globins has since been confirmed by a number of authors (2, 12–14). For example, the number of observed amino acid differences between the $\alpha$ and $\beta$ hemoglobin chains of man is approximately equal to that between the $\alpha$ chain of the carp and the $\beta$ chain of man (12). Table 1 lists the numbers of amino acid sites in these two sets of comparisons that can be interpreted from the code table as due to a minimum of 0, 1, and 2 nucleotide substitutions. Also, the number of gaps due to insertion and/or deletion is listed. Since the human and carp $\alpha$ chains differ from each other at roughly 50% of the amino acid sites, the data suggest that the two structural genes coding for the $\alpha$ and $\beta$ chains of hemoglobin have diverged independently of each other and to the same extent in the two lines since their origin by duplication which occurred possibly at the end of the Ordovician period. It is remarkable that mutant substitutions at gene loci coding for the $\alpha$ and $\beta$ chains have occurred at practically the same average rates in the two separate lines that have evolved independently over nearly a half billion years. From these comparisons, the rate of amino acid substitution/site per year turns out to be about $0.9 \times 10^{-9}$. On the other hand, from comparisons of the $\alpha$ hemoglobin chains among various mammalian species, we obtain roughly the rate $10^{-9}$ site per year which is in good agreement with the above estimate. Although local fluctuations no doubt occur, constancy rather than variation of the evolutionary rate distinguishes the process of molecular evolution. This is

TABLE 1. *Comparison of amino acid differences between $\alpha$ and $\beta$ hemoglobins*

| Type of change* | Human $\alpha$ vs. human $\beta$ | Carp $\alpha$ vs. human $\beta$ |
|---|---|---|
| 0 | 63 | 61 |
| 1 | 53 | 49 |
| 2 | 22 | 29 |
| Gap | 9 | 10 |
| Total | 147 | 149 |

* The numbers of amino acid sites that can be interpreted from the code table as due to a minimum of 0, 1, 2 nucleotide substitutions in two sets of comparisons involving the $\alpha$ and $\beta$ hemoglobin chains. The number of gaps is also listed for each comparison.

* Contribution no. 1000 from the National Institute of Genetics, Mishima, Shizuoka-ken, 411, Japan.

† *We dedicate this paper to Dr. Hitoshi Kihara the former director of our institute in honor of his 80th birthday anniversary. He was really far-sighted when he wrote, as early as 1947, in relation to his outstanding cytogenetical work on the origin of cultivated wheat, "The history of the earth is recorded in the layers of its crust; the history of all organisms is inscribed in the chromosomes" (original in Japanese, ref. 42). With this paper we also celebrate the 25th anniversary of the National Institute of Genetics.*

particularly noteworthy since it is well-known (15) that there are enormous differences among evolutionary rates at the organism level; some forms have evolved very rapidly while others have stayed essentially unchanged over hundreds of millions of years (especially in organisms known as living fossils). Approximate constancy of the evolutionary rate per year has also been noted in cytochrome *c* and fibrinopeptides (4, 16, 17) although each has its characteristic rate; i.e., the evolutionary rate of cytochrome *c* is about $^1/_3$ while that of fibrinopeptides is roughly 4 to about 9 times that of hemoglobin. Constancy of the evolutionary rate per year has also been noted in albumin evolution of primates (18).

Recently, some authors have questioned the concept of a molecular clock by emphasizing local variation of evolutionary rates. For example, Goodman and his associates (19) emphasize that the evolutionary rates of the hemoglobin $\alpha$ chain slowed down in higher primates. Their method is based on estimating hidden mutant substitutions with the so-called "maximum parsimony" method, and accepts time spans from paleontological studies. In our opinion, the validity of their method (particularly the maximum parsimony principle) has to be tested in several cases rather than being taken for granted. More recently, Langley and Fitch (20) performed a somewhat more reliable analysis on the variation of evolutionary rates among the branches of a phylogenetic tree involving simultaneously the evolution of the $\alpha$ and $\beta$ hemoglobins, cytochrome *c*, and fibrinopeptide A. They found that variation of evolutionary rates among branches ("legs") over proteins is significantly higher than expected by pure chance, with a $\chi^2$ value about 2.5 times its degree of freedom. Since the expected value of $\chi^2$ is equal to its degrees of freedom, their results mean that variation of evolutionary rate in terms of mutant substitutions among lines is about 2.5 times as large as that expected from chance fluctuations. Their estimation of the number of mutant substitutions is based on the assumption of minimum evolution, and it is likely that the estimation is biased in such a direction that lineages with more branches tend to show more hidden mutant substitutions. Yet, their results essentially agree with our previous analysis of the variation of evolutionary rates among lines using data on hemoglobins and cytochrome *c* (21). Namely, the observed variance of evolutionary rates among mammalian lines is roughly 1.5 to about 2.5 times the expected variance. However, the existing data indicate that, when averaged over a long period, the rate of evolution is remarkably uniform among different lineages, even though local fluctuations do occur.

We conclude, therefore, that constancy of evolutionary rate per year is valid as a first approximation. Such a constancy can be explained by the neutral mutation-random drift hypothesis if we assume that the rate of occurrence of neutral mutants is constant per year (4–6). Highly complicated and arbitrary sets of assumptions must be invoked regarding mutation, gene interaction, and ecological conditions as well as population size in order to explain the approximate constancy solely from the neo-Darwinian viewpoint. As predicted by one of us (12), it is likely that genes of "living fossils" in general have undergone essentially as many DNA base substitutions as corresponding genes in more rapidly evolving species. It is this constancy which makes the molecular data so useful and of such great potential value in constructing phylogenetic trees. Eventually, it will be possible to go far back into the history of life to clarify the early stage of evolution far beyond the capability of the traditional methods based on phenotypes.

(ii) *Functionally less important molecules or parts of a molecule evolve (in terms of mutant substitutions) faster than more important ones.* The rate of amino acid substitution has been estimated (with differing degrees of accuracy) for more than twenty different proteins as shown in Table 6-1 of Dayhoff (22). The highest rate is represented by fibrinopeptides ($9 \times 10^{-9}$/amino acid per year according to their estimation) while the lowest rate is that of histone IV ($0.006 \times 10^{-9}$). From this table it turns out that the median rate is $1.3 \times 10^{-9}$/amino acid per year. (represented by myoglobin in the table). This is not very different from $1.6 \times 10^{-9}$/amino acid per year which was estimated earlier by King and Jukes (2) as the average rate for seven proteins. Thus, hemoglobins show an evolutionary rate typical of those proteins that have been studied.

It is interesting to note that fibrinopeptides, the most rapidly evolving molecules, have little known function after they become separated from fibrinogen in the blood clot. The relationship between the functional importance (or more strictly, functional constraint) and the evolutionary rate has been beautifully explained by Dickerson (17) as follows. In fibrinopeptides, virtually any amino acid change (mutant substitution) that permits the peptides to be removed is "acceptable" to the species. Thus, the rate of evolutionary substitution of amino acids may be very near to the actual mutation rate. Hemoglobins, because they have a definite function of carrying oxygen and, so specifications for them are more restrictive than for fibrinopeptides, have a lower evolutionary rate. Cytochrome *c* interacts with cytochrome oxidase and reductase, both of which are much larger than it, and there is more functional constraint in cytochrome *c* than in hemoglobins. Thus cytochrome *c* has a lower evolutionary rate than hemoglobins. Histone IV binds to DNA in the nucleus, and is believed to control the expression of genetic information. It is quite probable that a protein so close to the genetic information storage system is highly specified with little evolutionary change over a billion years. Boyer *et al.* (23) reported that the $\delta$ chain of hemoglobin $A_2$ ($\alpha_2\delta_2$), which forms the minor component of adult hemoglobin, shows higher evolutionary rates and a higher level of polymorphism than the $\beta$ chain which forms the major component A ($\alpha_2\beta_2$). This appears to agree with the present principle that less constraint enables more rapid change.

The evolutionary rate differs not only between different molecules but also between different parts of one molecule. For example, in both the $\alpha$ and $\beta$ hemoglobin chains, the surface part of the molecule evolves nearly 10 times as fast as the functionally important heme pocket (7). In addition, two histidines binding to the heme are absolutely invariant throughout the entire history of vertebrate evolution extending nearly a half billion years (13). The Perutz model of hemoglobins (24) helps us greatly to interpret such observations in terms of structure and function of these molecules. More generally, if we consider the oil drop model of globular proteins (25), the inside of a molecule is filled with nonpolar (hydrophobic) amino acids, while the surface parts are occupied by polar (hydrophilic) amino acids. The functionally vital "active center" is located inside a crevice, and the rate of evolutionary substitutions of amino acids in this part is expected to be very low. On the other hand, the surface parts are usually not very critical in maintaining the function or the tertiary structure, and the evolutionary rates in these parts are expected to be much higher. Another interesting example is the middle segment (C) of the proinsulin molecule. This part is removed when the active insulin is formed, and it is now known that this part evolves at the rate $4.4 \times 10^{-9}/$ amino acid per year, which is roughly 10 times as fast as that of insulin (6, 17). An additional example is afforded by the recent report of Barnard *et al.* (26). According to them, sequence 15–24 of pancreatic ribonucleases evolves at a very high rate comparable to rapidly evolving parts of fibrinopeptides, and this "hypervariability" can be correlated with a lack of any contribution of this part either to the enzymatic activity or to the maintenance of structure required for the activity. Incidentally, their Table 3 listing frequencies of amino acids in hypervariable segments suggests that in such regions there might still exist some selective constraint in amino acid substitutions, so that not all of the mutations are tolerated.

All the observations in this section allow a very simple interpretation from the neutral mutation-random drift hypothesis. Namely, in a molecule or a part of a molecule which is functionally less important, the chance of a mutant being selectively neutral (or very slightly deleterious) is higher, and therefore it has a higher chance of being fixed in the population by random drift. On the other hand, from the neo-Darwinian view-point, we must assume that a rapidly evolving part has an important functional role and is undergoing very rapid adaptive improvements by accumulating many advantageous mutations. It may be argued that the smaller the effect of a mutational change, the higher the chance of it being beneficial as Fisher (27) said, and therefore observations in this section can also be explained by positive natural selection. However, if the selective advantage of a mutant becomes small, then the chance of its fixation in the population becomes correspondingly small. Thus, apart from the problem of validity of Fisher's statement when applied to molecular data, it may not necessarily follow that the smaller the effect, the higher the rate of mutant substitution by natural selection.

(iii) *Those mutant substitutions that disrupt less the existing structure and function of a molecule (conservative substitutions) occur more frequently in evolution than more disruptive ones.* The conservative nature of amino acid substitutions was earlier noted by Zuckerkandl and Pauling (9). They also noted that the code table itself is conservative in that single base substitution often leads to both the substitution of a similar amino acid as well as a synonymous substitution. Since then, the conservative nature of substitutions has been amply documented in evolutionary studies of proteins (17, 22, 28, 29). Clarke (30) treated this problem in quantitative terms by using Sneath's (31) measure of chemical similarity of amino acids and by considering the regression of the relative frequency of evolutionary substitutions on the similarity. His results confirm the well-known fact that chemically similar substitutions occur more frequently than dissimilar ones.

The principle of conservative substitution holds also for nucleotide substitutions. In their extensive study on the evolution of transfer RNA, Holmquist *et al.* (32) found that, among the mispairings in the helical regions, G·U or U·G pairs that do not interfere with helicity occur much more frequently than other forms of mispairing; of 68 observed "non-Watson-Crick pairs," 43 turned out to be either G·U or U·G. For each transfer RNA molecule, the total number of mispairings in helical regions is limited to one or two, suggesting that beyond such a small number, a mutation leading to an additional mispairing becomes highly deleterious and rejected (it is likely that even the mutation causing the first mispairing is deleterious, but it can be fixed by random drift due to very small effect, see ref. 8); only when one of the existing mispairings is closed by a mutant substitution, is the molecule ready to accept a new mutation through random drift and/or selection. This offers an excellent model of Fitch's concept of concomitantly variable codons or "covarions" (33); according to him, only 10% of codons in cytochrome *c* can accept mutations at any moment in the course of evolution. He also found (34) that the proportion of covarions is about 35% in the hemoglobin $\alpha$, but nearly 100% in fibrinopeptide A. A remarkable fact emerging from his analyses is that if the rate of amino acid substitution is calculated on the bases of covarions, cytochrome *c*, hemoglobin $\alpha$, and fibrinopeptide A are all evolving at about the same rate. Fitch's covarion idea, we believe, has a clearer meaning now in the light of selective constraints involved in the secondary and tertiary structure necessary for the function of the molecule.

Similarly, one might expect that synonymous substitutions causing no change in amino acids would occur more frequently in evolution than nucleotide substitutions leading to amino acid change. From studies of amino acid sequences of tryptophan synthetase A-chains of three bacterial species, *Escherichia coli, Salmonella typhimurium,* and *Aerobacter aerogenes,* in conjunction with the estimated nucleotide sequence differences among the corresponding structural genes (determined by mRNA·DNA hybridization, Li *et al.* (35) obtained results suggesting that synonymous codon differences in the gene for tryptophan synthetase A chain are quite common. According to their estimate, there are about as many base differences that do not alter the amino acid sequences as those that alter the sequences. It is possible, as Li *et al.* point out, that not every synonymous substitution is completely neutral with respect to natural selection. Some of them might be subject to selective elimination based on structural requirement (such as the one involved in forming the secondary structure of the RNA molecule). However, because synonymous substitutions, in general, must have a higher chance of being selectively neutral or only very slightly deleterious (other things being equal) than mis-sense substitutions, they have a greater chance of becoming fixed in the population by random drift. One prediction that we could therefore make is that the slower the evolutionary rate of a protein molecule, the higher the ratio of synonymous to mis-sense substitutions.

(iv) *Gene duplication must always precede the emergence of a gene having a new function.* The importance of gene duplication in evolution has been noted earlier by the great Drosophila workers of the Morgan school (see ref. 5). The crucial point pertinent here is that the existence of two copies of the same gene enables one of the copies to accumulate mutations and to eventually emerge as a new gene, while another copy retains the old function required by the species for survival through the transitional period. Shielded by the normal counterpart in the corresponding site of the duplicated DNA segment, mutations that would have been rejected before duplication can now accumulate, and through their accumulation, a stage is set for emergence of a new gene. The creative role which gene duplication plays in evolution has been much clarified by Ohno (36) in his stimulating book in which he considers new evidence based on modern molecular, cytological, and paleontological researches. Together with his recent paper (37), Ohno has made an important contribution to the modern evolutionary theory by bringing to light the remarkably conservative nature of mutant substitutions in evolution. Gene duplication, at the same time, must have caused a great deal of degeneration in duplicated DNA segments. This is because many mutations, which would have been definitely deleterious before duplication, become neutral or only very slightly deleterious after duplication, thus enabling them to spread in the population by random drift (38, 39).

(v) *Selective elimination of definitely deleterious mutants and random fixation of selectively neutral or very slightly deleterious mutants occur far more frequently in evolution than positive Darwinian selection of definitely advantageous mutants.* This is an extended form of the neutral mutation-random drift hypothesis, and is based on the thesis put forward by one of us (8) which argues that very slightly deleterious mutations as well as selectively neutral mutations play an important role in molecular evolution. Adaptive changes due to positive Darwinian selection no doubt occur at the molecular level, but we believe that definitely advantageous mutant substitutions are a minority when compared with a relatively large number of "non-Darwinian" type mutant substitutions, that is, fixations of mutant alleles in the population through the process of random drift of gene frequency. We emphasize that neutral or nearly neutral mutations should be considered not as a limit of selectively advantageous mutants but as a limit of deleterious mutants when the effect of mutation on fitness becomes small. In other words, mutational pressure causes evolutionary change whenever the negative-selection barrier is lifted. As an application of this principle, let us consider the evolutionary change of guinea pig insulin. Although the insulin (A and B segments) in general has a very low evolutionary rate (about $0.33 \times 10^{-9}$/amino acid per year), guinea pig insulin is exceptional in that it diverged very rapidly with the estimated rate of $5.3 \times 10^{-9}$/amino acid site per year (2). From the neo-Darwinian point of view, one might naturally consider such a rapid evolutionary change the result of adaptive change by natural selection. In fact, even King and Jukes (2) in their paper "Non-Darwinian Evolution" invoked "positive natural selection" to explain the rapid change. We suggest that guinea pig insulin lost its original selective constraint in the process of speciation. This allowed the accumulation of mutations which before would have been rejected. This inference is supported by a recent report of Blundell *et al.* (40) who studied the three-dimensional structure of insulin molecules. According to them, guinea pig insulin is accompanied by the loss of zinc in the islet cells (coinciding with the loss of usually invariant histidine B10). This suggests a drastic change in the tertiary structure. It is assumed then that, with the loss of the zinc constraint, mutations in guinea pig insulin started to accumulate at a very high rate approaching the rate in fibrinopeptides (the rate that might be called the fibrinopeptide limit).

When we consider the action of natural selection at the molecular level, we must keep in mind that higher order (i.e., secondary, tertiary, and quaternary) structures rather than the primary structure (i.e., amino acid sequence) are subject to selective constraint, usually in the form of negative selection, that is, elimination of functionally deleterious changes. The existence of selective constraint, often inferred from nonrandomness in amino acid or nucleotide sequences, does not contradict the neutral mutation-random drift hypothesis. Incidentally, it is interesting to note that the fibrinopeptide rate, when expressed in terms of nucleotide substitutions, is roughly equal to the rate of nucleotide substitution in the DNA of the mammalian genome (39). We note also that accumulation of very slightly deleterious mutants by random drift is essentially equivalent to the deterioration of environment, and definitely adaptive gene substitutions must occur from time to time to save the species from extinction.

Although clearly documented cases at the genic level are rather scarce, there is not a slightest doubt that the marvellous adaptations of all the living forms to their environments have been brought about by positive Darwinian selection. It is likely, however, that the ways in which mutations become advantageous are so opportunistic that no simple rules could be formulated to describe them. On the whole, mutations are disadvantageous, and, when a mutant is advantageous, it can be advantageous only under restricted conditions (41). We note also that difference in function at the molecular level, does not necessarily lead to effective natural selection at the level of individuals within a population.

In the past half century, with the rise of neo-Darwinism or more precisely, the synthetic theory of evolution, the claim that mutation is the main cause of evolution has completely been rejected. Instead, the orthodox view has been formed which maintains that the rate and direction of evolution are almost exclusively determined by positive natural selection. We believe that such a view has to be re-examined, particularly regarding evolutionary changes at the molecular level. We think that evolution by mutational pressure is a reality.

We thank Drs. T. H. Jukes and J. L. King for stimulating discussions which helped greatly to compose the manuscript. Especially, we are indebted to Dr. Jukes for critically reviewing the first draft and offering many suggestions for improvement. Thanks are also due to Drs. J. F. Crow and E. R. Dempster for reading the manuscript and offering suggestions for improving the presentation.

1. Kimura, M. (1968) *Nature* **217**, 624–626.
2. King, J. L. & Jukes, T. H. (1969) *Science* **164**, 788–798.
3. Crow, J. F. (1969) *Proc. XII Intern. Congr. Genetics (Tokyo)* **3**, 105–113.
4. Kimura, M. & Ohta, T. (1971) *J. Mol. Evolut.* **1**, 1–17.
5. Kimura, M. & Ohta, T. (1971) *Theoretical Aspects of Population Genetics* (Princeton University Press, Princeton, N.J.).
6. Kimura, M. & Ohta, T. (1972) *Proc. 6th Berkeley Symp. on Math. Stat. and Probability* **5**, 43–68.
7. Kimura, M. & Ohta, T. (1973) *Genetics (Sup.)* **73**, 19–35.
8. Ohta, T. (1973) *Nature* **246**, 96–98.
9. Zuckerkandl, E. & Pauling, L. (1965) in *Evolving Genes and Proteins*, eds. Bryson, V. & Vogel, H. J. (Academic Press, New York), pp. 97–166.
10. Ingram, V. M. (1961) *Nature* **189**, 704–708.
11. Jukes, T. H. (1963) *Advan. Biol. Med. Phys.* **9**, 1–41.
12. Kimura, M. (1969) *Proc. Nat. Acad. Sci. USA* **63**, 1181–1188.
13. Jukes, T. H. (1971) *J. Mol. Evolut.* **1**, 46–62.
14. Air, G. M., Thompson, E. O. P., Richardson, B. J. & Sharman, G. B. (1971) *Nature* **229**, 391–394.

15. Simpson, G. G. (1944) *Tempo and Mode in Evolution* (Columbia Univ. Press, New York).
16. Margoliash, E., Fitch, W. M. & Dickerson, R. E. (1968) *Brookhaven Symp. Biol.* **21,** 259–305.
17. Dickerson, R. E. (1971) *J. Mol. Evolut.* **1,** 26–45.
18. Sarich, V. M. & Wilson, A. C. (1967) *Proc. Nat. Acad. Sci. USA* **58,** 142–148.
19. Goodman, M., Barnabas, J., Matsuda, G. & Moore, G. W. (1971) *Nature* **233,** 604–613.
20. Langley, C. H. & Fitch, W. M. (1973) in *Genetic Structure of Populations* ed. Morton, N. E. (Univ. Press of Hawaii, Honolulu), pp. 246–262.
21. Ohta, T. & Kimura, M. (1971) *J. Mol. Evolut.* **1,** 18–25.
22. Dayhoff, M. O. (1972) *Atlas of Protein Sequence and Structure 1972* (National Biomedical Research Foundation, Silver Spring, Md.).
23. Boyer, S. H., Crosby, E. F., Thurmon, T. F., Noyes, A. N., Fuller, G. F., Leslie, S. E., Shepard, M. K. & Herndon, C. N. (1969) *Science* **166,** 1428–1431.
24. Perutz, M. F. & Lehman, H. (1968) *Nature* **219,** 902–909.
25. Dickerson, R. E. & Geis, I. (1969) *The Structure and Action of Proteins* (Harper & Row, New York, Evanston, London).
26. Barnard, E. A., Cohen, M. S., Gold, M. H. & Kim, Jae-Kyoung (1972) *Nature* **240,** 395–398.
27. Fisher, R. A. (1930) *The Genetical Theory of Natural Selection* (Clarendon Press, Oxford).
28. Epstein, C. J. (1967) *Nature* **215,** 355–359.
29. Lanks, K. W. & Kitchin, F. D. (1972) *Nature* **226,** 753–754.
30. Clarke, B. (1970) *Nature* **228,** 159–160.
31. Sneath, P. H. A. (1966) *Theoret. Biol.* **12,** 157–193.
32. Holmquist, R., Jukes, T. H. & Pangburn, S. (1973) *J. Mol. Biol.* **78,** 91–116.
33. Fitch, W. M. & Markowitz, E. (1970) *Biochem. Genet.* **4,** 579–593.
34. Fitch, W. M. (1972) in *Haematologie und Bluttransfusion* ed. Martin, H. (J. F. Lehmanns Verlag, Munich, Germany), pp. 199–215.
35. Li, S. L., Denney, R. M. & Yanofsky, C. (1973) *Proc. Nat. Acad. Sci. USA* **70,** 1112–1116.
36. Ohno, S. (1970) *Evolution by Gene Duplication* (Springer-Verlag, Berlin).
37. Ohno, S. (1973) *Nature* **244,** 259–262.
38. Nei, M. (1969) *Nature* **221,** 40–42.
39. Ohta, T. & Kimura, M. (1971) *Nature* **233,** 118–119.
40. Blundell, T. L., Cutfield, J. F., Cutfield, S. M., Dodson, E. J., Dodson, G. G., Hodgkin, D. C., Mercola, D. A. & Vijayan, M. (1971) *Nature* **231,** 506–511.
41. Ohta, T. (1972) *J. Mol. Evolut.* **1,** 305–314.
42. Kihara, H. (1947) *Ancestors of Common Wheat* (in Japanese) (Sōgensha, Tokyo).

# HOW GENES EVOLVE; A POPULATION GENETICIST'S VIEW

Motoo KIMURA

*National Institute of Genetics, Mishima (Japan).*

It is a great pleasure as well as an honor for me to be invited to give a lecture at the Collège de France, which Professor Monod in his book "Le Hasard et la Nécessité" referred to as "a fine and precious institution".

In the present talk, I would like to present, following Professor Ruffie's suggestion, my view of evolution as a theoretical population geneticist who is deeply involved in the search for the mechanism of molecular evolution.

Before I go into the main subject, let me review quickly a history of evolutionary theories (i. e., evolution of evolutionary theories) leading to the development of population genetics. I shall also present a short history of population genetics itself (ultimately leading to what I call "molecular population genetics"), and then a short summary of dynamics of mutant substitutions in a finite population. These are intended to serve as a preliminary to our discussion on the mechanism of evolution at the molecular level.

## FROM LAMARCK TO THE THREE SAVANTS

The fact that all the living things on the earth are the products of evolution rather than divine and inmutable creations was first clearly recognized by Jean-Baptiste Lamarck early in the last century. He also put forward, for the first time in biology, a systematic theory to explain how evolution has occurred.

Although his explanation assuming inheritance of acquired characters is now shown to be untenable, his thesis that organisms gradually change with time has since been amply confirmed. It is unfortunate, therefore, that the term Lamarckism has come to be used to represent his unsuccessful theory later shown to be wrong; I personally think that it would have been much nicer and more fair to this truly great French naturalist whose old age was rather tragic, if the term had been reserved to represent evolution, one of the most fundamental processes in the living world.

As everyone knows, the correct explanation of how evolution occurred, especially how organisms have become adapted to their environments was first supplied by Charles Darwin and Alfred Russel Wallace with the theory of natural selection or the survival of the fittest. Furthermore, Darwin elaborated his theory in his book "The Origin of Species" (1859) which has had unmeasurable influence not only on biology but also on human thought in general. We tend to forget that a storm of opposition and criticism once raged against the Darwinian view, for it is now so well established.

Contribution n° 1116 from the National Institute of Genetics, Mishima, Shizuoka-ken, 411, Japan.

Lecture delivered at the College de France on May 14th 1976.

KIMURA Motoo (1976). — How genes evolve ; a population geneticist's view. *Ann. Génét.*, *19*, n° 3, 153-168.

The rise of Mendelian genetics in this century has led to the elucidation of the mechanism of inheritance and the nature of heritable variations, which Darwin vainly struggled to know. The dawn of the Mendelian era, however, was stormy. Soon, a bitter conflict arose between the biometric school championed by Karl Pearson and W. F. R. Weldon and the Mendelian school led by William Bateson [see Provine, 1971]. The biometric school denied the generality of Mendelian inheritance and claimed, following Darwin, that evolution occurred continuously through natural selection acting on small variations. On the other hand, the Mendelians believed that all the heritable variations are discontinuous and therefore evolution occurs with descrete steps.

Although the Mendelian school won the battle, disharmony between Darwinism and Mendelism continued for more than a decade. In addition, the mutation theory put forward by Hugo de Vries at the beginning of this century claimed that evolution occurs by mutational leaps rather than by gradual natural selection. This theory soon became very popular among biologists and found many adherents at that time.

Eventually, population genetics was developed through the efforts to supply a genetical base to Darwin's theory of natural selection. It was a synthesis of Darwinism and Mendelism by the method of biometry. That this was achieved mainly by the works of the three savants, R. A. Fisher, J. B. S. Haldane and Sewall Wright, culminating in their definitive writings early in 1930's, is now widely recognized.

## DEVELOPMENT OF THEORETICAL POPULATION GENETICS

As a branch of genetics, population genetics has a long tradition of investigating the laws which govern the genetic composition of natural populations. The fundamental quantity which is used here is the gene frequency or the proportion of a given allele in the population.

Although early contributions to the field go back to Hardy (1908) and Weinberg (1908), whose names are remembered in biology by the Hardy-Weinberg principle, the systematic treatment of gene frequency changes in a population under various evolutionary factors such as mutation, migration and natural selection, was really started by Haldane in his series of papers, the first of which appeared in 1924. Haldane's treatments in these papers are largely deterministic in that random fluctuation of gene frequencies is disregarded, although he made the first calculation on the probability of gene fixation assuming simple conditions in 1927.

Taking random fluctuation into account and treating the processes of gene frequency changes as stochastic processes really started when Fisher and Wright presented their famous papers early in the 1930's [Fisher, 1930; Wright, 1931]. While providing epoch-making contributions to our understanding of the stochastic behavior of mutant alleles in a finite population, they arrived at diametrically opposed views concerning the role of sampling drift in evolution. Here again a controversy started. The issue this time is whether random genetic drift, that is, random fluctuation of gene frequencies caused by sampling of gametes in reproduction, has any important role to play in evolution.

Fisher came to the conclusion that the number of individuals making up a species is generally so large that the chance effect due to random sampling of gametes is negligible. He also thought that for most mutant alleles the product of the population size and the selection coefficient is unlikely to be restricted to the neighborhood of zero in the course of evolution, so that selectively neutral mutants must be extremely rare.

Wright arrived at an opposite conclusion, that random drift is a significant factor in evolution. Namely, random fluctuation of gene frequencies within subpopulations or demes followed by local selection creates the basis for interdemic selection, and through such a process, epistatic (i. e. non-additive, between-locus) gene interaction can be utilized for rapid evolution. Compared with this, a species consisting of one large panmictic population will soon reach an adaptive plateau by mass selection and it will end up with evolutionary stagnation. Thus a subdivided population structure is most favorable for rapid evolutionary progress of the species as a whole.

During the decade or two that followed, Wright's theory of evolution attracted much attention. Its popularity was greatly enhanced by its detailed exposition in Dobzhansky's book, "Genetics and the Origin of Species" (1937), which was widely read among biologists. In fact, it was an attractive idea (and I still think it is) to say that apparently non-adaptive characters separating related species were brought about through the process of random genetic drift. Terms such as "Sewall Wright effect" and "Wright drift" to designate such a process were at one time popular jargon.

Unfortunately, convincing evidence supporting its universal importance in speciation was very difficult to obtain, and gradually, mainly due to strong opposition by Fisher, Ford and their school in England [see for example, Fisher and Ford, 1950; Ford, 1965], the tide turned against Wright. At the same time, the idea of selectively neutral genes had become quite unpopular. In fact, in his influential book "Animal Species and Evolution" (1965), Ernst Mayr claims ; "...I consider it therefore exceedingly unlikely that any gene will remain selectively neutral for any length of time". He also says; "Selective neutrality can be excluded almost automatically whenever polymorphism or character clines are found in natural populations". I can attest that these represent then prevailing views, for I myself had essentially the same view early in the 1960's.

Although Wright's theory of evolution has been rather controversial, we should not overlook the fact that he has made many fundamental contributions, since his 1931 paper, to our understanding of the stochastic behavior of mutant alleles in finite populations. He has worked out gene frequency distributions under irreversible as well as reversible mutation [Wright, 1937, 1938, 1942, 1945, 1949; for review see Wright, 1969]. Ensuing development of the subject in terms of diffusion models [Kimura, 1964], i. e., by the use of the diffusion equation methods, owes much to his pioneering work.

In discussing the history of theoretical population genetics, I should not fail to mention the name of Gustave Malécot, one of the world's greatest mathematical geneticists. He has elucidated the probabilistic meaning of the inbreeding coefficient, and he has developed an elegant mathematical theory to treat the genetics of structured populations. Both are beautifully presented in his book published in 1948.

During the three decades that followed Wright's 1931 paper, the mathematical theory of population genetics, and the diffusion models in particular, grew gradually, but their applicability to the study of evolution had been quite limited until the era of molecular biology was ushered in. The main reason for this is that the population genetics theory is built on the concept of gene frequencies and the conventional studies of evolution are conducted at the phenotypic level, and there is no direct way of unambiguously connecting the two.

Such a limitation has been removed with the advent of molecular genetics, for it has become possible to study evolution at the molecular level. In particular, by comparing the direct products of genes such as proteins and RNA molecules, we can estimate the genetic differences between two distantly related organisms. From comparative studies of hemoglobins $\alpha$ and $\beta$, cytochrome c, and fibrinopeptides, for example, we can find out how many changes have taken place within the corresponding genes since man and horse diverged some 80 million years ago. Emile Zuckerkandl and Linus Pauling (1965), with their systematic and quantitative approach to the problem, pioneered the development of this new branch of science.

Thus, early in the last decade, the time was ripe to apply the mathematical theory of population genetics to the problems of molecular evolution in order to find out how genes evolve. One should have expected then that the principle of Darwinian selection or the survival of the fittest would be clearly shown to prevail at this most fundamental level of evolutionary change.

Contrary to such an expectation, what has emerged through the quantitative study is the view that random genetic drift rather than positive Darwinian selection prevails at the molecular level. My "neutral mutation-random drift hypothesis" [Kimura, 1968] and the "non-Darwinian" theory of evolution by Jack Lester King and Thomas Jukes (1969) both claim that the majority of mutant substitutions that we can observe at the molecular level are the results of random fixation of selectively neutral or nearly neutral mutations rather than by selective substitutions of definitely advantageous mutations.

These papers marked the beginning of a new controversy which has grown a great deal since then and which is still continuing today. The confrontation between the neutralists and the selectionists has been ably documented by Crow (1972), Calder (1973) and Lewontin (1974). At the same time, stimulated by various problems in population genetics and evolution at the molecular level, much progress has been made in the diffusion theory treating the behavior of mutant alleles in finite populations. This led me to organize a theoretical framework that might be called "molecular population genetics" [Kimura, 1971]. Let me add that for the development of this field, not only myself but also James F. Crow, Takeo Maruyama, Tomoko Ohta and Masatoshi Nei all have contributed significantly. Recently, an excellent book treating this subject was published by Nei (1975).

Throughout my talk, I shall use the term evolution to mean any cumulative change with time in the genetic constitution characteristic to the species; such a change may be progressive, nonadaptive and neutral, or even deteriorating.

## DYNAMICS OF MUTANT SUBSTITUTION AND THE RATE OF EVOLUTION

Except for self-replication, mutation is probably the most fundamental attribute of the genetic material. We should expect that in any natural population of appreciable size, a large number of mutants arise in each generation. Restricting our considerations to mutants produced by DNA base substitutions, consultation of the code table shows that roughly 3/4 or random base changes within a gene (cistron) lead to amino acid substitutions in the corresponding protein. Of these, roughly 1/3 could be detected by ordinary electrophoretic methods through their change in electric charges.

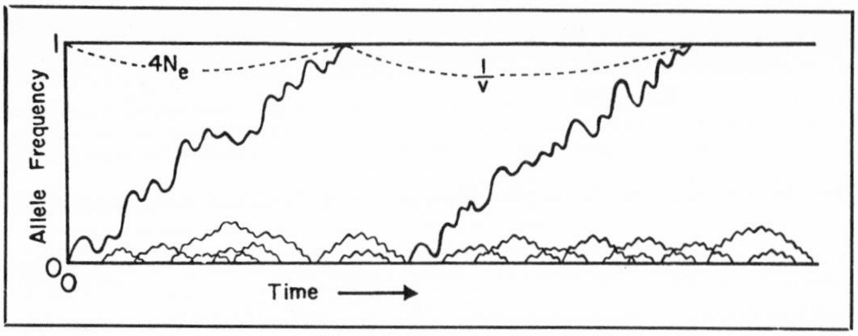


Fig. 1. — Diagram illustrating the behavior of mutant alleles following their appearance in a finite population.

Courses of change in the frequency of mutants destined for fixation are depicted by thick paths.

Since each gene consists of a large number (say, 1000) of nucleotide sites, the mutation rate per site must be very small, likely to be of the order of $10^{-8}$ or less per generation. On the other hand, since each site can be occupied by one of the four kinds of nucleotide bases, the total number of possible allelic states per locus must be astronomical [Kimura and Crow, 1964]. Thus each mutant is likely to be unique at the molecular level, and therefore represented only once at the moment of appearance.

From the standpoint of population genetics, the process of evolution consists of a series of gene substitutions. So, I shall discuss briefly the dynamics of mutant substitutions in a finite population. Figure 1 illustrates the course of change in the frequencies of mutant alleles following their appearance in a finite population. A great majority of such mutants, including those having small selective advantage, are lost from the population within a small number of generations (say, in 10 generations). Only a tiny fraction can spread to the whole population to reach fixation taking a very long time (say, 100,000 generations). In this figure, courses of change in the frequency of mutant alleles destined for fixation are depicted by thick paths. We can make these descriptions more quantitative. For example, if the mutants are selectively neutral and if each mutant is represented only once at the moment of appearance, then the probability of its eventual fixation is $1/(2N)$, where N is the actual size of the population. The average length of time which a mutant takes until fixation, excluding the cases of loss, is $4N_e$ generations [Kimura and Ohta, 1969], where $N_e$ is the effective population size. Note that in ordinary circumstances $N_e$ is roughly equal to the number of reproducing individuals during one generation. Still assuming neutral mutations, if a new allele is produced at a locus with the rate v per generation, then the average length of time between consecutive substitutions of alleles in the population is $1/v$ generations [Crow, 1969; Crow and Kimura, 1970].

If a mutant is advantageous, the probability of ultimate fixation is of course higher, but it was shown by Haldane (1927) that this probability is roughly twice the selective advantage if the advantage is small. As an example, let us suppose that the advantage is one per cent. Then, the chance is about two per cent that a mutant will eventually spread into the whole population. In the remaining 98 % of the cases, it will be lost by chance from the population without being used in evolution. Note that the overwhelming majority of advantageous mutations are likely to have only a slightly advantageous effect, for mutations with large effects tend to be disadvantageous. Thus, a large difference exists between the total number of advantageous mutants that have ever occurred in any species in the course of evolution and the number that have actually been incorporated into the species. An even greater difference exists for neutral mutants.

These considerations makes clear the basic distinction between "gene mutation" at the individual level and "mutant substitution" at the population level. Without keeping these two events conceptually distinct, there can be no meaningful discussions in population genetics. In the literature of molecular evolution, amino acid differences between homologous proteins of related species are usually ascribed simply to "accepted point mutations". But, we should not overlook the fact that each amino acid difference is the result of at least one mutant substitution in which a rare molecular mutant increased its frequency and finally spread to the whole species.

Let us denote by k the rate of mutant substitution per unit time (year, generation, etc.). This is the rate by which mutant genes are substituted one after another in the population in the course of time. Since each substitution may take a very long time, the rate should be measured over a long time period. The rate of substitution thus defined is independent of how quickly individual mutants spread into the population. What matters is the interval between two consecutive fixations (see fig. 2).

Consider a gene locus (or a part of it, such as a codon) and let v be the mutation rate for a particular class of mutants (such as a neutral class), and let u be the probability of ultimate fixation of an individual mutant. If we assume that each mutant is represented only once at the moment of appearance, then we have

VOLUME 19
N° 3 — 1976



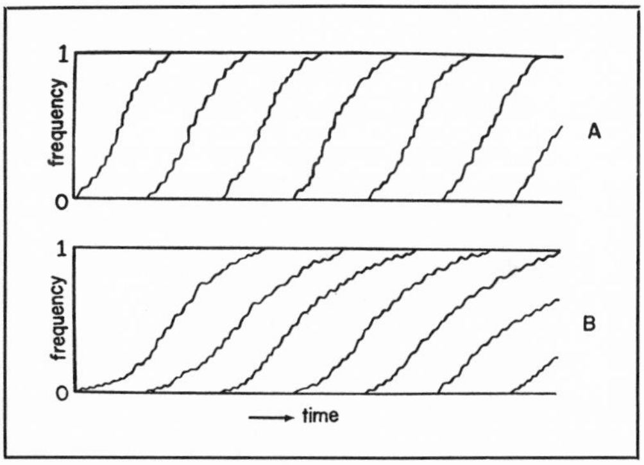


Fig. 2. — Patterns of mutant substitutions in evolution. Of the two cases illustrated, individual mutants increase much more rapidly within population A than B.

Yet the rate of mutant substitution is the same in these two cases, because the interval between two consecutive fixations is the same in A and B.

$$k = 2Nvu, \tag{1}$$

because in a population of N individuals, 2Nv new mutants appear in each generation, of which the fraction u eventually spread to the whole population.

For neutral mutants, the probability of fixation is equal to the initial frequency, i. e. $u = 1/(2N)$ so that

$$k = v. \tag{2}$$

In other words, the rate of mutant substitution in the population is equal to the rate of mutation per gamete. On the other hand, for definitely advantageous mutants with selective advantage s over the preexisting allele, it can be shown that $u = 2sN_e/N$ [Kimura, 1964], so that

$$k = 4N_e sv. \tag{3}$$

In this case, the rate is determined by the product of the effective population size, selective advantage, and mutation rate.

Then, what is the likely figure for the rate of mutant substitution in the actual process of evolution. Comparative studies of the protein sequences among related organisms give a valuable clue to answer this question. Comparison of hemoglobin α and β chains among vertebrates shows that the rate of amino acid substitution is approximately $10^{-9}$ per amino acid site per year, and this rate is quite uniform among diverse lines of vertebrate evolution when the rate is measured per year rather than per generation [Kimura, 1969]. Also, for cytochrome c, a rough constancy is found, although the rate is about 1/3 of the hemoglobin rate. At the moment we do not know for sure how representative these figures are for all cistrons, but King and Jukes (1969) obtained $1.6 \times 10^{-9}$ per amino acid site per year as the average of seven proteins. McLaughlin and Dayhoff (1972) list 21 estimates of the rate of amino acid substitution for various proteins. These values range from $9 \times 10^{-9}$ for fibrinopeptides down to $0.06 \times 10^{-9}$ for histone IV. It is possible that the frequency distribution of these rates does not follow a normal distribution, so the median is more suitable as a representative value than the arithmetic mean. It turned out, however, that the median rate is $1.3 \times 10^{-9}$ per amino acid site per year [Kimura, 1974] which is not very different from the average value ($1.6 \times 10^{-9}$) obtained by King and Jukes.

If we take $1.5 \times 10^{-9}$ as the representative rate per amino acid site, this corresponds roughly to the substitution rate of $0.7 \times 10^{-9}$ per nucleotide site in which synonymous mutations (that is, mutations which do not lead to amino acid change) are taken into account. These are very low rates; the substitution rate of $10^{-9}$ represents a process in which one mutant substitution occurs on the average during one billion ($10^9$) years. Considering the fact that the length of time since the origin of life on the earth is some four billion, it is likely that the average nucleotide has been replaced only a few times in its entire history (if we exclude the possibility that error rates were much higher when life was getting started).

On the other hand, if we extrapolate the figure per site to the total number of nucleotide sites making up the haploid chromosome set of mammals, the amazing conclusion is obtained that substitution of mutant sites within a population goes on at the rate of about 2.5 per year. This is because the total number of nucleotide sites per haploid chromosome set of mammals is very large, amounting to about 3.5 billion. Then, for mammals which take three years for one generation, the rate of mutant substitution amounts to at least half a dozen per generation. This is more than 2 thousand times higher than Haldane's famous estimate made in 1957, that, in standard-rate evolution, new genes might be substituted in the species at the average rate of once every 300 generations. Although a more conservative estimate for molecular evolution was used in my original paper on the subject [Kimura, 1968], the discrepancy between the extrapolated rate and Haldane's estimate was very large, and this led me to propose the hypothesis that a majority of amino acid (and nucleotide) substitutions in the course of evolution are the result of random fixation of selectively neutral or nearly neutral mutants rather than of "positive" Darwinian selection. I am not going to detail the rationale which led me to this idea, except to mention that it was based on the consideration of what was then known as Haldane's concept of the cost of natural selection [Haldane, 1957] and also on the very high frequency at which advantageous mutants must be assumed to occur if all the mutant substitutions were indeed due to positive Darwinian selection [Kimura, 1971].

My original argument was concerned with the rate of mutant substitutions at all the nucleotide sites of the mammalian genome. If the part of DNA that act as "genes" (i. e. translated into amino acids) is a small fraction of the total DNA, then, as pointed out by Crow (1972), the observed rates of amino acid substitutions do not contradict Haldane's concept of the cost. On the other hand, the actual rate of mutant substitutions per haploid genome appears to be even higher than estimated above. In fact, the rate of evolutionary divergence of non-repetitive DNA sequences as estimated by the hybridization technique [Kohne, 1970] is higher than that of most proteins (in terms of nucleotide substitutions), and is roughly equal to that of fibrinopeptide. This conforms with the neutral theory if we assume that a large fraction of DNA does not code for proteins and is therefore less subject to selective constraint [Ohta and Kimura, 1971]. Furthermore, there is evidence showing that total single-copy DNA evolves roughly twice as fast as DNA which is transcribed into poly(A)-containing (i. e. messenger) RNA in mouse and rat [Rosbash, Campo and Gummerson, 1975]. This is consistent with the view that the part of DNA which is transcribed (but not necessarily be translated) is subject to selective constraint stronger on the average than the single copy DNA but weaker than the part of DNA which is translated.

Before we discuss more fully the problems of molecular evolution, I would like to spend a few words on the rate of evolution in general. Traditionally, the rate of evolution has been considered at least at two different levels. One is the rate measured at the "organism" level, as extensively studied by Simpson (1944). For example, the line leading to the horse (*Equus*) from *Eohippus* went through eight successive genera during about 45 million years, giving approximately 0.18 genera per million years. This is within the range of "standard" rates which Simpson called "horotelic". It is well known among students of evolution that there are enormous differences among evolutionary rates. Some forms have evolved very rapidly while others have changed so slowly as to be hardly evolving at all.

Another is the rate measured at the level of quantitative characters such as body size and length of tooth as studied by Haldane (1949). For example, in the evolution of the horse, he found that tooth length changed on the average at the rate of about $4 \times 10^{-8}$ per year or 4 % per million years. He proposed the term *darwin* as a unit of evolutionary rate, representing a change in measurement at the rate of $10^{-6}$ per year. In these terms the horse rate is about 40 millidarwins. Also according to Haldane, the rate of evolutionary increase in body length of Dinosaurs during the mesozoic era is roughly half as large. On the other hand, it is known that in hominid evolution, cranial capacity increased by a factor of about 2 in 3 million years [Leakey, 1973; Pilbeam and Gould, 1974], so the rate is near 1/5 darwin, an example of rapid evolution.

As compared with these, an appropriate measure of evolutionary rate at the molecular level, as we have seen, is the rate of mutant substitution per site per year. Some years ago, I proposed [Kimura, 1969] the term *pauling* to represent the rate of substitution of $10^{-9}$ per amino acid site per year. In terms of this, the hemoglobin rate is roughly one *pauling,* and, this appears to represent a standard rate in molecular evolution.

## EVOLUTIONARY CHANGE AT THE MOLECULAR LEVEL

Although evolution as a fact was recognized by Lamarck more than 150 years ago, it is only during the last 10 years that we have started to uncover the process of evolutionary change at the molecular level, that is, inside the gene. Already, new observations have brought many enlightening results, but at the same time some puzzling ones.

Before molecular data became available, it looked as if the neo-Darwinian view, which claims that the rate and direction of evolution are determined almost exclusively by positive natural selection, had been firmly established. According to this view, selectively neutral genes are extremely rare if they ever exist, and random genetic drift has only a very minor role to play in evolution, except possibly through the "founder effect" when a few individuals are isolated to start a new colony. Every change in evolution is thought to be adaptive. Similarly, every mutational change that spreads through the species is believed to have some selective advantage. As far as phenotypes are concerned, treating evolution as an adaptive process had been very useful in the past, and, it was natural to expect that the same applies to evolution at the molecular level.

The picture of evolutionary change that has emerged from molecular studies, however, appears to be quite different from such an expectation. Some of the outstanding features disclosed are uniformity of the rates of amino acid substitutions per year for each protein among diverse lineages, the apparent randomness in the pattern of substitutions, and the very high overall rate, amounting to at least a half-dozen mutant nucleotide substitutions per genome per year when extrapolated to the haploid DNA material of mammals.

Recently, we have summarized [Kimura and Ohta, 1974] various observations in the following four empirical principles: [1] for each protein, the rate of evolution in terms of amino acid substitutions is

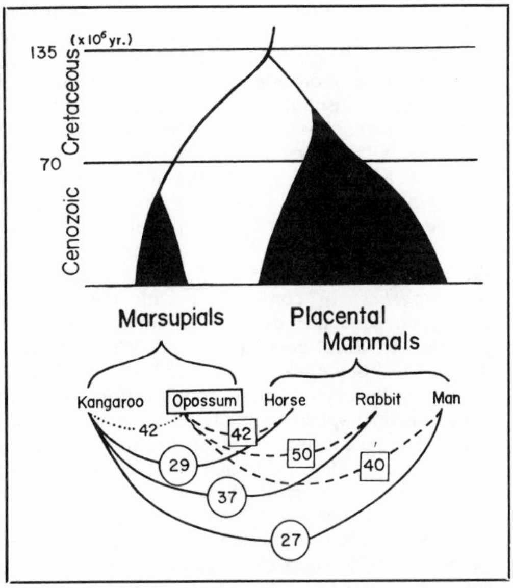


Fig. 3. — Phylogeny and hemoglobin divergence among some marsupials and placental mammals with special reference to the opossum. Numerals indicate the number of amino acid differences in the hemoglobin α chain between the animals compared.

approximately constant per year per site for various lines, as long as the function and tertiary structure of the molecule remain essentially unaltered. [2] Functionally less important molecules or parts of a molecule evolve (in terms of mutant substitutions) faster than more important ones. [3] Those mutant substitutions that disrupt less the existing structure and function of a molecule (conservative substitutions) occur more frequently in evolution than more disruptive ones. [4] Gene duplication must always precede the emergence of a gene having a new function.

Also, we have pointed out that all these empirical principles can be interpreted in a unified way by assuming that random fixation of selectively neutral or very slightly deleterious mutants occur far more frequently in evolution than selective substitution of definitely advantageous mutants. In this connection, neutral mutants should be considered not as a limit of selectively advantageous mutants but as a limit of deleterious mutants when the effect of mutation on fitness becomes small. In other words, mutational pressure causes evolutionary change whenever the negative-selection barrier is lifted.

As to the observed constancy per year of evolutionary rate for each protein among diverse lineages, it can be explained much more naturally by the neutral theory than by neo-Darwinian theory. Namely, under the neutral theory such a constancy can be explained by assuming that the rate of occurrence of neutral mutants (due to base substitutions) is constant per year. Needless to say, the validity of this assumption has to be tested. Actually, some doubt has been expressed by not a few geneticists if the mutation rate for neutral alleles is the same per year rather than per generation, for example, between man and Drosophila. However, if the two species to be compared have roughly the same generation time, such a problem does not arise. Note that for neutral mutants the rate of substitution in the population is equal to the mutation rate (see eq. 2). On the other hand, in order to explain the observed constancy by Darwinian substitutions of definitely advantageous mutants, a highly complicated and arbitrary set of assumptions has to be invoked regarding mutation rate, gene interaction, and ecological conditions as well as population size. This is apparent from eq. [3] in the previous section.

A few years ago, I proposed [Kimura, 1969] that if hemoglobins and other proteins show the same rate of amino acid substitutions in "living fossils" as in rapidly evolving species, this would support my neutral mutation theory. Recently, in order to test this proposal, Stenzel (1974) undertook to determine the sequence of the α chain of hemoglobin from a "living fossil", the Virginia opossum. Figure 3 illustrates the numbers of amino acid differences between Opossum, Kangaroo and a few mammalian α chains. These numbers are taken from Stenzel's Table 2. It is evident from the figure that the α chain of the opossum has undergone quite extensive evolutionary change despite the animal's very slow evolutionary change in phenotypes. In fact, the opossum α chain appears to have evolved more rapidly than any other α chains, even exceeding the expectations of the neutral mutation theory [Stenzel, 1974]. At any rate, if my hypothesis turns out to be correct, we should expect that every species, including those that have remained unchanged for a very long time at the phenotypic level, such as coelacanths, horseshoe crabs and *Lingula,* are undergoing constant change at the nucleotide level. In other words, underneath their unchanged morphology, and probably physiology as well, that have been kept remarkably constant by incessant action of "stabilizing" selection for hundred million years, a steady but great stream of almost neutral genetic variations has flowed through random genetic drift, transforming the nucleotide sequences of their genes tremendously.

Although the neutral theory emphasizes the role played by selectively neutral mutants, it does not claim that all mutations at the molecular level are neutral. It rather claims that the probability of a mutational change being neutral depends much on molecular constraint; if a molecule of a part of a molecule is functionally less important, or more strictly speaking, has less functional constraint, then

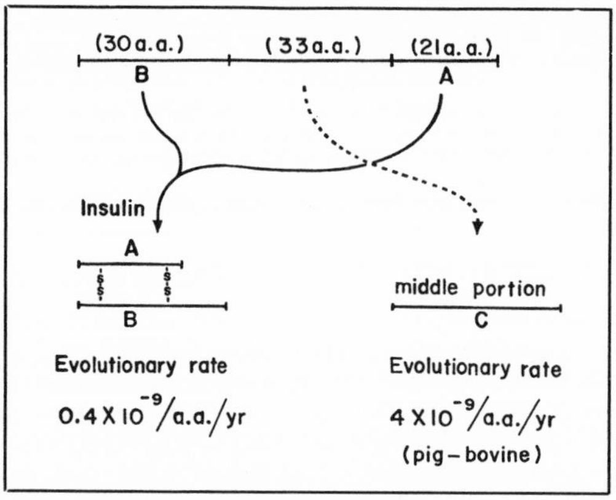


Fig. 4. — Diagram showing the process of insulin formation after the removal of the middle segment (C) from proinsulin. Evolutionary rates are those given by Kimura and Ohta (1972).

the probability of mutational change in it being selectively neutral is higher. This means that the rate of mutant substitutions due to random fixation of selectively neutral mutants is higher in a functionally less important molecule or part of a molecule. This accords with the observation that the fibrinopeptides and the middle segment (C) of the proinsulin molecule (see fig. 4) which are discarded when functionally active molecules are formed, have very high evolutionary rates [Dickerson, 1971; Kimura and Ohta, 1972].

Also, in hemoglobins, it is known [Perutz and Lehman, 1968] that the surface part generally has much less constraint in keeping the structure and function of the molecules than the heme pocket which is vitally important. We have shown that in both α and β hemoglobins the surface part evolves about 10 times as fast as the heme pocket [Kimura and Ohta, 1973a]. An alternative explanation that the higher evolutionary rates are due to higher mutation rates does not conform to observations. Particularly noteworthy is the relationship between evolutionary conservatism and spontaneous mutability of the two histidines that are linked to the heme [Kimura and Ohta, 1973a]. It has been reported [Jukes, 1971] that these two histidines in the globin molecule are invariant in vertebrate evolution. Yet, in human populations hemoglobin variants caused by substituting tyrosine for one of these histidine residues are known. Such variants (leading to hemoglobin M disease), although rare, are by no means excessively rare compared with other hemoglobin variants. Because of their highly deleterious effects in homozygous condition, this type of mutations has been rejected in vertebrate evolution.

Similarly, we should expect that "synonymous" or "silent" substitutions causing no change in amino acids occur more frequently (per site basis) than nucleotide substitutions leading to amino acid change. This follows from the premise that mutant substitutions which are less disruptive to structure and function of a molecule occur more frequently in evolution. Consultation of the genetic code table shows that roughly 1/4 of random nucleotide substitutions within a codon are synonymous, but evidence is accumulating that the actual fraction of synonymous change is higher. From studies of amino acid sequences of trytophan synthetase A-chains of synonymous change is higher. From A-chains of three bacterial species (*E. coli, S. typhimurium* and *A. aerogenes*) in conjunction with the estimated nucleotide sequence differences among the corresponding structural genes (using mRNA-DNA hybridization technique), Li et al. (1973) obtained results suggesting that synonymous codon differences in the gene for tryptophan synthetase A chain are quite common. According to their estimate, there are about as many base differences that do not alter the amino acid sequences as those that alter the sequences. There is another example of more pronounced occurrence of synonymous change. Robertson and Jeppsen (1972) report on three related RNA phages, MS2, R17, and f2. MS2 and R17 each differ from f2 by only one amino acid substitution in the coat protein cistron. Yet they differ in 15 to 20 base replacements. Also, there is evidence indicating that although histons III and IV are highly conserved in evolution, genes coding for them accumulate mutations significantly at the third positions; according to Farquhar and McCarthy (1973), DNA divergence, as estimated from hybridization experiments with sea urchin histone mRNA and mouse DNA, is approximately 8 %, of which only 0.5 % can be attributed to amino acid charges. One more example was reported by Salser et al. (1976); comparison of rabbit and human hemoglobin mRNA's indicates that substitution rate per nucleotide is some 10 times higher for mutations that do not alter the amino acid sequence than those that alter it. In this case synonymous changes per site turn out to be even higher than the substitution rate estimated from fibrinopeptide A.

It is possible, that not every synonymous substitution is completely neutral with respect to natural selection. Some of them might be subject to selective elimination based on structural requirement, such as the one involved in forming the secondary structure of the RNA molecule. However, synonymous substitutions in general must have a higher chance of being selectively neutral than mis-sense substitutions, and therefore have a higher chance of becoming fixed in the population by random drift.

On the other hand, in order to explain these observations by the neo-Darwinian theory, we must assume that a rapidly evolving part has some important function and is undergoing very rapid adaptive improvements by accumulating many advantageous mutations

In this connection it might be argued that the smaller the effect of a mutational change, the higher the chance of its being beneficial as Fisher (1930) showed, and therefore these observations can equally well be explained by positive natural selection. Fisher derived his result by considering the process of adaptive evolution of an organ, such as an eye, effected through a series of adjustments of its component parts. Fisher's treatment may be appropriate when we consider, in general terms, the process of adaptive change of a continuously varying phenotypic character which is determined by many loci. In this case, not only a smaller mutational effect has a higher chance of becoming beneficial, but also mutations with smaller effects may occur more frequently [Mukai, 1969] than those with larger effects. This is because more loci are available for minor mutations or every locus tends to produce such mutations. Thus gradual phenotypic change should characterize the process of adaptive evolution in agreement with the view of Darwin. This type of argument, however, does not apply when we consider evolutionary rate in terms of mutant substitutions per site. It is important to note that the rate of mutant substitutions in evolution is determined not only by the rate of production of mutants (per amino acid site) but also by their probability of fixation. If the selective advantage of a mutant becomes small, then the probability of its fixation becomes correspondingly small, as may be seen from the formula $u = 2sN_e/N$. In addition, there is no evidence for higher mutation rate *per site* at functionally less important parts of a molecule. These considerations show that resorting to Fisher (1930) is of no help to explain the observations in neo-Darwinian terms.

In my opinion, in order to gain deeper understanding of the mechanism of molecular evolution, it is much more rewarding to search for the relationship between the pattern of mutant substitutions on one hand and the tertiary structure and function of the molecule on the other, rather than trying to carry over concepts from statistical genetics which were once developed to analyse visible phenotypic characters. We are still unable to describe in concrete physico-chemical terms how molecular constraint becomes shifted as new mutants are substituted one by one in the course of evolution. This is particularly true for proteins, although Fitch's concept of concomitantly variable codons or "covarions" will serve as a useful guide. According to the statistical analysis of Fitch and Markowitz (1970), only 10 % of codons in cytochrome c can accept mutations at any moment in the course of evolution, and the positions of covarions inside the molecule change with time. Also, it was found [Fitch, 1971a] that the proportion of covarions is about 30 % in the hemoglobin α, but nearly 100 % in fibrinopeptide A. What is really interesting is his finding that if the

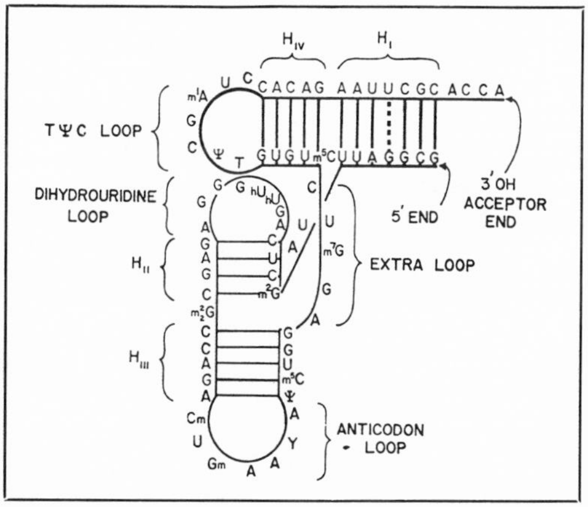


Fig. 5. — A model of the yeast phenylalanine tRNA showing four helica regions, $H_I$, $H_{II}$, $H_{III}$ and $H_{IV}$ (from Kim et al., 1973, slightly modified)

rate of amino acid substitutions is calculated on the bases of covarions, cytochrome c, hemoglobin α and fibrinopeptide A are all evolving at about the same rate. From the standpoint of the neutral theory, covarions can be interpreted in a natural way by assuming that they represent amino acid sites which can accept mutations by random genetic drift without impairing structure and function of the molecule. Fitch (1971b) has also been able to estimate the persistence of a covarion, that is, the probability that a codon which belongs to a covarion group remains so after a mutation becomes fixed at a different codon. His estimate of the persistence turns out to be less than 0.25 suggesting rapid turnover of covarions within one molecule.

I shall now show that in transfer RNA, shifting of molecular constraints can be explained in more concrete terms, especially at its helical regions. As is well-known, the secondary structure of all the tRNA molecules can be arranged in a clover leaf configuration, although the molecules are folded in more complex fashion to form the tertiary structure [Kim et al., 1974; Klug et al., 1974]. In figure 5, yeast phenylalanine tRNA is illustrated by taking into account the fact that its three-dimensional structure is L-shaped [Kim et al., 1973]. In general, the tRNA molecule contains four helical regions, as denoted by $H_I$, $H_{II}$, $H_{III}$, and $H_{IV}$ in figure 5. They correspond respectively to the amino acid acceptor, DHU (dihydrouridine), anticodon and TΨC stems. In these regions, bases are mostly paired forming Watson-Crick pairs, although one or two mispairs tend to develop, especially in the first helical region. It was suggested [Jukes, 1969; Holmquist et al., 1973] that in these regions the maintenance of double helical structure is their only requirement. This

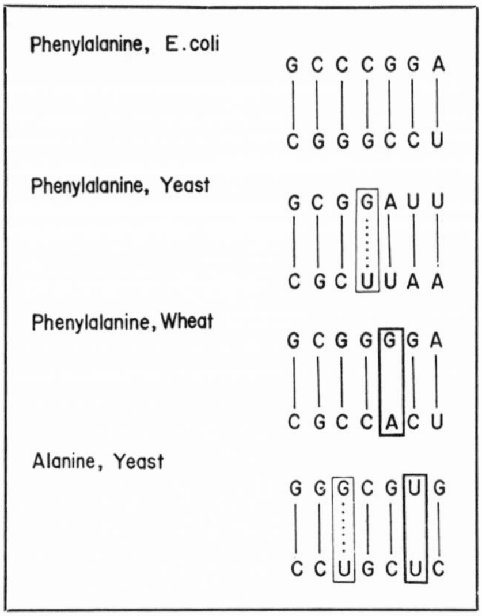


Fig. 6. — Examples showing 0, 1 and 2 nucleotide mispairings at the amino acid acceptor stem in the tRNA molecule.

means that individual bases may change freely in the course of evolution as long as the base substitutions are coupled in such a way that Watson-Crick pairs are extensively maintained in these regions. According to Holmquist et al. (1973), among the mispairings in the helical regions, G-U pairs (by this I mean G-U or U-G) that do not interfere with helicity occur much more frequently than other forms of mispairings. In fact, G-U pairs amounts to more than 60 % of all the "non-Watson-Crick pairs". It is interesting to note also that in these regions G-C pairs that have 3 hydrogen bonds occur much more frequently (69 %) than A-U pairs that have 2 hydrogen bonds. This suggests the importance of holding a tight double stranded structure. On the other hand, in the non-helical regions consisting of various loops, we find far fewer G-C pairs (39 % G + C), suggesting the opposite requirement of maintaining an open, unpaired structure necessary for molecular recognition. The majority of the "invariant sites" also occur in these loops.

Let us now examine in more detail the helical regions, particularly the amino acid acceptor stem (a. a. stem) consisting of seven base pairs. Figure 6 illustrates a few examples having 0, 1 and 2 mispairings. Of 43 tRNAs compiled by Holmquist et al. (1973), the average number of mispairings per a. a. stem is 0.65. Of 28 mispairs, the majority, that is 23, are G-U, other types such as G-A and U-U occuring much less frequently. Furthermore the incidence of a non G-U type mispairing appears to be independent of whether one G-U pair exists in another part of this stem or not. We can visualize the process of evolutionary change in this part of the molecule as an alternating sequence of opening and closing of Watson-Crick pairs. As pointed out by Ohta (1973), the first opening is likely to be very slightly deleterious, but a mutation causing such a change can spread in the population by random drift. It is possible that one mispairing, especially when it is G-U, has such a small deleterious effect that it is practically neutral. Thus, G-U serves as a most natural transitional state going from G-C to A-U or vice versa as pointed out by Jukes (1969). On the other hand, non-G-U type mispairings cause definite weakening of the bonding, and mutations leading to more than one such mispairing are eliminated from the population because of marked deleterious effects; only when the existing mispairing is closed by a mutant substitution, is the molecule ready to accept a new mutation. Even G-U pairs, when more than two accumulate in this region, appear to show a definitely deleterious effect. Thus, although there are seven base pairs (14 nucleotide sites) in this region, the effective number of variable sites at any moment in the course of evolution is smaller.

So far, we have considered nucleotide and amino acid substitutions in evolution. Recently, a fascinating but pazzling new phenomenon has been found from molecular studies. This is the "coincidental evolution" of multigene families [Hood L. et al., 1975]. I shall explain this phenomenon taking spacer regions associated with ribosomal RNA genes in eukaryotes as an example. In the African toad *Xenopus laevis,* the gene for 18S-28S rRNA is repeated about 450 times within a chromosome, each repeating unit consisting of 18S, 28S and spacer regions. Essentially the same number of repeats with the same structure are found in the related species *Xenopus mulleri.* However, when nucleotide sequences are compared between these two species, the 18S and 28S regions are shown to be identical, but the spacer regions have diverged significantly. Since the spacer has no known function, it is reasonable to assume that it is subject to less selective contraint and therefore mutations have accumulated more rapidly by random drift. What is puzzling, however, is that, within each species, sequences of spacer regions are essentially identical. How such intraspecific homogeneity of the spacer regions is maintained while they are evolving rapidly ?

An ingeneous explanation has been put forward by Smith (1974) and Black and Gibson (1974) assuming homologous but unequal crossing over. Figure 7 illustrates the process by which a multigene family within a chromosome becomes homogeneous under repreated "intrachromosomal" (i. e. between-sister-chromatid) unequal crossing over. As pointed out by Hood et al. (1975) as well as by Smith, the

VOLUME 19
No 3 — 1976 

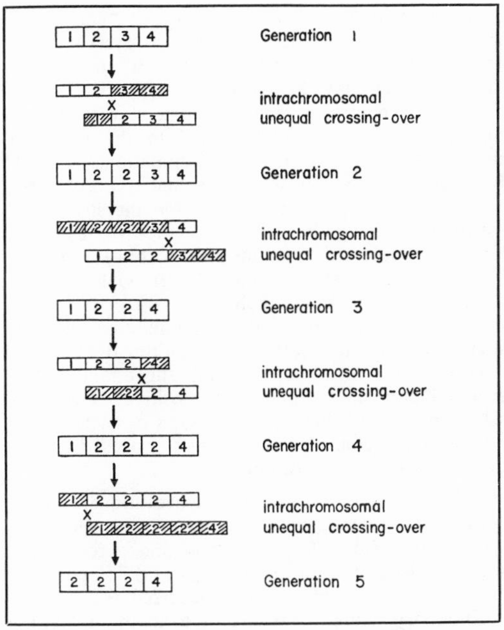


Fig. 7. — Diagram illustrating the process of crossover fixation in a multigene family.

process is analogous to random fixation of one of the alleles in a finite population, so that the term "crossover fixation" has been proposed to designate it. The average time until one of the variant copies of the family becomes fixed in a chromosome lineage is called the crossover fixation time. Smith, Black and Gibson performed simulation experiments to obtain some quantitative estimates regarding the number of crossovers that are needed to attain a certain level of homogeneity among the family members. It is possible that by suitably choosing a model for unequal crossing over, the theory of Kimura and Ohta (1969) on the average number of generations until fixation of a mutant allele within a finite population could be used to treat the problem analytically (*). It gives me some pleasure to note that here again the diffusion models appear to be useful to treat an important problem in molecular evolution.

## PROGRESS AND DEGENERATION IN EVOLUTION

The word evolution is not a monopoly of biology. Nowadays, it is often used to represent a historical development in the physical world such as "evolution" of stars and of the universe. However, biological evolution is distinguished from the rest in that it usually entails increased adaptation to the environment.

It is true that adaptive change brought about by positive Darwinian selection is the most important aspect of biological evolution. Through accumulation of such changes, progressively higher forms have emerged on the earth. What we call higher forms not only have more complex organizations but also can utilize a wider environment for their own benefit. This is why the notion of progress tends to be associated with the concept of evolution. Yet, we should not overlook the fact that a great deal of degeneration as well as neutral change has occurred in the course of evolution.

In the preceeding sections I have presented evidence suggesting that a majority of mutant substitutions that we observe at the molecular level are selectively neutral. This does not mean that adaptive changes do not occur at the level of information macromolecules. On the contrary, the marvellous function of molecular machineries on which life depends must be the products of positive Darwinian selection. However, most of the molecules that we are studying now, such as hemoglobins, cytochrome c, transfer RNAs, etc., must have had their essential designs perfected very far in the past; the changes thereafter have been mostly variations on the theme. This is clearly shown, for example, by the fact that all the globin molecules, including hemoglobins of an insect and a marine annelid worm have exactly the same tertiary structure characterized by the "myoglobin fold" [see Dickerson and Geis, 1969], despite an extensive change in the primary structure.

A majority of amino acid substitutions that we observe through comparative studies of vertebrate hemoglobins does not seem to have any obvious relationship to environmental conditions where the animals are placed. For example, man and gorilla differ by one amino acid in each of the $\alpha$ and $\beta$ hemoglobin chains, but it is doubtful if these differences have evolved in order to fit different lives of these two species. I note that the existence of functional differences among homologous proteins in different species, as detected under experimental conditions, is often associated with adaptive significance. Tomita and Riggs (1971) studied the oxygen affinity of mouse, human, and elephant hemoglobins, and found that the Bohr effect is larger for animals having smaller bodies and higher metabolic rates. They conclude that this finding does not support the neutral mutation-random drift hypothesis. I admit that hemoglobins of different species may have different function. But it is by no means clear that each of the hemoglobins has special qualities that fit it exactly to the need of the species, and that these needs must therefore be different from the exact

(*) Recently, a promising model and its analytical treatments have been presented by Tomoko Ohta (see *Nature,* 1976, *262,* No 5570.

needs of all other species in relation to ecological conditions. In addition, it is important to note that difference in function at the molecular level does not necessarily lead to effective natural selection at the level of individuals within a population.

From the standpoint of population genetics, positive Darwinian selection represents a process whereby advantageous mutants spread through the species. Considering their great importance in evolution, it is perhaps surprising that well established cases are so scarce; examples constantly being cited are industrial melanisms in moths and increase of DDT resistance in insects. On the other hand, examples abound showing that negative selection is at work to eliminate variants produced by mutation [see Kimura and Ohta, 1971]. As pointed out by Haldane (1959) elimination of deviants to keep the status quo (in the form of stabilizing or centripetal selection) is the most common type of natural selection. Also a remarkably conservative nature of natural selection has been brought to light by Ohno (1970) in his important discussion on the role of gene duplication in evolution. What has been revealed by recent studies of molecular evolution is again the prevalence of this type of natural selection.

At the same time, the possibility of mutational pressure directing the course of evolutionary change has been much enhanced; it appears that whenever a character becomes shielded from the direct action of natural selection, mutations start to accumulate in the population by random drift leading to degeneration in many cases. Note that mutation, being random in nature, has more chance of disabling the function of a gene than enhancing it. That mutational pressure is the main cause of degeneration of organs and characters that have ceased to be useful has been pointed out by several authors [Haldane, 1933; Muller, 1939; Wright, 1938; Brace, 1963].

Degeneration of eyes and loss of pigment of cave animals are conspicuous examples, although selectionist interpretations may also be possible.

If we adhere to the view that every change in evolution is adaptive, we must assume that degeneration of eyes itself is advantageous in the dark environment such as saving the animal from eye infections. However, cave animals in general have small population numbers and it is likely that random genetic drift plays a prominent role in them. Recently, Avise and Selander (1972) investigated 17 enzyme loci in cave populations of the characid fish *Astyanax mexicanus* together with their conspecific surface populations. In two cave populations that are isolated and having small population numbers (with $N_e$ probably 50 or less), genetic variability is either absent or very low, while in surface populations (with very large $N_e$) levels of genetic variability are quite high with average heterozygosity of 11.2 %.

As to the pattern of accumulation of "amorphic" mutants leading to degeneration of eyes in cave environment, it may be noted that in the first approximation this may follow the pattern of random fixation of neutral mutants. Namely, the rate of mutant substitutions (k) is equal to the mutation rate (v) but independent of the population size ($N_e$). It is possible, however, that the actual speed of degeneration is lower than expected from the strict neutral theory. This is because, as pointed out by Muller (1939), amorphic mutants tend to have undesirable pleiotropic effects which lower viability and fertility independent of their effect on eyes. Thus the real pattern follows the accumulation of very slightly deleterious mutants by random drift rather than strictly neutral ones. This suggests that the accumulation of modifiers by positive Darwinian selection should follow the process in order to remedy the undesirable pleotropic effects caused by fixation of slightly deleterious mutants.

Another interesting example of evolutionary degeneration of characters is the loss of ascorbic acid synthesizing ability found in man and a few other species. Recently, Jukes and King (1975) put forward a thesis that this loss had occurred as a neutral evolutionary change. According to them, the ability to synthesize ascorbic acid (vitamin C) is in general a characteristic of terrestrial vertebrates. However, besides the human, the ability is not present in monkeys, guinea pigs, fruit-eating bats and some passerine birds. These animals consume food rich in ascorbic acid, and it is assumed that the mutants leading to loss of the synthesising ability were neutral and became fixed in the species through random gene frequency drift. The alternative hypothesis is that the loss was adaptive and occurred by positive natural selection. Since many herbivorous vertebrate species which consume food high in ascorbic acid have retained the ability to synthesize it, Jukes and King's hypothesis appears to be more plausible. The observation that the loss occurred in species that are widely scattered in phylogeny also supports their hypothesis.

The evolutionary effect of "use or disuse" on the development or degeneration of characters is a subject which was once much discussed in relation to Lamarck's theory of evolution. The explanation of this phenomenon in terms of inheritance of acquired characters had been consistently disproved with the development of Mendelian genetics. The last blow to this explanation came with the rise of molecular genetics which has shown that information flows from nucleic acids (i. e., genes) to proteins (i. e., phenotypes) but never in the opposite direction. Thus phenotypic modifications of an individual caused by the environment during its lifetime are never be able to impinge on genes to change their nucleotide sequences correspondingly. At the same time, molecular genetics, in collaboration with population genetics has led us to reexamine the neo-Darwinian view which once seemed to be so well established. On the other hand, the old mutationism and Lamarckism, even if they contain some interesting ideas, are certainly wrong in the forms they were originally formulated, and we shall never go back to them.

I think that Haldane (1959) had a remarkable insight when he wrote:

« The history of science makes it almost certain that facts will be discovered which show that the theory of natural selection is not fully adequate to account for evolution. But the same history makes it extremely improbable that these facts will be in any way related to the criticisms at present made of it. The physics of Newton and Galileo have proved inadequate in several respects, and are being replaced by relativistic and quantum mechanics. These, however, are even further from the medieval physical theories than were the theories of Galileo and Newton... Darwinism will, I do not doubt, be modified. Like any other successful theory it will ultimately develop its own internal contradictions ».

## OUR PAST AND FUTURE

Our uniqueness as individuals stem from genetic variability inherent in human populations; we are born unequal as far as genetic constitution is concerned. We can regard ourselves as a random sample extracted from the immense gene pool that characterizes human being as a species. With the great amount of variability contained in the pool, it is utterly unlikely that two genetically identical individuals will ever appear in the entire history of man as a species, except for monozygotic twins. Recently, I have estimated [Kimura, 1974] that two unrelated individuals within a large human population differ from each other on the average by some million nucleotide sites.

It is known that the haploid chromosome set of man comprizes some 3.5 billion ($3.5 \times 10^9$) nucleotide sites, so with the world population of roughly 4.0 billion, man as a diploid species may be characterized by a set of $2.8 \times 10^{19}$ nucleotides, a number comparable to the total number of stars in the visible universe, and whose information content probably exceeds that of all the books man has ever written. Yet, when compressed, they can occupy the size of a pinhead, as H. J. Muller once remarked, for the nucleotide material of a haploid chromosome set would occupy only four cubic microns [see Muller, 1958].

Although man as a species is only a million or so years old, comprising a cumulative total of some $6.6 \times 10^{10}$ individuals from inception until the invention of agriculture [Deevy, 1960], a large part of his genetic heritage must have been formed during his evolution as a mammal (starting some 200 million years) or even as vertebrate (starting some 500 million years). During these stages of evolution, sexual reproduction must have played an important role as the means of speeding up evolutionary progress by bringing together into one individual two or more advantageous mutants that occurred separately in different individuals [Fisher, 1930; Muller, 1932; Crow and Kimura, 1965]. Extensive recombination of genes through sexual reproduction in the species has brought about the spectacular evolution of higher organisms since Cambrian times (500-600 million years ago).

Although a primitive form of recombination, or interchange of genetic material between individuals, might have occurred soon after the first appearance of self-replicating entities on the earth, it is likely that elaborate and full-fledged sexual processes became possible only with the appearance of eukaryotes. By eukaryotes, I mean organism with a true nucleus.

It is generally accepted that the earth was condensed from interstellar dust 4.6 billion years ago; this was followed by the formation of its crust, oceans, and atmosphere some half a billion years later. It is possible, as suggested by the existence of rocks in Greenland at least 3.7 billion years old [Moorbath et al., 1972], that the physical conditions for life to begin existed more than 4 billion years ago. If the reported microfossil "Eobacterium" in the figure 3 Cherts ($3.2 \times 10^9$ years old, see Barghoorn, 1971) is genuine, it is likely that life first started on the earth more than 3.5 billion years ago. The appearance of eukaryotes was, of course, much later.

Recently, we [Kimura and Ohta, 1973b] estimated the time of divergence of the eukaryotes and prokaryotes through comparative studies of 5S ribosomal RNA sequences coupled with those of cytochrome c. By prokaryotes I mean primitive forms having no true nucleus, i.e. bacteria and blue-green algae. On the other hand, eukaryotes are represented by higher forms such as plants and animals including yeasts and fungi. Figure 8 illustrates the results of our analysis; the eukaryote-prokaryotes divergence goes back to some 1.8 billion years ago. More extensive analysis made by Hori (1975) using more data on 5S rRNA sequences gave essentially the same results. It is likely therefore that the eukaryotes diverged from the prokaryotes nearly two billion years ago, thus opening up the way toward higher organisms, eventually culminating in the emergence of man.

Our past success as a species on the earth and our future prospect as an intelligent being certainly rest on our brains. So, I would like to make a few remarks on the evolution of brains in animals. The brains is a computer installed in an animal's body: as the central data processing machine, with input

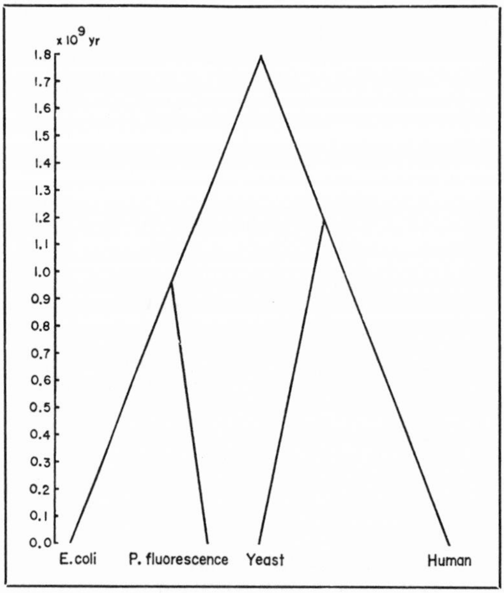


Fig. 8. — Eukaryote-prokaryote divergence as estimated through comparative study of 5S ribosomal RNA sequences.

coming through sense organs, with output directed toward efficient control of the body, the brain is of highest importance for survival and reproduction of vertebrates. Other things being equal, it is obvious that a better brain is more advantageous in the Darwinian sense. However, just as a better computer is more costly, a more developed brain and its efficient use make a slower rate of reproduction inevitable. At any rate, comparative anatomy of brains in vertebrates from the fish to man clearly shows that its structure has much developed. Using a computer terminology, the "hardware" has been improved great deal in the course of evolution.

Although uncomfortable to our ethical feeling, it is likely that the predatory life of animals throughout the long history of evolution was a real cause for the development of intelligence: in order that carnivores can catch their prey efficiently while herbivores can escape safely from their enemies, constant improvement of brain functions must have been required in both parties. So, literally, through struggle for existence, evolutionary progress of brains has been effected. Incidentally, a rapid increase of cranial capacity at the rate of 1/5 darwin in the last 3 million years of human evolution suggests that an exceptionally rapid improvement has occurred also in the brain function.

Thus, our perception of the outer world in terms of "space and time" is what has been created in the course of evolution as the best means of organizing the data from outside for the purpose useful to our survival. It seems to me that what we call a "scientific truth" or "law of nature" is, in a sense, a "subroutine program" that always works well for the purpose of simulation experiments inside our brains, if we again resort to computer terminologies.

As a social animal, and particularly after the rise of human civilization, exchange of informations between individuals has gradually grown in ever wider scale, like interconnection of more and more computers to perform greater and greater data processing and computation. In our time, with the help of the computers of man's own making, this is becoming an international scale. Now, we know that man, emerging from a simple self-replicating molecule, has gained enough ability to understand his own molecular make-up and to measure the size of the universe in which he is placed. I think that Muller (1960) was apt when he said that "the gene through its creation of man, has now, at long last, *found itself*". It was indeed a long way for the gene to do it !

According to a modern astronomical view, our universe is still young and it has countless eons to persist. Shall we be content to preserve ourselves as a superb example of living fossils on this tiny speck of the universe ? Or, shall we try with all our might, to improve ourselves to become supermen and to still higher forms, in order to expand into the wider part of the universe, and to show that life after all is not a meaningless episode ? Before very long, we shall have to decide ourselves.

## REFERENCES

AVISE J.C., SELANDER R.K. (1972). — Evolutionary genetics of cave-dwelling fishes of the genus *Astyanax*. *Evolution*, *26*, 1-19.

BARGHOORN E.S. (1971). — The oldest fossils. *Scientific American 224*, 5, 30-42.

BLACK John A., GIBSON D. (1974). — Neutral evolution and immunoglobulin diversity. *Nature*, *250*, 327-328.

BRACE C.L. (1963). — Structural reduction in evolution. *Amer. Nat. 97*, 39-49.

CALDER N. (1973). — The Life Game. BBC, London.

CROW J.F. (1969). — Molecular genetics and population genetics. Proc. XII Intern. Congr. Genetics (Tokyo), *3*, 105-113.

CROW J.F. (1972). — The dilemma of nearly neutral mutations : How important are they for evolution and human welfare ? *J. Hered.*, *63*, 306-316.

CROW J.F., KIMURA M. (1965). — Evolution in sexual and asexual populations. *Amer. Nat.*, *99*, 439-450.

CROW J.F., KIMURA M. (1970). — An Introduction to Population Genetics Theory. Harper and Row, New York.

DARWIN Charles (1859). — The Origin Species. John Murray, London.

DEEVEY Jr. E.S. (1960). — The human population. *Scientific American*, *203*, 3, 195-204.

DICKERSON R.E. (1971). — The structure of cytochrome c and the rates of molecular evolution. *J. molec. Evol.*, *1*, 26-45.

DICKERSON R.E., GEIS I. (1969). — The Structure and Action of Proteins. Harper & Row, New York, Evanston, London.

DOBZHANSKY Th. (1937). — Genetics and the Origin of Species. Columbia Univ. Press, New York.

FARQUHAR M.N., MCCARTHY B.J. (1973). — Evolutionary stability of the histone genes of sea urchins. *Biochemistry*, *12*, 4113-4122.

FISHER R.A. (1930). — The Genetical Theory of Natural Selection. Clarendon Press, Oxford.

FISHER R.A., FORD E.B. (1950). — The « Sewall Wright Effect ». *Heredity*, *4*, 117-119.

FITCH W.M. (1971a). — Evolutionary variability hemoglobins, pp. 201-215. In « Synthese, Struktur und Funktion des Hämoglobins » (Martin and Nowicki ed.), J. F. LEHMANNS Verlag, München.

FITCH W.M. (1971b). — Rate of change of concomitantly variable codons. *J. molec. Evol.*, *1*, 84-96.

FITCH W.M., MARKOWITZ E. (1970). — An improved method for determining codon variability in a gene and its application to the rate of fixation of mutations in evolution. *Biochem. Genet.*, *4*, 579-593.

FORD E.B. (1965). — Genetic Polymorphism. Faber and Faber, London.

HALDANE J.B.S. (1924). — A mathematical theory of natural and artificial selection, Part. I. *Trans. Camb. Phil. Soc.*, *23*, 19-41.

HALDANE J.B.S. (1927). — A mathematical theory of natural and artificial selection. Part. V. Selection and mutation. *Proc. Camb. Phil. Soc.*, *23*, 838-844.

HALDANE J.B.S. (1933). — The part played by recurrent mutation in evolution. *Amer. Nat.*, *67*, 5-19.

HALDANE J.B.S. (1949). — Suggestions as to quantitative measurement of rates of evolution. *Evolution*, *3*, 51-56.

HALDANE J.B.S. (1957). — The cost of natural selection. *J. Genet.*, *55*, 511-524.

HALDANE J.B.S. (1959). — Natural selection, in « Darwin's Biological Work » (Bell P.R. ed.), Cambridge University Press, Cambridge.

HARDY G.H. (1908). — Mendelian proportions in a mixed population. *Science*, *28*, 49-50.

HOLMQUIST R., JUKES T.H., PANGBURN S. (1973). — Evolution of transfer RNA. *J. molec. Biol.*, *78*, 91-116.

HOOD L., CAMPBELL J.H., ELGIN S.C.R. (1975). — The organization, expression and evolution of antibodies and other multigene families. *Ann. Rev. Genet.*, *9*, 305-353.

HORI H. (1975). — Evolution of 5s RNA. *J. molec. Evol.*, *7*, 75-86.

JUKES T.H. (1969). — Recent advances in studies of evolutionary relationships between proteins and nucleic acids. *Space Life Sciences*, *1*, 469-490.

JUJES T.H. (1971). — Comparisons of the polypeptide chains of globins. *J. molec. Evol.*, *1*, 46-62.

JUKES T.H., KING J.L. (1975). — Evolutionary loss of ascorbic acid synthesizing ability. *J. hum. Evol.*, *4*, 85-88.

KIM S.H., QUIGLEY G.J., SUDDATH F.L., MCPHERSON A., SNEDEN D., KIM J.J., WEINZIERL J., RICH A. (1973). — Three dimensional structure of yeast phenylalanine transfer RNA : folding of the polynucleotide chain. *Science*, *179*, 285-288.

KIM S.H., SUDDATH F.L., QUIGLEY G.J., MCPHERSON A., SUSSMAN J.L., WANG A.H.J., SEEMAN N.C., RICH A. (1974). Three-dimensional tertiary structure of yeast plenylalanine transfer RNA. *Science*, *185*, 435-440.

KIMURA M. (1964). — Diffusion models in population genetics. *J. appl. Probab.*, *1*, 177-232.

KIMURA M. (1968). — Evolutionary rate at the molecular level. *Nature*, *217*, 624-626.

KIMURA M. (1969). — The rate of molecular evolution considered from the standpoint of population genetics. *Proc. nat. Acad. Sci (Wash.)*, *63*, 1181-1188.

KIMURA M. (1971). — Theoretical foundation of population genetics at the molecular level. *Theoretical Population Biology*, *2*, 174-208.

KIMURA M. (1974). — Gene pool of higher organisms as a product of evolution. *Cold Spring Harbor Symp. Quant. Biol.*, *38*, 515-524.

KIMURA M., CROW J.F. (1964). — The number of alleles that can be maintained in a finite population. *Genetics*, *49*, 725-738.

KIMURA M., OHTA T. (1969). — The average number of generations until fixation of a mutant gene in a finite population. *Genetics*, *61*, 763-771.

KIMURA M., OHTA T. (1971). — Theoretical Aspects of Population Genetics. Princeton University Press, Princeton.

KIMURA M., OHTA T. (1972). — Population genetics, molecular biometry, and evolution. Proc. 6 th Berkeley Symp. on Math. Stat. and Probability, Vol. 5 (Darwinian, Neo-Darwinian, and Non-Darwinian Evolution), 43-68.

KIMURA M., OHTA T. (1973a). — Mutation and evolution at the molecular level. *Genetics*, *73*, Suppl., 19-35.

KIMURA M., OHTA T. (1973b). — Eukaryotes-prokaryotes divergence estimated by 5S ribosomal RNA sequences. *Nature New Biol.*, *243*, 199-200.

KIMURA M., OHTA T. (1974). — On some principles governing molecular evolution. *Proc. nat. Acad. Sci (Wash.)*, *71*, 2848-2852.

KING J.L., JUKES T.H. (1969). — Non-Darwinian evolution. *Science*, *164*, 788-798.

KLUG A., LADNER J., ROBERTUS J.D. (1974). — The structural geometry of co-ordinated base changes in transfer RNA. *J. molec. Biol.*, *89*, 511-516.

KOHNE D.E. (1970). — Evolution of higher-organism DNA. *Quart. Rev. Biophys.*, *3*, 327-375.

LEAKEY R.E.F. (1973). — Evidence for an advanced Plio-Pleistocene hominid from East Rudolf, Kenya. *Nature*, *242*, 447-450.

LEWONTIN R.C. (1974). — The Genetics Basis of Evolutionary Change. Columbia Univ. Press, New York and London.

LI S.L., DENNEY R.M., YANOFSKY C. (1973). — Nucleotide sequence divergence in the α-chain-structural genes of tryptophan synthetase from *Escherichia coli*, *Salmonella typhimurium* and *Aerobacter aerogenes*. *Proc. nat. Acad. Sci. (Wash.)*, *70*, 1112-1116.

MALÉCOT G. (1948). — Les Mathématiques de l'Hérédité. Masson et Cie, éd. Paris.

MAYR E. (1965). — Animal Species and Evolution. The Belknap Press of Harvard Univ. Press, Cambridge.

MCLAUGHLIN P.J., DAYHOFF M.O., (1972). — Evolution of species and proteins : a time scale. in « Atlas of Protein Sequence and Structure 1972 » (M.O. Dayhoff ed.) National Biomedical Research Foundation, Washington D.C., pp. 47-52.

MOORBATH S., O'NIONS R.K., PANKHURST R.J., GALE N.H., MCGREGOR V.R. (1972). — Further Rubidium-Strontium age determinations on the early Precambrian rocks of the Godthaab district, West Greenland. *Nature (Physical Science)*, *240*, 78-82.

MUKAI T. (1969). — Maintenance of polygenic and isoallelic variation in populations. Proc. XII Intern. Congr. Genetics (Tokyo), *3*, 293-308.

MULLER H.J. (1932). — Some genetic aspects of sex. *Amer. Nat.*, *66*, 118-138.

MULLER H.J. (1939). — Reversibility in evolution considered from the standpoint of genetics. *Biol. Rev.*, *14*, 261-280.

MULLER H.J. (1958). — Evolution by mutation. *Bull. Amer. Math. Soc.*, *64*, 137-160.

MULLER H.J. (1960). — Evolution and genetics. *Academia Nazionale dei Lincei*, *47*, 15-37.

NEI M. (1975). — Molecular Population Genetics and Evolution. North-Holland Publishing Co., Amsterdam, Oxford.

OHNO S. (1970). — Evolution by Gene Duplication. Springer-Verlag Berlin, Heidelberg, New York.

OHTA T. (1973). — Slightly deleterious mutant substitutions in evolution. *Nature*, *246*, 96-98.

OHTA T., KIMURA M. (1971). — Functional organization of genetic material as a product of molecular evolution. *Nature*, *233*, 118-119.

PERUTZ M.F., LEHMAN H. (1968). — Molecular pathology of human haemoglobin. *Nature*, *219*, 902-909.

PILBEAM D., GOULD S.J. (1974). — Size and scaling in human evolution. *Science*, *186*, 892-901.

PROVINE W.B. (1971). — The Origins of Theoretical Population Genetics. The Univ. of Chicago Press, Chicago and London.

ROBERTSON H.D., JEPPESEN P.G.N. (1972). — Extent of variation in three related bacteriophage RNA molecules. *J. molec. Biol.*, *68*, 417-428.

ROSBASH M., CAMPO M.S., GUMMERSON K.S. (1975). — Conservation of cytoplasmic poly (A) -containing RNA in mouse and rat. *Nature*, *258*, 682-686.

SALSER W., BOWEN S., BROWNE D., ADLI F.E., FEDOROFF N., FRY K., HEINDELL H., PADDOCK G., POON R., WALLACE B., WHITCOME P. (1976). — Investigation of the organization of mammalian chromosomes at the DNA sequence level. *Fed. Proc.*, *35*, 1, 23-35.

SIMPSON G.G. (1944). — Tempo and Mode in Evolution. Columbia Univ. Press, New York.

SMITH G.P. (1974). — Unequal crossover and the evolution of multigene families. *Cold Spring Harbor Symp. Quant. Biology*, *38*, 507-513.

STENZEL P. (1974). — Opossum Hb chain sequence and neutral mutation theory. *Nature*, *252*, 62-63.

TOMITA S., RIGGS A. (1971). — Studies of the interaction of 2, 3-diphosphoglycerate and carbon dioxide with hemoglobins from mouse, man, and elephant. *J. Biol. Chem.* *246*, 547-554.

WEINBERG W. (1908). — Über den Nachweis der Vererbung beim Menschen. *Jber.* Verein f. Vaterl. Naturk. Württem., *64*, 368-382.

WRIGHT S. (1931). — Evolution in Mendelian populations. *Genetics*, *16*, 97-159.

WRIGHT S. (1937). — The distribution of gene frequencies in populations. *Proc. nat. Acad. Sci. (Wash.)*, *23*, 307-320.

WRIGHT S. (1938). — The distribution of gene frequencies under irreversible mutation. *Proc. nat. Acad. Sci. (Wash.)*, *24*, 253-259.

WRIGHT S. (1942). — Statistical genetics and evolution. *Bull. Amer. Math. Soc.*, *48*, 223-246.

WRIGHT S. (1945). — The differential equation of the distribution of gene frequencies. *Proc. nat. Acad. Sci (Wash.)*, *31*, 382-389.

WRIGHT S. (1949). — Genetics of populations. *Encyclop. Brit.*, *10*, 111-112.

WRIGHT S. (1969). — Evolution and the Genetics of Populations. Vol. 2. The Theory of Gene Frequencies. University of Chicago Press, Chicago.

ZUCKERKANDL E., PAULING L. (1965). — Evolutionary divergence and convergence in proteins. in « Evolving Genes and Proteins » (Bryson, V. and H.J. Vogel ed.) Academic Press, New York, pp. 97-166.

# Preponderance of synonymous changes as evidence for the neutral theory of molecular evolution

ACCORDING to the neutral mutation–random drift hypothesis of molecular evolution and polymorphism[1,2], most mutant substitutions detected through comparative studies of homologous proteins (and the nucleotide sequences) are the results of random fixation of selectively neutral or nearly neutral mutations. This is in sharp contrast to the orthodox neo-Darwinian view that practically all mutant substitutions occurring within species in the course of evolution are caused by positive Darwinian selection[3–5]. This paper shows that by comparative studies of messenger RNA (mRNA) sequences reliable estimates can be obtained of the evolutionary rates (in terms of mutant substitutions) at the third positions of the codon, and that the estimates conform remarkably well with the framework of the neutral theory.

Salser *et al.*[6] have presented a comparison of homologous parts from the fragments of the human and rabbit haemoglobin mRNA sequences. Among 53 nucleotide positions that can be compared, there are six base differences, of which only one leads to amino acid difference. Their Table 5 shows that among 17 third nucleotide positions of the codon which can be compared between the two species, there are 5 nucleotide differences. To estimate the number of nucleotide substitutions per site ($K$) that have occurred in the course of evolution since the divergence of the rabbit and human lineages, I use the following formula which converts the observed difference into the estimate of evolutionary divergence

$$K = -\tfrac{3}{4}\log_e(1-4/3\,\lambda) \qquad (1)$$

In this formula, $\lambda$ is the observed fraction of the sites by which two homologous sequences differ from each other (for details, see ref. 7); $K$ includes superimposed and reverted mutant substitutions. The rate of nucleotide substitution per site per year can then be obtained by $k_{nuc} = K/(2T)$, where $T$ is the time (yr) since the divergence of the two lineages. Also, the standard error ($\sigma_K$) of $K$ may be computed using a formula given in ref. 7. Letting $\lambda = 5/17$ in equation (1), we obtain $K = 0.373$. The corresponding standard error ($\sigma_K$) is 0.182. Thus, taking $T = 8\times10^7$, we obtain $k_{nuc} = (2.3\pm1.1)\times10^{-9}$ (the present estimate should be statistically more reliable than the corresponding estimate given in ref. 6). This is a very high evolutionary rate, comparable with that of the fibrinopeptides. If we denote by $k_{aa}$ the rate of amino acid substitutions per site, it is known that for fibrinopeptides $k_{aa} = (4\sim9)\times10^{-9}$ (see refs 8, 9). This figure, when converted into the nucleotide substitution rate, is $k_{nuc} = (1.8\sim4.0)\times10^{-9}$. In addition, a direct comparison of human and rabbit fibrinopeptide A (using data given in ref. 8) gives $k_{aa} = 5.2\times10^{-9}$ leading to $k_{nuc} = 2.3\times10^{-9}$.

A similar but more interesting report comes from Grunstein *et al.*[10] who compared histone H4 (that is, histone IV) mRNA sequences of two sea urchin species *S. purpuratus* and *L. pictus*. It is well known that histone H4 is by far the most highly conserved protein (with $k_{aa} = 0.006\times10^{-9}$ (ref. 8)). Their report[10] now shows, however, that many synonymous changes (that is, the changes that do not lead to amino acid changes) have occurred in the gene coding for this protein in the course of evolution. From their Table 3, showing 84 nucleotide sites that can be compared between the messenger sequences of the two species, there are ten base differences, of which nine are located at the third positions of the codon. Actually, there are 27 third nucleotide positions which can be compared between these two species and there are nine nucleotide differences. Thus, we obtain $K = 0.440$ and $\sigma_K = 0.163$. According to Grunstein *et al.*, these sea urchins shared a common ancestor approximately $6\times10^7$ yr ago. Therefore, letting $T = 6\times10^7$, we obtain $k_{nuc} = (3.7\pm1.4)\times10^{-9}$ as an estimate of the average evolutionary rate at the third nucleotide position of the codon in the histone H4 gene. It is remarkable that, in this gene, synonymous mutant substitutions have occurred with the highest known rate, in spite of the fact that the amino acid substitutions have occurred in the corresponding protein with the lowest known rate.

How do these observations bear relation to the neutral theory? Let $k$ be the rate by which mutant genes are substituted in the species in evolution. If a certain fraction $f_0$ of the molecular mutants are selectively neutral and the rest are definitely deleterious (assuming that definitely advantageous mutations are negligible in frequency), then for neutral alleles, the rate of mutant substitutions is equal to the mutation rate[1]. Therefore

$$k = v_T f_0 \qquad (2)$$

where $v_T$ is the total mutation rate. Now, among the prominent features[11] of molecular evolution the following two are particularly noteworthy: (1) for a given protein, the rate of evolution is roughly constant per year and (2) the molecules or parts of molecules that are subject to less functional constraints evolve (in terms of mutant substitutions) faster.

As to the first feature, the neutralists assume that for a given molecule $f_0$ is roughly constant. Recently, some authors, such as Fitch and Langley[12], have emphasised the non-constancy, but their results show, in agreement with our earlier estimate[13], that the variance of evolutionary rates among lines is only about 2.6 times as large as that expected from chance fluctuations. It is possible that there is delicate fluctuation of intrinsic evolutionary rate around the mean, as caused by shifting of the molecular constraint as amino acids are substituted one after another in various parts of the molecule. This means that the probability $f_0$ of a mutation being neutral is not strictly constant but fluctuates around its characteristic mean. Also, as pointed out by Ohta[14], if nearly neutral but very slightly deleterious mutations are prevalent, the evolutionary rate fluctuates as the population goes through a series of bottlenecks. As to the difference of evolutionary rates among molecules or parts of molecules, the neutralists assume that the probability of a mutational change being neutral depends on functional constraints. Namely, the weaker the functional constraint, the larger the probability ($f_0$) of a random change being selectively neutral, with the result that $k$ in equation (2) gets larger. According to this explanation, the maximum evolutionary rate is attained when $f_0 = 1$, that is, when all the mutations are neutral. Now, the high evolutionary rates observed at the third position of the codon can be explained from the neutral theory by assuming that the majority of synonymous changes are selectively neutral. Note that roughly 2/3 of random nucleotide substitutions at the third position of the codon are synonymous. The possibility of synonymous nucleotide changes being selectively neutral was discussed extensively in an early paper[2] on the neutral theory. Of course, it is possible that not all synonymous mutations are completely neutral, but the possibility is very high that, on average, synonymous changes are subject to natural selection very much less than the mis-sense mutations. On the other hand, if we adhere to the selectionist position that practically all the mutant substitutions in evolution are caused by positive natural selection, there can be no upper limit to the evolutionary rate at the molecular level (as directly set by the mutation rate $v_T$), and there is no reason to believe that the rate of evolution is uniform even approximately for a given molecule among different species.

In my opinion, various observations suggest that as the functional constraint diminishes the rate of evolution converges to that of the synonymous substitutions. If this is valid, such a convergence (or plateauing) of molecular evolutionary rates will turn out to be strong supporting evidence for the neutral theory.

MOTOO KIMURA

*National Institute of Genetics,*
*Mishima, 411, Japan*

Received 22 December 1976; accepted 27 March 1977.

1 Kimura, M. *Nature* **217**, 624–626 (1968).
2 Kimura, M. *Genet. Res. Camb.* **11**, 247–269 (1968).
3 Fisher, R. A. *The Genetical Theory of Natural Selection* (Clarendon, Oxford, 1930).
4 Fisher, R. A. *Proc. R. Soc.* **B121**, 58–62 (1936).
5 Mayr, E. *Animal Species and Evolution* (Harvard University Press, Cambridge, Massachusetts, 1965).
6 Salser, W. *et al. Fedn Proc.* **35(1)**, 23–35 (1976).
7 Kimura, M. & Ohta, T. *J. molec. Evol.* **2**, 87–90 (1972).
8 Dayhoff, M. O. *Atlas of Protein Sequence and Structure 1972* (National Biomedical Research Foundation, Washington, D.C., 1972).
9 Barnard, E. A., Cohen, M. S., Gold, M. H. & Kim, J-K. *Nature* **240**, 395–398 (1972).
10 Grunstein, M., Schedl, P. & Kedes, L. *J. molec. Biol.* **104**, 351–369 (1976).
11 Kimura, M. & Ohta, T. *Proc. natn. Acad. Sci. U.S.A.* **71**, 2848–2852 (1974).
12 Fitch, W. M. & Langley, C. H. *Fedn Proc.* **35**, 2092–2097 (1976).
13 Ohta, T. & Kimura, M. *J. molec. Evol.* **1**, 18–25 (1971).
14 Ohta, T. *Nature* **252**, 351–354 (1974).

# Model of effectively neutral mutations in which selective constraint is incorporated

(molecular evolution/protein polymorphism/population genetics/neutral mutation theory)

MOTOO KIMURA

National Institute of Genetics, Mishima 411, Japan

*Contributed by Motoo Kimura, April 25, 1979*

**ABSTRACT** Based on the idea that selective neutrality is the limit when the selective disadvantage becomes indefinitely small, a model of neutral (and nearly neutral) mutations is proposed that assumes that the selection coefficient ($s'$) against the mutant at various sites within a cistron (gene) follows a $\Gamma$ distribution; $f(s') = \alpha^{\beta} e^{-\alpha s'} s'^{\beta-1}/\Gamma(\beta)$, in which $\alpha = \beta/\bar{s}'$ and $\bar{s}'$ is the mean selection coefficient against the mutants ($\bar{s}' > 0$; $1 \geq \beta > 0$). The mutation rate for alleles whose selection coefficients $s'$ lie in the range between 0 and $1/(2N_e)$, in which $N_e$ is the effective population size, is termed the effectively neutral mutation rate (denoted by $v_e$). Using the model of "infinite sites" in population genetics, formulas are derived giving the average heterozygosity ($\bar{h}_e$) and evolutionary rate per generation ($k_g$) in terms of mutant substitutions. It is shown that, with parameter values such as $\beta = 0.5$ and $\bar{s}' = 0.001$, the average heterozygosity increases much more slowly as $N_e$ increases, compared with the case in which a constant fraction of mutations are neutral. Furthermore, the rate of evolution per year ($k_1$) becomes constant among various organisms, if the generation span ($g$) in years is inversely proportional to $\sqrt{N_e}$ among them and if the mutation rate per generation is constant. Also, it is shown that we have roughly $k_g = v_e$. The situation becomes quite different if slightly advantageous mutations occur at a constant rate independent of environmental conditions. In this case, the evolutionary rate can become enormously higher in a species with a very large population size than in a species with a small population size, contrary to the observed pattern of evolution at the molecular level.

Among difficult questions that confront the neutral mutation theory purporting to treat quantitatively the evolution and variation at the molecular level, the following two are particularly acute. First, why the evolutionary rate in terms of mutant substitutions is roughly constant per year for each protein (such as hemoglobin $\alpha$; see refs. 1 and 2) among diverse lineages, even if the mutation rate appears to be constant per generation rather than per year. Secondary, why the observed level of the average heterozygosity stays mostly in a rather narrow range (between 0% and 20%; see ref. 3) among various species, even if their population sizes differ enormously.

The present paper proposes a model of neutral mutations in which selective constraint (negative selection) is incorporated, and shows that the model can go a long way toward solving these problems in the framework of the neutral mutation theory (4, 5). The model is based on the idea that selective neutrality is the limit when the selective disadvantage becomes indefinitely small (2). For the mathematical formulation of this idea, we must consider the distribution of the selection coefficients of new mutations at the neighborhood of strict neutrality (6, 7). Recently, Ohta (8) investigated a model in which the selection coefficients against the mutants follow an exponential distribution. From the standpoint of the neutral mutation theory, however, Ohta's model has a drawback in that it cannot accommodate enough mutations that behave effectively as neutral when the population size gets large. This difficulty can be overcome by assuming that the selection coefficients follow a $\Gamma$ distribution.

The publication costs of this article were defrayed in part by page charge payment. This article must therefore be hereby marked "*advertisement*" in accordance with 18 U. S. C. §1734 solely to indicate this fact.

## MODEL OF EFFECTIVELY NEUTRAL MUTATIONS

Let us assume that the frequency distribution of the selective disadvantage (denoted by $s'$) of mutants among different sites within a gene (cistron) follows the $\Gamma$ distribution

$$f(s') = \alpha^{\beta} e^{-\alpha s'} s'^{\beta-1}/\Gamma(\beta), \quad [1]$$

in which $\alpha = \beta/\bar{s}'$, $\bar{s}'$ is the mean selective disadvantage, and $\beta$ is a parameter such that $0 < \beta \leq 1$. If we measure the selective advantage in terms of Fisher's Malthusian parameter (9), $s'$ has the range $(0, \infty)$. On the other hand, if we measure it, as we shall do in this paper, in terms of conventional selection coefficient, the true range of $s'$ is restricted to the interval (0, 1). However, because we assume that $\bar{s}'$ is small, with a typical value of $10^{-3}$, $f(s')$ is negligible beyond $s' = 0.1$ so that we can take the entire positive axis as the range of integration without serious error. Note that in this formulation, we disregard beneficial mutants, and restrict our consideration only to deleterious and neutral mutations. Admittedly, this is an oversimplification, but as I shall show later, a model assuming that beneficial mutations also arise at a constant rate independent of environmental changes leads to unrealistic results.

Let us consider a diploid population of the effective size $N_e$ and denote by $v_e$ the effectively neutral mutation rate that is defined by the relationship

$$v_e = v \int_0^{1/(2N_e)} f(s')ds', \quad [2]$$

in which $v$ is the total mutation rate. For $2N_e\bar{s}' \gg 1$, Eq. 2 is approximated by

$$v_e = \frac{v}{\Gamma(1+\beta)} \left(\frac{\beta}{2N_e\bar{s}'}\right)^{\beta}. \quad [3]$$

Fig. 1 illustrates the distribution $f(s')$ for the case $\beta = 0.5$ and $\bar{s}' = 10^{-3}$. In this figure, the shaded area represents the fraction of effectively neutral mutations ($v_e/v$) when the effective population size ($N_e$) is 2500. This fraction becomes smaller as the population size increases. Note that even if the frequency of strictly neutral mutations (for which $s' = 0$) is zero in the present model, a large fraction of mutations can be effectively neutral if $\beta$ is small [note that $f(0) = \infty$ for $0 < \beta < 1$]. We may regard $\beta$ as representing the degree of physiological homeostasis, while $\bar{s}'$ represents the degree of functional constraint of the molecule. In the limiting situation $\beta \to 0$, all mutations become neutral. On the other hand, if $\beta = 1$, the model reduces to Ohta's model (8) for which $v_e/v \approx 1/(2N_e\bar{s}')$ when $2N_e\bar{s}' \gg 1$.

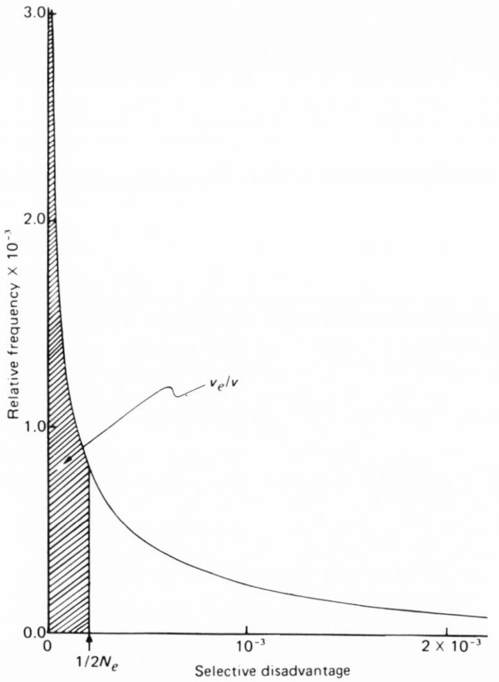


FIG. 1. Frequency distribution of selection coefficients among mutants at different sites within a cistron (gene). The shaded area represents the fraction of effectively neutral mutations. Parameter values assumed are $\beta = 0.5$ and $\bar{s}' = 0.001$. For details, see text.

## EVOLUTIONARY RATE

In order to calculate the rate of evolution in terms of mutant substitutions, we assume that the number of available sites (nucleotide or codon sites) for mutation is sufficiently large, while the mutation rate per site is very low so that whenever a mutation occurs it represents a new site in which no mutant forms are segregating within the population. This assumption is known as the model of infinite sites in population genetics. This model was originally formulated (10) with all the nucleotide sites of the genome in mind. The number of nucleotide sites making up a single gene is much smaller, being of the order of several hundreds. Nevertheless, we may apply the infinite site model to a gene locus as a reasonable approximation if the number of segregating sites per gene constitutes a small fraction. It is known (10) that under this model if $v$ is the total mutation rate and if all the mutations are neutral, the expected number of segregating sites is

$$I_1 = 4N_e v[\log_e(2N) + 1], \quad [4]$$

in which $N$ and $N_e$ are, respectively, the actual (apparent) and the effective sizes of the population. If the mutations are deleterious, the number of segregating sites is smaller. So, $I_1$ in Eq. 4 may be used to check if the infinite site model is appropriate. As a typical situation, we take $v = 2 \times 10^{-6}$, $N_e = 10^5$, and $N = 10^6$; then we get $I_1 = 12.4$. This constitutes a small fraction compared with several hundred, so the infinite site model may be applicable. However, as $N_e$ becomes larger, $I_1$ soon gets large enough so that the assumption of infinite sites becomes no longer valid. In such cases, the treatment gives overestimates particularly for the level of heterozygosity. The treatment may still be useful to obtain the upper limit to the heterozygosity.

With this precaution, we proceed to calculate the rate of evolution in terms of mutant substitutions. Let us assume that the mutant is semidominant in fitness so that selection coefficients against the mutant homo- and heterozygotes at a site are $2s'$ and $s'$, respectively. For such a mutant, the probability of eventual fixation in the population is given by

$$u = [1 - e^{2s'(N_e/N)}]/(1 - e^{4N_e s'}), \quad [5]$$

or, if $s'$ is small,

$$u = 2s'(N_e/N)/(e^{4N_e s'} - 1) \quad [6]$$

to a good approximation. For the rationale of Eq. 5 see ref. 11, particularly p. 426. Then the rate of mutant substitution per generation is

$$k_g = \int_0^\infty 2Nvuf(s')ds', \quad [7]$$

in which the subscript $g$ denotes that it refers to the rate per generation rather than per year. Eq. 7 is based on the consideration that the expected number of new mutants that arise in the population in each generation having selective disadvantage in the range $s' \sim s' + ds'$ is $2Nvf(s')ds'$, of which the fraction $u$ eventually reaches fixation in the population. Substituting Eqs. 1 and 6 in Eq. 7, we get, after some computation,

$$k_g = v\beta R^\beta \sum_{j=0}^{\infty} (j + 1 + R)^{-\beta-1}, \quad [8]$$

in which $R = \beta/(4N_e\bar{s}')$. In Fig. 2, $k_g$ is shown by a solid curve taking the effective population size ($N_e$) as the abscissa, and

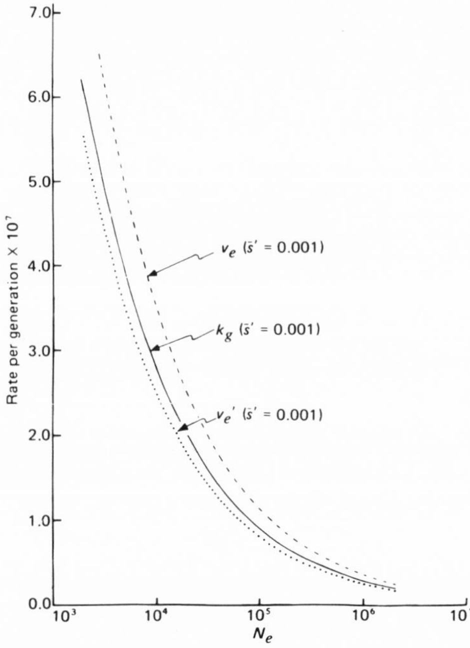


FIG. 2. Comparison between the evolutionary rate ($k_g$; ——) and the effectively neutral mutation rate ($v_e$; - - -). ·····, Mutation rate $v_e'$, the rate of occurrence of mutations whose selective disadvantage is less than $1/(4N_e)$. For all curves, $\beta = 0.5$.

*Proc. Natl. Acad. Sci. USA 76 (1979)*

assuming $v = 2 \times 10^{-6}$, $\beta = 0.5$, and $\bar{s}' = 0.001$. In the same figure, the effectively neutral mutation rate $v_e$ is plotted by a broken curve for the same set of parameters. Because

$$\sum_{j=0}^{\infty} (j + 1 + R)^{-\beta-1} \approx (1 + R)^{-\beta-1} + \int_{1.5}^{\infty} (\lambda + R)^{-\beta-1} d\lambda$$
$$= (1 + R)^{-\beta-1} + \beta^{-1}(1.5 + R)^{-\beta}, \quad [9]$$

Eq. 7 may be approximated by

$$k_g = vR^{\beta}[\beta\cdot(1 + R)^{-\beta-1} + (1.5 + R)^{-\beta}]. \quad [10]$$

This approximation gives about 17% overestimation for $\beta = 0.5$ and $N_e\bar{s}' \geqq 1$, but it is accurate enough for most practical purposes. From this we can easily show that, at the limit of either $4N_e\bar{s}' \rightarrow 0$ or $\beta \rightarrow 0$, we get

$$k_g = v, \quad [11]$$

which is a well-known result for strictly neutral mutations (4). We can also show that, for $4N_e\bar{s}' \gg 1$,

$$k_g \approx v[\beta^{\beta+1}/2^{\beta} + (\beta/3)^{\beta}]/(2N_e\bar{s}')^{\beta}. \quad [12]$$

Comparison of this with Eq. 3 suggests that, roughly speaking, we have

$$k_g \approx v_e. \quad [13]$$

Rough agreement of $k_g$ (solid curve) and $v_e$ (broken curve) may be seen in Fig. 2, in which $v_e'$ (the rate of occurrence of mutations whose $s'$ value is less than $1/4N_e$) is also plotted by a dotted curve for the same set of parameters as used for the other two curves. Thus, Eq. 13 may be regarded as an extension of Eq. 11. In the case of $\beta = 0.5$ as illustrated, the rate of evolution per generation is inversely proportional to $\sqrt{N_e}$ when $N_e\bar{s}'$ is large.

## MEAN HETEROZYGOSITY

Let $H_n$ be the expected number of heterozygous sites. Then, as shown in ref. 10 (see equation 15′ therein), we have, for a given value of $s'$,

$$H_n = \frac{8Nv}{-2s'}\left(u - \frac{1}{2N}\right), \quad [14]$$

in which $u$ is given by Eq. 6. This can also be expressed as

$$H_n = 8N_e v \frac{e^{4N_e s'} - 1 - 4N_e s'}{4N_e s'(e^{4N_e s'} - 1)}$$
$$= 8N_e v \sum_{i=1}^{\infty} \frac{(4N_e s')^i}{(i + 1)!} \sum_{j=1}^{\infty} e^{-4N_e s' j}. \quad [15]$$

Thus the mean number of heterozygous sites, when $s'$ follows the $\Gamma$ distribution 1, is

$$\bar{H}_n = \int_0^{\infty} H_n f(s')ds'$$
$$= 8N_e vR^{\beta} \sum_{i=1}^{\infty} \frac{\Gamma(i + \beta)}{\Gamma(\beta)(i + 1)!} \sum_{j=1}^{\infty} (j + R)^{-i-\beta}, \quad [16]$$

in which $R = \beta/(4N_e\bar{s}')$. Then, if we introduce the approximation

$$\sum_{j=1}^{\infty} (j + R)^{-i-\beta} \approx (1 + R)^{-i-\beta} + \int_{1.5}^{\infty} (\lambda + R)^{-i-\beta} d\lambda$$
$$= (1 + R)^{-i-\beta} + (1.5 + R)^{1-i-\beta}/(i + \beta - 1),$$

Eq. 16 becomes

$$\bar{H}_n = 8N_e vR^{\beta}\left[\sum_{i=1}^{\infty} \frac{\Gamma(i + \beta)}{\Gamma(\beta)(i + 1)!}(1 + R)^{-i-\beta} + \sum_{i=1}^{\infty} \frac{\Gamma(i + \beta - 1)}{\Gamma(\beta)(i + 1)!}(1.5 + R)^{1-i-\beta}\right]. \quad [17]$$

In the special case of $\beta = 1$, this reduces to

$$\bar{H}_n = \frac{8N_e v}{\bar{S}'}\left\{\frac{1}{1 + \bar{S}'} + \log_e(1 + \bar{S}') + \left(\frac{1}{2} + \frac{1}{\bar{S}'}\right)\log_e \frac{1 + 0.5\bar{S}'}{1 + 1.5\bar{S}'}\right\}, \quad [18]$$

in which $\bar{S}' = 4N_e\bar{s}'$.

In order to obtain a simpler expression for Eq. 17 for the case $0 < \beta < 1$, we start from the following formal expression.

$$\sum_{n=0}^{\infty} \frac{\alpha(\alpha - 1) \ldots (\alpha - n + 1)}{n!} x^n = (1 + x)^{\alpha}. \quad [19]$$

Letting $\alpha = -\beta$ and substituting $-xt$ for $x$ in this formula, and then integrating both sides of the resulting equation with respect to $t$ over the interval (0, 1), we get

$$\sum_{n=0}^{\infty} \frac{\beta(\beta + 1) \ldots (\beta + n - 1)}{(n + 1)!} x^{n+1} = \frac{1 - (1 - x)^{1-\beta}}{1 - \beta}. \quad [20]$$

Next, substituting $xt$ for $x$ in Eq. 20, and integrating both sides of the resulting equation with respect to $t$ over the interval (0, 1), followed by putting $n = i - 1$, we have

$$\sum_{i=1}^{\infty} \frac{\Gamma(i + \beta - 1)}{\Gamma(\beta)(i + 1)!} x^{i+1} = \frac{x}{1 - \beta} - \frac{1 - (1 - x)^{2-\beta}}{(1 - \beta)(2 - \beta)}.$$

Finally, if we substitute $x^{-1}$ for $x$ in this equation and then multiply $x^{2-\beta}$ through both sides we get

$$\sum_{i=1}^{\infty} \frac{\Gamma(i + \beta - 1)}{\Gamma(\beta)(i + 1)!} x^{1-i-\beta} = \frac{x^{1-\beta}}{1 - \beta} - \frac{x^{2-\beta} - (x - 1)^{2-\beta}}{(1 - \beta)(2 - \beta)}. \quad [21]$$

Going back to Eq. 20, we note that with a slight modification this may be expressed as

$$\sum_{i=0}^{\infty} \frac{\Gamma(i + \beta)}{\Gamma(\beta)(i + 1)!} x^i = \frac{x^{-1} - x^{-1}(1 - x)^{1-\beta}}{1 - \beta}.$$

Substituting $x^{-1}$ for $x$ in this equation, and after some rearrangements, we get

$$\sum_{i=1}^{\infty} \frac{\Gamma(i + \beta)}{\Gamma(\beta)(i + 1)!} x^{-i-\beta} = \frac{x^{1-\beta} - (x - 1)^{1-\beta}}{1 - \beta} - x^{-\beta}. \quad [22]$$

Applying Eq. 22 with $x = 1 + R$ and Eq. 21 with $x = 1.5 + R$ to the right hand side of Eq. 17, we obtain

$$\bar{H}_n = 8N_e vR^{\beta}\left[\frac{(1 + R)^{1-\beta} + (1.5 + R)^{1-\beta} - R^{1-\beta}}{1 - \beta} - (1 + R)^{-\beta} - \frac{(1.5 + R)^{2-\beta} - (0.5 + R)^{2-\beta}}{(1 - \beta)(2 - \beta)}\right], \quad [23]$$

in which $R = \beta/(4N_e\bar{s}')$. In the limiting situation either for $\beta \rightarrow 0$ or $\bar{s}' \rightarrow 0$, this equation reduces to $\bar{H}_n = 4N_e v$, which agrees with the result obtained in ref. 10 for strictly neutral mutations.

Let $\bar{h}_e$ be the expected heterozygosity of the gene under consideration. Then, assuming that different sites behave independently, we have

$$\bar{h}_e = 1 - e^{-\bar{H}_n}, \quad [24]$$

because this represents the probability that the gene is heterozygous at least in one of the sites. In Fig. 3, the expected heterozygosity is shown as a function of $N_e$ for various parameter values ($\beta$ and $\bar{s}'$) assuming $v = 2 \times 10^{-6}$. Note that as compared with the situation in which all the mutations are neutral ($\beta \rightarrow 0$), a case such as $\beta = 0.5$ and $\bar{s}' = 0.001$ is interesting because

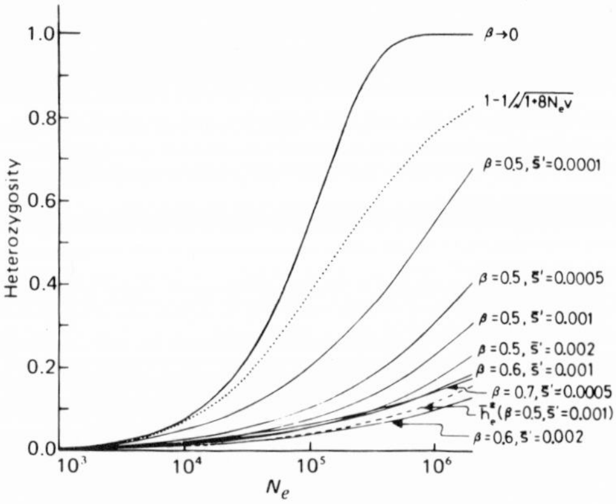


FIG. 3. The mean heterozygosity as a function of the effective population size for various combinations of parameters, $\beta$ and $\bar{s}'$, assuming the mutation rate $v = 2 \times 10^{-6}$.

of slow rate at which heterozygosity increases as $N_e$ gets large. In the same figure, the dotted line represents $\bar{h}_e = 1 - 1/(1 + 8N_ev)^{1/2}$, the heterozygosity expected under the stepwise mutation model (12). Similarly, the broken line represents $\bar{h}_e^* = 1 - 1/(1 + 8N_ev_e)^{1/2}$ in which $v_e$ is the effectively neutral mutation rate with $\beta = 0.5$ and $\bar{s}' = 0.001$. In all these cases the mutation rate $v = 2 \times 10^{-6}$ is assumed.

## SLIGHTLY ADVANTAGEOUS MUTATIONS

To make our analysis complete, let us investigate how the evolutionary rate is influenced by assuming that a certain fraction of mutations are advantageous. Let $v_a$ be the rate of occurrence of advantageous mutations, and assume that the selection coefficient $s$ for such a mutant follows a $\Gamma$ distribution with the mean $\bar{s}$ and the parameter $\gamma$;

$$f_a(s) = \alpha^{\gamma} e^{-\alpha s} s^{\gamma-1}/\Gamma(\gamma), \qquad [25]$$

in which $\alpha = \gamma/\bar{s}$ and $\gamma > 0$.

Noting that the probability of ultimate fixation of a single mutant with selective advantage $s$ (>0) is $u = 2s(N_e/N)/(1 - e^{-4N_es})$ (see ref. 11, p. 426), the rate of evolution due to advantageous mutations is

$$k_g = \int_0^{\infty} 2Nv_a u f_a(s) ds$$

$$= 4N_ev_a\bar{s}R^{\gamma+1} \sum_{j=0}^{\infty} (j + R)^{-\gamma-1}, \qquad [26]$$

in which $R = \gamma/(4N_e\bar{s})$. This can be approximated by

$$k_g = v_a\{4N_e\bar{s} + [\gamma/(6N_e\bar{s} + \gamma)]^{\gamma}\}. \qquad [27]$$

For $N_e\bar{s} \gg 1$, we have $k \approx 4N_e\bar{s}v_a$. This means that the rate of evolution can become enormously high in a very large population, $k_g$ being directly proportional to $N_e$, contrary to actual observations.

## DISCUSSION

The distribution of selection coefficients of new mutations at the neighborhood of strict neutrality was discussed by Crow (6) and King (7). However, it was Ohta (8) who investigated quantitatively the problem of "near neutrality" by assuming a specific mathematical form of the distribution. In Ohta's model, mutations are assumed to be deleterious and the selection coefficients against individual mutants follow an exponential distribution. With this model, she showed that the level of heterozygosity reaches an upper limit as $N_e$ increases, whereas the rate of evolution per generation ($k_g$) in terms of mutant substitutions is inversely proportional to the effective population size $N_e$. The present model assuming a $\Gamma$ distribution of selective disadvantages of mutants among different sites within a gene (or among different amino acid sites within a protein) has an advantage over Ohta's model in that it can accommodate a much larger fraction of effectively neutral mutations. Actually, Ohta's model corresponds to $\beta = 1$, while the assumption that all the mutations are neutral corresponds to $\beta \to 0$. It is likely that an intermediate parameter value such as $\beta = 0.5$ (as depicted in Fig. 1) may be more realistic to describe the typical situation observed in natural populations. In the case of $\beta = 0.5$, the rate of evolution per generation is inversely proportional to $\sqrt{N_e}$ if $N_e\bar{s}' \gg 1$; i.e., $k_g \propto 1/\sqrt{N_e}$. Thus the evolutionary rate per year is $k_1 \propto 1/(g\sqrt{N_e})$, in which $g$ is the generation span in years. If $g$ is inversely proportional to $\sqrt{N_e}$ among various organisms, then $g\sqrt{N_e}$ is constant, and therefore the evolutionary rate per year is constant, provided that the mutation rate $v$ per generation is constant (uniform) among them.

Note that in the present model those mutations that become fixed in the population by random drift in the course of evolution are restricted to effectively neutral mutations. The selective disadvantage of such mutants is at most of the order of $1/(2N_e)$, which means $10^{-5}$ or less in many mammals. The proportion of effectively neutral mutants decreases as the population size increases. This is why the heterozygosity increases much more slowly in the present model as compared with the conventional model of neutral mutations (see Fig. 3). The observations that the average heterozygosity is restricted in most organisms to the range 0% to 20% have been used repeatedly as evidence against the neutral mutation theory (see ref. 13). It is likely that this difficulty is resolved by the present model if we assume in addition that a population bottleneck occurs from time to time in all organisms in the course of evolution, reducing their effective population sizes substantially (14). Recently, Li (15, 16) investigated the amount of genetic variability maintained in a finite population using the $K$ allele model incorporating two or three classes of mutations including the neutral and slightly deleterious classes. Similarly, Maruyama and Kimura (17) used a stepwise mutation model incorporating two types of mutations, neutral and slightly deleterious, to investigate the same problem. The present model has a more desirable feature of incorporating a continuous spectrum of mutations conferring different fitnesses, but it does not take into account the limited detection ability of electrophoretic methods. At any rate, the present model can explain the observation made by Ohta (18) that in both *Drosophila* and humans the proportion of rare alleles is greater than what is expected under the assumption that all the mutations are strictly neutral. This observation, if valid, will greatly reduce the utility of Ewens' sampling theory (19), a point that was recently elaborated by Li (20). In the examples illustrated in Figs. 2 and 3, we assume the mutation rate $2 \times 10^{-6}$ per locus per generation. This value is based on the results reported by Mukai and Cockerham (21) for *Drosophila melanogaster* and by Nei (22) for humans and the Japanese macaque. Recently, higher estimates of mutation rates have been reported by Neel and Rothman (23) for tribal Ameridians.

As to the rate of molecular evolution, the present model with $v = 2 \times 10^{-6}$, $\beta = 0.5$, and $\bar{s}' = 0.001$, as illustrated in Fig. 2, seems to give realistic values; in mammals, $N_e = 10^5$ may be a representative effective population size for many species during evolution, and $k \approx 10^{-7}$, as shown in Fig. 2, is not very far from the typical rate, which is of the order $1.5 \times 10^{-7}$ per cistron per year (as represented by globins).

It is likely that the value of the parameter $\beta$ is smaller in mammals than in insects, because of higher physiological homeostasis in the mammals. The possibility of more mutations being neutral in higher forms such as mammals with advanced homeostasis has been suggested by Kondo (24). Low physiological homeostasis and frequent local extinction of colonies must be the main reason why the heterozygosity (or 1, minus the sum of squares of allelic frequencies in haploid organisms) does not go very high in organisms having immense apparent population sizes such as neotropical *Drosophila* (25) and *Escherichia coli* (26). The mathematical model proposed in this paper represents my attempt to make the neutral mutation theory more precise and realistic. The model assumes that molecular evolution and polymorphism are caused by random drift of very slightly deleterious but effectively neutral mutations. In this respect, the present theory resembles Ohta's theory of slightly deleterious mutations (27–29). But there are some important differences. Ohta (29) claims that, in very large populations, the stable mutation–selection balance will be realized with heterozygosity reaching the upper limit, while molecular evolution should have stopped or at least have slowed down. Then, fixation of mutants is mainly restricted to population bottlenecks at the time of speciation. On the other hand, I assume that, even in very large populations, alleles at intermediate frequencies, as often found in *Drosophila* species (see ref. 30, table II), represent effectively neutral mutations carried by random drift and that evolution by drift is unlikely to be stopped in these species. Finally, there is one biological problem that we have to consider. Under the present model, effectively neutral, but, in fact, very slightly deleterious mutants accumulate continuously in every species. The selective disadvantage of such mutants (in terms of an individual's survival and reproduction—i.e., in Darwinian fitness) is likely to be of the order of $10^{-5}$ or less, but with $10^4$ loci per genome coding for various proteins and each accumulating the mutants at the rate of $10^{-6}$ per generation, the rate of loss of fitness per generation may amount to $10^{-7}$ per generation. Whether such a small rate of deterioration in fitness constitutes a threat to the survival and welfare of the species (not to the individual) is a moot point, but this will easily be taken care of by adaptive gene substitutions that must occur from time to time (say once every few hundred generations).

I thank Dr. Tomoko Ohta and Professor Masao Kotani for stimulating discussions. Thanks are also due to Drs. J. F. Crow, Alan Robertson, and Roger Milkman, who read the first draft and made many useful suggestions. This is contribution no. 1252 from the National Institute of Genetics, Mishima, Shizuoka-ken 411, Japan.

1. Kimura, M. (1969) *Proc. Natl. Acad. Sci. USA* **63,** 1181–1188.
2. Kimura, M. & Ohta, T. (1974) *Proc. Natl. Acad. Sci. USA* **71,** 2848–2852.
3. Fuerst, P. A., Chakraborty, R. & Nei, M. (1977) *Genetics* **86,** 455–483.
4. Kimura, M. (1968) *Nature (London)* **217,** 624–626.
5. Kimura, M. & Ohta, T. (1971) *Nature (London)* **229,** 467–469.
6. Crow, J. F. (1972) in *Proceedings of the Sixth Berkeley Symposium on Mathematical Statistics and Probability,* eds. LeCam, L. M., Neyman, J. & Scott, E. L. (Univ. of California Press, Berkeley), Vol. 5, pp. 1–22.
7. King, J. L. (1972) in *Proceedings of the Sixth Berkeley Symposium on Mathematical Statistics and Probability,* eds. LeCam, L. M., Neyman, J. & Scott, E. L. (Univ. of California Press, Berkeley), Vol. 5, pp. 69–100.
8. Ohta, T. (1977) in *Molecular Evolution and Polymorphism,* ed. Kimura, M. (Natl. Inst. of Genet., Mishima, Japan), pp. 148–167.
9. Fisher, R. A. (1930) *The Genetical Theory of Natural Selection* (Clarendon, Oxford).
10. Kimura, M. (1969) *Genetics* **61,** 893–903.
11. Crow, J. F. & Kimura, M. (1970) *An Introduction to Population Genetics Theory* (Harper & Row, New York).
12. Ohta, T. & Kimura, M. (1973) *Genet. Res.* **22,** 201–204.
13. Lewontin, R. C. (1974) *The Genetic Basis of Evolutionary Change* (Columbia Univ. Press, New York).
14. Nei, M., Maruyama, T. & Chakraborty, R. (1975) *Evolution* **29,** 1–10.
15. Li, W.-H. (1977) *Proc. Natl. Acad. Sci. USA* **74,** 2509–2513.
16. Li, W.-H. (1978) *Genetics* **90,** 349–382.
17. Maruyama, T. & Kimura, M. (1978) *Proc. Natl. Acad. Sci. USA* **75,** 919–922.
18. Ohta, T. (1975) *Proc. Natl. Acad. Sci. USA* **72,** 3194–3196.
19. Ewens, W. J. (1972) *Theor. Popul. Biol.* **3,** 87–112.
20. Li, W.-H. (1979) *Genetics,* in press.
21. Mukai, T. & Cockerham, C. C. (1977) *Proc. Natl. Acad. Sci. USA* **74,** 2514–2517.
22. Nei, M. (1977) *Am. J. Hum. Genet.* **29,** 225–232.
23. Neel, J. V. & Rothman, E. D. (1978) *Proc. Natl. Acad. Sci. USA* **75,** 5585–5588.
24. Kondo, S. (1977) in *Molecular Evolution and Polymorphism,* ed. Kimura, M. (Natl. Inst. of Genet., Mishima, Japan), pp. 313–331.
25. Ayala, F. J., Powell, J. R., Tracey, M. L., Mourão, C. A. & Pérez-Salas, S. (1972) *Genetics* **70,** 113–139.
26. Milkman, R. (1973) *Science* **182,** 1024–1026.
27. Ohta, T. (1973) *Nature (London)* **246,** 96–98.
28. Ohta, T. (1974) *Nature (London)* **252,** 351–354.
29. Ohta, T. (1976) *Theor. Popul. Biol.* **10,** 254–275.
30. Selander, R. K. (1976) in *Molecular Evolution,* ed. Ayala, F. J. (Sinauer, Sunderland, MA), pp. 21–45.

# Selective constraint in protein polymorphism: Study of the effectively neutral mutation model by using an improved pseudosampling method

(population genetics/neutral theory of molecular evolution/polyallelic random drift)

MOTOO KIMURA AND NAOYUKI TAKAHATA

National Institute of Genetics, Mishima, 411 Japan

*Contributed by Motoo Kimura, September 30, 1982*

**ABSTRACT** To investigate the pattern of allelic distribution in enzyme polymorphism, with special reference to the relationship between the mean ($\overline{H}$) and the variance ($V_H$) of heterozygosity, we used the model of effectively neutral mutations involving multiple alleles in which selective disadvantage of mutant alleles follows a Γ distribution. A simulation method was developed that enables us to study efficiently the process of random drift in a multiallelic genetic system and that saves a great deal of computer time. It is an improved version of the pseudosampling-variable (PSV) method [Kimura, M. (1980) *Proc. Natl. Acad. Sci. USA* 77, 522–526] previously used to simulate random drift in a diallelic system. This method will be useful for simulating many models of population genetics that involve behavior of multiple alleles in a finite population. By using this method, it was shown that, as compared with the model of strictly neutral mutations, the present model gives the reduction of both $\overline{H}$ and $V_H$ and an excess of rare variant alleles. The results were discussed in the light of recent observations on protein polymorphism with special reference to the functional constraint of proteins involved.

From the standpoint of the neutral mutation–random drift hypothesis (1) (or the neutral theory, for short; see refs. 2 and 3 for review), protein polymorphism is a transient phase of molecular evolution (4); therefore, it is expected to be influenced by selective constraint in a similar way as in molecular evolution (5). In other words, the stronger the functional and structural constraints to which a protein is subject, the smaller the probability of an amino acid change by mutation being selectively neutral (i.e., not harmful) and, therefore, the lower the heterozygosity at its locus.

The observation that loci for substrate-specific enzymes are on the average less heterozygous than those for substrate-nonspecific enzymes (6) is consistent with this expectation. Furthermore, Yamazaki (7) and more recently Gojobori (8) showed that substrate specific enzymes not only have lower mean heterozygosity but also have smaller variance of heterozygosity among species than expected from the standard neutral infinite allele model (9).

In this paper, we intend to show that these observations can be explained by the model of effectively neutral mutations (10) in which selective constraint (negative selection) is incorporated. This model is based on the idea that selective neutrality is the limit when the selective disadvantage becomes indefinitely small (5) and is an extention of Ohta's model (11) in which the selection coefficients against the mutants follow an exponential distribution. These models are ultimately traced to Ohta's hypothesis (12, 13) that very slightly deleterious mutations as well as the strictly neutral mutations play an important role in molecular evolution and polymorphism.

One problem in our approach is that rigorous mathematical study of the model is rather difficult. Previously, Wright derived a general formula for the equilibrium distribution of multiallelic frequencies when mutation, selection, and random drift balance each other (see ref. 14, p. 394, for a review). Later, Watterson (15) and Li (16) developed a method for calculating moments of allele frequencies from Wright's formula. Theoretically, the formula and the method could be applied to treat the present model. It turned out, however, that when the possible number of selectively different alleles involved is large or the product of the effective population size and selection coefficient is large, or both are large, the application of this method is very difficult.

Here, we intend to present a method that can simulate the process of polyallelic random drift very efficiently. This method is an improved version of the pseudosampling method (17) and it contains a device (to be called a "telescoping" method) of sampling multiple alleles. As compared with other methods (18–20), it can treat a multiallelic system much more easily. Furthermore, it has the merit of enabling us to incorporate a suitable adjustment for rare alleles easily (see also ref. 21), which makes the simulation valid over all the combination of parameter values. This adjustment is particularly important when we try to obtain a satisfactory distribution for a whole domain of allele frequencies.

## Telescoping method of sampling multiple alleles

The essential idea underlying the PSV method is that we simulate the diffusion model of the process of gene frequency change by generating a sequence of some simple random numbers, rather than faithfully following binomial or multinomial process by drawing, in each generation, individual gametes from a finite population. For example, in the simplest case of two neutral alleles in a diploid population of size $N$, we generate in each generation a random number from the uniform distribution having the mean 0 and the variance $x(1-x)/(2N)$, where $x$ is the frequency of one of the alleles at a given moment. Then, this random number is added to $x$ to form the allelic frequency in the next generation. This saves computer time enormously and allows us to simulate easily a process of change in a very large population. Also, we can make many replicate trials without prohibitive computing time. Originally, the PSV method was intended to simulate the diffusion process itself rather than the discrete binomial sampling process ("Fisher–Wright model") for which the diffusion model is usually regarded as an approximation (17).

In the present paper, we shall be concerned with a multiallelic system and assume that there are $K$ possible allelic states ($K > 2$) at a locus in a random-mating diploid population of effective size $N_e$. Let $A_k$ be the $k$th allele ($k = 1, 2, \ldots, K$), whose frequency in the population before sampling is $x_k$. We first sam-

Abbreviation: PSV, pseudosampling variable.

ple allele $A_1$ so that its number follows binomial distribution

$$B(i_1|n, x_1) = \binom{n}{i_1} x_1^{i_1}(1 - x_1)^{n-i_1}, \qquad [1]$$

where $n = 2N_e$ is the total number of gametes that contributes to the next generation and $i_1$ is the number of gametes that contains allele $A_1$ after sampling. Actually, there is no need for the distribution to follow faithfully the binomial distribution, but we simply generate a uniform random number, $U_1$ with mean 0 and variance unity, and substitute the frequency of $A_1$ after sampling (corresponding to $i_1/n$) by

$$x_1' = x_1 + U_1 \sqrt{\{[x_1(1 - x_1)]/n\}}. \qquad [2]$$

One caution we need to take here is that when $A_1$ is represented by a very small number of individuals so that $nx_1$ is less than 3, say, then a more exact sampling procedure must be used (see below). A similar caution is also needed when $x_1$ is near unity, so that $n(1 - x_l)$ is small, say less than 3. Next, we sample $A_2$ so that the number of gametes ($i_2$) containing this allele follows the distribution $B(i_2|n_2, y_2)$, where $n_2 = n - i_1$ is the remainder of the gametes after $A_1$ is removed, and $y_2 = x_2/(1 - x_1)$. Likewise, $A_3$ is sampled so that the number of gametes ($i_3$) containing this allele follows the distribution $B(i_3|n_3, y_3)$, where $n_3 = n - i_1 - i_2$ and $y_3 = x_3/(1 - x_1 - x_2)$. A similar procedure is repeated until all $K - 1$ alleles are sampled. In general, if $i_k$ ($k = 1, 2, \ldots, K - 1$) is the number of $A_k$-bearing gametes in the sample, this number follows

$$B(i_k|n_k, y_k), \qquad [3]$$

where $n_k = n - i_1 - i_2 \ldots -i_{k-1}$ and $y_k = x_k/(1 - x_1 - x_2 \ldots -x_{k-1})$. (Note: $n_1 = n$ and $y_1 = x_1$ for $k = 1$). Then, it can be shown that

$$E\{i_k\} = nx_k, \qquad [4]$$
$$\mathrm{Var}\{i_k\} = nx_k(1 - x_k),$$

and

$$\mathrm{Cov}\{i_k i_l\} = -nx_k x_l, \ (k \neq l),$$

where $1 \leqq k \leqq K - 1$ and $1 \leqq l < K - 1$. Furthermore, if we let

$$y_k' = i_k/n_k, \qquad [5a]$$

then, for $k = 2, \ldots, K - 1$,

$$x_k' = i_k/n = y_k' n_k/n = y_k' \left(1 - \sum_{j=1}^{k-1} x_j'\right) \qquad [5b]$$

gives the frequency of $A_i$ after sampling of gametes.

Eqs. **1–5b** provide the procedure (telescoping method) for sampling $n$ or $2N_e$ gametes from the gene pool containing multiple alleles by simply repeating the binomial sampling procedure of Eq. **3**. Thus, the key point of the telescoping method is how to simulate Eq. **3** to generate $y_k'$. Unless $n_k$ is small, and more importantly, unless none of $n_k y_k$ and $n_k(1 - y_k)$ is very small, say less than 3, we can obtain $y_k'$ by

$$y_k' = y_k + U_k \sqrt{y_k(1 - y_k)/n_k}, \qquad [6a]$$

where $U_k$s are mutually independent arbitrary random variables, each with mean 0 and variance 1, and we usually find it convenient to choose them from the uniform distribution. Then from Eq. **5b**, we can obtain $x_k'$.

When $n_k$ is large but either one of $n_k y_k$ or $n_k(1 - y_k)$ is small, we obtain $y_k'$ by

$$y_k' = \eta_k/n_k \qquad \text{if } n_k y_k \text{ is small,}$$

or $\qquad [6b]$

$$y_k' = 1 - \zeta_k/n_k \quad \text{if } n_k(1 - y_k) \text{ is small,}$$

where $\eta_k$ and $\zeta_k$ are Poisson random numbers with the mean $n_k y_k$ and $n_k(1 - y_k)$, respectively (see also ref. 21).

The approximations given either by Eqs. **6a** or **6b** are satisfactory when $n_k$ is reasonably large, say larger than 20. If, on the other hand, $n_k \leq 20$, we must resort to a more exact procedure, and in this paper, we adopt the following method: we generate $n_k$ random variates that follow the uniform distribution in the range (0, 1), and count the number of such variates that happen to fall in the range (0, $y_k$). This number, $i_k$, is a binomial random number, so we let

$$y_i' = i_k/n_k, \quad (\text{if } n_k \leq 20), \qquad [6c]$$

from which we obtain $x_k'$ by using Eq. **5b**.

The advantage of the telescoping method in saving computer time comes from the fact that we can obtain each of the $y_k'$s (and therefore $x_k'$, the frequency of $A_k$, by using Eq. **5b**) by generating only one random number when $n_k$ is larger than 20. In addition, the accuracy and reliability are ensured by incorporating adjustments **6b** and **6c** for treating rare variant alleles. These adjustments are particularly pertinent if we note that, for $y_k \approx 0$ or 1, application of Eq. **6a** frequently produces $y_k'$ values either negative or larger than 1, thus creating a possibility of serious bias in a simulation experiment. Such adjustments were not incorporated in the treatments by Kimura (17) and by Maruyama and Takahata (22). Later, the problem was treated in a heuristic manner by Maruyama and Nei (19) and Takahata (20). Their heuristic treatment is concerned with allele frequency changes due to mutation; thus, in the absence of mutation, their simulation methods of random drift are not satisfactory for treating the changes of rare alleles.

**Model of effectively neutral mutations**

In the original formulation (10) of the model of effectively neutral mutations, it was assumed that the selective disadvantage of mutants among different sites within a gene (cistron) follows a Γ distribution. Here, we modify the model slightly and assume that there are $K$ possible allelic states ($A_1, A_2, \ldots, A_K$) at a locus, and that each allele mutates with the rate $v = u/(K - 1)$ to any one of the remaining $K - 1$ alleles (23). We consider the genic selection throughout this paper and designate the selection coefficient against $A_i$ by $s_i'$ so that the relative fitness of $A_i$ is $1 - s_i'$. We assume that a set of $s_i'$ values are a random sample extracted from the universe of $s'$ values that follow the Γ distribution

$$f(s') = \frac{\alpha^\beta}{\Gamma(\beta)} s'^{\beta-1} e^{-\alpha s'}, \qquad [7]$$

where $\alpha = \beta/\bar{s}'$, $\bar{s}'$ (>0) is the mean selective disadvantage, and $\beta$ is a parameter such that $0 < \beta \leq 1$. A set of values of $s_i'$s allocated to the entire allelic states are assumed not to change throughout each run of simulation experiments. Thus, we can treat selection deterministically once these values are decided.

Under the above assumptions of mutation and selection, the mean changes of the allelic frequencies per generation are

$$M(\delta x_i) = v - (u + v)x_i + (w_i - \bar{w})/\bar{w}, \qquad [8a]$$

where $w_i = 1 - s_i'$ and $\bar{w} = 1 - \Sigma_{j=1}^{K} s_j' x_j$ and $i = 1, 2, \ldots,$

$K$. In a population of effective size $N_e$, the changes of allelic frequencies due to random sampling of gametes have the variances

$$V(\delta x_i) = x_i(1 - x_i)/(2N_e), \quad [8b]$$

and covariances

$$W(\delta x_i \delta x_j) = -x_i x_j/(2N_e), \quad (i \neq j), \quad [8c]$$

where $i, j = 1, 2, \ldots, K - 1$.

**Simulation experiments**

In our simulation experiments, we tried various values of $\bar{s}'$ and $u$, while we mostly assumed $N_e = 10^4$, $K = 20$, and $\beta = 0.5$. In each run of computer simulation, we first generated $K$ selection coefficients ($s_i$'s) by drawing $K$ random variates from the $\Gamma$ distribution with given parameters $\beta$ and $\bar{s}'$. The change of $x_i$ due to mutation and selection was calculated deterministically by using Eq. **8a**, whereas the change due to random sampling of gametes was done by the telescoping method.

Starting from an arbitrary composition of allele frequencies, we discarded the first $4N_e$ generations to ensure that an equilibrium had been reached. Thereafter we observed the allele frequencies every $N_e/100$ generations. Each run was continued until the total number of observations reached 5,000, and various quantities of interest were obtained by taking the average over them. However, we must keep in mind that these quantities are still subject to statistical fluctuations, because $K$ selection coefficients are samples from a $\Gamma$ distribution. This caution is important particularly when $\bar{s}'$ is large and $K$ is small.

The unit of time was taken as one generation, and, as mentioned above, the effective population size ($N_e$) was kept constant, such as $10^4$, whereas we changed the mutation rate and selection coefficients a great deal for different runs. In our approach using the diffusion equation method, what matters are the products $N_e\bar{s}'$ and $N_e u$ rather than individual parameters $N_e$, $\bar{s}'$ and $u$. Therefore, such a change is equivalent to multiplying the diffusion equation by a constant factor and altering the time scale and $N_e$ but holding $u$ and $s_i'$ unchanged. For example, to simulate the case $u = 10^{-6}$, $N_e = 10^6$, and $\bar{s}' = 10^{-4}$, we may choose parameters such as $u = 10^{-4}$, $N_e = 10^4$, and $\bar{s}' = 10^{-2}$. However, such a correspondence is valid when $\bar{s}'$ is sufficiently small (If it is relatively large, say $\bar{s}' = 0.1$ or more, sampling of $s_i'$ values may cause significant differences). In order to carry out a computer simulation in a relatively short time, such a scaling is necessary, and this corresponds to measuring time in the unit of 100 generations, (see refs. 19 and 20). (We also checked the accuracy of our method by examining the distribution of allelic frequencies, and the results were satisfactory.)

## RESULTS AND DISCUSSION

The mean heterozygosity ($\bar{H}$ or $1 - \bar{F}$) obtained by the experiments is shown in Fig. 1. These results were obtained for a particular set of $s_i'$ values sampled from the $\Gamma$ distribution. To see the effect of a different set of $s_i'$ values on the extent of heterozygosity, we repeatedly simulated the case $\bar{s}' = 0.1$, $u = 0.025$, and $N_e = 10^4$ and also the case $N_e = \infty$ (no random drift), $\bar{s}' = 0.1$, and $u = 0.0002$. In both cases, the mean heterozygosity fluctuated to a large extent because of the difference in the set $s_i'$ drawn from a given $\Gamma$ distribution. This was due to the relatively large value of $\bar{s}'$. However, this sampling effect, coming from a different set of $s_i'$ values, was not very large when $\bar{s}' \leq 0.01$, even if $K = 20$. We checked this for the case $N_e = 10^4$, $u = 10^{-4}$, and $\bar{s}' = 0.01$ by performing five different runs, each with a different set of $s_i'$ values sampled from the same $\Gamma$ distribution. The results for the mean and the variance of homozygosity ($F$) together with their observed standard deviations

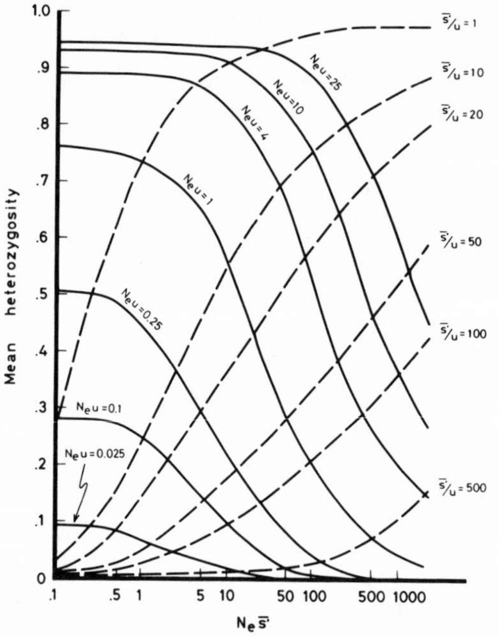


FIG. 1. The mean heterozygosity ($\bar{H}$) as a function of $N_e\bar{s}'$. The effectively neutral mutation model with $K = 20$ and $\beta = 0.5$ is assumed. The lines were drawn based on the results of simulation experiments. ——, Decrease of $\bar{H}$ with increasing $N_e\bar{s}'$ when $N_e u$ is kept constant; – – –, increase of $\bar{H}$ when $\bar{s}'/u$ is kept constant.

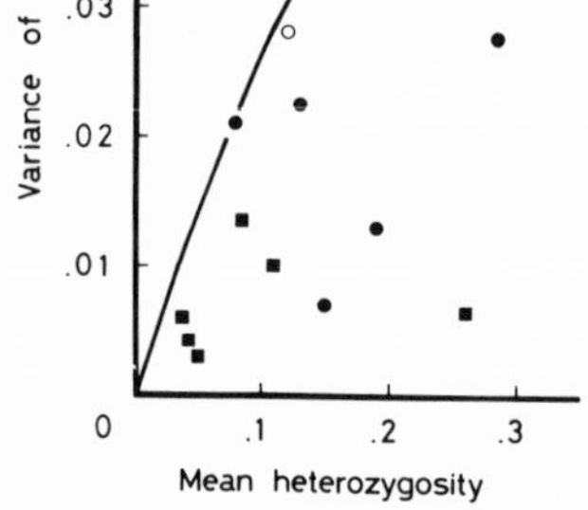

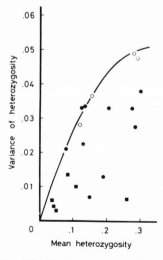

FIG. 2. The relationship between the mean and variance of heterozygosity. These simulation experiments assume that $K = 20$ and $\beta = 0.5$ (over the range of heterozygosity less than 0.3) and assess the intensity of mean selective disadvantage $\bar{s}'$ relative to the mutation rate $u$: ○, $\bar{s}' \simeq u$; ●, $\bar{s}' \simeq 10\,u$; ■, $\bar{s}' = 50$ or $100\,u$. The solid curve represents the theoretical relationship for strictly neutral mutations.

were as follows; $\bar{F} = 0.711 \pm 0.0847$ and $V_F = 0.0357 \pm 0.0067$.

It is obvious that $\bar{H}$ under effectively neutral mutations is always lower than $\bar{H}$ in the case of strictly neutral mutations for the same parameter $N_eu$. For a given $\bar{s}'/u$, $\bar{H}$ increases slowly with increasing $N_e$. However, as compared with the case in which all of the deleterious alleles have the same selection coefficient, the effect of selection on reducing genetic variation is relatively small for effectively neutral mutations; a large value of $\bar{s}'/u$ was required to reduce the amount of genetic variation significantly. Such low efficiency of negative selection may be understood by considering the situation where $N_e$ is indefinitely large so that the genetic variation is maintained by the deterministic balance between mutation and selection. In this case, the mean frequency of a deleterious mutant allele, $A_i$, is $u/s_i'$ so that the mean frequency per allele is $u/\tilde{s}$, where $\tilde{s}$ is the harmonic mean of $s_i'$ (excluding the most fit allele). Because $\tilde{s}$ is influenced much more by a small $s_i'$ than a large one, $\tilde{s}$ is always smaller than the arithmetic mean $\bar{s}'$.

It is known that the mean heterozygosity per individual at allozyme loci is restricted mostly to the range 0 – 0.3 in diverse species (24). If the mean heterozygosity in a large population having $N_eu = 1$ or more is maintained by effectively neutral mutations, it is required that $\bar{s}'/u$ is 100 or more (Fig. 1). If we take $u = 10^{-6}$ as a representative value, we must assume that $\bar{s}'$ is around $10^{-3}$ in order to explain the observed level of the mean heterozygosity by negative selection alone. A more plausible explanation is that the relevant effective size ($N_e$) for discussing the average heterozygosity is much smaller than the apparent population size due to a number of factors. Particularly, if the population goes through a sequence of bottlenecks in the course of evolution and if recovery from a reduced population size after each bottleneck is slow, the average heterozygosity will be much reduced. In nature, even if some species are distributed widely, covering an enormous area and comprising an immense number of individuals (as in some neotropical *Drosophila*), it is rather unlikely that they always have been so in the last millions of years and will continue to be so in the coming millions of years. Sooner or later, such a state would be disrupted by the process of speciation.

The variance of heterozygosity, $V_H$, is very close to that predicted from strictly neutral mutations (25) when $\bar{s}'/u$ is less than 10. A significant reduction of variance as compared with the case of strictly neutral mutations occurs where $\bar{s}'/u$ is much larger than 10 in a large population. In Fig. 2, the relationship between the mean and variance of heterozygosity is shown in cases where $\bar{s}'/u$ is about 1, 10, and in the range of 50–100, respectively.

The distribution of allele frequencies also was studied, although no illustrations are shown. When $4N_eu < 1$, the distribution is always U-shaped, and we can hardly distinguish the case of large $\bar{s}'/u$ from that of neutral mutations. When $4N_eu \geq 1$ and $\bar{s}'/u$ is large, the distribution has a peak at a frequency very close to 1, and also there appears an excess of rare alleles compared with neutral mutations (see also ref. 26). The peak becomes still closer to 1 as $\bar{s}'/u$ further increases, and the distribution becomes very much like U-shaped.

We also checked if $K = 20$ is sufficient to approximate the infinite allele model for our present purpose by calculating the mean number of segregating alleles after random sampling gametes. When $N_eu = 4$ and $N_e\bar{s}' = 1,000$, it was about 16, but most alleles have their frequencies less than 0.001. Therefore,

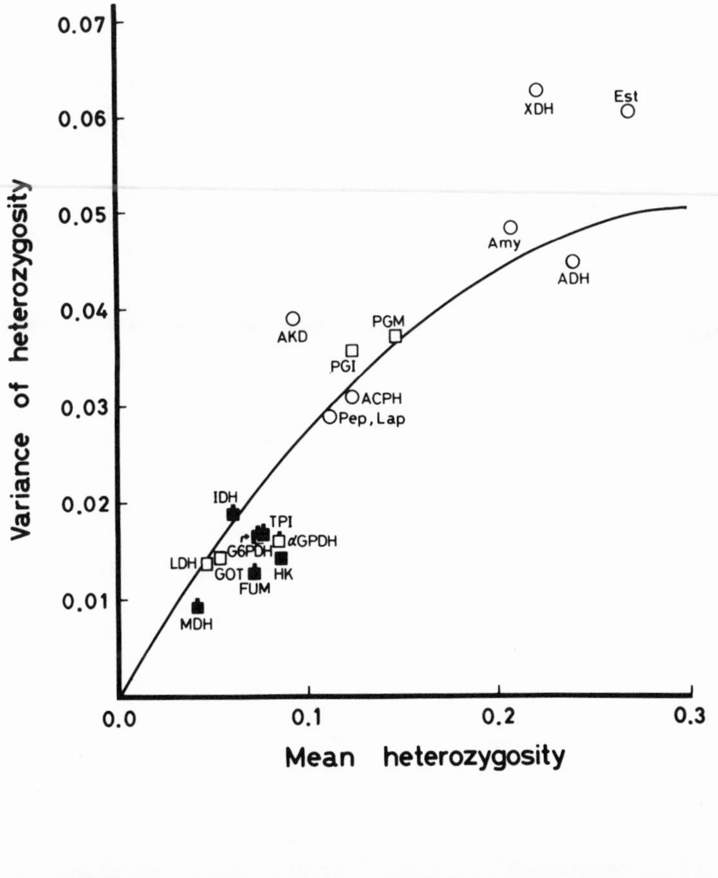


FIG. 3. The observed relationship between the variance of heterozygosity and the average heterozygosity. □, Substrate-specific enzymes; ■, substrate-specific enzymes concerned with main pathways but not with subpathways; ○, substrate-nonspecific enzymes. A short upright bar associated with these symbols means that the enzyme is concerned with only one pathway. The observed values are based on data compiled by Gojobori (8), and the 18 enzymes involved (from left to right) are as follows: MDH, malate dehydrogenase; LDH, lactate dehydrogenase; GOT, glutamate-oxaloacetate aminotransferase; IDH, isocitrate dehydrogenase; FUM, fumerase; G6PDH, glucose-6-phosphate dehydrogenase; TPI, triosephosphate isomerase; αGPDH, α-glycerophosphate dehydrogenase; HK, hexokinase; AKD, adenylate kinase; Pep. Lap, peptidase; ACPH, acid phosphatase; PGI, phosphoglucoisomerase; PGM, phosphoglucomutase; Amy, amylase; XDH, xanthine dehydrogenase; ADH, alcohol dehydrogenase; Est, esterase.

it is likely that, for any practical purpose, the mean and variance of heterozygosity thus obtained are largely independent of the number of $K$ if $K$ is 20 or more, although the expected number of rare alleles may be somewhat less as compared with the case $K = \infty$.

Finally, let us consider the bearing of the present simulation results, particularly those pertaining to the relationship between the mean and variance of heterozygosity (see Fig. 2) on the observed pattern of protein polymorphism. It has been shown by Nei and his associates (24, 27) that the observed relationship between the mean and variance of heterozygosity agrees fairly well with what is predicted from the neutral infinite allele model. Yamazaki (7) pointed out, however, using *Drosophila* data, that in substrate-specific enzymes, the variance of heterozygosity ($V_H$) tends to be lower than expected from the model of strictly neutral mutations. More recently, Gojobori (8) presented his analysis suggesting that enzymes that are subject to strong functional constraints show a similar tendency of lower-than-expected $V_H$. He analyzed data taken from 14 *Drosophila* species, 14 *Anolis* species, and 31 other species. Fig. 3 was constructed based on the results of his analysis presented in his tables I and II (8). In this figure, we did not include hemoglobin and transferrin, which are carrier proteins rather than enzymes.

The curve in Fig. 3 shows the theoretical relationship between the mean and the variance of heterozygosity under the infinite neutral alleles model. Enzymes represented by open circles (nonspecific, not restricted to a single main pathway) tend to have high average heterozygosity. Also, they tend to have higher than expected heterozygosity. On the other hand, those enzymes represented by solid squares each with an upright bar (substrate specific, restricted only to one main pathway) tend to have low heterozygosity. Furthermore, they tend to have lower variance than the theoretical curve. Of these two groups of enzymes, the latter must be subject to much stronger selective constraint than the former. Therefore, if the effectively neutral mutation model is applicable, enzymes in the latter group must have much larger $\bar{s}'$ values than those in the former group.

In comparing these observed values with the theoretical predictions given by the curve, we must keep in mind that this theoretical curve is based on the assumption that $4N_eu$ values are the same among different species for each enzyme locus. In reality, however, these values must be different from species to species because of the difference in the effective population size ($N_e$), even if the mutation rate ($u$) is the same among them for a given enzyme. Thus, the true theoretical curve should be raised. It is likely that, if this were done, open circles would fit roughly to the theoretical curve. Then, downward departure of solid squares from the theoretical curve would become even more pronounced. Comparison of these observations with experimental results in Fig. 2 supports the view that the very slightly deleterious but nearly neutral mutants are playing a significant role in the maintenance of polymorphism for substrate-specific enzymes.

This is contribution no. 1445 from the National Institute of Genetics, Mishima, Shizuoka-ken, 411 Japan. This work is supported in part by Grant-in-Aid 57120009 from the Ministry of Education, Science and Culture of Japan.

1. Kimura, M. (1968) *Nature (London)* **217**, 624–626.
2. Kimura, M. (1979) *Sci. Am.* **241(5)**, 94–104.
3. Kimura, M. (1982) in *Molecular Evolution, Protein Polymorphism and the Neutral Theory*, ed. Kimura, M. (Jpn. Sci. Soc. Press, Tokyo), pp. 3–56.
4. Kimura, M. & Ohta, T. (1971) *Nature (London)* **229**, 467–469.
5. Kimura, M. & Ohta, T. (1974) *Proc. Natl. Acad. Sci. USA* **71**, 2848–2852.
6. Gillespie, J. H. & Langley, C. H. (1974) *Genetics* **76**, 837–848.
7. Yamazaki, T. (1977) in *Molecular Evolution and Polymorphism*, ed. Kimura, M. (Natl. Inst. of Genet., Mishima, Japan), pp. 127–147.
8. Gojobori, T. (1982) in *Molecular Evolution, Protein Polymorphism and the Neutral Theory*, ed. Kimura, M. (Jpn. Sci. Soc. Press, Tokyo), pp. 137–148.
9. Kimura, M. & Crow, J. F. (1964) *Genetics* **49**, 725–738.
10. Kimura, M. (1979) *Proc. Natl. Acad. Sci. USA* **76**, 3440–3444.
11. Ohta, T. (1977) in *Molecular Evolution and Polymorphism*, ed. Kimura, M. (Natl. Inst. of Genet., Mishima, Japan), pp. 148–167.
12. Ohta, T. (1973) *Nature (London)* **246**, 96–98.
13. Ohta, T. (1974) *Nature (London)* **252**, 351–354.
14. Wright, S. (1969) *The Theory of Gene Frequencies*. Evolution and the Genetics of Populations (Univ. of Chicago Press, Chicago), Vol. 2.
15. Watterson, G. A. (1977) *Genetics* **85**, 789–814.
16. Li, W.-H. (1977) *Proc. Natl. Acad. Sci. USA* **74**, 2509–2513.
17. Kimura, M. (1980) *Proc. Natl. Acad. Sci. USA* **77**, 522–526.
18. Itoh, Y. (1979) *Inst. Stat. Math. Res. (Jpn.) Res. Memorandum* **154**, 1–20.
19. Maruyama, T. & Nei, M. (1981) *Genetics* **98**, 441–459.
20. Takahata, N. (1981) *Genetics* **98**, 427–440.
21. Pederson, D. G. (1973) *Biometrics* **29**, 814–821.
22. Maruyama, T. & Takahata, N. (1981) *Heredity* **46**, 49–57.
23. Kimura, M. (1968) *Genet. Res.* **11**, 247–269.
24. Fuerst, P. A., Chakraborty, R. & Nei, M. (1977) *Genetics* **86**, 455–483.
25. Stewart, F. M. (1976) *Theor. Popul. Biol.* **9**, 188–201.
26. Ohta, T. (1975) *Proc. Natl. Acad. Sci. USA* **72**, 3194–3196.
27. Nei, M., Fuerst, P. A. & Chakraborty, R. (1976) *Nature (London)* **262**, 491–493.

# Rare Variant Alleles in the Light of the Neutral Theory[1]

*Motoo Kimura*
National Institute of Genetics

Based on the neutral theory of molecular evolution and polymorphism, and particularly assuming "the model of infinite alleles," a method is proposed which enables us to estimate the fraction of selectively neutral alleles (denoted by $P_{neut}$) among newly arisen mutations. It makes use of data on the distribution of rare variant alleles in large samples together with information on the average heterozygosity. The formula proposed is $P_{neut} = [\bar{H}_e/(1 - \bar{H}_e)]\,[\log_e(2\bar{n}q)/\bar{n}_a(x < q)]$, where $\bar{n}_a(x < q)$ is the average number of rare alleles per locus whose frequency, $x$, is less than $q$; $\bar{n}$ is the average sample size used to count rare alleles; $\bar{H}_e$ is the average heterozygosity per locus; and $q$ is a small preassigned number such as $q = 0.01$. The method was applied to observations on enzyme and other protein loci in plaice, humans (European and Amerindian), Japanese monkeys, and fruit flies. Estimates obtained for them range from 0.064 to 0.21 with the mean and standard error $P_{neut} = 0.14 \pm 0.06$. It was pointed out that these estimates are consistent with the corresponding estimate $P_{neut}(\text{Hb}) = 0.14$ obtained independently based on the neutral theory and using data on the evolutionary rate of nucleotide substitutions in globin pseudogenes together with those in the normal globins.

## Introduction

During the past decade and a half, much attention has been paid to protein polymorphisms (and more recently, DNA polymorphism), and various statistical methods have been developed to analyze the data (see Kimura [1983], pp. 271–281, for review). There has also been much discussion, particularly in the form of the neutralist-selectionist controversy (Crow 1972, 1981; Calder 1973; Lewontin 1974; Harris 1976; Ruffié 1976; Selander 1976), regarding the mechanism by which molecular polymorphisms are maintained.

However, very little attention has been paid to rare variant alleles whose frequencies in the population are too low for them to be regarded as members of polymorphic systems. This is understandable, because such alleles do not make

1. Contribution no. 1488 from the National Institute of Genetics, Mishima, Shizuoka-ken, 411 Japan. Key words: protein polymorphism, population genetics, molecular evolution, neutral mutations.

Address for correspondence and reprints: Motoo Kimura, National Institute of Genetics, Yata 1,111 Mishima, 411, Japan.

*Mol. Biol. Evol.* 1(1):84–93.

any significant contribution to the genetic variability of the species, and also because they cannot be detected unless the sample size is unusually large.

In this paper I intend to show that such rare variant alleles can, nevertheless, supply valuable information on the mechanism by which polymorphism at the molecular level is maintained. In particular, I shall demonstrate that the neutral theory of protein polymorphism (Kimura 1968*a*, 1968*b*; Kimura and Ohta 1971; Kimura [1983] for review) can supply a theoretical basis to connect observations on rare variant alleles with those of polymorphic alleles. Furthermore, I shall endeavor to show that the result of data analysis fits well the larger picture of molecular evolution as seen from the standpoint of the neutral theory.

**Basic Theory**

Let us assume a random mating, diploid population of effective size $N_e$. Consider a particular locus and assume the infinite allele model (Kimura and Crow 1964), that is, assume that whenever mutation occurs it leads to an allele not already existing. Let $v$ be the mutation rate per locus per generation. I denote by $\Phi(x)$ the distribution of allelic frequencies under the assumption of an equilibrium in which mutational production of new alleles is balanced by random extinction of existing alleles. This distribution means that $\Phi(x)dx$ represents the expected number of alleles whose frequencies lie in the range between $x$ and $x + dx$. When all of the mutations are selectively neutral, it was shown by Kimura and Crow (1964) that

$$\Phi(x) = 4N_e v(1-x)^{4N_e v-1}x^{-1}. \tag{1}$$

Note that this distribution refers only to those alleles actually contained in the population ($x > 0$); although I assume there are an infinite number of possible alleles, only a limited number of them are present at any given moment in the population, and I do not include countless missing alleles. In the following, I shall use the letter $M$ to stand for $4N_e v$, so that the right-hand side of equation (1) becomes $M(1 - x)^{M-1}x^{-1}$.

The average value of the sum of squares of allelic frequencies or the average homozygosity is

$$\bar{H}_0 = \int_0^1 x^2\Phi(x)dx = 1/(M+1), \tag{2}$$

and, therefore, the average heterozygosity is

$$\bar{H}_e = 1 - \bar{H}_0 = M/(M+1). \tag{3}$$

This means that, if I know the value of $\bar{H}_e$ from observation, I can estimate the parameter $M$ by the formula

$$M = \bar{H}_e/(1-\bar{H}_e). \tag{4}$$

As emphasized by Nei (1975), a reliable estimate of the average heterozygosity of any species can only be obtained by averaging heterozygosities over a number of loci. Therefore, it is necessary to consider the possibility that mutation rates for neutral alleles differ among loci. To take such a variation into account, Nei et al. (1976) proposed an infinite allele model assuming that $M$ or $4N_e v$ among loci follows a gamma distribution with the mean $\bar{M}$ and the variance $V_M$. In this model, the relationship between the average heterozygosity and $\bar{M}$ is more complicated, but Nei (1975) derived a useful approximation formula,

$$\bar{H}_e = \frac{\bar{M}}{1 + \bar{M}} - \frac{V_M}{(1 + \bar{M})^3}, \tag{5}$$

which is valid unless $\alpha = \bar{M}^2/V_M$ is small and $\bar{M}$ is large. According to Nei et al. (1976), an appropriate value of $\alpha$ is about 1. Also, for a wide range of organisms, $\bar{M}$ seldom exceeds 0.3 (see, e.g., Nevo 1978), so this formula should have wide applicability. Note that under these circumstances $V_M$ is much smaller than $\bar{M}$, and therefore variation of $M$ among loci has a relatively small effect on $\bar{H}_e$, as easily seen by comparing equations (3) and (5).

One important point which I should note in estimating $M$ from $\bar{H}_e$ using equation (4) is that $M$ is determined largely by polymorphic alleles; rare alleles contribute very little to $\bar{H}_e$ and therefore to $M$.

Next, let us examine the occurrence of rare alleles whose frequencies are less than a certain small value $q$. Bearing in mind the standard practice of defining a polymorphic locus as one in which the most frequent allele does not exceed 99%, I find it appropriate to take $q = 0.01$.

It can be shown mathematically (see, e.g., Kimura 1983, p. 227) that, in the neighborhood of $x = 0$, the population behavior of alleles in general, including those having mild selective advantage or disadvantage, is essentially the same as that of selectively neutral mutants. Thus the average number of alleles per locus whose frequencies are less than $q$ is

$$\bar{n}_a(x<q) = \int_{1/(2n)}^{q} \Phi(x)dx \approx M \log_e(2nq), \tag{6}$$

where $n$ is the sample size. This formula is valid if $4N_e|s|q$ is small, where $|s|$ is the absolute value of the selection coefficient of a mutant allele. In this formula $M$ stands for $4N_e v$; however, $v$ here represents the mutation rate for practically all types of alleles, as pointed out by Nei (1977), and not just for selectively neutral alleles. In fact, he proposed the use of equation (6) for estimating the mutation rate for protein loci. I shall denote $M$ in this equation by $M_q$ in order to distinguish it from $M$ obtained by equation (4). If rare variants are scored at more than one locus with a large sample for each locus (the mean size being $\bar{n}$ per locus), and if $N_e$ is known, I can estimate the total mutation rate per locus by

$$v_{T(E)} = M_q/(4N_e), \tag{7}$$

where the subscript $E$ refers to electrophoretically detectable alleles, and

$$M_q = \bar{n}_a(x < q)/\log_e(2\bar{n}q). \tag{8}$$

However, if I denote by $v_{0(E)}$ the mutation rate per locus for selectively neutral (and electrophoretically detectable) alleles, then $M = 4N_e v_{0(E)}$, so that

$$v_{0(E)} = M/(4N_e). \tag{9}$$

Thus, I can estimate the fraction of neutral alleles among all the mutations that can be detected electrophoretically by

$$P_{\text{neut}} = v_{0(E)}/v_{T(E)} = M/M_q. \tag{10}$$

What is important here is that, even when the actual value of $N_e$ is not known, we can estimate the fraction of neutral alleles at the time of their occurrence by the ratio $M/M_q$, provided that the neutral theory is correct. This equation may be rewritten as

$$P_{\text{neut}} = \frac{\bar{H}_e}{1-\bar{H}_e} \cdot \frac{\log_e(2\bar{n}q)}{\bar{n}_a(x<q)}, \tag{10a}$$

where $\bar{H}_e$ is the mean heterozygosity per locus estimated by averaging over a number of polymorphic as well as monomorphic loci, $\bar{n}_a(x < q)$ is the number of rare variant alleles per locus, and $\bar{n}$ is the average sample size over loci used to count the rare alleles whose frequencies are less than $q$. As mentioned already, an appropriate value for $q$ is 0.01, although other values, such as $q = 0.005$, may be used.

Equations (6)–(10a) contain several assumptions and approximations. In particular, equation (6) is derived by assuming that the distribution of rare alleles in the sample is sufficiently close to that in the population. More accurate (but more complicated) formulas on the subject have been derived by Chakraborty (1981). Also, the use of the average sample size $\bar{n}$ in equation (8) requires that variation of the sample size among loci is relatively small. In the plaice data, $n$ varies around the mean = 1,956 with the standard deviation 508, so that use of the mean ($\bar{n}$ = 1,956) alone will not cause much error. It is hoped that these approximations are acceptable for the moment, and that, in the future, more extensive data will be analyzed with better statistical methods.

**Data Analysis**

As the first example of the application of the theory above to estimate the fraction of selectively neutral mutations, I shall use the data from Ward and Beardmore (1977) on protein variation in the plaice, *Pleuronectes platessa*, a marine flatfish. They screened electrophoretically detectable variation at 46 protein loci (39 enzyme and seven nonenzyme proteins), taking very large samples from the Bristol Channel population. This probably represents the most comprehensive investigation of protein variation in fishes. The mean heterozygosity per individual per locus turned out to be 0.102 ± 0.026. The sample sizes differ from locus to locus: they are more than 2,000 in 8 loci, between 2,000 and 500 in 9 loci, between 500 and 100 in 16 loci, but less than 100 in the remaining 13 loci. Using equation (4), I get $M = 0.114$ for $\bar{H}_e = 0.102$. If I use Nei's formula 5, then $\bar{M} = 0.128$ for $\alpha = 1$ and $\bar{H}_e = 0.102$, so the effect of variation of mutation rate among loci is rather small. Therefore, in the following, I shall use equation (4) to simplify our calculation.

In order to estimate $M_q$ from observations on rare variants, the sample size must be large. Therefore, I have chosen from the data of Ward and Beardmore (1977, tables 2 and 3) 11 loci for which the sample size per locus is larger than 1,000. The average sample size per locus for them turned out to be $\bar{n}$ = 1,956 ±

508. Of these 11 loci, 8 are polymorphic and 3 are monomorphic. The average heterozygosity of these 11 loci is 0.147, which is not very different from the average heterozygosity of 46 loci, that is, 0.102. Among these loci, 30 alleles are found whose frequencies in the sample are less than 0.01. Thus, $\bar{n}_a(x < 0.01) = 30/11 = 2.73$ per locus. Then, applying equation (8), where I assume $\bar{q} = 0.01$, I get $M_q = 2.73/\log_e(2 \times 19.56) = 0.744$. This leads to $v_{T(E)} = 0.744/(4N_e)$. Although the real value of $N_e$ is not known, if it is $10^6$, we have $v_{T(E)} = 1.86 \times 10^{-7}$. Finally, substituting $M = 0.114$ and $M_q = 0.744$ in equation (10), I get $P_{\text{neut}} = 0.15$. This means that one mutation out of 6.5 on the average is selectively neutral while the remaining 5.5 are too deleterious to contribute to protein polymorphism.

As the second example, I shall use the data on human populations of Harris et al. (1974), who reported the incidence of rare alleles determining electrophoretic variants at 43 enzyme loci in Europeans. From their table 1, I have chosen 26 loci for which the sample size is larger than 1,000. The average sample size for them is $\bar{n} = 4{,}058.04$. The average number of rare alleles per locus has turned out to be 49/26 or 1.88. Since Harris et al. defined rare alleles as those alleles whose individual frequency in the sample was less than 0.005, this corresponds to $q = 0.005$ of equation (8). Then, substituting $\bar{n}_a(x < q) = 1.88$, $\bar{n} = 4{,}058.04$, and $q = 0.005$ in this equation, I get $M_q = 0.508$. This gives an estimate for $4N_e v_{T(E)}$, where $v_{T(E)}$ is the total mutation rate per enzyme locus for electrophoretically detectable alleles. However, from equation (4), I obtain $M = 0.0718$ by assuming $\bar{H}_e = 0.067$, which is an approximate value for the average heterozygosity per locus due to common polymorphic alleles (Harris and Hopkinson 1972). Then from equation (10) the fraction of mutations that are selectively neutral among all electrophoretically detectable mutations is $P_{\text{neut}} = M/M_q = 0.14$. This value is very close to the corresponding estimate obtained for the plaice.

Extensive studies of rare variants in human populations have also been done by Neel and his associates on Amerindians, and valuable data have been obtained. I use the data presented in table 1 of Neel (1978), which lists the occurrence of rare variants at 28 loci in 21 Amerindian tribes. His definition of rare variant alleles corresponds to $q = 0.01$ in my terminology, and from his table I obtain $\bar{n}_a(x < q) = 1.29$ and $\bar{n} = 6{,}442.07$, giving $M_q = 0.266$. An interesting feature of his data is that some of the variants represent what he calls "private polymorphisms," that is, they are concentrated in a single or several related tribes where their frequencies are well above the minimum for a polymorphism. For example, an allele called YAN-2 at the albumin (Alb) locus is present in more than 6% of the members of the Yanomama tribe but absent in other tribes. We can calculate the value of $M$ using the average heterozygosity at 23 loci over 12 tribes as listed in table 5 of Neel (1978), where I find $\bar{H}_e = 0.054$. Thus, I obtain $M = 0.057$. Therefore, the fraction of neutral mutations, as estimated by $M/M_q$, turns out to be $P_{\text{neut}} = 0.21$, which is not very different from the corresponding value obtained for European populations (i.e., $P_{\text{neut}} = 0.14$).

The occurrence of rare variants is also reported in the Japanese macaque (*Macaca fuscata fuscata*) studied by Nozawa and his associates (see, e.g., Nozawa et al. 1982). They surveyed 32 independent protein loci and obtained 1.3% as the average heterozygosity, which is a rather low value even for mammals. Their extensive studies so far yield the following data (Nozawa, personal communication, 1981). The average number of rare variants per locus is $\bar{n}_a(x < 0.01) = 23/32 = 0.719$, and the average sample size is $\bar{n} = 1{,}609.9$. Thus I get $M_q =$

0.207 from equation (8). The observed average heterozygosity per locus is $\bar{H}_e$ = 0.013 ± 0.0014, from which I get $M$ = 0.0132. Using these values, I obtain $P_{neut}$ = $M/M_q$ = 0.064. This means that, roughly speaking, one mutation out of 16, on the average, is selectively neutral in Japanese monkeys. This is less than half as large as the corresponding value obtained for the plaice.

The census number of the total population of the Japanese macaque is estimated to be 20,000–70,000. It is also estimated that the effective population size is about one-third of its census number (cited from Nozawa et al. 1975). Following Nei (1977), if I assume $N_e = 2 \times 10^4$, I obtain $v_{T(E)} = M_q/(4N_e) = 2.6 \times 10^{-6}$. However, the mutation rate for neutral alleles is $v_{0(E)} = M/(4N_e) = 1.65 \times 10^{-7}$ per generation.

As the final example, I shall analyze the data from *Drosophila willistoni* group studied by Ayala and his associates (1974). The sample size per locus per species in this case is not as large as in the previous examples. Of the five species studied, only the *D. willistoni* data are extensive enough for the average sample size per locus to be larger than 500 (in terms of the gene number, i.e., $2n$), so I shall concentrate on this species. From their table 1, which lists allelic frequencies at 31 loci, I have chosen alleles whose frequencies are less than 1% ($q$ = 0.01). There are 85 such alleles, so that $\bar{n}_a(x < q)$ = 85/31 = 2.74. The average sample size per locus is $2\bar{n}$ = 568.06. From these values, I obtain $M_q$ = 1.60. The average heterozygosity per locus ($\bar{H}_e$) as listed in table 6 of Ayala et al. (1974) is 0.177, from which I obtain $M$ = 0.215. Therefore, the estimate for the fraction of neutral mutations among all electrophoretic mutations at the time of occurrence is $P_{neut}$ = $M/M_q$ = 0.13. This estimate is not very different from the corresponding estimates obtained for human populations, as well as for the plaice.

The results of analyses of the five examples above are summarized in table 1. The average of five $P_{neut}$ values turns out to be 0.14 ± 0.06.

## Discussion

From the standpoint of the neutral theory, the rare variant alleles are simply those alleles whose frequencies within a species happen to be in a low-frequency range (0, $q$), whereas polymorphic alleles are those whose frequencies happen to be in the higher-frequency range ($q$, 1 − $q$), where I arbitrarily take $q$ = 0.01. Both represent a phase of molecular evolution.

**Table 1**
**Proportion of Selectively Neutral Mutations at the Time of Occurrence among Electrophoretically Detectable Mutations ($P_{neut}$) Estimated from Five Data Sets**

| Organism | $\bar{H}_e$ | $M$ | $M_q$ | $P_{neut}$ |
|---|---|---|---|---|
| Plaice | | | | |
| (*Pleuronectes platessa*) | .102 | .114 | .744 | .15 |
| Human: | | | | |
| European | .067 | .072 | .509 | .14 |
| Amerindian | .054 | .057 | .266 | .21 |
| Japanese macaque | .013 | .013 | .207 | .064 |
| Fruit fly | | | | |
| (*Drosophila willistoni*) | .177 | .215 | 1.60 | .13 |

However, in contrast to polymorphic alleles, which are predominantly neutral, the rare variant alleles may include slightly deleterious and sometimes even definitely deleterious alleles in addition to selectively neutral ones. This means that rare variant alleles reflect the total mutation rate much more faithfully than polymorphic alleles. In this connection, Harris et al.'s (1974) observation is relevant. They found that "polymorphic" and "monomorphic" loci do not differ in the average heterozygosity for rare alleles if the placental alkaline phosphatase, an unusually variable locus, is excluded. This is easy to understand if we note that the intrinsic mutation rates ($v_{T(E)}$) at these two classes of loci may essentially be the same.

In the analysis above, I have estimated the fraction ($P_{\text{neut}}$) of selectively neutral mutations at the time of occurrence among mutations that can be detected by electrophoretic method, using data from the plaice, humans, the Japanese macaque, and the fruit fly (see table 1). It is remarkable that this fraction is relatively uniform among widely separated species with highly different average heterozygosities.

The present analysis is consistent with Ohta's (1975) finding on the excess of rare alleles: using *Drosophila* and human data, she noticed that the observed and theoretical distributions of allelic frequencies agree quite well under the neutral theory with respect to polymorphic alleles but that there is a marked excess of rare alleles in the observed distribution. Ohta (1976) went further and showed, using data on *D. willistoni,* that the excess of rare alleles is more pronounced in the substrate-specific enzymes than the substrate-nonspecific enzymes (see Ohta's [1976] table 3). Actually, if we apply the present method for each of these classes of enzymes separately (16 specific and 15 nonspecific enzymes in *D. willistoni*), we obtain $P_{\text{neut}} = 0.070$ for the substrate-specific group and $P_{\text{neut}} = 0.204$ for the substrate-nonspecific group. From these two $P_{\text{neut}}$ values, it is evident that the probability of a mutational change being selectively neutral is much smaller for the substrate-specific enzymes than for the nonspecific enzymes.

A similar calculation can be done using human data (Harris et al. 1974) following the classification of group I (substrate-specific) and group II (substrate-nonspecific) enzymes proposed by Gillespie and Langley (1974). For 13 loci of the group I enzymes, it turns out that $P_{\text{neut}} = 0.11$, and for 10 loci of the group II enzymes, $P_{\text{neut}} = 0.43$.

Previously, Gillespie and Langley (1974) showed that the average heterozygosity ($\bar{H}_e$) per locus is much lower for substrate-specific than for the nonspecific enzymes not only for *Drosophila* but also for the human and the mouse. These observations are compatible with the neutral theory if it is assumed that selective constraint (negative selection) is stronger for substrate-specific than nonspecific enzymes. This means that the probability of an amino acid change being not harmful, that is, selectively neutral, is smaller for the substrate-specific than for nonspecific enzymes, even if the total mutation rate per locus ($v_{T(E)}$) is the same for these two types of loci.

Finally, I would like to show that the present analysis on $P_{\text{neut}}$ is consistent with the results obtained from recent studies on the evolutionary rate of globin pseudogenes. As shown by Miyata and Yasunaga (1981) and Li et al. (1981), the evolutionary rate of nucleotide substitutions is very high for pseudogenes. This is easily understandable from the neutral theory, because pseudogenes can be regarded as "dead genes" which have been liberated from the constraint of negative selection, so that all the mutations in them become selectively neutral. Thus, pseudogenes accumulate mutational changes at the maximum speed as predicted by the neutral theory. This can be explained in more quantitative terms as follows: if I denote the fraction of neutral mutations by $f_0$ (which is determined by the degree of selective constraint), the rate of evolution in terms of mutant substitutions is

$$k = v_0 = f_0 v_T, \tag{11}$$

where $v_0$ is the neutral mutation rate and $v_T$ is the total mutation rate. Note that, under the neutral theory, the rate of evolution is equal to the mutation rate for neutral alleles (Kimura 1968*a*). As predicted by Kimura (1977), the maximum evolutionary rate is attained when $f_0 = 1$, and it is likely that pseudogenes indeed represent such a case.

If I adopt the estimates given in table 3 of Li et al. (1981), the average rate for the three globin pseudogenes, mouse $\psi\alpha3$, human $\psi\alpha1$, and rabbit $\psi\beta2$, is 4.6 $\times 10^{-9}$ substitutions per nucleotide site per year. However, the rates of nucleotide substitutions at the first, second, and third positions of the codons in the normal globin genes are $0.71 \times 10^{-9}$, $0.62 \times 10^{-9}$, and $2.64 \times 10^{-9}$, respectively (Li et al. 1981). In order to estimate the mutation rates $v_{T(E)}$ and $v_{0(E)}$ from these observed values, one needs to know what fraction of nucleotide changes at each of the three positions of the codon cause electrophoretically detectable amino acid changes. For this purpose, assume that electrophoretic mobility of a protein is determined solely by its net charge and that, among 20 amino acids, aspartic and glutamic acids are acidic and negatively charged, lysine and arginine are basic and positively charged, while the rest are electrically neutral. Then, from the standard code table, we find that the probability of a random nucleotide change causing an electrophoretically detectable amino acid change is about 0.28 for the first position, one-third for the second position, and only one-twelfth for the third position of the codon.

I also note that nucleotide changes always cause amino acid changes at the second position and predominantly so at the first position. However, at the third position, nucleotide changes cause amino acid changes in only some one-third of the cases, the rest being synonymous. Furthermore, in globins, the synonymous component of nucleotide substitutions has an evolutionary rate at least two or three times as high as the amino acid altering nucleotide substitutions (Jukes 1980; Kimura 1981), suggesting that the probability of a random nucleotide change being selectively neutral is much higher for the synonymous than for amino acid altering changes. There is also the phenomenon of nonrandom usage of synonymous codons (Grantham 1980; Ikemura 1981), and this, too, complicates the problem. For these reasons, I exclude the data from the codon's third position in the following calculation.

Then, using Li et al.'s (1981) estimates, I can compute the fraction of neutral mutations with respect to electrophoretically detectable changes in hemoglobin by the ratio $(0.71 \times 0.28 + 0.62/3)/(4.6 \times 0.28 + 4.6/3)$, which gives $P_{neut}(\text{Hb}) = 0.14$. Although I do not know the evolutionary rates in terms of amino acid substitutions of the various enzymes and other proteins used to estimate $P_{neut}$ in table 1, it is likely that their average evolutionary rate is not very different from the evolutionary rate of hemoglobin which is near the median of the evolutionary rates of proteins (Kimura 1974). Considering the many uncertainties involved in the process of estimating the fraction of neutral mutations, the agreement between the two independent estimates above, that is, $P_{\text{neut}} = 0.14 \pm 0.06$ for enzyme and other protein loci in the four organisms and $P_{\text{neut}}(\text{Hb}) = 0.14$ for hemoglobin in mammals, is impressive. I believe that this consistency strongly supports the neutral theory. I also think that a detailed study of rare variant alleles is just as important for understanding the mechanism of the maintenance of genetic variability as that of polymorphic alleles. It is hoped that more data on rare variants will be obtained for wild species whose ecologies, particularly the population sizes, are well known.

**Acknowledgment**

This work is supported in part by a grant-in-aid from the Japanese Ministry of Education, Science and Culture.

LITERATURE CITED

AYALA, F. J., M. L. TRACEY, L. G. BARR, J. F. MCDONALD, and S. PÉREZ-SALS. 1974 Genetic variation in natural populations of five *Drosophila* species and the hypothesis of the selective neutrality of protein polymorphisms. Genetics **77:**343–384.

CALDER, N. 1973. The life game. BBC, London.

CHAKRABORTY, R. 1981. Expected number of rare alleles per locus in a sample and estimation of mutation rates. Amer. J. Human Genet. **33:**481–484.

CROW, J. F. 1972. The dilemma of nearly neutral mutations: how important are they for evolution and human welfare? J. Hered. **63:**306–316.

———. 1981. The neutralist-selectionist controversy: an overview. Pp. 3–14 *in* E. B. HOOK, ed. Population and biological aspects of human mutation. Academic Press, New York.

GILLESPIE, J. H., and C. H. LANGLEY. 1974. A general model to account for enzyme variation in natural populations. Genetics **76:**837–848.

GRANTHAM, R. 1980. Workings of the genetic code. Trends Biochem. Sci. **5:**327–331.

HARRIS, H. 1976. Molecular evolution: the neutralist-selectionist controversy. Fed. Proc. **35:**2079–2082.

HARRIS, H., and D. A. HOPKINSON. 1972. Average heterozygosity per locus in man: an estimate based on the incidence of enzyme polymorphisms. Ann. Human Genet. **36:**9–20.

HARRIS, H., D. A. HOPKINSON, and E. B. ROBSON. 1974. The incidence of rare alleles determining electrophoretic variants: data on 43 enzyme loci in man. Ann. Human Genet. **37:**237–253.

IKEMURA, T. 1981. Correlation between the abundance of *Escherichia coli* transfer RNA and the occurrence of the respective codons in its protein genes. J. Mol. Biol. **146:**1–21.

JUKES, T. H. 1980. Silent nucleotide substitutions and the molecular evolutionary clock. Science **210:**973–978.

KIMURA, M. 1968*a*. Evolutionary rate at the molecular level. Nature **217:**624–626.

———. 1968*b*. Genetic variability maintained in a finite population due to mutational production of neutral and nearly neutral isoalleles. Genet. Res. **11:**247–269.

———. 1974. Gene pool of higher organisms as a product of evolution. Cold Spring Harbor Symp. Quant. Biol. **38:**515–524.

———. 1977. Preponderance of synonymous changes as evidence for the neutral theory of molecular evolution. Nature **267:**275–276.

———. 1981. Estimation of evolutionary distances between homologous nucleotide sequences. Proc. Natl. Acad. Sci. USA **78:**454–458.

———. 1983. The neutral theory of molecular evolution. Cambridge University Press, Cambridge.

KIMURA, M., and J. F. CROW. 1964. The number of alleles that can be maintained in a finite population. Genetics **49:**725–738.

KIMURA, M., and T. OHTA. 1971. Protein polymorphism as a phase of molecular evolution. Nature **229:**467–469.

LEWONTIN, R. C. 1974. The genetic basis of evolutionary change. Columbia University Press, New York.

LI, W.-H., T. GOJOBORI, and M. NEI. 1981. Pseudogenes as a paradigm of neutral evolution. Nature **292:**237–239.

MIYATA, T., and T. YASUNAGA. 1981. Rapidly evolving mouse α-globin-related pseudogene and its evolutionary history. Proc. Natl. Acad. Sci. USA **78:**450–453.

NEEL, J. V. 1978. Rare variants, private polymorphisms, and locus heterozygosity in Amerindian populations. Amer. J. Human Genet. **30:**465–490.

NEI, M. 1975. Molecular population genetics and evolution. North-Holland, Amsterdam.

———. 1977. Estimation of mutation rate from rare protein variants. Amer. J. Human Genet. **29:**225–232.

NEI, M., R. CHAKRABORTY, and P. A. FUERST. 1976. Infinite allele model with varying mutation rate. Proc. Natl. Acad. Sci. USA **73:**4164–4168.

NEVO, E. 1978. Genetic variation in natural populations: patterns and theory. Theoret. Pop. Biol. **13:**121–177.

NOZAWA, K., T. SHOTAKE, Y. KAWAMOTO, and Y. TANABE. 1982. Population genetics of Japanese monkeys. II. Blood protein polymorphisms and population structure. Primates **23:**252–271.

NOZAWA, K., T. SHOTAKE, and Y. OKURA. 1975. Blood protein polymorphisms and population structure of the Japanese macaque, *Macaca fuscata fuscata*. Pp. 225–241 *in* C. L. MARKERT, ed. Isozymes IV: genetics and evolution. Academic Press, New York.

OHTA, T. 1975. Statistical analyses of *Drosophila* and human protein polymorphisms. Proc. Natl. Acad. Sci. USA **72:**3194–3196.

———. 1976. Role of very slightly deleterious mutations in molecular evolution and polymorphism. Theoret. Pop. Biol. **10:**254–275.

RUFFIÉ, J. 1976. De la biologie à la culture. Flammarion, Paris.

SELANDER, R. K. 1976. Genetic variation in natural populations. Pp. 21–45 *in* F. J. AYALA, ed. Molecular evolution. Sinauer, Sunderland, Mass.

WARD, R. D., and J. A. BEADMORE. 1977. Protein variation in the plaice, *Pleuronectes platessa* L. Genet. Res. **30:**45–62.

MASATOSHI NEI, reviewing editor

Received June 22, 1983; revision received August 2, 1983.

# DNA and the neutral theory

By M. Kimura

*National Institute of Genetics, Mishima, 411 Japan*

The neutral theory claims that the great majority of evolutionary changes at the molecular (DNA) level are caused not by Darwinian selection but by random fixation of selectively neutral or nearly neutral mutants. The theory also asserts that the majority of protein and DNA polymorphisms are selectively neutral and that they are maintained in the species by mutational input balanced by random extinction. In conjunction with diffusion models (the stochastic theory) of gene frequencies in finite populations, it treats these phenomena in quantitative terms based on actual observations.

Although the theory has been strongly criticized by the 'selectionists', supporting evidence has accumulated over the years. Particularly, the recent outburst of DNA sequence data lends strong support to the theory both with respect to evolutionary base substitutions and DNA polymorphism, including rapid evolutionary base substitutions in pseudogenes. In addition, the observed pattern of synonymous codon choice can now be readily explained in the framework of this theory. I review these recent findings in the light of the neutral theory.

## Introduction

When the neutral theory of molecular evolution was proposed 17 years ago (Kimura 1968*a*), DNA data did not exist. All that was available was a small number of protein sequences, mostly those of haemoglobins and cytochrome *c*. In addition, electrophoretic data on enzyme polymorphism started to become available for a few organisms including the human and a species of fruit fly, which stimulated population geneticists to ponder over their implications.

Since then, enormous amounts of data have accumulated. Especially, during the last few years, we have witnessed an outburst of DNA sequence data. We are now much better informed about the rate and pattern of evolution, and the amount of intraspecific variability at the molecular level. As a result, we can make a much more detailed examination of the neutral theory, particularly with respect to the issues on which the so-called 'neutralist–selectionist controversy' has centred.

## The neutral theory

The neutral theory (or more precisely, the neutral-mutation–random-drift hypothesis) claims that the great majority of evolutionary changes at the molecular level are caused not by Darwinian selection acting on advantageous mutants, but by random fixation of selectively neutral or nearly neutral mutants. The theory does not deny the role of natural selection in determining the course of adaptive evolution, but it assumes that only a minute fraction of DNA changes are adaptive in nature.

The neutral theory also asserts that most of the intraspecific variability at the molecular level (including protein and DNA polymorphism) is essentially neutral, so that the majority of polymorphic alleles are maintained in the species by the balance between mutational input and random extinction. It regards protein and DNA polymorphisms as a transient phase of molecular evolution and rejects the notion that the majority of such polymorphisms are adaptive and maintained in the species by some form of balancing selection.

As a scientific hypothesis, the neutral theory has the advantage that its underlying assumptions are sufficiently simple that its population genetical consequences can be worked out by using suitable mutational models. This enables us to examine the theory in quantitative terms by comparing theoretical predictions with observations. In this enterprise the diffusion equation method, or 'diffusion model' as it is usually called (Kimura 1964), which treats the behaviour of mutant alleles as a stochastic process, and which at one time was regarded as being too theoretical to be of actual use, has proved to be extremely useful or even indispensable. This contrasts with the traditional evolutionary theories which rely almost exclusively on verbal, qualitative arguments.

Before I proceed, I would like to present a few formulae which are useful in discussing molecular evolution and variation from the standpoint of the neutral theory (for more details, see Kimura 1983*a*). If we denote by $k$ the rate of evolution in terms of mutant substitutions, and if we assume that this is caused by random fixation of selectively neutral mutants through random sampling drift, then

$$k = v_T f_0, \tag{1}$$

where $v_T$ is the total mutation rate and $f_0$ is the fraction of neutral mutants. In other words, under the neutral theory, the evolutionary rate is equal to the mutation rate to neutral alleles, provided that the same unit is used to measure both rates. Note that $k$ represents the rate at which molecular mutants are substituted one after another in the course of evolution within the lineage. Each of these events takes a long time, that is, four times the effective population size, as shown by Kimura & Ohta (1969). Advantageous mutations may occur, but the theory assumes that they are so rare as to be negligible. In this formulation, $1-f_0$ represents the fraction of definitely deleterious mutants which are eliminated from the population by negative selection and which do not contribute to evolution. Note that the neutral theory does not deny the occurrence of deleterious mutations. In fact, selective constraint due to such negative selection is a very important part of the neutralist explanation of some prominent features of molecular evolution, as I shall explain later.

Usually, the rate of evolution ($k$) is expressed by taking one year as the unit length of time, while mutation rate is often measured per generation. Accordingly, (1) may be modified so that

$$k = (v_T/g) f_0, \tag{2}$$

where $g$ is the generation span. In other words, $v_T/g$ is the total mutation rate per year. Next, let us consider intraspecific variability, and assume that, at a particular locus (or site), there are $n$ possible allelic states which are selectively equivalent (that is, neutral), and that mutation occurs with a equal rate in all directions. Then, at equilibrium in which the mutational input and random extinction of alleles balance each other, the average heterozygosity at this locus (or site) is

$$\overline{H}_e = 4N_e v_0/\{4N_e v_0 [n/(n-1)]+1\}, \tag{3}$$

where $N_e$ is the effective population size and $v_0$ is the mutation rate for neutral alleles (Kimura 1968*b*). For an individual nucleotide site, there are four DNA bases so that $n = 4$. On the other hand, for a gene as a whole, $n$ is so large that we may put $n = \infty$. This leads to the infinite allele model which was proposed by Kimura & Crow (1964), and which is suitable for treating enzyme polymorphisms.

At an individual nucleotide site, one of the four DNA bases is fixed for most of the time. This ensures the existence of species-specific or 'consensus' DNA sequences. Let us call a population 'monomorphic' with respect to a particular site if the total frequency of 'variant' or less frequent alleles is $q$ or less, where $q$ is a small positive number such as $q = 0.01$. Then the probability of polymorphism per site is

$$P_{\text{poly}} = 1 - 4C_1 q^{\alpha}, \tag{4}$$

where $\alpha = 4N_e v_{\text{nuc}}^{(0)}$, $C_1 = \Gamma(\alpha+\beta)/\{F(\alpha+1)\,\Gamma(\beta)\}$ and $\beta = \alpha/3$, in which $v_{\text{nuc}}^{(0)}$ is the neutral mutation rate per site and $\Gamma(\cdot)$ stands for the gamma function (see p. 197 of Kimura (1983*a*) for details). Since $\alpha$ is small, $P_{\text{poly}} = 1 - q^{\alpha}$ is sufficiently accurate for our purpose.

By comparing (3) and (4) with (1) or (2), we note that the heterozygosity and also the probability of polymorphism tend to be high at a site in which the rate of evolution is high, although the relationship involved is not a simple, linear one.

## Some misunderstandings

The neutral theory has been the target of a number of criticisms based on misunderstandings, so I shall try to discuss some of them.

First of all, genes involved in neutral evolution are not necessarily functionless as mistakenly suggested by some authors. By 'neutral evolution' I mean the cumulative genetic change caused by random drift under mutational pressure. What the neutral theory assumes is that mutant forms of each gene participating in neutral evolution are *selectively nearly equivalent*, namely, they can do the job equally well in terms of survival and reproduction of individuals. The fact that the protein and RNA molecules can tolerate many component substitutions without loss of their essential function, coupled with physiological homeostasis of organisms, is important in this context. Sometimes, neutral changes are called evolutionary 'noise', but I think this is a misnomer. Just as synonyms are not noise in language, it is not proper to regard the substitution of neutral alleles simply as noise or loss of genetic information. If the variants represent amino acid changes in a protein, this means that such changes are equally acceptable for the working of the protein in the body. Furthermore, this equality need not be exact; all that is required is that the resulting difference in fitness be small, say, for example, less than $1/(2N_e)$.

It is possible, and indeed likely, that the latitude for such interchangeability will increase as functional importance of a molecule or a part of one molecule decreases, and vice versa.

Sometimes, it is remarked that neutral alleles are by definition not relevant to adaptation, and therefore not biologically very important. I think that this is too short-sighted a view. Even if the so-called neutral alleles are selectively equivalent under a prevailing set of environmental conditions of a species, it is possible that some of them, when a new environmental condition is imposed, will become selected. Experiments suggesting this possibility have been reported by Dykhuizen & Hartl (1980) who called attention to the possibility that neutral alleles have a 'latent potential for selection'. I concur with them and believe that 'neutral mutations' can be the raw material for adaptive evolution.

## Synonymous and other silent substitutions

The first strong evidence for the neutral theory which emerged with the advent of DNA (or RNA) sequence data was the preponderance of synonymous change, namely, the observation that nucleotide substitutions within codons that do not cause amino acid changes occur at a much higher rate than amino-acid-altering substitutions.

A similar observation came with the discovery that most eukaryotic genes contain intervening sequences or 'introns' which are removed when the mature messenger RNA is formed and which therefore do not participate in protein formation. It was found that evolutionary nucleotide substitutions in introns are also high, with a rate comparable to the synonymous substitutions or even higher.

Since natural selection acts on the phenotype of the organism in the determination of which the structure and function of proteins play a decisive role, one should expect that silent mutations which do not cause amino acid changes in proteins, other things being equal, would be much less subject to natural selection than those that cause amino acid changes. Yet, it is the silent substitutions that really accumulate at a higher rate per site in evolution.

These observations are quite consistent with a similar observation made previously on proteins. For example, when active insulin is formed from proinsulin, the middle segment C of proinsulin is removed and discarded. It was found that for this peptide C the rate of evolution in terms of amino acid substitutions is several times as fast as that of insulin (see p. 159 of Kimura 1983*a*).

A general rule that has emerged through these observations is that *molecular changes that are less likely to be subject to natural selection occur more rapidly in evolution*. This empirical rule can readily be understood from the standpoint of the neutral theory, because such molecular changes have a higher chance of being selectively neutral (that is, a larger $f_0$ in (1)) and therefore neutral evolution occurs at a higher rate (that is, a larger $k$ in the same equation).

One interesting property of synonymous substitutions is that their rates are not only high in general but also are roughly equal to each other among different molecules.

Table 1 lists some estimates of the rate of evolutionary nucleotide substitutions for three molecules, presomatotropin (pregrowth hormone), β-globin and α-tubulin (for references to original data used for these estimations, see pp. 172–174 of Kimura (1983*a*).) In this table, $k_2$ stands for the rate of substitution per site per year at the second position of codons, taking $10^{-9}$ as the unit. Also, $k_S$ denotes the rate of synonymous substitutions per site, which was obtained by dividing the synonymous component $k_S'$ of the substitution rate at the third codon position by $\frac{2}{3}$. (For details, see p. 175 of Kimura (1983*a*); also readers may consult Kimura (1981*a*) and Kimura (1980) for details of the statistical methods used.) Since all the nucleotide changes at the second position and also the majority of changes at the first position of codons lead to amino acid changes, $2k_2$ roughly represents the rate of amino-acid-altering substitutions. It is interesting to note that in terms of amino acid substitutions, pregrowth hormone evolves more than a hundred times faster than α-tubulin, yet for synonymous changes, the former evolves only three or four times as fast as the latter.

Figure 1 illustrates the results of more extensive studies on this subject made by Miyata (1984); it is evident that the rates of synonymous substitutions are very close to each other even among proteins that differ widely in amino acid substitution rates. This suggests that the nature of the selective constraint is different for these two types of changes.

TABLE 1. ESTIMATES OF EVOLUTIONARY RATES

| protein (comparison) | evolutionary rate† $k_A$ | evolutionary rate† $k_S$ |
|---|---|---|
| presomatotropins (human versus rat) | 1.13±0.19 | 4.13±0.66 |
| β-globin (human versus mouse and human versus rabbit) | 0.59±0.12 | 2.48±0.48 |
| α-tubulin (chicken versus rat) | 0.008±0.005 | 1.18±0.13 |

† Units, $10^{-9}$ per site per year.

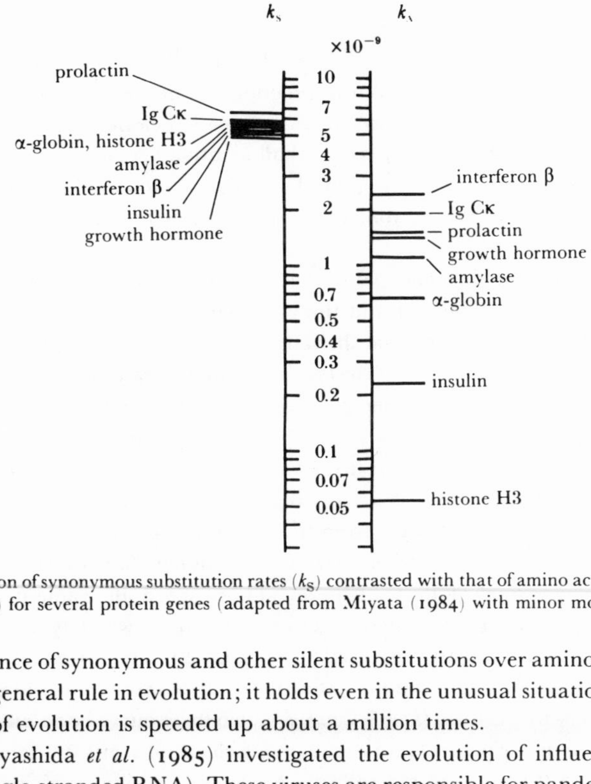


FIGURE 1. Distribution of synonymous substitution rates ($k_S$) contrasted with that of amino acid-altering substitution rates ($k_A$) for several protein genes (adapted from Miyata (1984) with minor modification).

A preponderance of synonymous and other silent substitutions over amino-acid-altering ones appears to be a general rule in evolution; it holds even in the unusual situation described below, where the rate of evolution is speeded up about a million times.

Recently, Hayashida *et al.* (1985) investigated the evolution of influenza A virus genes (consisting of single-stranded RNA). These viruses are responsible for pandemic influenza, and they can undergo radical antigenic changes. The authors compared nucleotide sequences of homologous genes among strains that were isolated in different years, and found that the genes evolve at extremely high rates in clock-like fashion. The average rate of silent substitutions estimated was $1.1 \times 10^{-2}$ per site per year, yet, compared with the corresponding rate of amino acid altering substitutions, it is only four or five times higher in this case. It is known that an outburst of pandemic influenza is caused by antigenic shift, but the number of amino acid sites responsible for surface antigens is quite limited. So, even in this case, neutral evolution predominates over adaptive evolution.

Gojobori & Yokoyama (1985) estimated the evolutionary rates of both the retroviral oncogene of Moloney murine sarcoma virus (*v-mos*$^{MO}$ gene), and its cellular homologue (*c-mos*$^{MO}$

gene). They found that the former evolves nearly 0.8 million times faster than the latter. For the former ($v$-$mos^{MO}$), the rate of nucleotide substitutions at the first, second and third codon positions are respectively $1.31 \times 10^{-3}$, $0.56 \times 10^{-3}$ and $2.06 \times 10^{-3}$ per site per year. These show clearly that even in this case the rate of nucleotide substitution is highest at the third position, indicating the preponderance of synonymous changes.

These examples show that neutral evolution can occur in RNA viruses at extraordinarily high rates. The rapid evolution is caused by correspondingly high mutation rates due to a lack of the proofreading enzymes that ensure accurate replication, as pointed out by Gojobori & Yokoyama (1985), coupled with the very high replication rates of these viruses.

## Evolutionary change in pseudogenes

The rapid evolutionary change observed in pseudogenes provided very strong support for the neutral theory in a previously unexpected way. Since I reviewed this topic extensively in my book (see pp. 178–183 of Kimura 1983*a*), I shall mention here only a few salient features. Generally speaking, a pseudogene is a region of DNA that shows definite homology with a known functional gene but has lost its ability to produce a functional product. It is sometimes called a 'dead gene'.

Comparison of pseudo globin genes with their normal counterparts revealed that base substitutions occurred at very high rates in the pseudogenes (Miyata & Yasunaga 1981; Li *et al.* 1981). What is really interesting is that the rates of substitutions are equally high in all three codon positions. According to Li *et al.* (1981), the estimated rate of substitutions in globin pseudogenes is $k_0 = 4.6 \times 10^{-9}$ per site per year. On the other hand, in normal globin genes, the estimated rates at the first, second and third codon positions are $k_1 = 0.71 \times 10^{-9}$, $k_2 = 0.62 \times 10^{-9}$ and $k_3 = 2.64 \times 10^{-9}$ respectively. In addition to base substitutions, pseudogenes accumulate deletions and additions also at very high rates in the course of evolution. It looks as if the pseudogenes have been liberated from the constraint of negative selection and are on the way to disintegration by accumulating various mutational changes (some of which must be highly damaging to normal globins) at the maximum speed allowable under mutational pressure and random drift. If this interpretation is correct, we may assume $f_0 = 1$ for pseudoglobins so that $k_0 = 4.6 \times 10^{-9}$ is equal to $v_T/g$ in (2). Then, we can obtain a rough estimate of the fraction of neutral mutations among newly arisen amino-acid-altering mutations in normal haemoglobins by the ratio $(k_1 + k_2)/(2k_0)$, which turns out to be about 0.14. Similarly, we can estimate the fraction of neutral mutations among electrophoretically detectable changes, and this gives $f_{0(E)} \approx 0.14$ (Kimura 1983*b*). Since the amino acid substitution rate ($k_{aa} \approx 1.2 \times 10^{-9}$) of haemoglobins is very close to the median value of amino acid substitution rates of various proteins (Kimura 1974), we may regard these $f_0$ values as the representative values for the fractions of neutral mutations ($P_{neut}$) among coding loci in the mammalian genome.

It is reassurig that $f_{0(E)} \approx 0.14$ thus obtained agrees quite well with the corresponding value ($P_{neut} = 0.14 \pm 0.06$) obtained by using data on the distribution of rare variant alleles together with that of polymorphic alleles which were detected by the electrophoretic method (Kimura 1983*b*).

Compared with protein polymorphism, data on DNA sequence polymorphism of nuclear genes are still very scanty. However, the study of Kreitman (1983) who determined DNA sequences of 11 cloned alcohol dehydrogenase (*Adh*) genes in *D. melanogaster*, together with

Bodmer & Ashburner's (1984) study on *Adh* sequences in *D. simulans* and a few other sibling species, show that synonymous (or silent) sites are much more polymorphic and evolve much faster than non-synonymous sites, in qualitative agreement with the neutral theory. A more recent and extensive study by Aquadro *et al.* (1985) on DNA sequence variation around the *Adh* gene region in natural populations of *D. melanogaster* revealed extensive variation due to base substitutions, insertions and deletions. Of particular interest is their finding that length variation due to unique sequence insertions or deletions and transposable element insertions are very common in this species.

## Unequal usage of synonymous codons as one aspect of selective constraint

It is now well known that synonymous codons are used quite unequally or in 'non-random' fashion in many genes of various organisms (see, for example, Grantham 1980; Grantham *et al.* 1980). In fact, non-random usage is the rule rather than the exception, and often this has been (and still is) quoted as evidence against the neutrality of synonymous base substitutions in evolution. At the beginning, this seemed to be a very puzzling and troublesome problem for the neutralists. Fortunately, however, it has become clear, mainly due to a series of papers by Ikemura (1980, 1981 *a,b*; see also Ikemura (1985) for review), that the unequal usage of synonymous codons is a result of selective constraint mainly caused by unequal availability of the cognate tRNA species in the cell at least in unicellular organisms such as *Escherichia coli* and yeast. He made the important discovery that among synonymous codons for an amino acid, the most frequently used codon invariably corresponds to the most abundant isoaccepting tRNA species. For example, there are six codons coding for leucine, but in *E. coli*, the codon CUG is used most frequently. This matches with the observation that among the cognate tRNA species for leucine in this organism, the one ($tRNA_1^{Leu}$) that recognizes this particular codon is the most abundant one.

On the other hand, in yeast, the codon UUG is used most frequently, and, at the same time the tRNA recognizing this codon is much more abundant than other tRNA species for leucine. The observation that the synonymous codon choice-pattern is similar among different genes within an organism also adds support to Ikemura's theory.

An additional rule that has emerged through recent studies by Ikemura and others is that more highly expressed genes tend to show stronger choice pattern among synonymous codons. In fact, Ikemura (1985) showed that a very strong correlation exists between the number of molecules per cell and the $F_{op}$ value, namely the fraction by which the 'optimal' codon (corresponding to the most abundant cognate tRNA species) is used.

These observations can readily be understood on the neutral theory by noting that in evolution synonymous substitutions are constrained in such a way that they do not deviate much from the established pattern of relative availability of isoaccepting tRNA species, since otherwise translational efficiency would be reduced in the cell thereby causing loss of Darwinian fitness. Such a constraint (negative selection) must be stronger for more highly expressed genes which must be more directly related to the survival and reproduction of individuals (or cells) than less highly expressed genes. Also, it has been found that essentially the same codon choice pattern is shared by a very wide group of related organisms. For example, the coding 'dialect' of *E. coli* is similar to those of other Enterobacteriaceae (Ikemura 1985).

I want to bring out one more observation that is highly pertinent to the neutralist–selectionist controversy. As shown by Miyata (1982) and Ikemura (1985), and as exemplified in table 2, *stronger choice among synonymous codons tends to slow down rather than accelerate the synonymous base substitutions in evolution.* If synonymous substitutions were caused by positive Darwinian selection, one should expect that the stronger choice would accelerate evolution rather than retard it, but the actual observations are the other way around.

TABLE 2. CODON CHOICE PATTERN AND EVOLUTIONARY RATE

| molecule (protein) | number of molecules† | $F_{op}$ | evolutionary distance $K_S$ | $K_{aa}$ |
|---|---|---|---|---|
| omp A | $3 \times 10^4$ | 0.92 | 0.18 | 0.07 |
| trp A | under 1000 | 0.61 | 1.34 | 0.15 |

† Number of protein molecules per genome for omp A has been measured in *E. coli* (Ikemura 1985), while that for trp A is an inferred value (T. Ikemura, personal communication).

Abbreviations: omp A, outer membrane major protein; trp A, tryptophan synthetase α; $F_{op}$, relative frequency of use of optimal codons (data, from Ikemura 1985); $K_S$, number of synonymous substitutions per nucleotide site; $K_{aa}$, number of amino acid substitutions per codon.

Evolutionary distances are estimated by comparing DNA sequences between *Escherichia coli* and *Salmonella typhimurium*. Data for trp A are from Nichols & Yanofsky (1979), and those for omp A are taken from Ikemura (1985).

## FEATURES OF MOLECULAR EVOLUTION CONTRASTED WITH THOSE OF PHENOTYPIC EVOLUTION

There are at least two features that distinguish molecular evolution from phenotypic evolution (table 3). Molecular evolution is characterized by (i) constancy in rate and (ii) conservatism in mode (see Kimura (1983*a*) for review), while phenotypic evolution exhibits (i) irregularity in rate and (ii) opportunism in mode (see, for example, Simpson 1949). These features can readily be understood if we assume that phenotypic evolution is largely controlled by positive Darwinian selection that brings about adaptation of organisms to their environment, while molecular evolution is mainly caused by random fixation of selectively neutral or nearly neutral mutants under mutation pressure.

TABLE 3. CONTRAST BETWEEN MOLECULAR AND PHENOTYIC EVOLUTION

| type of evolution | rate | mode |
|---|---|---|
| molecular | constant | conservative |
| phenotypic | irregular | opportunistic |

As to the first feature of molecular evolution, enough evidence has now accumulated indicating that the rate of evolution in terms of nucleotide substitutions is approximately constant among lines for a given type of gene or DNA region. I believe that the usefulness of the rate constancy hypothesis (or 'molecular clock' concept) in constructing phylogenetic trees attests its validity. The fact that the constancy is in terms of physical time (year) even among organisms with very different generation spans has often been cited as evidence against the neutral theory, because constancy of $v_T/g$ in (2) appears to be inconsistent with observations. In fact, traditional studies of mutations on visible and viability traits (including lethals) strongly suggest that the spontaneous mutation rate per generation, but not per year, is roughly equal among different animals (such as *Drosophila*, mouse and man) whose generation spans are very different. It now appears, however, that many of these 'mutations' are caused or controlled by transposons and insertion sequences (see, for example, Rubin 1983; Mukai & Yukuhiro 1983; Mackay 1984). On the other hand, no definite data are available at present to settle the issue whether the mutation rate for nucleotide substitutions (with which the neutral theory is concerned) is proportional to year or generation. Experimental studies on this subject are much needed.

The conservative nature of molecular evolution has now been well established; those mutant substitutions that are less disruptive to the existing structure and function of a molecule occur more frequently in evolution than more disruptive ones. This is easy to understand from the neutral theory, because the more conservative the mutational change, the more likely it is to turn out to be selectively neutral.

Some years back, we (Kimura & Ohta 1974) enumerated five principles which govern molecular evolution, including constancy in rate and conservatism in mode. One of the principles states that functionally less important molecules or parts of a molecule evolve (in terms of mutant substitutions) faster than more important ones. This too can readily be understood by the neutral theory, because the fraction of neutral mutations must be higher in less important molecules or parts of one molecule, that is, $f_0$ in (1) becomes larger in them.

When this principle, accompanied by its neutralist explanation, was first proposed, much opposition was voiced by 'selectionists', but I am glad to say that it has become a part of common sense among molecular biologists. It is now a routine practice among them, to search for various signals by comparing a relevant region of homologous DNA sequences of diverse organisms and to pick out a constant (consensus) pattern, but disregard variable ones as unimportant.

## Concluding remarks

Recent data on DNA sequences have strongly vindicated the neutral theory: I believe that evidence is now overwhelming that, at the molecular level, neutral evolution predominate over Darwinian evolution. Then, the question that immediately arises is why random fixation of selectively neutral or nearly neutral mutants prevails at the molecular level, even if Darwinian evolution by positive natural selection appears to be so important at the phenotypic level.

The answer to this question, I think, comes from the fact that the most common type of natural selection at the phenotypic level is stabilizing selection, to use Mather's (1953) teminology. Here it is important to note that natural selection acts directly on phenotype, but only secondarily on the molecular constitution of genes.

Unlike the type of natural selection which Darwin had in mind when he tried to explain evolution through the accumulation of small beneficial changes, stabilizing selection eliminates phenotypically extreme individuals and preserves those near the population mean. It acts to keep the *status quo*, rather than to produce a directional change. It was shown by Wright (1935) and Robertson (1956) that if genes are additive with respect to a quantitative character which is subject to stabilizing selection and if the mean and the optimum coincide, then the alleles involved behave as if negatively overdominant. Pursuing this problem further, I have shown (Kimura 1981 *b*) that extensive neutral evolution can occur under stabilizing selection if a large number of loci or sites are involved in a quantitative character. This applies, for example, to the situation in which each individual in a mammalian species is heterozygous on the average for one million nucleotide sites and the total selection intensity per individual is 50%. I believe that the concept of 'random drift under stabilizing selection' will help the selectionist camp to appreciate the merit of the neutral theory in understanding the mechanism of evolution and intraspecific variability at the molecular level. A similar view was presented by Milkman (1982) under the designation 'a unified selection theory'. More recently, he gave an interesting personal account on 'the rational resolution of the selectionist–neutralist controversy' along the lines I have outlined above (see pp. 328–334 of Milkman 1983).

Note that the selection involved here is not 'balancing selection' as routinely invoked by the selectionists. In fact, careful, large scale experiments by Mukai and his associates, using *Drosophila melanogaster* have produced no evidence for the three types of balancing selection, that is, overdominance, frequency-dependent selection and diversifying selection at work in protein polymorphism (see Mukai *et al.* (1982) for review).

I must add here that there is a possibility that a certain fraction of nucleotide sites (presumably a large fraction) produce no phenotypic effects at all, and therefore are completely neutral with respect to natural selection. On the other hand, a certain fraction (probably a very small fraction) of nucleotide or amino acid substitutions are definitely advantageous for the species in adapting to new environments, and therefore they are subjected to straightforward positive natural selection.

Recently, Perutz (1983) has made a detailed stereochemical examination of amino acid substitutions among vertebrate haemoglobins in relation to species adaptation. He concluded that adaptations leading to response to new chemical stimuli have evolved by only a few (one to five) amino acid substitutions in key positions, while most of the amino acid replacements between species are functionally neutral. He says that the evidence supports my neutral theory.

For me, a really encouraging aspect of the neutral theory is that its position becomes stronger with increasing data.

I would like to close my talk by quoting from Haldane (1959) who, a quarter of a century ago, wrote as follows in discussing Darwin's theory of evolution by natural selection.

> 'The history of science makes it almost certain that facts will be discovered which show that the theory of natural selection is not fully adequate to account for evolution. But the same history makes it extremely improbable that these facts will be in any way related to the criticisms at present made of it. The physics of Newton and Galileo have proved inadequate in several respects, and are being replaced by relativistic and quantum mechanics. These, however, are even further from the medieval physical theories than were the theories of Galileo and Newton. They were discovered because when the consequences of Newtonian physics were fully worked out, certain facts disagreed with them. It was not possible in Newton's time to guess at these discrepancies,.... Darwinism will, I do not doubt, be modified. Like any other successful theory it will ultimately develop its own internal contradictions.'

Contribution number 1615 from the National Institute of Genetics, Mishima, Shizuoka-Ken, 411 Japan.

## References


Aquadro, C. F., Deese, S. F., Bland, M. M., Langley, C. H. & Laurie-Ahlberg, C. C. 1985 Molecular population genetics of the alcohol dehydrogenase gene region of *Drosophila melanogaster*. *Genetics* (In the press.)

Bodmer, M. & Ashburner, M. 1984 Conservation and change in the DNA sequences coding for alcohol dehydrogenase in sibling species of *Drosophila*. *Nature, Lond.* **309**, 425–430.

Dykhuizen, D. & Hartl, D. L. 1980 Selective neutrality of 6PGD allozymes in *E. coli* and the effects of genetic background. *Genetics* **96**, 801–817.

Gojobori, T. & Yokoyama, S. 1985 Rates of evolution of the retroviral oncogene of Moloney murine sarcoma virus and of its cellular homologues. *Proc. natn. Acad. Sci. U.S.A.* **82**, 4198–4201.

Grantham, R. 1980 Workings of the genetic code. *Trends biochem. Sci.* **5**, 327–331.

Grantham, R., Gautier, C. & Gouy, M. 1980 Codon frequencies in 119 individual genes confirm consistent choices of degenerate bases according to genome type. *Nucl. Acids Res.* **8**, 1893–1912.

Haldane, J. B. S. 1959 Natural selection. In *Darwin's biological work* (ed. P. R. Bell), pp. 101–149. Cambridge University Press.

Hayashida, H., Toh, H., Kikuno, R. & Miyata, T. 1985 Evolution of influenza virus genes. *Molec. Biol. Evol.* **2**, 289–303.

Ikemura, T. 1980 The frequency of codon usage in *E. coli* genes: correlation with abundance of cognate tRNA. In *Genetics and evolution of RNA polymerase, tRNA and ribosomes* (ed. S. Osawa, H. Ozeki, H. Uchida & T. Yura), pp. 519–523. University of Tokyo Press.

Ikemura, T. 1981*a* Correlation between the abundance of *Escherichia coli* transfer RNAs and the occurrence of the respective codons in its protein genes. *J. molec. Biol.* **146**, 1–21.

Ikemura, T. 1981*b* Correlation between the abundance of *Escherichia coli* transfer RNAs and the occurrence of the respective codons in its protein genes: a proposal for a synonymous codon choice that is optimal for the *E. coli* translational system. *J. molec. Biol.* **151**, 389–409.

Ikemura, T. 1985 Codon usage and tRNA content in unicellular and multicellular organisms. *Molec. Biol. Evol.* **2**, 13–34.

Kimura, M. 1964 Diffusion models in population genetics. *J. appl. Prob.* **1**, 177–232.

Kimura, M. 1968*a* Evolutionary rate at the molecular level. *Nature, Lond.* **217**, 624–626.

Kimura, M. 1968*b* Genetic variability maintained in a finite population due to mutational production of neutral and nearly neutral isoalleles. *Genet. Res.* **11**, 247–269.

Kimura, M. 1974 Gene pool of higher organisms as a product of evolution. *Cold Spring Harb. Symp. quant. Biol.* **38**, 515–524.

Kimura, M. 1980 A simple method for estimating evolutionary rates of base substitutions through comparative studies of nucleotide sequences. *J. molec. Evol.* **16**, 111–120.

Kimura, M. 1981*a* Estimation of evolutionary distances between homologous nucleotide sequences. *Proc. natn. Acad. Sci. U.S.A.* **78**, 454–458.

Kimura, M. 1981*b* Possibility of extensive neutral evolution under stabilizing selection with special reference to non-random usage of synonymous codons. *Proc. natn. Acad. Sci. U.S.A.* **78**, 5773–5777.

Kimura, M. 1983*a* *The neutral theory of molecular evolution*. Cambridge University Press.

Kimura, M. 1983*b* Rare variant alleles in the light of the neutral theory. *Molec. Biol. Evol.* **1**, 84–93.

Kimura, M. & Crow, J. F. 1964 The number of alleles that can be maintained in a finite population. *Genetics* **49**, 725–738.

Kimura, M. & Ohta, T. 1969 The average number of generations until fixation of a mutant gene in a finite population. *Genetics* **61**, 763–771.

Kimura, M. & Ohta, T. 1974 On some principles governing molecular evolution. *Proc. natn. Acad. Sci. U.S.A.* **71**, 2848–2852.

Kreitman, M. 1983 Nucleotide polymorphism at the alcohol dehydrogenase locus of *Drosophila melanogaster*. *Nature, Lond.* **304**, 412–417.

Li, W.-H., Gojobori, T. & Nei, M. 1981 Pseudogenes as a paradigm of neutral evolution. *Nature, Lond.* **292**, 237–239.

Mackay, T. F. C. 1984 Jumping genes meet abdominal bristles: hybrid dysgenesis-induced quantitative variation in *Drosophila melanogaster*. *Genet. Res.* **44**, 231–237.

Mather, K. 1953 The genetical structure of populations. In *Symp. Soc. Exp. Biol, VII, Evolution*, pp. 66–95. Cambridge University Press.

Milkman, R. 1982 Toward a unified selection theory. In *Perspectives on evolution* (ed. R. Milkman), pp. 105-118. Sunderland, Massachusetts: Sinauer Associates.

Milkman, R. (ed.) 1983 *Experimental population genetics*, Benchmark Papers in Genetics, 13. Stroudsburg, Pennsylvania: Hutchinson Ross.

Miyata, T. 1982 Evolutionary changes and functional constraints in DNA sequences. In *Molecular evolution, protein polymorphism and the neutral theory* (ed. M. Kimura), pp. 233–266. Tokyo: Japan Scientific Societies Press. Berlin: Springer-Verlag.

Miyata, T. 1984 Evolution of DNA; dynamically evolving eukaryotic genes. In *Introduction to molecular evolutionary study* (ed. M. Kimura), pp. 56–90. Tokyo: Baifukan. [In Japanese.]
Miyata, T. & Yasunaga, T. 1981 Rapidly evolving mouse α-globin-related pseudogene and its evolutionary history. *Proc. natn. Acad. Sci. U.S.A.* **78**, 450–453.
Mukai, T., Yamaguchi, O., Kusakabe, S., Tachida, H., Matsuda, M., Ichinose, M. & Yoshimaru, H. 1982 Lack of balancing selection for protein polymorphisms. In *Molecular evolution, protein polymorphism and the neutral theory* (ed. M. Kimura), pp. 81–120. Tokyo: Japan Scientific Societies Press. Berlin: Springer-Verlag.
Mukai, T. & Yukuhiro, K. 1983 An extremely high rate of deleterious viability mutations in *Drosophila* possibly caused by transposons in non-coding regions. *Proc. Japan Acad.* B **59**, 316–319.
Nichols, B. P. & Yanofsky, C. 1979 Nucleotide sequences of *trp A* of *Salmonella typhimurium* and *Escherichia coli*: An evolutionary comparison. *Proc. natn. Acad. Sci. U.S.A.* **76**, 5244–5248.
Perutz, M. F. 1983 Species adaptation in a protein molecule. *Molec. Biol. Evol.* **1**, 1–28.
Robertson, A. 1956 The effect of selection against extreme deviants based on deviation or on homozygosis. *J. Genet.* **54**, 236–248.
Rubin, G. M. 1983 Dispersed repetitive DNAs in *Drosophila*. In *Mobile genetic elements* (ed. J. A. Shapiro), pp. 329–361. New York: Academic Press.
Simpson, G. G. 1949 *The meaning of evolution.* Yale University Press.
Wright, S. 1935 The analysis of variance and the correlations between relatives with respect to deviations from an optimum. *J. Genet.* **30**, 243–256.

# MAJOR PUBLICATIONS OF MOTOO KIMURA


Asterisk (*) indicates those reprinted here.

*1953 Kimura, M. "Stepping-stone" model of population. *Annual Report of National Institute of Genetics* 3:62–63.

*1954 Kimura, M. Process leading to quasi-fixation of genes in natural populations due to random fluctuation of selection intensities. *Genetics* 39:280–95.

*1955 Kimura, M. Solution of a process of random genetic drift with a continuous model. *Proceedings of the National Academy of Sciences USA* 41:144–50.

1955 Kimura, M. Random genetic drift in multi-allelic locus. *Evolution* 9:419–35.

*1955 Kimura, M. Stochastic processes and distribution of gene frequencies under natural selection. *Cold Spring Harbor Symposia on Quantitative Biology* 20:33–53.

1956 Kimura, M. Stochastic processes in population genetics. Ph.D. thesis (unpublished), University of Wisconsin.

1956 Kimura, M. Random genetic drift in a tri-allelic locus: Exact solution with a continuous model. *Biometrics* 12:57–66.

1956 Kimura, M. Rules for testing stability of a selective polymorphism. *Proceedings of the National Academy of Sciences USA* 42:336–40.

*1956 Kimura, M. A model of a genetic system which leads to closer linkage by natural selection. *Evolution* 10:278–87.

1956 Crow, J. F., and M. Kimura. Some genetic problems in natural populations. *Proceedings of the Third Berkeley Symposium on Mathematical Statistics and Probability* 4:1–22. University of California Press.

*1957 Kimura, M. Some problems of stochastic processes in genetics. *Annals of Mathematical Statistics* 28:882–901.

1958 Kimura, M. Theoretical basis for the study of inbreeding in man. (In Japanese with English summary.) *Japanese Journal of Human Genetics* 3:51–70.

*1958 Kimura, M. On the change of population fitness by natural selection. *Heredity* 12:145–67.

*1960 Kimura, M. Optimum mutation rate and degree of dominance as determined by the principle of minimum genetic load. *Journal of Genetics* 57:21–34.

1960 Kimura, M. Genetic load of a population and its significance in evolution. (In Japanese with English summary.) *Japanese Journal of Genetics* 35:7–33.

1960 Kimura, M. Relative applicability of the classical and the balance hypotheses to man. Especially with respect to quantitative characters. *Journal of Radiation Research* 1–2:155–64.

1960 Kimura, M. *Outline of Population Genetics* (in Japanese). Tokyo: Baifukan.

1961 Kimura, M. Some calculations on the mutational load. *Japanese Journal of Genetics* (suppl.) 36:179–90.

*1961 Kimura, M. Natural selection as the process of accumulating genetic information in adaptive evolution. *Genetical Research, Cambridge* 2:127–40.

*1961 Kimura, M., and H. Kayano. The maintenance of supernumerary chromosomes in wild populations of *Lilium callosum* by preferential segregation. *Genetics* 46:1699–1712.

*1962 Kimura, M. On the probability of fixation of mutant genes in a population. *Genetics* 47:713–19.

1962 Kimura, M. A suggestion on the experimental approach to the origin of supernumerary chromosomes. *American Naturalist* 96:319–20.

*1963 Kimura, M. A probability method for treating inbreeding systems, especially with linked genes. *Biometrics* 19:1–17.

1963 Kimura, M., and J. F. Crow. The measurement of effective population number. *Evolution* 17:279–88.

*1963 Kimura, M., and J. F. Crow. On the maximum avoidance of inbreeding. *Genetical Research, Cambridge* 4:399–415.

1963 Kimura, M., T. Maruyama, and J. F. Crow. The mutation load in small populations. *Genetics* 48:1303–12.

*1964 Kimura, M., and J. F. Crow. The number of alleles that can be maintained in a finite population. *Genetics* 49:725–38.

*1964 Kimura, M., and G. H. Weiss. The stepping stone model of population structure and the decrease of genetic correlation with distance. *Genetics* 49:561–76.

*1964 Kimura, M. Diffusion models in population genetics. *Journal of Applied Probability* 1:177–232.

1964 Crow, J. F., and M. Kimura. The theory of genetic loads. *Proceedings of the XI International Congress of Genetics,* pp. 495–506. The Hague, Netherlands.

*1965 Kimura, M. A stochastic model concerning the maintenance of genetic variability in quantitative characters. *Proceedings of the National Academy of Sciences USA* 54:731–36.

1965 Weiss, G. H., and M. Kimura. A mathematical analysis of the stepping stone model of genetic correlation. *Journal of Applied Probability* 2:129–49.

*1965 Kimura, M. Attainment of quasi linkage equilibrium when gene frequencies are changing by natural selection. *Genetics* 52:875–90.

*1965 Crow, J. F., and M. Kimura. Evolution in sexual and asexual populations. *American Naturalist* 99:439–50.

1966 Cavalli-Sforza, L. L., M. Kimura, and I. Barrai. The probability of consanguineous marriages. *Genetics* 54:37–60.

*1966 Kimura, M., and T. Maruyama. The mutational load with epistatic gene interactions in fitness. *Genetics* 54:1337–51.

*1967 Kimura, M. On the evolutionary adjustment of spontaneous mutation rates. *Genetical Research, Cambridge* 9:23–34.

*1968 Kimura, M. Evolutionary rate at the molecular level. *Nature* 217:624–26.

1968 Yasuda, N., and M. Kimura. A gene-counting method of maximum likelihood for estimating gene frequencies in ABO and ABO-like systems. *Annals of Human Genetics* 31:409–20.

*1968 Kimura, M. Genetic variability maintained in a finite population due to mutational production of neutral and nearly neutral isoalleles. *Genetical Research, Cambridge* 11:247–69.

1968 Kimura, M. Haldane's contributions to the mathematical theories of evolution and population genetics. In *Haldane and Modern Biology,* ed. K. R. Dronamraju, pp. 133–40. Baltimore: Johns Hopkins University Press.

1969 Ohta, T., and M. Kimura. Linkage disequilibrium due to random genetic drift. *Genetical Research, Cambridge* 13:47–55.

1969 Kimura, M., and J. F. Crow. Natural selection and gene substitution. *Genetical Research, Cambridge* 13:127–41.

1969 Kimura, M., and T. Maruyama. The substitutional load in a finite population. *Heredity* 24:101–14.

*1969 Kimura, M., and T. Ohta. The average number of generations until fixation of a mutant gene in a finite population. *Genetics* 61:763–71.

*1969 Kimura, M. The number of heterozygous nucleotide sites maintained in a finite population due to steady flux of mutations. *Genetics* 61:893–903.

1969 Crow, J. F., and M. Kimura. Evolution in sexual and asexual populations: A reply. *American Naturalist* 103:89–91.

*1969 Kimura, M. The rate of molecular evolution considered from the standpoint of population genetics. *Proceedings of the National Academy of Sciences USA* 63:1181–88.

1969 Ohta, T., and M. Kimura. Linkage disequilibrium at steady state determined by random genetic drift and recurrent mutation. *Genetics* 63:229–38.

1969 Kimura, M., and T. Ohta. The average number of generations until extinction of an individual mutant gene in a finite population. *Genetics* 63:701–9.

*1970 Kimura, M. The length of time required for a selectively neutral mutant to reach fixation through random frequency drift in a finite population. *Genetical Research, Cambridge* 15:131–33.

1970 Crow, J. F., and M. Kimura. *An Introduction to Population Genetics Theory.* New York: Harper and Row.

1970 Ohta, T., and M. Kimura. Statistical analysis of the base composition of genes using data on the amino acid composition of proteins. *Genetics* 64:387–95.

1970 Kimura, M. Stochastic processes in population genetics, with special reference to distribution of gene frequencies and probability of gene fixation. In *Mathematical Topics in Population Genetics,* ed. K. Kojima, pp. 178–209. Berlin: Springer-Verlag.

1970 Kimura, M., and T. Ohta. Probability of fixation of a mutant gene in a finite population when selective advantage decreases with time. *Genetics* 65:525–34.

1970 Kimura, M., and T. Ohta. Genetic loads at a polymorphic locus which is maintained by frequency-dependent selection. *Genetical Research, Cambridge* 16:145–50.

*1970 Ohta, T., and M. Kimura. Development of associative overdominance through linkage disequilibrium in finite populations. *Genetical Research, Cambridge* 16:165–77.

*1971 Kimura, M., and T. Ohta. Protein polymorphism as a phase of molecular evolution. *Nature* 229:467–69.

1971 Kimura, M. Theoretical foundation of population genetics at the molecular level. *Theoretical Population Biology* 2:174–208.

1971 Kimura, M., and T. Ohta. *Theoretical Aspects of Population Genetics.* Princeton: Princeton University Press.

1971 Kimura, M. Population genetics and human genetics, with special reference to variation and evolution at the molecular level. (In Japanese with English summary.) *Japanese Journal of Human Genetics* 16(1):1–14.

1971 Ohta, T., and M. Kimura. Amino acid composition of proteins as a product of molecular evolution. *Science* 174:150–53.

1971 Ohta, T., and M. Kimura. Functional organization of genetic material as a product of molecular evolution. *Nature* 233:118–19.

1971 Kimura, M., and T. Ohta. On the rate of molecular evolution. *Journal of Molecular Evolution* 1:1–17.

1971 Ohta, T., and M. Kimura. On the constancy of the evolutionary rate of cistrons. *Journal of Molecular Evolution* 1:18–25.

*1971 Ohta, T., and M. Kimura. Linkage disequilibrium between two segregating nucleotide sites under the steady flux of mutations in a finite population. *Genetics* 68:571–80.

1971 Ohta, T., and M. Kimura. Genetic load due to mutations with very small effects. *Japanese Journal of Genetics* 46:393–401.

*1971 Maruyama, T., and M. Kimura. Some methods for treating continuous stochastic processes in population genetics. *Japanese Journal of Genetics* 46:407–10.

1971 Kimura, M., and T. Maruyama. Pattern of neutral polymorphism in a geographically structured population. *Genetical Research, Cambridge* 18:125–31.

1971 Ohta, T., and M. Kimura. Behavior of neutral mutants influenced by associated overdominant loci in finite populations. *Genetics* 69:247–60.

1972 Crow, J. F., and M. Kimura. The effective number of a population with overlapping generations: A correction and further discussion. *American Journal of Human Genetics* 24:1–10.

1972 Kimura, M., and T. Ohta. Population genetics, molecular biometry, and evolution. Proceedings of the Sixth Berkeley Symposium on Mathematical Statistics and Probability. Vol. 5. In *Darwinian, Neo-Darwinian and Non-Darwinian Evolution,* pp. 43–68. Berkeley: University of California Press.

1972 Ohta, T., and M. Kimura. Fixation time of overdominant alleles influenced by random fluctuation of selection intensity. *Genetical Research, Cambridge* 20:1–7.

1972 Kimura M., and T. Ohta. On the stochastic model for estimation of mutational distance between homologous proteins. *Journal of Molecular Evolution* 2:87–90.

1973 Yasuda, N., and M. Kimura. A study of human migration in the Mishima district. *Annals of Human Genetics, London* 36:313–22.

*1973 Kimura, M., and T. Ohta. Mutation and evolution at the molecular level. *Genetics* (suppl.) 73:19–35.

*1973 Kimura, M., and T. Ohta. Eukaryotes-prokaryotes divergence estimated by 5S ribosomal RNA sequences. *Nature: New Biology* 243:199–200.

*1973 Kimura, M., and T. Ohta. The age of a neutral mutant persisting in a finite population. *Genetics* 75:199–212.

*1973 Ohta, T., and M. Kimura. A model of mutation appropriate to estimate the number of electrophoretically detectable alleles in a finite population. *Genetical Research, Cambridge* 22:201–4.

1973 Kimura, M. Gene pool of higher organisms as a product of evolution. *Cold Spring Harbor Symposia on Quantitative Biology* 38:515–24.

*1974 Maruyama, T., and M. Kimura. A note on the speed of gene frequency changes in reverse directions in a finite population. *Evolution* 28:161–63.

1974 Maruyama, T., and M. Kimura. Geographical uniformity of selectively neutral polymorphisms. *Nature* 249:30–32.

1974 Ohta, T., and M. Kimura. Simulation studies on electrophoretically detectable genetic variability in a finite population. *Genetics* 76:615–24.

*1974 Kimura, M., and T. Ohta. On some principles governing molecular evolution. *Proceedings of the National Academy of Sciences USA* 71:2848–52.

1974 Kimura, M., and T. Ohta. Probability of gene fixation in an expanding finite population. *Proceedings of the National Academy of Sciences USA* 71:3377–79.

1974 Kimura, M. Some models of allelic mutation in molecular population genetics. *Lectures on Mathematics in the Life Sciences* 7:1–23. American Mathematical Society, Providence, Rhode Island.

1974 Kimura, M. *Genetics and the Future of Man* (in Japanese), ed. M. Kimura. Tokyo: Baifukan.

1975 Ohta, T., and M. Kimura. Theoretical analysis of electrophoretically detectable polymorphisms: Models of very slightly deleterious mutations. *American Naturalist* 109:137–45.

*1975 Maruyama, T., and M. Kimura. Moments for sum of an arbitrary function of gene frequency along a stochastic path of gene frequency change. *Proceedings of the National Academy of Sciences USA* 72:1602–4.

1975 Kimura, M., and T. Ohta. Distribution of allelic frequencies in a finite population under stepwise production of neutral alleles. *Proceedings of the National Academy of Sciences USA* 72:2761–64.

1975 Kimura, M. An introduction to theoretical population genetics. In *Foundations of Human Genetics* (in Japanese). Iwanami Lecture Series in Modern Biology 6, ed. M. Kimura. Tokyo: Iwanami Shoten.

1975 Ohta, T., and M. Kimura. The effect of selected linked locus on heterozygosity of neutral alleles (the hitch-hiking effect). *Genetical Research, Cambridge* 25:313–26.

1975 Kimura, M. Mathematical contributions to population genetics. Proceedings of the XIII International Congress of Genetics. Part II. *Genetics* (suppl.) 79:91–100.

1976 Kimura, M. The origin of life and molecular evolution (in Japanese). Iwanami Lecture Series in Modern Biology 7, ed. M. Kimura and S. Kondo. Tokyo: Iwanami Shoten.

1976 Kimura, M. Random genetic drift prevails. *Trends in Biochemical Sciences* 1:N152–N154.

1976 Kimura, M. Population genetics and molecular evolution. *Johns Hopkins Medical Journal* 138:253–61.

*1976 Kimura, M. How genes evolve: A population geneticist's view. *Annales de Génétique* 19:153–68.

1976 Kimura, M., The neutral theory as a supplement to Darwinism. *Trends in Biochemical Sciences* 1:N248–N249.

*1977 Kimura, M. Preponderance of synonymous changes as evidence for the neutral theory of molecular evolution. *Nature* 267:275–76.

1977 Kimura, M. *Molecular Evolution and Polymorphism. Proceedings of the Second Taniguchi International Symposium on Biophysics,* ed. M. Kimura. Mishima, Japan: National Institute of Genetics.

1977 Kimura, M. Causes of evolution and polymorphism at the molecular level. *Proceedings of the Second Taniguchi International Symposium on Biophysics,* ed. M. Kimura, pp. 1–28. Mishima, Japan: National Institute of Genetics.

1977 Kimura, M., and T. Ohta. Further comments on "counterexamples to a neutralist hypothesis." *Journal of Molecular Evolution* 9:367–68.

1977 Kimura, M. Essai sur notre passé, notre présent et notre avenir: Le point de vue d'un généticien des populations. *Mémoires de l'Académie des Sciences, Inscriptions et Belles-Lettres de Toulouse* 139:103–7.

1977 Kimura, M. The neutral theory of molecular evolution and polymorphism. *Scientia* 112:687–707.

1977 Kimura, M. Genetic codes and the laws of evolution as the bases for our understanding of the biological nature of man. *Proceedings of the Fifth International Conference on the Unity of the Sciences,* pp. 621–30. Washington, D.C.

1978 Maruyama, T., and M. Kimura. Theoretical study of genetic variability, assuming stepwise production of neutral and very slightly deleterious mutations. *Proceedings of the National Academy of Sciences USA* 75:919–22.

*1978 Kimura, M. Change of gene frequencies by natural selection under population number regulation. *Proceedings of the National Academy of Sciences USA* 75:1934–37.

1978 Kimura, M., and T. Ohta. Stepwise mutation model and distribution of allelic frequencies in a finite population. *Proceedings of the National Academy of Sciences USA* 75:2868–72.

*1978 Kimura, M., and J. F. Crow. Effect of overall phenotypic selection on genetic change at individual loci. *Proceedings of the National Academy of Sciences USA* 75:6168–71.

*1979 Crow, J. F., and M. Kimura. Efficiency of truncation selection. *Proceedings of the National Academy of Sciences USA* 76:396–99.

*1979 Kimura, M., and J. L. King. Fixation of a deleterious allele at one of two "duplicate" loci by mutation pressure and random drift. *Proceedings of the National Academy of Sciences USA* 76:2858–61.

*1979 Kimura, M. Model of effectively neutral mutations in which selective constraint is incorporated. *Proceedings of the National Academy of Sciences USA* 76:3440–44.

*1979 Kimura, M., and T. Ohta. Population genetics of multigene family with special reference to decrease of genetic correlation with distance between gene members on a chromosome. *Proceedings of the National Academy of Sciences USA* 76:4001–5.

1979 Kimura, M. The neutral theory of molecular evolution. *Scientific American* 241, no. 5:98–126.

1979 Takahata, N., and M. Kimura. Genetic variability maintained in a finite population under mutation and autocorrelated random fluctuation of selection intensity. *Proceedings of the National Academy of Sciences USA* 76:5813–17.

1980 Kimura, M. Average time until fixation of a mutant allele in a finite population under continued mutation pressure: Studies by analytical, numerical, and pseudo-sampling methods. *Proceedings of the National Academy of Sciences USA* 77:522–26.

1980 Kimura, M. Contributions of population genetics to molecular evolutionary studies. In *Genetics and Evolution of RNA Polymerase, tRNA and Ribosomes,* ed S. Osawa, H. Ozeki, H. Uchida, and T. Yura, pp. 499–518. Tokyo: University of Tokyo Press.

*1980 Kimura, M. A simple method for estimating evolutionary rates of base substitutions through comparative studies of nucleotide sequences. *Journal of Molecular Evolution* 16:111–20.

*1980 Maruyama, T., and M. Kimura. Genetic variability and effective population size when local extinction and recolonization of subpopulations are frequent. *Proceedings of the National Academy of Sciences USA* 77:6710–14.

*1981 Kimura, M. Estimation of evolutionary distances between homologous nucleotide sequences. *Proceedings of the National Academy of Sciences USA* 78:454–58.

1981 Kimura, M. Was globin evolution very rapid in its early stages?: A dubious case against the rate-constancy hypothesis. *Journal of Molecular Evolution* 17:110–13.

1981 Kimura, M. Doubt about studies of globin evolution based on maximum parsimony codons and the augmentation procedure. *Journal of Molecular Evolution* 17:121–22.

1981 Ohta, T., and M. Kimura. Some calculations on the amount of selfish DNA. *Proceedings of the National Academy of Sciences USA* 78:1129–32.

1981 Kimura, M. Data on our evolutionary heritage. In *Data for Science and Technology,* ed. P. S. Glaeser, pp. 23–29. Oxford: Pergamon Press.

*1981 Kimura, M. Possibility of extensive neutral evolution under stabilizing selection with special reference to nonrandom usage of synonymous codons. *Proceedings of the National Academy of Sciences USA* 78:5773–77.

*1981 Takahata, N., and M. Kimura. A model of evolutionary base substitutions and its application with special reference to rapid change of pseudogenes. *Genetics* 98:641–57.

1982 Kimura, M., ed. *Molecular Evolution, Protein Polymorphism and the Neutral Theory.* Tokyo: Japan Scientific Societies Press; Berlin: Springer-Verlag.

1982 Kimura, M. The neutral theory as a basis for understanding the mechanism of evolution and variation at the molecular level. In *Molecular Evolution, Protein Polymorphism and the Neutral*

*Theory,* ed. M. Kimura, pp. 3–56. Tokyo: Japan Scientific Societies Press; Berlin: Springer-Verlag.

*1983 Kimura, M., and N. Takahata. Selective constraint in protein polymorphism: Study of the effectively neutral mutation model by using an improved pseudosampling method. *Proceedings of the National Academy of Sciences USA* 80:1048–52.

1983 Kimura, M. The neutral theory of molecular evolution. In *Evolution of Genes and Proteins,* ed. M. Nei and R. K. Koehn, pp. 208–33. Sunderland, Mass.: Sinauer.

*1983 Kimura, M. Diffusion model of intergroup selection, with special reference to evolution of an altruistic character. *Proceedings of the National Academy of Sciences USA* 80:6317–21.

1983 Kimura, M. *The Neutral Theory of Molecular Evolution.* Cambridge: Cambridge University Press.

*1983 Kimura, M. Rare variant alleles in the light of the neutral theory. *Molecular Biology and Evolution* 1:84–93.

1984 Kimura, M. Evolution of an altruistic trait through group selection as studied by the diffusion equation method. *IMA Journal of Mathematics Applied in Medicine and Biology* 1:1–15.

1984 Kimura, M. Neutral evolution is an inevitable process of change at the molecular level. In *Darwin a Barcelona,* ed. Josep Sancho i Valls, pp. 231–52. Promocions Publications Universitàries, Barcelona.

1984 Jukes T. H., and M. Kimura. Evolutionary constraints and the neutral theory. *Journal of Molecular Evolution* 21:90–92.

1985 Kimura, M. The neutral theory of molecular evolution. *New Scientist* 1464:41–46.

1985 Kimura, M. Natural selection and neutral evolution. In *What Darwin Began,* ed. L. R. Godfrey, pp. 73–93. Boston: Allyn and Bacon.

1985 Kimura, M. The role of compensatory neutral mutations in molecular evolution. *Journal of Genetics* 64:7–19.

*1985 Kimura, M. Diffusion models in population genetics with special reference to fixation time of molecular mutants under mutational pressure. In *Population Genetics and Molecular Evolution,* ed. T. Ohta and K. Aoki, pp. 19–39. Tokyo: Japan Scientific Societies Press; Berlin: Springer-Verlag.

1985 Kimura, M. Genes, populations, and molecules: A memoir. In *Population Genetics and Molecular Evolution,* ed. T. Ohta and K. Aoki, pp. 459–81. Tokyo: Japan Scientific Societies Press; Berlin: Springer-Verlag.

1985 Kimura, M. Neutralisme dans l'évolution. *Sciences et Avenir* (numéro spécial), pp. 29–34.

*1986 Kimura, M. DNA and the neutral theory. *Philosophical Transactions of the Royal Society of London,* series B, 312:343–54.

1986 Kimura, M. Diffusion model of population genetics incorporating group selection, with special reference to an altruistic trait.

In *Stochastic Processes and Their Applications,* ed. K. Ito and T. Hida, pp. 101–18. Lecture Notes in Mathematics. Berlin: Springer-Verlag.

1986 Kimura, M. Diffusion models of population genetics in the age of molecular biology. In *The Craft of Probabilistic Modelling,* ed. J. Gani and C. C. Heyde, pp. 150–65. New York: Springer-Verlag.

1987 Kimura, M. A stochastic model of compensatory neutral evolution. In *Stochastic Methods in Biology,* ed. M. Kimura, G. Kallianpur, and T. Hida, pp. 2–18. Berlin: Springer-Verlag.

*1987 Kimura, M. Molecular evolutionary clock and the neutral theory. *Journal of Molecular Evolution* 26:24–33.

1987 Kimura, M., G. Kallianpur, and T. Hida, eds. *Stochastic Methods in Biology.* Lecture Notes in Biomathematics 70. Berlin: Springer-Verlag.

1988 Kimura, M. Thirty years of population genetics with Dr. Crow. *Japanese Journal of Genetics* 63:1–10.

1988 Kimura, M. Natural selection and neutral evolution, with special reference to evolution and variation at the molecular level. In *L'Évolution dans sa réalité et ses diverses modalités,* ed. Fondation Singer-Polignac, pp. 269–84. Paris: Masson.

1989 Kimura, M. The neutral theory of molecular evolution and the world view of the neutralists. *Genome* 31:24–31.

1989 Kimura, M. Neutral theory. In *Evolution and Animal Breeding,* ed. W. G. Hill and T. F. C. Mackay, pp. 13–16. Wallingford: C.A.B. International.

1990 Kimura, M. Some models of neutral evolution, compensatory evolution, and the shifting balance process. *Theoretical Population Biology* 37:150–58.

1990 Kimura, M. The present status of the neutral theory. In *Population Biology of Genes and Molecules,* ed. N. Takahata and J. F. Crow, pp. 1–16. Tokyo: Baifukan.

1990 Gojobori, T., E. N. Moriyama, and M. Kimura. Molecular clock of viral evolution, and the neutral theory. *Proceedings of the National Academy of Sciences USA* 87:10015–18.

1991 Kimura, M. Recent development of the neutral theory viewed from the Wrightian tradition of theoretical population genetics. *Proceedings of the National Academy of Sciences USA* 88:5969–73.

1991 Kimura, M. Some recent data supporting the neutral theory. In *new Aspects of the Genetics of Molecular Evolution,* ed. M. Kimura and N. Takahata, pp. 3–14. Tokyo: Japan Scientific Societies Press; Berlin: Springer-Verlag.

1991 Kimura, M. The neutral theory of molecular evolution: A review of recent evidence. *Japanese Journal of Genetics* 66:367–86.

1991 Kimura, M., and N. Takahata, eds. *New Aspects of the Genetics of Molecular Evolution.* Tokyo: Japan Scientific Societies Press; Berlin: Springer-Verlag.

1991 Kimura, M. Neutral evolution. In *Evolution of Life,* ed. S. Osawa and T. Honjo, pp. 67–78. Tokyo: Springer-Verlag.

1993 Kimura, M. Retrospective of the last quarter century of the neutral theory. *Japanese Journal of Genetics* 68:521–28.

# REFERENCES


Akin, E. 1979. *The Geometry of Population Genetics.* In Lecture Notes in Biomathematics. Vol. 31. Berlin, Heidelberg, New York: Springer-Verlag.

Aoki, K. 1982. A condition for group selection to prevail over counteracting individual selection. *Evolution* 36:832–42.

Atwood, K. C., L. K. Schneider, and F. J. Ryan. 1951. Selective mechanisms in bacteria. *Cold Spring Harbor Symp. Quant. Biol.* 16:345–55.

Avery, P. J. 1977. The effect of random selection coefficients on populations of finite size—some particular models. *Genet. Res. Camb.* 29:97–112.

Bailey, G. S., R. T. M. Poulter, and P. A. Stockwell, 1978. Gene duplication in tetraploid fish: Model for gene silencing at unlinked duplicated loci. *Proc. Natl. Acad. Sci. USA* 75:5575–79.

Barton, N. H., and M. Turelli. 1987. Adaptive landscapes, genetic distance and the evaluation of quantitative characters. *Genet. Res. Camb.* 49:157–73.

———. 1989. Evolutionary quantitative genetics: How little do we know? *Ann. Rev. Genet.* 23:237–70.

Begun, D. J., and C. F. Aquadro. 1992. Levels of naturally occurring DNA polymorphism correlate with recombination rates in *D. melanogaster. Nature* 356:519–20.

Birky, C. W., Jr. 1983. Relaxed cellular controls and organelle heredity. *Science* 222: 468–75.

———. 1991. Evolution and population genetics of organelle genes: Mechanism and models. In *Evolution at the Molecular Level,* ed. R. K. Selander A. G. Clark, and T. S. Whittam, pp. 112–34. Sunderland, Mass.: Sinauer.

Bodmer, W. F., and J. Felsenstein. 1967. Linkage and selection theoretical analysis of the deterministic two locus random mating model. *Genetics* 57:237–65.

Boucher, W., and C. W. Cotterman. 1990. On the classification of regular systems of inbreeding. *J. Math. Biol.* 28:293–305.

Britten, R. J. 1986. Rates of DNA sequence evolution differ between taxonomic groups. *Science* 231:1393–98.

Brown, A. H. D., D. R. Marshall, and B. S. Weir. 1981. Current status of the charge state model for protein polymorphism. In *Genetic Studies of Drosophila Populations,* ed. J. B. Gibson and J. G. Oakeshott, pp. 15–43. Australian National University Press.

Bulmer, M. G. 1972. The genetic variability of polygenic characters under optimizing selection, mutation, and drift. *Genet. Res. Camb.* 19:17–25.

———. 1973. Geographical uniformity of protein polymorphisms. *Nature, Lond.* 241: 199–200.

———. 1980. *The Mathematical Theory of Quantitative Genetics.* Oxford: University of Oxford/Clarendon Press.

Cavalli-Sforza, L. L., and F. Conterio. 1960. Analysis of the fluctuation of the gene frequencies in the population of the Val Parma (in Italian). *Atti Associazione Genetica Italiana* 5:333–43.

Charlesworth, B., and D. Hartl. 1978. Population dynamics of the segregation distorter polymorphism of Drosophila melanogaster. *Genetics* 89:171–92.

Cockerham, C. C., and P. M. Burrows. 1980. Selection limits and strategies. *Proc. Natl. Acad. Sci. USA* 77:546–49.

Cotterman, C. W. 1940. A calculus for statistico-genetics. Ph.D. diss., Ohio State University, Columbus. In *Genetics and Social Structure,* ed. P. Ballonoff, pp. 157–272. Stroudsburg: Dowden, Hutchinson, and Ross.

Crow, J. F. 1972. The dilemma of nearly neutral mutations: How important are they for evolution and human welfare? *J. Heredity* 63:306–16.

———. 1979. Genes that violate Mendel's rules. *Scientific American* 240:134–46.

———. 1985. The neutrality-selection controversy in the history of evolution and population genetics. In *Population Genetics and Molecular Evolution,* ed. T. Ohta and K. Aoki, pp. 1–18. Tokyo: Japan Scientific Societies Press; Berlin: Springer-Verlag.

———. 1987. Motoo Kimura and molecular evolution. *Genetics* 116:183–84.

———. 1989a. The importance of recombination. In *The Evolution of Sex: An Examination of Current Ideas,* ed. R. E. Michod and B. R. Levin. Sunderland, Mass.: Sinauer.

———. 1989b. The infinite allele model. *Genetics* 121:631–34.

———. 1990a. Fisher's contributions to genetics and evolution. *Theor. Popul. Biol.* 38: 263–75.

———. 1990b. The third phase of Wright's theory of evolution. In *Population Biology of Genetics and Molecules,* ed. N. Takahata and J. F. Crow, pp. 93–103. Tokyo: Baifukan.

———. 1991. Why is Mendelian segregation so exact? *BioEssays* 13:305–12.

———. 1992. An advantage of sexual reproduction in a rapidly changing environment. *J. Hered.* 83:169–73.

Crow, J. F., and K. Aoki. 1984. Group selection for a polygenic behavioral trait: Estimating the degree of population subdivision. *Proc. Natl. Acad. Sci. USA* 81: 6073–77.

Crow, J. F., W. R. Engels, and C. Denniston. 1990. Phase three of Wright's shifting-balance theory. *Evolution* 44:233–47.

Crow, J. F., and M. Kimura. 1964. The theory of genetic loads. *Proceedings of the XI International Congress of Genetics,* pp. 495–506. The Hague, Netherlands.

———. 1970. *An Introduction to Population Genetics Theory.* New York: Harper and Row.

Dynkin, E. B. 1965. *Markov Processes.* Berlin, Gottingen, Heidelberg: Springer-Verlag.

Easteal, S. 1991. The relative rate of DNA evolution in primates. *Mol Evol. Biol.* 8:115.

Edwards, A. W. F. 1971. Fisher's fundamental theorem of natural selection and Wright's adaptive topography: A controversy revisited. *Biometrics* 27:762–63.

Eshel, I., and M. W. Feldman. 1982. On evolutionary genetic stability of the sex ratio. *Theor. Popul. Biol.* 11:410–24.

Ethier, S. N., and R. C. Griffiths. 1984. The infinitely-many-sites model as a measure-valued diffusion. Statistics Research Report, Monash University, Clayton, Victoria, Australia.

Ewens, W. J. 1963. The mean time for absorption in a process of genetic type. *J. Australia Math. Soc.* 3:375–83.

———. 1972. The sampling theory of selectively neutral alleles. *Theor. Popul. Biol.* 3:87–112.

———. 1979. *Mathematical Population Genetics.* Biomathematics, Vol. 9. Berlin, Heidelberg, New York: Springer-Verlag.

———. 1989. An interpretation and proof of the fundamental theorem of natural selection. *Theor. Popul. Biol.* 36:167–80.

Ewens, W. J., and W.-H. Li. 1980. Frequency spectra of neutral and deleterious alleles in a finite population. *J. Math. Biol.* 10:155–66.

Falconer, D. S. 1960. *Introduction to Quantitative Genetics.* Longman Scientific & Technical, copub. in the United States with John Wiley & Sons, New York.

Feldman, M. W. 1972. Selection for linkage modification. I: Random mating populations. *Theor. Popul. Biol.* 3:324–46.

Feldman, M. W., U. Liberman. 1986. An evolutionary reduction principle for genetic modifiers. *Proc. Natl. Acad. Sci. USA* 83:4824–27.

Feller, W. 1951. Diffusion processes in genetics. *Proc. Second Berkeley Symp. on Math. Stat. and Prob.*, pp. 227–46.

———. 1952. The parabolic differential equations and the associated semigroup of transformations. *Ann. Math.* 55:468–519.

Felsenstein, J. 1974. The evolutionary advantage of recombination. *Genetics* 78:737–56.

———. 1975. A pain in the torus: Some difficulties with models of isolation by distance. *Am. Nat.* 109:359–68.

———. 1976. The theoretical population genetics of variable selection and migration. *Ann. Rev. Genet.* 10:253–80.

———. 1981. Evolutionary trees from DNA sequences: A maximum likelihood approach. *J. Mol. Evol.* 17:368–76.

Felsenstin, J., and S. Yokoyama. 1976. The evolutionary advantage of recombination. II. Individual selection for recombination. *Genetics* 83:845–59. (Corrigendum 85: 372, 1977.)

Fisher, R. A. 1922. On the dominance ratio. *Proc. Roy. Soc. Edin.* 42:321–431.

———. 1930. *The Genetical Theory of Natural Selection,* 2d rev. ed. New York: Dover, 1958.

Fleming, W. H., and C.-H. Su. 1974. Some one-dimensional migration models in population genetics theory. *Theor. Popul. Biol.* 5:431–49.

Frank, S. A., and M. Slatkin. 1992. Fisher's fundamental theorem of natural selection. *Tree* 3:92–95.

Freese, E. 1962. On the evolution of the base composition of DNA. *J. Theor. Biol.* 3:82–101.

Frydenberg, O. 1963. Population studies of a lethal mutant in *Drosophila melanogaster.* I. Behaviour in populations with discrete generations. *Hereditas* 50:89–116.

Fuerst, P. A., R. Chakraborty, and M. Nei. 1977. Statistical studies on protein polymorphism in natural populations. I. Distribution of single locus heterozygosity. *Genetics* 86:455–83.

Fuerst, P. A., and R. E. Ferrell. 1980. The stepwise mutation model: An experimental evaluation utilizing hemoglobin variants. *Genetics* 94:185–201.

Gillespie, J. H. 1973. Polymorphism in random environments. *Theor. Popul. Biol.* 4:193–95.

———. 1984. The molecular clock may be an episodic clock. *Proc. Natl. Acad. Sci. USA* 81:8009–13.

———. 1986. Natural selection and the molecular clock. *Mol. Biol. Evol.* 3:138–55.

———. 1987. Molecular evolution and the neutral allele theory. In *Oxford Surveys in Evolutionary Biology,* ed. P. H. Harvey and L. Partridge, pp. 10–37. Oxford: Oxford University Press.

Gojobori, T., E. N. Moriyama, and M. Kimura. 1990. Molecular clock of viral evolution, and the neutral theory. *Proc. Natl. Acad. Sci. USA* 87:10015–18.

Graur, D., W. A. Hide, and W.-H. Li. 1991. Is the guinea-pig a rodent? *Nature, Lond.* 351:649–52.

Griffiths, R. C. 1980. Lines of descent in the diffusion approximation of neutral Wright-Fisher models. *Theor. Popul. Biol.* 17: 37–50.

Grunstein, M., P. Schedl, and L. Kedes. 1976. Sequence analysis and evolution of sea urchin (*Lytechinus pictus* and *Strongylocentrotus purpuratus*) histone H4 messenger RNAs. *J. Mol. Biol.* 104:351–69.

Haigh, J. 1978. The accumulation of deleterious genes in a population—Muller's ratchet. *Theor. Popul. Biol.* 14:251–67.

Haldane, J. B. S. 1927. A mathematical theory of natural and artificial selection, part V: Selection and mutation. *Proc. Camb. Philos. Soc.* 23:838–44.

———. 1937. The effect of variation on fitness. *Am. Nat.* 71:337–49.

———. 1957. The cost of natural selection. *J. Genet.* 55:511–24.

Haldane, J. B. S., and C. H. Waddington. 1931. Inbreeding and linkage. *Genetics* 16:357–74.

Harris, H. 1966. Enzyme polymorphism in man. *Proc. Roy. Soc. London,* ser. B, 164:298–310.

Hartl, D. L. 1980. Genetic dissection of segregation distortion. III. Unequal recovery of reciprocal recombinants. *Genetics* 96: 685–96.

Hasegawa, M., and H. Kishino. 1989. Confidence limits on the maximum-likelihood estimate of the hominoid tree from mitochondrial-DNA sequences. *Evolution* 43: 672–77.

Hedrick, P. W., T. S. Whittam, and P. Parham. 1991. Heterozygosity at individual amino acid sites: Extremely high levels for *HLA-A* and *-B* genes. *Proc. Natl. Acad. Sci. USA* 88:5897–5901.

Hill, W. G. 1982. Rates of change in quantitative traits from fixation of new mutations. *Proc. Natl. Acad. Sci. USA* 79:142–45.

———. 1990. Mutation, selection and quantitative genetic variation. In *Population Biology of Genes and Molecules,* ed. N. Takahata and J. F. Crow, pp. 219–32. Tokyo: Baifukan.

Hiraizumi, Y., and J. F. Crow. 1960. Heterozygous effects on viability, fertility, rate of development and longevity of Drosophila chromosomes that are lethal when homozygous. *Genetics* 45:1071–84.

Holmquist, R., M. Goodman, T. Conroy, and J. Czelunsniak. 1983. The spatial distribution of fixed mutations within genes coding for proteins. *J. Mol. Evol.* 19:437–48.

Hori, H., and S. Osawa. 1987. Origin and evolution of organisms as deducted from 5S ribosomal RNA sequences. *Mol. Biol. Evol.* 4:445–72.

Hubby, J. L., and R. C. Lewontin. 1966. A molecular approach to the study of genic heterozygosity in natural populations. I. The number of alleles at different loci in *Drosophila pseudoobscura. Genetics* 54: 577–94.

Hudson, R. R. 1983. Testing the constant rate neutral allele model with protein sequence data. *Evolution* 37:203–17.

Hudson, R. R., M. Kreitman, and M. Aguadé. 1987. A test of neutral molecular evolution based on nucleotide data. *Genetics* 116: 153–59.

Ikemura, T. 1985. Codon usage and tRNA content in unicellular and multicellular organisms. *Mol. Biol. Evol.* 2:13–34.

Ishii, K., H. Matsuda, Y. Iwasa, and A. Sasaki. 1989. Evolutionary stable mutation rate in a periodically changing environment. *Genetics* 121:163–74.

Jukes, T. H. 1991. Early development of the neutral theory. *Pers. in Biol. & Med.* 34:473–85.

Kaplan, N. L., R. R. Hudson, and C. H. Langley. 1989. The "hitchhiking effect" revisited. *Genetics* 123:887–99.

Karlin, S., and H. M. Taylor. 1981. *A Second Course in Stochastic Processes.* New York: Academic Press.

Kimura, M. 1955. Random genetic drift in multi-allelic locus. *Evolution* 9:419–35.

———. 1956. Random genetic drift in a triallelic locus: Exact solution with a continuous model. *Biometrics* 12:57–66.

———. 1960. Genetic load of a population and its significance in evolution. (In Japanese with English summary.) *Jap. J. Genet.* 35:7–33.

———. 1971. Theoretical foundation of population genetics at the molecular level. *Theor. Popul. Biol.* 2:174–208.

———. 1979. The neutral theory of molecular evolution. *Scientific American* 241, no. 5: 94–104.

———. 1982. The neutral theory as a basis for understanding the mechanism of evolution and variation at the molecular level. In *Molecular Evolution, Protein Polymorphism and the Neutral Theory,* ed. M. Kimura, pp. 3–56. Tokyo: Japan Scientific Societies Press; Berlin: Springer-Verlag.

———. 1983a. *The Neutral Theory of Molecular Evolution.* Cambridge: Cambridge University Press.

———. 1983b. The neutral theory of molecular evolution. In *Evolution of Genes and Proteins,* ed. M. Nei and R. K. Koehn, pp. 208–33. Sunderland, Mass.: Sinauer.

———. 1985a. Genes, populations and molecules: A memoir. In *Population Genetics and Molecular Evolution,* ed. T. Ohta and K. Aoki, pp. 459–81. Tokyo: Japan Scientific Societies Press; Berlin: Springer-Verlag.

———. 1985b. The role of compensatory neutral mutations in molecular evolution. *J. Genet.* 64:7–19.

———. 1989. The neutral theory of molecular evolution and the world view of the neutralists. *Genome* 31:24–31.

———. 1990. The present status of the neutral theory. In *Population Biology of Genes and Molecules*. ed. N. Takahata and J. F. Crow, pp. 1–16. Tokyo: Baifukan.

———. 1991a. The neutral theory of molecular evolution: A review of recent evidence. *Jap. J. Genet.* 66:367–86.

———. 1991b. Some recent data supporting the neutral theory. In *New Aspects of the Genetics of Molecular Evolution,* ed. M. Kimura and N. Takahata, pp. 3–14. Tokyo: Japan Scientific Societies Press; Berlin: Springer-Verlag.

Kimura, M., and T. Ohta. 1969. The average number of generations until extinction of an individual mutant gene in a finite population. *Genetics* 63:701–9.

———. 1971. *Theoretical Aspects of Population Genetics*. Princeton: Princeton University press.

King, J. L. 1967. Continuously distributed factors affecting fitness. *Genetics* 33:483–92.

King, J. L., and T. H. Jukes. 1969. Non-Darwinian evolution: Random fixation of selectively neutral mutations. *Science* 164:788–98.

Kingman, J. F. C. 1976. Coherent random walks arising in some genetical models. *Proc. Roy. Soc. A,* 351:19–31.

———. 1977. A note on multi-dimensional models of neutral mutation. *Theor. Popul. Biol.* 11:285–90.

———. 1982a. On the genealogy of large populations. *J. Appl. Prob.* A19:27–43.

———. 1982b. The coalescent. *Stochast. Proc. Appl.* 13:235–48.

Klein, J. 1986. *Natural History of the Major Histocompatibility Complex*. New York: John Wiley & Sons.

Kolmogorov, A. 1931. Über die analytischen Methoden in der Wahrscheinlichkeitsrechnung. *Math. Ann.* 104:415–58.

Kondrashov, A. S. 1984. Deleterious mutations as an evolutionary factor. 1. The advantage of recombination. *Genet. Res.* 44:199–214.

Kondrashov, A. S., and J. F. Crow. 1991. Haploidy or diploidy: Which is better? *Nature, Lond.* 351:314–15.

Lanave, C., G. Preparata, C. Saccone, and G. Serio. 1984. A new method for calculating evolutionary substitution rates. *J. Mol. Evol.* 20:86–93.

Lande, R. 1975. The maintenance of genetic variation by mutation in a polygenetic character with linked loci. *Genet. Res. Camb.* 26:221–35.

———. 1976. Natural selection and random genetic drift in phenotypic evolution. Evolution 30:314–34.

———. 1977. The influence of the mating system on the maintenance of genetic variability in polygenic characters. *Genetics* 86: 485–98.

Langley, C. H. 1990. The molecular population genetics of Drosophila. In *Population Biology of Genes and Molecules,* ed. N. Takahata and J. F. Crow, pp. 75–91. Tokyo: Baifukan.

Leigh, E. G. 1970. Natural selection and mutability. *Am. Nat.* 104:301–5.

Levikson, B. 1977. *The age distribution of Markov processes. J. Appl. Prob.* 14: 492–506.

Levin, B. R., and F. M. Stewart. 1976. Conditions for the existence of conjugationally transmitted plasmids in bacterial populations. *Genetics* 83:s45–s46.

Lewontin, R. C. 1974. *The Genetic Basis of Evolutionary Change*. New York and London: Columbia University Press.

———. 1991. Electrophoresis in the development of evolutionary genetics: Milestone or millstone? *Genetics* 128:657–62.

Lewontin, R. C., and J. L. Hubby. 1966. A molecular approach to the study of genic heterozygosity in natural populations. II. Amount of variation and degree of heterozygosity in natural populations of *Drosophila pseudoobscura. Genetics* 54:595–609.

Li, W.-H. 1976. A mixed model of mutation for electrophoretic identity of proteins within and between populations. *Genetics* 83:423–32.

———. 1980. Rate of gene silencing at duplicate loci: A theoretical study and interpretation of data from tetraploid fishes. *Genetics* 95:237–58.

Li, W.-H., C.-C. Luo, and C.-I. Wu. 1985. Evolution of DNA sequences. In *Molecular*

*Evolutionary Genetics,* ed. R. J. MacIntyre, pp. 1–96. New York: Plenum.

Li, W.-H., and M. Nei. 1977. Persistence of common alleles in two related species. *Genetics* 86:901–14.

Li, W.-H., and M. Tanimura. 1987. The molecular clock runs more slowly in man than in apes and monkeys. *Nature, Lond.* 326: 93–96.

Malécot, G. 1941. Étude mathématique des populations "mendéliennes." *Ann. Univ. Lyon Sci.* Sec. A, 4:45–60.

———. 1944. Sur un problème de probabilités en chaîne que pose la génétique. *C. R. Acad. Sci. Paris* 219:379–81.

———. 1948. *Les mathématiques de l'hérédité.* Paris: Masson.

———. 1951. Un traitement stochastique des problèmes linéaires (mutation, linkage, migration) en génétique de population. *Ann. Univ. Lyon Sci.* Sec. A, 14:79–117.

———. 1952. Les processus stochastiques et la méthode des fonctions génératrices ou caractéristiques. *Publ. Inst. Stat. Univ. Paris* 1: Fasc. 3, 1–16.

———. 1955. The decrease of relationship with distance. *Cold Spring Harbor Symp. Quant. Biol.* 20:52–53.

———. 1959. Les modèles stochastiques en génétique de population. *Publ. Inst. Stat. Univ. Paris* 8: Fasc. 3, 173–210.

Maruyama, T. 1970. Analysis of population structure. I. One-dimensional stepping stone models of finite length. *Ann. Hum. Genet.* 34:201–19.

———. 1974. The age of an allele in a finite population. *Genet. Res. Camb.* 23:137–43.

———. 1977. *Stochastic Problems in Population Genetics.* In Lecture Notes in Biomathematics. Vol. 17. Berlin, Heidelberg, New York: Springer-Verlag.

Maruyama, T., and M. Kimura. 1974. Geographical uniformity of selectively neutral polymorphisms. *Nature* 249:30–32.

Maynard Smith, J. 1968. Evolution in sexual and asexual populations. *Am. Nat.* 102: 469–73.

———. 1978. *The Evolution of Sex.* London, New York: Cambridge University Press.

———. 1989. *Evolutionary Genetics.* Oxford, New York, Tokyo: Oxford University Press.

Maynard Smith, J., and J. Haigh. 1974. The hitch-hiking effect of a favourable gene. *Genet. Res. Camb.* 23:23–35.

Medawar, P. B. 1979. *Advice to a Young Scientist.* New York: Basic Books.

Michod, R. E., and B. R. Levin, eds. 1989. *The Evolution of Sex: An Examination of Current Ideas.* Sunderland, Mass.: Sinauer.

Milkman, R. D. 1967. Heterosis as a major cause of heterozygosity in nature. *Genetics* 55:493–95.

———. 1978. Selection differentials and selection coefficients. *Genetics* 88:391–403.

Moran, P. A. P. 1975. Wandering distributions and the electrophoretic profile. *Theor. Popul. Biol.* 8:318–30.

———. 1976. A selective model for electrophoretic profiles in protein polymorphisms. *Genet. Res. Camb.* 28:47–53.

Mukai, T. 1964. The genetic structure of natural populations of *Drosophila melanogaster.* I. Spontaneous mutation rate of polygenes controlling viability. *Genetics* 50: 1–19.

———. 1965. Synergistic interaction between spontaneous mutant polygenes controlling viability in *Drosophila melanogaster. Annu. Rep. Nat. Inst. Genet.* 15:28–29.

———. 1990. Viability polygenes in populations of *Drosophila melanogaster.* In *Population Biology of Genetics and Molecules,* ed. N. Takahata and J. F. Crow, pp. 199–217. Tokyo: Baifukan.

Muller, H. J. 1914. A gene for the fourth chromosome of Drosophila. *J. Exp. Zool.* 17: 325–36.

———. 1932. Further studies on the nature and causes of gene mutations. *Proceedings of the Sixth International Congress of Genetics* 1:213–55.

———. 1950. Our road of mutations. *Amer. J. Hum. Genet.* 2:111–76.

———. 1958. Evolution by mutation. *Bull. Amer. Math. Soc.* 64:137–60.

Nagylaki, T. 1974a. Genetic structure of a population occupying a circular habitat. *Genetics* 78:777–89.

———. 1974b. The decay of genetic variability in geographically structured populations. *Proc. Natl. Acad. Sci. USA* 71:2932–36.

———. 1977. *Selection in One- and Two-*

*Locus Systems*. In Lecture Notes in Biomathematics. Vol. 15. Berlin, Heidelberg, New York: Springer-Verlag.

———. 1983. The robustness of neutral models of geographical variation. *Theor. Popul. Biol.* 24:268–94.

———. 1989a. Gustave Malécot and the transition from classical to modern population genetics. *Genetics* 122:253–68.

———. 1989b. Rate of evolution of a character without epistasis. *Proc. Natl. Acad. Sci. USA* 86:1910–13.

———. 1991. Error bounds for the fundamental and secondary theorems of natural selection. *Proc. Natl. Acad. Sci. USA* 88: 2402–6.

Narain, P. 1970. A note on the diffusion approximation for the variance of the number of generations until fixation of a neutral mutant gene. *Gent. Res. Camb.* 15:251–55.

Nei, M. 1969. Linkage modification and sex difference in recombination. *Genetics* 63: 681–99.

———. 1971. Fertility excess necessary for gene substitution in regulated populations. *Genetics* 68:169–84.

———. 1975. *Molecular Population Genetics and Evolution*. In North-Holland Research Monographs, Frontiers of Biology. Vol. 40. North-Holland: American Elsevier.

———. 1977. Estimation of mutation rate from rare protein variants. *Amer. J. Hum. Genet.* 29:225–32.

———. 1987. *Molecular Evolutionary Genetics*. New York: Columbia University Press.

Nei, M., P. A. Fuerst, and R. Chakraborty. 1976. Testing the neutral mutation hypothesis by distribution of single locus heterozygosity. *Nature, Lond.* 262:491–93.

Nei, M., and D. Graur. 1984. Extent of protein polymorphism and the neutral mutation theory. *Evol. Biol.* 17:73–118.

Nei, M., and R. K. Koehn. 1983. *Evolution of Genes and Proteins*. Sunderland, Mass.: Sinauer.

Nei, M., and A. A. K. Roychoudhury. 1973. Probability of fixation and mean fixation time of an overdominant mutation. *Genetics* 74:371–80.

Nei, M., and N. Takahata. 1993. Effective population size, genetic diversity, and coalescence time in subdivided population. *J. Mol. Evol.* 37:240–44.

Nei, M., and S. Yokoyama. 1976. Effects of random fluctuations of selection intensity on genetic variability in a finite population. *Jap. J. Genet.* 51:355–69.

Novick, A., and L. Szilard. 1950. Experiments with the chemostat on spontaneous mutations in bacteria. *Proc. Natl. Acad. Sci. USA* 36:708–19.

Ohnishi, O. 1977. Spontaneous and ethyl methanesulfonate-induced mutations controlling viability in *Drosophila melanogaster*. II. Homozygous effect of polygenic mutations. *Genetics* 87:529–45.

Ohno, S. 1970. *Evolution by Gene Duplication*. Berlin, Heidelberg, New York: Springer-Verlag.

Ohta, T. 1972. Fixation probability of a mutant influenced by random fluctuation of selection intensity. *Genet. Res. Camb.* 19: 33–38.

———. 1973. Slightly deleterious mutant substitutions in evolution. *Nature, Lond.* 246:96–98.

———. 1976. A simple model for treating the evolution of multigene families. *Nature, Lond.* 263:74–76.

———. 1977. Extension to the neutral mutation random drift hypothesis. In *Molecular Evolution and Polymorphism*, ed. M. Kimura, pp. 148–67. Mishima: National Institute of Genetics.

———. 1980. *Evolution and Variation of Multigene Families*. In Lecture Notes in Biomathematics. Vol. 37. Berlin, Heidelberg, New York: Springer-Verlag.

Ohta, T., and M. Kimura. 1971. On the constancy of the evolutionary rate of cistrons. *J. Mol. Evol.* 1:18–25.

Pamilo, P., M. Nei, and W.-H. Li. 1987. Accumulation of mutations in sexual and asexual populations. *Genet. Res. Camb.* 49: 135–46.

Popper, K. R. 1975. *The Logic of Scientific Discovery* (8th imp.). London: Hutchinson.

Price, G. R. 1972. Fisher's "Fundamental Theorem" made clear. *Ann. Hum. Genet.* 36: 129–40.

Provine, W. B. 1986. *Sewall Wright and Evolu-*

*tionary Biology.* Chicago: University of Chicago Press.

———. 1990. The neutral theory of molecular evolution in historical perspective. In *Population Biology of Genes and Molecules,* ed. N. Takahata and J. F. Crow, pp. 17–31. Tokyo: Baifukan.

Robertson, A. 1952. The effect of inbreeding on the variation due to recessive genes. *Genetics* 37:189–207.

Rutledge, R. A. 1970. The survival of epistatic gene complexes in subdivided populations. Ph.D. thesis in the Faculty of Pure Science, Columbia University.

Sandler, L., and Y. Hiraizumi. 1960. Meiotic drive in natural populations of *Drosophila melanogaster.* V. On the nature of the *SD* region. *Genetics* 45:1671–89.

Satta, Y., C. O'hUigin, N. Takahata, and J. Klein. 1993. The synonymous substitution rate of the major histocompatibility complex loci in primates. *Proc. Natl. Acad. Sci. USA* 90:7480–84.

Sawyer, S. 1977. On the past history of an allele now known to have frequency. *J. Appl. Prob.* 14:439–50.

Schmalhausen, I. I. 1958. Hereditary information and its transformations. *Proc. X International Congress of Genetics,* vol. 2, p. 253.

Selander, R. K., and B. R. Levin. 1980. Genetic diversity and structure in *Escherichia coli* populations. *Science* 210:545–47.

Shimakura, N. 1985. Existence and uniqueness of solutions for a diffusion model of intergroup selection. *J. Math. Kyoto Univ.* 25:775–88.

Simmons, M. J., and J. F. Crow. 1977. Mutations affecting fitness in Drosophila populations. *Annu. Rev. Genet.* 11:49–78.

Slatkin, M. 1977. Gene flow and genetic drift in a species subject to frequent local extinctions. *Theor. Popul. Biol.* 12:253–62.

———. 1985. Gene flow in natural populations. *Ann. Rev. Ecol. Syst.* 16:393–430.

———. 1987. Quantitative genetics of heterochrony. *Evolution* 41:799–811.

Stewart, I. 1992. Warning—handle with care! *Nature* 335:16–17.

Sueoka, N. 1962. On the genetic basis of variation and heterogeneity of DNA base composition. *Proc. Natl. Acad. Sci. USA* 48: 166–69.

Sved, J. A., T. E. Reed, and W. F. Bodmer. 1967. The number of balanced polymorphisms that can be maintained in a natural population. *Genetics* 55:469–81.

Tachida, H., and C. C. Cockerham. 1990. Evolution of neutral quantitative characters with gene interaction and mutation. In *Population Biology of Genes and Molecules,* ed. N. Takahata and J. F. Crow, pp. 233–49. Tokyo: Baifukan.

Tajima, F. 1983. Evolutionary relationship of DNA sequences in finite populations. *Genetics* 105:437–60.

———. 1989a. The effect of change in population size on DNA polymorphism. *Genetics* 123:597–601.

———. 1989b. Statistical method for testing the neutral mutation hypothesis. *Genetics* 123:585–95.

Takahata, N. 1980. Composite stepwise mutation model under the neutral mutation hypothesis. *J. Mol. Biol.* 15:13–20.

———. 1981. Genetic variability and rate of gene substitution in a finite population under mutation and fluctuating selection. *Genetics* 98:427–40.

———. 1982. Sexual recombination under the joint effects of mutation, selection, and random sampling drift. *Theor. Popul. Biol.* 22:258–77.

———. 1985. Population genetics of extranuclear genomes: A model and review. In *Population Genetics and Molecular Evolution,* ed. T. Ohta and K. Aoki, pp. 195–212. Tokyo: Japan Scientific Societies Press; Berlin: Springer-Verlag.

———. 1987. On the overdispersed molecular clock. *Genetics* 116:169–79.

———. 1990. A simple genealogical structure of strongly balanced allelic lines and trans-species evolution of polymorphism. *Proc. Natl. Acad. Sci. USA* 87:2419–23.

———. 1991a. A trend in population genetics theory. In *New Aspects of the Genetics of Molecular Evolution,* ed. M. Kimura and N. Takahata, pp. 27–47. Tokyo: Japan Sci-

entific Societies Press; Berlin: Springer-Verlag.
———. 1991b. Overdispersed molecular clock at the major histocompatibility complex loci. *Proc. R. Soc. Lond.* B, 243:13–18.
Takahata, N., and M. Kimura. 1979. Genetic variability maintained in a finite population under mutation and autocorrelated random fluctuation of selection intensity. *Proc. Natl. Acad. Sci. USA* 76:5813–17.
Takahata, N., and Maruyama, T. 1979. Polymorphism and loss of duplicate gene expression: A theoretical study with application to tetraploid fish. *Proc. Natl. Acad. Sci. USA* 76:4521–25.
Tavaré, S. 1984. Lines-of-descent and genealogical processes, and their applications in population genetics models. *Theor. Popul. Biol.* 26:119–64.
Temin, R. G., B. Ganetzky, P. A. Powers, T. W. Lyttle, S. Pimpinelli, P. Dimitri, C.-I. Wu, and Y. Hiraizumi. 1991. Segregation distortion in *Drosophila melanogaster:* Genetic and molecular analyses. *Am. Nat.* 137: 287–331.
Turelli, M. 1984. Heritable genetic variation via mutation-selection balance: Lerch's zeta meets the abdominal bristle. *Theor. Popul. Biol.* 25:138–93.
———. 1988. Population genetic models for polygenic variation and evolution. In *Proceedings of the Second International Conference on Quantitative Genetics,* ed. B. S. Weir, E. J. Eisen, M. M. Goodman, and G. Namkoong, pp. 601–18. Sunderland, Mass.: Sinauer.
Uyenoyama, M. K. 1979. Evolution of altruism under group selection in large and small populations in fluctuating environments. *Theor. Popul. Biol.* 15:58–85.
Uzzell, T., and K. W. Corbin. 1971. Fitting discrete probability distributions to evolutionary events. *Science* 172:1089–96.
Watson, J. D., N. H. Hopkins, J. W. Roberts, J. A. Steitz, and A. M. Weiner. 1987. *Molecular Biology of the Gene,* 4th ed. Menlo Park, Calif.: Benjamin/Cummings.
Watterson, G. A. 1962. Some theoretical aspects of diffusion theory in population genetics. *Ann. Math. Stat.* 33:939–57.
———. 1975. On the number of segregating sites in genetic models without recombination. *Theor. Popul. Biol.* 7:256–76.
———. 1976. Reversibility and the age of an allele. I. Moran's infinitely many neutral alleles model. *Theor. Popul. Biol.* 10:239–53.
———. 1983. On the time for gene silencing at duplicate loci. *Genetics* 105:745–66.
———. 1984. Lines-of-descent and the coalescent. *Theor. Popul. Biol.* 26:77–92.
———. 1987. Estimating the proportion of neutral mutants. *Gent. Res. Camb.* 501: 155–63.
Weir, B. S., E. J. Eisen, M. M. Goodman, and G. Namkoong. 1988. *Proceedings of the Second International Conference on Quantitative Genetics.* Sunderland, Mass.: Sinauer.
Weiss, G. H., and M. Kimura. 1965. A mathematical analysis of the stepping stone model of genetic correlation. *J. Appl. Prob.* 2:129–49.
Wilson, E. O. 1975. *Sociobiology.* Cambridge, Mass., and London: Belknap Press of Harvard University.
Wilson, A. C., S. S. Carlson, and T. J. White. 1977. Biochemical evolution. *Annu. Rev. Biochem.* 46:573–639.
Wimsatt, W. C. 1987. False models as means to truer theories. In *Neutral Models in Biology,* ed. M. H. Nitechi and A. Hoffman, pp. 23–55. Oxford: Oxford University Press.
Woese, C. R. 1991. The use of ribosomal RNA in reconstructing evolutionary relationships among bacteria. In *Evolution at the Molecular Level,* ed. R. K. Selander, A. G. Clark, and T. S. Whittam, pp. 1–24. Sunderland, Mass.: Sinauer.
Wright, S. 1918. On the nature of size factors. *Genetics* 3:367–74.
———. 1920. The relative importance of heredity and environment in determining the piebald pattern of guinea pigs. *Proc. Natl. Acad. Sci. USA* 6:320–32.
———. 1921. Systems of mating I: The biometric relation between parent and offspring. *Genetics* 6:111–23.
———. 1922. Coefficients of inbreeding and relationship. *Am. Nat.* 56:330–38.

———. 1931. Evolution ion Mendelian populations. *Genetics* 16:97–159.
———. 1940. Breeding structure of populations in relation to speciation. *Am. Nat.* 74:232–48.
———. 1943. Isolation by distance. *Genetics* 28:114–38.
———. 1948. On the roles of directed and random changes in gene frequency in the genetics of populations. *Evolution* 2:279–94.
———. 1950. Population structure as a factor in evolution. In *Moderne Biologie, Festschrift Zum 60. Geburtstag von Hans Nachtsheim,* ed. H. Gruehneberg and W. Ulrich, pp. 274–87. Berlin: F. W. Peter.
———. 1965. The interpretation of population structure by F-statistics with special regard to systems of mating. *Evolution* 19:395–420.
———. 1966. Polyallelic random drift in relation to evolution. *Proc. Natl. Acad. Sci. USA* 55:1074–81.
———. 1970. Random drift and the shifting balance theory of evolution. In *Mathematical Topics in Population Genetics,* ed. K. Kojima, pp. 1–31. Heidelberg: Springer-Verlag.
———. 1977. *Evolution and the Genetics of Populations,* vol. 3: *Experimental Results and Evolutionary Deductions.* Chicago: University of Chicago Press.
Wu, C.-I., and M. F. Hammer. 1991. Molecular evolution of ultraselfish genes of meiotic drive systems. In *Evolution at the Molecular Level,* ed. R. K. Selander, A. G. Clark, and T. S. Whittam, pp. 177–203. Sunderland, Mass.: Sinauer.
Wu, C.-I, and W.-H. Li. 1985. Evidence for higher rates of nucleotide substitution in rodents than in man. *Proc. Natl. Acad. Sci. USA* 82:1741–45.
Yamada, Y. 1985. Contribution of population genetics theory to animal breeding made by Dr. Motoo Kimura. In *Population Genetics and Molecular Evolution,* ed. T. Ohta and K. Aoki, pp. 455–57. Tokyo: Japan Scientific Societies Press; Berlin: Springer-Verlag.
Zuckerkandl, E., and L. Pauling. 1965. Molecules as documents of evolutionary history. *J. Theor. Biol.* 8:357–66.